HANDBOOKS IN OPERATIONS RESEARCH
AND MANAGEMENT SCIENCE
VOLUME 12

Handbooks in Operations Research and Management Science

Editors

J.K. Lenstra
Centrum voor Wiskunde en Informatica, Amsterdam

G.L. Nemhauser
Georgia Institute of Technology

Volume 12

ELSEVIER

Amsterdam – Boston – Heidelberg – London – New York – Oxford – Paris
San Diego – San Francisco – Singapore – Sydney – Tokyo

Discrete Optimization

Edited by

K. Aardal
Centrum voor Wiskunde en Informatica,
Amsterdam

G.L. Nemhauser
Georgia Institute of Technology

R. Weismantel
Otto-von-Guericke-University of Magdeburg

ELSEVIER
Amsterdam – Boston – Heidelberg – London – New York – Oxford – Paris
San Diego – San Francisco – Singapore – Sydney – Tokyo

ELSEVIER B.V.	ELSEVIER Inc.	ELSEVIER Ltd	ELSEVIER Ltd
Radarweg 29	525 B Street, Suite 1900	The Boulevard, Langford Lane	84 Theobalds Road
P.O. Box 211, 1000 AE	San Diego, CA 92101-4495	Kidlington, Oxford OX5 1GB	London WC1X 8RR
Amsterdam, The Netherlands	USA	UK	UK

First edition 2005

Library of Congress Cataloging in Publication Data
A catalog record is available from the Library of Congress.

British Library Cataloguing in Publication Data
A catalogue record is available from the British Library.

ISBN-10: 0-444-51507-0
ISBN-13: 978-0-444-51507-0
ISSN: 0927-0507

The paper used in this publication meets the requirements of ANSI/NISO Z39.48-1992 (Permanence of Paper).
Printed in The Netherlands.

Contents

Preface

Optimization was the subject of the first handbook of this series published in 1989. Two articles from that handbook, Polyhedral Combinatorics and Integer Programming, were on discrete optimization. Since then, there have been many very significant developments in the theory, methodology and applications of discrete optimization, enough to easily justify a full handbook on the subject. While such a handbook could not possibly be all-inclusive, we have chosen nine main topics that are representative of recent theoretical and algorithmic developments in the field. In addition to the nine papers that present recent results, there is an article on the early history of the field.

All of the articles in this handbook are written by authors who have made significant original contributions to their topics. We believe that the handbook will be a useful reference to experts in the field as well as to students and others who want to learn about discrete optimization. We also hope that these articles provide not only the current state of the art, but also a glimpse into future developments. Below we provide a brief introduction to the chapters of the handbook.

Besides being well known for his research contributions in combinatorial optimization, Lex Schrijver is a scholar of the history of the field, and we are very fortunate to have his article "On the history of combinatorial optimization (till 1960)". This article goes back to work of Monge in the 18th century on the assignment problem and presents six problem areas: assignment, transportation, maximum flow, shortest spanning tree, shortest path and traveling salesman.

The branch-and-cut algorithm of integer programming is the computational workhorse of discrete optimization. It provides the tools that have been implemented in commercial software such as CPLEX and Xpress MP that make it possible to solve practical problems in supply chain, manufacturing, telecommunications and many other areas. The article "Computational integer programming and cutting planes" by Armin Fügenschuh and Alexander Martin presents the key ingredients of these algorithms.

Although branch-and-cut based on linear programming relaxation is the most widely used integer programming algorithm, other approaches are needed to solve instances for which branch-and-cut performs poorly and to understand better the structure of integral polyhedra. The next three chapters discuss alternative approaches.

The article "The structure of group relaxations" by Rekha Thomas studies a family of polyhedra obtained by dropping certain nonnegativity restrictions on integer programming problems. Thomas surveys recent algebraic results obtained from the theory of Gröbner bases.

Although integer programming is NP-hard in general, it is polynomially solvable in fixed dimension. The article "Integer programming, lattices, and results in fixed dimension" by Karen Aardal and Friedrich Eisenbrand presents results in this area including algorithms that use reduced bases of integer lattices that are capable of solving certain classes of integer programs that defy solution by branch-and-cut.

Relaxation or dual methods, such as cutting plane algorithms, progressively remove infeasibility while maintaining optimality to the relaxed problem. Such algorithms have the disadvantage of possibly obtaining feasibility only when the algorithm terminates. Primal methods for integer programs, which move from a feasible solution to a better feasible solution, were studied in the 1960's but did not appear to be competitive with dual methods. However, recent development in primal methods presented in the article "Primal integer programming" by Bianca Spille and Robert Weismantel indicate that this approach is not just interesting theoretically but may have practical implications as well.

The study of matrices that yield integral polyhedra has a long tradition in integer programming. A major breakthrough occurred in the 1990's with the development of polyhedral and structural results and recognition algorithms for balanced matrices. Michele Conforti and Gérard Cornuéjols were two of the researchers who obtained these results and their article "Balanced matrices" is a tutorial on the subject.

Submodular function minimization generalizes some linear combinatorial optimization problems such as minimum cut and is one of the fundamental problems of the field that is solvable in polynomial time. The article "Submodular function minimization" by Tom McCormick presents the theory and algorithms of this subject.

In the search for tighter relaxations of combinatorial optimization problems, semidefinite programming provides a generalization of linear programming that can give better approximations and is still polynomially solvable. Monique Laurent and Franz Rendl discuss this subject in their article "Semidefinite programming and integer programming".

Many real world problems have uncertain data that is known only probabilistically. Stochastic programming treats this topic, but until recently it was limited, for computational reasons, to stochastic linear programs. Stochastic integer programming is now a high profile research area and recent developments are presented in the article "Algorithms for stochastic mixed-integer programming models" by Suvrajeet Sen.

Resource constrained scheduling is an example of a class of combinatorial optimization problems that is not naturally formulated with linear constraints so that linear programming based methods do not work well. The article

"Constraint programming" by Alexander Bockmayr and John Hooker presents an alternative enumerative approach that is complementary to branch-and-cut. Constraint programming, primarily designed for feasibility problems, does not use a relaxation to obtain bounds. Instead nodes of the search tree are pruned by constraint propagation, which tightens bounds on variables until their values are fixed or their domains are shown to be empty.

K. Aardal
G.L. Nemhauser
R. Weismantel

K. Aardal et al., Eds., *Handbooks in OR & MS, Vol. 12*

Chapter 1

On the History of Combinatorial Optimization (Till 1960)

Alexander Schrijver[1]

1 Introduction

As a coherent mathematical discipline, combinatorial optimization is relatively young. When studying the history of the field, one observes a number of independent lines of research, separately considering problems like optimum assignment, shortest spanning tree, transportation, and the traveling salesman problem. Only in the 1950's, when the unifying tool of linear and integer programming became available and the area of operations research got intensive attention, these problems were put into one framework, and relations between them were laid.

Indeed, linear programming forms the hinge in the history of combinatorial optimization. Its initial conception by Kantorovich and Koopmans was motivated by combinatorial applications, in particular in transportation and transshipment. After the formulation of linear programming as generic problem, and the development in 1947 by Dantzig of the simplex method as a tool, one has tried to attack about all combinatorial optimization problems with linear programming techniques, quite often very successfully.

A cause of the diversity of roots of combinatorial optimization is that several of its problems descend directly from practice, and instances of them were, and still are, attacked daily. One can imagine that even in very primitive (even animal) societies, finding short paths and searching (for instance, for food) is essential. A traveling salesman problem crops up when you plan shopping or sightseeing, or when a doctor or mailman plans his tour. Similarly, assigning jobs to men, transporting goods, and making connections, form elementary problems not just considered by the mathematician.

It makes that these problems probably can be traced back far in history. In this survey however we restrict ourselves to the mathematical study of these problems. At the other end of the time scale, we do not pass 1960, to keep size

[1]CWI, Kruislaan 413, 1098 SJ Amsterdam, The Netherlands, and Department of Mathematics, University of Amsterdam, Plantage Muidergracht 24, 1018 TV Amsterdam, The Netherlands.

in hand. As a consequence, later important developments, like Edmonds' work on matchings and matroids and Cook and Karp's theory of complexity (NP-completeness) fall out of the scope of this survey.

We focus on six problem areas, in this order: assignment, transportation, maximum flow, shortest tree, shortest path, and the traveling salesman problem.

2 The assignment problem

In mathematical terms, the assignment problem is: given an $n \times n$ 'cost' matrix $C=(c_{i,j})$, find a permutation π of $1,\ldots,n$ for which

$$\sum_{i=1}^{n} c_{i,\pi(i)} \tag{1}$$

is as small as possible.

Monge 1784

The assignment problem is one of the first studied combinatorial optimization problems. It was investigated by G. Monge [1784], albeit camouflaged as a continuous problem, and often called a transportation problem.

Monge was motivated by transporting earth, which he considered as the discontinuous, combinatorial problem of transporting molecules. There are two areas of equal acreage, one filled with earth, the other empty. The question is to move the earth from the first area to the second, in such a way that the total transportation distance is as small as possible. The total transportation distance is the distance over which a molecule is moved, summed over all molecules. Hence it is an instance of the assignment problem, obviously with an enormous cost matrix. Monge described the problem as follows:

> Lorsqu'on doit transporter des terres d'un lieu dans un autre, on a coutime de donner le nom de *Déblai* au volume des terres que l'on doit transporter, & le nom de *Remblai* à l'espace qu'elles doivent occuper après le transport.
>
> Le prix du transport d'une molécule étant, toutes choses d'ailleurs égales, proportionnel à son poids & à l'espace qu'on lui fait parcourir, & par conséquent le prix du transport total devant être proportionnel à la somme des produits des molécules multipliées chacune par l'espace parcouru, il s'ensuit que le déblai & le remblai étant donnés de figure & de position, il n'est pas indifférent que telle molécule du déblai soit transportée dans tel ou tel autre endroit du remblai, mais qu'il y a une

> certaine distribution à faire des molécules du premier dans le second, d'après laquelle la somme de ces produits sera la moindre possible, & le prix du transport total sera un *minimum*.[2]

Monge gave an interesting geometric method to solve this problem. Consider a line that is tangent to both areas, and move the molecule *m* touched in the first area to the position *x* touched in the second area, and repeat, till all earth has been transported. Monge's argument that this would be optimum is simple: if molecule *m* would be moved to another position, then another molecule should be moved to position *x*, implying that the two routes traversed by these molecules cross, and that therefore a shorter assignment exists:

> Étant données sur un même plan deux aires égales *ABCD*, & *abcd*, terminées par des contours quelconques, continus ou discontinus, trouver la route que doit suivre chaque molécule *M* de la premiere, & le point *m* où elle doit arriver dans la seconde, pour que tous les points étant semblablement transportés, ils replissent exactement la seconde aire, & que la somme des produits de chaque molécule multipliée par l'espace parcouru soit un *minimum*.
>
> Si par un point *M* quelconque de la première aire, on mène une droite *Bd*, telle que le segment *BAD* soit égal au segment *bad*, je dis que pour satisfaire à la question, il faut que toutes les molécules du segment *BAD*, soient portées sur le segment *bad*, & que par conséquent les molécules du segment *BCD* soient portées sur le segment égal *bcd*; car si un point *K* quelconque du segment *BAD*, étoit porté sur un point *k* de *bcd*, il faudroit nécessairement qu'un point égal *L*, pris quelque part dans *BCD*, fût transporté dans un certain point *l de bad*, ce qui ne pourroit pas se faire sans que les routes *Kk*, *Ll*, ne se coupassent entre leurs extrémités, & la somme des produits des molécules par les espaces parcourus ne seroit pas un *minimum*. Pareillement, si par un point *M′* infiniment proche du point *M*, on mène la droite *B′d′*, telle qu'on ait encore le segment *B′A′D′*, égal au segment *b′a′d′*, il faut pour que la question soit satisfaite, que les molécules du segment *B′A′D′* soient transportées sur *b′a′d′*. Donc toutes les molécules de l'élément *BB′D′D* doivent être transportées sur l'élément égal *bb′d′d*. Ainsi en divisant le déblai & le remblai en une infinité d'élémens par des droites qui coupent dans l'un & dans l'autre des

[2] When one must transport earth from one place to another, one usually gives the name of *Déblai* to the volume of earth that one must transport, & the name of *Remblai* to the space that they should occupy after the transport.

The price of the transport of one molecule being, if all the rest is equal, proportional to its weight & to the distance that one makes it covering, & hence the price of the total transport having to be proportional to the sum of the products of the molecules each multiplied by the distance covered, it follows that, the déblai & the remblai being given by figure and position, it makes difference if a certain molecule of the déblai is transported to one or to another place of the remblai, but that there is a certain distribution to make of the molecules from the first to the second, after which the sum of these products will be as little as possible, & the price of the total transport will be a *minimum*.

> segmens égaux entr'eux, chaque élément du déblai doit être porté sur l'élément correspondant du remblai.
>
> Les droites *Bd* & *B'd'* étant infiniment proches, il est indifférent dans quel ordre les molécules de l'élément *BB'D'D* se distribuent sur l'élément *bb'd'd*; de quelque manière en effet que se fasse cette distribution, la somme des produits des molécules par les espaces parcourus, est toujours la même, mais si l'on remarque que dans la pratique il convient de débleyer premièrement les parties qui se trouvent sur le passage des autres, & de n'occuper que les dernières les parties du remblai qui sont dans le même cas; la molécule *MM'* ne devra se transporter que lorsque toute la partie *MM'D'D* qui la précède, aura été transportée en *mm'd'd*; donc dans cette hypothèse, si l'on fait *mm'd'd*=*MM'D'D*, le point *m* sera celui sur lequel le point *M* sera transporté.[3]

Although geometrically intuitive, the method is however not fully correct, as was noted by Appell [1928]:

> Il est bien facile de faire la figure de manière que les chemins suivis par les deux parcelles dont parle Monge ne se croisent pas.[4]

(cf. Taton [1951]).

---

[3] Being given, in the same plane, two equal areas *ABCD* & *abcd*, bounded by arbitrary contours, continuous or discontinuous, find the route that every molecule *M* of the first should follow & the point *m* where it should arrive in the second, so that, all points being transported likewise, they fill precisely the second area & so that the sum of the products of each molecule multiplied by the distance covered, is *minimum*.

If one draws a straight line *Bd* through an arbitrary point *M* of the first area, such that the segment *BAD* is equal to the segment *bad*, I assert that, in order to satisfy the question, all molecules of the segment *BAD* should be carried on the segment *bad*, & hence the molecules of the segment *BCD* should be carried on the equal segment *bcd*; for, if an arbitrary point *K* of segment *BAD*, is carried to a point *k* of *bcd*, then necessarily some point *L* somewhere in *BCD* is transported to a certain point *l* in *bad*, which cannot be done without that the routes *Kk*, *Ll* cross each other between their end points, & the sum of the products of the molecules by the distances covered would not be a *minimum*. Likewise, if one draws a straight line *B'd'* through a point *M'* infinitely close to point *M*, in such a way that one still has that segment *B'A'D'* is equal to segment *b'a'd'*, then in order to satisfy the question, the molecules of segment *B'A'D'* should be transported to *b'a'd'*. So all molecules of the element *BB'D'D* must be transported to the equal element *bb'd'd*. Dividing the déblai & the remblai in this way into an infinity of elements by straight lines that cut in the one & in the other segments that are equal to each other, every element of the déblai must be carried to the corresponding element of the remblai.

The straight lines *Bd* & *B'd'* being infinitely close, it does not matter in which order the molecules of element *BB'D'D* are distributed on the element *bb'd'd*; indeed, in whatever manner this distribution is being made, the sum of the products of the molecules by the distances covered is always the same; but if one observes that in practice it is convenient first to dig off the parts that are in the way of others, & only at last to cover similar parts of the remblai; the molecule *MM'* must be transported only when the whole part *MM'D'D* that precedes it will have been transported to *mm'd'd*; hence with this hypothesis, if one has *mm'd'd*=*MM'D'D*, point *m* will be the one to which point *M* will be transported.

[4] It is very easy to make the figure in such a way that the routes followed by the two particles of which Monge speaks, do not cross each other.

Bipartite matching: Frobenius 1912-1917, Kőnig 1915-1931

Finding a largest matching in a bipartite graph can be considered as a special case of the assignment problem. The fundaments of matching theory in bipartite graphs were laid by Frobenius (in terms of matrices and determinants) and Kőnig. We briefly review their work.

In his article *Über Matrizen aus nicht negativen Elementen*, Frobenius [1912] investigated the decomposition of matrices, which led him to the following 'curious determinant theorem':

> *Die Elemente einer Determinante nten Grades seien n^2 unabhängige Veränderliche. Man setze einige derselben Null, doch so, daß die Determinante nicht identisch verschwindet. Dann bleibt sie eine irreduzible Funktion, außer wenn für einen Wert $m < n$ alle Elemente verschwinden, die m Zeilen mit $n - m$ Spalten gemeinsam haben.*[5]

Frobenius gave a combinatorial and an algebraic proof.

In a reaction to this, Dénes Kőnig [1915] realized that Frobenius' theorem can be equivalently formulated in terms of bipartite graphs, by introducing a now quite standard construction of associating a bipartite graph with a matrix $(a_{i,j})$: for each row index i there is a vertex v_i and for each column index j there is a vertex u_j, while vertices v_i and u_j are adjacent if and only if $a_{i,j} \neq 0$. With the help of this, Kőnig gave a proof of Frobenius' result.

According to Gallai [1978], Kőnig was interested in graphs, particularly bipartite graphs, because of his interest in set theory, especially cardinal numbers. In proving Schröder-Bernstein type results on the equicardinality of sets, graph-theoretic arguments (in particular: matchings) can be illustrative. This led Kőnig to studying graphs and their applications in other areas of mathematics.

On 7 April 1914, Kőnig had presented at the *Congrès de Philosophie mathématique* in Paris (cf. Kőnig [1916,1923]) the theorem that each regular bipartite graph has a perfect matching. As a corollary, Kőnig derived that the edge set of any regular bipartite graph can be decomposed into perfect matchings. That is, each k-regular bipartite graph is k-edge-colourable. Kőnig observed that these results follow from the theorem that the edge-colouring number of a bipartite graph is equal to its maximum degree. He gave an algorithmic proof of this.

In order to give an elementary proof of his result described above, Frobenius [1917] proved the following 'Hilfssatz', which now is a fundamental theorem in graph theory:

> II. *Wenn in einer Determinante nten Grades alle Elemente verschwinden, welche p ($\leq n$) Zeilen mit $n - p + 1$ Spalten gemeinsam haben, so verschwinden alle Glieder der entwickelten Determinante.*

[5] *Let the elements of a determinant of degree n be n^2 independent variables. One sets some of them equal to zero, but such that the determinant does not vanish identically. Then it remains an irreducible function, except when for some value $m < n$ all elements vanish that have m rows in common with $n - m$ columns.*

> *Wenn alle Glieder einer Determinante nten Grades verschwinden, so verschwinden alle Elemente, welche p Zeilen mit $n-p+1$ Spalten gemeinsam haben für $p=1$ oder 2, $\cdots$ oder n.*[6]

That is, if $A=(a_{i,j})$ is an $n \times n$ matrix, and for each permutation π of $\{1,\ldots,n\}$ one has $\prod_{i=1}^{n} a_{i,j}=0$, then for some p there exist p rows and $n-p+1$ columns of A such that their intersection is all-zero.

In other words, a bipartite graph $G=(V,E)$ with colour classes V_1 and V_2 satisfying $|V_1|=|V_2|=n$ has a perfect matching, if and only if one cannot select p vertices in V_1 and $n-p+1$ vertices in V_2 such that no edge is connecting two of these vertices.

Frobenius gave a short combinatorial proof (albeit in terms of determinants), and he stated that Kőnig's results follow easily from it. Frobenius also offered his opinion on Kőnig's proof method of his 1912 theorem:

> Die Theorie der Graphen, mittels deren Hr. KŐNIG den obigen Satz abgeleitet hat, ist nach meiner Ansicht ein wenig geeignetes Hilfsmittel für die Entwicklung der Determinantentheorie. In diesem Falle führt sie zu einem ganz speziellen Satze von geringem Werte. Was von seinem Inhalt Wert hat, ist in dem Satze II ausgesprochen.[7]

While Frobenius' result characterizes which bipartite graphs have a perfect matching, a more general theorem characterizing the maximum size of a matching in a bipartite graph was found by Kőnig [1931]:

> Páros körüljárású graphban az éleket kimerítő szögpontok minimális száma megegyezik a páronként közös végpontot nem tartalmazó élek maximális számával.[8]

In other words, the maximum size of a matching in a bipartite graph is equal to the minimum number of vertices needed to cover all edges.

This result can be derived from that of Frobenius [1917], and also from the theorem of Menger [1927] — but, as Kőnig detected, Menger's proof contains an essential hole in the induction basis — see Section 4. This induction basis is precisely the theorem proved by Kőnig.

---

[6] II. *If in a determinant of the nth degree all elements vanish that p ($\leq n$) rows have in common with $n-p+1$ columns, then all members of the expanded determinant vanish.*

If all members of a determinant of degree n vanish, then all elements vanish that p rows have in common with $n-p+1$ columns for $p=1$ or 2, $\cdots$ or n.

[7] The theory of graphs, by which Mr. KŐNIG has derived the theorem above, is to my opinion of little appropriate help for the development of determinant theory. In this case it leads to a very special theorem of little value. What from its contents has value, is enunciated in Theorem II.

[8] In an even circuit graph, the minimal number of vertices that exhaust the edges agrees with the maximal number of edges that pairwise do not contain any common end point.

Egerváry 1931

After the presentation by Kőnig of his theorem at the Budapest Mathematical and Physical Society on 26 March 1931, E. Egerváry [1931] found a weighted version of Kőnig's theorem. It characterizes the maximum *weight* of a matching in a bipartite graph, and thus applies to the assignment problem:

> *Ha az* $\|a_{ij}\|$ *n-edrendű matrix elemei adott nem negatív egész számok, úgy a*
>
> $$\lambda_i + \mu_j \geq a_{ij}, \quad (i, j = 1, 2, \ldots n), \quad (\lambda_i, \mu_j \textit{ nem negatív egész számok})$$
>
> *feltételek mellett*
>
> $$\text{min.} \sum_{k=1}^{n} (\lambda_k + \mu_k) = \text{max.}(a_{1\nu_1} + a_{2\nu_2} + \cdots + a_{n\nu_n}).$$
>
> *hol* $\nu_1, \nu_2, \ldots \nu_n$ *az* $1, 2, \ldots n$ *számok összes permutációit befutják.*[9]

The proof method of Egerváry is essentially algorithmic. Assume that the $a_{i,j}$ are integer. Let λ_i^*, μ_j^* attain the minimum. If there is a permutation ν of $\{1, \ldots, n\}$ such that $\lambda_i^* + \mu_{\nu_i}^* = a_{i,\nu_i}$ for all i, then this permutation attains the maximum, and we have the required equality. If no such permutation exists, by Frobenius' theorem there are subsets I, J of $\{1, \ldots, n\}$ such that

$$\lambda_i^* + \mu_j^* > a_{i,j} \quad \text{for all } i \in I, j \in J \tag{2}$$

and such that $|I| + |J| = n + 1$. Resetting $\lambda_i^* := \lambda_i^* - 1$ if $i \in I$ and $\mu_j^* := \mu_j^* + 1$ if $j \notin J$, would give again feasible values for the λ_i and μ_j, however with their total sum being decreased. This is a contradiction.

Egerváry's theorem and proof method formed, in the 1950's, the impulse for Kuhn to develop a new, fast method for the assignment problem, which he therefore baptized the *Hungarian method.* But first there were some other developments on the assignment problem.

[9] *If the elements of the matrix* $\|a_{ij}\|$ *of order n are given nonnegative integers, then under the assumption*

$$\lambda_i + \mu_j \geq a_{ij}, \quad (i, j = 1, 2, \ldots n), \quad (\lambda_i, \mu_j \textit{ nonnegative integers})$$

we have

$$\text{min.} \sum_{k=1}^{n} (\lambda_k + \mu_k) = \text{max.}(a_{1\nu_1} + a_{2\nu_2} + \cdots + a_{n\nu_n}).$$

where $\nu_1, \nu_2, \ldots \nu_n$ *run over all possible permutations of the numbers* $1, 2, \ldots n$.

Easterfield 1946

The first algorithm for the assignment problem might have been published by Easterfield [1946], who described his motivation as follows:

> In the course of a piece of organisational research into the problems of demobilisation in the R.A.F., it seemed that it might be possible to arrange the posting of men from disbanded units into other units in such a way that they would not need to be posted again before they were demobilised; and that a study of the numbers of men in the various release groups in each unit might enable this process to be carried out with a minimum number of postings. Unfortunately the unexpected ending of the Japanese war prevented the implications of this approach from being worked out in time for effective use. The algorithm of this paper arose directly in the course of the investigation.

Easterfield seems to have worked without knowledge of the existing literature. He formulated and proved a theorem equivalent to Kőnig's theorem and he described a primal-dual type method for the assignment problem from which Egerváry's result given above can be derived. Easterfield's algorithm has running time $O(2^n n^2)$. This is better than scanning all permutations, which takes time $\Omega(n!)$.

Robinson 1949

Cycle reduction is an important tool in combinatorial optimization. In a RAND Report dated 5 December 1949, Robinson [1949] reports that an 'unsuccessful attempt' to solve the traveling salesman problem, led her to the following cycle reduction method for the optimum assignment problem.

Let matrix $(a_{i,j})$ be given, and consider any permutation π. Define for all i, j a 'length' $l_{i,j}$ by: $l_{i,j} := a_{j,\pi(i)} - a_{i,\pi(i)}$ if $j \neq \pi(i)$ and $l_{i,\pi(i)} = \infty$. If there exists a negative-length directed circuit, there is a straightforward way to improve π. If there is no such circuit, then π is an optimal permutation. This clearly is a finite method, and Robinson remarked:

> I believe it would be feasible to apply it to as many as 50 points provided suitable calculating equipment is available.

The simplex method

A breakthrough in solving the assignment problem came when Dantzig [1951a] showed that the assignment problem can be formulated as a linear programming problem that automatically has an integer optimum solution. The reason is a theorem of Birkhoff [1946] stating that the convex hull of the permutation matrices is equal to the set of *doubly stochastic* matrices — nonnegative matrices in which each row and column sum is equal to 1.

Therefore, minimizing a linear functional over the set of doubly stochastic matrices (which is a linear programming problem) gives a permutation matrix, being the optimum assignment. So the assignment problem can be solved with the simplex method.

Votaw [1952] reported that solving a 10×10 assignment problem with the simplex method on the SEAC took 20 minutes. On the other hand, in his reminiscences, Kuhn [1991] mentioned the following:

> The story begins in the summer of 1953 when the National Bureau of Standards and other US government agencies had gathered an outstanding group of combinatorialists and algebraists at the Institute for Numerical Analysis (INA) located on the campus of the University of California at Los Angeles. Since space was tight, I shared an office with Ted Motzkin, whose pioneering work on linear inequalities and related systems predates linear programming by more than ten years. A rather unique feature of the INA was the presence of the Standards Western Automatic Computer (SWAC), the entire memory of which consisted of 256 Williamson cathode ray tubes. The SWAC was faster but smaller than its sibling machine, the Standards Eastern Automatic Computer (SEAC), which boasted a liquid mercury memory and which had been coded to solve linear programs.

According to Kuhn:

> the 10 by 10 assignment problem is a linear program with 100 nonnegative variables and 20 equation constraints (of which only 19 are needed). In 1953, there was no machine in the world that had been programmed to solve a linear program this large!

If 'the world' includes the Eastern Coast of the U.S.A., there seems to be some discrepancy with the remarks of Votaw [1952] mentioned above.

The complexity issue

The assignment problem has helped in gaining the insight that a finite algorithm need not be practical, and that there is a gap between exponential time and polynomial time.

Also in other disciplines it was recognized that while the assignment problem is a finite problem, there is a complexity issue. In an address delivered on 9 September 1949 at a meeting of the American Psychological Association at Denver, Colorado, Thorndike [1950] studied the problem of the 'classification' of personnel (being job assignment):

> The past decade, and particularly the war years, have witnessed a great concern about the classification of personnel and a vast expenditure of effort presumably directed towards this end.

He exhibited little trust in mathematicians:

> There are, as has been indicated, a finite number of permutations in the assignment of men to jobs. When the classification problem as formulated above was presented to a mathematician, he pointed to this fact and said that from the point of view of the mathematician there was no problem. Since the number of permutations was finite, one had only to try them all and choose the best. He dismissed the problem at that point. This is rather cold comfort to the psychologist, however, when one considers that only ten men and ten jobs mean over three and a half million permutations. Trying out all the permutations may be a mathematical solution to the problem, it is not a practical solution.

Thorndike presented three heuristics for the assignment problem, the *Method of Divine Intuition*, the *Method of Daily Quotas*, and the *Method of Predicted Yield*.

(Other heuristic and geometric methods for the assignment problem were proposed by Lord [1952], Votaw and Orden [1952], Törnqvist [1953], and Dwyer [1954] (the 'method of optimal regions').)

Von Neumann considered the complexity of the assignment problem. In a talk in the Princeton University Game Seminar on October 26, 1951, he showed that the assignment problem can be reduced to finding an optimum column strategy in a certain zero-sum two-person game, and that it can be found by a method given by Brown and von Neumann [1950]. We give first the mathematical background.

A zero-sum two-person game is given by a matrix A, the 'pay-off matrix'. The interpretation as a game is that a 'row player' chooses a row index i and a 'column player' chooses simultaneously a column index j. After that, the column player pays the row player $A_{i,j}$. The game is played repeatedly, and the question is what is the best strategy.

Let A have order $m \times n$. A *row strategy* is a vector $x \in \mathbb{R}^m_+$ satisfying $\mathbf{1}^\mathsf{T} x = 1$. Similarly, a *column strategy* is a vector $y \in \mathbb{R}^n_+$ satisfying $\mathbf{1}^\mathsf{T} y = 1$. Then

$$\max_x \min_j (x^\mathsf{T} A)_j = \min_y \max_i (Ay)_i, \tag{3}$$

where x ranges over row strategies, y over column strategies, i over row indices, and j over column indices. Equality (3) follows from LP duality.

It can be derived that the best strategy for the row player is to choose rows with distribution an optimum x in (3). Similarly, the best strategy for the column player is to choose columns with distribution an optimum y in (3). The average pay-off then is the value of (3).

The method of Brown [1951] to determine the optimum strategies is that each player chooses in turn the line that is best with respect to the distribution of the lines chosen by the opponent so far. It was proved by Robinson [1951] that this converges to optimum strategies. The method of Brown and von Neumann [1950] is a continuous version of this, and amounts to solving a system of linear differential equations.

Now von Neumann noted that the following reduces the assignment problem to the problem of finding an optimum column strategy. Let $C=(c_{i,j})$ be an $n \times n$ cost matrix, as input for the assignment problem. We may assume that C is positive. Consider the following pay-off matrix A, of order $2n \times n^2$, with columns indexed by ordered pairs (i,j) with $i,j=1,\ldots,n$. The entries of A are given by: $A_{i,(i,j)}:=1/c_{i,j}$ and $A_{n+j,(i,j)}:=1/c_{i,j}$ for $i,j=1,\ldots,n$, and $A_{k,(i,j)}:=0$ for all i,j,k with $k \neq i$ and $k \neq n+j$. Then any minimum-cost assignment, of cost γ say, yields an optimum column strategy y by: $y_{(i,j)}:=c_{i,j}/\gamma$ if i is assigned to j, and $y_{(i,j)}:=0$ otherwise. Any optimum column strategy is a convex combination of strategies obtained this way from optimum assignments. So an optimum assignment can in principle be found by finding an optimum column strategy.

According to a transcript of the talk (cf. von Neumann [1951,1953]), von Neumann noted the following on the number of steps:

> It turns out that this number is a moderate power of n, i.e., considerably smaller than the "obvious" estimate n! mentioned earlier.

However, no further argumentation is given.

In a Cowles Commission Discussion Paper of 2 April 1953, Beckmann and Koopmans [1953] noted:

> It should be added that in all the assignment problems discussed, there is, of course, the obvious brute force method of enumerating all assignments, evaluating the maximand at each of these, and selecting the assignment giving the highest value. This is too costly in most cases of practical importance, and by a method of solution we have meant a procedure that reduces the computational work to manageable proportions in a wider class of cases.

The Hungarian method: Kuhn 1955-1956, Munkres 1957

The basic combinatorial (nonsimplex) method for the assignment problem is the *Hungarian method*. The method was developed by Kuhn [1955b,1956], based on the work of Egerváry [1931], whence Kuhn introduced the name Hungarian method for it.

In an article "On the origin of the Hungarian method", Kuhn [1991] gave the following reminiscences from the time starting Summer 1953:

> During this period, I was reading Kőnig's classical book on the theory of graphs and realized that the matching problem for a bipartite graph

> on two sets of n vertices was exactly the same as an n by n assignment problem with all $a_{ij}=0$ or 1. More significantly, Kőnig had given a combinatorial algorithm (based on augmenting paths) that produces optimal solutions to the matching problem and its combinatorial (or linear programming) dual. In one of the several formulations given by Kőnig (p. 240, Theorem D), given an n by n matrix $A=(a_{ij})$ with all $a_{ij}=0$ or 1, the maximum number of 1's that can be chosen with no two in the same line (horizontal row or vertical column) is equal to the minimum number of lines that contain all of the 1's. Moreover, the algorithm seemed to be 'good' in a sense that will be made precise later. The problem then was: how could the general assignment problem be reduced to the 0-1 special case?
>
> Reading Kőnig's book more carefully, I was struck by the following footnote (p. 238, footnote 2): "... Eine Verallgemeinerung dieser Sätze gab Egerváry, Matrixok kombinatorius tulajdonságairól (Über kombinatorische Eigenschaften von Matrizen), Matematikai és Fizikai Lapok, *38*, 1931, S. 16-28 (ungarisch mit einem deutschen Auszug) ..." This indicated that the key to the problem might be in Egerváry's paper. When I returned to Bryn Mawr College in the fall, I obtained a copy of the paper together with a large Hungarian dictionary and grammar from the Haverford College library. I then spent two weeks learning Hungarian and translated the paper [1]. As I had suspected, the paper contained a method by which a general assignment problem could be reduced to a finite number of 0-1 assignment problems.
>
> Using Egerváry's reduction and Kőnig's maximum matching algorithm, in the fall of 1953 I solved several 12 by 12 assignment problems (with 3-digit integers as data) by hand. Each of these examples took under two hours to solve and I was convinced that the combined algorithm was 'good'. This must have been one of the last times when pencil and paper could beat the largest and fastest electronic computer in the world.

(Reference [1] is the English translation of the paper of Egerváry [1931].)

The method described by Kuhn is a sharpening of the method of Egerváry sketched above, in two respects: (i) it gives an (augmenting path) method to find either a perfect matching or sets I and J as required, and (ii) it improves the λ_i and μ_j not by 1, but by the largest value possible.

Kuhn [1955b] contented himself with stating that the number of iterations is finite, but Munkres [1957] observed that the method in fact runs in strongly polynomial time ($O(n^4)$).

Ford and Fulkerson [1956b] reported the following computational experience with the Hungarian method:

> The largest example tried was a 20×20 optimal assignment problem. For this example, the simplex method required well over an hour, the present method about thirty minutes of hand computation.

3 The transportation problem

The transportation problem is: given an $m \times n$ 'cost' matrix $C = (a_{i,j})$, a 'supply vector' $b \in \mathbb{R}^m_+$ and a 'demand' vector $d \in \mathbb{R}^n_+$, find a nonnegative $m \times n$ matrix $X = (x_{i,j})$ such that

$$\begin{aligned} &\text{(i)} \quad \sum_{j=1}^{n} x_{i,j} = b_i \text{ for } i = 1, \ldots, m, \\ &\text{(ii)} \quad \sum_{i=1}^{m} x_{i,j} = d_j \text{ for } j = 1, \ldots, n, \\ &\text{(iii)} \quad \sum_{i=1}^{m} \sum_{j=1}^{n} c_{i,j} x_{i,j} \text{ is as small as possible.} \end{aligned} \tag{4}$$

So the transportation problem is a special case of a linear programming problem.

Tolstoĭ 1930

An early study of the transportation problem was made by A.N. Tolstoĭ [1930]. He published, in a book on transportation planning issued by the National Commissariat of Transportation of the Soviet Union, an article called *Methods of finding the minimal total kilometrage in cargo-transportation planning in space*, in which he formulated and studied the transportation problem, and described a number of solution approaches, including the, now well-known, idea that an optimum solution does not have any negative-cost cycle in its residual graph.[10] He might have been the first to observe that the cycle condition is necessary for optimality. Moreover, he assumed, but did not explicitly state or prove, the fact that checking the cycle condition is also sufficient for optimality.

Tolstoĭ illuminated his approach by applications to the transportation of salt, cement, and other cargo between sources and destinations along the railway network of the Soviet Union. In particular, a, for that time large-scale, instance of the transportation problem was solved to optimality.

We briefly review the article here. Tolstoĭ first considered the transportation problem for the case where there are only two sources. He observed that in that case one can order the destinations by the difference between the distances to the two sources. Then one source can provide the destinations starting from the beginning of the list, until the supply of that source has been

[10] The *residual graph* has arcs from each source to each destination, and moreover an arc from a destination to a source if the transport on that connection is positive; the cost of the 'backward' arc is the negative of the cost of the 'forward' arc.

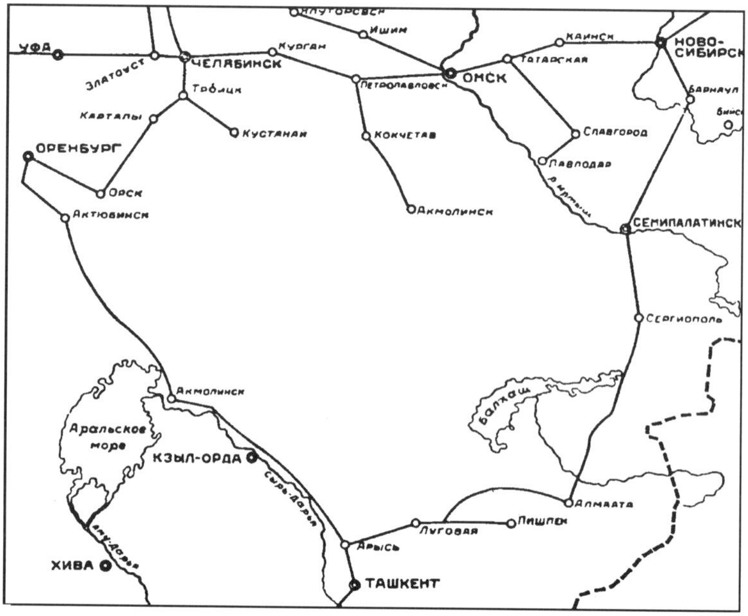

Figure 1. Figure from Tolstoĭ [1930] to illustrate a negative cycle.

used up. The other source supplies the remaining demands. Tolstoĭ observed that the list is independent of the supplies and demands, and hence it

> is applicable for the whole life-time of factories, or sources of production. Using this table, one can immediately compose an optimal transportation plan every year, given quantities of output produced by these two factories and demands of the destinations.

Next, Tolstoĭ studied the transportation problem in the case when all sources and destinations are along one circular railway line (cf. Figure 1), in which case the optimum solution is readily obtained by considering the difference of two sums of costs. He called this phenomenon *circle dependency*.

Finally, Tolstoĭ combined the two ideas into a heuristic to solve a concrete transportation problem coming from cargo transportation along the Soviet railway network. The problem has 10 sources and 68 destinations, and 155 links between sources and destinations (all other distances are taken to be infinite).

Tolstoĭ's heuristic also makes use of insight into the geography of the Soviet Union. He goes along all sources (starting with the most remote sources), where, for each source X, he lists those destinations for which X is the closest source or the second closest source. Based on the difference of the distances to the closest and second closest sources, he assigns cargo from X to the destinations, until the supply of X has been used up. (This obviously is equivalent to considering cycles of length 4.) In case Tolstoĭ foresees a negative-cost cycle in the residual graph, he deviates from this rule to avoid such a cycle. No backtracking occurs.

After 10 steps, when the transports from all 10 factories have been set, Tolstoĭ 'verifies' the solution by considering a number of cycles in the network, and he concludes that his solution is optimum:

> Thus, by use of successive applications of the method of differences, followed by a verification of the results by the circle dependency, we managed to compose the transportation plan which results in the minimum total kilometrage.

The objective value of Tolstoĭ's solution is 395,052 kiloton-kilometers. Solving the problem with modern linear programming tools (CPLEX) shows that Tolstoĭ's solution indeed is optimum. But it is unclear how sure Tolstoĭ could have been about his claim that his solution is optimum. Geographical insight probably has helped him in growing convinced of the optimality of his solution. On the other hand, it can be checked that there exist feasible solutions that have none of the negative-cost cycles considered by Tolstoĭ in their residual graph, but that are yet not optimum.

Later, Tolstoĭ [1939] described similar results in an article entitled *Methods of removing irrational transportations in planning* in the September 1939 issue of *Sotsialisticheskiĭ Transport*. The methods were also explained in the book *Planning Goods Transportation* by Pariĭskaya, Tolstoĭ, and Mots [1947].

According to Kantorovich [1987], there were some attempts to introduce Tolstoĭ's work by the appropriate department of the People's Commissariat of Transport.

Kantorovich 1939

Apparently unaware (by that time) of the work of Tolstoĭ, L.V. Kantorovich studied a general class of problems, that includes the transportation problem. The transportation problem formed the big motivation for studying linear programming. In his memoirs, Kantorovich [1987] wrote how questions from practice motivated him to formulate these problems:

> Once some engineers from the veneer trust laboratory came to me for consultation with a quite skilful presentation of their problems. Different productivity is obtained for veneer-cutting machines for different types of materials; linked to this the output of production of this group of machines depended, it would seem, on the chance factor of which group of raw materials to which machine was assigned. How could this fact be used rationally?
>
> This question interested me, but nevertheless appeared to be quite particular and elementary, so I did not begin to study it by giving up everything else. I put this question for discussion at a meeting of the

> mathematics department, where there were such great specialists as Gyunter, Smirnov himself, Kuz'min, and Tartakovskii. Everyone listened but no one proposed a solution; they had already turned to someone earlier in individual order, apparently to Kuz'min. However, this question nevertheless kept me in suspense. This was the year of my marriage, so I was also distracted by this. In the summer or after the vacation concrete, to some extent similar, economic, engineering, and managerial situations started to come into my head, that also required the solving of a maximization problem in the presence of a series of linear constraints.
>
> In the simplest case of one or two variables such problems are easily solved—by going through all the possible extreme points and choosing the best. But, let us say in the veneer trust problem for five machines and eight types of materials such a search would already have required solving about a billion systems of linear equations and it was evident that this was not a realistic method. I constructed particular devices and was probably the first to report on this problem in 1938 at the October scientific session of the Herzen Institute, where in the main a number of problems were posed with some ideas for their solution.
>
> The universality of this class of problems, in conjunction with their difficulty, made me study them seriously and bring in my mathematical knowledge, in particular, some ideas from functional analysis.
>
> What became clear was both the solubility of these problems and the fact that they were widespread, so representatives of industry were invited to a discussion of my report at the university.

This meeting took place on 13 May 1939 at the Mathematical Section of the Institute of Mathematics and Mechanics of the Leningrad State University. A second meeting, which was devoted specifically to problems connected with construction, was held on 26 May 1939 at the Leningrad Institute for Engineers of Industrial Construction. These meetings provided the basis of the monograph *Mathematical Methods in the Organization and Planning of Production* (Kantorovich [1939]).

According to the Foreword by A.R. Marchenko to this monograph, Kantorovich's work was highly praised by mathematicians, and, in addition, at the special meeting industrial workers unanimously evinced great interest in the work.

In the monograph, the relevance of the work for the Soviet system was stressed:

> I want to emphasize again that the greater part of the problems of which I shall speak, relating to the organization and planning of production, are connected specifically with the Soviet system of economy and in the

> majority of cases do not arise in the economy of a capitalist society. There the choice of output is determined not by the plan but by the interests and profits of individual capitalists. The owner of the enterprise chooses for production those goods which at a given moment have the highest price, can most easily be sold, and therefore give the largest profit. The raw material used is not that of which there are huge supplies in the country, but that which the entrepreneur can buy most cheaply. The question of the maximum utilization of equipment is not raised; in any case, the majority of enterprises work at half capacity.
>
> In the USSR the situation is different. Everything is subordinated not to the interests and advantage of the individual enterprise, but to the task of fulfilling the state plan. The basic task of an enterprise is the fulfillment and overfulfillment of its plan, which is a part of the general state plan. Moreover, this not only means fulfillment of the plan in aggregate terms (i.e. total value of output, total tonnage, and so on), but the certain fulfillment of the plan for all kinds of output; that is, the fulfillment of the assortment plan (the fulfillment of the plan for each kind of output, the completeness of individual items of output, and so on).

One of the problems studied was a rudimentary form of a transportation problem:

given: an $m \times n$ matrix $(c_{i,j})$;
find: an $m \times n$ matrix $(x_{i,j})$ such that:

$$\begin{aligned} &\text{(i)} \quad x_{i,j} \geq 0 \quad \text{for all } i, j; \\ &\text{(ii)} \quad \sum_{i=1}^{m} x_{i,j} = 1 \quad \text{for each } j = 1, \ldots, n; \\ &\text{(iii)} \quad \sum_{j=1}^{n} c_{i,j} x_{i,j} \quad \text{is independent of } i \text{ and is maximized.} \end{aligned} \tag{5}$$

Another problem studied by Kantorovich was 'Problem C' which can be stated as follows:

$$\begin{aligned} \text{maximize} \quad & \lambda \\ \text{subject to} \quad & \sum_{i=1}^{m} x_{i,j} = 1 && (j = 1, \ldots, n) \\ & \sum_{i=1}^{m} \sum_{j=1}^{n} c_{i,j,k} x_{i,j} = \lambda && (k = 1, \ldots, t) \\ & x_{i,j} \geq 0 && (i = 1, \ldots, m; j = 1, \ldots, n). \end{aligned} \tag{6}$$

The interpretation is: let there be n machines, which can do m jobs. Let there be one final product consisting of t parts. When machine i does job j, $c_{i,j,k}$ units of part k are produced ($k = 1, \ldots, t$). Now $x_{i,j}$ is the fraction of time machine i does job j. The number λ is the amount of the final product produced. "Problem C" was later shown (by H.E. Scarf, upon a suggestion by Kantorovich — see Koopmans [1959]) to be equivalent to the general linear programming problem.

Kantorovich outlined a new method to maximize a linear function under given linear inequality constraints. The method consists of determining dual variables ('resolving multipliers') and finding the corresponding primal solution. If the primal solution is not feasible, the dual solution is modified following prescribed rules. Kantorovich indicated the role of the dual variables in sensitivity analysis, and he showed that a feasible solution for Problem C can be shown to be optimal by specifying optimal dual variables.

The method resembles the simplex method, and a footnote in Kantorovich [1987] by his son V.L. Kantorovich suggests that Kantorovich had found the simplex method in 1938:

> In L.V. Kantorovich's archives a manuscript from 1938 is preserved on "Some mathematical problems of the economics of industry, agriculture, and transport" that in content, apparently, corresponds to this report and where, in essence, the simplex method for the machine problem is described.

Kantorovich gave a wealth of practical applications of his methods, which he based mainly in the Soviet plan economy:

> Here are included, for instance, such questions as the distribution of work among individual machines of the enterprise or among mechanisms, the correct distribution of orders among enterprises, the correct distribution of different kinds of raw materials, fuel, and other factors. Both are clearly mentioned in the resolutions of the 18th Party Congress.

He gave the following applications to transportation problems:

> Let us first examine the following question. A number of freights (oil, grain, machines and so on) can be transported from one point to another by various methods; by railroads, by steamship; there can be mixed methods, in part by railroad, in part by automobile transportation, and so on. Moreover, depending on the kind of freight, the method of loading, the suitability of the transportation, and the efficiency of the different kinds of transportation is different. For example, it is particularly advantageous to carry oil by water transportation if oil tankers are available, and so on. The solution of the problem of the distribution of a given freight flow over kinds of transportation, in order to complete the haulage plan in the shortest

time, or within a given period with the least expenditure of fuel, is possible by our methods and leads to Problems A or C.

Let us mention still another problem of different character which, although it does not lead directly to questions A, B, and C, can still be solved by our methods. That is the choice of transportation routes.

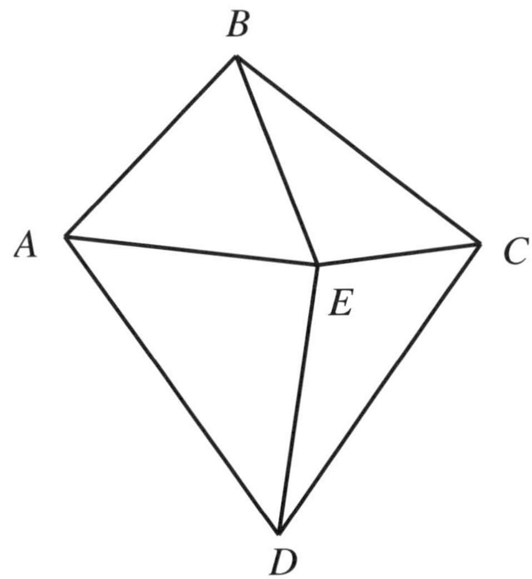

Let there be several points *A*, *B*, *C*, *D*, *E* (Fig. 1) which are connected to one another by a railroad network. It is possible to make the shipments from *B* to *D* by the shortest route *BED*, but it is also possible to use other routes as well: namely, *BCD*, *BAD*. Let there also be given a schedule of freight shipments; that is, it is necessary to ship from *A* to *B* a certain number of carloads, from *D* to *C* a certain number, and so on. The problem consists of the following. There is given a maximum capacity for each route under the given conditions (it can of course change under new methods of operation in transportation). It is necessary to distribute the freight flows among the different routes in such a way as to complete the necessary shipments with a minimum expenditure of fuel, under the condition of minimizing the empty runs of freight cars and taking account of the maximum capacity of the routes. As was already shown, this problem can also be solved by our methods.

As to the reception of his work, Kantorovich [1987] wrote in his memoirs:

The university immediately published my pamphlet, and it was sent to fifty People's Commissariats. It was distributed only in the

> Soviet Union, since in the days just before the start of the World War it came out in an edition of one thousand copies in all.
>
> The number of responses was not very large. There was quite an interesting reference from the People's Commissariat of Transportation in which some optimization problems directed at decreasing the mileage of wagons was considered, and a good review of the pamphlet appeared in the journal "The Timber Industry."
>
> At the beginning of 1940 I published a purely mathematical version of this work in Doklady Akad. Nauk [76], expressed in terms of functional analysis and algebra. However, I did not even put in it a reference to my published pamphlet—taking into account the circumstances I did not want my practical work to be used outside the country.
>
> In the spring of 1939 I gave some more reports—at the Polytechnic Institute and the House of Scientists, but several times met with the objection that the work used mathematical methods, and in the West the mathematical school in economics was an anti-Marxist school and mathematics in economics was a means for apologists of capitalism. This forced me when writing a pamphlet to avoid the term "economic" as much as possible and talk about the organization and planning of production; the role and meaning of the Lagrange multipliers had to be given somewhere in the outskirts of the second appendix and in the semi Aesopian language.

(Here reference [76] is Kantorovich [1940].)

Kantorovich mentions that the new area opened by his work played a definite role in forming the Leningrad Branch of the Mathematical Institute (LOMI), where he worked with M.K. Gavurin on this area. The problem they studied occurred to them by itself, but they soon found out that railway workers were already studying the problem of planning haulage on railways, applied to questions of driving empty cars and transport of heavy cargoes.

Kantorovich and Gavurin developed a method (the method of 'potentials'), which they wrote down in a paper 'Application of mathematical methods in questions of analysis of freight traffic'. This paper was presented in January 1941 to the mathematics section of the Leningrad House of Scientists, but according to Kantorovich [1987] there were political problems in publishing it:

> The publication of this paper met with many difficulties. It had already been submitted to the journal "Railway Transport" in 1940, but because of the dread of mathematics already mentioned it was not printed then either in this or in any other journal, despite the support of Academicians A.N. Kolmogorov and V.N. Obraztsov, a well-known transport specialist and first-rank railway General.

(The paper was finally published as Kantorovich and Gavurin [1949].) Kantorovich [1987] said that he fortunately made an abstract version of the problem, which was published as Kantorovich [1942]. In this, he considered the following generalization of the transportation problem.

Let R be a compact metric space, with two measures μ and μ'. Let $\mathcal{B}$ be the collection of measurable sets in R. A *translocation (of masses)* is a function $\Psi:\mathcal{B}\times\mathcal{B}\to\mathbb{R}_+$ such that for each $X\in\mathcal{B}$ the functions $\Psi(X,.)$ and $\Psi(.,X)$ are measures and such that

$$\Psi(X,R)=\mu(X) \text{ and } \Psi(R,X)=\mu'(X) \tag{7}$$

for each $X\in\mathcal{B}$.

Let a continuous function $r:R\times R\to\mathbb{R}_+$ be given. The value $r(x,y)$ represents the work necessary to transfer a unit mass from x to y. The *work* of a *translocation* Ψ is defined by:

$$\int_R\int_R r(x,y)\Psi(\mathrm{d}\mu,\mathrm{d}\mu'). \tag{8}$$

Kantorovich argued that, if there exists a translocation, then there exists a *minimal* translocation, that is, a translocation Ψ minimizing (8).

He called a translocation Ψ *potential* if there exists a function $p:R\to\mathbb{R}$ such that for all $x,y\in R$:

$$\begin{aligned}&\text{(i)}\quad |p(x)-p(y)|\le r(x,y);\\&\text{(ii)}\quad p(y)-p(x)=r(x,y) \text{ if } \Psi(U_x,U_y)>0\\&\qquad\text{for any neighbourhoods } U_x \text{ and } U_y \text{ of } x \text{ and } y.\end{aligned} \tag{9}$$

Kantorovich showed that a translocation Ψ is minimal if and only if it is potential. This framework applies to the transportation problem (when $m=n$), by taking for R the space $\{1,\ldots,n\}$, with the discrete topology. Kantorovich seems to assume that r satisfies the triangle inequality.

Kantorovich remarked that his method in fact is algorithmic:

> The theorem just demonstrated makes it easy for one to prove that a given mass translocation is or is not minimal. He has only to try and construct the potential in the way outlined above. If this construction turns out to be impossible, i.e. the given translocation is not minimal, he at least will find himself in the possession of the method how to lower the translocation work and eventually come to the minimal translocation.

Kantorovich gave the transportation problem as application:

> Problem 1. *Location of consumption stations with respect to production stations.* Stations $A_1,A_2,\cdots,A_m$, attached to a network of railways

> deliver goods to an extent of $a_1, a_2, \cdots, a_m$ carriages per day respectively. These goods are consumed at stations $B_1, B_2, \cdots, B_n$ of the same network at a rate of $b_1, b_2, \cdots, b_n$ carriages per day respectively ($\sum a_i = \sum b_k$). Given the costs $r_{i,k}$ involved in moving one carriage from station A_i to station B_k, assign the consumption stations such places with respect to the production stations as would reduce the total transport expenses to a minimum.

Kantorovich [1942] also gave a cycle reduction method for finding a minimum-cost transshipment (which is a *uncapacitated* minimum-cost flow problem). He restricted himself to symmetric distance functions.

Kantorovich's work remained unnoticed for some time by Western researchers. In a note introducing a reprint of the article of Kantorovich [1942], in *Management Science* in 1958, the following reassuring remark was made:

> It is to be noted, however, that the problem of determining an effective method of actually acquiring the solution to a specific problem is *not* solved in this paper. In the category of development of such methods we seem to be, currently, ahead of the Russians.

Hitchcock 1941

Independently of Kantorovich, the transportation problem was studied by Hitchcock and Koopmans.

Hitchcock [1941] might be the first giving a precise mathematical description of the problem. The interpretation of the problem is, in Hitchcock's words:

> When several factories supply a product to a number of cities we desire the least costly manner of distribution. Due to freight rates and other matters the cost of a ton of product to a particular city will vary according to which factory supplies it, and will also vary from city to city.

Hitchcock showed that the minimum is attained at a vertex of the feasible region, and he outlined a scheme for solving the transportation problem which has much in common with the simplex method for linear programming. It includes pivoting (eliminating and introducing basic variables) and the fact that nonnegativity of certain dual variables implies optimality. He showed that the *complementary slackness* condition characterizes optimality.

Hitchcock gave a method to find an initial basic solution of (4), now known as the *north-west rule*: set $x_{1,1} := \min\{a_1, b_1\}$; if the minimum is attained by a_1, reset $b_1 := b_1 - a_1$ and recursively find a basic solution $x_{i,j}$ satisfying $\sum_{j=1}^{n} x_{i,j} = a_i$ for each $i = 2, \ldots, m$ and $\sum_{i=2}^{m} x_{i,j} = b_j$ for each $j = 1, \ldots, n$; if the minimum is attained by b_1, proceed symmetrically. (The north-west rule was also described by Salvemini [1939] and Fréchet [1951] in a statistical context, namely in order to complete correlation tables given the marginal distributions.)

Hitchcock however seems to have overlooked the possibility of cycling of his method, although he pointed at an example in which some dual variables are negative while yet the primal solution is optimum.

Koopmans 1942-1948

Koopmans was appointed, in March 1942, as a statistician on the staff of the British Merchant Shipping Mission, and later the Combined Shipping Adjustment Board (CSAB), a British-American agency dealing with merchant shipping problems during the Second World War. Influenced by his teacher J. Tinbergen (cf. Tinbergen [1934]) he was interested in tanker freights and capacities (cf. Koopmans [1939]). Koopmans' wrote in August 1942 in his diary that, while the Board was being organized, there was not much work for the statisticians,

> and I had a fairly good time working out exchange ratio's between cargoes for various routes, figuring how much could be carried monthly from one route if monthly shipments on another route were reduced by one unit.

At the Board he studied the assignment of ships to convoys so as to accomplish prescribed deliveries, while minimizing empty voyages. According to the memoirs of his wife (Wanningen Koopmans [1995]), when Koopmans was with the Board,

> he had been appalled by the way the ships were routed. There was a lot of redundancy, no intensive planning. Often a ship returned home in ballast, when with a little effort it could have been rerouted to pick up a load elsewhere.

In his autobiography (published posthumously), Koopmans [1992] wrote:

> My direct assignment was to help fit information about losses, deliveries from new construction, and employment of British-controlled and U.S.-controlled ships into a unified statement. Even in this humble role I learned a great deal about the difficulties of organizing a large-scale effort under dual control—or rather in this case four-way control, military and civilian cutting across U.S. and U.K. controls. I did my study of optimal routing and the associated shadow costs of transportation on the various routes, expressed in ship days, in August 1942 when an impending redrawing of the lines of administrative control left me temporarily without urgent duties. My memorandum, cited below, was well received in a meeting of the Combined Shipping Adjustment Board (that I did not attend) as an explanation of the "paradoxes of shipping" which were always difficult to explain to higher authority. However, I have no knowledge of any systematic use of my ideas in the combined U.K.-U.S. shipping problems thereafter.

In the memorandum for the Board, Koopmans [1942] analyzed the sensitivity of the optimum shipments for small changes in the demands. In this memorandum (first published in Koopmans' Collected Works), Koopmans did not yet give a method to find an optimum shipment.

Further study led him to a 'local search' method for the transportation problem, stating that it leads to an optimum solution. Koopmans found these results in 1943, but, due to wartime restrictions, published them only after the war (Koopmans [1948], Koopmans and Reiter [1949a,1949b,1951]). Wanningen Koopmans [1995] writes that

> Tjalling said that it had been well received by the CSAB, but that he doubted that it was ever applied.

As Koopmans [1948] wrote:

> Let us now for the purpose of argument (since no figures of war experience are available) assume that one particular organization is charged with carrying out a world dry-cargo transportation program corresponding to the actual cargo flows of 1925. How would that organization solve the problem of moving the empty ships economically from where they become available to where they are needed? It seems appropriate to apply a procedure of trial and error whereby one draws tentative lines on the map that link up the surplus areas with the deficit areas, trying to lay out flows of empty ships along these lines in such a way that a minimum of shipping is at any time tied up in empty movements.

He gave an optimum solution for the following supplies and demands:

Net receipt of dry cargo in overseas trade, 1925
Unit: Millions of metric tons per annum

Harbour	Received	Dispatched	Net receipts
New York	23.5	32.7	−9.2
San Francisco	7.2	9.7	−2.5
St. Thomas	10.3	11.5	−1.2
Buenos Aires	7.0	9.6	−2.6
Antofagasta	1.4	4.6	−3.2
Rotterdam	126.4	130.5	−4.1
Lisbon	37.5	17.0	20.5
Athens	28.3	14.4	13.9
Odessa	0.5	4.7	−4.2
Lagos	2.0	2.4	−0.4
Durban	2.1	4.3	−2.2
Bombay	5.0	8.9	−3.9
Singapore	3.6	6.8	−3.2
Yokohama	9.2	3.0	6.2
Sydney	2.8	6.7	−3.9
Total	266.8	266.8	0.0

So Koopmans solved a 3×12 transportation problem.

Koopmans stated that if no improvement on a solution can be obtained by a cyclic rerouting of ships, then the solution is optimum. It was observed by Robinson [1950] that this gives a finite algorithm.

Koopmans moreover claimed that there exist *potentials* $p_1, \ldots, p_n$ and $q_1, \ldots, q_m$ such that $c_{i,j} \geq p_i - q_j$ for all i, j and such that $c_{i,j} = p_i - q_j$ for each i, j for which any optimum solution x has $x_{i,j} > 0$.

Koopmans and Reiter [1951] investigated the economic implications of the model and the method:

> For the sake of definiteness we shall speak in terms of the transportation of cargoes on ocean-going ships. In considering only shipping we do not lose generality of application since ships may be "translated" into trucks, aircraft, or, in first approximation, trains, and ports into the various sorts of terminals. Such translation is possible because all the above examples involve particular types of movable transportation equipment.

In a footnote they contemplate the application of graphs in economic theory:

> The cultural lag of economic thought in the application of mathematical methods is strikingly illustrated by the fact that linear graphs are making their entrance into transportation theory just about a century after they were first studied in relation to electrical networks, although organized transportation systems are much older than the study of electricity.

Linear programming and the simplex method 1949-1950

The transportation problem was pivotal in the development of the more general problem of linear programming. The simplex method, found in 1947 by G.B. Dantzig, extends the methods of Kantorovich, Hitchcock, and Koopmans. It was published in Dantzig [1951b]. In another paper, Dantzig [1951a] described a direct implementation of the simplex method as applied to the transportation problem.

Votaw and Orden [1952] reported on early computational results (on the SEAC), and claimed (without proof) that the simplex method is polynomial-time for the transportation problem (a statement refuted by Zadeh [1973]):

> As to computation time, it should be noted that for moderate size problems, say $m \times n$ up to 500, the time of computation is of the same order of magnitude as the time required to type the initial data. The computation time on a sample computation in which m and n were both 10 was 3 minutes. The time of computation can be shown by study of the computing method and the code to be proportional to $(m+n)^3$.

The new ideas of applying linear programming to the transportation problem were quickly disseminated, although in some cases applicability to practice was met by scepticism. At a Conference on Linear Programming in May 1954 in London, Land [1954] presented a study of applying linear programming to the problem of transporting coal for the British Coke Industry:

> The real crux of this piece of research is whether the saving in transport cost exceeds the cost of using linear programming.

In the discussion which followed, T. Whitwell of Powers Samas Accounting Machines Ltd remarked

> that in practice one could have one's ideas of a solution confirmed or, much more frequently, completely upset by taking a couple of managers out to lunch.

Alternative methods for the transportation problem were designed by Gleyzal [1955] (a primal-dual method), and by Ford and Fulkerson [1955, 1956a,1956b], Munkres [1957], and Egerváry [1958] (extensions of the Hungarian method for the assignment problem). It was also observed that the problem is a special case of the minimum-cost flow problem, for which several new algorithms were developed — see Section 4.

4 Menger's theorem and maximum flow

Menger's theorem 1927

Menger's theorem forms an important precursor of the max-flow min-cut theorem found in the 1950's by Ford and Fulkerson.

The topologist Karl Menger published his theorem in an article called *Zur allgemeinen Kurventheorie* (On the general theory of curves) (Menger [1927]) in the following form:

> Satz β. *Ist K ein kompakter regulär eindimensionaler Raum, welcher zwischen den beiden endlichen Mengen P und Q n-punktig zusammenhängend ist, dann enthält K n paarweise fremde Bögen, von denen jeder einen Punkt von P und einen Punkt von Q verbindet.*[11]

The result can be formulated in terms of graphs as: Let $G=(V,E)$ be an undirected graph and let $P,Q\subseteq V$. Then the maximum number of disjoint $P-Q$ paths is equal to the minimum cardinality of a set W of vertices such that each $P-Q$ path intersects W.

Menger's interest in this question arose from his research on what he called 'curves': a *curve* is a connected, compact topological space X with the property that for each $x\in X$, each neighbourhood of x contains a neighbourhood of x with totally disconnected boundary.

[11] Theorem β. *If K is a compact regular one-dimensional space which is n-point connected between the two finite sets P and Q, then K contains n disjoint curves, each of which connects a point in P and a point in Q.*

It was however noticed by Kőnig [1932] that Menger's proof of 'Satz β' is incomplete. Menger applied induction on $|E|$, where E is the edge set of the graph G. The basis of the induction is when P and Q contain all vertices. Menger overlooked that this constitutes a nontrivial case. It amounts to the theorem of Kőnig [1931] that in a bipartite graph $G=(V,E)$, the maximum size of a matching is equal to the minimum number of vertices needed to cover all edges. (According to Kőnig [1932], Menger informed him that he was aware of the hole in his proof.)

In his reminiscences on the origin of the 'n-arc theorem', Menger [1981] wrote:

> In the spring of 1930, I came through Budapest and met there a galaxy of Hungarian mathematicians. In particular, I enjoyed making the acquaintance of Dénes Kőnig, for I greatly admired the work on set theory of his father, the late Julius Kőnig — to this day one of the most significant contributions to the continuum problem — and I had read with interest some of Dénes' papers. Kőnig told me that he was about to finish a book that would include all that was known about graphs. I assured him that such a book would fill a great need; and I brought up my n-Arc Theorem which, having been published as a lemma in a curve-theoretical paper, had not yet come to his attention. Kőnig was greatly interested, but did not believe that the theorem was correct. "This evening," he said to me in parting, "I won't go to sleep before having constructed a counterexample." When we met again the next day he greeted me with the words, "A sleepless night!" and asked me to sketch my proof for him. He then said that he would add to his book a final section devoted to my theorem. This he did; and it is largely thanks to Kőnig's valuable book that the n-Arc Theorem has become widely known among graph theorists.

Variants of Menger's theorem 1927-1938

In a paper presented 7 May 1927 to the American Mathematical Society, Rutt [1927,1929] gave the following variant of Menger's theorem, suggested by Kline. Let $G=(V,E)$ be a planar graph and let $s,t\in V$. Then the maximum number of internally disjoint $s-t$ paths is equal to the minimum number of vertices in $V\setminus\{s,t\}$ intersecting each $s-t$ path.

In fact, the theorem follows quite easily from Menger's theorem by deleting s and t and taking for P and Q the sets of neighbours of s and t respectively. (Rutt referred to Menger and gave an independent proof of the theorem.)

This construction was also observed by Knaster [1930] who showed that, conversely, Menger's theorem would follow from Rutt's theorem for general (not necessarily planar) graphs. A similar theorem was published by Nöbeling [1932], using Menger's result.

A result implied by Menger's theorem was presented by Whitney [1932] on 28 February 1931 to the American Mathematical Society: a graph is n-connected if and only if any two vertices are connected by n internally disjoint paths. While referring to the papers of Menger and Rutt, Whitney gave a direct proof.

Other proofs of Menger's theorem were given by Hajós [1934] and Grünwald [1938] (=T. Gallai) — the latter gave an algorithmic proof similar to the flow-augmenting path method for finding a maximum flow of Ford and Fulkerson [1955].

Gallai observed, in a footnote, that the theorem also holds for directed graphs:

> Die ganze Betrachtung lässt sich auch bei orientierten Graphen durchführen und liefert dann eine Verallgemeinerung des Mengerschen Satzes.[12]

Maximum flow 1954

The maximum flow problem is: given a graph, with a 'source' vertex s and a 'terminal' vertex t specified, and given a capacity function c defined on its edges, find a flow from s to t subject to c, of maximum value.

In their basic paper *Maximal Flow through a Network* (published first as a RAND Report of 19 November 1954), Ford and Fulkerson [1954] mentioned that the maximum flow problem was formulated by T.E. Harris as follows:

> Consider a rail network connecting two cities by way of a number of intermediate cities, where each link of the network has a number assigned to it representing its capacity. Assuming a steady state condition, find a maximal flow from one given city to the other.

In their 1962 book *Flows in Networks*, Ford and Fulkerson [1962] give a more precise reference to the origin of the problem[13]:

> It was posed to the authors in the spring of 1955 by T.E. Harris, who, in conjunction with General F.S. Ross (Ret.), had formulated a simplified model of railway traffic flow, and pinpointed this particular problem as the central one suggested by the model [11].

Ford-Fulkerson's reference [11] is a secret report by Harris and Ross [1955] entitled *Fundamentals of a Method for Evaluating Rail Net Capacities*, dated 24 October 1955[14] and written for the US Air Force. At our request, the Pentagon downgraded it to 'unclassified' on 21 May 1999.

[12] The whole consideration lets itself carry out also for oriented graphs and then yields a generalization of Menger's theorem.

[13] There seems to be some discrepancy between the date of the RAND Report of Ford and Fulkerson (19 November 1954) and the date mentioned in the quotation (spring of 1955).

[14] In their book, Ford and Fulkerson incorrectly date the Harris-Ross report 24 October 1956.

In fact, the Harris-Ross report solves a relatively large-scale maximum flow problem coming from the railway network in the Western Soviet Union and Eastern Europe ('satellite countries'). Unlike what Ford and Fulkerson said, the interest of Harris and Ross was not to find a maximum flow, but rather a minimum cut ('interdiction') of the Soviet railway system. We quote:

> Air power is an effective means of interdicting an enemy's rail system, and such usage is a logical and important mission for this Arm.
>
> As in many military operations, however, the success of interdiction depends largely on how complete, accurate, and timely is the commander's information, particularly concerning the effect of his interdiction-program efforts on the enemy's capability to move men and supplies. This information should be available at the time the results are being achieved.
>
> The present paper describes the fundamentals of a method intended to help the specialist who is engaged in estimating railway capabilities, so that he might more readily accomplish this purpose and thus assist the commander and his staff with greater efficiency than is possible at present.

First, much attention is given in the report to modeling a railway network: taking each railway junction as a vertex would give a too refined network (for their purposes). Therefore, Harris and Ross proposed to take 'railway divisions' (organizational units based on geographical areas) as vertices, and to estimate the capacity of the connections between any two adjacent railway divisions. In 1996, Ted Harris remembered (Alexander [1996]):

> We were studying rail transportation in consultation with a retired army general, Frank Ross, who had been chief of the Army's Transportation Corps in Europe. We thought of modeling a rail system as a network. At first it didn't make sense, because there's no reason why the crossing point of two lines should be a special sort of node. But Ross realized that, in the region we were studying, the "divisions" (little administrative districts) should be the nodes. The link between two adjacent nodes represents the total transportation capacity between them. This made a reasonable and manageable model for our rail system. Problems about the effect of cutting links turned out to be linear programming, so we asked for help from George Dantzig and other LP specialists at Rand.

The Harris-Ross report stresses that specialists remain needed to make up the model (which is always a good strategy to get new methods accepted):

> The ability to estimate with relative accuracy the capacity of single railway lines is largely an art. Specialists in this field have no

> authoritative text (insofar as the authors are informed) to guide their efforts, and very few individuals have either the experience or talent for this type of work. The authors assume that this job will continue to be done by the specialist.

The authors next dispute the naive belief that a railway network is just a set of disjoint through lines, and that cutting them implies cutting the network:

> It is even more difficult and time-consuming to evaluate the capacity of a railway network comprising a multitude of rail lines which have widely varying characteristics. Practices among individuals engaged in this field vary considerably, but all consume a great deal of time. Most, if not all, specialists attack the problem by viewing the railway network as an aggregate of through lines.
>
> The authors contend that the foregoing practice does not portray the full flexibility of a large network. In particular it tends to gloss over the fact that even if every one of a set of independent through lines is made inoperative, there may exist alternative routings which can still move the traffic.
>
> This paper proposes a method that departs from present practices in that it views the network as an aggregate of railway operating divisions. All trackage capacities within the divisions are appraised, and these appraisals form the basis for estimating the capability of railway operating divisions to receive trains from and concurrently pass trains to each neighboring division in 24-hour periods.

Whereas experts are needed to set up the model, to solve it is routine (when having the 'work sheets'):

> The foregoing appraisal (accomplished by the expert) is then used in the preparation of comparatively simple work sheets that will enable relatively inexperienced assistants to compute the results and thus help the expert to provide specific answers to the problems, based on many assumptions, which may be propounded to him.

For solving the problem, the authors suggested applying the 'flooding technique', a heuristic described in a RAND Report of 5 August 1955 by A.W. Boldyreff [1955a]. It amounts to pushing as much flow as possible greedily through the network. If at some vertex a 'bottleneck' arises (that is, more trains arrive than can be pushed further through the network), the excess trains are returned to the origin. The technique does not guarantee optimality, but Boldyreff speculates:

> In dealing with the usual railway networks a single flooding, followed by removal of bottlenecks, should lead to a maximal flow.

Presenting his method at an ORSA meeting in June 1955, Boldyreff [1955b] claimed simplicity:

> The mechanics of the solutions is formulated as a simple game which can be taught to a ten-year-old boy in a few minutes.

The well-known flow-augmenting path algorithm of Ford and Fulkerson [1955], that does guarantee optimality, was published in a RAND Report dated only later that year (29 December 1955). As for the simplex method (suggested for the maximum flow problem by Ford and Fulkerson [1954]), Harris and Ross remarked:

> The calculation would be cumbersome; and, even if it could be performed, sufficiently accurate data could not be obtained to justify such detail.

The Harris-Ross report applied the flooding technique to a network model of the Soviet and Eastern European railways. For the data it refers to several secret reports of the Central Intelligence Agency (C.I.A.) on sections of the Soviet and Eastern European railway networks. After the aggregation of railway divisions to vertices, the network has 44 vertices and 105 (undirected) edges.

The application of the flooding technique to the problem is displayed step by step in an appendix of the report, supported by several diagrams of the railway network. (Also work sheets are provided, to allow for future changes in capacities.) It yields a flow of value 163,000 tons from sources in the Soviet Union to destinations in Eastern European 'satellite' countries (Poland, Czechoslovakia, Austria, Eastern Germany), together with a cut with a capacity of, again, 163,000 tons. (This cut is indicated as 'The bottleneck' in Figure 2 from the Harris-Ross report.) So the flow value and the cut capacity are equal, hence optimum.

The max-flow min-cut theorem

In the RAND Report of 19 November 1954, Ford and Fulkerson [1954] gave (next to defining the maximum flow problem and suggesting the simplex method for it) the max-flow min-cut theorem for undirected graphs, saying that the maximum flow value is equal to the minimum capacity of a cut separating source and terminal. Their proof is not constructive, but for planar graphs, with source and sink on the outer boundary, they give a polynomial-time, constructive method. In a report of 26 May 1955, Robacker [1955a] showed that the max-flow min-cut theorem can be derived also from the vertex-disjoint version of Menger's theorem.

As for the directed case, Ford and Fulkerson [1955] observed that the max-flow min-cut theorem holds also for directed graphs. Dantzig and Fulkerson [1955] showed, by extending the results of Dantzig [1951a] on integer solutions for the transportation problem to the maximum flow problem, that

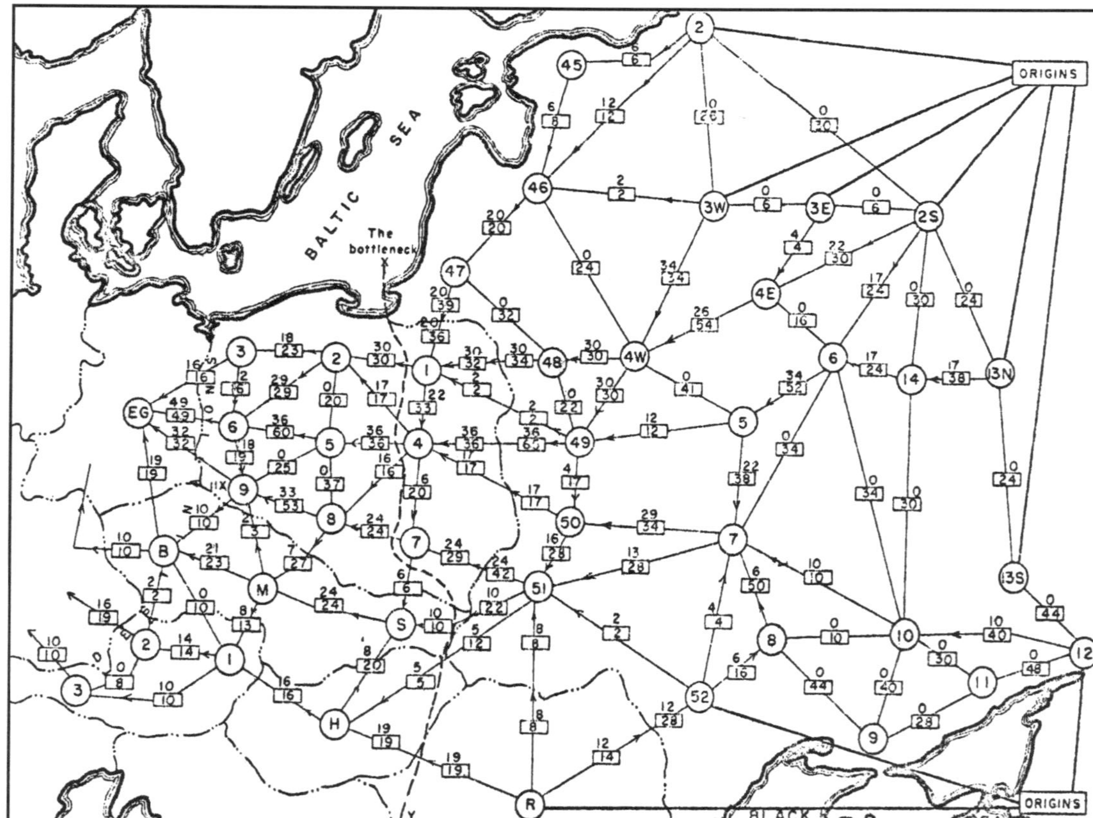

Figure 2. From Harris and Ross [1955]: Schematic diagram of the railway network of the Western Soviet Union and Eastern European countries, with a maximum flow of value 163,000 tons from Russia to Eastern Europe, and a cut of capacity 163,000 tons indicated as 'The bottleneck'.

if the capacities are integer, there is an integer maximum flow (the 'integrity theorem'). Hence, the arc-disjoint version of Menger's theorem for directed graphs follows as a consequence.

Also Kotzig gave the edge-disjoint version of Menger's theorem, but restricted to undirected graphs. In his dissertation for the degree of Academical Doctor, Kotzig [1956] defined, for any undirected graph G and any pair u, v of vertices of G, $\sigma_G(u, v)$ to be the minimum size of a $u-v$ cut. He stated:

> Veta 35. Nech G je l'ubovol'ný graf obsahujúci uzly $u \neq v$, o ktorých platí $\sigma_G(u, v) = k > 0$, potom existuje systém ciest $\{C_1, C_2, \ldots, C_k\}$ taký že každa cesta spojuje uzly u, v a žiadne dve rôzne cesty systému nemajú spoločnej hrany. Takýto systém ciest v G existuje len vtedy, keď je $\sigma_G(u, v) \geq k$.[15]

The proof method is to consider a minimal graph satisfying the cut condition, and next to orient it so as to make a directed graph in which each vertex (except u and v) has indegree equal to outdegree, while u has outdegree k and indegree 0. This then gives the paths.

Although the dissertation has several references to Kőnig's book, which contains the vertex-disjoint version of Menger's theorem, Kotzig did not link his result to that of Menger.

An alternative proof of the max-flow min-cut theorem was given by Elias, Feinstein, and Shannon [1956] ('manuscript received by the PGIT, July 11, 1956'), who claimed that the result was known by workers in communication theory:

> This theorem may appear almost obvious on physical grounds and appears to have been accepted without proof for some time by workers in communication theory. However, while the fact that this flow cannot be exceeded is indeed almost trivial, the fact that it can actually be achieved is by no means obvious. We understand that proofs of the theorem have been given by Ford and Fulkerson and Fulkerson and Dantzig. The following proof is relatively simple, and we believe different in principle.

The proof of Elias, Feinstein, and Shannon is based on a reduction technique similar to that used by Menger [1927] in proving his theorem.

Minimum-cost flows

The *minimum-cost* flow problem was studied, in rudimentary form, by Dantzig and Fulkerson [1954], in order to determine the minimum number

---

[15] Theorem 35. Let G be an arbitrary graph containing vertices $u \neq v$ for which $\sigma_G(u, v) = k > 0$, then there exists a system of paths $\{C_1, C_2, \ldots, C_k\}$ such that each path connects vertices u, v and no two distinct paths have an edge in common. Such a system of paths in G exists only if $\sigma_G(u, v) \geq k$.

of tankers to meet a fixed schedule. Similarly, Bartlett [1957] and Bartlett and Charnes [1957] gave methods to determine the minimum railway stock to run a given schedule.

It was noted by Orden [1955] and Prager [1957] that the minimum-cost flow problem is equivalent to the capacitated transportation problem.

A basic combinatorial minimum-cost flow algorithm was given (in disguised form) by Ford and Fulkerson [1957]. It consists of repeatedly finding a zero-length $s-t$ path in the residual graph, making lengths nonnegative by translating the cost with the help of a potential. If no zero-length path exists, the potential is updated. The complexity of this method was studied in a report by Fulkerson [1958].

5 Shortest spanning tree

The problem of finding the shortest spanning tree came up in several applied areas, like construction of road, energy and communication networks, and in the clustering of data in anthropology and taxonomy.

We refer to Graham and Hell [1985] for an extensive historical survey of shortest tree algorithms, with several quotes (with translations) from old papers. Our notes below have profited from their investigations.

Borůvka 1926

Borůvka [1926a] seems to be the first to consider the shortest spanning tree problem. His interest came from a question of the Electric Power Company of Western Moravia in Brno, at the beginning of the 1920's, asking for the most economical construction of an electric power network (see Borůvka [1977]).

Borůvka formulated the problem as follows:

> In dieser Arbeit löse ich folgendes Problem:
> Es möge eine Matrix der bis auf die Bedingungen $r_{\alpha\alpha}=0$, $r_{\alpha\beta}=r_{\beta\alpha}$ positiven und von einander verschiedenen Zahlen $r_{\alpha\beta}$ $(\alpha,\beta=1,2,\ldots,n;$ $n\geq 2)$ gegeben sein.
>
> Aus dieser ist eine Gruppe von einander und von Null verschiedener Zahlen auszuwählen, so dass
>
> 1° in ihr zu zwei willkürlich gewählten natürlichen Zahlen p_1, $p_2(\leq n)$ eine Teilgruppe von der Gestalt
>
> $$r_{p_1c_2}, r_{c_2c_3}, r_{c_3c_4}, \ldots r_{c_{q-2}c_{q-1}}, r_{c_{q-1}p_2}$$
>
> existiere,

2° die Summe ihrer Glieder kleiner sei als die Summe der Glieder irgendeiner anderen, der Bedingung 1° genügenden Gruppe von einander und von Null verschiedenen Zahlen.[16]

So Borůvka stated that the spanning tree found is the unique shortest. He assumed that all edge lengths are different.

As a method, Borůvka proposed *parallel merging*: connect each component to its nearest neighbouring component, and iterate. His description is somewhat complicated, but in a follow-up paper, Borůvka [1926b] gave an easier description of his method.

Jarník 1929

In a reaction to Borůvka's work, Jarník wrote on 12 February 1929 a letter to Borůvka in which he described a 'new solution of a minimal problem discussed by Mr. Borůvka.'

The 'new solution' amounts to *tree growing*: keep a tree on a subset of the vertices, and iteratively extend it by adding a shortest edge joining the tree with a vertex outside of the tree.

An extract of the letter was published as Jarník [1930]. We quote from the German summary:

a_1 ist eine beliebige unter den Zahlen $1, 2, \ldots, n$.

a_2 ist durch

$$r_{a_1,a_2} = \min_{\left(\substack{l = 1, 2, \ldots, n \\ l \neq a_1}\right)} r_{a_1,l}$$

definiert.

Wenn $2 \leq k < n$ und wenn $[a_1, a_2], \ldots, [a_{2k-3}, a_{2k-2}]$ bereits bestimmt sind, so wird $[a_{2k-1}, a_{2k}]$ durch

$$r_{a_{2k-1},a_{2k}} = \min r_{i,j},$$

[16] In this work, I solve the following problem:
A matrix may be given of positive distinct numbers $r_{\alpha\beta}$ ($\alpha, \beta = 1, 2, \ldots n$; $n \geq 2$), besides the conditions $r_{\alpha\alpha} = 0$, $r_{\alpha\beta} = r_{\beta\alpha}$.
From this, a group of numbers, different from each other and from zero, should be selected such that
1° for arbitrarily chosen natural numbers p_1, p_2 ($\leq n$) a subgroup of it exist of the form

$$r_{p_1 c_2}, r_{c_2 c_3}, r_{c_3 c_4}, \ldots r_{c_{q-2} c_{q-1}}, r_{c_{q-1} p_2},$$

2° the sum of its members be smaller than the sum of the members of any other group of numbers different from each other and from zero, satisfying condition 1°.

> definiert, wo i alle Zahlen $a_1, a_2, \ldots, a_{2k-2}$, j aber alle übrigen von den Zahlen $1, 2, \ldots, n$ durchläuft.[17]

(For a detailed discussion and a translation of the article of Jarník [1930] (and of Jarník and Kössler [1934] on the Steiner tree problem), see Korte and Nešetřil [2001].)

Parallel merging was also described by Choquet [1938] (without proof) and Florek, Łukaszewicz, Perkal, Steinhaus, and Zubrzycki [1951a,1951b]. Choquet gave as a motivation the construction of road systems:

> Étant donné n villes du plan, il s'agit de trouver un réseau de routes permettant d'aller d'une quelconque de ces villes à une autre et tel que:
>
> 1° la longueur globale du réseau soit minimum;
>
> 2° exception faite des villes, on ne peut partir d'aucun point dans plus de deux directions, afin d'assurer la sûreté de la circulation; ceci entraîne, par exemple, que lorsque deux routes semblent se croiser en un point qui n'est pas une ville, elles passent en fait l'une au-dessus de l'autre et ne communiquent pas entre elles en ce point, qu'on appellera faux-croisement.[18]

Choquet might be the first concerned with the complexity of the method:

> Le réseau cherché sera tracé après $2n$ opérations élémentaires au plus, en appelant opération élémentaire la recherche du continu le plus voisin d'un continu donné.[19]

---

[17] a_1 is an arbitrary one among the numbers $1, 2, \ldots, n$.
a_2 is defined by

$$r_{a_1,a_2} = \min_{\begin{pmatrix} l = 1, 2, \ldots, n \\ l \neq a_1 \end{pmatrix}} r_{a_1,l}.$$

If $2 \leq k < n$ and if $[a_1, a_2], \ldots, [a_{2k-3}, a_{2k-2}]$ are determined already, then $[a_{2k-1}, a_{2k}]$ is determined by

$$r_{a_{2k-1},a_{2k}} = \min r_{i,j},$$

where i runs through all numbers $a_1, a_2, \ldots a_{2k-2}$, j however through all remaining of the numbers $1, 2, \ldots, n$.

[18] Being given n cities of the plane, the point is to find a network of routes allowing to go from an arbitrary of these cities to another and such that:
1° the global length of the network be minimum;
2° except for the cities, one cannot depart from any point in more than two directions, in order to assure the certainty of the circulation; this entails, for instance, that when two routes seem to cross each other in a point which is not a city, they pass in fact one above the other and do not communicate among them in this point, which we shall call a false crossing.

[19] The network looked for will be traced after at most $2n$ elementary operations, calling the search for the continuum closest to a given continuum an elementary operation.

Florek *et al.* were motivated by clustering in anthropology, taxonomy, etc. They applied the method to:

> 1° the capitals of Poland's provinces, 2° two collections of excavated skulls, 3° 42 archeological finds, 4° the liverworts of Silesian Beskid mountains with forests as their background, and to the forests of Silesian Beskid mountains with the liverworts appearing in them as their background.

Shortest spanning trees 1956-1959

In the years 1956-1959 a number of papers appeared that again presented methods for the shortest spanning tree problem. Several of the results overlap, also with the earlier papers of Borůvka and Jarník, but also a few new and more general methods were given.

Kruskal [1956] was motivated by Borůvka's first paper and by the application to the traveling salesman problem, described as follows (where [1] is reference Borůvka [1926a]):

> Several years ago a typewritten translation (of obscure origin) of [1] raised some interest. This paper is devoted to the following theorem: If a (finite) connected graph has a positive real number attached to each edge (the *length* of the edge), and if these lengths are all distinct, then among the spanning trees (German: Gerüst) of the graph there is only one, the sum of whose edges is a minimum; that is, the shortest spanning tree of the graph is unique. (Actually in [1] this theorem is stated and proved in terms of the "matrix of lengths" of the graph, that is, the matrix $\|a_{ij}\|$ where a_{ij} is the length of the edge connecting vertices i and j. Of course, it is assumed that $a_{ij} = a_{ji}$ and that $a_{ii} = 0$ for all i and j.)
>
> The proof in [1] is based on a not unreasonable method of constructing a spanning subtree of minimum length. It is in this construction that the interest largely lies, for it is a solution to a problem (Problem 1 below) which on the surface is closely related to one version (Problem 2 below) of the well-known traveling salesman problem.
>
> Problem 1. Give a practical method for constructing a spanning subtree of minimum length.
>
> Problem 2. Give a practical method for constructing an unbranched spanning subtree of minimum length.
>
> The construction in [1] is unnecessarily elaborate. In the present paper I give several simpler constructions which solve Problem 1, and I show how one of these constructions may be used to prove the theorem of [1]. Probably it is true that any construction which solves Problem 1 may be used to prove this theorem.

Kruskal next described three algorithms: Construction A: choose iteratively the shortest edge that can be added so as not to create a circuit; Construction B: fix a nonempty set U of vertices, and choose iteratively the shortest edge leaving some component intersecting U; Construction A′: remove iteratively the longest edge that can be removed without making the graph disconnected.

In his reminiscences, Kruskal [1997] wrote about Borůvka's method:

> In one way, the method of construction was very elegant. In another way, however, it was unnecessarily complicated. A goal which has always been important to me is to find simpler ways to describe complicated ideas, and that is all I tried to do here. I simplified the construction down to its essence, but it seems to me that the idea of Professor Borůvka's method is still present in my version.

Another paper on the minimum spanning tree problem was published by Prim [1957], who was at Bell Laboratories, and who was motivated by the problem of finding a shortest telecommunication network:

> A problem of inherent interest in the planning of large-scale communication, distribution and transportation networks also arises in connection with the current rate structure for Bell System leased-line services.

He described the following algorithm: choose a component of the current forest, and connect it to the nearest other component. He observed that Kruskal's constructions A and B are special cases of this.

Prim noticed that in fact only the order of the lengths determines if a spanning tree is shortest:

> The *shortest spanning subtree* of a connected labelled graph also minimizes all increasing symmetric functions, and maximizes all decreasing symmetric functions, of the edge "lengths."

Prim preferred the tree growing method for computational reasons:

> This computational procedure is easily programmed for an automatic computer so as to handle quite large-scale problems. One of its advantages is its avoidance of checks for closed cycles and connectedness. Another is that it never requires access to more than two rows of distance data at a time — no matter how large the problem.

The implementation described by Prim has $O(n^2)$ running time.

A paper by Loberman and Weinberger [1957] gave minimizing wire connections as motivation:

> In the construction of a digital computer in which high-frequency circuitry is used, it is desirable and often necessary when making connections between terminals to minimize the total wire length in order to reduce the capacitance and delay-line effects of long wire leads.

They described two methods: tree growing and *forest merging*: keep a forest, and iteratively add a shortest edge connecting two components.

Only after they had designed their algorithms, Loberman and Weinberger discovered that their algorithms were given earlier by Kruskal [1956]:

> However, it is felt that the more detailed implementation and general proofs of the procedures justify this paper.

They next described how to implement Kruskal's method, in particular, how to merge forests. And, like Prim, they observed that the minimality of a spanning tree depends only on the order of the lengths, and not on their specific values:

> After the initial sorting into a list where the branches are of monotonically increasing length, the actual value of the length of any branch no longer appears explicitly in the subsequent manipulations. As a result, some other parameter such as the square of the length could have been used. More generally, the same minimum tree will persist for all variations in branch lengths that do not disturb the original relative order.

Dijkstra [1959] gave again the tree growing method, which he prefers (for computational reasons) to the methods given by Kruskal and Loberman and Weinberger (overlooking the fact that these authors also gave the tree growing method):

> The solution given here is to be preferred to the solution given by J.B. KRUSKAL [*1*] and those given by H. LOBERMAN and A. WEINBERGER [*2*]. In their solutions all the — possibly $\frac{1}{2}n(n-1)$ — branches are first of all sorted according to length. Even if the length of the branches is a computable function of the node coordinates, their methods demand that data for all branches are stored simultaneously.

(Dijkstra's references [1] and [2] are Kruskal [1956] and Loberman and Weinberger [1957]). Also Dijkstra described an $O(n^2)$ implementation.

Extension to matroids: Rado 1957

Rado [1957] noticed that the methods of Borůvka and Kruskal can be extended to finding a minimum-weight basis in a matroid. He first showed that if the elements of a matroid are linearly ordered by $<$, there is a unique minimal basis $\{b_1, \ldots, b_r\}$ with $b_1 < b_2 < \cdots < b_r$ such that for each $i = 1, \ldots, r$ all elements $s < b_i$ belong to span $(\{b_1, \ldots, b_{i-1}\})$. Rado derived that for any independent set $\{a_1, \ldots, a_k\}$ with $a_1 < \cdots < a_k$ one has $b_i \leq a_i$ for $i = 1, \ldots, k$. According to Rado, this 'leads to the result of' Borůvka [1926a] and Kruskal [1956].

6 Shortest path

Compared with other combinatorial optimization problems, like shortest spanning tree, assignment and transportation, mathematical research in the shortest path problem started relatively late. This might be due to the fact that the problem is elementary and relatively easy, which is also illustrated by the fact that at the moment that the problem came into the focus of interest, several researchers independently developed similar methods.

Yet, the problem has offered some substantial difficulties. For some considerable period heuristical, nonoptimal approaches have been investigated (cf. for instance Rosenfeld [1956], who gave a heuristic approach for determining an optimal trucking route through a given traffic congestion pattern).

Path finding, in particular searching in a maze, belongs to the classical graph problems, and the classical references are Wiener [1873], Lucas [1882] (describing a method due to C.P. Trémaux), and Tarry [1895] — see Biggs, Lloyd, and Wilson [1976]. They form the basis for depth-first search techniques.

Path problems were also studied at the beginning of the 1950's in the context of 'alternate routing', that is, finding a second shortest route if the shortest route is blocked. This applies to freeway usage (Trueblood [1952]), but also to telephone call routing. At that time making long-distance calls in the U.S.A. was automatized, and alternate routes for telephone calls over the U.S. telephone network nation-wide should be found automatically. Quoting Jacobitti [1955]:

> When a telephone customer makes a long-distance call, the major problem facing the operator is how to get the call to its destination. In some cases, each toll operator has two main routes by which the call can be started towards this destination. The first-choice route, of course, is the most direct route. If this is busy, the second choice is made, followed by other available choices at the operator's discretion. When telephone operators are concerned with such a call, they can exercise choice between alternate routes. But when operator or customer toll dialing is considered, the choice of routes has to be left to a machine. Since the "intelligence" of a machine is limited to previously "programmed" operations, the choice of routes has to be decided upon, and incorporated in, an automatic alternate routing arrangement.

Matrix methods for unit-length shortest path 1946-1953

Matrix methods were developed to study relations in networks, like finding the transitive closure of a relation; that is, identifying in a directed graph the pairs of points s, t such that t is reachable from s. Such methods were studied because of their application to communication nets (including neural nets) and to animal sociology (e.g. peck rights).

The matrix methods consist of representing the directed graph by a matrix, and then taking iterative matrix products to calculate the transitive closure. This was studied by Landahl and Runge [1946], Landahl [1947], Luce and Perry [1949], Luce [1950], Lunts [1950, 1952], and by A. Shimbel.

Shimbel's interest in matrix methods was motivated by their applications to neural networks. He analyzed with matrices which sites in a network can communicate to each other, and how much time it takes. To this end, let S be the 0,1 matrix indicating that if $S_{i,j}=1$ then there is direct communication from i to j (including $i=j$). Shimbel [1951] observed that the positive entries in S^t correspond to pairs between which there exists communication in t steps. An *adequate* communication system is one for which the matrix S^t is positive for some t. One of the other observations of Shimbel [1951] is that in an adequate communication system, the time it takes that all sites have all information, is equal to the minimum value of t for which S^t is positive. (A related phenomenon was observed by Luce [1950].)

Shimbel [1953] mentioned that the distance from i to j is equal to the number of zeros in the i,j position in the matrices $S^0, S^1, S^2, \dots, S^t$. So essentially he gave an $O(n^4)$ algorithm to find all distances in a directed graph with *unit lengths*.

Shortest-length paths

If a directed graph $D=(V,A)$ and a length function $l: A \rightarrow \mathbb{R}$ are given, one may ask for the distances and shortest-length paths from a given vertex s.

For this, there are two well-known methods: the 'Bellman-Ford method' and 'Dijkstra's method'. The latter one is faster but is restricted to nonnegative length functions. The former method only requires that there is no directed circuit of negative length.

The general framework for both methods is the following scheme, described in this general form by Ford [1956]. Keep a provisional distance function d. Initially, set $d(s):=0$ and $d(v):=\infty$ for each $v\neq s$. Next, iteratively,

$$\begin{array}{c}\text{choose an arc } (u, v) \text{ with } d(v) > d(u) + l(u, v) \\ \text{and reset } d(v) := d(u) + l(u, v).\end{array} \tag{10}$$

If no such arc exists, d is the distance function.

The difference in the methods is the rule by which the arc (u, v) with $d(v) > d(u)+l(u, v)$ is chosen. The Bellman-Ford method consists of considering all arcs consecutively and applying (10) where possible, and repeating this (at most $|V|$ rounds suffice). This is the method described by Shimbel [1955], Bellman [1958], and Moore [1959].

Dijkstra's method prescribes to choose an arc (u, v) with $d(u)$ smallest (then each arc is chosen at most once, if the lengths are nonnegative). This was described by Leyzorek, Gray, Johnson, Ladew, Meaker, Petry, and Seitz [1957] and Dijkstra [1959]. A related method, but slightly slower than Dijkstra's method when implemented, was given by Dantzig [1958], and chooses an arc (u, v) with $d(u)+l(u, v)$ smallest.

Parallel to this, a number of further results were obtained on the shortest path problem, including a linear programming approach and 'good characterizations'. We review the articles in a more or less chronological order.

Shimbel 1955

The paper of Shimbel [1955] was presented in April 1954 at the Symposium on Information Networks in New York. Extending his matrix methods for unit-length shortest paths, he introduced the following 'min-sum algebra':

> Arithmetic
>
> For any arbitrary real or infinite numbers x and y
>
> $$x + y \equiv \min(x, y) \quad \text{and}$$
> $$xy \equiv \text{the algebraic sum of } x \text{ and } y.$$

He transferred this arithmetic to the matrix product. Calling the distance matrix associated with a given length matrix S the 'dispersion', he stated:

> It follows trivially that S^k $k \geq 1$ is a matrix giving the shortest paths from site to site in S given that $k-1$ other sites may be traversed in the process. It also follows that for any S there exists an integer k such that $S^k = S^{k+1}$. Clearly, the dispersion of S (let us label it $D(S)$) will be the matrix S^k such that $S^k = S^{k+1}$.

This is equivalent to the Bellman-Ford method.

Although Shimbel did not mention it, one trivially can take $k \leq |V|$, and hence the method yields an $O(n^4)$ algorithm to find the distances between all pairs of points.

Shortest path as linear programming problem 1955-1957

Orden [1955] observed that the shortest path problem is a special case of a transshipment problem (=uncapacitated minimum-cost flow problem), and hence can be solved by linear programming. Dantzig [1957] described the following graphical procedure for the simplex method applied to this problem. Let T be a rooted spanning tree on $\{1, \ldots, n\}$, with root 1. For each $i = 1, \ldots, n$, let u_i be equal to the length of the path from 1 to i in T. Now if $u_j \leq u_i + d_{i,j}$ for all i, j, then for each i, the $1-i$ path in T is a shortest path.

If $u_j > u_i + d_{i,j}$, replace the arc of T entering j by the arc (i,j), and iterate with the new tree.

Trivially, this process terminates (as $\sum_{j=1}^{n} u_j$ decreases at each iteration, and as there are only finitely many rooted trees). Dantzig illustrated his method by an example of sending a package from Los Angeles to Boston. (Edmonds [1970] showed that this method may take exponential time.)

In a reaction to the paper of Dantzig [1957], Minty [1957] proposed an 'analog computer' for the shortest path problem:

> Build a string model of the travel network, where knots represent cities and string lengths represent distances (or costs). Seize the knot 'Los Angeles' in your left hand and the knot 'Boston' in your right and pull them apart. If the model becomes entangled, have an assistant untie and re-tie knots until the entanglement is resolved. Eventually one or more paths will stretch tight — they then are alternative shortest routes.
>
> Dantzig's 'shortest-route tree' can be found in this model by weighting the knots and picking up the model by the knot 'Los Angeles'.
>
> It is well to label the knots since after one or two uses of the model their identities are easily confused.

A similar method was proposed by Bock and Cameron [1958].

Ford 1956

In a RAND report dated 14 August 1956, Ford [1956] described a method to find a shortest path from P_0 to P_N, in a network with vertices $P_0, \ldots, P_N$, where l_{ij} denotes the length of an arc from i to j. We quote:

> Assign initially $x_0 = 0$ and $x_i = \infty$ for $i \neq 0$. Scan the network for a pair P_i and P_j with the property that $x_i - x_j > l_{ji}$. For this pair replace x_i by $x_j + l_{ji}$. Continue this process. Eventually no such pairs can be found, and x_N is now minimal and represents the minimal distance from P_0 to P_N.

So this is the general scheme described above ((10)). No selection rule for the arc (u, v) in (10) is prescribed by Ford.

Ford showed that the method terminates. It was shown however by Johnson [1973a,1973b,1977] that Ford's liberal rule can take exponential time.

The correctness of Ford's method also follows from a result given in the book *Studies in the Economics of Transportation* by Beckmann, McGuire, and Winsten [1956]: given a length matrix $(l_{i,j})$, the distance matrix is the unique matrix $(d_{i,j})$ satisfying

$$\begin{aligned} d_{i,i} &= 0 \text{ for all } i; \\ d_{i,k} &= \min_j (l_{i,j} + d_{j,k}) \text{ for all } i, k \text{ with } i \neq k. \end{aligned} \tag{11}$$

Good characterizations for shortest path 1956-1958

It was noticed by Robacker [1956] that shortest paths allow a theorem dual to Menger's theorem: the minimum length of a $P_0 - P_n$ path in a graph N is equal to the maximum number of pairwise disjoint $P_0 - P_n$ cuts. In Robacker's words:

> the maximum number of mutually disjunct cuts of N is equal to the length of the shortest chain of N from P_0 to P_n.

A related 'good characterization' was found by Gallai [1958]: A length function $l: A \to \mathbb{Z}$ on the arcs of a directed graph (V, A) does not give negative-length directed circuits, if and only if there is a function ('potential') $p: V \to \mathbb{Z}$ such that $l(u, v) \geq p(v) - p(u)$ for each arc (u, v).

Case Institute of Technology 1957

The shortest path problem was also investigated by a group of researchers at the Case Institute of Technology in Cleveland, Ohio, in the project *Investigation of Model Techniques*, performed for the Combat Development Department of the Army Electronic Proving Ground. In their *First Annual Report*, Leyzorek, Gray, Johnson, Ladew, Meaker, Petry, and Seitz [1957] presented their results.

First, they noted that Shimbel's method can be speeded up by calculating S^k by iteratively raising the current matrix to the square (in the min-sum matrix algebra). This solves the all-pairs shortest path problem in time $O(n^3 \log n)$.

Next, they gave a rudimentary description of a method equivalent to Dijkstra's method. We quote:

> (1) All the links joined to the origin, a, may be given an outward orientation. . . .
>
> (2) Pick out the link or links radiating from a, $a_{a\alpha}$, with the smallest delay. . . . Then it is impossible to pass from the origin to any other node in the network by any "shorter" path than $a_{a\alpha}$. Consequently, the minimal path to the general node α is $a_{a\alpha}$.
>
> (3) All of the other links joining α may now be directed outward. Since $a_{a\alpha}$ must necessarily be the minimal path to α, there is no advantage to be gained by directing any other links toward α. . . .
>
> (4) Once α has been evaluated, it is possible to evaluate immediately all other nodes in the network whose minimal values do not exceed the value of the second-smallest link radiating from the origin. Since the minimal values of these nodes are less than the values of the second-smallest, third-smallest, and all other links radiating directly from the origin, only the smallest link, $a_{a\alpha}$, can form a part of the minimal path

> to these nodes. Once a minimal value has been assigned to these nodes, it is possible to orient all other links except the incoming link in an outward direction.
>
> (5) Suppose that all those nodes whose minimal values do not exceed the value of the second-smallest link radiating from the origin have been evaluated. Now it is possible to evaluate the node on which the second-smallest link terminates. At this point, it can be observed that if conflicting directions are assigned to a link, in accordance with the rules which have been given for direction assignment, that link may be ignored. It will not be a part of the minimal path to either of the two nodes it joins. ...
>
> Following these rules, it is now possible to expand from the second-smallest link as well as the smallest link so long as the value of the third-smallest link radiating from the origin is not exceeded. It is possible to proceed in this way until the entire network has been solved.

(In this quotation we have deleted sentences referring to figures.)

Bellman 1958

After having published several papers on dynamic programming (which is, in some sense, a generalization of shortest path methods), Bellman [1958] eventually focused on the shortest path problem by itself, in a paper in the *Quarterly of Applied Mathematics.* He described the following 'functional equation approach' for the shortest path problem, which is the same as that of Shimbel [1955].

There are N cities, numbered $1, \ldots, N$, every two of which are linked by a direct road. A matrix $T=(t_{i,j})$ is given, where $t_{i,j}$ is time required to travel from i to j (not necessarily symmetric). Find a path between 1 and N which consumes minimum time.

Bellman remarked:

> Since there are only a finite number of paths available, the problem reduces to choosing the smallest from a finite set of numbers. This direct, or enumerative, approach is impossible to execute, however, for values of N of the order of magnitude of 20.

He gave a "functional equation approach"

> The basic method is that of successive approximations. We choose an initial sequence $\{f_i^{(0)}\}$, and then proceed iteratively, setting
>
> $$f_i^{(k+1)} = \operatorname*{Min}_{j \neq i}(t_{ij} + f_j^{(k)}], \quad i = 1, 2, \cdots, N-1,$$
> $$f_N^{(k+1)} = 0,$$
>
> for $k = 0, 1, 2 \cdots,$.

As initial function $f_i^{(0)}$ Bellman proposed (upon a suggestion of F. Haight) to take $f_i^{(0)} = t_{i,N}$ for all i. Bellman noticed that, for each fixed i, starting with this choice of $f_i^{(0)}$ gives that $f_i^{(k)}$ is monotonically nonincreasing in k, and stated:

> It is clear from the physical interpretation of this iterative scheme that at most $(N-1)$ iterations are required for the sequence to converge to the solution.

Since each iteration can be done in time $O(N^2)$, the algorithm takes time $O(N^3)$. As for the complexity, Bellman said:

> It is easily seen that the iterative scheme discussed above is a feasible method for either hand or machine computation for values of N of the order of magnitude of 50 or 100.

In a footnote, Bellman mentioned:

> *Added in proof* (*December* 1957): After this paper was written, the author was informed by Max Woodbury and George Dantzig that the particular iterative scheme discussed in Sec. 5 had been obtained by them from first principles.

Dantzig 1958

The paper of Dantzig [1958] gives an $O(n^2 \log n)$ algorithm for the shortest path problem with nonnegative length function. It consists of choosing in (10) an arc with $d(u) + l(u, v)$ as small as possible. Dantzig assumed

> (a) that one can write down without effort for each node the arcs leading to other nodes in increasing order of length and (b) that it is no effort to ignore an arc of the list if it leads to a node that has been reached earlier.

He mentioned that, beside Bellman, Moore, Ford, and himself, also D. Gale and D.R. Fulkerson proposed shortest path methods, 'in informal conversations'.

Dijkstra 1959

Dijkstra [1959] gave a concise and clean description of 'Dijkstra's method', yielding an $O(n^2)$-time implementation. Dijkstra stated:

> The solution given above is to be preferred to the solution by L.R. FORD [*3*] as described by C. BERGE [*4*], for, irrespective of the number of branches, we need not store the data for all branches simultaneously but only those for the branches in sets I and II, and this number is

> always less than n. Furthermore, the amount of work to be done seems to be considerably less.

(Dijkstra's references [3] and [4] are Ford [1956] and Berge [1958].)

Dijkstra's method is easier to implement (as an $O(n^2)$ algorithm) than Dantzig's, since we do not need to store the information in lists: in order to find a next vertex v minimizing $d(v)$, we can just scan all vertices.

Moore 1959

At the International Symposium on the Theory of Switching at Harvard University in April 1957, Moore [1959] of Bell Laboratories, presented a paper "The shortest path through a maze":

> The methods given in this paper require no foresight or ingenuity, and hence deserve to be called algorithms. They would be especially suited for use in a machine, either a special-purpose or a general-purpose digital computer.

The motivation of Moore was the routing of toll telephone traffic. He gave algorithms A, B, C, and D.

First, Moore considered the case of an undirected graph $G=(V,E)$ with no length function, in which a path from vertex A to vertex B should be found with a minimum number of edges. Algorithm A is: first give A label 0. Next do the following for $k=0,1,\ldots$: give label $k+1$ to all unlabeled vertices that are adjacent to some vertex labeled k. Stop as soon as vertex B is labeled.

> If it were done as a program on a digital computer, the steps given as single steps above would be done serially, with a few operations of the computer for each city of the maze; but, in the case of complicated mazes, the algorithm would still be quite fast compared with trial-and-error methods.

In fact, a direct implementation of the method would yield an algorithm with running time $O(m)$. Algorithms B and C differ from A in a more economical labeling (by fewer bits).

Moore's algorithm D finds a shortest route for the case where each edge of the graph has a nonnegative length. This method is a refinement of Bellman's method described above: (i) it extends to the case that not all pairs of vertices have a direct connection; that is, if there is an underlying graph $G=(V,E)$ with length function; (ii) at each iteration only those $d_{i,j}$ are considered for which u_i has been decreased at the previous iteration.

The method has running time $O(nm)$. Moore observed that the algorithm is suitable for parallel implementation, yielding a decrease in running time bound to $O(n\Delta(G))$, where $\Delta(G)$ is the maximum degree of G. Moore concluded:

> The origin of the present methods provides an interesting illustration of the value of basic research on puzzles and games. Although such research is often frowned upon as being frivolous, it seems plausible that these algorithms might eventually lead to savings of very large sums of money by permitting more efficient use of congested transportation or communication systems. The actual problems in communication and transportation are so much complicated by timetables, safety requirements, signal-to-noise ratios, and economic requirements that in the past those seeking to solve them have not seen the basic simplicity of the problem, and have continued to use trial-and-error procedures which do not always give the true shortest path. However, in the case of a simple geometric maze, the absence of these confusing factors permitted algorithms *A*, *B*, and *C* to be obtained, and from them a large number of extensions, elaborations, and modifications are obvious.
>
> The problem was first solved in connection with Claude Shannon's maze-solving machine. When this machine was used with a maze which had more than one solution, a visitor asked why it had not been built to always find the shortest path. Shannon and I each attempted to find economical methods of doing this by machine. He found several methods suitable for analog computation, and I obtained these algorithms. Months later the applicability of these ideas to practical problems in communication and transportation systems was suggested.

Among the further applications of his method, Moore described the example of finding the fastest connections from one station to another in a given railroad timetable. A similar method was given by Minty [1958].

In May 1958, Hoffman and Pavley [1959] reported, at the Western Joint Computer Conference in Los Angeles, the following computing time for finding the distances between all pairs of vertices by Moore's algorithm (with nonnegative lengths):

> It took approximately three hours to obtain the minimum paths for a network of 265 vertices on an IBM *704*.

7 The traveling salesman problem

The *traveling salesman problem* (TSP) is: given n cities and their intermediate distances, find a shortest route traversing each city exactly once. Mathematically, the traveling salesman problem is related to, in fact generalizes, the question for a Hamiltonian circuit in a graph. This question goes back to Kirkman [1856] and Hamilton [1856,1858] and was also studied by Kowalewski [1917a,1917b] — see Biggs, Lloyd, and Wilson [1976]. We restrict our survey to the traveling salesman problem in its general form.

The mathematical roots of the traveling salesman problem are obscure. Dantzig, Fulkerson, and Johnson [1954] say:

> It appears to have been discussed informally among mathematicians at mathematics meetings for many years.

A 1832 manual

The traveling salesman problem has a natural interpretation, and Müller-Merbach [1983] detected that the problem was formulated in a 1832 manual for the successful traveling salesman, *Der Handlungsreisende — wie er sein soll und was er zu thun hat, um Aufträge zu erhalten und eines glücklichen Erfolgs in seinen Geschäften gewiß zu sein — von einem alten Commis-Voyageur*[20] [1832]. (Whereas the politically correct nowadays prefer to speak of the traveling salesperson problem, the manual presumes that the 'Handlungsreisende' is male, and it warns about the risks of women in or out of business.)

The booklet contains no mathematics, and formulates the problem as follows:

> Die Geschäfte führen die Handlungsreisenden bald hier, bald dort hin, und es lassen sich nicht füglich Reisetouren angeben, die für alle vorkommende Fälle passend sind; aber es kann durch eine zweckmäßige Wahl und Eintheilung der Tour, manchmal so viel Zeit gewonnen werden, daß wir es nicht glauben umgehen zu dürfen, auch hierüber einige Vorschriften zu geben. Ein Jeder möge so viel davon benutzen, als er es seinem Zwecke für dienlich hält; so viel glauben wir aber davon versichern zu dürfen, daß es nicht wohl thunlich sein wird, die Touren durch Deutschland in Absicht der Entfernungen und, worauf der Reisende hauptsächlich zu sehen hat, des Hin- und Herreisens, mit mehr Oekonomie einzurichten. Die Hauptsache besteht immer darin: so viele Orte wie möglich mitzunehmen, ohne den nämlichen Ort zweimal berühren zu müssen.[21]

The manual suggests five tours through Germany (one of them partly through Switzerland). In Figure 3 we compare one of the tours with a shortest

[20] "The traveling salesman — how he should be and what he has to do, to obtain orders and to be sure of a happy success in his business — by an old traveling salesman."

[21] Business brings the traveling salesman now here, then there, and no travel routes can be properly indicated that are suitable for all cases occurring; but sometimes, by an appropriate choice and arrangement of the tour, so much time can be gained, that we don't think we may avoid giving some rules also on this. Everybody may use that much of it, as he takes it for useful for his goal; so much of it however we think we may assure, that it will not be well feasible to arrange the tours through Germany with more economy in view of the distances and, which the traveler mainly has to consider, of the trip back and forth. The main point always consists of visiting as many places as possible, without having to touch the same place twice.

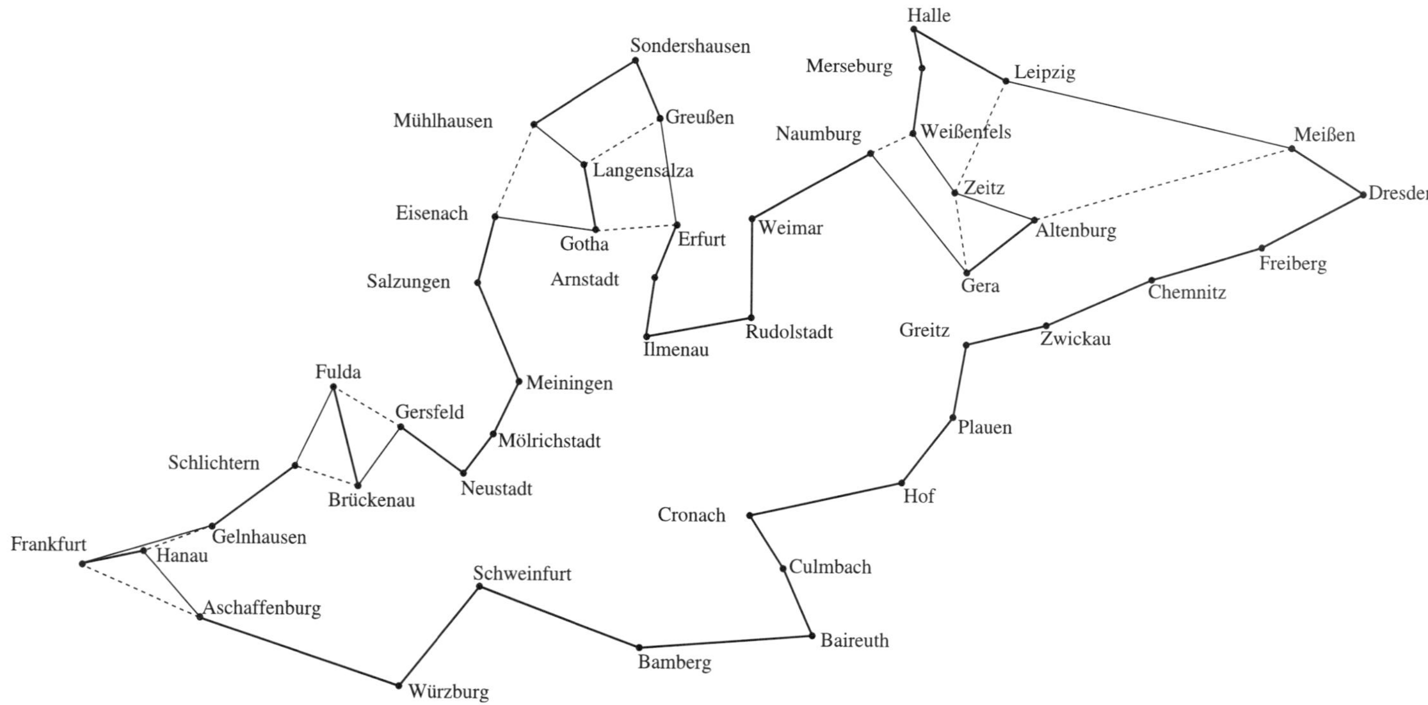

Figure 3. A tour along 45 German cities, as described in the 1832 traveling salesman manual, is given by the unbroken (bold and thin) lines (1285 km). A shortest tour is given by the unbroken bold and by the dashed lines (1248 km). We have taken geodesic distances — taking local conditions into account, the 1832 tour might be optimum.

tour, found with 'modern' methods. (Most other tours given in the manual do not qualify for 'die Hauptsache' as they contain subtours, so that some places are visited twice.)

Menger's Botenproblem 1930

K. Menger seems to be the first mathematician to have written about the traveling salesman problem. The root of his interest is given in his paper Menger [1928b]. In this, he studies the *length* $l(C)$ of a simple curve C in a metric space S, which is, by definition,

$$l(C) := \sup \sum_{i=1}^{n-1} \operatorname{dist}(x_i, x_{i+1}), \tag{12}$$

where the supremum ranges over all choices of $x_1, \ldots, x_n$ on C *in the order determined by* C. What Menger showed is that we may relax this to finite subsets X of C and minimize over all possible orderings of X. To this end he defined, for any finite subset X of a metric space, $\lambda(X)$ to be the shortest length of a path through X (in graph terminology: a *Hamiltonian path*), and he showed that

$$l(C) = \sup_X \lambda(X), \tag{13}$$

where the supremum ranges over all finite subsets X of C. It amounts to showing that for each $\varepsilon > 0$ there is a finite subset X of C such that $\lambda(X) \geq l(C) - \varepsilon$.

Menger [1929a] sharpened this to:

$$l(C) = \sup_X \kappa(X), \tag{14}$$

where again the supremum ranges over all finite subsets X of C, and where $\kappa(X)$ denotes the minimum length of a *spanning tree* on X.

These results were reported also in Menger [1930]. In a number of other papers, Menger [1928a,1929b,1929a] gave related results on these new characterizations of the length function.

The parameter $\lambda(X)$ clearly is close to the practical application of the traveling salesman problem. This relation was mentioned explicitly by Menger in the session of 5 February 1930 of his *mathematisches Kolloquium* in Vienna (organized at the desire of some students). According to the report in Menger [1931a,1932], he first asked if a further relaxation is possible by replacing $\kappa(X)$ by the minimum length of an (in current terminology) *Steiner tree* connecting X — a spanning tree on a superset of X in S. (So Menger toured along some basic combinatorial optimization problems.) This problem was solved for Euclidean spaces by Mimura [1933].

Next Menger posed the traveling salesman problem, as follows:

> Wir bezeichnen als *Botenproblem* (weil diese Frage in der Praxis von jedem Postboten, übrigens auch von vielen Reisenden zu lösen ist) die Aufgabe, für endlichviele Punkte, deren paarweise Abstände bekannt sind, den kürzesten die Punkte verbindenden Weg zu finden. Dieses Problem ist natürlich stets durch endlichviele Versuche lösbar. Regeln, welche die Anzahl der Versuche unter die Anzahl der Permutationen der gegebenen Punkte herunterdrücken würden, sind nicht bekannt. Die Regel, man solle vom Ausgangspunkt erst zum nächstgelegenen Punkt, dann zu dem diesem nächstgelegenen Punkt gehen usw., liefert im allgemeinen nicht den kürzesten Weg.[22]

So Menger asked for a shortest Hamiltonian path through the given points. He was aware of the complexity issue in the traveling salesman problem, and he knew that the now well-known nearest neighbour heuristic might not give an optimum solution.

Harvard, Princeton 1930-1934

Menger spent the period September 1930-February 1931 as visiting lecturer at Harvard University. In one of his seminar talks at Harvard, Menger presented his results on lengths of arcs and shortest paths through finite sets of points quoted above. According to Menger [1931b], a suggestion related to this was given by Hassler Whitney, who at that time did his Ph.D. research in graph theory at Harvard. This paper however does not mention if the practical interpretation was given in the seminar talk.

The year after, 1931-1932, Whitney was a National Research Council Fellow at Princeton University, where he gave a number of seminar talks. In a seminar talk, he mentioned the problem of finding the shortest route along the 48 States of America.

There are some uncertainties in this story. It is not sure if Whitney spoke about the 48 States problem during his 1931-1932 seminar talks (which talks he did give), or later, in 1934, as is said by Flood [1956] in his article on the traveling salesman problem:

> This problem was posed, in 1934, by Hassler Whitney in a seminar talk at Princeton University.

[22] We denote by *messenger problem* (since in practice this question should be solved by each postman, anyway also by many travelers) the task to find, for finitely many points whose pairwise distances are known, the shortest route connecting the points. Of course, this problem is solvable by finitely many trials. Rules which would push the number of trials below the number of permutations of the given points, are not known. The rule that one first should go from the starting point to the closest point, then to the point closest to this, etc., in general does not yield the shortest route.

That memory can be shaky might be indicated by the following two quotes. Dantzig, Fulkerson, and Johnson [1954] remark:

> Both Flood and A.W. Tucker (Princeton University) recall that they heard about the problem first in a seminar talk by Hassler Whitney at Princeton in 1934 (although Whitney, recently queried, does not seem to recall the problem).

However, when asked by David Shmoys, Tucker replied in a letter of 17 February 1983 (see Hoffman and Wolfe [1985]):

> I cannot confirm or deny the story that I heard of the TSP from Hassler Whitney. If I did (as Flood says), it would have occurred in 1931-32, the first year of the old Fine Hall (now Jones Hall). That year Whitney was a postdoctoral fellow at Fine Hall working on Graph Theory, especially planarity and other offshoots of the 4-color problem. ... I was finishing my thesis with Lefschetz on *n*-manifolds and Merrill Flood was a first year graduate student. The Fine Hall Common Room was a very lively place — 24 hours a day.

(Whitney finished his Ph.D. at Harvard University in 1932.)

Another uncertainty is in which form Whitney has posed the problem. That he might have focused on finding a shortest route along the 48 states in the U.S.A., is suggested by the reference by Flood, in an interview on 14 May 1984 with Tucker [1984], to the problem as the "48 States Problem of Hassler Whitney". In this respect Flood also remarked:

> I don't know who coined the peppier name 'Traveling Salesman Problem' for Whitney's problem, but that name certainly has caught on, and the problem has turned out to be of very fundamental importance.

TSP, Hamiltonian paths, and school bus routing

Flood [1956] mentioned a number of connections of the TSP with Hamiltonian games and Hamiltonian paths in graphs, and continues:

> I am indebted to A.W. Tucker for calling these connections to my attention, in 1937, when I was struggling with the problem in connection with a schoolbus routing study in New Jersey.

In the following quote from the interview by Tucker [1984], Flood referred to school bus routing in a different state (West Virginia), and he mentioned the involvement in the TSP of Koopmans, who spent 1940-1941 at the Local Government Surveys Section of Princeton University ("the Princeton Surveys"):

> Koopmans first became interested in the "48 States Problem" of Hassler Whitney when he was with me in the Princeton Surveys,

> as I tried to solve the problem in connection with the work by Bob Singleton and me on school bus routing for the State of West Virginia.

1940

In 1940, some papers appeared that study the traveling salesman problem, in a different context. They seem to be the first containing mathematical results on the problem.

In the American continuation of Menger's *mathematisches Kolloquium*, Menger [1940] returned to the question of the shortest path through a given set of points in a metric space, followed by investigations of Milgram [1940] on the shortest Jordan curve that covers a given, not necessarily finite, set of points in a metric space. As the set may be infinite, a shortest curve need not exist.

Fejes [1940] investigated the problem of a shortest curve through n points in the unit square. In consequence of this, Verblunsky [1951] showed that its length is less than $2 + \sqrt{2.8n}$. Later work in this direction includes Few [1955] and Beardwood, Halton, and Hammersley [1959].

Lower bounds on the expected value of a shortest path through n random points in the plane were studied by Mahalanobis [1940] in order to estimate the cost of a sample survey of the acreage under jute in Bengal. This survey took place in 1938 and one of the major costs in carrying out the survey was the transportation of men and equipment from one survey point to the next. He estimated (without proof) the minimum length of a tour along n random points in the plane, for Euclidean distance:

> It is also easy to see in a general way how the journey time is likely to behave. Let us suppose that n sampling units are scattered at random within any given area; and let us assume that we may treat each such sample unit as a geometrical point. We may also assume that arrangements will usually be made to move from one sample point to another in such a way as to keep the total distance travelled as small as possible; that is, we may assume that the path traversed in going from one sample point to another will follow a straight line. In this case it is easy to see that the mathematical expectation of the total length of the path travelled in moving from one sample point to another will be $(\sqrt{n} - 1/\sqrt{n})$. The cost of the journey from sample to sample will therefore be roughly proportional to $(\sqrt{n} - 1/\sqrt{n})$. When n is large, that is, when we consider a sufficiently large area, we may expect that the time required for moving from sample to sample will be roughly proportional to $\sqrt{n}$, where n is the total number of samples in the given area. If we consider the journey time per sq. mile, it will be roughly proportional to $\sqrt{y}$, where y is the density of number of sample units per sq. mile.

This research was continued by Jessen [1942], who estimated empirically a similar result for l_1-distance (Manhattan distance), in a statistical investigation of a sample survey for obtaining farm facts in Iowa:

> If a route connecting y points located at random in a fixed area is minimized, the total distance, D, of that route is[23]
>
> $$D = d\left(\frac{y-1}{\sqrt{y}}\right)$$
>
> where d is a constant.
>
> This relationship is based upon the assumption that points are connected by direct routes. In Iowa the road system is a quite regular network of mile square mesh. There are very few diagonal roads, therefore, routes between points resemble those taken on a checker-board. A test wherein several sets of different members of points were located at random on an Iowa county road map, and the minimum distance of travel from a given point on the border of the county through all the points and to an end point (the county border nearest the last point on route), revealed that
>
> $$D = d\sqrt{y}$$
>
> works well. Here y is the number of randomized points (border points not included). This is of great aid in setting up a cost function.

Marks [1948] gave a proof of Mahalanobis' bound. In fact he showed that $\sqrt{\frac{1}{2}A}(\sqrt{n} - 1/\sqrt{n})$ is a lower bound, where A is the area of the region. Ghosh [1949] showed that asymptotically this bound is close to the expected value, by giving a heuristic for finding a tour, yielding an upper bound of $1.27\sqrt{An}$. He also observed the complexity of the problem:

> After locating the n random points in a map of the region, it is very difficult to find out *actually* the shortest path connecting the points, unless the number n is very small, which is seldom the case for a large-scale survey.

TSP, transportation, and assignment

As is the case for many other combinatorial optimization problems, the RAND Corporation in Santa Monica, California, played an important role in the research on the TSP. Hoffman and Wolfe [1985] write that

> John Williams urged Flood in 1948 to popularize the TSP at the RAND Corporation, at least partly motivated by the purpose of

[23] At this point, Jessen referred in a footnote to Mahalanobis [1940].

> creating intellectual challenges for models outside the theory of games. In fact, a prize was offered for a significant theorem bearing on the TSP. There is no doubt that the reputation and authority of RAND, which quickly became the intellectual center of much of operations research theory, amplified Flood's advertizing.

At RAND, researchers considered the idea of transferring the successful methods for the transportation problem to the traveling salesman problem. Flood [1956] mentioned that this idea was brought to his attention by Koopmans in 1948. In the interview with Tucker [1984], Flood remembered:

> George Dantzig and Tjallings Koopmans met with me in 1948 in Washington, D.C., at the meeting of the International Statistical Institute, to tell me excitedly of their work on what is now known as the linear programming problem and with Tjallings speculating that there was a significant connection with the Traveling Salesman Problem.

(This meeting was in fact held 6–18 September 1947.)

The issue was taken up in a RAND Report by Julia Robinson [1949], who, in an 'unsuccessful attempt' to solve the traveling salesman problem, considered, as a relaxation, the assignment problem, for which she found a cycle reduction method. The relation is that the assignment problem asks for an optimum permutation, and the TSP for an optimum *cyclic* permutation.

Robinson's RAND report might be the earliest mathematical reference using the term 'traveling salesman problem':

> The purpose of this note is to give a method for solving a problem related to the traveling salesman problem. One formulation is to find the shortest route for a salesman starting from Washington, visiting all the state capitals and then returning to Washington. More generally, to find the shortest closed curve containing n given points in the plane.

Flood wrote (in a letter of 17 May 1983 to E.L. Lawler) that Robinson's report stimulated several discussions on the TSP with his research assistant at RAND, D.R. Fulkerson, during 1950-1952.[24]

It was noted by Beckmann and Koopmans [1952] that the TSP can be formulated as a quadratic assignment problem, for which however no fast methods are known.

Dantzig, Fulkerson, and Johnson 1954

Fundamental progress on the traveling salesman was made in a seminal paper by the RAND researchers Dantzig, Fulkerson, and Johnson

[24] Fulkerson started at RAND only in March 1951.

[1954] — according to Hoffman and Wolfe [1985] 'one of the principal events in the history of combinatorial optimization'. The paper introduced several new methods for solving the traveling salesman problem that are now basic in combinatorial optimization. In particular, it shows the importance of *cutting planes* for combinatorial optimization.

By a theorem of Birkhoff [1946], the convex hull of the $n \times n$ permutation matrices is precisely the set of doubly stochastic matrices — nonnegative matrices with all row and column sums equal to 1. In other words, the convex hull of the permutation matrices is determined by:

$$x_{i,j} \geq 0 \text{ for all } i, j; \qquad \sum_{j=1}^{n} x_{i,j} = 1 \text{ for all } i; \quad \sum_{i=1}^{n} x_{i,j} = 1 \text{ for all } j. \tag{15}$$

This makes it possible to solve the assignment problem as a linear programming problem. It is tempting to try the same approach to the traveling salesman problem. For this, one needs a description in linear inequalities of the *traveling salesman polytope* — the convex hull of the *cyclic* permutation matrices. To this end, one may add to (15) the following *subtour elimination constraints*:

$$\sum_{i \in I, j \notin I} x_{i,j} \geq 1 \quad \text{for each } I \subseteq \{1, \ldots, n\} \quad \text{with } \emptyset \neq I \neq \{1, \ldots, n\}. \tag{16}$$

However, while these inequalities are enough to cut off the noncyclic permutation matrices from the polytope of doubly stochastic matrices, they yet do not yield all facets of the traveling salesman polytope (if $n \geq 5$), as was observed by Heller [1953a]: there exist doubly stochastic matrices, of any order $n \geq 5$, that satisfy (16) but are not a convex combination of cyclic permutation matrices.

The inequalities (16) can nevertheless be useful for the TSP, since we obtain a lower bound for the optimum tour length if we minimize over the constraints (15) and (16). This lower bound can be calculated with the simplex method, taking the (exponentially many) constraints (16) as *cutting planes* that can be added during the process when needed. In this way, Dantzig, Fulkerson, and Johnson were able to find the shortest tour along cities chosen in the 48 U.S. states and Washington, D.C. Incidentally, this is close to the problem mentioned by Julia Robinson in 1949 (and maybe also by Whitney in the 1930's).

The Dantzig-Fulkerson-Johnson paper does not give an algorithm, but rather gives a tour and proves its optimality with the help of the subtour elimination constraints. This work forms the basis for most of the later work on large-scale traveling salesman problems.

Early studies of the traveling salesman polytope were made by Heller [1953a,1953b,1955a,1956b,1955b,1956a], Kuhn [1955a], Norman [1955], and Robacker [1955b], who also made computational studies of the probability that a random instance of the traveling salesman problem needs the constraints (16) (cf. Kuhn [1991]). This made Flood [1956] remark on the intrinsic complexity of the traveling salesman problem:

> Very recent mathematical work on the traveling-salesman problem by I. Heller, H.W. Kuhn, and others indicates that the problem is fundamentally complex. It seems very likely that quite a different approach from any yet used may be required for succesful treatment of the problem. In fact, there may well be no general method for treating the problem and impossibility results would also be valuable.

Flood mentioned a number of other applications of the traveling salesman problem, in particular in machine scheduling, brought to his attention in a seminar talk at Columbia University in 1954 by George Feeney.

Other work on the traveling salesman problem in the 1950's was done by Morton and Land [1955] (a linear programming approach with a 3-exchange heuristic), Barachet [1957] (a graphic solution method), Bock [1958], Croes [1958] (a heuristic), and Rossman and Twery [1958]. In a reaction to Barachet's paper, Dantzig, Fulkerson, and Johnson [1959] showed that their method yields the optimality of Barachet's (heuristically found) solution.

Acknowledgements. I thank Sasha Karzanov for his efficient help in finding Tolstoĭ's and several other papers in the (former) Lenin Library in Moscow, Irina V. Karzanova for accurately providing me with an English translation of Tolstoĭ's 1930 paper, Alexander Rosa for sending me a copy of Kotzig's thesis and for providing me with translations of excerpts of it, András Frank and Tibor Jordán for translating parts of Hungarian articles, Adri Steenbeek and Bill Cook for finding the shortest traveling salesman tour along the 45 German towns from the 1832 manual, Karin van Gemert and Wouter Mettrop at CWI's Library for providing me with bibliographic information and copies of numerous papers, Alfred B. Lehman for giving me copies of old reports of the Case Institute of Technology, Jan Karel Lenstra for giving me copies of letters of Albert Tucker to David Shmoys and of Merrill M. Flood to Eugene L. Lawler on TSP history, Alan Hoffman and David Williamson for helping me to understand Gleyzal's paper on transportation, Steve Brady (RAND) and Dick Cottle for their help in obtaining classical RAND Reports, Kim H. Campbell and Joanne McLean at Air Force Pentagon for declassifying the Harris-Ross report, Richard Bancroft and Gustave Shubert at RAND Corporation for their mediation in this, Bruno Simeone for sending me Salvemini's paper, and Truus Wanningen Koopmans for imparting to me her "Stories and Memories" and quotations from the diary of Tj.C. Koopmans.

References

[1996] K.S. Alexander, A conversation with Ted Harris, *Statistical Science* 11 (1996) 150–158.

[1928] P. Appell, *Le problème géométrique des déblais et remblais*, [Mémorial des Sciences Mathématiques XXVII], Gauthier-Villars, Paris, 1928.

[1957] L.L. Barachet, Graphic solution to the traveling-salesman problem, *Operations Research* 5 (1957) 841–845.

[1957] T.E. Bartlett, An algorithm for the minimum number of transport units to maintain a fixed schedule, *Naval Research Logistics Quarterly* 4 (1957) 139–149.

[1957] T.E. Bartlett, A. Charnes, [Cyclic scheduling and combinatorial topology: assignment and routing of motive power to meet scheduling and maintenance requirements]. Part II Generalization and analysis, *Naval Research Logistics Quarterly* 4 (1957) 207–220.

[1959] J. Beardwood, J.H. Halton, J.M. Hammersley, The shortest path through many points, *Proceedings of the Cambridge Philosophical Society* 55 (1959) 299–327.

[1952] M. Beckmann, T.C. Koopmans, *A Note on the Optimal Assignment Problem*, Cowles Commission Discussion Paper: Economics 2053, Cowles Commission for Research in Economics, Chicago, Illinois, [October 30] 1952.

[1953] M. Beckmann, T.C. Koopmans, *On Some Assignment Problems*, Cowles Commission Discussion Paper: Economics No. 2071, Cowles Commission for Research in Economics, Chicago, Illinois, [April 2] 1953.

[1956] M. Beckmann, C.B. McGuire, C.B. Winsten, *Studies in the Economics of Transportation*, Cowles Commission for Research in Economics, Yale University Press, New Haven, Connecticut, 1956.

[1958] R. Bellman, On a routing problem, *Quarterly of Applied Mathematics* 16 (1958) 87–90.

[1958] C. Berge, *Théorie des graphes et ses applications*, Dunod, Paris, 1958.

[1976] N.L. Biggs, E.K. Lloyd, R.J. Wilson, *Graph Theory 1736–1936*, Clarendon Press, Oxford, 1976.

[1946] G. Birkhoff, Tres observaciones sobre el algebra lineal, *Revista Facultad de Ciencias Exactas, Puras y Aplicadas Universidad Nacional de Tucuman, Serie A (Matematicas y Fisica Teorica)* 5 (1946) 147–151.

[1958] F. Bock, An algorithm for solving "travelling-salesman" and related network optimization problems [abstract], *Operations Research* 6 (1958) 897.

[1958] F. Bock, S. Cameron, Allocation of network traffic demand by instant determination of optimum paths [paper Presented at the 13th National (6th Annual) Meeting of the Operations Research Society of America, Boston, Massachusetts, 1958], *Operations Research* 6 (1958) 633–634.

[1955a] A.W. Boldyreff, *Determination of the Maximal Steady State Flow of Traffic through a Railroad Network*, Research Memorandum RM-1532, The RAND Corporation, Santa Monica, California, [5 August] 1955, [Published in *Journal of the Operations Research Society of America* 3 (1955) 443–465].

[1955b] A.W. Boldyreff, The gaming approach to the problem of flow through a traffic network [abstract of lecture presented at the Third Annual Meeting of the Society, New York, June 3–4, 1955], *Journal of the Operations Research Society of America* 3 (1955) 360.

[1926a] O. Borůvka, O jistém problému minimálním [Czech, with German summary; On a minimal problem], *Práce Moravské Přírodovědecké Společnosti Brno [Acta Societatis Scientiarum Naturalium Moravi[c]ae]* 3 (1926) 37–58.

[1926b] O. Borůvka, Příspěvek k řešení otázky ekonomické stavby elektrovodných sítí [Czech; Contribution to the solution of a problem of economical construction of electrical networks], *Elektrotechnický Obzor* 15:10 (1926) 153–154.

[1977] O. Borůvka, Několik vzpomínek na matematický život v Brně, *Pokroky Matematiky, Fyziky, a Astronomie* 22 (1977) 91–99.

[1951] G.W. Brown, Iterative solution of games by fictitious play, in: *Activity Analysis of Production and Allocation — Proceedings of a Conference* (Proceedings Conference on Linear Programming, Chicago, Illinois, 1949; Tj.C. Koopmans, ed.), Wiley, New York, 1951, pp. 374–376.

[1950] G.W. Brown, J. von Neumann, Solutions of games by differential equations, in: *Contributions to the Theory of Games* (H.W. Kuhn, A.W. Tucker, eds.) [Annals of Mathematics Studies 24], Princeton University Press, Princeton, New Jersey, 1950, pp. 73–79.

[1938] G. Choquet, Étude de certains réseaux de routes, *Comptes Rendus Hebdomadaires des Séances de l'Académie des Sciences* 206 (1938) 310–313.

[1832] ["ein alter Commis-Voyageur"], *Der Handlungsreisende — wie er sein soll und was er zu thun hat, um Aufträge zu erhalten und eines glücklichen Erfolgs in seinen Geschäften gewiß zu sein — Von einem alten Commis-Voyageur*, B.Fr. Voigt, Ilmenau, 1832 [reprinted: Verlag Bernd Schramm, Kiel, 1981].

[1958] G.A. Croes, A method for solving traveling-salesman problems, *Operations Research* 6 (1958) 791–812.

[1951a] G.B. Dantzig, Application of the simplex method to a transportation problem, in: *Activity Analysis of Production and Allocation — Proceedings of a Conference* (Proceedings Conference on Linear Programming, Chicago, Illinois, 1949; Tj.C. Koopmans, ed.), Wiley, New York, 1951, pp. 359–373.

[1951b] G.B. Dantzig, Maximization of a linear function of variables subject to linear inequalities, in: *Activity Analysis of Production and Allocation — Proceedings of a Conference* (Proceedings Conference on Linear Programming, Chicago, Illinois, 1949; Tj. C. Koopmans, ed.), Wiley, New York, 1951, pp. 339–347.

[1957] G.B. Dantzig, Discrete-variable extremum problems, *Operations Research* 5 (1957) 266–277.

[1958] G.B. Dantzig, *On the Shortest Route through a Network*, Report P-1345, The RAND Corporation, Santa Monica, California, [April 12] 1958 [Revised April 29, 1959] [published in *Management Science* 6 (1960) 187–190].

[1954] G.B. Dantzig, D.R. Fulkerson, *Notes on Linear Programming: Part XV — Minimizing the Number of Carriers to Meet a Fixed Schedule*, Research Memorandum RM-1328, The RAND Corporation, Santa Monica, California, [24 August] 1954 [published in *Naval Research Logistics Quarterly* 1 (1954) 217–222].

[1956] G.B. Dantzig, D.R. Fulkerson, *On the Max Flow Min Cut Theorem of Networks*, Research Memorandum RM-1418, The RAND Corporation, Santa Monica, California, [1 January] 1955 [revised: Research Memorandum RM-1418-1 (= Paper P-826), The RAND Corporation, Santa Monica, California, [15 April] 1955 [published in: *Linear Inequalities and Related Systems* (H.W. Kuhn, A.W. Tucker, eds.) [Annals of Mathematics Studies 38], Princeton University Press, Princeton, New Jersey, 1956, pp. 215–221]].

[1954] G. Dantzig, R. Fulkerson, S. Johnson, *Solution of a Large Scale Traveling Salesman Problem*, Paper P-510, The RAND Corporation, Santa Monica, California, [12 April] 1954 [published in *Journal of the Operations Research Society of America* 2 (1954) 393–410].

[1959] G.B. Dantzig, D.R. Fulkerson, S.M. Johnson, *On a Linear-Programming-Combinatorial Approach to the Traveling-Salesman Problem: Notes on Linear Programming and Extensions-Part 49*, Research Memorandum RM-2321, The RAND Corporation, Santa Monica, California, 1959 [published in *Operations Research* 7 (1959) 58–66].

[1959] E.W. Dijkstra, A note on two problems in connexion with graphs, *Numerische Mathematik* 1 (1959) 269–271.

[1954] P.S. Dwyer, Solution of the personnel classification problem with the method of optimal regions, *Psychometrika* 19 (1954) 11–26.

[1946] T.E. Easterfield, A combinatorial algorithm, *The Journal of the London Mathematical Society* 21 (1946) 219–226.

[1970] J. Edmonds, *Exponential growth of the simplex method for shortest path problems*, manuscript [University of Waterloo, Waterloo, Ontario], 1970.

[1931] J. Egerváry, Matrixok kombinatorius tulajdonságairól [Hungarian, with German summary], *Matematikai és Fizikai Lapok* 38 (1931) 16–28. [English translation [by H.W. Kuhn]: On combinatorial properties of matrices, Logistics Papers, George Washington University, issue 11 (1955), paper 4, pp. 1–11].

[1958] E. Egerváry, Bemerkungen zum Transportproblem, *MTW Mitteilungen* 5 (1958) 278–284.

[1956] P. Elias, A. Feinstein, C.E. Shannon, A note on the maximum flow through a network, *IRE Transactions on Information Theory* IT-2 (1956) 117–119.

[1940] L. Fejes, Über einen geometrischen Satz, *Mathematische Zeitschrift* 46 (1940) 83–85.

[1955] L. Few, The shortest path and the shortest road through *n* points, *Mathematika [London]* 2 (1955) 141–144.

[1956] M.M. Flood, The traveling-salesman problem, *Operations Research* 4 (1956) 61–75 [also in: *Operations Research for Management — Volume II Case Histories, Methods, Information Handling* (J.F. McCloskey, J.M. Coppinger, eds.), Johns Hopkins Press, Baltimore, Maryland, 1956, pp. 340–357].

[1951a] K. Florek, J. Łukaszewicz, J. Perkal, H. Steinhaus, S. Zubrzycki, Sur la liaison et la division des points d'un ensemble fini, *Colloquium Mathematicum* 2 (1951) 282–285.

[1951b] K. Florek, J. Łukaszewicz, J. Perkal, H. Steinhaus, S. Zubrzycki, Taksonomia Wrocławska [Polish, with English and Russian summaries], *Przeglad Antropologiczny* 17 (1951) 193–211.

[1956] L.R. Ford, Jr, *Network Flow Theory*, Paper P-923, The RAND Corporation, Santa Monica, California, [August 14] 1956.

[1954] L.R. Ford, D.R. Fulkerson, *Maximal Flow through a Network*, Research Memorandum RM-1400, The RAND Corporation, Santa Monica, California, [19 November] 1954 [published in *Canadian Journal of Mathematics* 8 (1956) 399–404].

[1955] L.R. Ford, Jr, D.R. Fulkerson, *A Simple Algorithm for Finding Maximal Network Flows and an Application to the Hitchcock Problem*, Research Memorandum RM-1604, The RAND Corporation, Santa Monica, California, [29 December] 1955 [published in *Canadian Journal of Mathematics* 9 (1957) 210–218].

[1956a] L.R. Ford, Jr, D.R. Fulkerson, *A Primal Dual Algorithm for the Capacitated Hitchcock Problem* [Notes on Linear Programming: Part XXXIV], Research Memorandum RM-1798 [ASTIA Document Number AD 112372], The RAND Corporation, Santa Monica, California, [September 25] 1956 [published in *Naval Research Logistics Quarterly* 4 (1957) 47–54].

[1956b] L.R. Ford, Jr, D.R. Fulkerson, *Solving the Transportation Problem* [Notes on Linear Programming — Part XXXII], Research Memorandum RM-1736, The RAND Corporation, Santa Monica, California, [June 20] 1956 [published in *Management Science* 3 (1956-57) 24–32].

[1957] L.R. Ford, Jr, D.R. Fulkerson, *Construction of Maximal Dynamic Flows in Networks*, Paper P-1079 [= Research Memorandum RM-1981], The RAND Corporation, Santa Monica, California, [May 7,] 1957 [published in *Operations Research* 6 (1958) 419–433].

[1962] L.R. Ford, Jr, D.R. Fulkerson, *Flows in Networks*, Princeton University Press, Princeton, New Jersey, 1962.

[1951] M. Fréchet, Sur les tableaux de corrélation dont les marges sont données, *Annales de l'Université de Lyon, Section A, Sciences Mathématiques et Astronomie* (3) 14 (1951) 53–77.

[1912] F.G. Frobenius, Über Matrizen aus nicht negativen Elementen, *Sitzungsberichte der Königlich Preußischen Akademie der Wissenschaften zu Berlin* (1912) 456–477 [reprinted in: *Ferdinand Georg Frobenius, Gesammelte Abhandlungen*, Band III (J.-P. Serre, ed.), Springer, Berlin, 1968, pp. 546–567].

[1917] G. Frobenius, Über zerlegbare Determinanten, *Sitzungsberichte der Königlich Preußischen Akademie der Wissenschaften zu Berlin* (1917) 274–277 [reprinted in: *Ferdinand*

Georg Frobenius, Gesammelte Abhandlungen, Band III (J.-P. Serre, ed.), Springer, Berlin, 1968, pp. 701–704].

[1958] D.R. Fulkerson, *Notes on Linear Programming: Part XLVI – Bounds on the Primal-Dual Computation for Transportation Problems*, Research Memorandum RM-2178, The RAND Corporation, Santa Monica, California, 1958.

[1958] T. Gallai, Maximum-minimum Sätze über Graphen, *Acta Mathematica Academiae Scientiarum Hungaricae* 9 (1958) 395–434.

[1978] T. Gallai, The life and scientific work of Dénes Kőnig (1884–1944), *Linear Algebra and Its Applications* 21 (1978) 189–205.

[1949] M.N. Ghosh, Expected travel among random points in a region, *Calcutta Statistical Association Bulletin* 2 (1949) 83–87.

[1955] A. Gleyzal, An algorithm for solving the transportation problem, *Journal of Research National Bureau of Standards* 54 (1955) 213–216.

[1985] R.L. Graham, P. Hell, On the history of the minimum spanning tree problem, *Annals of the History of Computing* 7 (1985) 43–57.

[1938] T. Grünwald, Ein neuer Beweis eines Mengerschen Satzes, *The Journal of the London Mathematical Society* 13 (1938) 188–192.

[1934] G. Hajós, Zum Mengerschen Graphensatz, *Acta Litterarum ac Scientiarum Regiae Universitatis Hungaricae Francisco-Josephinae, Sectio Scientiarum Mathematicarum [Szeged]* 7 (1934–35) 44–47.

[1856] W.R. Hamilton, Memorandum respecting a new system of roots of unity (the Icosian calculus), *Philosophical Magazine* 12 (1856) 446.

[1858] W.R. Hamilton, On a new system of roots of unity, *Proceedings of the Royal Irish Academy* 6 (1858) 415–416.

[1955] T.E. Harris, F.S. Ross, *Fundamentals of a Method for Evaluating Rail Net Capacities*, Research Memorandum RM-1573, The RAND Corporation, Santa Monica, California, [October 24,] 1955.

[1953a] I. Heller, On the problem of shortest path between points. I [abstract], *Bulletin of the American Mathematical Society* 59 (1953) 551.

[1953b] I. Heller, On the problem of shortest path between points. II [abstract], *Bulletin of the American Mathematical Society* 59 (1953) 551–552.

[1955a] I. Heller, Geometric characterization of cyclic permutations [abstract], *Bulletin of the American Mathematical Society* 61 (1955) 227.

[1955b] I. Heller, Neighbor relations on the convex of cyclic permutations, *Bulletin of the American Mathematical Society* 61 (1955) 440.

[1956a] I. Heller, Neighbor relations on the convex of cyclic permutations, *Pacific Journal of Mathematics* 6 (1956) 467–477.

[1956b] I. Heller, On the travelling salesman's problem, in: *Proceedings of the Second Symposium in Linear Programming* (Washington, D.C., 1955; H.A. Antosiewicz, ed.), Vol. 2, National Bureau of Standards, U.S. Department of Commerce, Washington, D.C., 1956, pp. 643–665.

[1941] F.L. Hitchcock, The distribution of a product from several sources to numerous localities, *Journal of Mathematics and Physics* 20 (1941) 224–230.

[1959] W. Hoffman, R. Pavley, Applications of digital computers to problems in the study of vehicular traffic, in: *Proceedings of the Western Joint Computer Conference* (Los Angeles, California, 1958), American Institute of Electrical Engineers, New York, 1959, pp. 159–161.

[1985] A.J. Hoffman, P. Wolfe, History, in: *The Traveling Salesman Problem — A Guided Tour of Combinatorial Optimization* (E.L. Lawler, J.K. Lenstra, A.H.G., Rinnooy Kan, D.B. Shmoys, eds.), Wiley, Chichester, 1985, pp. 1–15.

[1955] E. Jacobitti, Automatic alternate routing in the 4A crossbar system, *Bell Laboratories Record* 33 (1955) 141–145.

[1930] V. Jarník, O jistém problému minimálním (Z dopisu panu O. Borůvkovi) [Czech; On a minimal problem (from a letter to Mr Borůvka)] *Práce Moravské Přírodovědecké Společnosti Brno [Acta Societatis Scientiarum Naturalium Moravicae]* 6 (1930-31) 57–63.

[1934] V. Jarník, M. Kössler, O minimálních grafech, obsahujíciích *n* daných bodů, *Časopis pro Pěstování Matematiky a Fysiky* 63 (1934) 223–235.

[1942] R.J. Jessen, *Statistical Investigation of a Sample Survey for Obtaining Farm Facts*, Research Bulletin 304, Iowa State College of Agriculture and Mechanic Arts, Ames, Iowa, 1942.

[1973a] D.B. Johnson, A note on Dijkstra's shortest path algorithm, *Journal of the Association for Computing Machinery* 20 (1973) 385–388.

[1973b] D.B. Johnson, *Algorithms for Shortest Paths*, Ph.D. Thesis [Technical Report CU-CSD-73-169, Department of Computer Science], Cornell University, Ithaca, New York, 1973.

[1977] D.B. Johnson, Efficient algorithms for shortest paths in sparse networks, *Journal of the Association for Computing Machinery* 24 (1977) 1–13.

[1939] L.V. Kantorovich, *Matematicheskie metody organizatsii i planirovaniia proizvodstva* [Russian], Publication House of the Leningrad State University, Leningrad, 1939 [reprinted (with minor changes) in: *Primenenie matematiki v èkonomicheskikh issledovaniyakh* [Russian; Application of Mathematics in Economical Studies] (V.S. Nemchinov, ed.), Izdatel'stvo Sotsial'no-Èkonomicheskoĭ Literatury, Moscow, 1959, pp. 251–309] [English translation: Mathematical methods of organizing and planning production, *Management Science* 6 (1959-60) 366–422 [also in: *The Use of Mathematics in Economics* (V.S. Nemchinov, ed.), Oliver and Boyd, Edinburgh, 1964, pp. 225–279]].

[1940] L.V. Kantorovich, An effective method for solving some classes of extremal problems [in Russian], *Doklady Akademii Nauk SSSR* 28 (1940) 212–215.

[1942] L.V. Kantorovich, O peremeshchenii mass [Russian]. *Doklady Akademii Nauk SSSR* 37:7-8 (1942) 227–230 [English translation: On the translocation of masses, *Comptes Rendus (Doklady) de l'Académie des Sciences de l'U.R.S.S.* 37 (1942) 199–201 [reprinted: *Management Science* 5 (1958) 1–4]].

[1987] L.V. Kantorovich, Moĭ put' v nauke (Predpolagavshiĭsya doklad v Moskovskom matematicheskom obshchestve) [Russian; My journey in science (proposed report to the Moscow Mathematical Society)], *Uspekhi Matematicheskikh Nauk* 42:2 (1987) 183–213 [English translation: *Russian Mathematical Surveys* 42:2 (1987) 233–270 [reprinted in: *Functional Analysis, Optimization, and Mathematical Economics, A Collection of Papers Dedicated to the Memory of Leonid Vital'evich Kantorovich* (L.J. Leifman, ed.), Oxford University Press, New York, 1990, pp. 8–45]; also in: *L.V. Kantorovich Selected Works Part I* (S.S. Kutateladze, ed.), Gordon and Breach, Amsterdam, 1996, pp. 17–54].

[1949] L.V. Kantorovich, M.K. Gavurin, Primenenie matematicheskikh metodov v voprosakh analiza gruzopotokov [Russian; The application of mathematical methods to freight flow analysis], in: *Problemy povysheniya effectivnosti raboty transporta* [Russian; Collection of Problems of Raising the Efficiency of Transport Performance], Akademiia Nauk SSSR, Moscow-Leningrad, 1949, pp. 110–138.

[1856] T.P. Kirkman, On the representation of polyhedra, *Philosophical Transactions of the Royal Society of London Series A* 146 (1856) 413–418.

[1930] B. Knaster, Sui punti regolari nelle curve di Jordan, in: *Atti del Congresso Internazionale dei Matematici* [Bologna 3–10 Settembre 1928] Tomo II, Nicola Zanichelli, Bologna, [1930], pp. 225–227.

[1915] D. Kőnig, Vonalrendszerek és determinánsok [Hungarian; Line systems and determinants], *Mathematikai és Természettudományi Értesitő* 33 (1915) 221–229.

[1916] D. Kőnig, Graphok és alkalmazásuk a determinánsok és a halmazok elméletére [Hungarian], *Mathematikai és Természettudományi Értesitő* 34 (1916) 104–119 [German translation: Über Graphen und ihre Anwendung auf Determinantentheorie und Mengenlehre, *Mathematische Annalen* 77 (1916) 453–465].

[1923] D. Kőnig, Sur un problème de la théorie générale des ensembles et la théorie des graphes [Communication faite, le 7 avril 1914, au Congrès de Philosophie mathématique à Paris], *Revue de Métaphysique et de Morale* 30 (1923) 443–449.

[1931] D. Kőnig, Graphok és matrixok [Hungarian; Graphs and matrices], *Matematikai és Fizikai Lapok* 38 (1931) 116–119.

[1932] D. Kőnig, Über trennende Knotenpunkte in Graphen (nebst Anwendungen auf Determinanten und Matrizen), *Acta Litterarum ac Scientiarum Regiae Universitatis Hungaricae Francisco-Josephinae, Sectio Scientiarum Mathematicarum [Szeged]* 6 (1932-34) 155–179.

[1939] T. Koopmans, *Tanker Freight Rates and Tankship Building — An Analysis of Cyclical Fluctuations*, Publication Nr 27, Netherlands Economic Institute, De Erven Bohn, Haarlem, 1939.

[1942] Tj.C. Koopmans, Exchange ratios between cargoes on various routes (non-refrigerating dry cargoes), Memorandum for the Combined Shipping Adjustment Board, Washington, D.C., 1942, 1–12 [first published in: *Scientific Papers of Tjalling C. Koopmans*, Springer, Berlin, 1970, pp. 77–86].

[1948] Tj.C. Koopmans, Optimum utilization of the transportation system, in: *The Econometric Society Meeting* (Washington, D.C., 1947; D.H. Leavens, ed.) [Proceedings of the International Statistical Conferences — Volume V], 1948, pp. 136–146 [reprinted in: *Econometrica* 17 (Supplement) (1949) 136–146] [reprinted in: *Scientific Papers of Tjalling C. Koopmans*, Springer, Berlin, 1970, pp. 184–193].

[1959] Tj.C. Koopmans, A note about Kantorovich's paper, "Mathematical methods of organizing and planning production", *Management Science* 6 (1959-1960) 363–365.

[1992] Tj.C. Koopmans, [autobiography] in: *Nobel Lectures Including Presentation Speeches and Laureates' Biographies — Economic Sciences 1969—1980* (A. Lindbeck, ed.), World Scientific, Singapore, 1992, pp. 233–238.

[1949a] T.C. Koopmans, S. Reiter, *Allocation of Resources in Production, I*, Cowles Commission Discussion Paper. Economics: No. 264, Cowles Commission for Research in Economics, Chicago, Illinois, [May 4] 1949.

[1949b] T.C. Koopmans, S. Reiter, *Allocation of Resources in Production II Application to Transportation*, Cowles Commission Discussion Paper: Economics: No. 264A, Cowles Commission for Research in Economics, Chicago, Illinois, [May 19] 1949.

[1951] Tj.C. Koopmans, S. Reiter, A model of transportation, in: *Activity Analysis of Production and Allocation — Proceedings of a Conference* (Proceedings Conference on Linear Programming, Chicago, Illinois, 1949; Tj.C. Koopmans, ed.), Wiley, New York, 1951, pp. 222–259.

[2001] B. Korte, J. Nešetřil, Vojtěch Jarník's work in combinatorial optimization, *Discrete Mathematics* 235 (2001) 1–17.

[1956] A. Kotzig, *Súvislosť a Pravidelná Súvislosť Konečných Grafov* [Slovak; Connectivity and Regular Connectivity of Finite Graphs], Academical Doctorate Dissertation, Vysoká Škola Ekonomická, Bratislava, [September] 1956.

[1917a] A. Kowalewski, Topologische Deutung von Buntordnungsproblemen, *Sitzungsberichte Kaiserliche Akademie der Wissenschaften in Wien Mathematisch-naturwissenschaftliche Klasse Abteilung IIa* 126 (1917) 963–1007.

[1917b] A. Kowalewski, W.R. Hamilton's, Dodekaederaufgabe als Buntordnungsproblem, *Sitzungsberichte Kaiserliche Akademie der Wissenschaften in Wien Mathematisch-naturwissenschaftliche Klasse Abteilung IIa* 126 (1917) 67–90.

[1956] J.B. Kruskal, Jr, On the shortest spanning subtree of a graph and the traveling salesman problem, *Proceedings of the American Mathematical Society* 7 (1956) 48–50.

[1997] J.B. Kruskal, A reminiscence about shortest spanning subtrees, *Archivum Mathematicum (Brno)* 33 (1997) 13–14.

[1955a] H.W. Kuhn, On certain convex polyhedra [abstract], *Bulletin of the American Mathematical Society* 61 (1955) 557–558.

[1955b] H.W. Kuhn, The Hungarian method for the assignment problem, *Naval Research Logistics Quarterly* 2 (1955) 83–97.

[1956] H.W. Kuhn, Variants of the Hungarian method for assignment problems, *Naval Research Logistics Quarterly* 3 (1956) 253–258.

[1991] H.W. Kuhn, On the origin of the Hungarian method, in: *History of Mathematical Programming — A Collection of Personal Reminiscences* (J.K. Lenstra, A.H.G. Rinnooy Kan, A. Schrijver, eds.), CWI, Amsterdam and North-Holland, Amsterdam, 1991, pp. 77–81.

[1954] A.H. Land, A problem in transportation, in: *Conference on Linear Programming May 1954* (London, 1954) , Ferranti Ltd., London, 1954, pp. 20–31.

[1947] H.D. Landahl, A matrix calculus for neural nets: II, *Bulletin of Mathematical Biophysics* 9 (1947) 99–108.

[1946] H.D. Landahl, R. Runge, Outline of a matrix algebra for neural nets, *Bulletin of Mathematical Biophysics* 8 (1946) 75–81.

[1957] M. Leyzorek, R.S. Gray, A.A. Johnson, W.C. Ladew, S.R. Meaker, Jr, R.M. Petry, R.N. Seitz, *Investigation of Model Techniques — First Annual Report — 6 June 1956 – 1 July 1957 — A Study of Model Techniques for Communication Systems*, Case Institute of Technology, Cleveland, Ohio, 1957.

[1957] H. Loberman, A. Weinberger, Formal procedures for connecting terminals with a minimum total wire length, *Journal of the Association for Computing Machinery* 4 (1957) 428–437.

[1952] F.M. Lord, Notes on a problem of multiple classification, *Psychometrika* 17 (1952) 297–304.

[1882] É. Lucas, *Récréations mathématiques, deuxième édition*, Gauthier-Villars, Paris, 1882–1883.

[1950] R.D. Luce, Connectivity and generalized cliques in sociometric group structure, *Psychometrika* 15 (1950) 169–190.

[1949] R.D. Luce, A.D. Perry, A method of matrix analysis of group structure, *Psychometrika* 14 (1949) 95–116.

[1950] A.G. Lunts, Prilozhen ie matrichnoĭ bulevskoĭ algebry k analizu i sintezu releĭno-kontaktiykh skhem [Russian; Application of matrix Boolean algebra to the analysis and synthesis of relay-contact schemes], *Doklady Akademii Nauk SSSR (N.S.)* 70 (1950) 421–423.

[1952] A.G. Lunts, Algebraicheskie metody analiza i sinteza kontaktiykh skhem [Russian; Algebraic methods of analysis and synthesis of relay contact networks], *Izvestiya Akademii Nauk SSSR, Seriya Matematicheskaya* 16 (1952) 405–426.

[1940] P.C. Mahalanobis, A sample survey of the acreage under jute in Bengal, *Sankhyā* 4 (1940) 511–530.

[1948] E.S. Marks, A lower bound for the expected travel among m random points, *The Annals of Mathematical Statistics* 19 (1948) 419–422.

[1927] K. Menger, Zur allgemeinen Kurventheorie, *Fundamenta Mathematicae* 10 (1927) 96–115.

[1928a] K. Menger, Die Halbstetigkeit der Bogenlänge, *Anzeiger — Akademie der Wissenschaften in Wien — Mathematisch-naturwissenschaftliche Klasse* 65 (1928) 278–281.

[1928b] K. Menger, Ein Theorem über die Bogenlänge, *Anzeiger — Akademie der Wissenschaften in Wien — Mathematisch-naturwissenschaftliche Klasse* 65 (1928) 264–266.

[1929a] K. Menger, Eine weitere Verallgemeinerung des Längenbegriffes, *Anzeiger — Akademie der Wissenschaften in Wien — Mathematisch-naturwissenschaftliche Klasse* 66 (1929) 24–25.

[1929b] K. Menger, Über die neue Definition der Bogenlänge, *Anzeiger — Akademie der Wissenschaften in Wien — Mathematisch-naturwissenschaftliche Klasse* 66 (1929) 23–24.

[1930] K. Menger, Untersuchungen über allgemeine Metrik. Vierte Untersuchung. Zur Metrik der Kurven, *Mathematische Annalen* 103 (1930) 466–501.

[1931a] K. Menger, Bericht über ein mathematisches Kolloquium, *Monatshefte für Mathematik und Physik* 38 (1931) 17–38.

[1931b] K. Menger, Some applications of point-set methods, *Annals of Mathematics* (2) 32 (1931) 739–760.

[1932] K. Menger, Eine neue Definition der Bogenlänge, *Ergebnisse eines Mathematischen Kolloquiums* 2 (1932) 11–12.

[1940] K. Menger, On shortest polygonal approximations to a curve, *Reports of a Mathematical Colloquium* (2) 2 (1940) 33–38.

[1981] K. Menger, On the origin of the n-arc theorem, *Journal of Graph Theory* 5 (1981) 341–350.

[1940] A.N. Milgram, On shortest paths through a set, *Reports of a Mathematical Colloquium* (2) 2 (1940) 39–44.

[1933] Y. Mimura, Über die Bogenlänge, *Ergebnisse eines Mathematischen Kolloquiums* 4 (1933) 20–22.

[1957] G.J. Minty, A comment on the shortest-route problem, *Operations Research* 5 (1957) 724.

[1958] G.J. Minty, A variant on the shortest-route problem, *Operations Research* 6 (1958) 882–883.

[1784] G. Monge, Mémoire sur la théorie des déblais et des remblais. *Histoire de l'Académie Royale des Sciences* [année 1781. Avec les Mémoires de Mathématique & de Physique, pour la même Année] (2e partie) (1784) [*Histoire:* 34–38, *Mémoire:*] 666–704.

[1959] E.F. Moore, The shortest path through a maze, in: *Proceedings of an International Symposium on the Theory of Switching, 2–5 April 1957, Part II* [The Annals of the Computation Laboratory of Harvard University Volume XXX] (H. Aiken, ed.), Harvard University Press, Cambridge, Massachusetts, 1959, pp. 285–292.

[1955] G. Morton, A. Land, A contribution to the 'travelling-salesman' problem, *Journal of the Royal Statistical Society Series B* 17 (1955) 185–194.

[1983] H. Müller-Merbach, Zweimal travelling Salesman, *DGOR-Bulletin* 25 (1983) 12–13.

[1957] J. Munkres, Algorithms for the assignment and transportation problems, *Journal of the Society for Industrial and Applied Mathematics* 5 (1957) 32–38.

[1951] J. von Neumann, *The Problem of Optimal Assignment and a Certain 2-Person Game*, unpublished manuscript, [October 26] 1951.

[1953] J. von Neumann, A certain zero-sum two-person game equivalent to the optimal assignment problem, in: *Contributions to the Theory of Games Volume II* (H.W. Kuhn, A.W. Tucker, eds.) [Annals of Mathematics Studies 28], Princeton University Press, Princeton, New Jersey, 1953, pp. 5–12 [reprinted in: *John von Neumann, Collected Works, Vol. VI* (A.H. Taub, ed.), Pergamon Press, Oxford, 1963, pp. 44–49].

[1932] G. Nöbeling, Eine Verschärfung des *n*-Beinsatzes, *Fundamenta Mathematicae* 18 (1932) 23–38.

[1955] R.Z. Norman, On the convex polyhedra of the symmetric traveling salesman problem [abstract], *Bulletin of the American Mathematical Society* 61 (1955) 559.

[1955] A. Orden, The transhipment problem, *Management Science* 2 (1955-56) 276–285.

[1947] Z.N. Pariĭskaya, A.N. Tolstoĭ, A.B. Mots, *Planirovanie Tovarnykh Perevozok — Metody Opredeleniya Ratsionaljiykh Puteĭ Tovarodvizheniya* [Russian; Planning Goods Transportation — Methods of Determining Efficient Routes of Goods Traffic], Gostorgizdat, Moscow, 1947.

[1957] W. Prager, A generalization of Hitchcock's transportation problem, *Journal of Mathematics and Physics* 36 (1957) 99–106.

[1957] R.C. Prim, Shortest connection networks and some generalizations, *The Bell System Technical Journal* 36 (1957) 1389–1401.

[1957] R. Rado, Note on independence functions, *Proceedings of the London Mathematical Society* (3) 7 (1957) 300–320.

[1955a] J.T. Robacker, *On Network Theory*, Research Memorandum RM-1498, The RAND Corporation, Santa Monica, California, [May 26,] 1955.

[1955b] J.T. Robacker, *Some Experiments on the Traveling-Salesman Problem*, Research Memorandum RM-1521, The RAND Corporation, Santa Monica, California, [28 July] 1955.

[1956] J.T. Robacker, *Min-Max Theorems on Shortest Chains and Disjoint Cuts of a Network*, Research Memorandum RM-1660, The RAND Corporation, Santa Monica, California, [12 January] 1956.

[1949] J. Robinson, *On the Hamiltonian Game (A Traveling Salesman Problem)*, Research Memorandum RM-303, The RAND Corporation, Santa Monica, California, [5 December] 1949.

[1950] J. Robinson, *A Note on the Hitchcock-Koopmans Problem*, Research Memorandum RM-407, The RAND Corporation, Santa Monica, California, [15 June] 1950.

[1951] J. Robinson, An iterative method of solving a game. *Annals of Mathematics* 54 (1951) 296–301 [reprinted in: *The Collected Works of Julia Robinson* (S. Feferman, ed.), American Mathematical Society, Providence, Rhode Island, 1996, pp. 41–46].

[1956] L. Rosenfeld, Unusual problems and their solutions by digital computer techniques, in: *Proceedings of the Western Joint Computer Conference* (San Francisco, California, 1956), The American Institute of Electrical Engineers, New York, 1956, pp. 79–82.

[1958] M.J. Rossman, R.J. Twery, A solution to the travelling salesman problem by combinatorial programming [abstract], *Operations Research* 6 (1958) 897.

[1927] N.E. Rutt, Concerning the cut points of a continuous curve when the arc curve, *ab*, contains exactly *n* independent arcs [abstract], *Bulletin of the American Mathematical Society* 33 (1927) 411.

[1929] N.E. Rutt, Concerning the cut points of a continuous curve when the arc curve, *AB*, contains exactly *N* independent arcs, *American Journal of Mathematics* 51 (1929) 217–246.

[1939] T. Salvemini, Sugl'indici di omofilia, *Supplemento Statistico* 5 (Serie II) (1939) [= Atti della Prima Riunione Scientifica della Società Italiana di Statistica, Pisa, 1939] 105–115 [English translation: On the indexes of homophilia, in: *Tommaso Salvemini — Scritti Scelti*, Cooperativa Informazione Stampa Universitaria, Rome, 1981, pp. 525–537].

[1951] A. Shimbel, Applications of matrix algebra to communication nets, *Bulletin of Mathematical Biophysics* 13 (1951) 165–178.

[1953] A. Shimbel, Structural parameters of communication networks, *Bulletin of Mathematical Biophysics* 15 (1953) 501–507.

[1955] A. Shimbel, Structure in communication nets, in: *Proceedings of the Symposium on Information Networks* (New York, 1954), Polytechnic Press of the Polytechnic Institute of Brooklyn, Brooklyn, New York, 1955, pp. 199–203.

[1895] G. Tarry, Le problème des labyrinths. *Nouvelles Annales de Mathématiques* (3) (14) (1895) 187–190 [English translation in: N.L. Biggs, E.K. Lloyd, R.J. Wilson, *Graph Theory 1736–1936*, Clarendon Press, Oxford, 1976, pp. 18–20].

[1951] R. Taton, *L'Œuvre scientifique de Monge*, Presses universitaires de France, Paris, 1951.

[1950] R.L. Thorndike, The problem of the classification of personnel, *Psychometrika* 15 (1950) 215–235.

[1934] J. Tinbergen, Scheepsruimte en vrachten, *De Nederlandsche Conjunctuur* (1934) maart 23–35.

[1930] A.N. Tolstoĭ, Metody nakhozhdeniya naimen'shego summovogo kilometrazha pri planirovanii perevozok v prostranstve [Russian; Methods of finding the minimal total kilometrage in cargo-transportation planning in space], in: *Planirovanie Perevozok, Sbornik pervyĭ* [Russian; Transportation Planning, Volume I], Transpechat' NKPS [TransPress of the National Commissariat of Transportation], Moscow, 1930, pp. 23–55.

[1939] A. Tolstoĭ, Metody ustraneniya neratsional'nykh perevozok pri planirovanii [Russian; Methods of removing irrational transportation in planning], *Sotsialisticheskiĭ Transport* 9 (1939) 28–51 [also published as 'pamphlet': *Metody ustraneniya neratsional'nykh perevozok pri sostavlenii operativnykh planov* [Russian; Methods of Removing Irrational Transportation in the Construction of Operational Plans], Transzheldorizdat, Moscow, 1941].

[1953] L. Törnqvist, *How to Find Optimal Solutions to Assignment Problems*, Cowles Commission Discussion Paper: Mathematics No. 424, Cowles Commission for Research in Economics, Chicago, Illinois, [August 3] 1953.

[1952] D.L. Trueblood, The effect of travel time and distance on freeway usage, *Public Roads* 26 (1952) 241–250.

[1984] Albert Tucker, Merrill Flood (with Albert Tucker) — This is an interview of Merrill Flood in San Francisco on 14 May 1984, in: *The Princeton Mathematics Community in the 1930s — An Oral-History Project* [located at Princeton University in the Seeley G. Mudd Manuscript Library web at the URL: `http://www.princeton.edu/mudd/math`], Transcript Number 11 (PMC11), 1984.

[1951] S. Verblunsky, On the shortest path through a number of points, *Proceedings of the American Mathematical Society* 2 (1951) 904–913.

[1952] D.F. Votaw, Jr, Methods of solving some personnel-classification problems, *Psychometrika* 17 (1952) 255–266.

[1952] D.F. Votaw, Jr, A. Orden, The personnel assignment problem, in: *Symposium on Linear Inequalities and Programming* [Scientific Computation of Optimum Programs, Project SCOOP, No. 10] (Washington, D.C., 1951; A. Orden, L. Goldstein, eds.), Planning Research Division, Director of Management Analysis Service, Comptroller, Headquarters U.S. Air Force, Washington, D.C., 1952, pp. 155–163.

[1995] T. Wanningen Koopmans, *Stories and Memories*, type set manuscript, [May] 1995.

[1932] H. Whitney, Congruent graphs and the connectivity of graphs. *American Journal of Mathematics* 54 (1932) 150–168 [reprinted in: *Hassler Whitney Collected Works Volume I* (J. Eells, D. Toledo, eds.), Birkhäuser, Boston, Massachusetts, 1992, pp. 61–79].

[1873] Chr. Wiener, Ueber eine Aufgabe aus der Geometria situs, *Mathematische Annalen* 6 (1873) 29–30.

[1973] N. Zadeh, A bad network problem for the simplex method and other minimum cost flow algorithms, *Mathematical Programming* 5 (1973) 255–266.

K. Aardal et al., Eds., *Handbooks in OR & MS, Vol. 12*

Chapter 2

Computational Integer Programming and Cutting Planes

Armin Fügenschuh and Alexander Martin

Abstract

The study and solution of mixed-integer programming problems is of great interest, because they arise in a variety of mathematical and practical applications. Today's state-of-art software packages for solving mixed-integer programs based on linear programming include preprocessing, branch-and-bound, and cutting planes techniques. The main purpose of this article is to describe these components and recent developments that can be found in many solvers. Besides linear programming based relaxation methods we also discuss Langrangean, Dantzig–Wolfe and Benders' decomposition and their interrelations.

1 Introduction

The study and solution of linear mixed integer programs lies at the heart of discrete optimization. Various problems in science, technology, business, and society can be modeled as linear mixed integer programming problems and their number is tremendous and still increasing. This handbook, for instance, documents the variety of ideas, approaches and methods that help to solve mixed integer programs, since there is no unique method that solves them all, see also the surveys Aardal, Weismantel, and Wolsey (2002); Johnson, Nemhauser, and Savelsbergh (2000); Marchand, Martin, Weismantel, and Wolsey (2002). Among the currently most successful methods are linear programming (LP, for short) based branch-and-bound algorithms where the underlying linear programs are possibly strengthened by cutting planes. For example, most commercial mixed integer programming solvers, see Sharda (1995), or special purpose codes for problems like the traveling salesman problem are based on this method.

The purpose of this chapter is to describe the main ingredients of today's (commercial or research oriented) solvers for integer programs. We assume the reader to be familiar with basics in linear programming and polyhedral theory, see for instance Chvátal (1983) or Padberg (1995).

Consider an integer program or more general a mixed integer program (MIP) in the form

$$
\begin{aligned}
z_{\text{MIP}} = \min \quad & c^T x \\
\text{s.t.} \quad & Ax \begin{Bmatrix} \le \\ = \end{Bmatrix} b \\
& l \le x \le u \\
& x \in \mathbb{Z}^N \times \mathbb{R}^C,
\end{aligned}
\tag{1}
$$

where $A \in \mathbb{Q}^{M \times (N \cup C)}$, $c \in \mathbb{Q}^{N \cup C}$, $b \in \mathbb{Q}^M$. Here, M, N and C are nonempty, finite, ordered sets with N and C disjoint. Without loss of generality, we may assume that the elements of N and C are represented by numbers, i.e., $N = \{1, \ldots, p\}$ and $C = \{p+1, \ldots, n\}$. The vectors $l \in (\mathbb{Q} \cup \{-\infty\})^{N \cup C}$, $u \in (\mathbb{Q} \cup \{\infty\})^{N \cup C}$ are called *lower* and *upper bounds* on x, respectively. A variable x_j, $j \in N \cup C$, is *unbounded from below* (*above*), if $l_j = -\infty$ ($u_j = \infty$). An integer variable $x_j \in \mathbb{Z}$ with $l_j = 0$ and $u_j = 1$ is called *binary*. In the following four cases we also use other notions for (1):

linear program or LP, if $N = \emptyset$,

integer program or IP, if $C = \emptyset$,

binary mixed integer program, $0-1$ *mixed integer program* or BMIP, if all variables x_j, $j \in N$, are binary,

binary integer program, $0-1$ *integer program* or BIP, if (1) is a BMIP with $C = \emptyset$.

Usually, (1) models a problem arising in some application and the formulation for modeling this problem is not unique. In fact, for the same problem various formulations might exist and the first question is how to select an appropriate formulation. This issue will be discussed in Section 2. Very often however, we do not have our hands on the problem itself but just get the problem formulation as given in (1). In this case, we must extract all relevant information for the solution process from the constraint matrix A, the right-hand side vector b and the objective function c, i.e., we have to perform a structure analysis. This is usually part of the so-called *preprocessing phase* of mixed integer programming solvers and will also be discussed in Section 2. Thereafter, we have a problem, still in the format of (1), but containing more information about the inherit structure of the problem. Secondly, preprocessing also tries to discover and eliminate redundant information from a MIP solver's point of view.

From a complexity point of view mixed integer programming problems belong to the class of $\mathcal{NP}$-hard problems (Garey and Johnson, 1979) which makes it unlikely that efficient, i.e., polynomial time, algorithms for their solution exist. The route one commonly follows to solve an $\mathcal{NP}$-hard problem like (1) to optimality is to attack it from two sides. First, one considers the dual side and determines a lower bound on the objective function by relaxing the problem. The common basic idea of relaxation methods is to get rid of some part of the problem that causes difficulties. The methods differ in their choice of which part to delete and in the way to reintroduce the deleted part. The most commonly used approach is to relax the integrality constraints to obtain a linear program and reintroduce the integrality by adding cutting planes. This will be the main focus of Section 3. In addition, we will discuss in this section other relaxation methods that delete parts of the constraints and/or variables. Second, we consider the primal side and try to find some good feasible solution in order to determine an upper bound. Unfortunately, very little is done in this respect in general mixed integers solvers, an issue that will be discussed in Section 4.3.

If we are lucky the best lower and upper bounds coincide and we have solved the problem. If not, we have to resort to some enumeration scheme, and the one that is mostly used in this context is the branch-and-bound method. We will discuss branch-and-bound strategies in Section 4 and we will see that they have a big influence on the solution time and quality.

Needless to say that the way described above is not the only way to solve (1), but it is definitely the most used, and often among the most successful. Other approaches include semidefinite programming, combinatorial relaxations, basis reduction, Gomory's group approach, test sets and optimal primal algorithms, see the various articles in this handbook.

2 Formulations and structure analysis

The first step in the solution of an integer program is to find a "right" formulation. Right formulations are of course not unique and they strongly depend on the solution method one wants to use to solve the problem. The method we mainly focus on in this chapter is LP based branch-and-bound. The criterion for evaluating formulations that is mostly used in this context is the tightness of the LP relaxation. If we drop the integrality condition on the variables $x_1, \ldots, x_p$ in problem (1), we obtain the so-called *linear programming relaxation*, or *LP relaxation* for short:

$$\begin{aligned} z_{\mathrm{LP}} = \min \quad & c^T x \\ \text{s.t.} \quad & Ax \left\{ \begin{matrix} \leq \\ = \end{matrix} \right\} b \\ & l \leq x \leq u \\ & x \in \mathbb{R}^n. \end{aligned} \tag{2}$$

For the solution of (2) we have either polynomial (ellipsoid and interior point) or computationally efficient (interior point and simplex) algorithms at hand.

To problem (1) we associate the polyhedron $P_{\text{MIP}} := \text{conv}\{x \in \mathbb{Z}^p \times \mathbb{R}^{n-p}: Ax \leq b\}$, i.e., the convex hull of all feasible points for (1). A proof for P_{MIP} being a polyhedron can be found, for instance, in Nemhauser and Wolsey (1988) and Schrijver (1986). In the same way we define the associated polyhedron of problem (2) by $P_{\text{LP}} := \{x \in \mathbb{R}^n : Ax \leq b\}$. Of course, $P_{\text{MIP}} \subseteq P_{\text{LP}}$ and $z_{\text{LP}} \leq z_{\text{MIP}}$, so P_{LP} is a relaxation of P_{MIP}. The crucial requirement in the theory of solving general mixed integer problems is a sufficiently good understanding of the underlying polyhedra in order to tighten this relaxation.

Very often a theoretical analysis is necessary to decide which formulation is superior. There are no general rules such as: "the fewer the number of variables and/or constraints the better the formulation." In the following we discuss as an example a classical combinatorial optimization problem, the Steiner tree problem, which underpins the statement that fewer variables are not always better.

Given an undirected graph $G=(V,E)$ and a node set $T \subseteq V$, a *Steiner tree for T in G* is a subset $S \subseteq E$ of the edges such that $(V(S), S)$ contains a path between s and t for all $s, t \in T$, where $V(S)$ denotes the set of nodes incident to an edge in S. In other words, a Steiner tree is an edge set S that spans T. (Note that by our definition, a Steiner tree might contain circles, in contrast to the usual meaning of the notion *tree* in graph theory.) The *Steiner tree problem* is to find a minimal Steiner tree with respect to some given edge costs $c_e \geq 0$, $e \in E$.

A canonical way to formulate the Steiner tree problem as an integer program is to introduce, for each edge $e \in E$, a variable x_e indicating whether e is in the Steiner tree ($x_e = 1$) or not ($x_e = 0$). Consider the integer program

$$\begin{array}{lll} z_u := \min & c^T x & \\ & x(\delta(W)) \geq 1, & \text{for all } W \subset V,\ W \cap T \neq \emptyset, \\ & & (V \backslash W) \cap T \neq \emptyset, \\ & 0 \leq x_e \leq 1, & \text{for all } e \in E, \\ & x \text{ integer}, & \end{array} \tag{3}$$

where $\delta(X)$ denotes the cut induced by $X \subseteq V$, i.e., the set of edges with one end node in X and one its complement, and $x(F) := \sum_{e \in F} x_e$, for $F \subseteq E$. The first inequalities are called (*undirected*) *Steiner cut inequalities* and the inequalities $0 \leq x_e \leq 1$ *trivial inequalities*. It is easy to see that there is a one-to-one correspondence between Steiner trees in G and 0/1 vectors satisfying the undirected Steiner cut inequalities. Hence, (3) models the Steiner tree problem correctly.

Another way to model the Steiner tree problem is to consider the problem in a directed graph. We replace each edge $\{u, v\} \in E$ by two directed arcs (u, v) and (v, u). Let A denote this set of arcs and $D=(V, A)$ the resulting digraph. We choose some terminal $r \in T$, which will be called the *root*. A *Steiner arborescence* (*rooted at r*) is a set of arcs $S \subseteq A$ such that $(V(S), S)$ contains a directed path from r to t for all $t \in T \backslash \{r\}$. Obviously, there is a one-to-one correspondence between (undirected) Steiner trees in G and Steiner arborscences in D which contain at most one of two directed arcs (u, v), (v, u). Thus, if we choose arc costs $\vec{c}_{(u,v)} := \vec{c}_{(v,u)} := c_{\{u,v\}}$, for $\{u, v\} \in E$, the Steiner tree problem can be solved by finding a minimal Steiner arborescence with respect to $\vec{c}$. Note that there is always an optimal Steiner arborescence which does not contain an arc and its anti-parallel counterpart, since $\vec{c} \geq 0$. Introducing variables y_a for $a \in A$ with the interpretation $y_a := 1$, if arc a is in the Steiner arborescence, and $y_a := 0$ otherwise, we obtain the integer program

$$\begin{array}{lll} z_d := \min & \vec{c}^T y & \\ & y(\delta^+(W)) \geq 1, & \text{for all } W \subset V, r \in W, \\ & & (V \backslash W) \cap T \neq \emptyset, \\ & 0 \leq y_a \leq 1, & \text{for all } a \in A, \\ & y \text{ integer}, & \end{array} \tag{4}$$

where $\delta^+(X) := \{(u, v) \in A | u \in X, v \in V \backslash X\}$ for $X \subset V$, i.e., the set of arcs with tail in X and head in its complement. The first inequalities are called *(directed) Steiner cut inequalities* and $0 \leq y_a \leq 1$ are the *trivial inequalities*. Again, it is easy to see that each 0/1 vector satisfying the directed Steiner cut inequalities corresponds to a Steiner arborescence, and conversely, the incidence vector of each Steiner arborescence satisfies (4). Which of the two models (3) and (4) should be used to solve the Steiner tree problem in graphs?

At first glance, (3) is preferable to (4), since it contains only half the number of variables and the same structure of inequalities. However, it turns out that the optimal value z_d of the LP relaxation of the directed model (4) is greater than or equal to the corresponding value z_u of the undirected formulation (3). Even if the undirected formulation is tightened by the so-called Steiner partition inequalities, this relation holds (Chopra and Rao, 1994). This is astonishing, since the separation problem of the Steiner partition inequalities is difficult ($\mathcal{NP}$-hard), see Grötschel, Monma, and Stoer (1992), whereas the directed Steiner cut inequalities can be separated in polynomial time by max flow computations. Finally, the disadvantage of the directed model that the number of variables is doubled is not really a bottleneck. Since we are minimizing a nonnegative objective function, the variable of one of the two anti-parallel arcs will usually be at its lower bound. If we solve the LP relaxations by the simplex algorithm, it would rarely let this variables enter the basis. Thus, the directed model is much better than the undirected model, though it contains more variables. And in fact, most state-of-the-art solvers for the Steiner tree problem in graphs use formulation (4) or one that is equivalent to (4), see Koch, Martin, and Voß (2001) for further references.

The Steiner tree problem shows that it is not easy to find a tight problem formulation and that often a nontrivial analysis is necessary to come to a good decision.

Once we have decided on some formulation we face the next step, that of eliminating redundant information in (1). This so-called preprocessing step is very important, in particular, if we have no influence on the formulation step discussed above. In this case it is not only important to eliminate redundant information, but also to perform a structure analysis to extract as much information as possible from the constraint matrix. We will give a nontrivial example concerning block diagonal matrices at the end of this section. Before we come to this point let us briefly sketch the main steps that are usually performed within preprocessing. Most of these options are drawn from Andersen and Andersen (1995), Bixby (1994), Crowder, Johnson, and Padberg (1983), Hoffman and Padberg (1991), Savelsbergh (1994), Suhl and Szymanski (1994). We denote by $s_i \in \{\leq, =\}$ the sense of row l, i.e., (1) reads $\min\{c^T x : Ax\, s b, l \leq x \leq u, x \in \mathbb{Z}^N \times \mathbb{R}^C\}$. We consider the following cases:

Duality fixing. Suppose there is some column j with $c_j \geq 0$ that satisfies $a_{ij} \geq 0$ if $s_i =$ '$\leq$', and $a_{ij} = 0$ if $s_i =$ '$=$' for $i \in M$. If $l_j > -\infty$, we can fix column j to its lower bound. If $l_j = -\infty$ the problem is unbounded or infeasible. The same arguments apply to some column j with $c_j \leq 0$. Suppose $a_{ij} \leq 0$ if $s_i =$ '$\leq$', $a_{ij} = 0$ if $s_i =$ '$=$' for $i \in M$. If $u_j < \infty$, we can fix column j to its upper bound. If $u_j = \infty$ the problem is unbounded or infeasible.

Forcing and dominated rows. Here, we exploit the bounds on the variables to detect so-called forcing and dominated rows. Consider some row i and let

$$
\begin{aligned}
L_i &= \sum_{j \in P_i} a_{ij} l_j + \sum_{j \in N_i} a_{ij} u_j \\
U_i &= \sum_{j \in P_i} a_{ij} u_j + \sum_{j \in N_i} a_{ij} l_j
\end{aligned}
\tag{5}
$$

where $P_i = \{j : a_{ij} > 0\}$ and $N_i = \{j : a_{ij} < 0\}$. Obviously, $L_i \leq \sum_{j=1}^{n} a_{ij} x_j \leq U_i$. The following cases might come up:

1. Infeasible row:

   (a) $s_i =$ '$=$,' and $L_i > b_i$ or $U_i < b_i$
   (b) $s_i =$ '$\leq$,' and $L_i > b_i$

   In these cases the problem is infeasible.

2. Forcing row:

   (a) $s_i =$ '$=$,' and $L_i = b_i$ or $U_i = b_i$
   (b) $s_i =$ '$\leq$,' and $L_i = b_i$

   Here, all variables in P_i can be fixed to their lower (upper) bound and all variables in N_i to their upper (lower) bound when $L_i = b_i (U_i = b_i)$. Row i can be deleted afterwards.

3. Redundant row:

   (a) $s_i =$ '$\leq$,' and $U_i < b_i$.

This row bound analysis can also be used to strengthen the lower and upper bounds of the variables. Compute for each variable x_j

$$\bar{u}_{ij} = \begin{cases} (b_i - L_i)/a_{ij} + l_j, & \text{if } a_{ij} > 0 \\ (b_i - U_i)/a_{ij} + l_j, & \text{if } a_{ij} < 0 \text{ and } s_i = `=' \\ (L_i - U_i)/a_{ij} + l_j, & \text{if } a_{ij} < 0 \text{ and } s_i = `\leq' \end{cases}$$

$$\bar{l}_{ij} = \begin{cases} (b_i - U_i)/a_{ij} + u_j, & \text{if } a_{ij} > 0 \text{ and } s_i = `=' \\ (L_i - U_i)/a_{ij} + u_j, & \text{if } a_{ij} > 0 \text{ and } s_i = `\leq' \\ (b_i - L_i)/a_{ij} + u_j, & \text{if } a_{ij} < 0 \end{cases}$$

Let $\bar{u}_j = \min_i \bar{u}_{ij}$ and $\bar{l}_j = \max_i \bar{l}_{ij}$. If $\bar{u}_j \leq u_j$ and $\bar{l}_j \geq l_j$, we speak of an *implied free variable*. The simplex method might benefit from not updating the bounds but treating variable x_j as a free variable (note, setting the bounds of j to $-\infty$ and $+\infty$ will not change the feasible region). Free variables will commonly be in the basis and are thus useful in finding a starting basis. For mixed integer programs however, it is better in general to update the bounds by setting $u_j = \min\{u_j, \bar{u}_j\}$ and $l_j = \max\{l_j, \bar{l}_j\}$, because the search region of the variable within an enumeration scheme is reduced. In case x_j is an integer (or binary) variable we round u_j down to the next integer and l_j up to the next integer. As an example consider the following inequality (taken from mod015 from the `Miplib`[1]):

$$-45x_6 - 45x_{30} - 79x_{54} - 53x_{78} - 53x_{102} - 670x_{126} \leq -443$$

Since all variables are binary, $L_i = -945$ and $U_i = 0$. For $j = 126$ we obtain $\bar{l}_{ij} = (-443 + 945)/-670 + 1 = 0.26$. After rounding up it follows that x_{126} must be one.

Note that with these new lower and upper bounds on the variables it might pay to recompute the row bounds L_i and U_i, which again might result in tighter bounds on the variables.

Coefficients reduction. The row bounds in (5) can also be used to reduce the absolute value of coefficients of binary variables. Consider some row i with $s_i = `\leq'$ and let x_j be a binary variable with $a_{ij} \neq 0$.

$$\text{If} \begin{cases} a_{ij} < 0,\ U_i + a_{ij} < b_i, & \text{set} \quad \bar{a}'_{ij} = b_i - U_i, \\ a_{ij} > 0,\ U_i - a_{ij} < b_i, & \text{set} \quad \begin{cases} \bar{a}'_{ij} = U_i - b_i, \\ \bar{b}_i = U_i - a_{ij}, \end{cases} \end{cases} \tag{6}$$

[1] `Miplib` is a public available test set of real-world mixed integer programming problems (Bixby, Ceria, McZeal, and Savelsbergh, 1998).

where $\bar{a}'_{ij}$ denotes the new reduced coefficient. Consider the following inequality from example p0033 in the `Miplib`:

$$-230x_{10} - 200x_{16} - 400x_{17} \leq -5$$

All variables are binary, $U_i = 0$, and $L_i = -830$. We have $U_i + a_{i,10} = -230 < -5$ and we can reduce $a_{i,10}$ to $b_i - U_i = -5$. The same can be done for the other coefficients, and we obtain the inequality

$$-5x_{10} - 5x_{16} - 5x_{17} \leq -5$$

Note that the operation of reducing coefficients to the value of the right-hand side can also be applied to integer variables if all variables in this row have negative coefficients and lower bound zero. In addition, we may compute the greatest common divisor of the coefficients and divide all coefficients and the right-hand side by this value. In case all involved variables are integer (or binary) the right-hand side can be rounded down to the next integer. In our example, the greatest common divisor is 5, and dividing by that number we obtain the set covering inequality

$$-x_{10} - x_{16} - x_{17} \leq -1.$$

Aggregation. In mixed integer programs very often equations of the form

$$a_{ij}x_j + a_{ik}x_k = b_i$$

appear for some $i \in M$, k, $j \in N \cup C$. In this case, we may replace one of the variables, x_k say, by

$$\frac{b_i - a_{ij}x_j}{a_{ik}}. \tag{7}$$

In case x_k is binary or integer, the substitution is only possible, if the term (7) is guaranteed to be binary or integer as well. If this is true or x_k is a continuous variable, we aggregate the two variables. The new bounds of variables x_j are $l_j = \max\{l_j, (b_i - a_{ik}l_k)/a_{ij}\}$ and $u_j = \min\{u_j, (b_i - a_{ik}u_k)/a_{ij}\}$ if $a_{ik}/a_{ij} < 0$, and $l_j = \max\{l_j, (b_i - a_{ik}u_k)/a_{ij}\}$ and $u_j = \min\{u_j, (b_i - a_{ik}l_k)/a_{ij}\}$ if $a_{ik}/a_{ij} > 0$.
Of course, aggregation can also be applied to equations whose support is greater than two. However, this might cause additional fill (i.e., nonzero coefficients) in the matrix A that increases computer memory demand and lowers the computational speed of the simplex algorithm. Hence, aggregation is usually restricted to constraints and columns with small support.

Disaggregation. Disaggregation of columns, to our knowledge, is not an issue in preprocessing of mixed integer programs, since this usually blows up

the solution space. It is however applied in interior point algorithms for linear programs, because dense columns result in dense blocks in the Cholesky decomposition and are thus to be avoided (Gondzio, 1997).
On the other hand, disaggregation of rows is an important issue for mixed integer programs. Consider the following inequality (taken from the `Miplib`-problem p0282)

$$x_{85} + x_{90} + x_{95} + x_{100} + x_{217} + x_{222} + x_{227} + x_{232} - 8x_{246} \leq 0 \quad (8)$$

where all variables involved are binary. The inequality says that whenever one of the variables x_i with $i \in S := \{85, 90, 95, 100, 217, 222, 227, 232\}$ is one, x_{246} must also be one. This fact can also be expressed by replacing (8) by the following eight inequalities:

$$x_i - x_{246} \leq 0 \quad \text{for all } i \in S. \quad (9)$$

This formulation is tighter in the following sense: Whenever any variable in S is one, x_{246} is forced to one as well, which is not guaranteed in the original formulation. On the other hand, one constraint is replaced by many (in our case 8) inequalities, which might blow up the constraint matrix. However, within a cutting plane procedure, see the next section, this problem is not really an issue, because the inequalities in (9) can be generated on demand.

Probing. Probing is sometimes used in general mixed integer programming codes, see, for instance, Savelsbergh (1994), Suhl and Szymanski (1994). The idea is to set some binary variable temporarily to zero or one and try to deduce further or stronger inequalities from that. These implications can be expressed in inequalities as follows:

$$\begin{aligned}
(x_j = 1 \Rightarrow x_i = \alpha) &\Rightarrow \begin{cases} x_i \geq l_i + (\alpha - l_i)x_j \\ x_i \leq u_i - (u_i - \alpha)x_j \end{cases} \\
(x_j = 0 \Rightarrow x_i = \alpha) &\Rightarrow \begin{cases} x_i \geq \alpha - (\alpha - l_i)x_j \\ x_i \leq \alpha + (u_i - \alpha)x_j. \end{cases}
\end{aligned} \quad (10)$$

As an example, suppose we set variable x_{246} temporarily to zero in (8). This implies that $x_i = 0$ for all $i \in S$. Applying (10) we deduce the inequality

$$x_i \leq 0 + (1 - 0)x_{246} = x_{246}$$

for all $i \in S$, which is exactly (9). For further aspects of probing refer to Atamtürk, Nemhauser, and Savelsbergh (2000), where probing is used for the construction of conflict graphs to strengthen the LP relaxation, Johnson, Nemhauser, and Savelsbergh (2000), where probing is applied to improve the coefficients of the given inequalities, and Savelsbergh (1994), where a comprehensive study of probing is provided.

Besides the cases described, there are trivial ones like empty rows, empty, infeasible, and fixed columns, parallel rows and singleton rows or columns that we refrain from discussing here. One hardly believes at this point that such examples or some of the above cases really appear in mixed integer programming formulations, because better formulations are straight-forward to derive. But such formulations do indeed come up and mixed integer programming solvers must be able to handle them. Reasons for their existence are that formulations are often made by nonexperts or are sometimes generated automatically by some matrix generating program.

In general, all these tests are iteratively applied until all of them fail. Typically, preprocessing is applied only once at the beginning of the solution procedure, but sometimes it pays to run the preprocessing routine more often on different nodes in the branch-and-bound phase, see Section 4. There is always the question of the break even point between the running time for preprocessing and the savings in the solution time for the whole problem. There is no unified answer to this question. It depends on the individual problem, when intensive preprocessing pays and when not. Martin (1998), for instance, performs some computational tests for the instances in the `Miplib`. His results show that preprocessing reduces the problem sizes in terms of number of rows, columns, and nonzeros by around 10% on average. The time spent in preprocessing is negligible (below one per mill). It is interesting to note that for some problems presolve is indispensable for their solution. For example, problem fixnet6 in the `Miplib` is an instance, on which most solvers fail without preprocessing, but with presolve the instance turns out to be very easy. Further results on this subject can be found in Savelsbergh (1994).

Observe also that the preprocessing steps discussed so far consider just one single row or column at a time. The question comes up, whether one could gain something by looking at the structure of the matrix as a whole. This is a topic of computational linear algebra where one tries on one side to speed-up algorithms for matrices in special forms and on the other hand tries to develop algorithms that detect certain forms after reordering columns and/or rows. It is interesting to note that the main application area in this field are matrices arising from PDE systems. Very little has been done in connection with mixed integer programs. In the following we discuss one case, which shows that there might be more potential for MIPs.

Consider a matrix in a so-called bordered block diagonal form as depicted in Fig. 1. Suppose the constraint matrix of (1) has such a form and suppose in addition that there are just a few or even no coupling constraints. In the latter case the problem decomposes into a number of blocks many independent problems, which can be solved much faster than the original problem. Even if

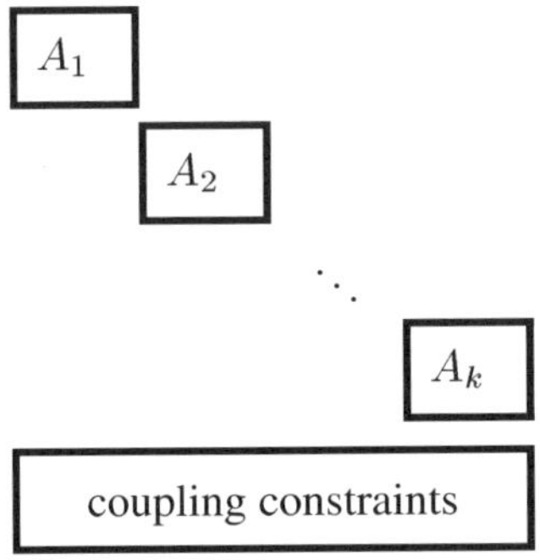

Fig. 1. Matrix in bordered block diagonal form.

there are coupling constraints this structure might help for instance to derive new cutting planes. The question arises whether MIPs have such a structure, possibly after reordering columns and rows? There are some obvious cases, where the matrix is already in this form (or can be brought into it), such as multi-commodity flow problems, multiple knapsack problems or other packing problems. But, there are problems where a bordered block diagonal form is hidden in the problem formulation (1) and can only be detected after reordering columns and rows. Borndörfer, Ferreira, and Martin (1998) have analyzed this question and checked whether matrices from MIPs can be brought into this form. They have tested various instances, especially problems whose original formulation is not in bordered block diagonal form, and it turns out that many problems have indeed such a form. Even more, the heuristics developed for detecting such a form are fast enough to be incorporated into preprocessing of a MIP solver. Martin and Weismantel (Martin, 1998; Martin and Weismantel, 1998) have developed cutting planes that exploit bordered block diagonal form and the computational results for this class of cutting planes are very promising. Of course, this is just a first step of exploiting special structures of MIP matrices and more needs to be done in this direction.

3 Relaxations

In obtaining good or optimal solutions of (1) one can approach it in two different ways: from the *primal* side by computing feasible solutions (mostly by heuristics) or from the dual side by determining good lower bounds. This is done by relaxing the problem. We consider three different types of relaxation ideas. The first and most common is to relax the integrality constraints and to find cutting planes that strengthen the resulting LP relaxation. This is the topic of Section 3.1. In Section 3.2 we sketch further well-known approaches, Lagrangean relaxation as well as Dantzig–Wolfe and Benders' decomposition.

The idea of these approaches is to delete part of the constraint matrix and reintroduce it into the problem either in the objective function or via column generation or cutting planes, respectively.

3.1 Cutting planes

The focus of this section is on describing cutting planes that are used in general mixed integer programming solvers. Mainly, we can classify cutting planes generating algorithms in two groups: one is exploiting the structure of the underlying mixed integer program, the other not. We first take a closer look on the latter group, in which we find the so-called Gomory cuts, mixed integer rounding cuts and lift-and-project cuts.

Suppose we want to solve the mixed integer program (1), where we assume for simplicity that we have no equality constraints and that $N=\{1,\ldots,p\}$ and $C=\{p+1,\ldots,n\}$. Note that if $x^*=(x_1^*,\ldots,x_n^*)$ is an optimal solution of (2) and x^* is in $\mathbb{Z}^p\times\mathbb{R}^{n-p}$, then it is already an optimal solution of (1) and we are done. But this is unlikely to happen after just solving the relaxation. It is more realistic to expect that some (or even all) of the variables $x_1^*,\ldots,x_p^*$ are not integral. In this case there exists at least one inequality $a^Tx\leq\beta$ that is feasible for P_{MIP} but not satisfied by x^*. From a geometric point of view, x^* is cut off by the hyperplane $a^Tx\leq\beta$ and therefore $a^Tx\leq\beta$ is called a *cutting plane*. The problem of determining whether x^* is in P_{MIP} and if not of finding such a cutting plane is called the *separation problem*. If we find a cutting plane $a^Tx\leq\beta$, we add it to the problem (2) and obtain

$$\begin{array}{ll}\min & c^Tx\\ \text{s.t.} & Ax\leq b\\ & a^Tx\leq\beta\\ & x\in\mathbb{R}^n,\end{array}\tag{11}$$

which strengthens (2) in the sense that $P_{\text{LP}}\subset P_{\text{LP}^1}\subseteq P_{\text{MIP}}$, where $P_{\text{LP}^1}:=\{x: Ax\leq b, a^Tx\leq\beta\}$ is the associated polyhedron of (11). Note that the first inclusion is strict by construction.

The process of solving (11) and finding a cutting plane is now iterated until the solution is in $\mathbb{Z}^p\times\mathbb{R}^{n-p}$ (this will be the optimal solution of (1)). Let us summarize the cutting plane algorithm discussed so far:

Algorithm 1. (*Cutting plane*)

1. *Let* $k:=0$ *and* LP^0 *the linear programming relaxation of the mixed integer program* (1).
2. *Solve* LP^k. *Let* $\tilde{x}^k$ *be an optimal solution.*
3. *If* $\tilde{x}^k$ *is in* $\mathbb{Z}^p\times\mathbb{R}^{n-p}$, *stop;* $\tilde{x}^k$ *is an optimal solution of* (1).

4. *Otherwise, find a linear inequality, that is satisfied by all feasible mixed integer points of* (1), *but not by* $\tilde{x}^k$.
5. *Add this inequality to* LP^k *to obtain* LP^{k+1}.
6. *Increase k by one and go to Step* 2.

The remaining of this section is devoted to the question on how to find good cutting planes.

3.1.1 Gomory integer cuts

We start with the pure integer case, i.e., $p=n$ in problem (1). The cutting plane algorithm we present in the sequel is based on simple integer rounding and makes use of information given by the simplex algorithm. Hereto we transform the problem into the standard form by adding slack variables and substituting unbounded variables $x_i := x_i^+ - x_i^-$ by two variables $x_i^+, x_i^- \geq 0$ that are bounded from below. Summing up, we turned (1) into a problem with the following structure:

$$\begin{array}{ll} \min & c^T x \\ \text{s.t.} & Ax = b \\ & x \in \mathbb{Z}_+^n, \end{array} \tag{12}$$

with $A \in \mathbb{Z}^{m \times n}$ and $b \in \mathbb{Z}^m$. (Note that this A, c and x may differ from those in (1).) We denote the associated polyhedron by $P_{\text{IP}}^{\text{St}} := \text{conv}\{x \in \mathbb{Z}_+^n : Ax = b\}$. Let x^* be an optimal solution of the LP relaxation of (12). We partition x^* into two subvectors x_B^* and x_N^*, where $B \subset \{1, \ldots, n\}$ is a basis of A, i.e., A_B nonsingular (regular), with

$$x_B^* = A_B^{-1} b - A_B^{-1} A_N x_N^* \geq 0 \tag{13}$$

and $x_N^* = 0$ for the nonbasic variables where $N = \{1, \ldots, n\} \backslash B$. (Note that this N completely differs from the N used in (1).) If x^* is integral, we have found an optimal solution of (12). Otherwise, at least one of the values in x_B^* must be fractional. So we choose $i \in B$ such that $x_i^* \notin \mathbb{Z}$. From (13) we get the following expression for the i-th variable of x_B:

$$A_{i\cdot}^{-1} b = \sum_{j \in N} A_{i\cdot}^{-1} A_{\cdot j} x_j + x_i, \tag{14}$$

where $A_{i\cdot}^{-1}$ denotes the i-th row of A^{-1} and $A_{\cdot j}$ the j-th column of A, respectively. We set $\bar{b}_i := A_{i\cdot}^{-1} b$ and $\bar{a}_{ij} := A_{i\cdot}^{-1} A_{\cdot j}$ for short. Since $x_j \geq 0$ for all j,

$$x_i + \sum_{j \in N} \lfloor \bar{a}_{ij} \rfloor x_j \leq x_i + \sum_{j \in N} \bar{a}_{ij} x_j = \bar{b}_i. \tag{15}$$

We can round down the right-hand side, since x is assumed to be integral and nonnegative, and thus the left-hand side in (15) is integral. So we obtain

$$x_i + \sum_{j \in N} \lfloor \bar{a}_{ij} \rfloor x_j \leq \lfloor \bar{b}_i \rfloor. \tag{16}$$

This inequality is valid for all integral points of $P_{\text{IP}}^{\text{St}}$, but it cuts off x^*, since $x_i^* = \bar{b}_i \notin \mathbb{Z}$, $x_j^* = 0$ for all $j \in N$ and $\lfloor \bar{b}_i \rfloor < \bar{b}_i$. Furthermore, all values of (16) are integral. After introducing another slack variable we add it to (12) still fulfilling the requirement that all values in the constraint matrix, the right-hand side and the new slack variable have to be integral. Named after their inventor, inequalities of this type are called *Gomory cuts* (Gomory, 1958, 1960). Gomory showed that an integer optimal solution is found after repeating these steps a finite number of times.

3.1.2 Gomory mixed integer cuts

The previous approach of generating valid inequalities fails if both integer and continuous variables are present. It fails, because rounding down the right-hand side may cut off some feasible points of $P_{\text{MIP}}^{\text{St}} := \text{conv}\{x \in \mathbb{Z}_+^p \times \mathbb{R}_+^{n-p} : Ax = b\}$, if x cannot be assumed to be integral. For the general mixed-integer case, we describe three different methods to obtain valid inequalities. They are all more or less based on the following disjunctive argument.

Lemma 3.2. *Let P and Q be two polyhedra in $\mathbb{R}_+^n$ and $a^T x \leq \alpha$, $b^T x \leq \beta$ valid inequalities for P and Q respectively. Then*

$$\sum_{i=1}^{n} \min(a_i, b_i) x_i \leq \max(\alpha, \beta)$$

is valid for conv $(P \cup Q)$.

We start again with a mixed integer problem in standard form, but this time with $p < n$, i.e.,

$$\begin{aligned} \min \quad & c^T x \\ \text{s.t.} \quad & Ax = b \\ & x \in \mathbb{Z}_+^p \times \mathbb{R}_+^{n-p}. \end{aligned} \tag{17}$$

Let $P_{\text{MIP}}^{\text{St}}$ be the convex hull of all feasible solutions of (17). Consider again (14), where B is a basis, x_i, $i \in B$, is an integer variable and $\bar{b}_i, \bar{a}_{ij}$ are defined accordingly. We divide the set N of nonbasic variables in $N^+ := \{j \in N : \bar{a}_{ij} \geq 0\}$ and $N^- := N \backslash N^+$. As we already mentioned, every feasible x of (17) satisfies $x_B = A_B^{-1} b - A_B^{-1} A_N x_N$, hence

$$\bar{b}_i - \sum_{j \in N} \bar{a}_{ij} x_j \in \mathbb{Z}$$

and there exists $k \in \mathbb{Z}$ such that

$$\sum_{j \in N} \bar{a}_{ij} x_j = f(\bar{b}_i) + k, \tag{18}$$

where $f(\alpha) := \alpha - \lfloor \alpha \rfloor$ for $\alpha \in \mathbb{R}$. In order to apply the disjunctive argument, we distinguish the following two cases, $\sum_{j \in N} \bar{a}_{ij} x_j \geq 0$ and $\sum_{j \in N} \bar{a}_{ij} x_j \leq 0$. In the first case

$$\sum_{j \in N^+} \bar{a}_{ij} x_j \geq f(\bar{b}_i)$$

follows. In the second case we get

$$\sum_{j \in N^-} \bar{a}_{ij} x_j \leq f(\bar{b}_i) - 1 \leq 0$$

or, equivalently,

$$-\frac{f(\bar{b}_i)}{1 - f(\bar{b}_i)} \sum_{j \in N^-} \bar{a}_{ij} x_j \geq f(\bar{b}_i).$$

Now we apply the disjunctive argument to the disjunction $P := P^{\text{St}}_{\text{MIP}} \cap \{x : \sum_{j \in N} \bar{a}_{ij} x_j \geq 0\}$ and $Q := P^{\text{St}}_{\text{MIP}} \cap \{x : \sum_{j \in N} \bar{a}_{ij} x_j \leq 0\}$. Because of $\max(\bar{a}_{ij}, 0) = \bar{a}_{ij}$ for $j \in N^+$ and $\max(-f(\bar{b}_i)/(1 - f(\bar{b}_i))\bar{a}_{ij}, 0) = -f(\bar{b}_i)/(1 - f(\bar{b}_i))\bar{a}_{ij}$ for $j \in N^-$ we obtain by applying Lemma 3.2 the valid inequality for $P^{\text{St}}_{\text{MIP}}$

$$\sum_{j \in N^+} \bar{a}_{ij} x_j - \frac{f(\bar{b}_i)}{1 - f(\bar{b}_i)} \sum_{j \in N^-} \bar{a}_{ij} x_j \geq f(\bar{b}_i), \tag{19}$$

which cuts off x^*, since all nonbasic variables are zero. It is possible to strengthen inequality (19) in the following way. Observe that the derivation does not change, if we add integer multiples to those variables x_j, $j \in N$, that are integral (only the value of k in (18) might change). By doing this we may put the coefficient of each integer variable x_j either in the set N^+ or N^-. If we put it in N^+, the derivation of the inequality yields $\bar{a}_{ij}$ as coefficient for x_j. Thus the best possible coefficient after adding integer multiples is $f(\bar{a}_{ij})$, the difference between the right-hand and left-hand side in (19) is now as small as possible. In N^- the final coefficient is $-f(\bar{b}_i)/(1 - f(\bar{b}_i))\bar{a}_{ij}$, so the smallest difference is achieved by the factor $f(\bar{b}_i)(1 - f(\bar{a}_{ij}))/(1 - f(\bar{b}_i))$. We still have the freedom to select between N^+ and N^-. We obtain the best possible coefficients by using $\min(f(\bar{a}_{ij}), f(\bar{b}_i)(1 - f(\bar{a}_{ij}))/(1 - f(\bar{b}_i)))$. Putting all this together yields Gomory's mixed integer cut (Gomory, 1960):

$$\sum_{\substack{j\in N, j\le p \\ f(\bar{a}_{ij})\le f(\bar{b}_i)}} f(\bar{a}_{ij})x_j + \sum_{\substack{j\in N, j\le p \\ f(\bar{a}_{ij})> f(\bar{b}_i)}} \frac{f(\bar{b}_i)(1-f(\bar{a}_{ij}))}{1-f(\bar{b}_i)} x_j + \sum_{j\in N^+, j>p} \bar{a}_j x_j - \sum_{j\in N^-, j>p} \frac{f(\bar{b}_i)}{1-f(\bar{b}_i)} \bar{a}_j x_j \ge f(\bar{b}_i). \tag{20}$$

Gomory (1960) showed that an algorithm based on iteratively generated inequalities of this type solves (1) after a finite number of steps, if the objective function value $c^T x$ is integer for all $x \in \mathbb{Z}_+^p \times \mathbb{R}_+^{n-p}$ with $Ax = b$.

In the derivation of Gomory's mixed integer cuts we followed the original path of Gomory (1960). Having mixed-integer-rounding cuts at hand, we can give another proof for their validity in just one single line at the end of the next section.

Though Gomory's mixed integer cuts have been known since the sixties, their computational breakthrough came in the nineties with the paper by Balas, Ceria, Cornuéjols, and Natraj (1996). In the meantime they are incorporated in many MIP solvers, see, for instance Bixby, Fenelon, Guand, Rothberg, and Wunderling (1999). Note that Gomory's mixed integer cuts can always be applied, as the separation problem for the optimal LP solution is easy. However, adding these inequalities might cause numerical difficulties, see the discussion in Padberg (2001).

3.1.3 Mixed-integer-rounding cuts

We start developing the idea of this kind of cutting planes by considering the subset $X := \{(x, y) \in \mathbb{Z} \times \mathbb{R}_+ : x - y \le b\}$ of $\mathbb{R}^2$ with $b \in \mathbb{R}$. We define two disjoint subsets $P := \text{conv}(X \cap \{(x, y) : x \le \lfloor b \rfloor\})$ and $Q := \text{conv}(X \cap \{(x, y) : x \ge \lfloor b \rfloor + 1\})$ of conv(X). For P the inequalities $x - \lfloor b \rfloor \le 0$ and $0 \le y$ are valid and therefore every linear combination of them is also valid. Hence, if we multiply them by $1 - f(b)$ and 1 respectively, we obtain

$$(x - \lfloor b \rfloor)(1 - f(b)) \le y.$$

For Q we scale the valid inequalities $-(x - \lfloor b \rfloor) \le -1$ and $x - y \le b$ with weights $f(b)$ and 1 to get

$$(x - \lfloor b \rfloor)(1 - f(b)) \le y.$$

Now the disjunctive argument, Lemma 3.2, implies that $(x - \lfloor b \rfloor)(1 - f(b)) \le y$, or equivalently:

$$x - \frac{1}{1 - f(b)} y \le \lfloor b \rfloor \tag{21}$$

is valid for $\text{conv}(P \cup Q) = \text{conv}(X)$.

From this basic situation we change now to more general settings. Consider the mixed integer set $X := \{(x,y) \in \mathbb{Z}^p_+ \times \mathbb{R}_+ : a^T x - y \leq b\}$ with $a \in \mathbb{R}^p$ and $b \in \mathbb{R}$. We define a partition of $\{1,\ldots,n\}$ by $N^1 := \{i \in \{1,\ldots,n\} : f(a_i) \leq f(b)\}$ and $N^2 := \{1,\ldots,n\} \backslash N^1$. With this setting we obtain

$$\sum_{i \in N^1} \lfloor a_i \rfloor x_i + \sum_{i \in N^2} a_i x_i - y \leq a^T x - y \leq b.$$

Now let $w := \sum_{i \in N^1} \lfloor a_i \rfloor x_i + \sum_{i \in N^2} \lceil a_i \rceil x_i \in \mathbb{Z}$ and $z := y + \sum_{i \in N^2} (1 - f(a_i)) x_i \geq 0$, then we obtain (remark that $\lceil a_i \rceil - \lfloor a_i \rfloor \leq 1$)

$$\begin{aligned} w - z &= \sum_{i \in N^1} \lfloor a_i \rfloor x_i + \sum_{i \in N^2} \lceil a_i \rceil x_i - \sum_{i \in N^2} (1 - a_i + \lfloor a_i \rfloor) x_i - y \\ &\leq \sum_{i \in N^1} \lfloor a_i \rfloor x_i + \sum_{i \in N^2} a_i x_i - y \leq b. \end{aligned}$$

and (21) yields

$$w - \frac{1}{1 - f(b)} z \leq \lfloor b \rfloor.$$

Substituting w and z gives

$$\sum_{i \in N^1} \lfloor a_i \rfloor x_i + \sum_{i \in N^2} \left(\lceil a_i \rceil - \frac{1 - f(a_i)}{1 - f(b)} \right) x_i - \frac{1}{1 - f(b)} y \leq \lfloor b \rfloor.$$

Easy computation shows that this is equivalent to

$$\sum_{i=1}^{p} \left(\lfloor a_i \rfloor + \frac{\max(0, f(a_i) - f(b))}{1 - f(b)} \right) x_i - \frac{1}{1 - f(b)} y \leq \lfloor b \rfloor.$$

Thus we have shown that this is a valid inequality for conv(X), the *mixed integer rounding* (*MIR*) *inequality*.
From MIR inequalities one can easily derive Gomory's mixed integer cuts. Consider the set $X := \{(x, y^-, y^+) \in \mathbb{Z}^p_+ \times \mathbb{R}^2_+ | a^T x + y^+ - y^- = b\}$, then $a^T x - y^- \leq b$ is valid for X and the computations shown above now yield

$$\sum_{i=1}^{p} \left(\lfloor a_i \rfloor + \frac{\max(0, f(a_i) - f(b))}{1 - f(b)} \right) x_i - \frac{1}{1 - f(b)} y^- \leq \lfloor b \rfloor$$

as a valid inequality. Subtracting $a^T x + y^+ - y^- = b$ gives Gomory's mixed integer cut.

Nemhauser and Wolsey (1990) discuss MIR inequalities in a more general setting. They prove that MIR inequalities provide a complete description for any mixed 0–1 polyhedron. Marchand and Wolsey (Marchand, 1998; Marchand and Wolsey, 2001) show the computational merits of MIR inequalities in solving general mixed integer programs.

3.1.4 Lift-and-project cuts

The cuts presented here only apply to 0–1 mixed integer problems. The idea of "lift and project" is to find new inequalities not in the original space but in a higher dimensional (lifting). By projecting these inequalities back to the original space tighter inequalities can be obtained. In literature many different ways to lift and to project back can be found (Balas, Ceria, and Cornuéjols, 1993; Bienstock and Zuckerberg, 2003; Lasserre, 2001; Lovász and Schrijver, 1991; Sherali and Adams, 1990). The method we want to review in detail is due to Balas et al. (1993, 1996). It is based on the following observation:

Lemma 3.3. *If* $\alpha + a^T x \geq 0$ *and* $\beta + b^T x \geq 0$ *are valid for a polyhedron* P, *then* $(\alpha + a^T x)(\beta + b^T x) \geq 0$ *is also valid for* P.

We consider a 0–1 program in the form of (1) having w.l.o.g. no equality constraints, in which the system $Ax \leq b$ already contains the trivial inequalities $0 \leq x_i \leq 1$ for all $i \in \{1, \ldots, p\}$. The following steps give an outline of the lift-and-project procedure:

Algorithm 4. (*Lift-and-project*)

1. *Choose an index* $j \in \{1, \ldots, p\}$.
2. *Multiply each inequality of* $Ax \leq b$ *once by* x_j *and once by* $1 - x_j$ *giving the new (nonlinear) system*:

$$\begin{aligned} (Ax)x_j &\leq bx_j \\ (Ax)(1 - x_j) &\leq b(1 - x_j) \end{aligned} \tag{22}$$

3. *Lifting: replace* $x_i x_j$ *by* y_i *for* $i \in \{1, \ldots, n\} \setminus \{j\}$ *and* x_j^2 *by* x_j. *The resulting system of inequalities is again linear and finite and the set of its feasible points* $L_j(P)$ *is therefore a polyhedron.*
4. *Projection: project* $L_j(P)$ *back to the original space by eliminating all variables* y_i. *Call the resulting polyhedron* P_j.

In Balas et al. (1993) it is proven that $P_j = \text{conv}(P \cap \{x \in \mathbb{R}^n : x_j \in \{0, 1\})$, i.e., the j-th component of each vertex of P_j is either zero or one. Moreover, it is shown that a repeated application of Algorithm 4 on the first p variables yields

$$((P_1)_2 \ldots)_p = \text{conv}(P \cap \{x \in \mathbb{R}^n : x_1, \ldots, x_p \in \{0, 1\}\}) = P_{\text{MIP}}.$$

In fact, this result does not depend on the order in which one applies lift-and-project. Every permutation of $\{1, \ldots, p\}$ yields P_{MIP}.

The crucial step we did not describe up to now is how to carry out the projection (Step 4). As $L_j(P)$ is a polyhedron, there exists matrices D, B and a vector d such that $L_j(P) = \{(x, y) : Dx + By \leq d\}$. Thus we can describe the (orthogonal-) projection of $L_j(P)$ onto the x-space by

$$P_j = \{x \in \mathbb{R}^n : (u^T D)x \leq u^T d \quad \text{for all } u \geq 0, u^T B = 0\}.$$

Now that we are back in our original problem space, we can start finding valid inequalities by solving the following linear program for a given fractional solution x^* of the underlying mixed integer problem:

$$\begin{aligned} \max \quad & u^T(Dx^* - d) \\ \text{s.t.} \quad & u^T B = 0 \\ & u \in \mathbb{R}^n_+. \end{aligned} \tag{23}$$

The set $C := \{u \in \mathbb{R}^n_+ : u^T B = 0\}$ in which we are looking for the optimum is a pointed polyhedral cone. The optimum is either 0, if the variable x_j is already integral, or the linear program is unbounded (infinity). In the latter case let $u^* \in C$ be an extreme ray of the cone in which direction the linear program (23) is unbounded. Then u^* will give us the cutting plane $(u^*)^T Dx \leq (u^*)^T d$ that indeed cuts off x^*.

Computational experiences with lift-and-project cuts to solve real-world problems are discussed in Balas et al. (1993, 1996).

3.1.5 *Knapsack inequalities*

The cutting planes discussed so far have one thing common: they do not make use of the special structures of the given problem. In this section we want to generate valid inequalities by investigating the underlying combinatorial problem. The inequalities that are generated in this way are usually stronger in the sense that one can prove that they induce high-dimensional faces, often facets, of the underlying polyhedron.

We start again with the pure integer case. A knapsack problem is a 0–1 integer problem with just one inequality $a^T x \leq \beta$. Its polytope, the 0–1 knapsack polytope, is the following set of points:

$$P_K(N, a, \beta) := \text{conv}\left\{x \in \{0, 1\}^N : \sum_{j \in N} a_j x_j \leq \beta\right\}$$

with a finite set N, weights $a \in \mathbb{Z}^N_+$ and some capacity $\beta \in \mathbb{Z}_+$.

Observe that each inequality of a 0–1 program gives rise to a 0–1 knapsack polytope. And thus each valid inequality known for the knapsack polytope can be used to strengthen the 0–1 program. In the sequel we derive some known inequalities for the 0–1 knapsack polytope that are also useful for solving general 0–1 integer problems.

Cover inequalities. A subset $C \subseteq N$ is called a *cover* if $\sum_{j \in C} a_j > \beta$, i.e., the sum of the weights of all items in C is bigger than the capacity of the knapsack. To each cover, we associate the cover inequality

$$\sum_{j \in C} x_j \leq |C| - 1,$$

a valid inequality for $P_K(N, a, \beta)$. If the underlying cover C is *minimal*, i.e., $C \subset N$ is a cover and for every $s \in C$ we have $\sum_{j \in S \setminus \{s\}} a_j \leq \beta$, the inequality defines a facet of $P_K(C, a, \beta)$, i.e., the dimension of the face that is induced by the inequality is one less than the dimension of the polytope. Nonminimal cover only give faces, but not facets. Indeed, if a cover is not minimal, the corresponding cover inequality is superfluous, because it can be expressed as a sum of minimal cover inequalities and some upper bound constraints. Minimal cover inequalities might be strengthened by a technique called lifting that we present in detail in the next section.

(1, *k*)-Configuration inequalities. Padberg (1980) introduced this class of inequalities. Let $S \subset N$ be a set of items that fits into the knapsack, $\sum_{j \in S} a_j \leq \beta$, and suppose there is another item $z \in N \setminus S$ such that $\tilde{S} \cup \{z\}$ is a minimal cover for every $\tilde{S} \subset S$ with cardinality $|\tilde{S}| = k$. Then (S, z) is called a $(1, k)$-*configuration*. We derive the following inequality

$$\sum_{j \in S} x_j + (|S| - k + 1)x_z \leq |S|,$$

which we call $(1, k)$-*configuration* inequality. They are connected to minimal cover inequalities in the following way: a minimal cover S is a $(1, |S| - 1)$-configuration and a $(1, k)$-configuration with respect to $(S, \{z\})$ with $k = |S|$ is a minimal cover. Moreover, one can show that $(1, k)$-configuration inequalities define facets of $P_K(S \cup \{z\}, a, \beta)$.

Extended weight inequalities. Weismantel (1997) generalized minimal cover and $(1, k)$-configuration inequalities. He introduced extended weight inequalities which include both classes of inequalities as special cases. Denote $a(T) := \sum_{j \in T} a_j$ and consider a subset $T \subset N$ such that $a(T) < \beta$. With $r := \beta - a(T)$, the inequality

$$\sum_{i \in T} a_i x_i + \sum_{i \in N \setminus T} \max(a_i - r, 0) x_i \leq a(T) \tag{24}$$

is valid for $P_K(N, a, \beta)$. It is called a *weight inequality with respect to T*. The name *weight inequality* reflects the fact that the coefficients of the items in T equal their original weights and the number $r := \beta - a(T)$ corresponds to the remaining capacity of the knapsack when $x_j = 1$ for all $j \in T$. There is a natural way to extend weight inequalities by (i) replacing

the original weights of the items by relative weights and (ii) using the method of sequential lifting that we outline in Section 3.1.8.
Let us consider a simple case by associating a weight of one to each of the items in T. Denote by S the subset of $N \backslash T$ such that $a_j \geq r$ for all $j \in S$. For a chosen permutation $\pi_1, \ldots, \pi_{|S|}$ of S we apply sequential lifting, see Section 3.1.8, and obtain lifting coefficients w_j, $j \in S$ such that

$$\sum_{j \in T} x_j + \sum_{j \in S} w_j x_j \leq |T|,$$

is a valid inequality for $P_K(N, a, \beta)$, called the (*uniform*) *extended weight inequality*. They already generalize minimal cover and $(1, k)$-configuration inequalities and can be generalized themselves to inequalities with arbitrary weights in the starting set T, see Weismantel (1997).

The separation of minimal cover inequalities is widely discussed in the literature. The complexity of cover separation has been investigated in Ferreira (1994), Gu, Nemhauser, and Savelsbergh (1998), Klabjan, Nemhauser, Tovey (1998), whereas algorithmic and implementation issues are treated among others in Crowder, Johnson, and Padberg (1983), Gu, Nemhauser, and Savelsbergh (1998), Hoffman and Padberg (1991), Van Roy and Wolsey (1987), Zemel (1989). The ideas and concepts suggested to separate cover inequalities basically carry over to extended weight inequalities. Typical features of a separation algorithm for cover inequalities are: fix all variables that are integers, find a cover (in the extended weight case some subset T) usually by some greedy-type heuristics, and lift the remaining variables sequentially.

Cutting planes derived from knapsack relaxations can sometimes be strengthened if special ordered set (SOS) inequalities $\sum_{j \in Q} x_j \leq 1$ for some $Q \subseteq N$ are available. In connection with a knapsack inequality these constraints are also called *generalized upper bound constraints* (*GUBs*). It is clear that by taking the additional SOS constraints into account stronger cutting planes may be derived. This possibility has been studied in Crowder, Johnson, and Padberg (1983), Gu, Nemhauser, and Savelsbergh (1998), Johnson and Padberg (1981), Nemhauser and Vance (1994), Wolsey (1990).

From pure integer knapsack problems we switch now to mixed 0–1 knapsacks, where some continuous variables appear. As we will see, the concept of covers is also useful in this case to describe the polyhedral structure of the associated polytopes. Consider the mixed 0–1 knapsack set

$$P_S(N, a, \beta) = \left\{(x, s) \in \{0, 1\}^N \times \mathbb{R}_+ : \sum_{j \in N} a_j x_j - s \leq \beta\right\}$$

with nonnegative coefficients, i.e., $a_j \geq 0$ for $j \in N$ and $\beta \geq 0$.

Now let $C \subset N$ be a cover and $\lambda := \sum_{j \in C} a_j - \beta > 0$. Marchand and Wolsey (1999) showed that the inequality

$$\sum_{j \in C} \min(a_j, \lambda) x_j - s \leq \sum_{j \in C} \min(a_j, \lambda) - \lambda \tag{25}$$

is valid for $P_S(N, a, \beta)$. Moreover, this inequality defines a facet of $P_S(C, a, \beta)$. This result marks a contrast to the pure 0–1 knapsack case, where only minimal covers induce facets. Computational aspects of these inequalities are discussed in Marchand (1998), Marchand and Wolsey (1999).

Cover inequalities also appear in other contexts. In Ceria, Cordier, Marchand, and Wolsey (1998) cover inequalities are derived for the knapsack set with general integer variables. Unfortunately, in this case, the resulting inequalities do not define facets of the convex hull of the knapsack set restricted to the variables defining the cover. More recently, the notion of cover has been used to define families of valid inequalities for the complementarity knapsack set (de Farias, Johnson, and Nemhauser, 2002). By lifting continous variables new inequalities are developed in Richard, de Farias, and Nemhauser (2001) that extended (25). Atamtürk (2001) studies the convex hull of feasible solutions for a single constraint taken from a mixed-integer programming problem. No sign restrictions are imposed on the coefficients and the variables are not necessarily bounded, thus mixed 0–1 knapsacks are contained as a special case. It is still possible to obtain strong valid inequalities that may be useful for general mixed-integer programming.

3.1.6 Flow cover inequalities

From (mixed) knapsack problems with only one inequality we now turn to more complex polyhedral structures. Consider within a capacitated network flow problem, some node with a set of ingoing arcs N. Each inflow arc $j \in N$ has a capacity a_j. By y_j we denote the (positive) flow that is actually on arc $j \in N$. Moreover, the total inflow (i.e., sum of all flows on the arcs in N) is bounded by $b \in \mathbb{R}_+$. Then the (flow) set of all feasible points of this problem is given by

$$X = \left\{ (x, y) \in \{0, 1\}^N \times \mathbb{R}_+^N : \sum_{j \in N} y_j \leq b,\ y_j \leq a_j x_j,\ \forall j \in N \right\}. \tag{26}$$

We want to demonstrate how to use the mixed knapsack inequality (25) to derive new inequalities for the polyhedron conv(X). Let $C \subset N$ be a cover for the knapsack in X, i.e., C is a subset of N satisfying $\lambda := \sum_{j \in C} a_j - b > 0$ (usually covers for flow problems are called *flow covers*). From $\sum_{j \in N} y_j \leq b$ we obtain

$$\sum_{j \in C} a_j x_j - \sum_{j \in C} s_j \leq b,$$

by discarding all y_j for $j \in N \backslash C$ and replacing y_j by $a_j x_j - s_j$ for all $j \in C$, where $s_j \geq 0$ is a slack variable. Using the mixed knapsack inequality (25), we have that the following inequality is valid for X:

$$\sum_{j \in C} \min(a_j, \lambda) x_j - \sum_{j \in C} s_j \leq \sum_{j \in C} \min(a_j, \lambda) - \lambda,$$

or equivalently, substituting $a_j x_j - y_j$ for s_j,

$$\sum_{j \in C} (y_j + \max(a_j - \lambda, 0)(1 - x_j)) \leq b. \tag{27}$$

It was shown by Padberg, Van Roy, and Wolsey (1985) that this last inequality, called *flow cover inequality*, defines a facet of conv(X), if $\max_{j \in C} a_j > \lambda$.

Flow models have been extensively studied in the literature. Various generalizations of the flow cover inequality (27) have been derived for more complex flow models. In Van Roy and Wolsey (1986), a family of flow cover inequalities is described for a general single node flow model containing variable lower and upper bounds. Generalizations of flow cover inequalities to lot-sizing and capacitated facility location problems can also be found in Aardal, Pochet, and Wolsey (1995) and Pochet (1998). Flow cover inequalities have been used successfully in general purpose branch-and-cut algorithms to tighten formulations of mixed integer sets (Atamtürk, 2002; Gu et al., 1999, 2000; Van Roy and Wolsey, 1987).

3.1.7 *Set packing inequalities*

The study of set packing polyhedra plays a prominent role in combinatorial optimization and integer programming. Suppose we are given a set $X := \{1, \ldots, m\}$ and a finite system of subsets $X_1, \ldots, X_n \subseteq X$. For each j we have a real number c_j representing the gain for the use of X_j. In the *set packing problem* we ask for a selection $N \subseteq \{1, \ldots, n\}$ such that $X_i \cap X_j = \emptyset$ for all $i, j \in N$ with $i \neq j$ and $\sum_{j \in N} c_j$ is maximal. We can model this problem by introducing incidence vectors $a_j \in \{0, 1\}^m$ for each X_j, $j \in \{1, \ldots, n\}$, where $a_{ij} = 1$ if and only if $i \in X_j$. This defines a matrix $A := (a_{ij}) \in \{0,1\}^{m \times n}$. For the decision which subset we put into the selection N we introduce a vector $x \in \{0,1\}^n$, with $x_j = 1$ if and only if $j \in N$. With this definition we can state the set packing problem as the flowing 0–1 integer program:

$$\begin{array}{ll} \max & c^T x \\ \text{s.t.} & Ax \leq \mathbb{1} \\ & x \in \{0, 1\}^n. \end{array} \tag{28}$$

This problem is important not only from a theoretical but from a computational point of view: set packing problems often occur as subproblems in (mixed) integer problems. Hence a good understanding of 0–1 integer programs with 0–1 matrices can substantially speed up the solution process of general mixed integer problems including such substructures.

In the sequel we study the *set packing polytope* $P(A) := \text{conv}\{x \in \{0,1\}^n : Ax \leq \mathbb{1}\}$ associated to A. An interpretation of this problem in a graph theoretic sense is helpful to obtain new valid inequalities that strengthens the LP relaxation of (28). The *column intersection graph* $G(A) = (V, E)$ of $A \in \{0,1\}^{m \times n}$ consists of n nodes, one for each column with edges (i,j) between two nodes i and j if and only if their corresponding columns in A have a common nonzero entry in some row. There is a one-to-one correspondence between 0–1 feasible solutions and stable sets in $G(A)$, where a stable set S is a subset of nodes such that $(i,j) \notin E$ for all $i,j \in S$. Consider a feasible vector $x \in \{0,1\}^n$ with $Ax \leq \mathbb{1}$, then $S = \{i \in N : x_i = 1\}$ is a stable set in $G(A)$ and vice versa, each stable set in $G(A)$ defines a feasible 0–1 solution x via $x_i = 1$ if and only if $i \in S$. Observe that different matrices A, A' have the same associated polyhedron if and only if their corresponding intersection graphs coincide. It is therefore customary to study $P(A)$ via the graph G and denote the set packing polytope and the stable set polytope, respectively, by $P(G)$. Without loss of generality we can assume that G is connected.

What can we say about $P(G)$? The following observations are immediate:

(i) $P(G)$ is full dimensional.
(ii) $P(G)$ is lower monotone, i.e., if $x \in P(G)$ and $y \in \{0,1\}^n$ with $0 \leq y \leq x$ then $y \in P(G)$.
(iii) The nonnegativity constraints $x_j \geq 0$ induce facets of $P(G)$.

It is a well-known fact that $P(G)$ is completely described by the nonnegative constraints (iii) and the edge-inequalities $x_i + x_j \leq 1$ for $(i,j) \in E$ if and only if G is bipartite, i.e., there exists a partition (V_1, V_2) of the nodes V such that every edge has one node in V_1 and one in V_2. If G is not bipartite, then it contains odd cycles. They give rise to the following *odd cycle inequality*

$$\sum_{j \in V_C} x_j \leq \frac{|V_C| - 1}{2},$$

where $V_C \subseteq V$ is the set of nodes of cycle $C \subset E$ of odd cardinality. This inequality is valid for $P(G)$ and defines a facet of $P((V_C, E_{V_C}))$ if and only if C is an odd hole, i.e., an odd cycle without chords (Padberg, 1973). This class of inequalities can be separated in polynomial time using an algorithm based on the computation of shortest paths, see Lemma 9.1.11 in Grötschel, Lovász, and Schrijver (1988) for details.

A *clique* (C, E_C) in a graph $G = (V, E)$ is a subset of nodes and edges such that for every pair $i,j \in C, i \neq j$ there exists an edge $(i,j) \in E_C$. From a clique (C, E_C) we obtain the *clique inequality*

$$\sum_{j \in C} x_j \leq 1,$$

which is valid for $P(G)$. It defines a facet of $P(G)$ if and only if the clique is maximal (Fulkerson, 1971; Padberg, 1973). A clique (C, E_C) is said to be *maximal* if every $i \in V$ with $(i,j) \in E$ for all $j \in C$ is already contained in C. In contrast to the class of odd cycle inequalities, the separation of clique inequalities is difficult ($\mathcal{NP}$-hard), see Theorem 9.2.9 in Grötschel, Lovász, and Schrijver (1988). But there exists a larger class of inequalities, called *orthonormal representation* (OR) *inequalities*, that includes the clique inequalities and can be separated in polynomial time (Grötschel et al., 1988). Beside odd cycle, clique and OR-inequalities there are many other inequalities known for the stable set polytope. Among these are blossom, odd antihole, and web, wedge inequalities and many more. Borndörfer (1998) gives a survey on these constraints including a discussion on their separability.

3.1.8 *Lifted inequalities*

The lifting technique is a general approach that has been used in a wide variety of contexts to strengthen valid inequalities. A field for its application is the reuse of inequalities within branch-and-bound, see Section 4, where some inequality that is only valid under certain variable fixings is made globally valid by applying lifting. Assume for simplicity that all integer variables are 0–1. Consider an arbitrary polytope $P \subseteq [0,1]^N$ and let $L \subset N$. Suppose we have an inequality

$$\sum_{j \in L} w_j x_j \leq w_0, \tag{29}$$

which is valid for $P_L := \operatorname{conv}(P \cap \{x : x_j = 0 \ \forall\, j \in N \backslash L\})$. We investigate the lifting of a variable x_j that has been set to 0, setting x_j to 1 is similar. The *lifting problem* is to find lifting coefficients w_j for $j \in N \backslash L$ such that

$$\sum_{j \in N} w_j x_j \leq w_0 \tag{30}$$

is valid for P. Ideally we would like inequality (3) to be "strong," i.e., if inequality (29) defines a face of high dimension of P_L, we would like the inequality (30) to define a face of high dimension of P as well.

One way of obtaining coefficients $(w_j)_{j \in N \backslash L}$ is to apply *sequential lifting:* lifting coefficients w_j are calculated one after another. That is we determine an ordering of the elements of $N \backslash L$ that we follow in computing the coefficients. Let $k \in N \backslash L$ be the first index in this sequence. The coefficient w_k is computed for a given $k \in N \backslash L$ so that

$$w_k x_k + \sum_{j \in L} w_j x_j \leq w_0 \tag{31}$$

is valid for $P_{L \cup \{k\}}$.

We explain the main idea of lifting on the knapsack polytope: $P := P_K(N, a, \beta)$. It is easily extended to more general cases. Define the *lifting function* as the solution of the following 0–1 knapsack problem:

$$\begin{aligned}\Phi_L(u) := \min \quad & w_0 - \sum_{j\in L} w_j x_j \\ \text{s.t.} \quad & \sum_{j\in L} a_j x_j \leq \beta - u, \\ & x \in \{0, 1\}^L.\end{aligned}$$

We set $\Phi_L(u) := +\infty$ if $\{x \in \{0,1\}^L : \sum_{j\in L} a_j x_j \leq \beta - u\} = \emptyset$. Then inequality (31) is valid for $P_{L\cup\{k\}}$ if $w_k \leq \Phi_L(a_k)$, see Padberg (1975), Wolsey (1975). Moreover, if $w_k = \Phi_L(a_k)$ and (29) defines a face of dimension t of P_L, then (31) defines a face of $P_{L\cup\{k\}}$ of dimension at least $t+1$.

If one now intends to lift a second variable, then it becomes necessary to update the function Φ_L. Specifically, if $k \in N\backslash L$ was introduced first with a lifting coefficient w_k, then the lifting function becomes

$$\begin{aligned}\Phi_{L\cup\{k\}}(u) := \min \quad & w_0 - \sum_{j\in L\cup\{k\}} w_j x_j \\ \text{s.t.} \quad & \sum_{j\in L\cup\{k\}} a_j x_j \leq \beta - u, \\ & x \in \{0, 1\}^{L\cup\{k\}},\end{aligned}$$

so in general for fixed u, function Φ_L can decrease as more variables are lifted in. As a consequence, lifting coefficients depend on the order in which variables are lifted and therefore different orders of lifting often lead to different valid inequalities.

One of the key questions to be dealt with when implementing such a lifting approach is how to compute lifting coefficients w_j. To perform "exact" sequential lifting (i.e., to compute at each step the lifting coefficient given by the lifting function), we have to solve a sequence of integer programs. In the case of the lifting of variables for the 0–1 knapsack set this can be done efficiently using a dynamic programming approach based on the following recursion formula:

$$\Phi_{L\cup\{k\}}(u) = \min(\Phi_L(u), \Phi_L(u + a_k) - \Phi_L(a_k)).$$

Using such a lifting approach, facet-defining inequalities for the 0–1 knapsack polytope have been derived (Balas, 1975; Balas and Zemel, 1978; Hammer, Johnson, and Peled, 1975; Padberg, 1975; Wolsey, 1975) and embedded in a branch-and-bound framework to solve particular types of 0–1 integer programs to optimality (Crowder et al., 1983).

We now take a look on how to apply the idea of lifting to the more complex polytope associated to the flow problem discussed in Section 3.1.6. Consider the set

$$X' = \left\{ (x, y) \in \{0, 1\}^{L \cup \{k\}} \times \mathbb{R}_+^{L \cup \{k\}} : \sum_{j \in L \cup \{k\}} y_j \leq b, y_j \leq a_j x_j, j \in L \cup \{k\} \right\}.$$

Note that with $(x_k, y_k) = (0, 0)$, this reduces to the flow set, see (26)

$$X = \left\{ (x, y) \in \{0, 1\}^L \times \mathbb{R}_+^L : \sum_{j \in L} y_j \leq b, y_j \leq a_j x_j, j \in L \right\}.$$

Now suppose that the inequality

$$\sum_{j \in L} w_j x_j + \sum_{j \in L} v_j y_j \leq w_0$$

is valid and facet-defining for conv(X). As before, let

$$\begin{aligned} \Psi_L(u) = \min \quad & w_0 - \sum_{j \in L} w_j x_j - \sum_{j \in L} v_j y_j \\ \text{s.t.} \quad & \sum_{j \in L} y_j \leq b - u \\ & y_j \leq a_j x_j, \ j \in L \\ & (x, y) \in \{0, 1\}^L \times \mathbb{R}_+^L. \end{aligned}$$

Now the inequality

$$\sum_{j \in L} w_j x_j - \sum_{j \in L} v_j y_j + w_k x_k + v_k y_k \leq w_0$$

is valid for conv(X') if and only if $w_k + v_k u \leq \Psi_L(u)$ for all $0 \leq u \leq a_k$, ensuring that all feasible points with $(x_k, y_k) = (1, u)$ satisfy the inequality.

The inequality defines a facet if the affine function $w_k + v_k u$ lies below the function $\Psi_L(u)$ in the interval $[0, a_k]$ and touches it in two points different from $(0, 0)$, thereby increasing the number of affinely independent tight points by the number of new variables. In theory, "exact" sequential lifting can be applied to derive valid inequalities for any kind of mixed integer set. However, in practice, this approach is only useful to generate valid inequalities for sets for which one can associate a lifting function that can be evaluated efficiently.

Gu et al. (1999) showed how to lift the pair (x_k, y_k) when y_k has been fixed to a_k and x_k to 1.

Lifting is applied in the context of set packing problems to obtain facets from odd-hole inequalities (Padberg, 1973). Other uses of sequential lifting can be found in Ceria et al. (1998) where the lifting of continuous and integer variables is used to extend the class of lifted cover inequalities to a mixed knapsack set with general integer variables. In Martin (1998), Martin and Weismantel (1998) lifting is applied to define (lifted) feasible set inequalities for an integer set defined by multiple integer knapsack constraints. Generalizations of the lifting procedure where more than one variable is lifted simultaneously (so-called *sequence-independent lifting*) can be found for instance in Atamtürk (2001) and Gu et al. (2000).

3.2 Further relaxations

In the preceding section we have simplified the mixed integer program by relaxing the integrality constraints and by trying to force the integrality of the solution by adding cutting planes. In the methods we are going to discuss now we keep the integrality constraints, but relax part of the constraint matrix that causes difficulties.

3.2.1 Lagrangean relaxation

Consider again (1). The idea of Lagrangean relaxation is to delete part of the constraints and reintroduce them into the problem by putting them into the objective function attached with some penalties. Split A and b into two parts

$$A = \begin{pmatrix} A_1 \\ A_2 \end{pmatrix} \quad \text{and} \quad b = \begin{pmatrix} b_1 \\ b_2 \end{pmatrix},$$

where $A_1 \in \mathbb{Q}^{m_1 \times n}$, $A_2 \in \mathbb{Q}^{m_2 \times n}$, $b_1 \in \mathbb{Q}^{m_1}$, $b_2 \in \mathbb{Q}^{m_2}$ with $m_1 + m_2 = m$. Then, assuming all equality constraints are divided into two inequalities each, (1) takes the form

$$\begin{aligned} z_{\text{MIP}} := \min \quad & c^T x \\ \text{s.t.} \quad & A_1 x \leq b_1 \\ & A_2 x \leq b_2 \\ & x \in \mathbb{Z}^p \times \mathbb{R}^{n-p}. \end{aligned} \tag{32}$$

Consider for some fixed $\lambda \in \mathbb{R}_+^{m_1}$ the following function

$$\begin{aligned} L(\lambda) = \min \quad & c^T x - \lambda^T (b_1 - A_1 x) \\ \text{s.t.} \quad & x \in P^2, \end{aligned} \tag{33}$$

where $P^2 = \{x \in \mathbb{Z}^p \times \mathbb{R}^{n-p} : A_2 x \le b_2\}$. $L(\cdot)$ is called the *Lagrangean function.* The evaluation of this function for a given λ is called the *Lagrangean subproblem*. Obviously, $L(\lambda)$ is a lower bound on z_{MIP}, since for any feasible solution $\bar{x}$ of (32) we have

$$c^T \bar{x} \ge c^T \bar{x} - \lambda^T (b_1 - A_1 \bar{x}) \ge \min_{x \in P^2} c^T x - \lambda^T (b_1 - A_1 x) = L(\lambda).$$

Since this holds for each $\lambda \ge 0$ we conclude that

$$\max_{\lambda \ge 0} L(\lambda) \tag{34}$$

yields a lower bound of z_{MIP}. (34) is called *Lagrangean dual.* Let λ^* be an optimal solution to (34). The questions remain, how good is $L(\lambda^*)$ and how to compute λ^*. The following equation provides an answer to the first question:

$$L(\lambda^*) = \min\{c^T x : A_1 x \le b_1, x \in \text{conv}(P^2)\}. \tag{35}$$

A proof of this result can be found for instance in Nemhauser and Wolsey (1988) and Schrijver (1986). Since

$$\begin{aligned} \{x \in \mathbb{R}^n : Ax \le b\} &\supseteq \{x \in \mathbb{R}^n : A_1 x \le b_1, x \in \text{conv}(P^2)\} \\ &\supseteq \text{conv}\{x \in \mathbb{Z}^p \times \mathbb{R}^{n-p} : Ax \le b\} \end{aligned}$$

we conclude from (35) that

$$z_{\text{LP}} \le L(\lambda^*) \le z_{\text{MIP}}. \tag{36}$$

Furthermore, $z_{\text{LP}} = L(\lambda^*)$ for all objective functions $c \in \mathbb{R}^n$ if

$$\{x \in \mathbb{R}^n : A_2 x \le b_2\} = \text{conv}\{x \in \mathbb{Z}^p \times \mathbb{R}^{n-p} : A_2 x \le b_2\}.$$

It remains to discuss how to compute $L(\lambda^*)$. From a theoretical point of view it can be shown using the polynomial equivalence of separation and optimization that $L(\lambda^*)$ can be determined in polynomial time, if $\min\{\tilde{c}^T x : x \in \text{conv}(P^2)\}$ can be computed in polynomial time for any objective function $\tilde{c}$, see for instance (Schrijver, 1986). In practice, $L(\lambda^*)$ is determined by applying subgradient methods. The function $L(\lambda)$ is piecewise linear, concave and bounded from above. Consider for some fixed $\lambda^0 \in \mathbb{R}_+^{m_1}$ an optimal solution x^0 for (33). Then, $g^0 := A_1 x^0 - b_1$ is a *subgradient* for L and λ^0, i.e.,

$$L(\lambda) - L(\lambda^0) \le (g^0)^T (\lambda - \lambda^0),$$

since

$$\begin{aligned} L(\lambda) - L(\lambda^0) &= c^T x^\lambda - \lambda^T(b_1 - A_1 x^\lambda) - (c^T x^0 - (\lambda^0)^T(b_1 - A_1 x^0)) \\ &\le c^T x^0 - \lambda^T(b_1 - A_1 x^0) - (c^T x^0 - (\lambda^0)^T(b_1 - A_1 x^0)) \\ &= (g^0)^T(\lambda - \lambda^0). \end{aligned}$$

Hence, for λ^* we have $(g^0)^T(\lambda^* - \lambda^0) \ge L(\lambda^*) - L(\lambda^0) \ge 0$. In order to find λ^* this suggests to start with some λ^0, compute

$$x^0 \in \operatorname{argmin}\{c^T x - (\lambda^0)^T(b_1 - A_1 x) : x \in P^2\}$$

and determine iteratively $\lambda^0, \lambda^1, \lambda^2, \ldots$ by setting $\lambda^{k+1} = \lambda^k + \mu^k g^k$, where $g^k := A_1 x^k - b_1$, and μ^k is some step length to be specified. This iterative method is the essence of the *subgradient method*. Details and refinements of this method can be found among others in Nemhauser and Wolsey (1988) and Zhao and Luh (2002).

Of course, the quality of the Lagrangean relaxation strongly depends on the set of constraints that is relaxed. On one side, we must compute (33) for various values of λ and thus it is necessary to compute $L(\lambda)$ fast. Therefore one may want to relax as many (complicated) constraints as possible. On the other hand, the more constraints are relaxed the worse the bound $L(\lambda^*)$ will get, see Lemaréchal and Renaud (2001). Therefore, one always must find a compromise between these two conflicting goals.

3.2.2 Dantzig–Wolfe decomposition

The idea of decomposition methods is to decouple a set of constraints (variables) from the problem and treat them at a superordinate level, often called the *master problem*. The resulting residual subordinate problem can often be solved more efficiently. Decomposition methods now work alternately on the master and subordinate problem and iteratively exchange information to solve the original problem to optimality. In this section we discuss two well known examples of this approach, Dantzig–Wolfe decomposition and Benders' decomposition. We will see that as in the case of Lagrangean relaxation these methods also delete part of the constraint matrix. But instead of reintroducing this part in the objective function, it is now reformulated and reintroduced into the constraint system.

Let us start with Dantzig–Wolfe decomposition (Dantzig and Wolfe, 1960) and consider again (32), where we assume for the moment that $p = 0$, i.e., a linear programming problem. Consider the polyhedron $P^2 = \{x \in \mathbb{R}^n : A_2 x \le b_2\}$. It is a well known fact about polyhedra that there exist vectors $v_1, \ldots, v_k$ and $e_1, \ldots, e_l$ such that $P^2 = \operatorname{conv}(\{v_1, \ldots, v_k\}) + \operatorname{cone}(\{e_1, \ldots, e_l\})$. In other words, $x \in P^2$ can be written in the form

$$x = \sum_{i=1}^{k} \lambda_i v_i + \sum_{j=1}^{l} \mu_j e_j \tag{37}$$

with $\lambda_1, \ldots, \lambda_k \geq 0$, $\sum_{i=1}^{k} \lambda_i = 1$ and $\mu_1, \ldots, \mu_l \geq 0$. Substituting for x from (37) we may write (32) as

$$\begin{aligned} \min \quad & c^T\left(\sum_{i=1}^{k}\lambda_i v_i + \sum_{j=1}^{l}\mu_j e_j\right) \\ \text{s.t.} \quad & A_1\left(\sum_{i=1}^{k}\lambda_i v_i + \sum_{j=1}^{l}\mu_j e_j\right) \leq b_1 \\ & \sum_{i=1}^{k}\lambda_i = 1 \\ & \lambda \in \mathbb{R}_+^k, \mu \in \mathbb{R}_+^l, \end{aligned}$$

which is equivalent to

$$\begin{aligned} \min \quad & \sum_{i=1}^{k}(c^T v_i)\lambda_i + \sum_{j=1}^{l}(c^T e_j)\mu_j \\ \text{s.t.} \quad & \sum_{i=1}^{k}(A_1 v_i)\lambda_i + \sum_{j=1}^{l}(A_1 e_j) \leq b_1 \\ & \sum_{i=1}^{k}\lambda_i = 1 \\ & \lambda \in \mathbb{R}_+^k, \mu \in \mathbb{R}_+^l. \end{aligned} \tag{38}$$

Problem (38) is called the *master problem* of (32). Comparing formulations (32) and (38) we see that we reduced the number of constraints from m to m_1, but obtained $k+l$ variables instead of n. $k+l$ might be large compared to n, in fact even exponential (consider for example the unit cube in $\mathbb{R}^n$ with $2n$ constraints and 2^n vertices) so that there seems to be at first sight no gain in using formulation (38). However, we can use the simplex algorithm for the solution of (38). For ease of exposition abbreviate (38) by $\min\{w^T\eta : D_\eta = d, \eta \geq 0\}$ with $D \in \mathbb{R}^{(m_1+1)\times(k+l)}$, $d \in \mathbb{R}^{m_1+1}$. Recall that the simplex algorithm starts with a (feasible) basis $B \subseteq \{1, \ldots, k+l\}$, $|B| = m_1 + 1$, with D_B nonsingular and the corresponding (feasible) solution $\eta_B^* = D_B^{-1}d$ and $\eta_N^* = 0$, where $N = \{1, \ldots, k+l\} \backslash B$. Observe that $D_B \in \mathbb{R}^{(m_1+1)\times(m_1\times 1)}$ is (much) smaller than a basis for the original system (32) and that only a fraction of variables ($m_1 + 1$ out of $k+l$) are possibly nonzero. In addition, on the way to an optimal solution the only operation within the simplex method that involves all columns is the pricing step, where it is checked whether the reduced costs $w_N - \tilde{y}^T D_N$ are nonnegative with $\tilde{y} \leq 0$ being the solution of $y^T D_B = w_B$. The nonnegativity of the reduced costs can be verified via the following linear program:

$$\begin{array}{ll} \min & (c^T - \bar{y}^T A_1)x \\ \text{s.t.} & A_2 x \le b_2 \\ & x \in \mathbb{R}^n, \end{array} \tag{39}$$

where $\bar{y}$ are the first m_1 components of the solution of $\tilde{y}$. The following cases might come up:

(i) Problem (39) has an optimal solution $\tilde{x}$ with $(c^T - \bar{y}^T A_1)\tilde{x} < \tilde{y}_{m_1+1}$. In this case, $\tilde{x}$ is one of the vectors v_i, $i \in \{1, \dots, k\}$, with corresponding reduced cost

$$w_i - \tilde{y}^T D_{\cdot i} = c^T v_i - \tilde{y}^T \begin{pmatrix} A_1 v_i \\ 1 \end{pmatrix} = c^T v_i - \bar{y}^T A_1 v_i - \tilde{y}_{m_1+1} < 0.$$

In other words, $\binom{A_1 v_i}{1}$ is the entering column within the simplex algorithm.

(ii) Problem (39) is unbounded.
Here we obtain a feasible extreme ray e^* with $(c^T - \bar{y}^T A_1)e^* < 0$. e^* is one of the vectors e_j, $j \in \{1, \dots, l\}$. It yields a column $\binom{A_1 e_j}{0}$ with reduced cost

$$w_{k+j} - D_{\cdot(k+j)} = c^T e_j - \tilde{y}^T \begin{pmatrix} A_1 e_j \\ 0 \end{pmatrix} = c^T e_j - \bar{y}^T (A_1 e_j) < 0.$$

That is, $\binom{A_1 e_j}{0}$ is the entering column.

(iii) Problem (39) has an optimal solution $\tilde{x}$ with $(c^T - \bar{y}^T A_1)^T \tilde{x} \ge \tilde{y}_{m_1+1}$. In this case we conclude using the same arguments as in (i) and (ii) that $w_i - \tilde{y}^T D_{\cdot i} \ge 0$ for all $i = 1, \dots, k+l$ proving that x^* is an optimal solution for the master problem (38).

Observe that the whole problem (32) is decomposed into two problems, i.e., (38) and (39), and the approach iteratively works on the master level (38) and the subordinate level (39). The procedure starts with some feasible solution for (38) and generates new promising columns on demand by solving (39). Such procedures are commonly called *column generation* or *delayed column generation algorithms*.

The approach can also be extended to general integer programs with some caution. In this case problem (39) turns from a linear to an integer linear program. In addition, we have to guarantee in (37) that all feasible integer solutions x of (32) can be generated by (integer) linear combinations of the vectors $v_1, \dots, v_k$ and $e_1, \dots, e_l$, where

$$\operatorname{conv}(\{x \in \mathbb{Z}^n : Ax \le b\}) = \operatorname{conv}(\{v_1, \dots, v_k\}) + \operatorname{cone}(\{e_1, \dots, e_l\}).$$

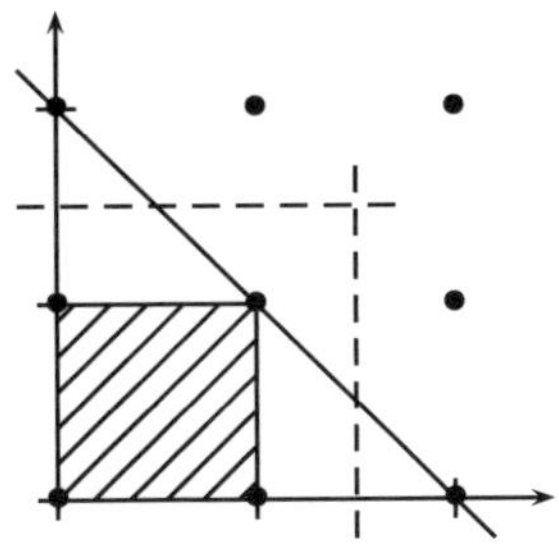

Fig. 2. Extending Dantzig–Wolfe decomposition to integer programs.

It is not sufficient to require λ and μ to be integer. Consider as a counterexample

$$A_1 = \begin{pmatrix} 1 & 0 \\ 0 & 1 \end{pmatrix}, \; b_1 = \begin{pmatrix} 1.5 \\ 1.5 \end{pmatrix} \quad \text{and} \quad A_2 = (1, 1), \; b_2 = 2$$

and the problem

$$\max\{x_1 + x_2 : A_1 x \leq b_1, A_2 x \leq b_2, x \in \{0, 1, 2\}^2\}$$

Then

$$P^2 = \text{conv}\left(\left\{\begin{pmatrix} 0 \\ 0 \end{pmatrix}, \begin{pmatrix} 2 \\ 0 \end{pmatrix}, \begin{pmatrix} 0 \\ 2 \end{pmatrix}\right\}\right),$$

see Fig. 2, but the optimal solution $\binom{1}{1}$ of the integer program is not an integer linear combination of the vertices of P^2. However, when all variables are 0–1, this difficulty does not occur, since any 0–1 solution of the LP relaxation of some binary MIP is always a vertex of that polyhedron. And in fact, column generation algorithms are not only used for the solution of large linear programs, but especially for large 0–1 integer programs. Of course, the Dantzig–Wolfe decomposition for linear or 0–1 integer programs is just one type of column generation algorithm. Others solve the subordinate problem not via general linear or integer programming techniques, but use combinatorial or explicit enumeration algorithms. Furthermore, the problem is often not modeled via (32), but directly as in (38). This is, for instance, the case when the set of feasible solutions have a rather complex description by linear inequalities, but these constraints can easily be incorporated into some enumeration scheme.

3.2.3 Benders' decomposition

Let us finally turn to *Benders' decomposition* (Benders, 1962). Benders' decomposition also deletes part of the constraint matrix, but in contrast to

Dantzig–Wolfe decomposition, where we delete part of the constraints and reintroduce them via column generation, we now delete part of the variables and reintroduce them via cutting planes. In this respect, Benders' decomposition is the same as Dantzig–Wolfe decomposition applied to the dual as we will see in detail in Section 3.2.4. Consider again (1) and write it in the form

$$\begin{array}{ll} \min & c_1^T x_1 + c_2^T x_2 \\ \text{s.t.} & A_1 x_1 + A_2 x_2 \le b \\ & x_1 \in \mathbb{R}^{n_1}, x_2 \in \mathbb{R}^{n_2}, \end{array} \tag{40}$$

where $A=[A_1, A_2] \in \mathbb{R}^{m\times n}$, $A_1 \in \mathbb{R}^{m\times n_1}$, $A_2 \in \mathbb{R}^{m\times n_2}$, c_1, $x_1 \in \mathbb{R}^{n_1}$, c_2, $x_2 \in \mathbb{R}^{n_2}$ with $n_1+n_2=n$. Note that we have assumed for ease of exposition the case of a linear program. We will see, however, that what follows is still true if $x_1 \in \mathbb{Z}^{n_1}$. Our intention is to get rid of the variables x_2. These variables prevent (40) from being a pure integer program in case $x_1 \in \mathbb{Z}^{n_1}$. Also in the linear programming case they might be the origin for some difficulties, see the applications in Section 4.4. One well known approach to get rid of variables is projection, see also the lift-and-project cuts in Section 3.1. In order to apply projection we must slightly reformulate (40) to

$$\begin{array}{lr} \min & z \\ \text{s.t.} & -z + c_1^T x_1 + c_2^T x_2 \le 0 \\ & A_1 x_1 + A_2 x_2 \le b \\ & z \in \mathbb{R}, x_1 \in \mathbb{R}^{n_1}, x_2 \in \mathbb{R}^{n_2}. \end{array} \tag{41}$$

Now, (41) is equivalent to

$$\begin{array}{ll} \min & z \\ \text{s.t.} & -uz + uc_1^T x_1 + v^T A_1 x_1 \le v^T b \\ & z \in \mathbb{R}, x_1 \in \mathbb{R}^{n_1}, \\ & \begin{pmatrix} u \\ v \end{pmatrix} \in C, \end{array} \tag{42}$$

where

$$C = \left\{ \begin{pmatrix} u \\ v \end{pmatrix} \in \mathbb{R}^{m+1} : v^T A_2 + uc_2^T = 0, u \ge 0, v \ge 0 \right\}.$$

C is a pointed polyhedral cone, thus there exist vectors

$$\begin{pmatrix} \bar{u}_1 \\ \bar{v}_s \end{pmatrix}, \ldots, \begin{pmatrix} \bar{u}_s \\ \bar{v}_s \end{pmatrix}$$

such that

$$C = \text{cone}\left(\left\{\begin{pmatrix}\bar{u}_1\\ \bar{v}_1\end{pmatrix}, \ldots, \begin{pmatrix}\bar{u}_s\\ \bar{v}_s\end{pmatrix}\right\}\right).$$

These extreme rays can be rescaled such that $\bar{u}_i$ is zero or one. Thus

$$C = \text{cone}\left(\left\{\begin{pmatrix}0\\ v_k\end{pmatrix} : k \in K\right\}\right) + \text{cone}\left(\left\{\begin{pmatrix}1\\ v_j\end{pmatrix} : j \in J\right\}\right)$$

with $K \cup J = \{1, \ldots, s\}$ and $K \cap J = \emptyset$. With this description of C, (42) can be restated as

$$\begin{array}{lllll} \min & z & & & \\ \text{s.t.} & -z \leq c_1^T x_1 + v_j^T(b - A_1 x_1) & \text{for all} & j \in J, & \\ & 0 \leq \qquad\qquad v_k^T(b - A_1 x_1) & \text{for all} & k \in K, & \\ & z \in \mathbb{R}, x_1 \in \mathbb{R}^{n_1}. & & & \end{array} \tag{43}$$

Problem (43) is called *Benders' master problem*. Benders' master problem has just $n_1 + 1$ variables instead of $n_1 + n_2$ variables in (40), or in case $x_1 \in \mathbb{Z}^{n_1}$ we have reduced the mixed integer program (40) to an almost pure integer program (43) with one additional continuous variable z. However, (43) contains an enormous number of constraints, in general exponentially many in n. To get around this problem, we solve Benders' master problem by cutting plane methods, see Section 3.1. We start with a small subset of extreme rays of C (possibly the empty set) and optimize (43) just over this subset. We obtain an optimal solution x^*, z^* of the relaxed problem and we must check whether this solution satisfies all other inequalities in (43). This can be done via the following linear program

$$\begin{array}{ll} \min & v^T(b - A_1 x_1^*) + u(z^* - c_1^T x_1^*) \\ \text{s.t.} & \begin{pmatrix}u\\ v\end{pmatrix} \in C. \end{array} \tag{44}$$

Problem (44) is called the *Benders' subproblem*. It is feasible, since $\binom{0}{0} \in C$, and (44) has an optimal solution value of zero or it is unbounded. In the first case, x_1^*, z^* satisfies all inequalities in (43) and we have solved (43) and thus (40). In the latter case we obtain an extreme ray $\binom{u^*}{v^*}$ from (44) with $(v^*)^T(b - A_1 x_1^*) + u^*(z^* - c_1^T x_1^*) < 0$ which after rescaling yields a cut for (43) violated by x_1^*, z^*. We add this cut to Benders' master problem (43) and iterate.

3.2.4 Connections between the approaches

At first sight, Lagrangean relaxation, Dantzig–Wolfe, and Benders' decomposition seem to be completely different relaxation approaches. However, they are strongly related as we will shortly outline in the following. Consider once again (39) which for some fixed $\bar{y} \leq 0$ can be rewritten as

$$\begin{array}{llll} \min & (c^T - \bar{y}^T A_1)x = \min & c^T x + \bar{y}^T(b_1 - A_1 x) - \bar{y}^T b_1 \\ \text{s.t.} & x \in P^2 & \text{s.t.} \quad x \in P^2 \\ & & = L(-\bar{y}) - \bar{y}^T b, \end{array}$$

that is, (33) and (39) are the same problems up to the constant $-\bar{y}^T b$. Even further, by replacing P^2 by $\text{conv}(\{v_1, \ldots, v_k\}) + \text{cone}(\{e_1, \ldots, e_l\})$ we see that (38) coincides with the right-hand side in (35) and thus with $L(\lambda^*)$. In other words, both Dantzig–Wolfe and Lagrangean relaxation compute the same bound. The only differences are that for updating the dual variables, i.e., λ in the Lagrangean relaxation and $\bar{y}$ in Dantzig–Wolfe, in the first case subgradient methods whereas in the latter linear programming techniques are applied. Other ways to compute λ^* are provided by the bundle method based on quadratic programming (Hiriart-Urruty and Lemarechal, 1993), and the analytic center cutting plane method that is based on an interior point algorithm (Goffin and Vial, 2002).

Similarly, Benders' decomposition is the same as that applied by Dantzig–Wolfe to the dual of (40). To see this, consider its dual

$$\begin{array}{ll} \max & -y^T b \\ \text{s.t.} & -y^T A_1 = c_1^T \\ & -y^T A_2 = c_2^T \\ & y \geq 0. \end{array} \tag{45}$$

Now reformulate $\bar{P}^2 = \{y \in \mathbb{R}^{n_2} : y^T A_2 = -c_2^T, y \leq 0\}$ by $\bar{P}^2 = \text{conv}(\{v_j : j \in J\}) + \text{cone}(\{v_k : k \in K\})$, where K, J and v_l, $l \in K \cup J$ are exactly those from (43), and rewrite (45) as

$$\begin{array}{ll} \max & \sum_{j \in J} (-v_j^T b)\lambda_j + \sum_{k \in K} (-v_k^T b)\mu_k \\ \text{s.t.} & \sum_{j \in J} (-v_j^T A_1)\lambda_j + \sum_{k \in K} (-v_k^T A_1)\mu_k = c_1^T \\ & \sum_{j \in J} \lambda_j = 1 \\ & \lambda \in \mathbb{R}_+^J, \mu \in \mathbb{R}_+^K. \end{array} \tag{46}$$

Now from Section 3.2.2 we conclude that (46) is the master problem from (45). Finally, dualizing (46) yields

$$\begin{aligned} \min\quad & c_1^T x_1 + z \\ \text{s.t.}\quad & v_i^T(b - A_1 x_1) \geq -z \ \forall j \in J \\ & v_k^T(b - A_1 x_1) \geq 0 \quad \forall k \in K, \end{aligned}$$

which is equivalent to (43), that is to the Benders' master problem of (40). In other words, Benders' and Dantzig–Wolfe decomposition yield the same bound, which by our previous discussion is also equivalent to the Lagrangean dual (34).

4 Branch-and-bound strategies

Branch-and-bound algorithms for mixed integer programming use a "divide and conquer" strategy to explore the set of all feasible mixed integer solutions. But instead of exploring the whole feasible set, they make use of lower and upper bounds and therefore avoid surveying certain (large) parts of the space of feasible solutions. Let $X := \{x \in \mathbb{Z}^p \times \mathbb{R}^{n-p} \colon Ax \leq b\}$ be the set of feasible mixed integer solutions of problem (1). If it is too difficult to compute

$$\begin{aligned} z_{\text{MIP}} = \min\quad & c^T x \\ \text{s.t.}\quad & x \in X \end{aligned}$$

(for instance with a cutting plane approach), we can split X into a finite number of subsets $X_1, \ldots, X_k \subset X$ such that $\cup_{j=1}^k X_j = X$ and then try to solve separately each of the subproblems

$$\begin{aligned} \min\quad & c^T x \\ \text{s.t.}\quad & x \in X_j, \quad \forall j = 1, \ldots, k. \end{aligned}$$

Later, we compare the optimal solutions of the subproblems and choose the best one. Each subproblem might be as difficult as the original problem, so one tends to solve them by the same method, i.e., splitting the subproblems again into further sub-subproblems. The (fast-growing) list of all subproblems is usually organized as a tree, called a *branch-and-bound tree*. Since this tree of subproblems looks like a family tree, one usually says that a *father* or *parent* problem is split into two or more *son* or *child* problems. This is the branching part of the branch-and-bound method.

For the bounding part of this method we assume that we can efficiently compute a lower bound b_{X_j} of subproblem X_j, i.e., $b_{X_j} \leq \min_{x \in X_j} c^T x$. In the case of mixed integer programming, this lower bound can be obtained by using any relaxation method discussed in Section 3. In the following, suppose we have chosen the LP relaxation method by relaxing the integrality constraints. In Section 4.4 we give references if one of the other relaxation methods is applied within branch-and-bound. For the ease of explanation we assume in the sequel that the LP relaxation has a finite optimum. It occasionally happens in the course of the branch-and-bound algorithm that the optimal solution $\tilde{x}_{X_{\bar{j}}}$ of the LP relaxation of a subproblem $X_{\bar{j}}$ is also a feasible mixed integer point, i.e., it lies in X. This allows us to maintain an upper bound $U := c^T \tilde{x}_{X_{\bar{j}}}$ on the optimal solution value z_{MIP} of X, as $z_{\text{MIP}} \leq U$. Having a good upper bound U is crucial in a branch-and-bound algorithm, because it keeps the branching tree small: suppose the solution of the LP relaxation of some other subproblem X_j satisfies $b_{X_j} \geq U$. Then subproblem X_j and further sub-subproblems derived from X_j need not be considered further, because the optimal solution of this subproblem cannot be better than the best feasible solution $\tilde{x}_{X_{\bar{j}}}$ corresponding to U. The following algorithm summarizes the whole procedure:

Algorithm 5. (*Branch-and-bound*)

1. *Let L be the list of unsolved problems. Initialize L with* (1). *Set* $U := +\infty$ *as upper bound.*
2. *Choose an unsolved problem* X_j *from the list L and delete it from L.*
3. *Compute the lower bound* b_{X_j} *by solving the linear programming relaxation. If problem* X_j *is infeasible, go to Step* 2 *until the list is empty. Otherwise, let* $\tilde{x}_{X_j}$ *be an optimal solution and set* $b_{X_j} := c^T \tilde{x}_{X_j}$.
4. *If* $\tilde{x}_{X_j} \in \mathbb{Z}^p \times \mathbb{R}^{n-p}$, *problem* X_j *is solved and we found a feasible solution of* X_j; *if* $U > b_{X_j}$ *set* $U := b_{X_j}$ *and delete all subproblems* X_i *with* $b_{X_i} \geq U$ *from the list L.*
5. *If* $\tilde{x}_{X_j} \notin \mathbb{Z}^p \times \mathbb{R}^{n-p}$, *split problem* X_j *into subproblems and add them to the list L.*
6. *Go to Step* 2 *until L is empty.*

Each (sub)problem X_j in the list L corresponds to a node in the branch-and-bound tree, where the unsolved problems are the leaves of the tree and the node that corresponds to the entire problem (1) is the root.

As crucial as finding a good upper bound is to find a good lower bound. Sometimes the LP relaxation turns out to be weak, but can be strengthened by adding cutting planes as discussed in Section 3.1. This combination of finding cutting planes and branch-and-bound leads to a hybrid algorithm called a *branch-and-cut algorithm*.

Algorithm 6. (*Branch-and-cut*)

1. *Let L be the list of unsolved problems. Initialize L with* (1). *Set* $U := +\infty$ *as upper bound.*
2. *Choose an unsolved problem* X_j *from the list L and delete it from L.*

3. *Compute the lower bound* b_{X_j} *by solving the linear programming relaxation. If problem* X_j *is infeasible, go to Step* 2 *until the list is empty. Let* $\tilde{x}_{X_j}$ *be an optimal solution and set* $b_{X_j} := c^T\tilde{x}_{X_j}$.
4. *If* $\tilde{x}_{X_j} \in \mathbb{Z}^p \times \mathbb{R}^{n-p}$, *problem* X_j *is solved and we found a feasible solution of* X_j; *if* $U > b_{X_j}$ *set* $U := b_{X_j}$ *and delete all subproblems* X_i *with* $b_{X_i} \geq U$ *from the list L.*
5. *If* $\tilde{x}_{X_j} \notin \mathbb{Z}^p \times \mathbb{R}^{n-p}$, *look for cutting planes and add them to the linear relaxation.*
6. *Go to Step* 3 *until no more violated inequalities can be found or violated inequalities have too little impact on improving the lower bound.*
7. *Split problem* X_j *into subproblems and add them to the list L.*
8. *Go to Step* 2 *until L is empty.*

In the general outline of the above branch-and-cut algorithm, there are two steps in the branch-and-bound part that leave some choices. In Step 2 of Algorithm 6 we have to select the next problem (node) from the list of unsolved problems to work on next, and in Step 7 we must decide on how to split the problem into subproblems. Usually this split is performed by choosing a variable $\tilde{x}_j \notin \mathbb{Z}$, $1 \leq j \leq p$, from an optimal solution of some subproblem X_k from the list of open problems and creating two subproblems: one with the additional bound $x_j \leq \lfloor \tilde{x}_j \rfloor$ and the other with $x_j \geq \lceil \tilde{x}_j \rceil$. Popular strategies are to branch on a variable that is closest to 0.5 and to choose a node with the worst dual bound, i.e., a problem $\tilde{j}$ from the list of open problems with $b_{X_{\tilde{j}}} = \min_j b_{X_j}$. In this section we briefly discuss some more alternatives that outperform the standard strategies. For a comprehensive study of branch-and-bound strategies we refer to Land and Powell (1979), Linderoth and Savelsbergh (1999), Achterberg, Koch, and Martin (2005), and the references therein.

4.1 Node selection

In this section we discuss three different strategies to select the node to be processed next, see Step 3 of Algorithm 6.

Best first search (bfs). Here, a node is chosen with the worst dual bound, i.e., a node with lowest lower bound, since we are minimizing in (1). The goal is to improve the dual bound. However, if this fails early in the solution process, the branch-and-bound tree tends to grow considerably resulting in large memory requirements.

Depth first search (dfs). This rule chooses some node that is "deepest" in the branch-and-bound tree, i.e., whose path to the root is longest. The advantages are that the tree tends to stay small, since one of the two sons is always processed next, if the node can not be fathomed. This fact also implies that the linear programs from one node to the next are very similar, usually the difference is just the change of one variable bound and thus the reoptimization goes fast. The main disadvantage is that the dual bound basically stays untouched during the solution process resulting in bad solution guarantees.

Best projection. When selecting a node the most important question is, where are the good (optimal) solutions hidden in the branch-and-bound tree? In other words, is it possible to guess at some node whether it contains a better solution? Of course, this is not possible in general. But, there are some rules that evaluate the nodes according to the potential of having a better solution. One such rule is *best projection*. The earliest reference we found for this rule is a paper of Mitra (1973) who gives the credit to J. Hirst. Let $z(p)$ be the dual bound of some node p, $z(\text{root})$ the dual bound of the root node, $\bar{z}_{\text{IP}}$ the value of the current best primal solution, and $s(p)$ the sum of the infeasibilities at node p, i.e., $s(p) = \sum_{i \in N} \min\{\bar{x}_i - \lfloor \bar{x}_i \rfloor, \lceil \bar{x}_i \rceil - \bar{x}_i\}$, where $\bar{x}$ is the optimal LP solution of node p and N the set of all integer variables. Let

$$\varrho(p) = z(p) + \frac{\bar{z}_{\text{IP}} - z(\text{root})}{s(\text{root})} \cdot s(p). \tag{47}$$

The term $(\bar{z}_{\text{IP}} - z(\text{root})/s(\text{root}))$ can be viewed as a measure for the change in the objective function per unit decrease in infeasibility. The *best projection* rule selects the node that minimizes $\varrho(\cdot)$.

The computational tests in Martin (1998) show that *dfs* finds by far the largest number of feasible solutions. This indicates that feasible solutions tend to lie deep in the branch-and-bound tree. In addition, the number of simplex iterations per LP is on an average much smaller (around one half) for *dfs* than for *bfs* or *best projection*. This confirms our statement that reoptimizing a linear program is fast when just one variable bound is changed. However, the *dfs* strategy does not take the dual bound into account. For many more difficult problems the dual bound is not improved resulting in very bad solution guarantees compared to the other two strategies. *Best projection* and *bfs* are doing better in this respect. There is no clear winner between the two, sometimes *best projection* outperforms *bfs*, but on average *bfs* is the best. Linderoth and Savelsbergh (1999) compare further node selection strategies and come to a similar conclusion that there is no clear winner and that a sophisticated MIP solver should allow many different options for node selection.

4.2 *Variable selection*

In this section we discuss rules on how to split a problem into subproblems, if it could not be fathomed in the branch-and-bound tree, see Step 7 of

Algorithm 6. The only way to split a problem within an LP based branch-and-bound algorithm is to branch on linear inequalities in order to keep the property of having an LP relaxation at hand. The easiest and most common inequalities are *trivial inequalities*, i.e., inequalities that split the feasible interval of a singleton variable. To be more precise, if j is some variable with a fractional value $\bar{x}_j$ in the current optimal LP solution, we obtain two subproblems, one by adding the trivial inequality $x_j \leq \lfloor \bar{x}_j \rfloor$ (called the *left subproblem* or *left son*) and one by adding the trivial inequality $x_j \geq \lceil \bar{x}_j \rceil$ (called the *right subproblem* or *right son*). This rule of branching on trivial inequalities is also called *branching on variables*, because it actually does not require the addition of an inequality, but only a change of the bounds of variable j. Branching on more complicated inequalities or even splitting the problem into more than two subproblems are rarely incorporated into general solvers, but turn out to be effective in special cases, see, for instance, Borndörfer et al. (1998), Clochard and Naddef (1993), Naddef (2002). In the following we present three variable selection rules.

Most infeasibility. This rule is to choose a variable that is closest to 0.5. The heuristic reason behind this choice is that this is a variable where the least tendency can be recognized to which "side" (up or down) the variable should be rounded. The hope is that a decision on this variable has the greatest impact on the LP relaxation.

Pseudo-costs. This is a more sophisticated rule in the sense that it keeps a history of the success of the variables on which one has already branched. To introduce this rule, which goes back to (Benichou et al., 1971), we need some notation. Let $\mathcal{P}$ denote the set of all problems (nodes) except the root node that have already been solved in the solution process. Initially, this set is empty. $\mathcal{P}^+$ denotes the set of all right sons, and $\mathcal{P}^-$ the set of all left sons, where $\mathcal{P} = \mathcal{P}^+ \cup \mathcal{P}^-$. For some problem $p \in \mathcal{P}$ let

$f(p)$ be the father of problem p.
$v(p)$ be the variable that has been branched on to obtain problem p from the father $f(p)$.
$x(p)$ be the optimal solution of the final linear program at node p.
$z(p)$ be the optimal objective function value of the final linear program at node p.

The *up pseudo-cost* of variable $j \in N$ is

$$\Phi^+(j) = \frac{1}{|P_j^+|} \sum_{p \in P_j^+} \frac{z(p) - z(f(p))}{\lceil x_{v(p)}(f(p)) \rceil - x_{v(p)}(f(p))}, \tag{48}$$

where $P_j^+ \subseteq \mathcal{P}^+$. The *down pseudo-cost* of variable $j \in N$ is

$$\Phi^-(j) = \frac{1}{|P_j^-|} \sum_{p \in P_j^-} \frac{z(p) - z(f(p))}{x_{\upsilon(p)}(f(p)) - \lfloor x_{\upsilon(p)}(f(p)) \rfloor}, \tag{49}$$

where $P_j^- \subseteq \mathcal{P}^-$. The terms

$$\frac{z(p) - z(f(p))}{\lceil x_{\upsilon(p)}(f(p)) \rceil - x_{\upsilon(p)}(f(p))} \quad \text{and} \quad \frac{z(p) - z(f(p))}{x_{\upsilon(p)}(f(p)) - \lfloor x_{\upsilon(p)}(f(p)) \rfloor},$$

respectively, measure the change in the objective function per unit decrease of infeasibility of variables j. There are many suggestions made on how to choose the sets P_j^+ and P_j^-, for a survey see Linderoth and Savelsbergh (1999). To name one possibility, following the suggestion of Eckstein (1994) one could choose $P_j^+ := \{p \in \mathcal{P}^+ : \upsilon(p) = j\}$ and $P_j^- := \{p \in \mathcal{P}^- : \upsilon(p) = j\}$, if j has already been considered as a branching variable, otherwise set $P_j^+ := \mathcal{P}^+$ and $P_j^- := \mathcal{P}^-$. It remains to discuss how to weight the *up* and *down pseudo-costs* against each other to obtain the final *pseudo-costs* according to which the branching variable is selected. Here one typically sets

$$\Phi(j) = \alpha_j^+ \Phi^+(j)(\lceil \bar{x}_j \rceil - x_j) + \alpha_j^- \Phi^-(j)(\bar{x}_j - \lfloor \bar{x}_j \rfloor), \tag{50}$$

where α_j^+, α_j^- are positive scalars. A variable that maximizes (50) is chosen to be the next branching variable. As formula (50) shows, the rule takes the earlier success of the variables into account when deciding on the next branching variable. The weakness of this approach is that at the very beginning there is no information available, and $\Phi(\cdot)$ is almost identical for all variables. Thus, at the beginning where the branching decisions are usually the most critical the pseudo-costs take no effect. An attempt is made to overcome this drawback in the following rule.

Strong branching. The idea of *strong branching*, invented by CPLEX (ILOG CPLEX Division, 1997), see also Applegate, Bixby, Chvátal, and Cook (1995), is before actually branching on some variable to test whether it indeed gives some progress. This testing is done by fixing the variable temporarily to its up and down value, i.e., to $\lceil \bar{x}_j \rceil$ and $\lfloor \bar{x}_j \rfloor$ if $\bar{x}_j$ is the fractional LP value of variable j, performing a certain fixed number of dual simplex iterations for each of the two settings, and measuring the progress in the objective function value. The testing is done, of course, not only for one variable but for a certain set of variables. Thus, the parameters of *strong branching* to be specified are the size of the candidate set, the maximum number of dual simplex iterations to be performed on each candidate variable, and a criterion according to which the candidate set is selected. Needless to say that each MIP solver has its own parameter settings, all are of heuristic nature and that their justifications are based only on experimental results.

Computational experience in Martin (1998) show that branching on a *most infeasible* variable is by far the worst, measured in CPU time, in solution quality as well as in the number of branch-and-bound nodes. Using *pseudo-costs* gives much better results. The power of *pseudo-costs* becomes particularly apparent if the number of already solved branch-and-bound nodes is large. In this case the function $\Phi(\cdot)$ properly represents the variables that are qualified for branching. In addition, the time necessary to compute the *pseudo-costs* is basically for free. The statistics change when looking at *strong branching*. *Strong branching* is much more expensive than the other two strategies. This comes as no surprise, since in general the average number of dual simplex iterations per linear program is very small (for the `Miplib`, for instance, below 10 on average). Thus, the testing of a certain number of variables (even if it is small) in *strong branching* is relatively expensive. On the other hand, the number of branch-and-bound nodes is much smaller (around one half) compared to the *pseudo-costs* strategy. This decrease, however, does not completely compensate the higher running times for selecting the variables in general. Thus, *strong branching* is normally not used as a default strategy, but can be a good choice for some hard instances. A similar report is given in Linderoth and Savelsbergh (1999), where Linderoth and Savelsbergh conclude that there is no branching rule that clearly dominates the others, though pseudo-cost strategies are essential to solve many instances.

The latter strategy is refined in several aspects in Linderoth and Savelsbergh (1999), Achterberg, Koch, and Martin (2005) to hybrid methods where the advantages of pseudo-cost and strong branching are put together. The basic idea is to use strong branching at the very beginning when pseudo-costs contain no or only poor information and switches to the much faster pseudo-cost strategy later in the solution process.

4.3 Further aspects

In this section we discuss some additional issues that can be found in basically every state-of-the-art branch-and-cut implementation.

LP management. The method that is commonly used to solve the LPs within a branch-and-cut algorithm is the dual simplex algorithm, because an LP basis stays dual feasible when adding cutting planes. There are fast and robust linear programming solvers available, see, for instance, DASH Optimization (2001) and ILOG CPLEX Division (2000). Nevertheless, one major aspect in the design of a branch-and-cut algorithm is to control the size of the linear programs. To this end, inequalities are often assigned an "age" (at the beginning the age is set to 0). Each time the inequality is not tight at the current LP solution, the age is increased by one. If the inequality gets too old, i.e., the age exceeds a certain limit, the inequality is eliminated from the LP. The value for this "age limit" varies from application to application. Another issue of LP management concerns the questions: When should an inequality be added to the LP? When is an inequality considered to be "violated"? And, how many and which inequalities should be added? The answers to these questions again depend on the applications. It is clear that one always makes sure that no redundant inequalities are added to the linear program. A commonly used data structure in this context is the *pool*. Violated inequalities that are added to the LP are stored in this data structure. Also inequalities that are eliminated from the LP are restored in the pool. Reasons for the pool are to reconstruct the LPs when switching from one node in the branch-and-bound tree to another and to keep inequalities that were "expensive" to separate for an easier access in the ongoing solution process.

Heuristics. Raising the lower bound using cutting planes is one important aspect in a branch-and-cut algorithm, finding good feasible solutions early to enable fathoming of branches of the search-tree is another. Primal heuristics strongly depend on the application. A very common way to find feasible solutions for general mixed integer programs is to "plunge" from time to time at some node of the branch-and-bound tree, i.e., to dive deeper into the tree and look for feasible solutions. This plunging is done by alternating rounding/fixing some variables and solving linear programs, until all the variables are fixed, the LP is infeasible, a feasible solution has been found, or the LP value exceeds the current best solution. This rounding heuristic can be detached from the regular branch-and-bound enumeration phase or considered within the global enumeration phase. The complexity and the sensitivity to the change of the LP solutions influences the frequency with which the heuristics are called. Some more information on this topic can be found, for instance, in Bixby, Fenelon, Guand, Rothberg, and Wunderling (1998), Cordier, Marchand, Laundy, and Wolsey (1999), Martin (1998).
Some ideas that go beyond this general approach of rounding and fixing variables can be found in Balas, Ceria, Dawande, Margot, and Pataki (2001), Balas and Martin (1980), Fischetti and Lodi (2002). Balas et al. (2001) observe that an LP solution consisting solely of slack variables must be integer and thus try to pivot in slack variables into the optimal LP solution to derive feasible integer solutions. In Balas et al. (2001) 0–1 solutions are generated by doing local search in a more sophisticated manner. Very recently, a new idea was proposed by Fischetti and Lodi (2002). Instead of fixing certain variables, they branch on the constraint that any new solution must have at least or at most a certain number of fixings in common with the current best solution. The computational

results show that with this branching rule very fast good feasible solutions are obtained.

Reduced cost fixing. The idea is to fix variables by exploiting the reduced costs of the current optimal LP solution. Let $\bar{z}=c^T\bar{x}$ be the objective function value of the current LP solutions, z^{IP} be an upper bound on the value of the optimal solution, and $d=(d_i)_{i=1,\ldots,n}$ the corresponding reduced cost vector. Consider a nonbasic variable x_i of the current LP solution with finite lower and upper bounds l_i and u_i, and nonzero reduced cost d_i. Set $\delta=(z^{\mathrm{IP}}-\bar{z}/|d_i|)$, rounded down in case x_j is a binary or an integer variable. Now, if x_i is currently at its lower bound l_i and $l_i+\delta< u_i$, the upper bound of x_i can be reduced to $l_i+\delta$. In case x_i is at its upper bound u_i and $u_i-\delta>l_i$, the lower bound of variable x_i can be increased to $u_i-\delta$. In case the new bounds l_i and u_i coincide, the variable can be fixed to its bounds and removed from the problem. This strengthening of the bounds is called *reduced cost fixing*. It was originally applied for binary variables (Crowder et al., 1983), in which case the variable can always be fixed if the criterion applied. There are problems where by the reduced cost criterion many variables can be fixed, see, for instance, (Ferreira, Martin, and Weismantel, 1996). Sometimes, further variables can be fixed by logical implications, for example, if some binary variable x_i is fixed to one by the reduced cost criterion and it is contained in an SOS constraint (i.e., a constraint of the form $\sum_{j\in J}x_j\leq 1$ with nonnegative variables x_j), all other variables in this SOS constraint can be fixed to zero.

4.4 Other relaxation methods within branch-and-bound

We have put our emphasis up to now on branch-and-cut algorithms where we investigated the LP-relaxation in combination with the generation of cutting planes. Of course the bounding within branch-and-bound algorithms could also be obtained by any other relaxation method discussed in Section 3.2.

Dantzig–Wolfe decomposition or delayed column generation in connection with branch-and-bound is commonly called *branch-and-price algorithm*. Branch-and-price algorithms have been successfully applied for instance in airline crew scheduling, vehicle routing, public mass transport, or network design, to name just a few. An outline of recent developments, practical applications, and implementation details of branch-and-price can be found for instance in Barnhart, Johnson, Nemhauser, Savelsbergh, and Vance (1998), Lübbecke and Desrosiers (2002), Savelsbergh (2001), Vanderbeck (1999), (2000). Of course, also integer programs with bordered block diagonal form, see Fig. 1, nicely fit into this context. In contrast to Lagrangean relaxation, see below, where the coupling constraints are relaxed, Dantzing–Wolfe decomposition keeps these constraints in the master problem and relaxes the constraints of the blocks having the advantage that (39) decomposes into independent problems, one for each block.

Lagrangean relaxation is very often used if the underlying linear programs of (1) are just too big to be solved directly and even the relaxed problems in (33) are still large (Löbel, 1997, 1998). Often the relaxation can be done in a way that the evaluation of (33) can be solved combinatorially. In the following we give some applications where this method has been successfully applied and a good balance between these two opposite objectives can be found.

Consider the traveling salesman problem where we are given a set of nodes $V=\{1,\dots,n\}$ and a set of edges E. The nodes are the cities and the edges are pairs of cities that are connected. Let $c_{(i,j)}$ for $(i,j)\in E$ denote the traveling time from city i to city j. The traveling salesman problem (TSP) now asks for a tour that starts in city 1, visits every other city exactly once, returns to city 1 and has minimal travel time. We can model this problem by the following 0–1 integer program. The binary variable $x_{(i,j)}\in\{0,1\}$ equals 1 if city j is visited right after city i is left, and equals 0 otherwise, that is $x\in\{0,1\}^E$. The equations

$$\sum_{\{i:(i,j)\in E\}} x_{(i,j)} = 2 \quad \forall j \in V$$

(*degree constraints*) ensure that every city is entered and left exactly once, respectively. To eliminate subtours, for any $U\subset V$ with $2\leq|U|\leq|V|-1$, the constraints

$$\sum_{\{(i,j)\in E:i,j\in U\}} x_{(i,j)} \leq |U| - 1$$

have to be added. By relaxing the degree constraints in the integer programming formulation for the traveling salesman problem, we are left with a spanning tree problem, which can be solved fast by the greedy algorithm. A main advantage of this TSP relaxation is that for the evaluation of (33) combinatorial algorithms are at hand and no general LP or IP solution techniques must be used. Held and Karp (1971) proposed this approach in the seventies and they solved instances that could not be solved with any other method at that time.

Other examples where Lagrangean relaxation is used are multicommodity flow problems arising for instance in vehicle scheduling or scenario decompositions of stochastic mixed integer programs. In fact, the latter two applications fall into a class of problems where the underlying matrix has bordered block diagonal form, see Fig. 1. If we relax the coupling constraints within a Lagrangean relaxation, the remaining matrix decomposes into k independent blocks. Thus, $L(\lambda)$ is the sum of k individual terms that can be determined separately. Often each single block A_i models a network flow

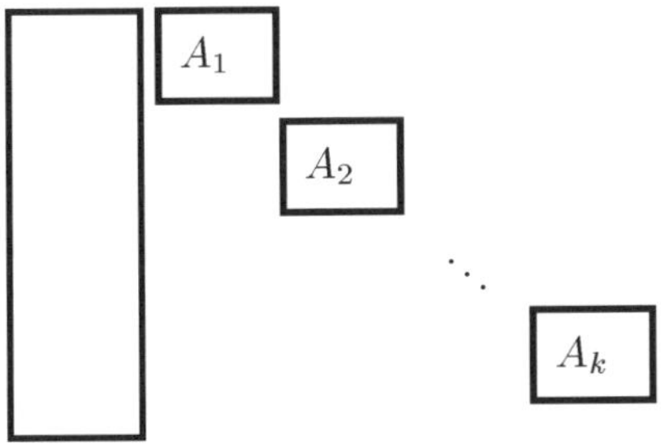

Fig. 3. Matrix in bordered block diagonal form with coupling variables.

problem, a knapsack problem or the like and can thus be solved using special purpose combinatorial algorithms.

The *volume algorithm* presented in Barahona and Anbil (2000) is a promising new algorithm also based on Lagrangean-type relaxation. It was successfully integrated in a branch-and-cut framework to solve some difficult instances of combinatorial optimization problems (Barahona and Ladanyi, 2001).

Benders' decomposition is very often implicitly used within cutting plane algorithms, see for instance the derivation of lift-and-project cuts in Section 3.1. Other applications areas are problems whose constraint matrix has bordered block diagonal form, where we have coupling variable instead of coupling constraints, see Fig. 3, i.e., the structure of the constraint matrix is the transposed of the structure of the constraint matrix in Fig. 1. Such problems appear, for instance, in stochastic integer programming (Sherali and Fraticelli, 2002). Benders' decomposition is attractive in this case, because Benders' subproblem decomposes into k independent problems.

5 Final remarks

In this chapter we have described the state-of-the-art in solving general mixed integer programs where we put our emphasis on the branch-and-cut method.

In Section 2 we explained in detail preprocessing techniques and some ideas used in structure analysis. These are however just two steps, though important, in answering the question on how information that is inhered in a problem can be carried over to the MIP solver. The difficulty is that the only "language" that MIP solvers understand and in which information can be transmitted are linear inequalities: The MIP solver gets as input some formulation as in (1). But such a formulation might be worse than others as we have seen for the Steiner tree problem in Section 2 and there is basically no way to reformulate (3) into (4) if no additional information like "this is a Steiner tree problem" is given. In other words, there are further tools necessary in order to transmit such information. Modeling languages like AMPL

(Fourer, Gay, and Kernighan, 1993) or ZIMPL (Koch, 2001) are going in this direction, but more needs to be done.

In Section 3 we described several relaxation methods where we mainly concentrated on cutting planes. Although the cutting plane method is among the most successful to solve general mixed integer programs, it is not the only one and there is pressure of competition from various sides like semi-definite programming, Gomory's group approach, basis reduction or primal approaches, see the various chapters in this handbook. We explained the most frequently used cutting planes within general MIP solvers, Gomory cuts, mixed integer rounding cuts, lift-and-project cuts as well as knapsack and set packing cutting planes. Of course, there are more and the interested reader will find a comprehensive survey in Marchand et al. (2002).

Finally, we discussed the basic strategies used in enumerating the branch-and-bound tree. We have seen that they have a big influence on the performance. A bit disappointing from a mathematical point of view is that these strategies are only evaluated computationally and that there is no theoretical proof that tells that one strategy is better than another.

All in all, mixed integer programming solvers have become much better during the last years. Their success lies in the fact that they gather more and more knowledge from the solution of special purpose problems and incorporate it into their codes. This process will and must continue to push the frontier of solvability further and further.

5.1 Software

The whole chapter was about the features of current mixed integer programming solvers. So we do not want to conclude without mentioning some of them. Due to the rich variety of applications and problems that can be modeled as mixed integer programs, it is not in the least surprising that many codes exist and not just a few of them are business oriented. In many cases, free trial versions of the software products mentioned below are available for testing. From time to time, the INFORMS newsletter OR/MS Today gives a survey on currently available commercial linear and integer programming solvers, see for instance Sharda (1995).

The following list shows software where we know that it has included many of the aspects that are mentioned in this chapter:

ABACUS, developed at the University of Cologne (Thienel, 1995), provides a branch-and-cut framework mainly for combinatorial optimization problems,

bc-opt, developed at CORE (Cordier et al., 1999), is very strong for mixed 0–1 problems,

CPLEX, developed at Incline Village (Bixby et al., 1998; ILOG CPLEX Division, 2000), is one of the currently best commercial codes,

`LINDO` and `LINGO` are commercial codes developed at Lindo Systems Inc. (1997) used in many real-world applications,

`MINTO`, developed at Georgia Institute of Technology (Nemhauser, Savelsbergh, and Sigismondi, 1994), is excellent in cutting planes and has included basically all the mentioned cutting planes and more,

`MIPO`, developed at Columbia University (Balas et al., 1996), is very good in lift-and-project cuts,

`OSL`, developed at IBM Corporation (Wilson, 1992), is now available with COIN, an open source Computational Infrastructure for Operations Research (COIN, 2002),

`SIP`, developed at Darmstadt University of Technology and ZIB, is the software of one of the authors,

`SYMPHONE`, developed at Cornell University and Lehigh University (Ralphs, 2000), has its main focus on providing a parallel framework,

`XPRESS-MP`, developed at DASH (DASH Optimization, 2001), is also one of the best commercial codes.

References

Aardal, K., Y. Pochet, L. A. Wolsey (1995). Capacitated facility location: valid inequalities and facets. *Mathematics of Operations Research* 20, 562–582.

Aardal, K., R. Weismantel, L. A. Wolsey (2002). Non-standard approaches to integer programming. *Discrete Applied Mathematics* 123/124, 5–74.

Achterberg, T., T. Koch, A. Martin (2005). Branching Rules Revisited, *Operation Research Letters* 33, 42–54.

Andersen, E. D., K. D. Andersen (1995). Presolving in linear programming. *Mathematical Programming* 71, 221–245.

Applegate, D., R. E. Bixby, V. Chvátal, W. Cook (March, 1995). Finding cuts in the TSP. Technical Report 95-05, DIMACS.

Atamtürk, A. (2003). On the facets of the mixed-integer knapsack polyhedron. *Mathematical Programming* 98, 145–175.

Atamtürk, A. (2004). Sequence independent lifting for mixed integer programming. *Operations Research* 52, 487–490.

Atamtürk, A. (2002). On capacitated network design cut-set polyhedral. *Mathematical Programming* 92, 425–437.

Atamtürk, A., G. L. Nemhauser, M. W. P. Savelsbergh (2000). Conflict graphs in integer programming. *European Journal of Operations Research* 121, 40–55.

Balas, E. (1975). Facets of the knapsack polytope. *Mathematical Programming* 8, 146–164.

Balas, E., S. Ceria, G. Cornuéjols (1993). A lift-and-project cutting plane algorithm for mixed 0–1 programs. *Mathematical Programming* 58, 295–324.

Balas, E., S. Ceria, G. Cornuéjols (1996). Mixed 0–1 programming by lift-and-project in a branch-and-cut framework. *Management Science* 42, 1229–1246.

Balas, E., S. Ceria, G. Cornuéjols, N. Natraj (1996). Gomory cuts revisited. *Operations Research Letters* 19, 1–9.

Balas, E., S. Ceria, M. Dawande, F. Margot, G. Pataki (2001). OCTANE: a new heuristic for pure 0–1 programs. *Operations Research* 49, 207–225.

Balas, E., R. Martin (1980). Pivot and complement: a heuristic for 0–1 programming. *Management Science* 26, 86–96.

Balas, E., E. Zemel (1978). Facets of the knapsack polytope from minimal covers. *SIAM Journal on Applied Mathematics* 34, 119–148.

Barahona, F., L. Ladanyi (2001). Branch and cut based on the volume algorithm: Steiner trees in graphs and max-cut. Technical Report RC22221, IBM.

Barahona, F., Ranga Anbil (2000). The volume algorithm: producing primal solutions with a subgradient method. *Mathematical Programming* 87(3), 385–399.

Barnhart, C., E. L. Johnson, G. L. Nemhauser, M. W. P. Savelsbergh, P. H. Vance (1998). Branch-and-price: column generation for huge integer programs. *Operations Research* 46, 316–329.

Benders, J. F. (1962). Partitioning procedures for solving mixed variables programming. *Numerische Mathematik* 4, 238–252.

Benichou, M., J. M. Gauthier, P. Girodet, G. Hentges, G. Ribiere, O. Vincent (1971). Experiments in mixed-integer programming. *Mathematical Programming* 1, 76–94.

Bienstock, D., M. Zuckerberg (2003). Subset algebra lift operators for 0–1 integer programming. Technical Report CORC Report 2002-01, Columbia University, New York.

Bixby, R. E. (1994). *Lectures on Linear Programming*. Rice University, Houston, Texas, Spring.

Bixby, R. E., S. Ceria, C. McZeal, M. W. P. Savelsbergh (1998). An updated mixed integer programming library: MIPLIB 3.0. Paper and Problems are available at WWW Page: http://www.caam.rice.edu/~ bixby/miplib/miplib.html.

Bixby, R. E., M. Fenelon, Z. Guand, E. Rothberg, R. Wunderling (1999). MIP: theory and practice closing the gap. Technical Report, ILOG Inc., Paris, France.

Borndörfer, R. (1998). *Aspects of Set Packing, Partitioning, and Covering*. Shaker, Aachen.

Borndörfer, R., C. E. Ferreira, A. Martin (1998). Decomposing matrices into blocks. *SIAM Journal on Optimization* 9, 236–269.

Ceria, S., C. Cordier, H. Marchand, L. A. Wolsey (1998). Cutting planes for integer programs with general integer variables. *Mathematical Programming* 81, 201–214.

Chopra, S., M. R. Rao (1994). The Steiner tree problem I: formulations, compositions and extension of facets. *Mathematical Programming* 64(2), 209–229.

Clochard, J. M., D. Naddef (1993). Using path inequalities in a branch-and-cut code for the symmetric traveling salesman problem, in: L. A. Wolsey, G. Rinaldi (eds.), *Proceedings on the Third IPCO Conference* 291–311.

COIN (2002). A COmputational INfrastructures for Operations Research. URL: http://www124.ibm.com/developerworks/opensource/coin.

Cordier, C., H. Marchand, R. Laundy, L. A. Wolsey (1999). bc – opt: a branch-and-cut code for mixed integer programs. *Mathematical Programming* 86, 335–354.

Crowder, H., E. Johnson, M. W. Padberg (1983). Solving large-scale zero-one linear programming problems. *Operations Reserch* 31, 803–834.

Dantzig, G. B., P. Wolfe (1960). Decomposition principle for linear programs. *Operations Research* 8, 101–111.

DASH Optimization (2001). Blisworth House, Church Lane, Blisworth, Northants NN7 3BX, UK. *XPRESS-MP Optimisation Subroutine Library*, Information available at URL http://www.dash.co.uk.

de Farias, I. R., E. L. Johnson, G. L. Nemhauser (2002). Facets of the complementarity knapsack polytope. *Mathematics of Operations Research*, 27, 210–226.

Eckstein, J. (1994). Parallel branch-and-bound algorithms for general mixed integer programming on the CM-5. *SIAM Journal on Optimization* 4, 794–814.

Ferreira, C. E. (1994). *On Combinatorial Optimization Problems Arising in Computer System Design*. PhD thesis, Technische Universität, Berlin.

Ferreira, C. E., A. Martin, R. Weismantel (1996). Solving multiple knapsack problems by cutting planes. *SIAM Journal on Optimization* 6, 858–877.

Fischetti, M., A. Lodi (2002). Local branching. *Mathematical Programming* 98, 23–47.

Fourer, R., D. M. Gay, B. W. Kernighan (1993). *AMPL: A Modeling Language for Mathematical Programming*. Duxbury Press/Brooks/Cole Publishing Company.

Fulkerson, D. R. (1971). Blocking and anti-blocking pairs of polyhedra. *Mathematical Programming* 1, 168–194.

Garey, M. R., D. S. Johnson (1979). *Computers and Intractability: A Guide to the Theory of NP-Completeness*. W. H. Freeman and Company, New York.

Goffin, J. L., J. P. Vial (1999). Convex nondifferentiable optimization: a survey focused on the analytic center cutting plane method. Technical Report 99.02, Logilab, Universite de Geneve. To appear in *Optimization Methods and Software*.

Gomory, R. E. (1958). Outline of an algorithm for integer solutions to linear programs. *Bulletin of the American Society* 64, 275–278.

Gomory, R. E. (1960). An algorithm for the mixed integer problem. Technical Report RM-2597, The RAND cooperation.

Gomory, R. E. (1960). Solving linear programming problems in integers, in: R. Bellman M. Hall (eds.), *Combinatorial Analysis, Proceedings of Symposia in Applied Mathematics* Vol. 10, Providence RI.

Gondzio, J. (1997). Presolve analysis of linear programs prior to apply an interior point method. *INFORMS Journal on Computing* 9, 73–91.

Grötschel, M., L. Lovász, A. Schrijver (1988). *Geometric Algorithms and Combinatorial Optimization*. Springer.

Grötschel, M., C. L. Monma, M. Stoer (1992). Computational results with a cutting plane algorithm for designing communication networks with low-connectivity constraints. *Operations Research* 40, 309–330.

Gu, Z., G. L. Nemhauser, M. W. P. Savelsbergh (1998). Cover inequalities for 0–1 linear programs: complexity. *INFORMS Journal on Computing* 11, 117–123.

Gu, Z., G. L. Nemhauser, M. W. P. Savelsbergh (1998). Cover inequalities for 0–1 linear programs: computation. *INFORMS Journal on Computing* 10, 427–437.

Gu, Z., G. L. Nemhauser, M. W. P. Savelsbergh (1999). Lifted flow cover inequalities for mixed 0–1 integer programs. *Mathematical Programming* 85, 439–468.

Gu, Z., G. L. Nemhauser, M. W. P. Savelsbergh (2000). Sequence independent lifting in mixed integer programming. *Journal on Combinatorial Optimization* 4, 109–129.

Hammer, P. L., E. Johnson, U. N. Peled (1975). Facets of regular 0–1 polytopes. *Mathematical Programming* 8, 179–206.

Held, M., R. Karp (1971). The traveling-salesman problem and minimum spanning trees: part II. *Mathematical Programming* 1, 6–25.

Hiriart-Urruty, J. B., C. Lemarechal. (1993). Convex analysis and minimization algorithms, part 2: advanced theory and bundle methods. *Grundlehren der Mathematischen Wissenschaften*. Springer-Verlag, Vol. 306.

Hoffman, K. L., M. W. Padberg (1991). Improved LP-representations of zero-one linear programs for branch-and-cut. *ORSA Journal on Computing* 3, 121–134.

ILOG CPLEX Division (1997). 889 Alder Avenue, Suite 200, Incline Village, NV 89451, USA. *Using the CPLEX Callabel Library*, Information available at URL http://www.cplex.com.

ILOG CPLEX Division (2000). 889 Alder Avenue, Suite 200, Incline Village, NV 89451, USA. *Using the CPLEX Callabel Library*, Information available at URL http://www.cplex.com.

Johnson, E., M. W. Padberg (1981). A note on the knapsack problem with special ordered sets. *Operations Research Letters* 1, 18–22.

Johnson, E. L., G. L. Nemhauser, M. W. P. Savelsbergh (2000). Progress in linear programming based branch-and-bound algorithms: an exposition. *INFORMS Journal on Computing* 12, 2–23.

Klabjan, D., G. L. Nemhauser, C. Tovey (1998). The complexity of cover inequality separation. *Operations Research Letters* 23, 35–40.

Koch, T. (2001). ZIMPL user guide. Technical Report Preprint 01-20, Konrad-Zuse-Zentrum Für Informationstechnik Berlin.

Koch, T., A. Martin, S. Voß (2001). SteinLib: an updated library on Steiner tree problems in graphs, in: D.-Z. Du, X. Cheng (eds.), *Steiner Tress in Industries*, Kluwer, 285–325.

Land, A., S. Powell (1979). Computer codes for problems of integer programming. *Annals of Discrete Mathematics* 5, 221–269.

Lasserre, J. B. (2001). An explicit exact SDP relaxation for nonlinear 0–1 programs, in: K. Aardal, A. M. H. Gerards (eds.), *Lecture Notes in Computer Science*, 293–303.

Lemaréchal, C., A. Renaud (2001). A geometric study of duality gaps, with applications. *Mathematical Programming* 90, 399–427.

Linderoth, J. T., M. W. P. Savelsbergh (1999). A computational study of search strategies for mixed integer programming. *INFORMS Journal on Computing* 11, 173–187.

Lindo Systems Inc. (1997). *Optimization Modeling with LINDO*. See web page: http://www.lindo.com.

Löbel, A. (1997). *Optimal Vehicle Scheduling in Public Transit*. PhD thesis, Technische Universität Berlin.

Löbel, A. (1998). Vehicle scheduling in public transit and lagrangean pricing. *Management Science* 12(44), 1637–1649.

Lovász, L., A. Schrijver (1991). Cones of matrices and set-functions and 0–1 optimization. *SIAM Journal on Optimization* 1, 166–190.

Lübbecke, J. E., Jacques Desrosiers (2002). Selected topics in column generation. Technical Report, Braunschweig University of Technology, Department of Mathematical Optimization.

Marchand, H. (1998). *A Polyhedral Study of the Mixed Knapsack Set and its Use to Solve Mixed Integer Programs*. PhD thesis, Université Catholique de Louvain, Louvain-la-Neuve, Belgium.

Marchand, H., A. Martin, R. Weismantel, L. A. Wolsey (2002). Cutting planes in integer and mixed integer programming. *Discrete Applied Mathematics* 123/124, 391–440.

Marchand, H., L. A. Wolsey (1999). The 0–1 knapsack problem with a single continuous variable. *Mathematical Programming* 85, 15–33.

Marchand, H., L. A. Wolsey (2001). Aggregation and mixed integer rounding to solve MIPs. *Operations Research* 49, 363–371.

Martin, A. (1998). Integer programs with block structure. Habilitations-Schrift, Technische Universität Berlin, Available as ZIB-Preprint SC-99-03, see www.zib.de.

Martin, A., R. Weismantel (1998). The intersection of knapsack polyhedra and extensions, in: R. E., Bixby, E. A. Boyd, R. Z. Ríos-Mercado (eds.), *Integer Programming and Combinatorial Optimization*, Proceedings of the 6th IPCO Conference, 243–256.

Mitra, G. (1973). Investigations of some branch and bound strategies for the solution of mixed integer linear programs. *Mathematical Programming* 4, 155–170.

Naddef, D. (2002). Polyhedral theory and branch-and-cut algorithms for the symmetric tsp. in: G. Gutin, A. Punnen (eds.), *The Traveling Salesman Problem and its Variations*. Kluwer.

Nemhauser, G. L., M. W. P. Savelsbergh, G. Minto, C. Sigismondi (1994). MINTO a mixed integer optimizer. *Operations Research Letters* 15, 47–58.

Nemhauser, G. L., P. H. Vance (1994). Lifted cover facets of the 0–1 knapsack Polytope with GUB constraints. *Operations Research Letters* 16, 255–263.

Nemhauser, G. L., L. A. Wolsey (1988). *Integer and Combinatorial Optimization*. Wiley.

Nemhauser, G. L., L. A. Wolsey (1990). A recursive procedure to generate all cuts for 0–1 mixed integer programs. *Mathematical Programming* 46, 379–390.

Padberg, M. W. (1973). On the facial structure of set packing polyhedra. *Mathematical Programming* 5, 199–215.

Padberg, M. W. (1975). A note on zero-one programming. *OR* 23(4), 833–837.

Padberg, M. W. (1980). $(1, k)$-Configurations and facets for packing problems. *Mathematical Programming* 18, 94–99.

Padberg, M. W. (1995). *Linear Optimization and Extensions*. Springer.

Padberg, M. W. (2001). Classical cuts for mixed-integer programming and branch-and-cut. *Mathematical Methods of OR* 53, 173–203.

Pedberg, M. W., T. J. Van Roy, L. A. Wolsey (1985). Valid inequalities for fixed charge problems. *Operations Research* 33, 842–861.

Pochet, Y. (1988). Valid inequalities and separation for capacitated economic lot-sizing. *Operations Research Letters* 7, 109–116.

Ralphs, T. K. (September, 2000). *SYMPHONY Version* 2.8 *User's Manual*. Information available at http://www.lehigh.edu/inime/ralphs.htm.

Richard, J. P., I. R. de Farias, G. L. Nemhauser (2001). Lifted inequalities for 0–1 mixed integer programming: basic theory and algorithms. *Lecture Notes in Computer Science*.

Van Roy, T. J., L. A. Wolsey (1986). Valid inequalities for mixed 0–1 programs. *Discrete Applied Mathematics* 4, 199–213.

Van Roy, T. J., L. A. Wolsey (1987). Solving mixed integer programming problems using automatic reformulation. *Operations Research* 35, 45–57.

Vasek Chvátal (1983). *Linear Programming*. W. H. Freeman and Company.

Savelsbergh, M. W. P. (1994). Preprocessing and probing for mixed integer programming problems. *ORSA J. on Computing* 6, 445–454.

Savelsbergh, M. W. P. (2001). Branch-and-price: integer programming with column generation, in: P. Pardalos, C. Flouda (eds.), *Encylopedia of Optimization*, Kluwer.

Schrijver, A. (1986). *Theory of Linear and Integer Programming*. Wiley, Chichester.

Sharda, R. (1995). Linear programming solver software for personal computers: 1995 report. *OR/MS Today* 22(5), 49–57.

Sherali, H., W. Adams (1990). A hierarchy of relaxations between the continuous and convex hull representations for zero-one programming problems. *SIAM Journal of Discrete Mathematics* 3, 411–430.

Sherali, H. D., B. M. P. Fraticelli (2002). A modification of Benders' decomposition algorithm for discrete subproblems: an approach for stochastic programs with integer recourse. *Journal of Global Optimization* 22, 319–342.

Suhl, U. H., R. Szymanski (1994). Supernode processing of mixed-integer models. *Computational Optimization and Applications* 3, 317–331.

Thienel, S. (1995). *ABACUS A Branch-And-Cut System*. PhD thesis, Universität zu Köln.

Vanderbeck, F. (1999). Computational study of a column generation algorithm for bin packing and cutting stock problems. *Mathematical Programming* 46, 565–594.

Vanderbeck, F. (2000). On Dantzig–Wolfe decomposition in integer programming and ways to perform branching in a branch-and-price algorithm. *Operations Research* 48(1), 111–128.

Weismantel, R. (1997). On the 0/1 knapsack polytope. *Mathematical Programming* 77(1), 49–68.

Wilson, D. G. (1992). A brief introduction to the ibm optimization subroutine library. *SIAG/OPT Views and News* 1, 9–10.

Wolsey, L. A. (1975). Faces of linear inequalities in 0–1 variables. *Mathematical Programming* 8, 165–178.

Wolsey, L. A. (1990). Valid inequalities for 0–1 knapsacks and MIPs with generalized upper bound constraints. *Discrete Applied Mathematics* 29, 251–261.

Zemel, E. (1989). Easily computable facets of the knapsack polytope. *Mathematics of Operations Research* 14, 760–764.

Zhao, X., P. B. Luh (2002). New bundle methods for solving Lagrangian relaxation dual problems. *Journal of Optimization Theory and Applications* 113(2), 373–397.

K. Aardal et al., Eds., *Handbooks in OR & MS, Vol. 12*

Chapter 3

The Structure of Group Relaxations

Rekha R. Thomas

Department of Mathematics, University of Washington, Box 354350, Seattle, Washington 98195, USA
E-mail: thomas@math.washington.edu

Abstract

This article is a survey of new results on the structure of group relaxations in integer programming that have come from the algebraic study of integer programs via the theory of Gröbner bases. We study all bounded group relaxations of all integer programs in the infinite family of programs arising from a fixed coefficient matrix and cost vector. The programs in the family are classified by the set of indices of the nonnegativity constraints that can be relaxed in a maximal group relaxation that solves each problem. A highlight of the theory is the "chain theorem" which proves that these sets come in saturated chains. We obtain a natural invariant of the infinite family of integer programs called its arithmetic degree. We also characterize all families of integer programs that can be solved by their Gomory relaxations. The article is self contained and assumes no familiarity with algebraic techniques.

1 Introduction

Group relaxations of integer programs were introduced by Ralph Gomory in the 1960s (Gomory, 1965, 1969). Given a general integer program of the form

$$\text{minimize}\{c \cdot x\colon\ Ax = b,\ x \geq 0, \text{integer}\}, \tag{1}$$

its group relaxation is obtained by dropping nonnegativity restrictions on all the basic variables in the optimal solution of its linear relaxation. In this article, we survey recent results on group relaxations obtained from the algebraic study of integer programming using *Gröbner bases* of *toric ideals* (Sturmfels, 1995). No knowledge of these methods is assumed, and the exposition is self-contained and hopefully accessible to a person familiar with the traditional methods of integer programming. For the reader who might be interested in the algebraic origins, motivations and counterparts of the described results, we have included brief comments in the last section. These comments are numbered in the style of footnotes and organized as paragraphs in Section 8. While they offer a more complete picture of the theory to those familiar with commutative algebra, they are not necessary for the continuity of the article.

For the sake of brevity, we will bypass a detailed account of the classical theory of group relaxations. A short expository account can be found in Schrijver (1986, §24.2), and a detailed set of lecture notes on this topic in Johnson (1980). We give a brief synopsis of the essentials based on the recent survey article by Aardal, Weismantel, and Wolsey (2002) and refer the reader to any of the above sources for further details and references on the classical theory of group relaxations.

Assuming that all data in (1) are integral and that A_B is the optimal basis of the linear relaxation of (1), Gomory's group relaxation of (1) is the problem

$$\text{minimize}\,\{\tilde{c}_N \cdot x_N\colon\, A_B^{-1}A_N x_N \equiv A_B^{-1}b\,(\text{mod}\,1),\ x_N \geq 0,\ \text{integer}\} \qquad (2)$$

Here B and N are the index sets for the basic and nonbasic columns of A corresponding to the optimal solution of the linear relaxation of (1). The vector x_N denotes the nonbasic variables and the cost vector $\tilde{c}_N = c_N - c_B A_B^{-1} A_N$ where $c = (c_B, c_N)$ is partitioned according to B and N. The notation $A_B^{-1}A_N x_N \equiv A_B^{-1}b$ (mod 1) indicates that $A_B^{-1}A_N x_N - A_B^{-1}b$ is a vector of integers. Problem (2) is called a "group relaxation" of (1) since it can be written in the canonical form

$$\text{minimize}\left\{\tilde{c}_N \cdot x_N\colon \sum_{j\in N} g_j x_j \equiv g_0(\text{mod } G),\, x_N \geq 0,\ \text{integer}\right\} \qquad (3)$$

where G is a finite abelian group and $g_j \in G$. Problem (3) can be viewed as a shortest path problem in a graph on $|G|$ nodes which immediately furnishes algorithms for solving it. Once the optimal solution x_N^* of (2) is found, it can be uniquely *lifted* to a vector $x^* = (x_B^*, x_N^*) \in \mathbb{Z}^n$ such that $Ax^* = b$. If $x_B^* \geq 0$ then x^* is the optimal solution of (1). Otherwise, $c \cdot x^*$ is a lower bound for the optimal value of (1). Several strategies are possible when the group relaxation fails to solve the integer program. See Bell and Shapiro (1977), Gorry, Northup, and Shapiro (1973), Nemhauser and Wolsey (1988) and Wolsey (1973) for the work in this direction. A particular idea according to Wolsey (1971), that is very relevant for this chapter, is to consider the *extended* group relaxations of (1). These are all the possible group relaxations of (1) obtained by dropping nonnegativity restrictions on all possible subsets of the basic variables x_B in the optimum of the linear relaxation of (1). Gomory's group relaxation (2) of (1) and (1) itself are therefore among these extended group relaxations. If (2) does not solve (1), then one could resort to other extended relaxations to solve the problem. At least one of these extended group relaxations (in the worst case (1) itself) is guaranteed to solve the integer program (1).

The convex hull of the feasible solutions to (2) is called the *corner polyhedron* (Gomory, 1967). A major focus of Gomory and others who worked on group relaxations was to understand the polyhedral structure of the corner polyhedron. This was achieved via the *master polyhedron* of the group G (Gomory, 1969) which is the convex hull of the set of points

$$\left\{ z: \sum_{g \in G} g z_g \equiv g_0 (\mathrm{mod}\ G),\ z \geq 0, \text{ integer} \right\}.$$

Facet-defining inequalities for the master polyhedron provide facet inequalities of the corner polyhedron (Gomory, 1969). As remarked in Aardal et al. (2002), this landmark paper (Gomory, 1969) introduced several of the now standard ideas in polyhedral combinatorics like projection onto faces, subadditivity, master polytopes, using automorphisms to generate one facet from another, lifting techniques and so on. See Gomory and Johnson (1972) for further results on generating facet inequalities. Recent results on the facets of master polyhedra and cutting planes can be found in Aráoz, Evans, Gomory, and Johnson (2003), Evans, Gomory, and Johnson (2003), and Gomory and Johnson (2003).

In the algebraic approach to integer programming, one considers the entire family of integer programs of the form (1) as the right hand side vector b varies. Definition 2.6 defines a set of group relaxations for each program in this family. Each relaxation is indexed by a face of a simplicial complex called a *regular triangulation* (Definition 2.1). This complex encodes all the optimal bases of the linear programs arising from the coefficient matrix A and cost vector c (Lemma 2.3). The main result of Section 2 is Theorem 2.8 which states that the group relaxations in Definition 2.6 are precisely all the bounded group relaxations of all programs in the family. In particular, they include all the extended group relaxations of all programs in the family and typically contain more relaxations for each program. This theorem is proved via a particular reformulation of the group relaxations which is crucial for the rest of the paper. This and other reformulations are described in Section 2.

The most useful group relaxations of an integer program are the "least strict" ones among all those that solve the program. By this we mean that any further relaxation of nonnegativity restrictions will result in group relaxations that do not solve the problem. The faces of the regular triangulation indexing all these special relaxations for all programs in the family are called the *associated sets* of the family (Definition 3.1). In Section 3, we develop tools to study associated sets. This leads to Theorem 3.11 which characterizes associated sets in terms of *standard pairs* and *standard polytopes*. Theorem 3.12 shows that one can "read off" the "least strict" group relaxations that solve a given integer program in the family from these standard pairs.

The results in Section 3 lead to an important invariant of the family of integer programs being studied called its *arithmetic degree*. In Section 4 we discuss the relevance of this invariant and give a bound for it based on a result of Ravi Kannan (Theorem 4.8). His result builds a bridge between our methods and those of Kannan, Lenstra, Lovasz, Scarf and others that use the geometry of numbers in integer programming.

Section 5 examines the structure of the poset of associated sets. The main result in this section is the *chain theorem* (Theorem 5.2) which shows that associated sets occur in saturated chains. Theorem 5.4 bounds the length of a maximal chain.

In Section 6 we define a particular family of integer programs called a *Gomory family*, for which all associated sets are maximal faces of the regular triangulation. Theorem 6.2 gives several characterizations of Gomory families. We show that this notion generalizes the classical notion of *total dual integrality* in integer programming Schrijver (1986, §22). We conclude in Section 7 with constructions of Gomory families from matrices whose columns form a Hilbert basis. In particular, we recast the existence of a Gomory family as a *Hilbert cover* problem. This builds a connection to the work of Sebö (1990), Bruns and Gubeladze (1999) and Firla and Ziegler (1999) on *Hilbert partitions* and *covers* of polyhedral cones. We describe the notions of *super* and Δ-*normality* both of which give rise to Gomory families (Theorems 7.8 and Theorems 7.15).

The majority of the material in this chapter is a translation of algebraic results from Hoşten and Thomas (1999a,b, 2003), Sturmfels (1995, §8 and §12.D), Sturmfels, Trung, and Vogel (1995) and Sturmfels, Weismantel, and Ziegler (1995). The translation has sometimes required new definitions and proofs. Kannan's theorem in Section 4 has not appeared elsewhere.

We will use the letter $\mathbb{N}$ to denote the set of nonnegative integers, $\mathbb{R}$ to denote the real numbers and $\mathbb{Z}$ for the integers. The symbol $P \subseteq Q$ denotes that P is a subset of Q, possibly equal to Q, while $P \subset Q$ denotes that P is a proper subset of Q.

2 Group relaxations

Throughout this chapter, we fix a matrix $A \in \mathbb{Z}^{d \times n}$ of rank d, a cost vector $c \in \mathbb{Z}^n$ and consider the family $\mathrm{IP}_{A,c}$ of all integer programs

$$\mathrm{IP}_{A,c}(b) := \text{minimize}\,\{c \cdot x\colon\ Ax = b,\ x \in \mathbb{N}^n\}$$

as b varies in the semigroup $\mathbb{N}A := \{Au\colon u \in \mathbb{N}^n\} \subseteq \mathbb{Z}^d$. This family is precisely the set of all feasible integer programs with coefficient matrix A and cost vector c. The semigroup $\mathbb{N}A$ lies in the intersection of the d-dimensional polyhedral cone $\text{cone}(A) := \{Au: u \geq 0\} \subseteq \mathbb{R}^d$ and the d-dimensional lattice $\mathbb{Z}A := \{Au: u \in \mathbb{Z}^n\} \subseteq \mathbb{Z}^d$. For simplicity, we will assume that cone(A) is pointed and that $\{u \in \mathbb{R}^n: Au = 0\}$, the kernel of A, intersects the nonnegative orthant of $\mathbb{R}^n$ only at the origin. This guarantees that all programs in $\text{IP}_{A,c}$ are bounded. In addition, the cost vector c will be assumed to be *generic* in the sense that each program in $\text{IP}_{A,c}$ has a unique optimal solution. The algebraic study of integer programming shows that all cost vectors in $\mathbb{R}^n$ except those on (parts of) a finite number of hyperplanes, are generic for the family $\text{IP}_{A,c}$ (Sturmfels and Thomas, 1997). Hence, the genericity assumption on c is almost always satisfied. In fact all cost vectors can be made generic by breaking ties with a fixed total order on $\mathbb{N}^n$ such as the lexicographic order. Geometrically, this has the effect of perturbing a nongeneric c to a vector that no longer lies on one of the forbidden hyperplanes, while keeping the optimal solution of the programs in $\text{IP}_{A,c}$ unchanged.

The *linear relaxation* of $\text{IP}_{A,c}(b)$ is the linear program

$$\text{LP}_{A,c}(b) := \text{minimize}\{c \cdot x: Ax = b, \ \ x \geq 0\}.$$

We denote by $\text{LP}_{A,c}$ the family of all linear programs of the form $\text{LP}_{A,c}(b)$ as b varies in cone(A). These are all the feasible linear programs with coefficient matrix A and cost vector c. Since all data are integral and all programs in $\text{IP}_{A,c}$ are bounded, all programs in $\text{LP}_{A,c}$ are bounded as well.

In the classical definitions of group relaxations of $\text{IP}_{A,c}(b)$, one assumes knowledge of the optimal basis of the linear relaxation $\text{LP}_{A,c}(b)$. In the algebraic set up, we define group relaxations for all members of $\text{IP}_{A,c}$ at one shot and, analogously to the classical setting, assume that the optimal bases of all programs in $\text{LP}_{A,c}$ are known. This information is carried by a *polyhedral complex* called the *regular triangulation* of cone(A) with respect to c.

A polyhedral complex Δ is a collection of polyhedra called *cells* (or *faces*) of Δ such that:

(i) every face of a cell of Δ is again a cell of Δ and,
(ii) the intersection of any two cells of Δ is a common face of both.

The set-theoretic union of the cells of Δ is called the *support* of Δ. If Δ is not empty, then the empty set is a cell of Δ since it is a face of every polyhedron. If all the faces of Δ are cones, we call Δ a cone *complex*.

For $\sigma \subseteq \{1, \ldots, n\}$, let A_σ be the submatrix of A whose columns are indexed by σ, and let cone (A_σ) denote the cone generated by the columns of A_σ. The *regular subdivision* Δ_c of cone(A) is a cone complex with support cone(A) defined as follows.

Definition 2.1. For $\sigma \subseteq \{1, \ldots, n\}$, cone($A_\sigma$) is a face of the *regular subdivision* Δ_c of cone(A) if and only if there exists a vector $y \in \mathbb{R}^d$ such that $y \cdot a_j = c_j$ for all $j \in \sigma$ and $y \cdot a_j < c_j$ for all $j \notin \sigma$.

The regular subdivision Δ_c can be constructed geometrically as follows. Consider the cone in $\mathbb{R}^{d+1}$ generated by the lifted vectors $(a_i^t, c_i) \in \mathbb{R}^{d+1}$ where a_i is the ith column of A and c_i is the ith component of c. The *lower* facets of this lifted cone are all those facets whose normal vectors have a negative $(d+1)$th component. Projecting these lower facets back onto cone(A) induces the regular subdivision Δ_c of cone(A) [see Billera, Filliman, and Sturmfels (1990)]. Note that if the columns of A span an affine hyperplane in $\mathbb{R}^d$, then Δ_c can also be seen as a subdivision of conv(A), the $(d-1)$-dimensional convex hull of the columns of A.

The genericity assumption on c implies that Δ_c is in fact a triangulation of cone(A) [see Sturmfels and Thomas (1997)]. We call Δ_c the *regular triangulation* of cone(A) with respect to c. For brevity, we may also refer to Δ_c as the regular triangulation of A with respect to c. Using σ to label cone(A_σ), Δ_c is usually denoted as a set of subsets of $\{1, \ldots, n\}$. Since Δ_c is a complex of simplicial cones, it suffices to list just the maximal elements (with respect to inclusion) in this set of sets. By definition, every one dimensional face of Δ_c is of the form cone(a_i) for some column a_i of A. However, not all cones of the form cone(a_i), a_i a column of A, need appear as a one dimensional cell of Δ_c.

Example 2.2. (i) Let

$$A = \begin{bmatrix} 1 & 1 & 1 & 1 \\ 0 & 1 & 2 & 3 \end{bmatrix}$$

and $c = (1, 0, 0, 1)$. The four columns of A are the four dark points in Fig. 1 labeled by their column indices $1, \ldots, 4$. Figure 1(a) shows the cone generated by the lifted vectors $(a_i^t, c_i) \in \mathbb{R}^3$. The rays generated by the lifted vectors have the same labels as the points that were lifted. Projecting the lower facets of this lifted cone back onto cone(A), we get the regular triangulation Δ_c of cone(A) shown in Fig. 1(b). The same triangulation is shown as a triangulation of conv(A) in Fig. 1(c). The faces of the triangulation Δ_c are $\{1,2\}, \{2,3\}, \{3,4\}, \{1\}, \{2\}, \{3\}, \{4\}$ and $\emptyset$. Using only the maximal faces, we may write $\Delta_c = \{\{1, 2\}, \{2,3\}, \{3, 4\}\}$.

(ii) For the A in (i), cone(A) has four distinct regular triangulations as c varies. For instance, the cost vector $c' = (0,1,0,1)$ induces the regular triangulation $\Delta_{c'} = \{\{1,3\}, \{3,4\}\}$ shown in Fig. 2(b) and (c). Notice that $\{2\}$ is not a face of $\Delta_{c'}$.

(iii) If

$$A = \begin{bmatrix} 1 & 3 & 2 & 1 \\ 0 & 1 & 2 & 3 \end{bmatrix}$$

and $c = (1,0,0,1)$, then $\Delta_c = \{\{1,2\}, \{2,3\}, \{3,4\}\}$. However, in this case, Δ_c can only be seen as a triangulation of cone(A) and not of conv(A). □

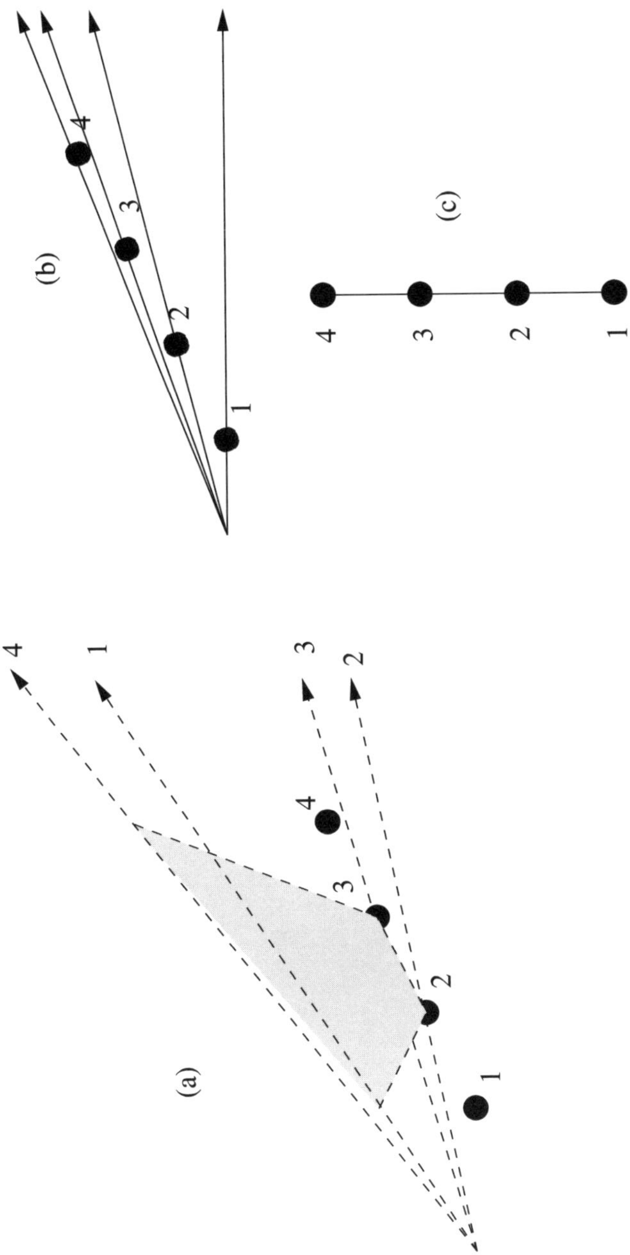

Fig. 1. Regular triangulation Δ_c for $c=(1,0,0,1)$ (Example 2.2 (i)).

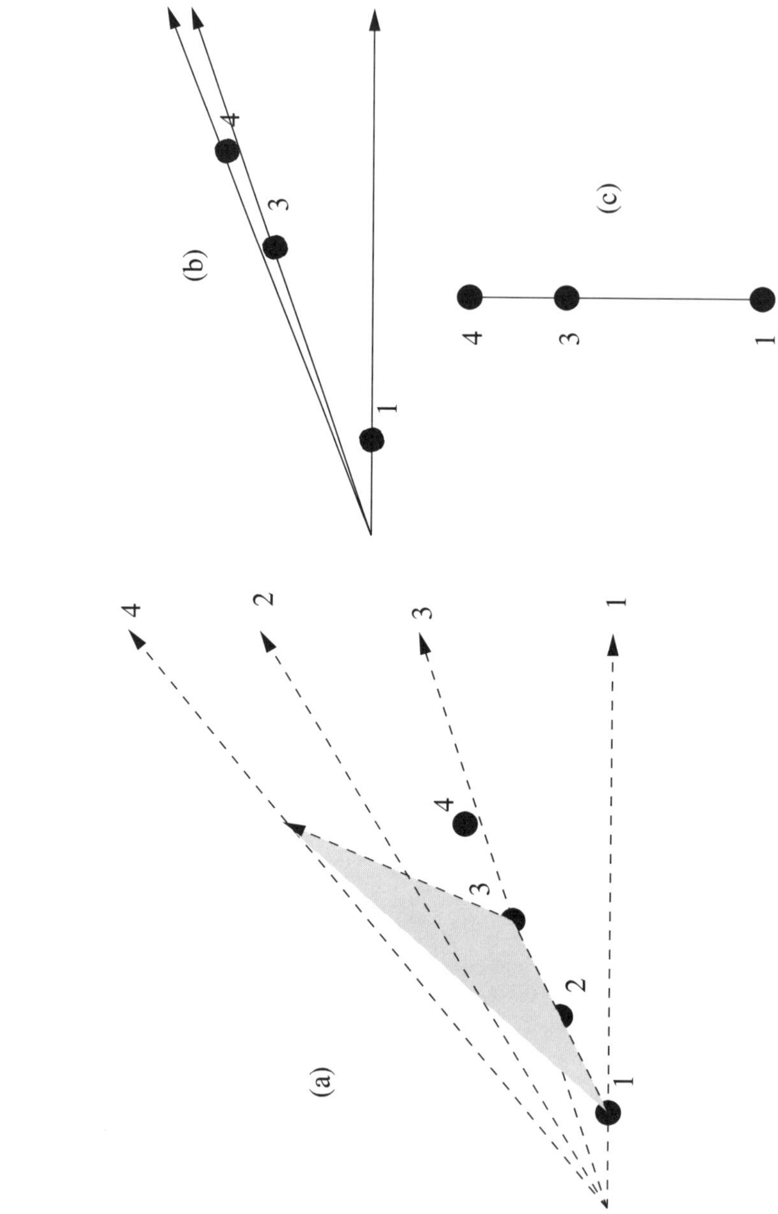

Fig. 2. Regular triangulation $\Delta_{c'}$ for $c' = (0, 1, 0, 1)$ (Example 2.2 (ii)).

For a vector $x \in \mathbb{R}^n$, let $\text{supp}(x) = \{i: x_i \neq 0\}$ denote the *support* of x. The significance of regular triangulations for linear programming is summarized in the following proposition.

Proposition 2.3. [Sturmfels and Thomas (1997, Lemma 1.4)] *An optimal solution of* $\text{LP}_{A,c}(b)$ *is any feasible solution* x^* *such that* $\text{supp}(x^*) = \tau$ *where* τ *is the smallest face of the regular triangulation* Δ_c *such that* $b \in \text{cone}(A_\tau)$.

Proposition 2.3 implies that $\sigma \subseteq \{1, \ldots, n\}$ is a maximal face of Δ_c if and only if A_σ is an optimal basis for all $\text{LP}_{A,c}(b)$ with b in $\text{cone}(A_\sigma)$. For instance, in Example 2.2 (i), if $b = (4, 1)^t$ then the optimal basis of $\text{LP}_{A,c}(b)$ is $[a_1, a_2]$ where as if $b = (2, 2)^t$, then the optimal solution of $\text{LP}_{A,c}(b)$ is degenerate and either $[a_1, a_2]$ or $[a_2, a_3]$ could be the optimal basis of the linear program. (Recall that a_i is the ith column of A.) All programs in $\text{LP}_{A,c}$ have one of $[a_1, a_2]$, $[a_2, a_3]$ or $[a_3, a_4]$ as its optimal basis.

Given a polyhedron $P \subset \mathbb{R}^n$ and a face F of P, the *normal* cone of F at P is the cone $N_P(F) := \{\omega \in \mathbb{R}^n: \omega \cdot x' \geq \omega \cdot x, \text{ for all } x' \in F \text{ and } x \in P\}$. The normal cones of all faces of P form a cone complex in $\mathbb{R}^n$ called the *normal fan* of P.

Proposition 2.4. *The regular triangulation* Δ_c *of cone*(A) *is the normal fan of the polyhedron* $P_c := \{y \in \mathbb{R}^d: yA \leq c\}$.

Proof. The polyhedron P_c is the feasible region of maximize $\{y \cdot b: yA \leq c, y \in \mathbb{R}^d\}$, the dual program to $\text{LP}_{A,c}(b)$. The support of the normal fan of P_c is $\text{cone}(A)$, since this is the polar cone of the recession cone $\{y \in \mathbb{R}^d: yA \leq 0\}$ of P_c. Suppose b is any vector in the interior of a maximal face $\text{cone}(A_\sigma)$ of Δ_c. Then by Proposition 2.3, $\text{LP}_{A,c}(b)$ has an optimal solution x^* with support σ. By complementary slackness, the optimal solution y to the dual of $\text{LP}_{A,c}(b)$ satisfies $y \cdot a_j = c_j$ for all $j \in \sigma$ and $y \cdot a_j \leq c_j$ otherwise. Since σ is a maximal face of Δ_c, $y \cdot a_j < c_j$ for all $j \notin \sigma$. Thus y is unique, and $\text{cone}(A_\sigma)$ is contained in the normal cone of P_c at the vertex y. If b lies in the interior of another maximal face $\text{cone}(A_\tau)$ then y', (the dual optimal solution to $\text{LP}_{A,c}(b)$) satisfies $y' \cdot A_\tau = c_\tau$ and $y' \cdot A_{\bar{\tau}} < c_{\bar{\tau}}$ where $\tau \neq \sigma$. As a result, y' is distinct from y, and each maximal cone in Δ_c lies in a distinct maximal cone in the normal fan of P_c. Since Δ_c and the normal fan of P_c are both cone complexes with the same support, they must therefore coincide. □

Example 2.2 continued. Figure 3(a) shows the polyhedron P_c for Example 2.2 (i) with all its normal cones. The normal fan of P_c is drawn in Fig. 3(b). Compare this fan with that in Fig. 1(b). □

Corollary 2.5. *The polyhedron* P_c *is simple if and only if the regular subdivision* Δ_c *is a triangulation of cone*(A).

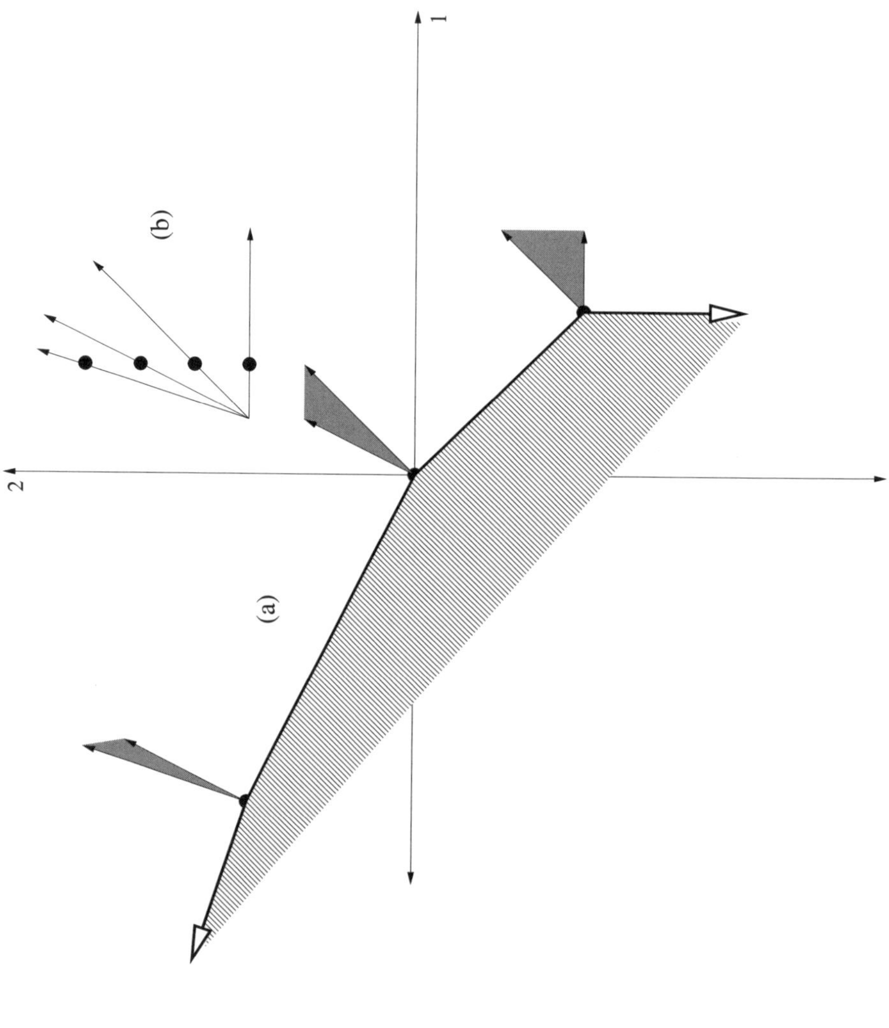

Fig. 3. The polyhedron P_c and its normal fan for Example 2.2 (i).

Regular triangulations were introduced by Gel'fand, Kapranov, and Zelevinsky (1994) and have various applications. They have played a central role in the algebraic study of integer programming (Sturmfels, 1995; Sturmfels and Thomas, 1997), and we use them now to define *group relaxations* of $IP_{A,c}(b)$.

A subset τ of $\{1,\ldots,n\}$ partitions $x=(x_1,\ldots,x_n)$ as x_τ and $x_{\bar{\tau}}$ where x_τ consists of the variables indexed by τ and $x_{\bar{\tau}}$ the variables indexed by the complementary set $\bar{\tau}$. Similarly, the matrix A is partitioned as $A=[A_\tau, A_{\bar{\tau}}]$ and the cost vector as $c=(c_\tau, c_{\bar{\tau}})$. If σ is a maximal face of Δ_c, then A_σ is nonsingular and $Ax=b$ can be written as $x_\sigma = A_\sigma^{-1}(b - A_{\bar{\sigma}}x_{\bar{\sigma}})$. Then $c\cdot x = c_\sigma(A_\sigma^{-1}(b - A_{\bar{\sigma}}x_{\bar{\sigma}})) + c_{\bar{\sigma}}x_{\bar{\sigma}} = c_\sigma A_\sigma^{-1}b + (c_{\bar{\sigma}} - c_\sigma A_\sigma^{-1}A_{\bar{\sigma}})x_{\bar{\sigma}}$. Let $\tilde{c}_{\bar{\sigma}} := c_{\bar{\sigma}} - c_\sigma A_\sigma^{-1}A_{\bar{\sigma}}$ and, for any face τ of σ, let $\tilde{c}_{\bar{\tau}}$ be the extension of $\tilde{c}_{\bar{\sigma}}$ to a vector in $\mathbb{R}^{|\bar{\tau}|}$ by adding zeros.

We now define a group relaxation of $IP_{A,c}(b)$ with respect to each face τ of Δ_c.

Definition 2.6. The group relaxation of the integer program $IP_{A,c}(b)$ with respect to the face τ of Δ_c is the program:

$$G^\tau(b) = \text{minimize}\{\tilde{c}_{\bar{\tau}}\cdot x_{\bar{\tau}}\colon A_\tau x_\tau + A_{\bar{\tau}}x_{\bar{\tau}} = b,\ x_{\bar{\tau}} \geq 0, (x_\tau, x_{\bar{\tau}}) \in \mathbb{Z}^n\}.$$

Equivalently, $G^\tau(b) = \text{minimize}\{\tilde{c}_{\bar{\tau}}\cdot x_{\bar{\tau}}\colon A_{\bar{\tau}}x_{\bar{\tau}} \equiv b \pmod{\mathbb{Z}A_\tau},\ x_{\bar{\tau}} \geq 0, \text{integer}\}$ where $\mathbb{Z}A_\tau$ is the lattice generated by the columns of A_τ. Suppose $x_{\bar{\tau}}^*$ is an optimal solution to the latter formulation. Since τ is a face of Δ_c, the columns of A_τ are linearly independent, and therefore the linear system $A_\tau x_\tau + A_{\bar{\tau}}x_{\bar{\tau}}^* = b$ has a unique solution. Solving this system for x_τ, the optimal solution $x_{\bar{\tau}}^*$ of $G^\tau(b)$ can be uniquely lifted to the solution $(x_\tau^*, x_{\bar{\tau}}^*)$ of $Ax=b$. The formulation of $G^\tau(b)$ in Definition 2.6 shows that x_τ^* is an integer vector. The group relaxation $G^\tau(b)$ solves $IP_{A,c}(b)$ if and only if x_τ^* is also nonnegative.

The group relaxations of $IP_{A,c}(b)$ from Definition 2.6 contain among them the classical group relaxations of $IP_{A,c}(b)$ found in the literature. The program $G^\sigma(b)$, where A_σ is the optimal basis of the linear relaxation $LP_{A,c}(b)$, is precisely Gomory's group relaxation of $IP_{A,c}(b)$ (Gomory, 1965). The set of relaxations $G^\tau(b)$ as τ varies among the subsets of this σ are the *extended* group relaxations of $IP_{A,c}(b)$ defined by Wolsey (1971). Since $\emptyset \in \Delta_c$, $G^\emptyset(b) = IP_{A,c}(b)$ is a group relaxation of $IP_{A,c}(b)$, and hence $IP_{A,c}(b)$ will certainly be solved by one of its extended group relaxations. However, it is possible to construct examples where a group relaxation $G^\tau(b)$ solves $IP_{A,c}(b)$, but $G^\tau(b)$ is neither Gomory's group relaxation of $IP_{A,c}(b)$ nor one of its nontrivial extended Wolsey relaxations (see Example 4.2). Thus, Definition 2.6 typically creates more group relaxations for each program in $IP_{A,c}$ than in the classical situation. This has the obvious advantage that it increases the chance that $IP_{A,c}(b)$ will be solved by some nontrivial relaxation, although one may have to keep track of many more relaxations for each program. In Theorem 2.8, we will prove that Definition 2.6 is the best possible in the sense that the relaxations of $\mathrm{IP}_{A,c}(b)$ defined there are precisely all the bounded group relaxations of the program.

The goal in the rest of this section is to describe a useful reformulation of the group problem $G^\tau(b)$ which is needed in the rest of the chapter and in the proof of Theorem 2.8. Given a sublattice Λ of $\mathbb{Z}^n$, a cost vector $w \in \mathbb{R}^n$ and a vector $v \in \mathbb{N}^n$, the *lattice program* defined by this data is

$$\text{minimize}\{w \cdot x\colon x \equiv v \pmod{\Lambda},\ \ x \in \mathbb{N}^n\}.$$

Let $\mathcal{L}$ denote the $(n-d)$-dimensional saturated lattice $\{x \in \mathbb{Z}^n\colon Ax = 0\} \subseteq \mathbb{Z}^n$ and u be a feasible solution of the integer program $\mathrm{IP}_{A,c}(b)$. Since $\mathrm{IP}_{A,c}(b) =$ minimize$\{c \cdot x\colon Ax = b(=Au),\ x \in \mathbb{N}^n\}$ can be rewritten as minimize$\{c \cdot x\colon x - u \in \mathcal{L},\ x \in \mathbb{N}^n\}$, $\mathrm{IP}_{A,c}(b)$ is equivalent to the lattice program

$$\text{minimize}\{c \cdot x\colon x \equiv u \pmod{\mathcal{L}},\ \ x \in \mathbb{N}^n\}.$$

For $\tau \in \Delta_c$, let π_τ be the projection map from $\mathbb{R}^n \to \mathbb{R}^{|\bar{\tau}|}$ that kills all coordinates, indexed by τ. Then $\mathcal{L}_\tau := \pi_\tau(\mathcal{L})$ is a sublattice of $\mathbb{Z}^{|\bar{\tau}|}$ that is isomorphic to $\mathcal{L}$: Clearly, $\pi_\tau : \mathcal{L} \to \mathcal{L}_\tau$ is a surjection. If $\pi_\tau(v) = \pi_\tau(v')$ for $v, v' \in \mathcal{L}$, then $A_\tau v_\tau + A_{\bar{\tau}} v_{\bar{\tau}} = 0 = A_\tau v'_\tau + A_{\bar{\tau}} v'_{\bar{\tau}}$, implies that $A_\tau(v_\tau - v'_\tau) = 0$. Then $v_\tau = v'_\tau$ since the columns of A_τ are linearly independent. Using this fact, $G^\tau(b)$ can also be reformulated as a lattice program:

$$\begin{aligned} G^\tau(b) &= \text{minimize}\{\tilde{c}_{\bar{\tau}} \cdot x_{\bar{\tau}}\colon A_\tau x_\tau + A_{\bar{\tau}} x_{\bar{\tau}} = b,\ \ x_{\bar{\tau}} \geq 0, (x_\tau, x_{\bar{\tau}}) \in \mathbb{Z}^n\} \\ &= \text{minimize}\{\tilde{c}_{\bar{\tau}} \cdot x_{\bar{\tau}}\colon (x_\tau, x_{\bar{\tau}})^t - (u_\tau, u_{\bar{\tau}})^t \in \mathcal{L},\ \ x_{\bar{\tau}} \in \mathbb{N}^{|\bar{\tau}|}\} \\ &= \text{minimize}\{\tilde{c}_{\bar{\tau}} \cdot x_{\bar{\tau}}\colon x_{\bar{\tau}} - u_{\bar{\tau}} \in \mathcal{L}_\tau,\ \ x_{\bar{\tau}} \in \mathbb{N}^{|\bar{\tau}|}\} \\ &= \text{minimize}\{\tilde{c}_{\bar{\tau}} \cdot x_{\bar{\tau}}\colon x_{\bar{\tau}} \equiv \pi_\tau(u) \pmod{\mathcal{L}_\tau},\ \ x_{\bar{\tau}} \in \mathbb{N}^{|\bar{\tau}|}\}. \end{aligned}$$

Lattice programs were shown to be solved by Gröbner bases in Sturmfels et al. (1995). Theorem 5.3 in Sturmfels et al. (1995) gives a geometric interpretation of these Gröbner bases in terms of corner polyhedra. This article was the first to make a connection between the theory of group relaxations and commutative algebra [see Sturmfels et al. (1995, §6)]. Special results are possible when the sublattice Λ is of finite index. In particular, the associated Gröbner bases are easier to compute.

Since the $(n-d)$-dimensional lattice $\mathcal{L} \subset \mathbb{Z}^n$ is isomorphic to $\mathcal{L}_\tau \subset \mathbb{Z}^{|\bar{\tau}|}$ for $\tau \in \Delta_c$, $\mathcal{L}_\tau$ is of finite index if and only if τ is a maximal face of Δ_c. Hence, by the last sentence of the previous paragraph, the group relaxations $G^\sigma(b)$ as σ varies over the maximal faces of Δ_c are the easiest to solve among all group relaxations of $\mathrm{IP}_{A,c}(b)$. They contain among them Gomory's group relaxation of $\mathrm{IP}_{A,c}$. We give these relaxations a collective name.

Definition 2.7. The group relaxations $G^{\sigma}(b)$ of $\mathrm{IP}_{A,c}(b)$, as σ varies among the maximal faces of Δ_c, are called the *Gomory relaxations* of $\mathrm{IP}_{A,c}(b)$.

It is useful to reformulate $G^{\tau}(b)$ once again as follows. Let $B \in \mathbb{Z}^{n \times (n-d)}$ be any matrix such that the columns of B generate the lattice $\mathcal{L}$, and let u be a feasible solution of $\mathrm{IP}_{A,c}(b)$ as before. Then

$$\begin{aligned} IP_{A,c}(b) &= \text{minimize}\{c \cdot x\colon x - u \in \mathcal{L},\ x \in \mathbb{N}^n\} \\ &= \text{minimize}\{c \cdot x\colon x = u - Bz,\ x \geq 0,\ z \in \mathbb{Z}^{n-d}\}. \end{aligned}$$

The last problem is equivalent to minimize$\{c(u - Bz) : Bz \leq u,\ z \in \mathbb{Z}^{n-d}\}$ and, therefore $\mathrm{IP}_{A,c}(b)$ is equivalent to the problem

$$\text{minimize}\{(-cB) \cdot z\colon Bz \leq u,\ z \in \mathbb{Z}^{n-d}\}. \tag{4}$$

There is a bijection between the set of feasible solutions of (4) and the set of feasible solutions of $\mathrm{IP}_{A,c}(b)$ via the map $z \mapsto u - Bz$. In particular, $0 \in \mathbb{R}^{n-d}$ is feasible for (4) and it is the pre-image of u under this map.

If $B^{\bar{\tau}}$ denotes the $|\bar{\tau}| \times (n-d)$ submatrix of B obtained by deleting the rows indexed by τ, then $\mathcal{L}_{\tau} = \pi_{\tau}(\mathcal{L}) = \{B^{\bar{\tau}} z : z \in \mathbb{Z}^{n-d}\}$. Using the same techniques as above, $G^{\tau}(b)$ can be reformulated as

$$\text{minimize}\{(-\tilde{c}_{\bar{\tau}} B^{\bar{\tau}}) \cdot z\colon B^{\bar{\tau}} z \leq \pi_{\tau}(u),\ z \in \mathbb{Z}^{n-d}\}.$$

Since $\tilde{c}_{\bar{\tau}} = \pi_{\tau}(c - c_{\sigma} A_{\sigma}^{-1} A)$ for any maximal face σ of Δ_c containing τ and the support of $c - c_{\sigma} A_{\sigma}^{-1} A$ is contained in $\bar{\tau}$, $\tilde{c}_{\bar{\tau}} B^{\bar{\tau}} = (c - c_{\sigma} A_{\sigma}^{-1} A) B = cB$ since $AB = 0$. Hence $G^{\tau}(b)$ is equivalent to

$$\text{minimize}\{(-cB) \cdot z\colon B^{\bar{\tau}} z \leq \pi_{\tau}(u),\ z \in \mathbb{Z}^{n-d}\}. \tag{5}$$

The feasible solutions to (4) are the lattice points in the rational polyhedron $P_u := \{z \in \mathbb{R}^{n-d}\colon Bz \leq u\}$, and the feasible solutions to (5) are the lattice points in the relaxation $P_u^{\bar{\tau}} := \{z \in \mathbb{R}^{n-d}\colon B^{\bar{\tau}} z \leq \pi_{\tau}(u)\}$ of P_u obtained by deleting the inequalities indexed by τ. In theory, one could define group relaxations of $\mathrm{IP}_{A,c}(b)$ with respect to any $\tau \subseteq \{1, \ldots, n\}$. The following theorem illustrates the completeness of Definition 2.6.

Theorem 2.8. *The group relaxation $G^{\tau}(b)$ of $IP_{A,c}(b)$ has a finite optimal solution if and only if $\tau \subseteq \{1, \ldots, n\}$ is a face of Δ_c.*

Proof. Since all data are integral it suffices to prove that the linear relaxation

$$\text{minimize}\{(-cB) \cdot z\colon z \in P_u^{\bar{\tau}}\}$$

is bounded if and only if $\tau \in \Delta_c$.

If τ is a face of Δ_c then there exists $y \in \mathbb{R}^d$ such that $yA_\tau = c_\tau$ and $yA_{\bar{\tau}} < c_{\bar{\tau}}$. Using the fact that $A_\tau B^\tau + A_{\bar{\tau}} B^{\bar{\tau}} = 0$ we see that $cB = c_\tau B^\tau + c_{\bar{\tau}} B^{\bar{\tau}} = yA_\tau B^\tau + c_{\bar{\tau}} B^{\bar{\tau}} = y(-A_{\bar{\tau}} B^{\bar{\tau}}) + c_{\bar{\tau}} B^{\bar{\tau}} = (c_{\bar{\tau}} - yA_{\bar{\tau}}) B^{\bar{\tau}}$. This implies that cB is a positive linear combination of the rows of $B^{\bar{\tau}}$ since $c_{\bar{\tau}} - yA_{\bar{\tau}} > 0$. Hence cB lies in the polar of $\{z \in \mathbb{R}^{n-d}: B^{\bar{\tau}} z \leq 0\}$ which is the recession cone of $P_u^{\bar{\tau}}$ proving that the linear program minimize$\{(-cB) \cdot z: z \in P_u^{\bar{\tau}}\}$ is bounded.

The linear program minimize$\{(-cB) \cdot z: z \in P_u^{\bar{\tau}}\}$ is feasible since 0 is a feasible solution. If it is bounded as well then minimize$\{c_\tau x_\tau + c_{\bar{\tau}} x_{\bar{\tau}}: A_\tau x_\tau + A_{\bar{\tau}} x_{\bar{\tau}} = b, x_{\bar{\tau}} \geq 0\}$ is feasible and bounded. As a result, the dual of the latter program maximize$\{y \cdot b: yA_\tau = c_\tau, yA_{\bar{\tau}} \leq c_{\bar{\tau}}\}$ is feasible. This shows that a superset of τ is a face of Δ_c which implies that $\tau \in \Delta_c$ since Δ_c is a triangulation. □

3 Associated sets

The group relaxation $G^\tau(b)$ (seen as (5)) solves the integer program $\mathrm{IP}_{A,c}(b)$ (seen as (4)) if and only if both programs have the same optimal solution $z^* \in \mathbb{Z}^{n-d}$. If $G^\tau(b)$ solves $\mathrm{IP}_{A,c}(b)$ then $G^{\tau'}(b)$ also solves $\mathrm{IP}_{A,c}(b)$ for every $\tau' \subset \tau$ since $G^{\tau'}(b)$ is a *stricter* relaxation of $\mathrm{IP}_{A,c}(b)$ (has more nonnegativity restrictions) than $G^\tau(b)$. For the same reason, one would expect that $G^\tau(b)$ is easier to solve than $G^{\tau'}(b)$. Therefore, the most useful group relaxations of $\mathrm{IP}_{A,c}(b)$ are those indexed by the *maximal* elements in the subcomplex of Δ_c consisting of all faces τ such that $G^\tau(b)$ solves $\mathrm{IP}_{A,c}(b)$. The following definition isolates such relaxations.

Definition 3.1. A face τ of the regular triangulation Δ_c is an *associated set* of $\mathrm{IP}_{A,c}$ (or is associated to $\mathrm{IP}_{A,c}$) if for some $b \in \mathbb{N}A$, $G^\tau(b)$ solves $\mathrm{IP}_{A,c}(b)$ but $G^{\tau'}(b)$ does not for all faces τ' of Δ_c such that $\tau \subset \tau'$.

The associated sets of $\mathrm{IP}_{A,c}$ carry all the information about all the group relaxations needed to solve the programs in $\mathrm{IP}_{A,c}$. In this section we will develop tools to understand these sets. We start by considering the set $\mathcal{O}_c \subset \mathbb{N}^n$ of all the optimal solutions of all programs in $\mathrm{IP}_{A,c}$. A basic result in the algebraic study of integer programming is that $\mathcal{O}_c$ is an *order ideal* or *down set* in $\mathbb{N}^n$, i.e., if $u \in \mathcal{O}_c$ and $v \leq u$, $v \in \mathbb{N}^n$, then $v \in \mathcal{O}_c$. One way to prove this is to show that the complement $\mathcal{N}_c := \mathbb{N}^n \backslash \mathcal{O}_c$ has the property that if $v \in \mathcal{N}_c$ then $v + \mathbb{N}^n \subseteq \mathcal{N}_c$. Every lattice point in $\mathbb{N}^n$ is a feasible solution to a unique program in $\mathrm{IP}_{A,c}$ ($u \in \mathbb{N}^n$ is feasible for $\mathrm{IP}_{A,c}(Au)$). Hence, $\mathcal{N}_c$ is the set of all nonoptimal solutions of all programs in $\mathrm{IP}_{A,c}$. A set $P \subseteq \mathbb{N}^n$ with the property that $p + \mathbb{N}^n \subseteq P$ whenever $p \in P$ has a finite set of minimal elements. Hence there exists $\alpha_1, \ldots, \alpha_t \in \mathcal{N}_c$ such that

$$\mathcal{N}_c = \bigcup_{i=1}^{t} (\alpha_i + \mathbb{N}^n).$$

As a result, $\mathcal{O}_c$ is completely specified by the finitely many "generators" $\alpha_1, \ldots, \alpha_t$ of its complement $\mathcal{N}_c$. See Thomas (1995) for proofs of these assertions. Recall that the cost vector c of $\mathrm{IP}_{A,c}$ was assumed to be generic in the sense that each program in $\mathrm{IP}_{A,c}$ has a unique optimal solution. This implies that there is a bijection between the lattice points of $\mathcal{O}_c$ and the semigroup $\mathbb{N}A$ via the map $\phi_A : \mathcal{O}_c \to \mathbb{N}A$ such that $u \mapsto Au$. The inverse of ϕ_A sends a vector $b \in \mathbb{N}A$ to the optimal solution of $\mathrm{IP}_{A,c}(b)$.

Example 3.2. Consider the family of knapsack problems:

$$\text{minimize}\{10000x_1 + 100x_2 + x_3 \ : \ 2x_1 + 5x_2 + 8x_3 = b, \ (x_1, x_2, x_3) \in \mathbb{N}^3\}$$

as b varies in the semigroup $\mathbb{N}[2\ 5\ 8]$. The set $\mathcal{N}_c$ is generated by the vectors

$$(0, 8, 0), (1, 0, 1), (1, 6, 0), (2, 4, 0), (3, 2, 0), \text{ and } (4, 0, 0)$$

which means that $\mathcal{N}_c = ((0,\ 8,\ 0) + \mathbb{N}^3) \cup \cdots \cup ((4,\ 0,\ 0) + \mathbb{N}^3)$. Figure 4 is a picture of $\mathcal{N}_c$ (created by Ezra Miller). The white points are its generators. One can see that $\mathcal{O}_c$ consists of finitely many points of the form $(p, q, 0)$ where $p \geq 1$ and the eight "lattice lines" of points $(0, i, *)$, $i = 0, \ldots, 7$. □

The most fundamental open question concerning $\mathcal{O}_c$ is the following.

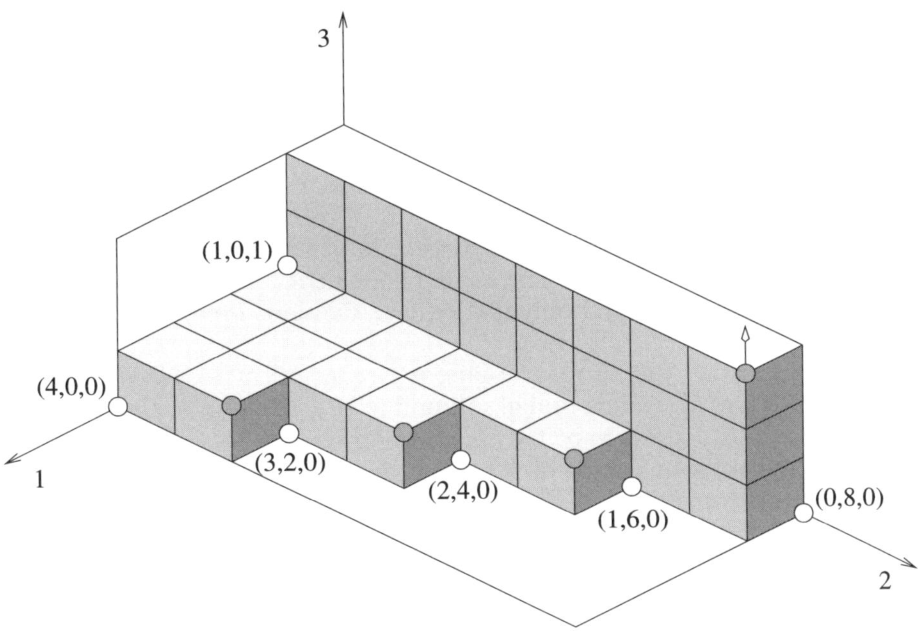

Fig. 4. The set of nonoptimal solutions $\mathcal{N}_c$ for Example 3.2.

Problem 3.3. *Characterize the order ideals in $\mathbb{N}^n$ that arise as $\mathcal{O}_c$ for a family of integer programs $IP_{A,c}$ where $A \in \mathbb{Z}^{d\times n}$ and $c \in \mathbb{Z}^n$ is generic.*

Several necessary conditions for an order ideal to be $\mathcal{O}_c$ are known, of which the *chain property* explained in Section 5 is the most sophisticated thus far. For the purpose of computations, it is most effective (as of now) to think of $\mathcal{N}_c$ and $\mathcal{O}_c$ algebraically.[1] These sets carry all of the information concerning the family $\text{IP}_{A,c}$ – the minimal test set (Gröbner basis) of the family, complete information on the group relaxations needed to solve all programs in the family, and precise sensitivity information for $\text{IP}_{A,c}$ to variations in the cost function c. The Gröbner bases approach to integer programming allows $\mathcal{N}_c$ (and thus $\mathcal{O}_c$) to be calculated via the Buchberger algorithm for Gröbner bases. Besides this, $\mathcal{O}_c$ can also be constructed by repeated calls to an integer programming oracle (Hoşten and Thomas, 1999b). This second method is yet to be implemented and tested seriously. The following problem remains important. Recent work by Deloera *et al.* has shown how to store $\mathcal{O}_c$ efficiently.

We will now describe a certain decomposition of the set $\mathcal{O}_c$ which in turn will shed light on the associated sets of $\text{IP}_{A,c}$. For $u \in \mathbb{N}^n$, consider $Q_u := \{z \in \mathbb{R}^{n-d}\colon Bz \le u, (-cB)\cdot z \le 0\}$ and its relaxation $Q_u^{\bar\tau} := \{z \in \mathbb{R}^{n-d}\colon B^{\bar\tau} z \le \pi_\tau(u), (-cB)\cdot z \le 0\}$ where $B, B^{\bar\tau}$ are as in (4) and (5) and $\tau \in \Delta_c$. By Theorem 2.8, both Q_u and $Q_u^{\bar\tau}$ are polytopes. Notice that if $\pi_\tau(u) = \pi_\tau(u')$ for two distinct vectors $u, u' \in \mathbb{N}^n$, then $Q_u^{\bar\tau} = Q_{u'}^{\bar\tau}$.

Lemma 3.5. (*i*) *A lattice point u is in $\mathcal{O}_c$ if and only if $Q_u \cap \mathbb{Z}^{n-d} = \{0\}$.*
(*ii*) *If $u \in \mathcal{O}_c$, then the group relaxation $G^\tau(Au)$ solves the integer program $IP_{A,c}(Au)$ if and only if $Q_u^{\bar\tau} \cap \mathbb{Z}^{n-d} = \{0\}$.*

Proof. (i) The lattice point u belongs to $\mathcal{O}_c$ if and only if u is the optimal solution to $\text{IP}_{A,c}(Au)$ which is equivalent to $0 \in \mathbb{Z}^{n-d}$ being the optimal solution to the reformulation (4) of $\text{IP}_{A,c}(Au)$. Since c is generic, the last statement is equivalent to $Q_u \cap \mathbb{Z}^{n-d} = \{0\}$. The second statement follows from (i) and the fact that (5) solves (4) if and only if they have the same optimal solution. □

In order to state the current results, it is convenient to assume that the vector u in (4) and (5) is the optimal solution to $\text{IP}_{A,c}(b)$. For an element $u \in \mathcal{O}_c$ and a face τ of Δ_c let $S(u,\tau)$ be the *affine* semigroup $u + \mathbb{N}(e_i\colon i \in \tau) \subseteq \mathbb{N}^n$ where e_i denotes the ith unit vector of $\mathbb{R}^n$. Note that $S(u,\tau)$ is not a semigroup if $u \ne 0$, but is a translation of the semigroup $\mathbb{N}(e_i\colon i \in \tau)$. We use the adjective *affine* here as in an *affine subspace* which is not a subspace but the translation of one. Note that if $v \in S(u,\tau)$, then $\pi_\tau(v) = \pi_\tau(u)$.

[1] See [A1] in Section 8.

Lemma 3.6. *For $u \in \mathcal{O}_c$ and a face τ of Δ_c, the affine semigroup $S(u, \tau)$ is contained in $\mathcal{O}_c$ if and only if $G^\tau(Au)$ solves $IP_{A,c}(Au)$.*

Proof. Suppose $S(u, \tau) \subseteq \mathcal{O}_c$. Then by Lemma 3.5 (i), for all $v \in S(u, \tau)$,

$$Q_v = \{z \in \mathbb{R}^{n-d}: B^\tau z \leq \pi_{\bar{\tau}}(v),\ B^{\bar{\tau}} z \leq \pi_\tau(u),\ (-cB) \cdot z \leq 0\} \cap \mathbb{Z}^{n-d} = \{0\}.$$

Since $\pi_{\bar{\tau}}(v)$ can be any vector in $\mathbb{N}^{|\tau|}$, $Q_u^{\bar{\tau}} \cap \mathbb{Z}^{n-d} = \{0\}$. Hence, by Lemma 3.5 (ii), $G^\tau(Au)$ solves $\text{IP}_{A,c}(Au)$.

If $v \in S(u, \tau)$, then $\pi_\tau(u) = \pi_\tau(v)$, and hence $Q_u^{\bar{\tau}} = Q_v^{\bar{\tau}}$. Therefore, if $G^\tau(Au)$ solves $\text{IP}_{A,c}(Au)$, then $\{0\} = Q_u^{\bar{\tau}} \cap \mathbb{Z}^{n-d} = Q_v^{\bar{\tau}} \cap \mathbb{Z}^{n-d}$ for all $v \in S(u, \tau)$. Since $Q_v^{\bar{\tau}}$ is a relaxation of Q_v, $Q_v \cap \mathbb{Z}^{n-d} = \{0\}$ for all $v \in S(u, \tau)$ and hence by Lemma 3.5 (i), $S(u, \tau) \subseteq \mathcal{O}_c$. □

Lemma 3.7. *For $u \in \mathcal{O}_c$ and a face τ of Δ_c, $G^\tau(Au)$ solves $IP_{A,c}(Au)$ if and only if $G^\tau(Av)$ solves $IP_{A,c}(Av)$ for all $v \in S(u, \tau)$.* □

Proof. If $v \in S(u, \tau)$ and $G^\tau(Au)$ solves $\text{IP}_{A,c}(Au)$, then as seen before, $\{0\} = Q_u^{\bar{\tau}} \cap \mathbb{Z}^{n-d} = Q_v^{\bar{\tau}} \cap \mathbb{Z}^{n-d}$ for all $v \in S(u, \tau)$. By Lemma 3.5 (ii), $G^\tau(Av)$ solves $\text{IP}_{A,c}(Av)$ for all $v \in S(u, \tau)$. The converse holds for the trivial reason that $u \in S(u, \tau)$.

Corollary 3.8. *For $u \in \mathcal{O}_c$ and a face τ of Δ_c, the affine semigroup $S(u, \tau)$ is contained in $\mathcal{O}_c$ if and only if $G^\tau(Av)$ solves* $\text{IP}_{A,c}(Av)$ *for all $v \in S(u, \tau)$.*

Since $\pi_\tau(u)$ determines the polytope $Q_u^{\bar{\tau}} = Q_v^{\bar{\tau}}$ for all $v \in S(u, \tau)$, we could have assumed that $\text{supp}(u) \subseteq \bar{\tau}$ in Lemmas 3.6 and 3.7.

Definition 3.9. For $\tau \in \Delta_c$ and $u \in \mathcal{O}_c$, (u, τ) is called an *admissible pair* of $\mathcal{O}_c$ if

(i) the support of u is contained in $\bar{\tau}$, and
(ii) $S(u, \tau) \subseteq \mathcal{O}_c$ or equivalently, $G^\tau(Av)$ solves $\text{IP}_{A,c}(Av)$ for all $v \in S(u, \tau)$.

An admissible pair (u, τ) is a *standard pair* of $\mathcal{O}_c$ if the affine semigroup $S(u,\tau)$ is not properly contained in $S(v, \tau')$ where (v, τ') is another admissible pair of $\mathcal{O}_c$.

Example 3.2 continued. From Fig. 4, one can see that the standard pairs of $\mathcal{O}_c$ are as:

$((1,0,0),\emptyset)$	$((1,3,0),\emptyset)$		$((0,0,0),\{3\})$
$((2,0,0),\emptyset)$	$((2,3,0),\emptyset)$		$((0,1,0),\{3\})$
$((3,0,0),\emptyset)$	$((1,4,0),\emptyset)$		$((0,2,0),\{3\})$
$((1,1,0),\emptyset)$	$((1,5,0),\emptyset)$	and	$((0,3,0),\{3\})$
$((2,1,0),\emptyset)$			$((0,4,0),\{3\})$
$((3,1,0),\emptyset)$			$((0,5,0),\{3\})$
$((1,2,0),\emptyset)$			$((0,6,0),\{3\})$
$((2,2,0),\emptyset)$			$((0,7,0),\{3\})$

□

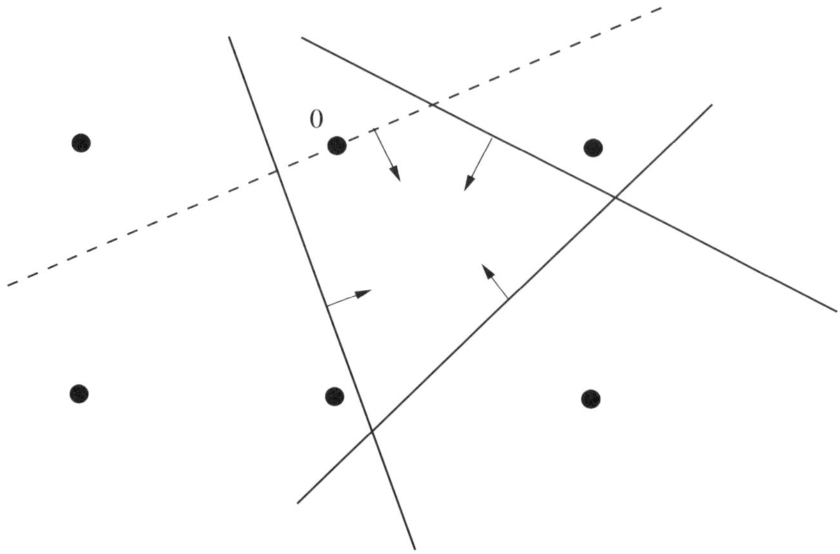

Fig. 5. A standard polytope.

Definition 3.10. For a face τ of Δ_c and a lattice point $u \in \mathbb{N}^n$, we say that the polytope $Q_u^{\bar{\tau}}$ is a *standard polytope* of $\mathrm{IP}_{A,c}$ if $Q_u^{\bar{\tau}} \cap \mathbb{Z}^{n-d} = \{0\}$ and every relaxation of $Q_u^{\bar{\tau}}$ obtained by removing an inequality in $B^{\bar{\tau}} z \leq \pi_\tau(u)$ contains a nonzero lattice point.

Figure 5 is a diagram of a standard polytope $Q_u^{\bar{\tau}}$. The dashed line is the boundary of the half space $(-cB) \cdot z \leq 0$ while the other lines are the boundaries of the halfspaces given by the inequalities in $B^{\bar{\tau}} z \leq \pi_\tau(u)$. The origin is the only lattice point in the polytope and if any inequality in $B^{\bar{\tau}} z \leq \pi_\tau(u)$ is removed, a lattice point will enter the relaxation.

We re-emphasize that if $Q_u^{\bar{\tau}}$ is a standard polytope, then $Q_{u'}^{\bar{\tau}}$ is the same standard polytope if $\pi_\tau(u) = \pi_\tau(u')$. Hence the same standard polytope can be indexed by infinitely many $u \in \mathbb{N}^n$. We now state the main result of this section which characterizes associated sets in terms of standard pairs and standard polytopes.

Theorem 3.11. *The following statements are equivalent:*

(i) *The admissible pair* (u, τ) *is a standard pair of* $\mathcal{O}_c$.
(ii) *The polytope* $Q_u^{\bar{\tau}}$ *is a standard polytope of* $IP_{A,c}$.
(iii) *The face* τ of Δ_c *is associated to* $IP_{A,c}$.

Proof. (i)$\Leftrightarrow$(ii): The admissible pair (u, τ) is standard if and only if for every $i \in \bar{\tau}$, there exists some positive integer m_i and a vector $v \in S(u, \tau)$ such that $v + m_i e_i \in \mathcal{N}_c$. (If this condition did not hold for some $i \in \bar{\tau}$, then

$(u', \tau \cup \{i\})$ would be an admissible pair of $\mathcal{O}_c$ such that $S(u', \tau \cup \{i\})$ contains $S(u, \tau)$ where u' is obtained from u by setting the ith component of u to zero. Conversely, if the condition holds for an admissible pair then the pair is standard.) Equivalently, for each $i \in \bar{\tau}$, there exists a positive integer m_i and a $v \in S(u, \tau)$ such that $Q^{\bar{\tau}}_{v+m_ie_i} = Q^{\bar{\tau}}_{u+m_ie_i}$ contains at least two lattice points. In other words, the removal of the inequality indexed by i from the inequalities in $B^{\bar{\tau}}z \leq \pi_\tau(u)$ will bring an extra lattice point into the corresponding relaxation of $Q^{\bar{\tau}}_u$. This is equivalent to saying that $Q^{\bar{\tau}}_u$ is a standard polytope of $\mathrm{IP}_{A,c}$.

(i)$\Leftrightarrow$(iii): Suppose (u, τ) is a standard pair of $\mathcal{O}_c$. Then $S(u, \tau) \subseteq \mathcal{O}_c$ and $G^\tau(Au)$ solves $\mathrm{IP}_{A,c}(Au)$ by Lemma 3.6. Suppose $G^{\tau'}(Au)$ solves $\mathrm{IP}_{A,c}(Au)$ for some face $\tau' \in \Delta_c$ such that $\tau \subset \tau'$. Lemma 3.6 then implies that $S(u, \tau')$ lies in $\mathcal{O}_c$. This contradicts the fact that (u, τ) was a standard pair of $\mathcal{O}_c$ since $S(u, \tau)$ is properly contained in $S(\hat{u}, \tau')$ corresponding to the admissible pair $(\hat{u}, \tau')$ where $\hat{u}$ is obtained from u by setting $u_i = 0$ for all $i \in \tau' \backslash \tau$.

To prove the converse, suppose τ is associated to $\mathrm{IP}_{A,c}$. Then there exists some $b \in \mathbb{N}A$ such that $G^\tau(b)$ solves $\mathrm{IP}_{A,c}(b)$ but $G^{\tau'}(b)$ does not for all faces τ' of Δ_c containing τ. Let u be the unique optimal solution of $\mathrm{IP}_{A,c}(b)$. By Lemma 3.6, $S(u, \tau) \subseteq \mathcal{O}_c$. Let $\hat{u} \in \mathbb{N}^n$ be obtained from u by setting $u_i = 0$ for all $i \in \tau$. Then $G^\tau(A\hat{u})$ solves $\mathrm{IP}_{A,c}(A\hat{u})$ since $Q^{\bar{\tau}}_u = Q^{\bar{\tau}}_{\hat{u}}$. Hence $S(\hat{u}, \tau) \subseteq \mathcal{O}_c$ and $(\hat{u}, \tau)$ is an admissible pair of $\mathcal{O}_c$. Suppose there exists another admissible pair (w, σ) such that $S(\hat{u}, \tau) \subset S(w, \sigma)$. Then $\tau \subseteq \sigma$. If $\tau = \sigma$ then $S(\hat{u}, \tau)$ and $S(w, \sigma)$ are both orthogonal translates of $\mathbb{N}(e_i : i \in \tau)$ and hence $S(\hat{u}, \tau)$ cannot be properly contained in $S(w, \sigma)$. Therefore, τ is a proper subset of σ which implies that $S(\hat{u}, \sigma) \subseteq \mathcal{O}_c$. Then, by Lemma 3.6, $G^\sigma(A\hat{u})$ solves $\mathrm{IP}_{A,c}(A\hat{u})$ which contradicts that τ was an associated set of $\mathrm{IP}_{A,c}$. □

Example 3.2 continued. In Example 3.2 we can choose B to be the 3×2 matrix

$$B = \begin{bmatrix} -1 & 4 \\ 2 & 0 \\ -1 & -1 \end{bmatrix}.$$

The standard polytope defined by the standard pair $((1,0,0), \emptyset)$ is hence

$$\{(z_1, z_2) \in \mathbb{R}^2 \colon -z_1 + 4z_2 \leq 1,\ 2z_1 \leq 0,\ -z_1 - z_2 \leq 0,\ 9801z_1 - 40001z_2 \leq 0\}$$

while the standard polytope defined by the standard pair $((0,2,0), \{3\})$ is:

$$\{(z_1, z_2) \in \mathbb{R}^2 \colon -z_1 + 4z_2 \leq 0,\ 2z_1 \leq 2,\ 9801z_1 - 40001z_2 \leq 0\}.$$

The associated sets of $\mathrm{IP}_{A,c}$ in this example are $\emptyset$ and $\{3\}$. There are twelve quadrangular and eight triangular standard polytopes for this family of knapsack problems. □

Standard polytopes were introduced in Hoşten and Thomas (1999a), and the equivalence of parts (i) and (ii) of Theorem 3.11 was proved in Hoşten and Thomas (1999a, Theorem 2.5). Under the linear map $\phi_A: \mathbb{N}^n \to \mathbb{N}A$ where $u \mapsto Au$, the affine semigroup $S(u, \tau)$ where (u, τ) is a standard pair of $\mathcal{O}_c$ maps to the affine semigroup $Au + \mathbb{N}A_\tau$ in $\mathbb{N}A$. Since every integer program in $\text{IP}_{A,c}$ is solved by one of its group relaxations, $\mathcal{O}_c$ is covered by the affine semigroups corresponding to its standard pairs. We call this cover and its image in $\mathbb{N}A$ under ϕ_A the *standard pair decompositions* of $\mathcal{O}_c$ and $\mathbb{N}A$, respectively. Since standard pairs of $\mathcal{O}_c$ are determined by the standard polytopes of $\text{IP}_{A,c}$, the standard pair decomposition of $\mathcal{O}_c$ is unique. The terminology used above has its origins in Sturmfels et al. (1995) which introduced the *standard pair decomposition* of a *monomial ideal*. The specialization to integer programming appear in Hoşten and Thomas (1999a,b) and Sturmfels (1995, §12.D). The following theorem shows how the standard pair decomposition of $\mathcal{O}_c$ dictates which group relaxations solve which programs in $\text{IP}_{A,c}$.

Theorem 3.12. *Let v be the optimal solution of the integer program $IP_{A,c}(b)$. Then the group relaxation $G^\tau(Av)$ solves $IP_{A,c}(Av)$ if and only if there is some standard pair (u, τ') of $\mathcal{O}_c$ with $\tau \subseteq \tau'$ such that v belongs to the affine semigroup $S(u, \tau')$.*

Proof. Suppose v lies in $S(u, \tau')$ corresponding to the standard pair (u, τ') of $\mathcal{O}_c$. Then $S(v, \tau') \subseteq \mathcal{O}_c$ which implies that $G^{\tau'}(Av)$ solves $\text{IP}_{A,c}(Av)$ by Lemma 3.6. Hence $G^\tau(Av)$ also solves $\text{IP}_{A,c}(Av)$ for all $\tau \subseteq \tau'$.

To prove the converse, suppose τ' is a maximal element in the subcomplex of all faces τ of Δ_c such that $G^\tau(Av)$ solves $\text{IP}_{A,c}(Av)$. Then τ' is an associated set of $\text{IP}_{A,c}$. In the proof of (iii) $\Rightarrow$ (i) in Theorem 3.11, we showed that $(\hat{v}, \tau')$ is a standard pair of $\mathcal{O}_c$ where $\hat{v}$ is obtained from v by setting $v_i = 0$ for all $i \in \tau'$. Then $v \in S(\hat{v}, \tau')$. □

Example 3.2 continued. The eight standard pairs of $\mathcal{O}_c$ of the form $(*, \{3\})$, map to the eight affine semigroups:

$$\mathbb{N}[8],\ (5 + \mathbb{N}[8]),\ (10 + \mathbb{N}[8]),\ (15 + \mathbb{N}[8]),\ (20 + \mathbb{N}[8]),$$

$$(25 + \mathbb{N}[8]),\ (30 + \mathbb{N}[8]) \text{ and } (35 + \mathbb{N}[8])$$

contained in $\mathbb{N}A = \mathbb{N}[2, 5, 8] \subset \mathbb{N}$. For all right hand side vectors b in the union of these sets, the integer program $\text{IP}_{A,c}(b)$ can be solved by the group relaxation $G^{\{3\}}(b)$. The twelve standard pairs of the from $(*, \emptyset)$ map to the remaining finitely many points

$$2, 4, 6, 7, 9, 11, 12, 14, 17, 19, 22 \text{ and } 27$$

of $\mathbb{N}[2, 5, 8]$. If b is one of these points, then $\mathrm{IP}_{A,c}(b)$ can only be solved as the full integer program. In this example, the regular triangulation $\Delta_c = \{\{3\}\}$. Hence $G^{\{3\}}(b)$ is a Gomory relaxation of $\mathrm{IP}_{A,c}(b)$. □

For most $b \in \mathbb{N}A$, the program $\mathrm{IP}_{A,c}(b)$ is solved by one of its Gomory relaxations, or equivalently, by Theorem 3.12, the optimal solution v of $\mathrm{IP}_{A,c}(b)$ lies in $S(*, \sigma)$ for some standard pair $(*, \sigma)$ where σ is a maximal face of Δ_c. For mathematical versions of this informal statement (see Sturmfels (1995, Proposition 12.16) and Gomory (1965, Theorems 1 and 2). Roughly speaking, these right hand sides are away from the boundary of cone(A). (This was seen in Example 3.2 above, where for all but twelve right hand sides, $\mathrm{IP}_{A,c}(b)$ was solvable by the Gomory relaxation $G^{\{3\}}(b)$. Further, these twelve right hand sides were toward the boundary of cone(A), the origin in this one-dimensional case.) For the remaining right hand sides, $\mathrm{IP}_{A,c}(b)$ can only be solved by $G^{\tau}(b)$ where τ is a lower dimensional face of Δ_c – possibly even the empty face. An important contribution of the approach described here is the identification of the minimal set of group relaxations needed to solve *all* programs in the family $\mathrm{IP}_{A,c}$ and of the particular relaxations necessary to solve any given program in the family.

4 Arithmetic degree

For an associated set τ of $\mathrm{IP}_{A,c}$ there are only finitely many standard pairs of $\mathcal{O}_c$ indexed by τ since there are only finitely many standard polytopes of the form $Q_u^{\bar{\tau}}$. Borrowing terminology from Sturmfels et al. (1995), we call the number of standard pairs of the form $(\cdot, \tau)$ the *multiplicity* of τ in $\mathcal{O}_c$ (abbreviated as *mult*(τ)). The total number of standard pairs of $\mathcal{O}_c$ is called the *arithmetic degree* of $\mathcal{O}_c$. Our main goal in this section is to provide bounds for these invariants of the family $\mathrm{IP}_{A,c}$ and discuss their relevance. We will need the following interpretation from Section 3.

Corollary 4.1. *The multiplicity of the face τ of Δ_c in $\mathcal{O}_c$ is the number of distinct standard polytopes of $IP_{A,c}$ indexed by $\bar{\tau}$, and the arithmetic degree of $\mathcal{O}_c$ is the total number of standard polytopes of $IP_{A,c}$.*

Proof. This result follows from Theorem 3.11. □

Example 3.2 continued. The multiplicity of the associated set $\{3\}$ is eight while the empty set has multiplicity twelve. The arithmetic degree of $\mathcal{O}_c$ is hence twenty. □

If the standard pair decomposition of $\mathcal{O}_c$ is known, then we can solve all programs in $\mathrm{IP}_{A,c}$ by solving (arithmetic degree) – many *linear systems* as follows. For a given $b \in \mathbb{N}A$ and a standard pair (u, τ), consider the linear system

$$A_{\bar{\tau}}\pi_{\tau}(u) + A_{\tau}x = b, \quad \text{or equivalently,} \quad A_{\tau}x = b - A_{\bar{\tau}}\pi_{\tau}(u). \quad (6)$$

As τ is a face of Δ_c, the columns of A_τ are linearly independent and the linear system (6) can be solved uniquely for x. Since the optimal solution of $\mathrm{IP}_{A,c}(b)$ lies in $S(w, \sigma)$ for some standard pair (w, σ) of $\mathcal{O}_c$, at least one nonnegative and integral solution for x will be found as we solve the linear systems (6) obtained by varying (u, τ) over all the standard pairs of $\mathcal{O}_c$. If the standard pair (u, τ) yields such a solution v, then $(\pi_\tau(u), v)$ is the optimal solution of $\mathrm{IP}_{A,c}(b)$. This preprocessing of $\mathrm{IP}_{A,c}$ has the same flavor as Kannan (1993). The main result in Kannan (1993) is that given a coefficient matrix $A \in \mathbb{R}^{m \times n}$ and cost vector c, there exists *floor functions* $f_1, \ldots, f_k : \mathbb{R}^m \to \mathbb{Z}^n$ such that for a right hand side vector b, the optimal solution of the corresponding integer program is the one among $f_1(b), \ldots, f_k(b)$ that is feasible and attains the best objective function value. The crucial point is that this algorithm runs in time bounded above by a polynomial in the length of the data for fixed n and j, where j is the affine dimension of the space of right hand sides. In our situation, the preprocessing involves solving (arithmetic-degree)-many linear systems. Given this, it is interesting to bound the arithmetic degree of $\mathcal{O}_c$.

The second equation in (6) suggests that one could think of the first arguments u in the standard pairs (u, τ) of $\mathcal{O}_c$ as "correction vectors" that need to be applied to find the optimal solutions of programs in $\mathrm{IP}_{A,c}$. Thus the arithmetic degree of $\mathcal{O}_c$ is the total number of correction vectors that are needed to solve all programs in $\mathrm{IP}_{A,c}$. The multiplicities of associated sets give a finer count of these correction vectors, organized by faces of Δ_c. If the optimal solution of $\mathrm{IP}_{A,c}(b)$ lies in the affine semigroup $S(w, \sigma)$ given by the standard pair (w, σ) of $\mathcal{O}_c$, then w is a correction vector for this b as well as all other b's in $(Aw + \mathbb{N}A_\sigma)$. One obtains all correction vectors for $\mathrm{IP}_{A,c}$ by solving the (arithmetic degree)-many integer programs with right hand sides Au for all standard pairs (u, τ) of $\mathcal{O}_c$. See Wolsey (1981) for a similar result from the classical theory of group relaxations.

In Example 3.2, $\Delta_c = \{\{3\}\}$ and both its faces $\{3\}$ and $\emptyset$ are associated to $\mathrm{IP}_{A,c}$. In general, not all faces of Δ_c need be associated sets of $\mathrm{IP}_{A,c}$ and the poset of associated sets can be quite complicated. (We will study this poset in Section 5.) Hence, for $\tau \in \Delta_c$, $\mathrm{mult}(\tau) = 0$ unless τ is an associated set of $\mathrm{IP}_{A,c}$. We will now prove that all maximal faces of Δ_c are associated sets of $\mathrm{IP}_{A,c}$. Further, if σ is a maximal face of Δ_c then $\mathrm{mult}(\sigma)$ is the absolute value of $\det(A_\sigma)$ divided by the g.c.d. of the maximal minors of A. This g.c.d. is nonzero since A has full row rank. If the columns of A span an affine hyperplane, then the absolute value of $\det(A_\sigma)$ divided by the g.c.d. of the maximal minors of A is called the *normalized volume* of the face σ in Δ_c. We first give a nontrivial example.

Example 4.2. Consider the rank three matrix

$$A = \begin{bmatrix} 5 & 0 & 0 & 2 & 1 & 0 \\ 0 & 5 & 0 & 1 & 4 & 2 \\ 0 & 0 & 5 & 2 & 0 & 3 \end{bmatrix}$$

and the generic cost vector $c = (21, 6, 1, 0, 0, 0)$. The first three columns of A generate cone(A) which is simplicial. The regular triangulation

$$\Delta_c = \{\{1, 3, 4\}, \{1, 4, 5\}, \{2, 5, 6\}, \{3, 4, 6\}, \{4, 5, 6\}\}$$

is shown in Fig. 6 as a triangulation of conv(A). The six columns of A have been labeled by their column indices. The arithmetic degree of $\mathcal{O}_c$ in this example is 70. The following table shows all the standard pairs organized by associated sets and the multiplicity of each associated set. Note that all maximal faces of Δ_c are associated to $\text{IP}_{A,c}$. The g.c.d. of the maximal minors of A is five. Check that mult(σ) is the normalized volume of σ whenever σ is a maximal face of Δ_c.

Observe that the integer program $\text{IP}_{A,c}(b)$ where $b = A(e_1 + e_2 + e_3)$ is solved by $G^\tau(b)$ with $\tau = \{1, 4, 5\}$. By Proposition 2.3, Gomory's relaxation of $\text{IP}_{A,c}(b)$ is indexed by $\sigma = \{4, 5, 6\}$ since b lies in the interior of the face cone(A_σ) of Δ_c.

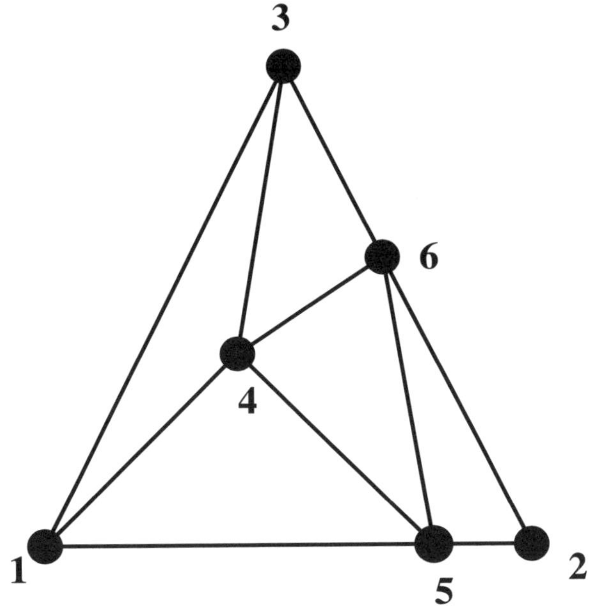

Fig. 6. The regular triangulation Δ_c for Example 4.2.

τ	Standard pairs $(\cdot, \tau)$	Mult (τ)
$\{1,3,4\}$	$(0, \cdot), (e_5, \cdot), (e_6, \cdot), (e_5+e_6, \cdot), (2e_6, \cdot)$	5
$\{1,4,5\}$	$(0, \cdot), (e_2, \cdot), (e_3, \cdot), (e_6, \cdot), (e_2+e_3, \cdot), (2e_2, \cdot),$ $(3e_2, \cdot), (2e_2+e_3, \cdot)$	8
$\{2,5,6\}$	$(0, \cdot), (e_3, \cdot), (2e_3, \cdot)$	3
$\{3,4,6\}$	$(0, \cdot), (e_5, \cdot), (2e_5, \cdot), (3e_5, \cdot)$	4
$\{4,5,6\}$	$(0, \cdot), (e_3, \cdot), (2e_3, \cdot), (3e_3, \cdot), (4e_3, \cdot)$	5
$\{1,4\}$	$(e_3+2e_5+e_6, \cdot), (2e_3+2e_5+e_6, \cdot), (2e_3+2e_5, \cdot),$ $(2e_3+3e_5, \cdot), (2e_3+4e_5, \cdot)$	5
$\{1,5\}$	$(e_2+e_6, \cdot), (2e_2+e_6, \cdot), (3e_2+e_6, \cdot)$	3
$\{2,5\}$	$(e_3+e_4, \cdot), (e_4, \cdot), (2e_4, \cdot)$	3
$\{3,4\}$	$(e_2, \cdot), (e_1+e_2, \cdot), (e_1+2e_5, \cdot), (e_1+2e_5+e_6, \cdot), (e_2+e_5, \cdot),$	5
$\{3,6\}$	$(e_2, \cdot), (e_2+e_5, \cdot)$	2
$\{4,5\}$	$(e_2+2e_3, \cdot), (e_2+3e_3, \cdot), (2e_2+2e_3, \cdot), (3e_2+e_3, \cdot), (4e_2, \cdot)$	5
$\{5,6\}$	$(e_2+3e_3, \cdot)$	1
$\{1\}$	$(e_2+e_3+e_6, \cdot), (e_2+e_3+e_5+e_6, \cdot), (e_2+2e_6, \cdot),$ $(e_2+e_3+2e_6, \cdot), (2e_2+2e_6, \cdot), (e_2+e_3+2e_5+e_6, \cdot)$	6
$\{3\}$	$(e_1+e_2+e_6, \cdot), (e_1+e_2+2e_6, \cdot)$	2
$\{4\}$	$(e_1+e_2+2e_3+e_5, \cdot), (e_1+e_2+2e_3+2e_5, \cdot),$ $(e_1+e_2+2e_3+3e_5, \cdot), (e_1+e_2+2e_3+4e_5, \cdot),$ $(e_1+3e_3+3e_5, \cdot), (e_1+3e_3+4e_5, \cdot)$	6
$\{\emptyset\}$	$(e_1+e_2+2e_3+e_5+e_6, \cdot), (e_1+e_2+2e_3+2e_5+e_6, \cdot),$ $(e_1+2e_2+e_3+e_6, \cdot), (e_1+2e_2+e_3+e_5+e_6, \cdot),$ $(e_1+2e_2+e_3+2e_5+e_6, \cdot), (e_1+2e_2+e_3+2e_6, \cdot),$ $(e_1+3e_2+2e_6, \cdot)$	7
	Arithmetic degree	70

However, neither this relaxation nor any nontrivial extended relaxation solves $\mathrm{IP}_{A,c}(b)$ since the optimal solution $e_1+e_2+e_3$ is not covered by any standard pair $(\cdot, \tau)$ where τ is a nonempty subset of $\{4,5,6\}$. □

Theorem 4.3. *For a set $\sigma \subseteq \{1,\ldots,n\}$, $(0,\sigma)$ is a standard pair of $\mathcal{O}_c$ if and only if σ is a maximal face of Δ_c.*

Proof. If σ is a maximal face of Δ_c, then by Definition 2.1, there exists $y \in \mathbb{R}^d$ such that $yA_\sigma = c_\sigma$ and $yA_{\bar\sigma} < c_{\bar\sigma}$. Then $p = c_{\bar\sigma} - yA_{\bar\sigma} > 0$ and $pB^{\bar\sigma} = (c_{\bar\sigma} - yA_{\bar\sigma})B^{\bar\sigma} = c_{\bar\sigma}B^{\bar\sigma} - yA_{\bar\sigma}B^{\bar\sigma} = c_{\bar\sigma}B^{\bar\sigma} + yA_\sigma B^\sigma = c_{\bar\sigma}B^{\bar\sigma} + c_\sigma B^\sigma = cB$. Hence there is a positive dependence relation among $(-cB)$ and the rows of $\mathrm{B}^{\bar\sigma}$. Since σ is a maximal face of Δ_c, $|\det(A_\sigma)| \neq 0$. However, $|\det(B^{\bar\sigma})| = |\det(A_\sigma)|$ which implies that $|\det(B^{\bar\sigma})| \neq 0$. Therefore, $(-cB)$ and the rows of $B^{\bar\sigma}$ span $\mathbb{R}^{n-d}$ positively. This implies that $Q_0^{\bar\sigma} = \{z \in \mathbb{R}^{n-d}: B^{\bar\sigma}z \leq 0,\ (-cB)\cdot z \leq 0\}$ is a polytope consisting of just the origin. If any inequality defining this simplex is dropped, the resulting relaxation is unbounded as only $n-d$ inequalities would remain. Hence $Q_0^{\bar\sigma}$ is a standard polytope of $\mathrm{IP}_{A,c}$ and by Theorem 3.11, $(0, \sigma)$ is a standard pair of $\mathcal{O}_c$.

Conversely, if $(0, \sigma)$ is a standard pair of $\mathcal{O}_c$ then $Q_0^{\bar\sigma}$ is a standard polytope of $\mathrm{IP}_{A,c}$. Since every inequality in the definition of $Q_0^{\bar\sigma}$ gives a halfspace containing the origin and $Q_0^{\bar{\sigma}}$ is a polytope, $Q_0^{\bar{\sigma}} = \{0\}$. Hence there is a positive linear dependence relation among $(-cB)$ and the rows of B^{σ}. If $|\bar{\sigma}| > n-d$, then $Q_0^{\bar{\sigma}}$ would coincide with the relaxation obtained by dropping some inequality from those in $B^{\bar{\sigma}}z \leq 0$. This would contradict that $Q_0^{\bar{\sigma}}$ was a standard polytope and hence $|\sigma| = d$ and σ is a maximal face of Δ_c. □

Corollary 4.4. *Every maximal face of Δ_c is an associated set of $IP_{A,c}$.*

For Theorem 4.5 and Corollary 4.6 below we assume that the g.c.d. of the maximal minors of A is one which implies that $\mathbb{Z}A = \mathbb{Z}^d$.

Theorem 4.5. *If σ is a maximal face of Δ_c then the multiplicity of σ in $\mathcal{O}_c$ is* $|\det(A_\sigma)|$.

Proof. Consider the full dimensional lattice $\mathcal{L}_\sigma = \pi_\sigma(\mathcal{L}) = \{B^{\bar{\sigma}}z : z \in \mathbb{Z}^{n-d}\}$ in $\mathbb{Z}^{n-d}$. Since the g.c.d. of the maximal minors of A is assumed to be one, the lattice $\mathcal{L}_\sigma$ has index $|\det(B^{\bar{\sigma}})| = |\det(A_\sigma)|$ in $\mathbb{Z}^{n-d}$. Since $\mathcal{L}_\sigma$ is full dimensional, it has a strictly positive element which guarantees that each equivalence class of $\mathbb{Z}^{n-d}$ modulo $\mathcal{L}_\sigma$ has a nonnegative member. This implies that there are $|\det(A_\sigma)|$ distinct equivalence classes of $\mathbb{N}^{n-d}$ modulo $\mathcal{L}_\sigma$. Recall that if u is a feasible solution to $\text{IP}_{A,c}(b)$ then

$$G^{\sigma}(b) = \text{minimize}\left\{\tilde{c}_{\bar{\sigma}} \cdot x_{\bar{\sigma}} : x_{\bar{\sigma}} \equiv u_{\bar{\sigma}} \ (\text{mod}\ \mathcal{L}_\sigma),\ \ x_{\bar{\sigma}} \in \mathbb{N}^{n-d}\right\}.$$

Since there are $|\det(A_\sigma)|$ equivalence classes of $\mathbb{N}^{n-d}$ modulo $\mathcal{L}_\sigma$, there are $|\det(A_\sigma)|$ distinct group relaxations indexed by σ. The optimal solution of each program becomes the right hand side vector of a standard polytope (simplex) of $\text{IP}_{A,c}$ indexed by σ. Since no two optimal solutions are the same (as they come from different equivalence classes of $\mathbb{N}^{n-d}$ modulo $\mathcal{L}_\sigma$), there are precisely $|\det(A_\sigma)|$ standard polytopes of $\text{IP}_{A,c}$ indexed by σ. □

Corollary 4.6. *The arithmetic degree of $\mathcal{O}_c$ is bounded below by the sum of the absolute values of $\det(A_\sigma)$ as σ varies among the maximal faces of Δ_c.*

Theorem 4.5 gives a precise bound on the multiplicity of a maximal associated set of $\text{IP}_{A,c}$, which in turn provides a lower bound for the arithmetic degree of $\mathcal{O}_c$ in Corollary 4.6. No exact result like Theorem 4.5 is known when τ is a lower dimensional associated set of $\text{IP}_{A,c}$. Such bounds would provide a bound for the arithmetic degree of $\mathcal{O}_c$. The reader interested in the algebraic origins of some of the above results may consult the notes [A2] in Section 8. We close this section with a first attempt at bounding the arithmetic degree of $\mathcal{O}_c$ (under certain nondegeneracy assumptions). This result is due to Ravi Kannan, and its simple arguments are along the lines of proofs in Kannan (1992) and Kannan, Lovász, and Scarf (1990).

Suppose $S \in \mathbb{Z}^{m\times n}$ and $u \in \mathbb{N}^m$ are fixed and $K_u := \{x \in \mathbb{R}^n: Sx \le u\}$ is such that $K_u \cap \mathbb{Z}^n = \{0\}$ and the removal of any inequality defining K_u will bring in a nonzero lattice point into the relaxation. Let $s^{(i)}$ denote the ith row of S, $M := \max||s^{(i)}||_1$ and $\Delta_k(S)$ and $\delta_k(S)$ be the maximum and minimum absolute values of the $k \times k$ subdeterminants of S. We will assume that $\delta_n(S) \neq 0$ which is a nondegeneracy condition on the data. We assume this set up in Theorem 4.8 and Lemmas 4.9 and 4.10.

Definition 4.7. If K is a convex set and v a nonzero vector in $\mathbb{R}^n$, the width of K along v, denoted as $\text{width}_v(K)$ is $\max\{v\cdot x: x\in K\} - \min\{v \cdot x: x \in K\}$.

Note that $\text{width}_v(K)$ is invariant under translations of K.

Theorem 4.8. *If K_u is as above then $0 \le u_i \le 2M(n+2)\frac{\Delta_n(S)}{\delta_n(S)}$.*

Lemma 4.9. *If K_u is as above then for some t, $1 \le t \le m$, $\text{width}_{s^{(t)}}(K_u) \le M(n+2)$.*

Proof. Clearly, K_u is bounded since otherwise there would be a nonzero lattice point on an unbounded edge of K_u due to the integrality of all data. Suppose $\text{width}_{s^{(t)}}(K_u) > M(n+2)$ for all rows $s^{(t)}$ of S. Let p be the center of gravity of K_u. Then by a property of the center of gravity, for any $x \in K_u$, $(1/(n+1))$th of the vector from p to the reflection of x about p is also in K_u, i.e., $(1+\frac{1}{n+1})p - \frac{1}{n+1}x \in K_u$. Fix i, $1 \le i \le m$ and let x_0 minimize $s^{(i)} \cdot x$ over K_u. By the definition of width, we then have $u_i - s^{(i)} \cdot x_0 > M(n+2)$ which implies that

$$s^{(i)} \cdot x_0 < u_i - M(n+2). \tag{7}$$

Now $s^{(i)} \cdot ((1+\frac{1}{n+1})p - \frac{1}{n+1}x_0) \le u_i$ implies that

$$s^{(i)} \cdot p \le u_i\left(\frac{n+1}{n+2}\right) + \frac{s^{(i)} \cdot x_0}{n+2} \tag{8}$$

Combining (7) and (8) we get

$$s^{(i)} \cdot p < u_i - M \tag{9}$$

Let $q = \lfloor p \rfloor$ be the vector obtained by rounding down all components of p. Then $p = q + r$ where $0 \le r_j < 1$ for all $j = 1, \ldots, n$, and by (9), $s^{(i)} \cdot (q+r) < u_i - M$ which leads to $s^{(i)} \cdot q + (s^{(i)} \cdot r + M) < u_i$. Since $M = \max||s^{(i)}||_1$,

$$-M \le s^{(i)} \cdot r \le M. \tag{10}$$

and hence, $s^{(i)} \cdot q < u_i$. Repeating this argument for all rows of S, we get that $q \in K_u$. Similarly, if $q' = \lceil p \rceil$ is the vector obtained by rounding up all components of p, then $p = q' - r$ where $0 \leq r_j < 1$ for all $j = 1, \ldots, n$. Then (9) implies that $s^{(i)} \cdot (q' - r) < u_i - M$ which leads to $s^{(i)} \cdot q' + (M - s^{(i)} \cdot r) < u_i$. Again by Eq. (10), $s^{(i)} \cdot q' < u_i$ and hence $q' \in K_u$. Since $q \neq q'$, at least one of them is nonzero which contradicts that $K_u \cap \mathbb{Z}^n = \{0\}$. □

Lemma 4.10. *For any two rows* $s^{(i)}$, $s^{(j)}$ of S, *width*$_{s^{(i)}}(K_u) \leq 2\frac{\Delta_n(S)}{\delta_n(S)}$ *width*$_{s^{(j)}}(K_u)$.

Proof. Without loss of generality we may assume that $j = n + 1$. Since K_u is bounded, width$_{s^{(j)}}(K_u)$ is finite. Suppose the minimum of $s^{(j)} \cdot x$ over K_u is attained at v. Since translations leave the quantities in the lemma invariant, we may prove the lemma for the body $K_{u'}$ obtained by translating K_u by $-v$. Now $s^{(j)} \cdot x$ is minimized over $K_{u'}$ at the origin. By LP duality, there are n linearly independent constraints among the m defining $K_{u'}$ such that the minimum of $s^{(n+1)} \cdot x$ subject to just these n constraints is attained at 0. After renumbering the inequalities if necessary, assume these n constraints are the first n. Let

$$D = \{x \colon s^{(l)} \cdot x \leq u'_l, \ \ l = 1, 2, \ldots, n + 1\}$$

where of course $u'_1 = u'_2 = \cdots = u'_n = 0$. Then by the above, D is a bounded simplex.

Since D contains $K_{u'}$, it suffices to show that for each i,

$$\text{width}_{s^{(i)}}(D) \leq 2\left(\frac{\Delta_n(S)}{\delta_n(S)}\right)\text{width}_{s^{(n+1)}}(K_{u'}) = 2\left(\frac{\Delta_n(S)}{\delta_n(S)}\right)u'_{n+1} \tag{11}$$

We show that for each vertex q of D, $|s^{(i)} \cdot q| \leq (\frac{\Delta_n(S)}{\delta_n(S)})u'_{n+1}$ which will prove (11). This is clearly true for $q = 0$. Without loss of generality assume that vertex q satisfies $s^{(l)} \cdot q = u'_l$ for $l = 2, 3, \ldots, n + 1$. Since the determinant of the submatrix of S consisting of the rows $s^{(2)}, \ldots, s^{(n+1)}$ is not zero, for any i there exists rationals λ_l such that $s^{(i)} = \sum_{l=2}^{n+1} \lambda_l s^{(l)}$. By Cramer's rule, $|\lambda_l| \leq (\frac{\Delta_n(S)}{\delta_n(S)})$. Therefore, $s^{(i)} \cdot q = \sum_{l=2}^{n+1} \lambda_l s^{(l)} \cdot q = \sum_{l=2}^{n+1} \lambda_l u'_l = \lambda_{i+1} u'_{n+1}$ since $u'_l = 0$ for $l = 2, \ldots, n$. This proves that

$$|s^{(i)} \cdot q| = |\lambda_{n+1} u'_{n+1}| = |\lambda_{n+1}| u'_{n+1} \leq \left(\frac{\Delta_n(S)}{\delta_n(S)}\right)u'_{n+1}$$ □

Proof of Theorem 4.8. From Lemmas 4.9 and 4.10 it follows that for any i, $1 \leq i \leq m$, width$_{s^{(i)}}(K_u) \leq 2(\frac{\Delta_n(S)}{\delta_n(S)})M(n+2) = 2M(n+2)(\frac{\Delta_n(S)}{\delta_n(S)})$. Since $0 \in K_u$, $\min\{s^{(i)} \cdot x \colon x \in K_u\} \leq 0$ while $\max\{s^{(i)} \cdot x \colon x \in K_u\} = u_i$. Therefore, $u_i = u_i - 0 \leq$ width$_{s^{(i)}}(K_u)$ and hence, $0 \leq u_i \leq 2M(n+2)(\frac{\Delta_n(S)}{\delta_n(S)})$ for all $1 \leq i \leq m$. □

Reverting back to our set up, let $\bar{B}=\left[\begin{smallmatrix} B \\ -cB \end{smallmatrix}\right]$. Suppose K_u is the standard polytope $Q_u^{\bar{\tau}}$. By Theorem 4.8, $0 \leq u_i \leq 2M(n-d+2)(\frac{\Delta_n(\bar{B})}{\delta_n(\bar{B})})$.

Corollary 4.11. *If no maximal minor of $\bar{B}$ is zero, then the arithmetic degree of $\mathcal{O}_c$ is at most* $(2M(n-d+2)(\frac{\Delta_n(\bar{B})}{\delta_n(\bar{B})^n}))$.

The above arguments do not use the condition that the removal of an inequality from K_u will bring in a lattice point into the relaxation. Further, the bound is independent of the number of facets of K_u, and Corollary 4.11 is straightforward. Thus, further improvements may be possible with more effort. However, these proofs provide a first bound for arithmetic degree and have the nice feature that they build a bridge to techniques from the geometry of numbers that have played a central role in theoretical integer programming in the work of Kannan, Lenstra, Lovász, Scarf and others. See Lovász (1989) for a survey.

Problem 4.12. *Is it possible to find improved bounds for the multiplicities of associated sets and the arithmetic degree of $\mathcal{O}_c$?*

5 The Chain theorem

We now examine the structure of the poset of associated sets of $\mathrm{IP}_{A,c}$ which we denote as Assets($\mathrm{IP}_{A,c}$). All elements of Assets($\mathrm{IP}_{A,c}$) are faces of the regular triangulation Δ_c and the partial order is set inclusion. Theorem 4.3 provides a first result.

Corollary 5.1. *The maximal elements of Assets($IP_{A,c}$) are the maximal faces of Δ_c.*

Example 4.2 continued. The lower dimensional associated sets of this example (except the empty set) are the thick faces of Δ_c shown in Fig. 7. □

Despite the seemingly chaotic structure of Assets($\mathrm{IP}_{A,c}$) beyond its maximal elements, it has an important structural property that we now explain. (The proof of Theorem 5.2 relies on simple geometric ideas. Following the seemingly technical arguments with a picture could be very helpful. For algebraic comments on Theorem 5.2 see [A3] in Section 8.)

Theorem 5.2 [The Chain theorem]. *If $\tau \in \Delta_c$ is an associated set of $IP_{A,c}$ and $|\tau| < d$ then there exists a face $\tau' \in \Delta_c$ that is also an associated set of $IP_{A,c}$ with the property that $\tau \subset \tau'$ and $|\tau' \backslash \tau| = 1$.*

Proof. Since τ is an associated set of $\mathrm{IP}_{A,c}$, by Theorem 3.11, $\mathcal{O}_c$ has a standard pair of the form (v,τ) and $Q_v^{\bar{\tau}} = \{z \in \mathbb{R}^{n-d}: B^{\bar{\tau}}z \leq \pi_\tau(v), (-cB)\cdot z \leq 0\}$ is

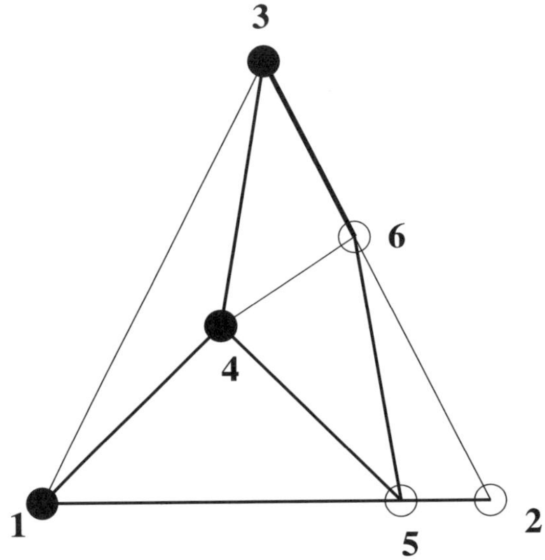

Fig. 7. Lower dimensional associated sets of Example 4.2 except the empty set.

a standard polytope of $IP_{A,c}$. Since $|\tau| < d$, τ is not a maximal face of Δ_c and hence by Theorem 4.3, $v \neq 0$. For each $i \in \bar{\tau}$, let R^i be the relaxation of $Q_v^{\bar{\tau}}$ obtained by removing the ith inequality $b_i \cdot z \leq v_i$ from $B^{\bar{\tau}} z \leq \pi_\tau(v)$, i.e.,

$$R^i := \{z \in \mathbb{R}^{n-d}: B^{\bar{\tau}\setminus\{i\}} z \leq \pi_{\tau\cup\{i\}}(v),\ (-cB)\cdot z \leq 0\}.$$

Let $E^i := R^i \setminus Q_v^{\bar{\tau}}$. Clearly, $E^i \cap Q_v^{\bar{\tau}} = \emptyset$, and, since the removal of $b_i \cdot z \leq v_i$ introduces at least one lattice point into R^i, $E^i \cap \mathbb{Z}^{n-d} \neq \emptyset$. Let z_i^* be the optimal solution to minimize$\{(-cB)\cdot z\colon z \in E^i \cap \mathbb{Z}^{n-d}\}$ if the program is bounded. This integer program is always feasible since $E^i \cap \mathbb{Z}^{n-d} \neq \emptyset$, but it may not have a finite optimal value. However, there exists at least one $i \in \bar{\tau}$ for which the above integer program is bounded. To see this, pick a maximal simplex $\sigma \in \Delta_c$ such that $\tau \subset \sigma$. The polytope $\{z \in \mathbb{R}^{n-d}: B^{\bar{\sigma}} z \leq \pi_\sigma(v),\ (-cB)\cdot z \leq 0\}$ is a simplex and hence bounded. This polytope contains all E^i for $i \in \sigma\setminus\tau$, and hence all these E^i are bounded and have finite optima with respect to $(-cB)\cdot z$. We may assume that the inequalities in $B^{\bar{\tau}} z \leq \pi_\tau(v)$ are labeled so that the finite optimal values are ordered as $(-cB)\cdot z_1^* \geq (-cB)\cdot z_2^* \geq \cdots \geq (-cB)\cdot z_p^*$ where $\{1, 2, \ldots, p\} \subseteq \bar{\tau}$.

Claim. *Let* $N^1 := \{z \in \mathbb{R}^{n-d}: B^{\bar{\tau}\setminus\{1\}} z \leq \pi_{\tau\cup\{1\}}(v),\ (-cB)\cdot z \leq (-cB)\cdot z_1^*\}$. *Then* z_1^* *is the unique lattice point in* N^1 *and the removal of any inequality from* $B^{\bar{\tau}/\{1\}} z \leq \pi_{\tau\cup\{1\}}(v)$ *will bring in a new lattice point into the relaxation.*

Proof. Since z_1^* lies in R^1, $0=(-cB)\cdot 0 \geq (-cB)\cdot z_1^*$. However, $0>(-cB)\cdot z_1^*$ since otherwise, both z_1^* and 0 would be optimal solutions to minimize$\{(-cB)\cdot z: z\in R^1\}$ contradicting that c is generic. Therefore,

$$\begin{aligned} N^1 &= R^1 \cap \{z \in \mathbb{R}^{n-d}: (-cB)\cdot z \leq (-cB)\cdot z_1^*\} \\ &= (E^1 \cup Q_v^{\bar{\tau}}) \cap \{z \in \mathbb{R}^{n-d}: (-cB)\cdot z \leq (-cB)\cdot z_1^*\} \\ &= (E^1 \cap \{z \in \mathbb{R}^{n-d}: (-cB)\cdot z \leq (-cB)\cdot z_1^*\}) \\ &\quad \cup (Q_v^{\bar{\tau}} \cap \{z \in \mathbb{R}^{n-d}: (-cB)\cdot z \leq (-cB)\cdot z_1^*\}). \end{aligned}$$

Since c is generic, z_1^* is the unique lattice point in the first polytope and the second polytope is free of lattice points. Hence z_1^* is the unique lattice point in N^1. The relaxation of N^1 got by removing $b_j\cdot z \leq v_j$ is the polyhedron $N^1 \cup (E^j \cap \{z\in\mathbb{R}^{n-d}: (-cB)\cdot z \leq (-cB)\cdot z_1^*\})$ for $j\in\bar{\tau}$ and $j\neq 1$. Either this is unbounded, in which case there is a lattice point z in this relaxation such that $(-cB)\cdot z_1^* \geq (-cB)\cdot z$, or (if $j\leq p$) we have $(-cB)\cdot z_1^* \geq (-cB)\cdot z_j^*$ and z_j^* lies in this relaxation. ◇

Translating N^1 by $-z_1^*$ we get $Q_{v'}^{\bar{\tau}\setminus\{1\}} := \{z\in\mathbb{R}^{n-d}: (-cB)\cdot z\leq 0, B^{\bar{\tau}\setminus\{1\}}z \leq v'\}$ where $v'=\pi_{\tau\cup\{1\}}(v)-B^{\bar{\tau}\setminus\{1\}}z_1^*\geq 0$ since z_1^* is feasible for all inequalities except the first one. Now $Q_{v'}^{\bar{\tau}\setminus\{1\}}\cap\mathbb{Z}^{n-d}=\{0\}$, and hence $(v',\tau\cup\{1\})$ is a standard pair of $\mathcal{O}_c$. □

Example 4.2 continued. The empty set is associated to $\mathrm{IP}_{A,c}$ and $\emptyset\subset\{1\}\subset\{1,4\}\subset\{1,4,5\}$ is a saturated chain in Assets($\mathrm{IP}_{A,c}$) that starts at the empty set. □

Since the elements of Assets($\mathrm{IP}_{A,c}$) are faces of Δ_c, a maximal face of which is a d-element set, the length of a maximal chain in Assets($\mathrm{IP}_{A,c}$) is at most d. We denote the maximal length of a chain in Assets($\mathrm{IP}_{A,c}$) by length(Assets($\mathrm{IP}_{A,c}$)). When $n-d$ (the corank of A) is small compared to d, length(Assets($\mathrm{IP}_{A,c}$)) has a stronger upper bound than d. We use the following result of Bell and Scarf to prove the bound.

Theorem 5.3. [Schrijver (1986, Corollary 16.5a)] *Let $Ax\leq b$ be a system of linear inequalities in n variables, and let $c\in\mathbb{R}^n$. If max $\{c\cdot x: Ax\leq b, x\in Z^n\}$ is a finite number, then max $\{c\cdot x: Ax\leq b, x\in Z^n\}=$max $\{c\cdot x: A'x\leq b', x\in Z^n\}$ for some subsystem $A'x\leq b'$ of $Ax\leq b$ with at most 2^n-1 inequalities.*

Theorem 5.4. *The length of a maximal chain in the poset of associated sets of $\mathrm{IP}_{A,c}$ is at most min$(d, 2^{n-d}-(n-d+1))$.*

Proof. As seen earlier, length(Assets($\mathrm{IP}_{A,c}$)) $\leq d$. If v lies in $\mathcal{O}_c$, then the origin is the optimal solution to the integer program minimize$\{(-cB)\cdot z: Bz\leq v,$

$z \in \mathbb{Z}^{n-d}\}$. By Theorem 5.3, we need at most $2^{n-d}-1$ inequalities to describe the same integer program which means that we can remove at least $n-(2^{n-d}-1)$ inequalities from $Bz \leq v$ without changing the optimum. Assuming that the inequalities removed are indexed by τ, $Q_v^{\bar{\tau}}$ will be a standard polytope of $\mathrm{IP}_{A,c}$. Therefore, $|\tau| \geq n - (2^{n-d}-1)$. This implies that the maximal length of a chain in Assets($\mathrm{IP}_{A,c}$) is at most $d-(n-(2^{n-d}-1)) = 2^{n-d}-(n-d+1)$. □

Corollary 5.5. *The cardinality of an associated set of $IP_{A,c}$ is at least $max(0, n-(2^{n-d}-1))$.*

Corollary 5.6. *If $n-d=2$, then length(Assets($\mathrm{IP}_{A,c}$)) ≤ 1.*

Proof. In this situation, $2^{n-d}-(n-d+1)=4-(4-2+1)=4-3=1$. □

We conclude this section with a family of examples for which length(Assets($\mathrm{IP}_{A,c}$)) $=2^{n-d}-(n-d+1)$. This is adapted from Hoşten and Thomas (1999, Proposition 3.9) which was modeled on a family of examples from Peeva and Sturmfels (1998).

Proposition 5.7. *For each $m > 1$, there is an integer matrix A of corank m and a cost vector $\mathrm{c} \in \mathbb{Z}^n$ where $n=2^m-1$ such that length(Assets($IP_{A,c}$)) $=2^m-(m+1)$.*

Proof. Given $m > 1$, let $B' = (b_{ij}) \in \mathbb{Z}^{(2^m-1)\times m}$ be the matrix whose rows are all the $\{1,-1\}$-vectors in $\mathbf{R}^m$ except $v=(-1,-1,\ldots,-1)$. Let $B \in \mathbb{Z}^{(2^m+m-1)\times m}$ be obtained by stacking B' on top of $-I_m$ where I_m is the $m \times m$ identity matrix. Set $n=2^m+m-1$, $d=2^m-1$ and $A'=[I_d|B'] \in \mathbb{Z}^{d\times n}$. By construction, the columns of B span the lattice $\{u \in \mathbb{Z}^n: A'u=0\}$. We may assume that the first row of B' is $(1,1,\ldots,1) \in \mathbb{R}^m$. Adding this row to all other rows of A' we get $A \in \mathbb{N}^{d\times n}$ with the same row space as A'. Hence the columns of B are also a basis for the lattice $\{u \in \mathbb{Z}^n: Au=0\}$. Since the rows of B span $\mathbb{Z}^m$ as a lattice, we can find a cost vector $c \in \mathbb{Z}^n$ such that $(-cB)=v$.

For each row b_i of B' set $r_i := |\{b_{ij}: b_{ij}=1\}|$, and let r be the vector of all r_is. By construction, the polytope $Q := \{z \in \mathbb{R}^m: B'z \leq r, -(cB)\cdot z \leq 0\}$ has no lattice points in its interior, and each of its 2^m facets has exactly one vertex of the unit cube in $\mathbb{R}^m$ in its relative interior. If we let $w_i = r_i - 1$, then the polytope $\{z \in \mathbb{R}^m: B'z \leq w, -(cB)\cdot z \leq 0\}$ is a standard polytope $Q_u^{\bar{\tau}}$ of $\mathrm{IP}_{A,c}$ where $\tau = \{d+1, d+2, \ldots, d+m=n\}$ and $w=\pi_\tau(u)$. Since a maximal face of Δ_c is a $d=(2^m-1)$-element set and $|\tau|=m$, Theorem 5.2 implies that length(Assets($\mathrm{IP}_{A,c}$)) $\geq 2^m-1-m=2^m-(m+1)$. However, by Theorem 5.4, length(Assets($\mathrm{IP}_{A,c}$)) $=\min(2^m-1, 2^m-(m+1))=2^m-(m+1)$ since $m > 1$ by assumption. □

Example 5.8. If we choose $m=3$ then $n=2^m+m-1=10$ and $d=2^m-1=7$. Constructing B' and A as in Proposition 5.7, we get

$$B' = \begin{bmatrix} 1 & 1 & 1 \\ -1 & 1 & 1 \\ 1 & -1 & 1 \\ 1 & 1 & -1 \\ -1 & -1 & 1 \\ -1 & 1 & -1 \\ 1 & -1 & -1 \end{bmatrix} \quad \text{and} \quad A = \begin{bmatrix} 1 & 0 & 0 & 0 & 0 & 0 & 0 & 1 & 1 & 1 \\ 1 & 1 & 0 & 0 & 0 & 0 & 0 & 0 & 2 & 2 \\ 1 & 0 & 1 & 0 & 0 & 0 & 0 & 2 & 0 & 2 \\ 1 & 0 & 0 & 1 & 0 & 0 & 0 & 2 & 2 & 0 \\ 1 & 0 & 0 & 0 & 1 & 0 & 0 & 0 & 0 & 2 \\ 1 & 0 & 0 & 0 & 0 & 1 & 0 & 0 & 2 & 0 \\ 1 & 0 & 0 & 0 & 0 & 0 & 1 & 2 & 0 & 0 \end{bmatrix}$$

The vector $c=(11, 0, 0, 0, 0, 0, 0, 10, 10, 10)$ satisfies $(-cB)=(-1, -1, -1)$. The associated sets of $\text{IP}_{A,c}$ along with their multiplicities are given below.

τ	Multiplicity	τ	Multiplicity
{4,5,6,7,8,9,10}*	4	{2,3,7,8,9,10}	2
{1,5,6,7,8,9,10}	4	{5,6,7,8,9,10}*	1
{3,4,6,7,8,9,10}	4	{4,5,6,7,8,9}	1
{2,3,4,6,7,9,10}	2	{2,4,7,8,9,10}	2
{2,3,4,7,8,9,10}	4	{1,5,7,8,9,10}	1
{3,4,5,6,7,8,10}	2	{2,3,4,8,9,10}	1
{2,3,4,5,6,7,10}	1	{4,5,7,8,9,10}	2
{2,4,5,6,7,9,10}	2	{2,5,6,7,9,10}	1
{2,3,6,7,9,10}	1	{4,5,6,8,9,10}	2
{3,4,5,6,8,10}	1	{1,5,6,8,9,10}	1
{2,4,5,7,9,10}	1	{3,4,6,8,9,10}	2
{1,6,7,8,9,10}	1	{6,7,8,9,10}*	1
{3,5,6,7,8,10}	1	{7,8,9,10}*	1
{3,6,7,8,9,10}	2	{8,9,10}*	1

The elements in the unique maximal chain in $\text{Assets}(\text{IP}_{A,c})$ are marked with a $*$ and $\text{length}(\text{Assets}(\text{IP}_{A,c}))=2^3-(3+1)=4$ as predicted by Proposition 5.7. □

6 Gomory integer programs

Recall from Definition 2.7 that a group relaxation $G^\sigma(b)$ of $\text{IP}_{A,c}(b)$ is called a Gomory relaxation if σ is a maximal face of Δ_c. As discussed in Section 2, these relaxations are the easiest to solve among all relaxations of $\text{IP}_{A,c}(b)$. Hence it is natural to ask under what conditions on A and c would all programs in $\text{IP}_{A,c}$ be solvable by Gomory relaxations. We study this question in this section. The majority of the results here are taken from Hoşten and Thomas (2003).

Definition 6.1. The family of integer programs $\mathrm{IP}_{A,c}$ is a **Gomory family** if, for *every* $b \in \mathbb{N}A$, $\mathrm{IP}_{A,c}(b)$ is solved by a group relaxation $G^{\sigma}(b)$ where σ is a *maximal* face of the regular triangulation Δ_c.

Theorem 6.2. *The following conditions are equivalent:*

(i) *$IP_{A,c}$ is a Gomory family.*
(ii) *The associated sets of $IP_{A,c}$ are precisely the maximal faces of Δ_c.*
(iii) *$(*, \tau)$ is a standard pair of $\mathcal{O}_c$ if and only if τ is a maximal face of Δ_c.*
(iv) *All standard polytopes of $IP_{A,c}$ are simplices.*

Proof. By Definition 6.1, $\mathrm{IP}_{A,c}$ is a Gomory family if and only if for all $b \in \mathbb{N}A$, $\mathrm{IP}_{A,c}(b)$ can be solved by one of its Gomory relaxations. By Theorem 3.12, this is equivalent of saying that every $u \in \mathcal{O}_c$ lies in some $S(*, \sigma)$ where σ is a maximal face of Δ_c and $(*, \sigma)$ a standard pair of $\mathcal{O}_c$. Definition 3.1 then implies that all associated sets of $\mathrm{IP}_{A,c}$ are maximal faces of Δ_c. By Theorem 4.3, every maximal face of Δ_c is an associated set of $\mathrm{IP}_{A,c}$ and hence (i)$\Leftrightarrow$(ii). The equivalence of statements (ii), (iii), and (iv) follow from Theorem 3.11. □

If c is a generic cost vector such that for a triangulation Δ of cone(A), $\Delta = \Delta_c$, then we say that Δ *supports* the order ideal $\mathcal{O}_c$ and the family of integer programs $\mathrm{IP}_{A,c}$. No regular triangulation of the matrix A in Example 4.2 supports a Gomory family. Here is a matrix with a Gomory family.

Example 6.3. Consider the 3×6 matrix

$$A = \begin{bmatrix} 1 & 0 & 1 & 1 & 1 & 1 \\ 0 & 1 & 1 & 1 & 2 & 2 \\ 0 & 0 & 1 & 2 & 3 & 4 \end{bmatrix}.$$

In this case, cone(A) has 14 distinct regular triangulations and 48 distinct sets $\mathcal{O}_c$ as c varies among all generic cost vectors. Ten of these triangulations support Gomory families; one for each triangulation. For instance, if $c = (0, 0, 1, 1, 0, 3)$, then

$$\Delta_c = \{\sigma_1 = \{1, 2, 5\},\ \sigma_2 = \{1, 4, 5\},\ \sigma_3 = \{2, 5, 6\},\ \sigma_4 = \{4, 5, 6\}\}$$

and $\mathrm{IP}_{A,c}$ is a Gomory family since the standard pairs of $\mathcal{O}_c$ are:

$$(0, \sigma_1),\ (e_3, \sigma_1),\ (e_4, \sigma_1),\ (0, \sigma_2),\ (0, \sigma_3),\ \text{and}\ (0, \sigma_4).$$ □

The algebraic approach to integer programming allows one to compute all down sets $\mathcal{O}_c$ of a fixed matrix A as c varies among the set of generic cost vectors. See Huber and Thomas (2000), Sturmfels (1995), and Sturmfels and Thomas (1997) for details. The software package TiGERS is custom-tailored for this purpose. The above example as well as many of the remaining examples in this chapter were done using TiGERS. See **[A4]** in Section 8 for comments on the algebraic equivalent of a Gomory family.

We now compare the notion of a Gomory family to the classical notion of *total dual integrality* [Schrijver (1986, §22)]. It will be convenient to assume that $\mathbb{Z}A = \mathbb{Z}^d$ for these results.

Definition 6.4. The system $yA \leq c$ is **totally dual integral (TDI)** if $\mathrm{LP}_{A,c}(b)$ has an integral optimal solution for each $b \in \mathrm{cone}(A) \cap \mathbb{Z}^d$.

Definition 6.5. The regular triangulation Δ_c is *unimodular* if $\mathbb{Z}A_\sigma = \mathbb{Z}^d$ for every maximal face $\sigma \in \Delta_c$.

Example 6.6. The regular triangulation in Example 2.2 (i) is unimodular while those in Example 2.2 (ii) and (iii) are not. □

Lemma 6.7. *The system* $yA \leq c$ *is TDI if and only if the regular triangulation* Δ_c *is unimodular.*

Proof. The regular triangulation Δ_c is the normal fan of P_c by Proposition 2.4, and it is unimodular if and only if $\mathbb{Z}A_\sigma = \mathbb{Z}^d$ for every maximal face $\sigma \in \Delta_c$. This is equivalent to every $b \in \mathrm{cone}(A_\sigma) \cap \mathbb{Z}^d$ lying in $\mathbb{N}A_\sigma$ for every maximal face σ of Δ_c. By Lemma 2.3, this happens if and only if $\mathrm{LP}_{A,c}(b)$ has an integral optimum for all $b \in \mathrm{cone}(A) \cap \mathbb{Z}^d$. □

For an algebraic algorithm to check TDI-ness see **[A5]** in Section 8.

Theorem 6.8. *If* $yA \leq c$ *is TDI then* $IP_{A,c}$ *is a Gomory family.*

Proof. By Theorem 4.3, $(0, \sigma)$ is a standard pair of $\mathcal{O}_c$ for every maximal face σ of Δ_c. Lemma 6.7 implies that $\mathrm{cone}(A_\sigma)$ is unimodular (i.e., $\mathbb{Z}A_\sigma = \mathbb{Z}^d$), and therefore $\mathbb{N}A_\sigma = \mathrm{cone}(A_\sigma) \cap \mathbb{Z}^d$ for every maximal face σ of Δ_c. Hence the semigroups $\mathbb{N}A_\sigma$ arising from the standard pairs $(0, \sigma)$ as σ varies over the maximal faces of Δ_c cover $\mathbb{N}A$. Therefore the only standard pairs of $\mathcal{O}_c$ are $(0, \sigma)$ as σ varies over the maximal faces of Δ_c. The result then follows from Theorem 6.2. □

When $yA \leq c$ is TDI, the multiplicity of a maximal face of Δ_c in $\mathcal{O}_c$ is one (from Theorem 4.5). By Theorem 6.8, no lower dimensional face of Δ_c is associated to $\mathrm{IP}_{A,c}$. While this is sufficient for $\mathrm{IP}_{A,c}(b)$ to be a Gomory family, it is far from necessary. TDI-ness guarantees *local integrality* in the sense that $\mathrm{LP}_{A,c}(b)$ has an integral optimum for every integral b in cone(A). In contrast, if $IP_{A,c}$ is a Gomory family, the linear optima of the programs in $LP_{A,c}$ may not be integral.

If A is *unimodular* (i.e., $\mathbb{Z}A_\sigma = \mathbb{Z}^d$ for every nonsingular maximal submatrix A_σ of A), then the feasible regions of the linear programs in $LP_{A,c}$ have integral vertices for each $b \in \text{cone}(A) \cap \mathbb{Z}^d$, and $yA \leq c$ is TDI *for all* c. Hence if A is unimodular, then $IP_{A,c}$ is a Gomory family for all generic cost vectors c. However, just as integrality of the optimal solutions of programs in $LP_{A,c}$ is not necessary for $IP_{A,c}$ to be a Gomory family, unimodularity of A is not necessary for $IP_{A,c}$ to be a Gomory family *for all* c.

Example 6.9. Consider the seven by twelve integer matrix

$$A = \begin{bmatrix} 1 & 0 & 0 & 0 & 0 & 0 & 1 & 1 & 1 & 1 & 1 & 0 \\ 0 & 1 & 0 & 0 & 0 & 0 & 1 & 1 & 0 & 0 & 0 & 1 \\ 0 & 0 & 1 & 0 & 0 & 0 & 1 & 0 & 1 & 0 & 0 & 1 \\ 0 & 0 & 0 & 1 & 0 & 0 & 0 & 1 & 0 & 1 & 0 & 0 \\ 0 & 0 & 0 & 0 & 1 & 0 & 0 & 0 & 1 & 0 & 1 & 0 \\ 0 & 0 & 0 & 0 & 0 & 1 & 0 & 0 & 0 & 1 & 1 & 1 \\ 0 & 0 & 0 & 0 & 0 & 0 & 1 & 1 & 1 & 1 & 1 & 1 \end{bmatrix}$$

of rank seven. The maximal minors of A have absolute values zero, one and two and hence A is not unimodular. This matrix has 376 distinct regular triangulations supporting 418 distinct order ideals $\mathcal{O}_c$ (computed using TiGERS). In each case, the standard pairs of $\mathcal{O}_c$ are indexed by just the maximal simplices of the regular triangulation Δ_c that supports it. Hence $IP_{A,c}$ is a Gomory family for all generic c. □

The above discussion shows that $IP_{A,c}$ being a Gomory family is more general than $yA \leq c$ being TDI. Similarly, $IP_{A,c}$ being a Gomory family for all generic c is more general than A being a unimodular matrix.

7 Gomory families and Hilbert bases

As we just saw, unimodular matrices or more generally, unimodular regular triangulations lead to Gomory families. A common property of unimodular matrices and matrices A such that cone(A) has a unimodular triangulation is that the columns of A form a *Hilbert basis* for cone(A), i.e., $\mathbb{N}A = \text{cone(A)} \cap \mathbb{Z}^d$ (assuming $\mathbb{Z}A = \mathbb{Z}^d$).

Definition 7.1. A $d \times n$ integer matrix A is *normal* if the semigroup $\mathbb{N}A$ equals $\text{cone}(A) \cap \mathbb{Z}^d$.

The reason for this (highly over used) terminology here is that if the columns of A form a Hilbert basis, then the zero set of the toric ideal I_A (called a *toric variety*) is a *normal* variety. See Sturmfels (1995, Chapter 14) for more details. We first note that if A is not normal, then $\text{IP}_{A,c}$ need not be a Gomory family for any cost vector c.

Example 7.2. The matrix

$$A = \begin{bmatrix} 1 & 1 & 1 & 1 \\ 0 & 1 & 3 & 4 \end{bmatrix}$$

is not normal since $(1,2)^t$ which lies in $\text{cone}(A) \cap \mathbb{Z}^2$ cannot be written as a nonnegative integer combination of the columns of A. This matrix gives rise to 10 distinct order ideals $\mathcal{O}_c$ supported on its four regular triangulations $\{\{1, 4\}\}$,$\{\{1, 2\},\{2, 4\}\}$,$\{\{1, 3\},\{3, 4\}\}$ and $\{\{1, 2\}, \{2, 3\},\{3, 4\}\}$. Each $\mathcal{O}_c$ has at least one standard pair that is indexed by a lower dimensional face of Δ_c.

The matrix in Example 4.2 is also not normal and has no Gomory families. While we do not know whether normality of A is sufficient for the existence of a generic cost vector c such that $\text{IP}_{A,c}$ is a Gomory family, we will now show that under certain additional conditions, normal matrices do give rise to Gomory families.

Definition 7.3. A $d \times n$ integer matrix A is Δ-normal if cone(A) has a triangulation Δ such that for every maximal face $\sigma \in \Delta$, the columns of A in cone(A_σ) form a Hilbert basis.

Remark 7.4. If A is Δ-normal for some triangulation Δ, then it is normal. To see this note that every lattice point in cone(A) lies in cone(A_σ) for some maximal face $\sigma \in \Delta$. Since A is Δ-normal, this lattice point also lies in the semigroup generated by the columns of A in cone(A_σ) and hence in $\mathbb{N}A$.

Observe that A is Δ-normal with respect to all the unimodular triangulations of cone(A). Hence triangulations Δ with respect to which A is Δ-normal generalize unimodular triangulations of cone(A).

Problem 7.5. *Are there known families of integer programs whose coefficient matrices are normal or Δ-normal but not unimodular? Are there known Gomory families of integer programs in the literature (not arising from unimodular matrices)?*

Examples 7.6 and 7.7 show that the set of matrices where cone(A) has a unimodular triangulation is a proper subset of the set of Δ-normal matrices which in turn is a proper subset of the set of normal matrices.

Example 7.6. Examples of normal matrices with no unimodular triangulations can be found in Bouvier and Gonzalez-Springberg (1994) and Firla and Ziegler (1999). If cone(A) is simplicial for such a matrix, A will be Δ-normal with respect to its coarsest (regular) triangulation Δ consisting of the single maximal face with support cone(A). For instance, consider the following example taken from Firla and Ziegler (1999):

$$A = \begin{bmatrix} 1 & 0 & 0 & 1 & 1 & 1 & 1 & 1 \\ 0 & 1 & 0 & 1 & 1 & 2 & 2 & 2 \\ 0 & 0 & 1 & 1 & 2 & 2 & 3 & 3 \\ 0 & 0 & 0 & 1 & 2 & 3 & 4 & 5 \end{bmatrix}$$

Here cone(A) has 77 regular triangulations and no unimodular triangulations. Since cone(A) is a simplicial cone generated by a_1, a_2, a_3 and a_8, A is Δ-normal with respect to its coarsest regular triangulation $\Delta = \{\{1, 2, 3, 8\}\}$.

Example 7.7. There are normal matrices A that are not Δ-normal with respect to any triangulation of cone(A). To see such an example, consider the following modification of the matrix in Example 7.6 that appears in Sturmfels (1995, Example 13.17):

$$A = \begin{bmatrix} 0 & 1 & 0 & 0 & 1 & 1 & 1 & 1 & 1 \\ 0 & 0 & 1 & 0 & 1 & 1 & 2 & 2 & 2 \\ 0 & 0 & 0 & 1 & 1 & 2 & 2 & 3 & 3 \\ 0 & 0 & 0 & 0 & 1 & 2 & 3 & 4 & 5 \\ 1 & 1 & 1 & 1 & 1 & 1 & 1 & 1 & 1 \end{bmatrix}$$

This matrix is again normal and each of its nine columns generate an extreme ray of cone(A). Hence the only way for this matrix to be Δ-normal for some Δ would be if Δ is a unimodular triangulation of cone(A). However, there are no unimodular triangulations of this matrix.

Theorem 7.8. *If A is Δ-normal for some regular triangulation Δ then there exists a generic cost vector $c \in \mathbb{Z}^n$ such that $\Delta = \Delta_c$ and $IP_{A,c}$ is a Gomory family.*

Proof. Without loss of generality we can assume that the columns of A in cone(A_σ) form a minimal Hilbert basis for every maximal face σ of Δ. If there were a redundant element, the smaller matrix obtained by removing this column from A would still be Δ-normal.

For a maximal face $\sigma \in \Delta$, let $\sigma_{\text{in}} \subset \{1, \ldots, n\}$ be the set of indices of all columns of A lying in cone(A_σ) that are different from the columns of A_σ. Suppose $a_{i_1}, \ldots, a_{i_k}$ are the columns of A that generate the one dimensional faces of Δ, and $c' \in \mathbb{R}^n$ a cost vector such that $\Delta = \Delta_{c'}$. We modify c' to obtain a new cost vector $c \in \mathbb{R}^n$ such that $\Delta = \Delta_c$ as follows. For $j = 1, \ldots, k$, let $c_{i_j} := c'_{i_j}$. If $j \in \sigma_{\text{in}}$ for some maximal face $\sigma \in \Delta$, then $a_j = \sum_{i \in \sigma} \lambda_i a_i$, $0 \leq \lambda_i < 1$

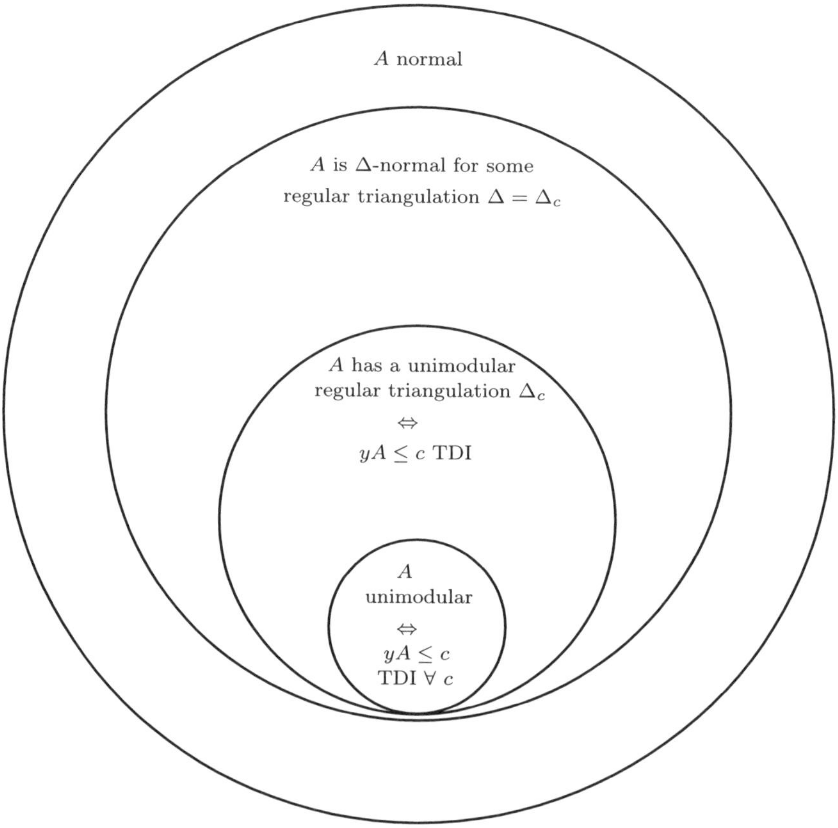

Fig. 8. Inclusions of sets of matrices.

and we define $c_j := \sum_{i \in \sigma} \lambda_i c_i$. Hence, for all $j \in \sigma_{\text{in}}$, $(a_j^t, c_j) \in \mathbb{R}^{d+1}$ lies in $C_\sigma := \text{cone}((a_i^t, c_i): i \in \sigma) = \text{cone}((a_i^t, c_i'): i \in \sigma)$ which was a facet of $C = \text{cone}((a_i^t, c_i'): i = 1, \ldots, n)$. If $y \in \mathbb{R}^d$ is a vector as in Definition 2.1 showing that σ is a maximal face of $\Delta_{c'}$ then $y \cdot a_i = c_i$ for all $i \in \sigma \cup \sigma_{\text{in}}$ and $y \cdot a_j < c_j$ otherwise. Since $\text{cone}(A_\sigma) = \text{cone}(A_{\sigma \cup \sigma_{\text{in}}})$, we conclude that $\text{cone}(A_\sigma)$ is a maximal face of Δ_c.

If $b \in \mathbb{N}A$ lies in $\text{cone}(A_\sigma)$ for a maximal face $\sigma \in \Delta_c$, then $\text{IP}_{A,c}(b)$ has at least one feasible solution u with support in $\sigma \cup \sigma_{\text{in}}$ since A is Δ-normal. Further, $(b^t, c \cdot u) = ((Au)^t, c \cdot u)$ lies in C_σ and all feasible solutions of $\text{IP}_{A,c}(b)$ with support in $\sigma \cup \sigma_{\text{in}}$ have the same cost value by construction. Suppose $v \in \mathbb{N}^n$ is any feasible solution of $\text{IP}_{A,c}(b)$ with support not in $\sigma \cup \sigma_{\text{in}}$. Then $c \cdot u < c \cdot v$ since $(a_i^t, c_i) \in C_\sigma$ if and only if $i \in \sigma \cup \sigma_{\text{in}}$ and C_σ is a lower facet of C. Hence the optimal solutions of $\text{IP}_{A,c}(b)$ are precisely those feasible solutions with support in $\sigma \cup \sigma_{\text{in}}$. The vector b can be expressed as $b = b' + \sum_{i \in \sigma} z_i a_i$ where $z_i \in \mathbb{N}$ are unique and $b' \in \{\sum_{i \in \sigma} \lambda_i a_i : 0 \leq \lambda_i < 1\} \cap \mathbb{Z}^d$ is also unique. The vector $b' = \sum_{j \in \sigma_{\text{in}}} r_j a_j$ where $r_j \in \mathbb{N}$.

Setting $u_i = z_i$ for all $i \in \sigma$, $u_j = r_j$ for all $j \in \sigma_{\text{in}}$ and $u_k = 0$ otherwise, we obtain all feasible solutions u of $\text{IP}_{A,c}(b)$ with support in $\sigma \cup \sigma_{\text{in}}$.

If there is more than one such feasible solution, then c is not generic. In this case, we can perturb c to a generic cost vector $c'' = c + \varepsilon\omega$ by choosing $1 \gg \varepsilon > 0$, $\omega_j \ll 0$ whenever $j = i_1, \ldots, i_k$ and $\omega_j = 0$ otherwise. Suppose $u_1, \ldots, u_t$ are the optimal solutions of the integer programs $\text{IP}_{A,c''}(b')$ where $b' \in \{\sum_{i \in \sigma} \lambda_i a_i \colon 0 \leq \lambda_i < 1\} \cap \mathbb{Z}^d$. (Note that $t = |\{\sum_{i \in \sigma} \lambda_i a_i \colon 0 \leq \lambda_i < 1\} \cap \mathbb{Z}^d|$ is the index of $\mathbb{Z}A_\sigma$ in $\mathbb{Z}A$.) The support of each such u_i is contained in σ_{in}. For any $b \in \text{cone}(A_\sigma) \cap \mathbb{Z}^d$, the optimal solution of $\text{IP}_{A,c''}(b)$ is hence $u = u_i + z$ for some $i \in \{1, \ldots, t\}$ and $z \in \mathbb{N}^n$ with support in σ. This shows that $\mathbb{N}A$ is covered by the affine semigroups $\phi_A(S(u_i, \sigma))$ where σ is a maximal face of Δ and u_i as above for each σ. By construction, the corresponding admissible pairs (u_i, σ) are all standard for $\mathcal{O}_{c''}$. Since all data is integral, $c'' \in \mathbb{Q}^n$ and hence can be scaled to lie in $\mathbb{Z}^n$. Renaming c'' as c, we conclude that $\text{IP}_{A,c}$ is a Gomory family. □

Corollary 7.9. *Let A be a normal matrix such that cone(A) is simplicial, and let Δ be the coarsest triangulation whose single maximal face has support cone(A). Then there exists a cost vector $c \in \mathbb{Z}^n$ such that $\Delta = \Delta_c$ and $IP_{A,c}$ is a Gomory family.*

Example 7.10. Consider the normal matrix in Example 6.3. Here cone(A) is generated by the first, second and sixth columns of A and hence A is Δ-normal with respect to the regular triangulation $\{\{1, 2, 6\}\}$. There are 13 distinct sets $\mathcal{O}_c$ supported on Δ. Among the 13 corresponding families of integer programs, only one is a Gomory family. A representative cost vector for this $\text{IP}_{A,c}$ is $c = (0, 0, 4, 4, 1, 0)$. The standard pair decomposition of $\mathcal{O}_c$ is the one constructed in Theorem 7.8. The affine semigroups $S(\cdot, \sigma)$ from this decomposition are:

$$S(0, \sigma), \quad S(e_3, \sigma), \quad S(e_4, \sigma), \quad \text{and} \quad S(e_5, \sigma).$$

Note that A is not Δ-normal with respect to the regular triangulation supporting the Gomory family $\text{IP}_{A,c}$ in Example 6.3. The columns of A in cone(A_{σ_1}) are the columns of A_{σ_1} and a_3. The vector $(1, 2, 2)$ is in the minimal Hilbert basis of cone(A_{σ_1}) but is not a column of A. This example shows that a regular triangulation Δ of cone(A) can support a Gomory family even if A is not Δ-normal. The Gomory families in Theorem 7.8 have a very special standard pair decomposition. □

Problem 7.11. *If $A \in \mathbb{Z}^{d \times n}$ is a normal matrix, does there exist a generic cost vector $c \in \mathbb{Z}^n$ such that $IP_{A,c}$ is a Gomory family?*

While we do not know the answer to this question, we will now show that stronger results are possible for small values of d.

Theorem 7.12. *If $A \in \mathbb{Z}^{d\times n}$ is a normal matrix and $d \leq 3$, then there exists a generic cost vector $c \in \mathbb{Z}^n$ such that $IP_{A,c}$ is a Gomory family.*

Proof. It is known that if $d \leq 3$ then cone(A) has a regular unimodular triangulation Δ_c (Sebö, 1990). The result then follows from Corollary 6.8. □

Before we proceed, we rephrase Problem 7.11 in terms of covering properties of cone(A) and $\mathbb{N}A$ along the lines of Bouvier and Gonzalez-Springberg (1994), Bruns and Gubeladze (1999), Bruns, Gubeladze, Henk, Martin, and Weismantel (1999), Firla and Ziegler (1999) and Sebö (1990). To obtain the same set up as in these articles we assume in this section that A is normal and the columns of A form the unique minimal Hilbert basis of cone(A). Using the terminology in Bruns and Gubeladze (1999), the *free Hilbert cover* problem asks whether there exists a covering of $\mathbb{N}A$ by semigroups $\mathbb{N}A_\tau$ where the columns of A_τ are linearly independent. The *unimodular Hilbert cover* problem asks whether cone(A) can be covered by full dimensional unimodular subcones cone(A_τ) (i.e., $\mathbb{Z}A_\tau = \mathbb{Z}^d$), while the stronger *unimodular Hilbert partition* problem asks whether cone(A) has a unimodular triangulation. (Note that if cone(A) has a unimodular Hilbert cover or partition using subcones cone(A_τ), then $\mathbb{N}A$ is covered by the semigroups $\mathbb{N}A_\tau$.) All these problems have positive answers if $d \leq 3$ since cone(A) admits a unimodular Hilbert partition in this case (Bouvier and Gonzalez-Springberg, 1994; Sebö, 1990). Normal matrices (with $d=4$) such that cone(A) has no unimodular Hilbert partition can be found in Bouvier and Gonzalez-Springberg (1994) and Firla and Ziegler (1999). Examples (with $d=6$) that admit no free Hilbert cover and hence no unimodular Hilbert cover can be found in Bruns and Gubeladze (1999) and Bruns et al. (1999).

When $yA \leq c$ is TDI, the standard pair decomposition of $\mathbb{N}A$ induced by c gives a unimodular Hilbert partition of cone(A) by Theorem 6.7. An important difference between Problem 7.11 and the Hilbert cover problems is that *affine* semigroups cannot be used in Hilbert covers. Moreover, affine semigroups that are allowed in standard pair decompositions come from integer programming. If there are no restrictions on the affine semigroups that can be used in a cover, $\mathbb{N}A$ can always be covered by full dimensional affine semigroups: for any triangulation Δ of cone(A) with maximal subcones cone(A_σ), the affine semigroups $b+\mathbb{N}A_\sigma$ cover $\mathbb{N}A$ as b varies in $\{\sum_{i\in\sigma}\lambda_i a_i\colon 0 \leq \lambda_i < 1\} \cap \mathbb{Z}^d$ and σ varies among the maximal faces of the triangulation. A *partition* of $\mathbb{N}A$ derived from this idea can be found in Stanley (1982, Theorem 5.2). We recall the notion of *supernormality* introduced in Hoşten, Maclagan, and Sturmfels (forthcoming).

Definition 7.13. A matrix $A \in \mathbb{Z}^{d\times n}$ is *supernormal* if for every submatrix A' of A, the columns of A that lie in cone(A') form a Hilbert basis for cone(A').

Proposition 7.14. *For $A \in \mathbb{Z}^{d \times n}$, the following are equivalent:*

(i) *A is supernormal,*
(ii) *A is Δ-normal for every regular triangulation Δ of cone(A),*
(iii) *Every triangulation of cone(A) in which all columns of A generate one dimensional faces is unimodular.*

Proof. The equivalence of (i) and (iii) was established in Hosten, Maclagan, and Sturmfels (forthcoming), Proposition 3.1. Definition 7.13 shows that (i) $\Rightarrow$ (ii). Hence we just need to show that (ii) $\Rightarrow$ (i). Suppose that A is Δ-normal for every regular triangulation of cone(A). In order to show that A is supernormal we only need to check submatrices A' where the dimension of cone(A') is d. Choose a cost vector c with $c_i \gg 0$ if the ith column of A does not generate an extreme ray of cone(A'), and $c_i = 0$ otherwise. This gives a polyhedral subdivision of cone(A) in which cone(A') is a maximal face. There are standard procedures that will refine this subdivision to a regular triangulation Δ of cone(A). Let T be the set of maximal faces σ of Δ such that cone(A_σ) lies in cone(A'). Since A is Δ-normal, the columns of A that lie in cone(A_σ) form a Hilbert basis for cone(A_σ) for each $\sigma \in T$. However, since their union is the set of columns of A that lie in cone(A'), this union forms a Hilbert basis for cone(A'). □

It is easy to catalog all Δ-normal and supernormal matrices, of the type considered in this chapter, for small values of d. We say that the matrix A is *graded* if its columns span an affine hyperplane in $\mathbb{R}^d$. If $d = 1$, cone(A) has n triangulations $\{\{i\}\}$ each of which has the unique maximal subcone cone(a_i) whose support is cone(A). If we assume that $a_1 \leq a_2 \leq \cdots \leq a_n$, then A is normal if and only if either $a_1 = 1$, or $a_n = -1$. Also, A is normal if and only if it is supernormal. If $d = 2$ and the columns of A are ordered counterclockwise around the origin, then A is normal if and only if $\det(a_i, a_{i+1}) = 1$ for all $i = 1, \ldots, n-1$. Such an A is supernormal since it is Δ-normal for every triangulation Δ – the Hilbert basis of a maximal subcone of Δ is precisely the set of columns of A in that subcone. If $d = 3$ then as mentioned before, cone(A) has a unimodular triangulation with respect to which A is Δ-normal. However, not every such A needs to be supernormal: the matrix in Example 6.3 is not Δ-normal for the Δ supporting the Gomory family in that example. If $d = 3$ and A is graded, then without loss of generality we can assume that the columns of A span the hyperplane $x_1 = 1$. If A is normal as well, then its columns are precisely all the lattice points in the convex hull of A. Conversely, every graded normal A with $d = 3$ arises this way – its columns are all the lattice points in a polygon in $\mathbb{R}^2$ with integer vertices. In particular, every triangulation of cone(A) that uses all the columns of A is unimodular. Hence, by Proposition 7.14, A is supernormal, and therefore Δ-normal for any triangulation of A.

Theorem 7.15. *Let $A \in \mathbb{Z}^{d \times n}$ be a normal matrix of rank d.*

(i) *If $d=1,2$ or A is graded and $d=3$, every regular triangulation of cone(A) supports at least one Gomory family.*
(ii) *If $d=2$ and A is graded, every regular triangulation of cone(A) supports exactly one Gomory family.*
(iii) *If $d=3$ and A is not graded, or if $d=4$ and A is graded, then not all regular triangulations of cone(A) may support a Gomory family. In particular, A may not be Δ-normal with respect to every regular triangulation.*

Proof. (i) If $d=1,2$ or A is graded and $d=3$, A is supernormal and hence by Proposition 7.14 and Theorem 7.8, every regular triangulation of cone(A) supports at least one Gomory family.
(ii) If $d=2$ and A is graded, then we may assume that

$$A = \begin{bmatrix} 1 & 1 & 1 & \dots & 1 \\ 0 & 1 & 2 & \dots & n-1 \end{bmatrix}$$

In this case, A is supernormal and hence every regular triangulation Δ of cone(A) supports a Gomory family by Theorem 7.8. Suppose the maximal cones of Δ, in counter-clockwise order, are $C_1, \dots, C_r$. Assume the columns of A are labeled such that $C_i = \text{cone}(a_{i-1}, a_i)$ for $i=1,\dots,r$, and the columns of A in the interior of C_i are labeled in counter-clockwise order as $b_{i1}, \dots, b_{ik_i}$. Hence the n columns of A from left to right are:

$$a_0, b_{11}, \dots, b_{1k_1}, a_1, b_{21}, \dots, a_{r-1}, b_{r1}, \dots, b_{rk_r}, a_r.$$

Indexing the columns of A by their labels, the maximal faces of Δ are $\sigma_i = \{i-1, i\}$ for $i=1,\dots,r$. Let e_i be the unit vector of $\mathbb{R}^n$ indexed by the true column index of a_i in A and e_{ij} be the unit vector of $\mathbb{R}^n$ indexed by the true column index of b_{ij} in A. Since the columns of A form a minimal Hilbert basis of cone(A), e_i is the unique solution to $\text{IP}_{A,c}(a_i)$ for all c and e_{ij} is the unique solution to $\text{IP}_{A,c}(b_{ij})$ for all c. Hence the standard pairs of Theorem 7.8 are $(0, \sigma_i)$ and (e_{ij}, σ_i) or $i=1,\dots,r$ and $j=1,\dots,k_i$.

Suppose Δ supports a second Gomory family $\text{IP}_{A,\omega}$. Then every standard pair of $\mathcal{O}_w$ is also of the form $(*, \sigma_i)$ for $\sigma_i \in \Delta$, and r of them are $(0, \sigma_i)$ for $i=1,\dots,r$. The remaining standard pairs are of the form (e_{ij}, σ_k). To see this, consider the semigroups in $\mathbb{N}A$ arising from the standard pairs of $\mathcal{O}_w$. The total number of standard pairs of $\mathcal{O}_c$ and $\mathcal{O}_w$ are the same. Since the columns of A all lie on $x_1=1$, no two b_{ij}s can be covered by a semigroup coming from the same standard pair and none of them are covered by a semigroup $(0, \sigma_i)$. We show that if (e_{ij}, σ_k) is a standard pair of $\mathcal{O}_w$ then $k=i$ and thus $\mathcal{O}_w = \mathcal{O}_c$.

If $r=1$, the standard pairs of $\mathcal{O}_w$ are $(0,\sigma_1),(e_{11},\sigma_1),\ldots,(e_{1k_1},\sigma_1)$ as in Theorem 7.8. If $r>1$, consider the last cone $C_r=\text{cone}(a_{r-1},a_r)$. If a_{r-1} is the second to last column of A, then C_r is unimodular and the semigroup from $(0,\sigma_r)$ covers $C_r\cap\mathbb{Z}^2$. The subcomplex comprised of $C_1,\ldots,C_{r-1}$ is a regular triangulation Δ' of cone(A') where A' is obtained by dropping the last column of A. Since A' is a normal graded matrix with $d=2$ and Δ' has less than r maximal cones, the standard pairs supported on Δ' are as in Theorem 7.8 by induction. If a_{r-1} is not the second to last column of A then b_{rk_r}, the second to last column of A is in the Hilbert basis of C_r but is not a generator of C_r. So $\mathcal{O}_w$ has a standard pair of the form (e_{rk_r},σ_i). If $\sigma_i\neq\sigma_r$, then the lattice point $b_{rk_r}+a_r$ cannot be covered by the semigroup from this or any other standard pair of $\mathcal{O}_w$. Hence $\sigma_i=\sigma_r$. By a similar argument, the remaining standard pairs indexed by σ_r are $(e_{r(k_r-1)},\sigma_r),\ldots,(e_{r1},\sigma_r)$ along with $(0,\sigma_r)$. These are precisely the standard pairs of $\mathcal{O}_c$ indexed by σ_r. Again we are reduced to considering the subcomplex comprised of $C_1,\ldots,C_{r-1}$ and by induction, the remaining standard pairs of $\mathcal{O}_w$ are as in Theorem 7.8.

(iii) The 3×6 normal matrix A of Example 6.3 has 10 distinct Gomory families supported on 10 out of the 14 regular triangulations of cone(A). Furthermore, the normal matrix

$$A=\begin{bmatrix}1&1&1&1&1&1&1\\1&0&1&1&1&1&0\\0&1&2&2&1&1&0\\0&0&4&3&2&1&0\end{bmatrix}$$

has 11 distinct Gomory families supported on 11 out of its 19 regular triangulations. □

8 Algebraic notes

[A1]: A monomial x^u in the polynomial ring $S:=k[x_1,\ldots,x_n]$ is a product $x^u=x_1^{u_1}x_2^{u_2}\ldots x_n^{u_n}$ where $u=(u_1,\ldots,u_n)\in\mathbb{N}^n$. We assume that k is a field, say the set of rational numbers. For a scalar $k_u\in k$ and a monomial x^u in S, we call k_ux^u a *term* of S. A *polynomial* $f=\sum k_ux^u$ in S is a combination of finitely many terms in S. A subset I of S is an *ideal* of S if (1) I is closed under addition, i.e., $f,\ g\in I\Rightarrow f+g\in I$ and (2) if $f\in I$ and $g\in S$ then $fg\in I$. We say that I is generated by the polynomials $f_1,\ldots,f_t$, denoted as $I=\langle f_1,\ldots,f_t\rangle$, if $I=\{\sum_{i=1}^t f_ig_i\colon g_i\in S\}$. By Hilbert's basis theorem, every ideal in S has a finite generating set. An ideal M in S is called a *monomial ideal* if it is generated by monomials, i.e., $M=\langle x^{v_1},\ldots,x^{v_t}\rangle$ for monomials $x^{v_1},\ldots,x^{v_t}$ in S. The monomials that do not lie in M are called the *standard monomials* of M. The *cost* of a term k_ux^u with respect to a vector $c\in\mathbb{R}^n$ is the dot product $c\cdot u$. The *initial term* of a polynomial $f=\sum k_ux^u\in S$ with respect to c, denoted as $\text{in}_c(f)$, is the sum of all terms in f of maximal cost. For any ideal $I \subset S$, the *initial ideal* of I with respect to c, denoted as $\text{in}_c(I)$, is the ideal generated by all the initial terms $\text{in}_c(f)$ of all polynomials f in I. These concepts come from the theory of *Gröbner bases* for polynomial ideals. See Cox, Little, and O'Shea (1996) for an introduction.

The *toric ideal* of the matrix A, denoted as I_A, is the *binomial* ideal in S defined as:

$$I_A := \langle x^u - x^v \colon u, v \in \mathbb{N}^n \text{ and } Au = Av \rangle.$$

Toric ideals provide the link between integer programming and Gröbner basis theory. See Sturmfels (1995) and Thomas (1997) for an introduction to this area of research. This connection yields the following basic facts that we state without proofs.

Lemma 8.1. [Sturmfels (1995)]

(i) *If c is generic, then the initial ideal $\text{in}_c(I_A)$ is a monomial ideal.*
(ii) *A lattice point u is nonoptimal for the integer program $IP_{A,c}(Au)$, or equivalently, $u \in \mathcal{N}_c$, if and only if x^u lies in the initial ideal $\text{in}_c(I_A)$. In other words, a lattice point u lies in $\mathcal{O}_c$ if and only if x^u is a standard monomial of $\text{in}_c(I_A)$.*
(iii) *The* **reduced Gröbner basis** *$\mathcal{G}_c$ of I_A with respect to c is the unique minimal test set for the family of integer programs $IP_{A,c}$.*
(iv) *If u is a feasible solution of $IP_{A,c}(b)$, and x^{u^*} is the unique* **normal form** *of x^u with respect to $\mathcal{G}_c$, then u^* is the optimal solution of $IP_{A,c}(b)$.*

The above lemma provides an algorithm to compute $\mathcal{O}_c$ algebraically as the lattice points in it are the exponents of the standard monomials of $\text{in}_c(I_A)$. This initial ideal is a byproduct of the computation of the reduced Gröbner basis of I_A with respect to the cost vector c. Gröbner bases of polynomial ideals can be computed by Buchberger's algorithm (Cox et al., 1996).

Example 3.2 continued. In this example, the toric ideal $I_A = \langle x_1^4 - x_3, x_2^2 - x_1x_3 \rangle$ and its initial ideal with respect to the cost vector $c = (10000, 100, 1)$ is

$$\text{in}_c(I_A) = \langle x_2^8,\ x_1x_3,\ x_1x_2^6,\ x_1^2x_2^4,\ x_1^3x_2^2,\ x_1^4 \rangle.$$

Note that the exponent vectors of the generators of $\text{in}_c(I_A)$ are the generators of $\mathcal{N}_c$. □

[A2]: A *primary* ideal J in $k[x_1, \ldots, x_n]$ is a proper ideal such that $fg \in J$ implies either $f \in J$ or $g^t \in J$ for some positive integer t. A *prime* ideal J of $k[x_1, \ldots, x_n]$ is a proper ideal such that $fg \in J$ implies that either $f \in J$ or $g \in J$.

A *primary decomposition* of an ideal I in $k[x_1, \ldots, x_n]$ is an expression of I as a finite intersection of *primary ideals* in $k[x_1, \ldots, x_n]$. Lemma 3.3 in Sturmfels (1995) shows that every monomial ideal M in $k[x_1, \ldots, x_n]$ admits a primary decomposition into irreducible primary ideals that are indexed by the standard pairs of M. The *radical* of an ideal $I \subset k[x_1, \ldots, x_n]$ is the ideal $\sqrt{I} := \{f \in S : f^t \in I, \text{ for some positive integer } t\}$. Radicals of primary ideals are prime. The radicals of the primary ideals in a minimal primary decomposition of an ideal I are called as the *associated primes* of I. This list of primes ideals is independent of the primary decomposition of the ideal. The minimal elements among the associated primes of I are called the *minimal primes* of I while the others are called as the *embedded primes* of I. The minimal primes of I are precisely the defining ideals of the isolated components of the *variety* of I while the embedded primes cut out embedded subvarieties in the isolated components. See a textbook in commutative algebra like Eisenbud (1994) for more details.

A face τ of Δ_c is an associated set of $\text{IP}_{A,c}$ if and only if the monomial prime ideal $p_\tau := \langle x_j : j \notin \tau \rangle$ is an associated prime of the ideal $\text{in}_c(I_A)$. Further, p_σ is a minimal prime of $\text{in}_c(I_A)$ if and only if σ is a maximal face of Δ_c. Hence the lower dimensional associated sets of $\text{IP}_{A,c}$ index the embedded primes of $\text{in}_c(I_A)$. The standard pair decomposition of a monomial ideal was introduced in Sturmfels et al. (1995) to study its associated primes. The multiplicity of an associated prime p_τ of $\text{in}_c(I_A)$ is an algebraic invariant of $\text{in}_c(I_A)$, and Sturmfels et al. shows that this is exactly the number of standard pairs indexed by τ. Similarly, the arithmetic degree of $\text{in}_c(I_A)$ is a refinement of the geometric notion of *degree* and Sturmfels et al. shows that this number is the total number of standard pairs of $\text{in}_c(I_A)$. These connections explain our choice of terminology. Theorem 4.3 is a translation of the specialization of Lemma 3.5 in Sturmfels et al. (1995) to toric initial ideals. We refer the interested reader to Sturmfels (1995, §8 and §12.D) and Sturmfels et al. (1995, §3) for the algebraic connections. Theorem 4.5 is a staple result of toric geometry and also follows from Gomory (1965, Theorem 1). It is proved via the algebraic technique of *localization* in Sturmfels (1995), Theorem 8.8. Bounds on the arithmetic degree of a general monomial ideal in terms of its dimension and minimal generators can be found in Sturmfels et al. (1995, Theorem 3.1). One hopes that stronger bounds are possible for toric initial ideals.

[A3]: In algebraic language, the chain theorem says that the associated primes of $\text{in}_c(I_A)$ occur in saturated chains. This was proved in Hoşten and Thomas (1999, Theorem 3.1). When the cost vector c is not generic, $\text{in}_c(I_A)$ is no longer a monomial ideal, and its associated primes need not come in saturated chains. See Hoşten and Thomas (1999, Remark 3.3) for such an example. Algebraically, Problem 3.3 asks for a characterization of all monomial ideals that can appear as a monomial initial ideal of a toric ideal. Theorem 5.2 imposes the necessary condition that the associated primes of such a monomial ideal has to come in saturated chains. Unfortunately, this is not sufficient. See Miller, Sturmfels, and Yanagawa (2000) for another class of monomial ideals that also have the chain property.

[A4]: Algebraically, $IP_{A,c}$ is a Gomory family if and only if the initial ideal $in_c(I_A)$ has no embedded primes and hence Theorem 6.2 is a characterization of toric initial ideals without embedded primes. A sufficient condition for an ideal in $k[x_1, \ldots, x_n]$ not to have embedded primes is that it is *Cohen-Macaulay* (Eisenbud, 1994). In general, Cohen-Macaulayness is not necessary for an ideal to be free of embedded primes. However, empirical evidence seemed to suggest for a while that for toric initial ideals, Cohen-Macaulayness might be equivalent of being free of embedded primes. A counterexample to this was found recently by Laura Matusevich.

[A5]: Corollary 8.4 in Strumfels (1995) shows that Δ_c is unimodular if and only if the monomial ideal $in_c(I_A)$ is generated by *square-free* monomials. Hence, by computing $in_c(I_A)$, one can determine whether $yA \leq c$ is TDI. Such computations can be carried out on computer algebra systems like CoCoA Cocoa 4.1 or MACAULAY 2 (Grayson and Stillman) for moderately sized examples. See Sturmfels (1995) for algorithms. Standard pair decompositions of monomial ideals can be computed with MACAULAY 2 (Hoşten and Smith, 2002).

Acknowledgment

This research was partially supported by NSF grant DMS-0100141.

References

Aardal, K., R. Weismantel, L. Wolsey (2002). Non-standard approaches to integer programming. *Discrete Appl. Math.*, 123, no. 1–3:5–74, Workshop on Discrete optimization DO'99 Piscataway, NJ.

Aráoz, J., L. Evans, R. E. Gomory, E. L. Johnson (2003). Cyclic group and knapsack facets. *Math. Programming, Series B* 96, 377–408.

Bell, D., J. Shapiro (1977). A convergent duality theory for integer programming. *Journal of Operations Research* 25, 419–434.

Billera, L. J., P. Filliman, B. Sturmfels (1990). Constructions and complexity of secondary polytopes. *Advances in Mathematics* 83, 155–179.

Bouvier, C., G. Gonzalez-Springberg (1994). Système generateurs minimal, diviseurs essentiels et G-dèsingularizations de vari'et'es toriques. *Tôhoku Math. Journal* 46, 125–149.

Bruns, W., J. Gubeladze (1999). Normality and covering properties of affine semigroups. *Journal für die reine und angewandte Mathematik* 510, 161–178.

Bruns, W., J. Gubeladze, M. Henk, A. Martin, R. Weismantel (1999). A counterexample to an integer analogue of caratheodory's theorem. *Journal für die reine und angewandte Mathematik* 510, 179–185.

Cocoa 4.1. Available from ftp://cocoa.dima.unige.it/cocoa.

Cox, D., J. Little, D. O'Shea (1996). *Ideals, Varieties, and Algorithms*, 2nd edition, Springer-Verlag, New York.

Eisenbud, D. (1994). *Commutative Algebra with a View Towards Algebraic Geometry*. Springer Graduate Texts in Mathematics.

Evans, L., R. E. Gomory, E. L. Johnson (2003). Corner polyhedra and their connection with cutting planes. *Math. Programming, Series B* 96, 321–339.

Firla, R., G. Ziegler (1999). Hilbert bases, unimodular triangulations, and binary covers of rational polyhedral cones. *Discrete and Computational Geometry* 21, 205–216.

Gel'fand, I. M., M. Kapranov, A. Zelevinsky (1994). *Multidimensional Determinants, Discriminants and Resultants*, Birkhäuser, Boston.

Gomory, R. E. (1965). On the relation between integer and noninteger solutions to linear programs. *Proceedings of the National Academy of Sciences* 53, 260–265.

Gomory, R. E. (1967). Faces of an integer polyhedron. *Proceedings of the National Academy of Sciences* 57, 16–18.

Gomory, R. E. (1969). Some polyhedra related to combinatorial problems. *Linear Algebra and its Applications* 2, 451–558.

Gomory, R. E., E. L. Johnson (1972). Some continuous functions related to corner polyhedra. *Mathematical Programming* 3, 23–85.

Gomory, R. E., E. L. Johnson (2003). T-space and cutting planes. *Math. Programming, Series B* 96, 341–375.

Gorry, G., W. Northup, J. Shapiro (1973). Computational experience with a group theoretic integer programming algorithm. *Mathematical Programming* 4, 171–192.

Grayson, D., M. Stillman, Macaulay 2, a software system for research in algebraic geometry. Available at http://www.math.uiuc.edu/Macaulay2.

Hoşten, S., D. Maclagan, B. Sturmfels. Supernormal vector configurations. *J. Algebraic Combinatorics*. To appear.

Hoşten, S., G. Smith (2002). Monomial ideals, in: D. Eisenbud, D. Grayson, M. Stillman, B. Sturmfels (eds.), *Mathematical Computations with Macaulay 2*, Springer Verlag, New York, pp. 73–100.

Hoşten, S., R. R. Thomas (1999a). The associated primes of initial ideals of lattice ideals. *Mathematical Research Letters* 6, 83–97.

Hoşten, S., R. R. Thomas (1999b). Standard pairs and group relaxations in integer programming. *Journal of Pure and Applied Algebra* 139, 133–157.

Hoşten, S., R. R. Thomas (2003). Gomory integer programs. *Math. Programming, Series B* 96, 271–292.

Huber, B., R. R. Thomas (2000). Computing Gröbner fans of toric ideals. *Experimental Mathematics* 9, 321–331.

Johnson, E. L. (1980). *Integer Programming: Facets, Subadditivity, and Duality for Group and Semi-group Problems*. SIAM CBMS Regional Conference Series in Applied Mathematics No. 32, Philadelphia.

Kannan, R. (1992). Lattice translates of a polytope and the Frobenius problem. *Combinatorica* 12, 161–177.

Kannan, R. (1993). Optimal solution and value of parametric integer programs, in: G. Rinaldi, L. Wolsey (eds.), *Proceedings of the Third IPCO Conference*, pp. 11–21.

Kannan, R., L. Lovaśz, H. E. Scarf (1990). Shapes of polyhedra. *Mathematics of Operations Research* 15, 364–380.

Lovász, L. (1989). Geometry of numbers and integer programming, in: M. Iri, K. Tanebe (eds.), *Mathematical Programming: Recent Developments and Applications*, Kluwer Academic Press, pp. 177–210.

Miller, E. N., B. Sturmfels, K. Yanagawa (2000). Generic and cogeneric monomial ideals. *Journal of Symbolic Computation* 29, 691–708.

Nemhauser, G., L. Wolsey (1988). *Integer and Combinatorial Optimization*, Wiley, New York.

Peeva, I., B. Sturmfels (1998). Syzygies of codimension 2 lattice ideals. *Mathematische Zeitschrift* 229, 163–194.

Schrijver, A. (1986). *Theory of Linear and Integer Programming*, Wiley-Interscience Series in Discrete Mathematics and Optimization, New York.

Sebö, A. (1990). Hilbert bases, Caratheodory's theorem and combinatorial optimization, in: R. Kannan, W. Pulleyblank (eds.), *Integer Programming and Combinatorial Optimization*, Mathematical Programming Society. University of Waterloo Press, Waterloo, pp. 431–456.

Stanley, R. P. (1982). Linear diophantine equations and local cohomology. *Inventiones Math.* 68, 175–193.

Sturmfels, B. (1995). *Gröbner Bases and Convex Polytopes*, American Mathematical Society, Providence, RI.

Sturmfels, B., R. R. Thomas (1997). Variation of cost functions in integer programming. *Mathematical Programming* 77, 357–387.

Sturmfels, B., N. Trung, W. Vogel (1995). Bounds on projective schemes. *Mathematische Annalen* 302, 417–432.

Sturmfels, B., R. Weismantel, G. Ziegler (1995). Gröbner bases of lattices, corner polyhedra and integer programming. *Beiträge zur Algebra und Geometrie* 36, 281–298.

Thomas, R. R. (1995). A geometric Buchberger algorithm for integer programming. *Mathematics of Operations Research* 20, 864–884.

Thomas, R. R. (1997). Applications to integer programming. Applications of Computational Algebraic Geometry. in: D. Cox, B. Sturmfels (eds.), *AMS Proceedings of Symposia in Applied Mathematics* 53, 119–142.

TiGERS. Available from http://www.math.washington.edu/~ thomas/programs.html.

Wolsey, L. (1971). Extensions of the group theoretic approach in integer programming. *Management Science* 18, 74–83.

Wolsey, L. (1973). Generalized dynamic programming methods in integer programming. *Mathematical Programming* 4, 222–232.

Wolsey, L. (1981). The b-hull of an integer program. *Discrete Applied Mathematics* 3, 193–201.

K. Aardal et al., Eds., *Handbooks in OR & MS, Vol. 12*

Chapter 4

Integer Programming, Lattices, and Results in Fixed Dimension

Karen Aardal and Friedrich Eisenbrand

Abstract

We review and describe several results regarding integer programming problems in fixed dimension. First, we describe various lattice basis reduction algorithms that are used as auxiliary algorithms when solving integer feasibility and optimization problems. Next, we review three algorithms for solving the integer feasibility problem. These algorithms are based on the idea of branching on lattice hyperplanes, and their running time is polynomial in fixed dimension. We also briefly describe an algorithm, based on a different principle, to count integer points in an integer polytope. We then turn the attention to integer optimization. Again, we describe three algorithms: binary search, a linear algorithm for a fixed number of constraints, and a randomized algorithm for a varying number of constraints. The topic of the next part of our chapter is how to use lattice basis reduction in problem reformulation. Finally, we review cutting plane results when the dimension is fixed.

1 Introduction

Integer programming problems have offered, and are still offering, many challenging theoretical and computational questions. We consider two integer programming problems. Given is a set of rational linear inequalities $\boldsymbol{Ax} \leq \boldsymbol{d}$. The first problem is the *integer feasibility problem*: Does there exist an integer vector $\boldsymbol{x}$ satisfying $\boldsymbol{Ax} \leq \boldsymbol{d}$? The second problem is the *integer optimization problem*: Determine an integer vector $\boldsymbol{x}$ that satisfies $\boldsymbol{Ax} \leq \boldsymbol{d}$, and also maximizes or minimizes a given linear function $\boldsymbol{c}^T\boldsymbol{x}$.

The feasibility problem was proved to be NP-complete in 1976, but an interesting complexity question remained: Is the feasibility problem solvable in polynomial time if the number of variables, i.e., the number of components of $\boldsymbol{x}$, is fixed? The predominantly used algorithm, branch-and-bound, is not a polynomial time algorithm in fixed dimension, but in 1983 H.W. Lenstra, Jr. developed an algorithm with a polynomial running time if the dimension is fixed. His algorithm is based on results from number theory; in particular on properties of lattices and lattice bases. Since then we have seen several results built on knowledge about lattices, and also many other results for integer programming problems in fixed dimension.

In our chapter we will illustrate some of these results. Since lattices and lattice bases play an important role we will present three algorithms for finding "good" lattice bases in Section 3. In this section we also review algorithms to compute a shortest vector of a lattice. In Section 4 we focus on the integer feasibility problem and describe three algorithms built on the fundamental result that if a polytope does not contain an integer vector, then there exists a nonzero integer direction in which the polytope is intersected by at most $f(n)$ so-called lattice hyperplanes, where $f(n)$ is a function depending on the dimension n only. The integer optimization problem is treated in Section 5. Again three algorithms are described; first binary search, second a more involved algorithm that solves the problem in linear time when the number of constraints is fixed, and finally a randomized algorithm which reduces the dependence of the complexity on the number of constraints. In Section 6 we take another view of solving integer feasibility problems. Here we try to construct a lattice in which we can prove that solutions to the considered problems are short vectors in that lattice. Solutions, if they exist, can then be found by considering bases of the lattice in which the basis vectors are short. Finally, in Section 7 we review various results regarding cutting planes if, again, the dimension is fixed. Even though little explicit use is made of lattices in this section, the results tie in well with the results discussed in Sections 4–6, and address several complexity questions that are naturally raised in the context of integer programming in a fixed dimension.

2 Notation and basic definitions

To make our chapter more accessible we present some basic notation and definitions in the following two subsections.

2.1 *Numbers, vectors, matrices, and polyhedra*

The set of real (integer, rational) numbers is denoted by $\mathbb{R}$ ($\mathbb{Z}$, $\mathbb{Q}$). If we require nonnegativity we use the notation $\mathbb{R}_{\geq 0}$, $\mathbb{Z}_{\geq 0}$, and $\mathbb{Q}_{\geq 0}$ respectively. The set of natural numbers is denoted by $\mathbb{N}$ and if we consider positive natural numbers we use the notation $\mathbb{N}_{>0}$. When we write $\boldsymbol{x}_j$ we mean the j-th vector in a sequence of vectors. The i-th component of a vector $\boldsymbol{x}$ will be denoted by x_i, and the i-th component of the vector $\boldsymbol{x}_j$ is written x_j^i. The *Euclidean length* of a vector $\boldsymbol{x} \in \mathbb{R}^n$ is denoted by $\|\boldsymbol{x}\|$ and is computed as $\|\boldsymbol{x}\| = \sqrt{\boldsymbol{x}^T\boldsymbol{x}}$, where $\boldsymbol{x}^T$ is the transpose of the vector $\boldsymbol{x}$. An $m \times n$ matrix $\boldsymbol{A}$ has columns $(\boldsymbol{a}_1, \ldots, \boldsymbol{a}_n)$, and element (i, j) of $\boldsymbol{A}$ is denoted by a_{ij}. We use $(\boldsymbol{c})^{(m\times n)}$ to denote an $m \times n$ matrix in which all elements are equal to c. The $n \times n$ identity matrix is denoted by $\boldsymbol{I}^{(n)}$, and when it is clear from the context the superscripts of $(\boldsymbol{c})^{(m\times n)}$ and $\boldsymbol{I}^{(n)}$ are dropped. Given an $m \times n$ matrix $\boldsymbol{A}$, the inequality

$$\sqrt{\det(\boldsymbol{A}^T\boldsymbol{A})} \leq \|\boldsymbol{a}_1\| \cdots \|\boldsymbol{a}_n\| \tag{1}$$

is known as the *Hadamard inequality*. An integer nonsingular matrix $\boldsymbol{U}$ is *unimodular* if $\det(\boldsymbol{U}) = \pm 1$. A matrix of full row rank is said to be in *Hermite Normal Form*, (HNF), if it has the form $(\boldsymbol{C}, (\boldsymbol{0})^{(m\times(n-m))})$, where $\boldsymbol{C}$ is a lower triangular nonnegative $m \times m$ matrix in which the unique row maxima can be found along the diagonal. A rational $m \times n$ matrix $\boldsymbol{A}$ of full row rank has a unique Hermite normal form, $\mathrm{HNF}(\boldsymbol{A}) = (\boldsymbol{C}, (\boldsymbol{0})^{(m\times(n-m))}) = \boldsymbol{AU}$, where $\boldsymbol{U}$ is unimodular.

We use the notation $\lfloor x \rfloor$ and $\lceil x \rceil$ for the round down and round up of the number x. We define $\lceil x \rfloor := \lceil x - \frac{1}{2} \rceil$. The *size* of an integer z is the number $\mathrm{size}(z) = 1 + \lceil \log_2(|z| + 1) \rceil$. Likewise, the size of a matrix $\boldsymbol{A} \in \mathbb{Z}^{m\times n}$ is the number of bits needed to encode $\boldsymbol{A}$, i.e., $\mathrm{size}(\boldsymbol{A}) = mn + \sum_{i,j} \mathrm{size}(a_{ij})$, see [99, p. 29].

A *polyhedron* P is a set of vectors of the form $P = \{\boldsymbol{x} \in \mathbb{R}^n \mid \boldsymbol{Ax} \le \boldsymbol{d}\}$, for some matrix $\boldsymbol{A} \in \mathbb{R}^{m\times n}$ and some vector $\boldsymbol{d} \in \mathbb{R}^m$. We write $P = P(\boldsymbol{A}, \boldsymbol{d})$. If P is given as $P(\boldsymbol{A}, \boldsymbol{d})$, then $\mathrm{size}(P) = \mathrm{size}(\boldsymbol{A}) + \mathrm{size}(\boldsymbol{d})$. The polyhedron $P = P(\boldsymbol{A}, \boldsymbol{d})$ is *rational* if both $\boldsymbol{A}$ and $\boldsymbol{d}$ can be chosen to be rational. If P is bounded, then P is called a *polytope*. The *integer hull* P_I of a polyhedron P is the convex hull of the integer vectors in P. If P is rational, then P_I is a rational polyhedron again. The *dimension* of P is the dimension of the affine hull of P.

A *rational halfspace* is a set of the form $H = \{\boldsymbol{x} \in \mathbb{R}^n \mid \boldsymbol{c}^T\boldsymbol{x} \le \delta\}$, for some non-zero vector $\boldsymbol{c} \in \mathbb{Q}^n$ and some $\delta \in \mathbb{Q}$. The halfspace H is then denoted by $(\boldsymbol{c}^T\boldsymbol{x} \le \delta)$. The corresponding *hyperplane*, denoted by $(\boldsymbol{c}^T\boldsymbol{x} = \delta)$, is the set $\{\boldsymbol{x} \in \mathbb{R}^n \mid \boldsymbol{c}^T\boldsymbol{x} = \delta\}$. A rational half space always has a representation in which the components of $\boldsymbol{c}$ are relatively prime integers. That is, we can choose $\boldsymbol{c} \in \mathbb{Z}^n$ with $\gcd(c_1, \ldots, c_n) = 1$.

An inequality $\boldsymbol{c}^T\boldsymbol{x} \le \delta$ is called *valid* for a polyhedron P, if $(\boldsymbol{c}^T\boldsymbol{x} \le \delta) \supseteq P$. A *face* of P is a set of the form $F = (\boldsymbol{c}^T\boldsymbol{x} = \delta) \cap P$, where $\boldsymbol{c}^T\boldsymbol{x} \le \delta$ is valid for P. The inequality $\boldsymbol{c}^T\boldsymbol{x} \le \delta$ is a *face-defining* inequality for F. Clearly F is a polyhedron. If $P \supset F \supset \emptyset$, then F is called *proper*. A maximal (inclusion wise) proper face of P is called a *facet* of P, i.e., a proper face F is a facet if and only if $\dim(F) = \dim(P) - 1$. If the face-defining inequality $\boldsymbol{c}^T\boldsymbol{x} \le \delta$ defines a facet of P, then $\boldsymbol{c}^T\boldsymbol{x} \le \delta$ is a *facet-defining* inequality. A proper face of P of dimension 0 is called a *vertex* of P. A vertex $\boldsymbol{v}$ of $P(\boldsymbol{A}, \boldsymbol{d})$ is uniquely determined by a subsystem $\boldsymbol{A}^{\boldsymbol{v}}\boldsymbol{x} \le \boldsymbol{d}^{\boldsymbol{v}}$ of $\boldsymbol{Ax} \le \boldsymbol{d}$, where $\boldsymbol{A}^{\boldsymbol{v}}$ is nonsingular and $\boldsymbol{v} = (\boldsymbol{A}^{\boldsymbol{v}})^{-1}\boldsymbol{d}^{\boldsymbol{v}}$. If P is full-dimensional, then P has a unique (up to scalar multiplication) minimal set of inequalities defining P, which correspond to the facets of P. A polytope P can be described as the convex hull of its vertices. A *d-simplex* is a polytope, which is the convex hull of $d + 1$ affinely independent points.

Let $P \subseteq \mathbb{R}^n$ be a rational polyhedron. The *facet complexity* of P is the smallest number φ satisfying

- $\varphi \ge n$, and
- there exists a system $\boldsymbol{Ax} \le \boldsymbol{d}$ of rational linear inequalities defining P such that each inequality in $\boldsymbol{Ax} \le \boldsymbol{d}$ has size at most φ.

The *vertex complexity* of P is the smallest number ν, such that there exist rational vectors $\boldsymbol{q}_1, \ldots, \boldsymbol{q}_k$, $\boldsymbol{c}_1, \ldots, \boldsymbol{c}_t$, each of size at most ν, with

$$P = \text{conv}(\{\boldsymbol{q}_1, \ldots, \boldsymbol{q}_k\}) + \text{cone}(\{\boldsymbol{c}_1, \ldots, \boldsymbol{c}_t\}).$$

Let $P \subseteq \mathbb{R}^n$ be a rational polyhedron of facet complexity φ and vertex complexity ν. Then (see Schrijver [99])

$$\nu \le 4n^2\varphi \text{ and } \varphi \le 4n^2\nu. \tag{2}$$

We refer to Nemhauser and Wolsey [85] and Schrijver [99] for further basics on the topics treated in this subsection.

2.2 Lattices and lattice bases

Let $\boldsymbol{b}_1, \ldots, \boldsymbol{b}_l$ be linearly independent vectors in $\mathbb{R}^n$. The set

$$L = \left\{ \boldsymbol{x} \in \mathbb{R}^n \mid \boldsymbol{x} = \sum_{j=1}^{l} \lambda_j \boldsymbol{b}_j,\ \lambda_j \in \mathbb{Z},\ 1 \le j \le l \right\} \tag{3}$$

is called a *lattice*. The set of vectors $\{\boldsymbol{b}_1, \ldots, \boldsymbol{b}_l\}$ is called a *lattice basis*. The vectors of a lattice L form an additive group, i.e., $\boldsymbol{0} \in L$, and if $\boldsymbol{x}$ belongs to L, so does $-\boldsymbol{x}$, and if $\boldsymbol{x}, \boldsymbol{y} \in L$, then $\boldsymbol{x} \pm \boldsymbol{y} \in L$. Moreover, the group L is *discrete*, i.e., there exists a real number $r > 0$ such that the n-dimensional ball with radius r, centered at the origin, does not contain any other element from L except the origin.

The *rank* of L, rk L, is equal to the dimension of the Euclidean vector space generated by a basis of L. The rank of the lattice L in Expression (3) is l, and we have $l \le n$. If $l = n$ we call the lattice *full-dimensional*. Let $\boldsymbol{B} = (\boldsymbol{b}_1, \ldots, \boldsymbol{b}_l)$. If we want to emphasize that we are referring to a lattice L that is generated by the basis $\boldsymbol{B}$, then we use the notation $L(\boldsymbol{B})$. Two matrices $\boldsymbol{B}_1$, $\boldsymbol{B}_2 \in \mathbb{R}^{n \times l}$ are bases of the same lattice $L \subseteq \mathbb{R}^n$, if and only if $\boldsymbol{B}_1 = \boldsymbol{B}_2\boldsymbol{U}$ for some $l \times l$ unimodular matrix $\boldsymbol{U}$. The *shortest nonzero vector* in the lattice L is denoted by SV(L) or SV($L(\boldsymbol{B})$).

We will frequently make use of *Gram-Schmidt orthogonalization*. The Gram-Schmidt process derives orthogonal vectors $\boldsymbol{b}_j^*$, $1 \le j \le l$, from linearly independent vectors $\boldsymbol{b}_j$, $1 \le j \le l$. The vectors $\boldsymbol{b}_j^*$, $1 \le j \le l$, and the real numbers μ_{jk}, $1 \le k < j \le l$, are determined from $\boldsymbol{b}_j$, $1 \le j \le l$, by the recursion

$$\boldsymbol{b}_1^* = \boldsymbol{b}_1$$

$$\boldsymbol{b}_j^* = \boldsymbol{b}_j - \sum_{k=1}^{j-1} \mu_{jk} \boldsymbol{b}_k^*, \quad 2 \le j \le l,$$

where

$$\mu_{jk} = \frac{\boldsymbol{b}_j^T \boldsymbol{b}_k^*}{\|\boldsymbol{b}_k^*\|^2}, \quad 1 \le k < j \le l.$$

The Gram-Schmidt process yields a factorization of the matrix $(\boldsymbol{b}_1, \ldots, \boldsymbol{b}_n)$ as

$$(\boldsymbol{b}_1, \ldots, \boldsymbol{b}_n) = (\boldsymbol{b}_1^*, \ldots, \boldsymbol{b}_n^*) \cdot \boldsymbol{R}, \tag{4}$$

where $\boldsymbol{R}$ is the matrix

$$\boldsymbol{R} = \begin{pmatrix} 1 & \mu_{21} & \cdots & \mu_{n1} \\ 0 & 1 & \cdots & \mu_{n2} \\ \cdots & \cdots & \cdots & \cdots \\ 0 & \cdots & 0 & 1 \end{pmatrix} \tag{5}$$

The vector $\boldsymbol{b}_j^*$ is the projection of $\boldsymbol{b}_j$ on the orthogonal complement of $\sum_{k=1}^{j-1} \mathbb{R}\,\boldsymbol{b}_k = \{\sum_{k=1}^{j-1} m_k \boldsymbol{b}_k : m_k \in \mathbb{R}, 1 \le k \le j-1\}$, i.e., $\boldsymbol{b}_j^*$ is the component of $\boldsymbol{b}_j$ orthogonal to the real subspace spanned by $\boldsymbol{b}_1, \ldots, \boldsymbol{b}_{j-1}$. Thus, any pair $\boldsymbol{b}_i^*$, $\boldsymbol{b}_k^*$ of the Gram-Schmidt vectors are mutually orthogonal. The multiplier μ_{jk} gives the length, relative to $\boldsymbol{b}_k^*$, of the component of the vector $\boldsymbol{b}_j$ in direction $\boldsymbol{b}_k^*$. The multiplier μ_{jk} is equal to zero if and only if $\boldsymbol{b}_j$ is orthogonal to $\boldsymbol{b}_k^*$. Notice that the Gram-Schmidt vectors corresponding to $\boldsymbol{b}_1, \ldots, \boldsymbol{b}_l$ do not in general belong to the lattice generated by $\boldsymbol{b}_1, \ldots, \boldsymbol{b}_l$, but they do span the same real vector space as $\boldsymbol{b}_1, \ldots, \boldsymbol{b}_l$.

Let W be the vector space spanned by the lattice L, and let B_W be an orthonormal basis for W. The *determinant of the lattice* L, $d(L)$, is defined as the absolute value of the determinant of any nonsingular linear transformation $W \rightarrow W$ that maps B_W onto a basis of L. Below we give three different formulae for computing $d(L)$. Let $\boldsymbol{B} = (\boldsymbol{b}_1, \ldots, \boldsymbol{b}_l)$ be a basis for the lattice $L \subset \mathbb{R}^n$, with $l \le n$, and let $\boldsymbol{b}_1^*, \ldots, \boldsymbol{b}_l^*$ be the vectors obtained from applying the Gram-Schmidt orthogonalization procedure to $\boldsymbol{b}_1, \ldots, \boldsymbol{b}_l$.

$$d(L) = \|\boldsymbol{b}_1^*\| \cdot \|\boldsymbol{b}_2^*\| \cdot \cdots \cdot \|\boldsymbol{b}_l^*\|,$$

$$d(L) = \sqrt{\det(\boldsymbol{B}^T \boldsymbol{B})},$$

$$d(L) = \lim_{r \rightarrow \infty} \frac{|\{\boldsymbol{x} \in L : \|\boldsymbol{x}\| < r\}|}{\text{vol}(B_l(r))}, \tag{6}$$

where $\text{vol}(B_l(r))$ is the volume of the l-dimensional ball with radius r. If L is full-dimensional, then $d(L(\boldsymbol{B}))$ can be interpreted as the volume of the parallelepiped $\sum_{j=1}^{n} [0, 1)\, \boldsymbol{b}_j$. In this case the determinant of the lattice can be computed straightforwardly as $d(L(\boldsymbol{B})) = |\det(\boldsymbol{B})|$. The determinant of $\mathbb{Z}^n$ is equal to one. It is clear from Expression (6) that the determinant of a lattice depends only on the lattice and not on the choice of basis, see also Section 3. We will often use Hadamard's inequality (1) to bound the determinant of the lattice, i.e.,

$$d(L(\boldsymbol{B})) = \sqrt{\det(\boldsymbol{B}^T\boldsymbol{B})} \leq \|\boldsymbol{b}_1\| \cdot \cdots \cdot \|\boldsymbol{b}_l\|, \tag{7}$$

where equality holds if and only if the basis $\boldsymbol{B}$ is orthogonal.

A convex set $K \in \mathbb{R}^n$ is *symmetric about the origin* if $\boldsymbol{x} \in K$ implies that $-\boldsymbol{x} \in K$. We will refer to the following theorem by Minkowski later in the chapter.

Theorem 1 (Minkowski's convex body theorem [83]). *Let K be a compact convex set in $\mathbb{R}^n$ of volume $vol(K)$ that is symmetric about the origin. Let m be an integer an let L be a lattice of determinant $d(L)$. Suppose that $vol(K) \geq m2^n d(L)$. Then K contains at least m pairs of points $\pm\boldsymbol{x}_j$, $1 \leq j \leq m$ that are distinct from each other and from the origin.*

Let L be a full-dimensional lattice in $\mathbb{R}^n$. Its *dual lattice* L^* is defined as

$$L^* = \{\boldsymbol{x} \in \mathbb{R}^n \mid \boldsymbol{x}^T\boldsymbol{y} \in \mathbb{Z} \text{ for all } \boldsymbol{y} \in L\}.$$

For a lattice L and its dual we have $d(L) = d(L^*)^{-1}$.

For more details about lattices, see e.g. Cassels [22], Grötschel, Lovász, and Schrijver [55], and Schrijver [99].

3 Lattice basis reduction

In several of the sections in this chapter we will use representations of lattices using bases that consist of vectors that are short and nearly orthogonal. In Section 3.1 we motivate why short lattice vectors are interesting objects, and we describe the basic principle of obtaining a new basis from a known basis of a given lattice. In Section 3.2 we describe Lovász' basis reduction algorithm, and some variants. The first vector in a Lovász-reduced basis is an approximation of the shortest non-zero lattice vector. In Section 3.3 we introduce Korkine-Zolotareff-reducedness and present Kannan's algorithm for computing the shortest non-zero lattice vector. We also discuss the complexity status of the shortest and closest lattice vector problem. In Section 3.4 we describe the generalized basis reduction algorithm by Lovász and Scarf, which uses a polyhedral norm instead of the Euclidean norm as in Lovász' algorithm. Finally, in Section 3.5 we discuss fast basis reduction algorithms in the bit model.

3.1 Reduced bases, an informal introduction

A lattice of rank at least two has infinitely many bases. Some of these bases are more useful than others, and in the applications we consider in this chapter we use bases whose elements are "nearly orthogonal". Such bases are called *reduced*. There are several definitions of reducedness, and some of them will be discussed in the following sections. Having a reduced basis makes it possible to obtain important bounds on both algorithmic running times and quality of solutions when lattice representations are used in integer programming and related areas. The study of reduced bases appears as early as in work by Gauß [49], Hermite [59], Minkowski [82], and Korkine and Zolotareff [72].

In many applications it becomes essential to determine the shortest nonzero vector in a lattice. In the following we motivate why an "almost orthogonal basis" helps us to find this vector. Suppose that $L \subseteq \mathbb{R}^n$ is generated by the basis $\boldsymbol{b}_1, \ldots, \boldsymbol{b}_n$ and assume that the vectors $\boldsymbol{b}_j$ are pairwise orthogonal. Consider a nonzero element $\boldsymbol{v} = \sum_{j=1}^n \lambda_j \boldsymbol{b}_j$ of the lattice, where $\lambda_j \in \mathbb{Z}$ for $j = 1, \ldots, n$. One has

$$\begin{aligned}\|\boldsymbol{v}\|^2 &= \left(\sum_{j=1}^n \lambda_j \boldsymbol{b}_j\right)^T \left(\sum_{j=1}^n \lambda_j \boldsymbol{b}_j\right) \\ &= \sum_{j=1}^n \lambda_j^2 \|\boldsymbol{b}_j\|^2 \\ &\geq \min\{\|\boldsymbol{b}_j\|^2 \mid j = 1, \ldots, n\},\end{aligned}$$

where the last inequality follows from the fact that the λ_j are integers and not all of them are zero. Therefore the shortest vector of L is the shortest vector of the basis $\boldsymbol{b}_1, \ldots, \boldsymbol{b}_n$.

How do we determine the shortest vector of L if the basis $\boldsymbol{b}_1, \ldots, \boldsymbol{b}_n$ is not orthogonal but "almost orthogonal"? The Gram-Schmidt orthogonalization procedure, see Section 2.2, computes pairwise orthogonal vectors $\boldsymbol{b}_1^*, \ldots, \boldsymbol{b}_n^*$ and an upper triangular matrix $R \in \mathbb{R}^{n \times n}$ whose diagonal entries are all one such that

$$(\boldsymbol{b}_1, \ldots, \boldsymbol{b}_l) = \left(\boldsymbol{b}_1^*, \ldots, \boldsymbol{b}_l^*\right) \cdot \boldsymbol{R}$$

holds. Furthermore one has $\|\boldsymbol{b}_j\| \geq \|\boldsymbol{b}_j^*\|$ for $j = 1, \ldots, n$. This implies the Hadadmard inequality (7): $d(L) = \|\boldsymbol{b}_1^*\| \cdots \|\boldsymbol{b}_n^*\| \leq \|\boldsymbol{b}_1\| \cdots \|\boldsymbol{b}_n\|$, where equality holds if and only if the $\boldsymbol{b}_1, \ldots, \boldsymbol{b}_n$ are pairwise orthogonal. The number $c = \|\boldsymbol{b}_1\| \cdots \|\boldsymbol{b}_n\| / d(L)$ is called the *orthogonality defect* of the lattice basis $\boldsymbol{b}_1, \ldots, \boldsymbol{b}_n$. By "almost orthogonal" we mean that the orthogonality defect of a reduced basis is bounded by a constant that depends on the dimension n of the lattice only.

How does the orthogonality defect c come into play if one is interested in the shortest vector of a lattice? Again, consider a vector $\boldsymbol{v} = \sum_{j=1}^n \lambda_j \boldsymbol{b}_j$ of the lattice L generated by the basis $\boldsymbol{b}_1, \ldots, \boldsymbol{b}_n$ with orthogonality defect c.

We now argue that if $\boldsymbol{v}$ is a shortest vector, then $|\lambda_j| \leq c$ for all j. This means that, with a reduced basis at hand, one only has to enumerate all $(2c+1)^n$ vectors $(\lambda_1, \ldots, \lambda_n)$ with $|\lambda_j| \leq c$, compute the corresponding vector $\boldsymbol{v} = \sum_{j=1}^{n} \lambda_j \boldsymbol{b}_j$, and choose the shortest among them.

So suppose that one of the λ_j has absolute value strictly larger than c. Since the orthogonality defect is invariant under permutation of the basis vectors, we can assume that $j=n$. Consider the Gram-Schmidt orthogonalization $\boldsymbol{b}_1^*, \ldots, \boldsymbol{b}_n^*$ of $\boldsymbol{b}_1, \ldots, \boldsymbol{b}_n$. Since $\|\boldsymbol{b}_j^*\| \leq \|\boldsymbol{b}_j\|$ and since $\|\boldsymbol{b}_1\| \cdots \|\boldsymbol{b}_n\| \leq c\|\boldsymbol{b}_1^*\| \cdots \|\boldsymbol{b}_n^*\|$ one has $\|\boldsymbol{b}_n\| \leq c\|\boldsymbol{b}_n^*\|$ and thus

$$\begin{aligned} \|\boldsymbol{v}\| &= \left\| \lambda_n \boldsymbol{b}_n + \sum_{j=1}^{n-1} \lambda_j \boldsymbol{b}_j \right\| \\ &= \|\lambda_n \boldsymbol{b}_n^* + \boldsymbol{u}\|, \end{aligned}$$

where $\boldsymbol{u}$ is a vector in the subspace generated by $\boldsymbol{b}_1, \ldots, \boldsymbol{b}_{n-1}$. Since $\boldsymbol{u}$ and $\boldsymbol{b}_n^*$ are orthogonal we obtain

$$\begin{aligned} \|\boldsymbol{v}\| &= |\lambda_n| \, \|\boldsymbol{b}_n^*\| + \|\boldsymbol{u}\| \\ &> \|\boldsymbol{b}_n\|, \end{aligned}$$

which shows that $\boldsymbol{v}$ is not a shortest vector. Thus, a shortest vector of L can be computed from a basis with orthogonality defect c in $O(c^{2n+1})$ steps.

In the following sections we present various reduction algorithms, and we begin with Lovász' algorithm that produces a basis with orthogonality defect bounded by $2^{n(n-1)/4}$. Lovász' algorithm runs in polynomial time in varying dimension. This implies that a shortest vector in a lattice can be computed from a Lovász-reduced basis by enumerating $(2 \cdot 2^{n(n-1)/4} + 1)^n = 2^{O(n^3)}$ candidates, and thus in polynomial time if the dimension is fixed.

Before discussing specific basis reduction algorithms, we describe the basic operations that are used to go from one lattice basis to another.

The following operations on a matrix are called *elementary column operations*:

- exchanging two columns,
- multiplying a column by -1,
- adding an integer multiple of one column to another column.

It is well known that a unimodular matrix can be derived from the identity matrix by elementary column operations.

To go from one basis to another is conceptually easy; given a basis $\boldsymbol{B}$ we just multiply $\boldsymbol{B}$ by a unimodular matrix, or equivalently, we perform a series of elementary column operations on $\boldsymbol{B}$, to obtain a new basis. The key question is of course how to do this efficiently such that the new basis is reduced according to the definition of reducedness we are using. In the following subsections we will describe some basis reduction algorithms, and highlight results relevant to integer programming.

3.2 Lovász' basis reduction algorithm

In Lovász' [75] basis reduction algorithm the length of the vectors are measured using the Euclidean length, and the Gram-Schmidt vectors corresponding to the current basis are used as a reference for checking whether the basis vectors are nearly orthogonal. Let $L \subset \mathbb{R}^n$ be a lattice, and let $\boldsymbol{b}_1, \ldots, \boldsymbol{b}_l, l \leq n$, be the current basis vectors for L. The vectors $\boldsymbol{b}_j^*, 1 \leq j \leq l$, and the numbers $\mu_{jk}, 1 \leq k < j \leq l$ result from the Gram-Schmidt process as described in Section 2.2. A basis $\boldsymbol{b}_1, \boldsymbol{b}_2, \ldots, \boldsymbol{b}_l$ is called *reduced* in the sense of Lovász if

$$|\mu_{jk}| \leq \frac{1}{2} \quad \text{for } 1 \leq k < j \leq l, \tag{8}$$

$$\|\boldsymbol{b}_j^* + \mu_{j,j-1}\boldsymbol{b}_{j-1}^*\|^2 \geq \frac{3}{4}\|\boldsymbol{b}_{j-1}^*\|^2 \quad \text{for } 1 < j \leq l. \tag{9}$$

The constant $\frac{3}{4}$ in inequality (9) is arbitrarily chosen and can be replaced by any fixed real number $\frac{1}{4} < y < 1$. In a practical implementation, one chooses a constant close to one. Below we explain why vectors satisfying Conditions (8) and (9) are relatively short and nearly orthogonal.

Condition (8) is satisfied in two cases. The first case, see Figure 1(a), is if $\boldsymbol{b}_j$ is almost orthogonal to $\boldsymbol{b}_k^*$. Then, clearly, if we project $\boldsymbol{b}_j$ on $\boldsymbol{b}_k^*$, the absolute value of the length of this projection is going to be short relative to the length of $\boldsymbol{b}_k^*$. The second possibility for (8) to be satisfied, see Figure 1(b), is if $\boldsymbol{b}_j$ is short relative to $\boldsymbol{b}_k^*$. Even if $\boldsymbol{b}_j$ and $\boldsymbol{b}_k^*$ are not close to being orthogonal, the length of the projection of $\boldsymbol{b}_j$ on $\boldsymbol{b}_k^*$ will still be small relative to the length of $\boldsymbol{b}_k^*$. If we would accept this case we would also accept a basis in which $\|\boldsymbol{b}_1\| \gg \|\boldsymbol{b}_2\| \gg \cdots \gg \|\boldsymbol{b}_l\|$, and where the vectors are far from being orthogonal.

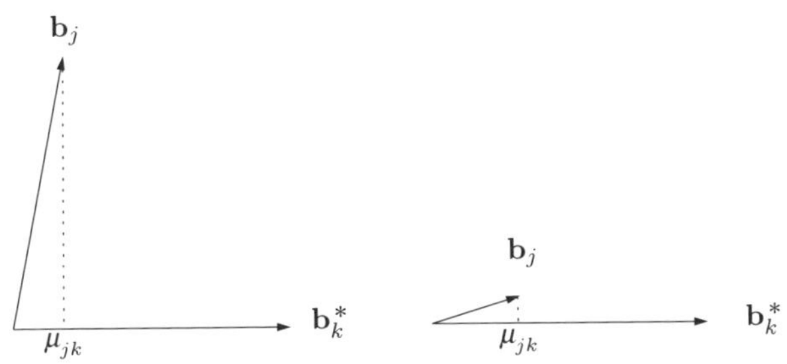

(a) $\mu_{jk} < 1/2$, since $\mathbf{b}_j$ is almost orthogonal to $\mathbf{b}_k^*$.

(b) $\mu_{jk} < 1/2$ since $\mathbf{b}_j$ is short compared to $\mathbf{b}_k^*$. If $k = j - 1$, Condition (9) is violated.

Figure 1. Cases for which Condition (8) are satisfied.

To prevent this, Condition (9) is enforced. Here we relate to the interpretation of the Gram-Schmidt vectors above, and notice that the vectors $\boldsymbol{b}_j^* + \mu_{j,j-1}\boldsymbol{b}_{j-1}^*$ and $\boldsymbol{b}_{j-1}^*$ are the projections of $\boldsymbol{b}_j$ and $\boldsymbol{b}_{j-1}$ on the orthogonal complement of $\sum_{k=1}^{j-2}\mathbb{R}\,\boldsymbol{b}_k$. Consider the case where $k=j-1$, i.e., suppose that $\boldsymbol{b}_j$ is short compared to $\boldsymbol{b}_{j-1}^*$, which implies that $\boldsymbol{b}_j^*$ is short compared to $\boldsymbol{b}_{j-1}^*$ as $\|\boldsymbol{b}_j^*\| \leq \|\boldsymbol{b}_j\|$. Suppose we *interchange* $\boldsymbol{b}_j$ and $\boldsymbol{b}_{j-1}$. Then the new $\boldsymbol{b}_{j-1}^*$ will be the vector $\boldsymbol{b}_j^* + \mu_{j,j-1}\boldsymbol{b}_{j-1}^*$, which will be short compared to the old $\boldsymbol{b}_{j-1}^*$, i.e., Condition (9) will be violated.

Given a basis $\boldsymbol{b}_1, \ldots, \boldsymbol{b}_n$ one can apply a sequence of elementary column operations to obtain a basis satisfying (8) in the following way. Recall (see (4)) that the Gram-Schmidt process yields a factorization of the matrix $(\boldsymbol{b}_1, \ldots, \boldsymbol{b}_n)$ as $(\boldsymbol{b}_1, \ldots, \boldsymbol{b}_n) = (\boldsymbol{b}_1^*, \ldots, \boldsymbol{b}_n^*) \cdot \boldsymbol{R}$ (see(4)), where $\boldsymbol{R}$ is upper triangular, with all diagonal entries being equal to one. By subtracting integer multiples of column $\boldsymbol{r}_i$ from the columns $\boldsymbol{r}_{i+1}, \ldots, \boldsymbol{r}_n$, one can achieve that the elements $R(i, j)$ for $i<j$ are at most 1/2 in absolute value. By doing so for $i=n-1, \ldots, 1$, in that order, one obtains a matrix $\boldsymbol{R}'$, which is upper triangular, with all diagonal elements equal to one, and all the elements above the diagonal being at most 1/2 in absolute value. This yields a new basis $(\boldsymbol{b}_1', \ldots, \boldsymbol{b}_n') = (\boldsymbol{b}_1^*, \ldots, \boldsymbol{b}_n^*) \cdot \boldsymbol{R}'$, which satisfies (8). The replacement of the basis $(\boldsymbol{b}_1, \ldots, \boldsymbol{b}_n)$ by $(\boldsymbol{b}_1', \ldots, \boldsymbol{b}_n')$ is called *size reduction*. Notice that the Gram-Schmidt orthogonalization of $(\boldsymbol{b}_1', \ldots, \boldsymbol{b}_n')$ is given by $(\boldsymbol{b}_1^*, \ldots, \boldsymbol{b}_n^*) \cdot \boldsymbol{R}'$.

If Condition (9) is violated for a certain index j, then the vectors $\boldsymbol{b}_j$ and $\boldsymbol{b}_{j-1}$ are *interchanged* to prevent us from accepting a basis with long nonorthogonal vectors as described in the previous paragraph. Lovász' basis reduction algorithm now performs size reductions and interchangings until the basis satisfies (8) and (9).

Algorithm 1 (Lovász' algorithm).

1. While Conditions (8) and (9) are not satisfied
   (a) Perform size reduction on the basis
   (b) If j is an index which violates (9), then interchange basis elements $j-1$ and j.

The key to the termination argument of Lovász' algorithm is the following potential function $\Phi(\boldsymbol{b}_1, \ldots, \boldsymbol{b}_n)$ of a lattice basis $\boldsymbol{B} = (\boldsymbol{b}_1, \ldots, \boldsymbol{b}_n)$, $\boldsymbol{b}_j \in \mathbb{Z}^n$, $1 \leq j \leq n$,

$$\Phi(B) = \|\boldsymbol{b}_1^*\|^{2n} \|\boldsymbol{b}_2^*\|^{2(n-1)} \cdots \|\boldsymbol{b}_1^*\|^{2}.$$

The potential of an integer lattice basis is always an integer. Furthermore, an *interchange* step in Lovász' algorithm decreases the potential by a factor of 3/4 or a smaller number. Thus, if $\boldsymbol{B}_1$ and $\boldsymbol{B}_2$ are two subsequent bases after an interchange step in Lovász' algorithm, then

$$\Phi(\boldsymbol{B}_2) \leq \frac{3}{4}\Phi(\boldsymbol{B}_1).$$

The potential of the input basis $\boldsymbol{B}$ can be bounded by $\Phi(\boldsymbol{B}) \leq (\|\boldsymbol{b}_1\| \cdots \|\boldsymbol{b}_n\|)^{2n}$. Therefore, the number of iterations of Lovász' algorithm is bounded by $O(n(\log\|\boldsymbol{b}_1\| + \cdots + \|\boldsymbol{b}_n\|))$. In order to conclude that Lovász' algorithm runs in polynomial time, one has further to show that the binary encoding lengths of the rational numbers representing the basis and the Gram-Schmidt orthogonalization remain polynomial in the input. For this, we refer to [75], where the following running time bound is given.

Theorem 2 ([75]). *Let $L \subseteq \mathbb{Z}^n$ be a lattice with basis $\boldsymbol{b}_1, \ldots, \boldsymbol{b}_n$, and let $\beta \in \mathbb{R}$, $\beta \geq 2$, be such that $\|\boldsymbol{b}_j\|^2 \leq \beta$ for $1 \leq j \leq n$. Then the number of arithmetic operations needed by the basis reduction algorithm as described in [75] is $O(n^4 \log \beta)$, and the integers on which these operations are performed each have binary length $O(n \log \beta)$.*

In terms of bit operations, Theorem 2 implies that Lovász' basis reduction algorithm has a running time of $O(n^6(\log \beta)^3)$ using classical algorithms for addition and multiplication.

Example 1. Here we give an example of an initial and a reduced basis for a given lattice. Let L be the lattice generated by the vectors

$$\boldsymbol{b}_1 = \begin{pmatrix} 4 \\ 1 \end{pmatrix} \quad \boldsymbol{b}_2 = \begin{pmatrix} 1 \\ 1 \end{pmatrix}.$$

The Gram-Schmidt vectors are $\boldsymbol{b}_1^* = \boldsymbol{b}_1$ and $\boldsymbol{b}_2^* = \boldsymbol{b}_2 - \mu_{21}\boldsymbol{b}_1^* = (1, 1)^T - \frac{5}{17}\boldsymbol{b}_1^* = \frac{1}{17}(-3, 12)^T$, see Figure 2a. Condition (8) is satisfied since $\boldsymbol{b}_2$ is short relative to $\boldsymbol{b}_1^*$. However, Condition (9) is violated, so we exchange $\boldsymbol{b}_1$ and $\boldsymbol{b}_2$, giving

$$\boldsymbol{b}_1 = \begin{pmatrix} 1 \\ 1 \end{pmatrix} \quad \boldsymbol{b}_2 = \begin{pmatrix} 4 \\ 1 \end{pmatrix}.$$

We now have $\boldsymbol{b}_1^* = \boldsymbol{b}_1$, $\mu_{21} = \frac{5}{2}$ and $\boldsymbol{b}_2^* = \frac{1}{2}(3, -3)^T$, see Figure 2b.

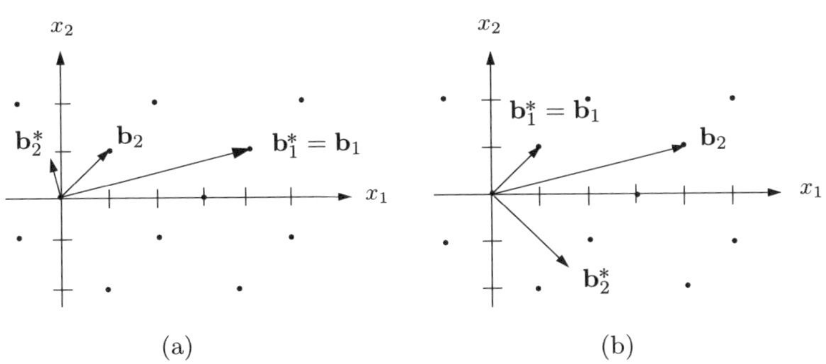

Figure 2.

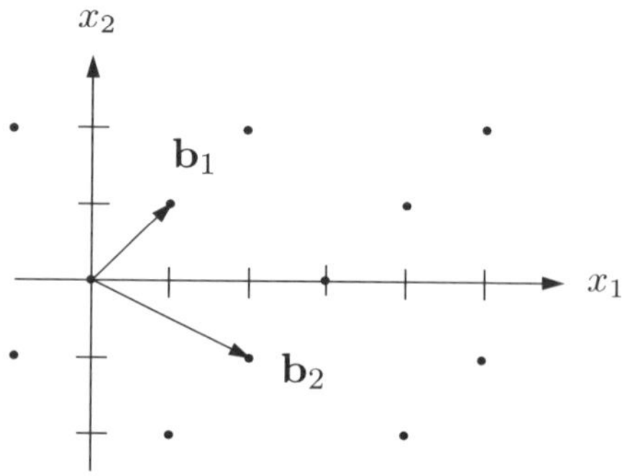

Figure 3. The reduced basis.

Condition (8) is now violated, so we replace $\boldsymbol{b}_2$ by $\boldsymbol{b}_2 - 2\boldsymbol{b}_1 = (2, -1)^T$. Conditions (8) and (9) are satisfied for the resulting basis

$$\boldsymbol{b}_1 = \begin{pmatrix} 1 \\ 1 \end{pmatrix} \quad \boldsymbol{b}_2 = \begin{pmatrix} 2 \\ -1 \end{pmatrix},$$

and hence this basis is reduced, see Figure 3. □

Next we will present some useful bounds on reduced basis vectors.

Proposition 1 ([75]). *Let $\boldsymbol{b}_1, \ldots, \boldsymbol{b}_n$ be a reduced basis for the lattice $L \subset \mathbb{R}^n$. Then,*

$$d(L) \leq \Pi_{j=1}^{n} ||\boldsymbol{b}_j|| \leq c_1 \cdot d(L), \tag{10}$$

where $c_1 = 2^{n(n-1)/4}$.

The first inequality in (10) is Hadamard's inequality (7) that holds for any basis of L. Recall that we refer to the ratio $\Pi_{j=1}^{n} \|\boldsymbol{b}_j\|/d(L)$ as the orthogonality defect. Hermite [58] proved that each lattice $L \subset \mathbb{R}^n$ has a basis $\boldsymbol{b}_1, \ldots, \boldsymbol{b}_n$ such that $\Pi_{j=1}^{n} \|\boldsymbol{b}_j\|/d(L) \leq c(n)$, where $c(n)$ is a constant depending only on n. The upper bound in (10) implies that the orthogonality defect of a Lovász-reduced basis is bounded from above by c_1. Better constants than c_1 are possible, but the question is then whether the basis can be obtained in polynomial time.

A consequence of Proposition 1 is that if we consider a basis that satisfies (10), and if $\boldsymbol{b}_n$ is the longest of the basis vectors, then the distance of $\boldsymbol{b}_n$ to the hyperplane generated by the basis vectors $\boldsymbol{b}_1, \ldots, \boldsymbol{b}_{n-1}$ is not too small as stated in the following corollary.

Corollary 1 ([76]). *Assume that $\boldsymbol{b}_1, \ldots, \boldsymbol{b}_n$ is a basis such that (10) holds, and that, after possible reordering, $\|\boldsymbol{b}_n\| = \max_{1 \leq j \leq n}\{\|\boldsymbol{b}_j\|\}$. Let $H = \sum_{j=1}^{n-1} \mathbb{R}\, \boldsymbol{b}_j$ and* *let h be the distance of basis vector* $\boldsymbol{b}_n$ *to H. Then*

$$c_1^{-1} \cdot \|\boldsymbol{b}_n\| \leq h \leq \|\boldsymbol{b}_n\|, \tag{11}$$

where $c_1 = 2^{n(n-1)/4}$.

Proof: Let $L' = \sum_{j=1}^{n-1} \mathbb{Z}\boldsymbol{b}_j$. We have

$$d(L) = h \cdot d(L'). \tag{12}$$

Expressions (10) and (12) give

$$\Pi_{j=1}^{n} \|\boldsymbol{b}_j\| \leq c_1 \cdot d(L) = c_1 \cdot h \cdot d(L') \leq c_1 \cdot h \cdot \Pi_{j=1}^{n-1} \|\boldsymbol{b}_j\|, \tag{13}$$

where the first inequality follows from the second inequality of (10), and where the last inequality follows from the first inequality of (10). From (13) we obtain $h \geq c_1^{-1} \|\boldsymbol{b}_n\|$. From the definition of h we have $h \leq \|\boldsymbol{b}_n\|$, and this bound holds with equality if and only if the vector $\boldsymbol{b}_n$ is orthogonal to H. □

The lower bound on h given in Corollary 1 plays a crucial role in the algorithm of H. W. Lenstra, Jr., which is described in Section 4.1.

Proposition 2 ([75]). *Let* $L \subset \mathbb{R}^n$ *be a lattice with reduced basis* $\boldsymbol{b}_1, \ldots, \boldsymbol{b}_n \in \mathbb{R}^n$. *Let* $\boldsymbol{x}_1, \ldots, \boldsymbol{x}_t \in L$ *be linearly independent. Then we have*

$$\|\boldsymbol{b}_1\|^2 \leq 2^{n-1} \|\boldsymbol{x}\|^2 \quad \textit{for all } \boldsymbol{x} \in L, \boldsymbol{x} \neq \boldsymbol{0}, \tag{14}$$

$$\|\boldsymbol{b}_j\|^2 \leq 2^{n-1} \max\{\|\boldsymbol{x}_1\|^2, \|\boldsymbol{x}_2\|^2, \ldots, \|\boldsymbol{x}_t\|^2\} \quad \textit{for } 1 \leq j \leq t. \tag{15}$$

Inequality (14) implies that the first reduced basis vector $\boldsymbol{b}_1$ is an approximation of the shortest nonzero vector in L.

Just as the first basis vector is an approximation of the shortest vector of the lattice (14), the other basis vectors are approximations of the *successive minima* of the lattice. The j-th successive minimum of $\| \ \|$ on L is the smallest positive value ν_j such that there exists j linearly independent elements of the lattice L in the ball of radius ν_j centered at the origin.

Proposition 3 ([75]). *Let* $\nu_1, \ldots, \nu_l$ *denote the successive minima of* $\| \ \|$ *on L, and let* $\boldsymbol{b}_1, \ldots, \boldsymbol{b}_l$ *be a reduced basis for L. Then*

$$2^{(1-j)/2} \nu_j \leq \|\boldsymbol{b}_j\| \leq 2^{(l-1)/2} \nu_j \quad \textit{for } 1 \leq j \leq l.$$

In recent years several new variants of Lovász' basis reduction algorithm have been developed and a number of variants for implementation have been suggested. We mention a few below, and recommend the paper by Schnorr and Euchner [93] for a more detailed overview. Schnorr [91] extended Lovász' algorithm to a family of polynomial time algorithms that, given $\epsilon > 0$, finds a non-zero vector in an n-dimensional lattice that is no longer than $(1+\epsilon)^n$ times the length of the shortest vector in the lattice. The degree of the polynomial that bounds the running time of the family of algorithms increases as ϵ goes to zero. Seysen [101] developed an algorithm in which the intermediate integers that are produced are no larger than the input integers. Seysen's algorithm performs well particularly on lower-dimensional lattices. Schnorr and Euchner [93] discuss the possibility of computing the Gram-Schmidt vectors using floating point arithmetic while keeping the basis vectors in exact arithmetic in order to improve the practical performance of the algorithm. The drawback of this approach is that the basis reduction algorithm might become unstable. They propose a floating point version with good stability, but cannot prove that the algorithm always terminates. Their computational study indicates that their version is stable on instances of dimension up to 125 having input numbers of bit length as large as 300. Our experience is that one can use basis reduction for problems of larger dimensions if the input numbers are smaller, but once the dimension reaches about 300–400, basis reduction will be slow. Another version considered by Schnorr and Euchner is basis reduction *with deep insertions*. Here, they allow for a vector $\boldsymbol{b}_k$ to be swapped with a vector with lower index than $k-1$. Schnorr [91], [92] also developed a variant of Lovász' algorithm in which not only two vectors are interchanged during the reduction process, but where blocks $\boldsymbol{b}_j, \boldsymbol{b}_{j+1}, \ldots, \boldsymbol{b}_{j+\beta-1}$ of β consecutive vectors are transformed so as to minimize the j-th Gram-Schmidt vector $\boldsymbol{b}_j^*$. This so called *block reduction* produces shorter basis vectors but needs more computing time. The shortest vector $\boldsymbol{b}_j^*$ in a block of size β is determined by complete enumeration of all short lattice vectors. Schnorr and Hörner [94] develop and analyze a rule for pruning this enumeration process.

For the reader interested in using a version of Lovász' basis reduction algorithm there are some useful libraries available on the Internet. Two of them are LiDIA - a C++ Library for Computational Number Theory [77] and NTL - a Library for doing Number Theory, developed by V. Shoup [102].

3.3 Korkine-Zolotareff reduction and fast algorithms for the shortest vector problem

As we have mentioned in Section 3.1, one can compute a shortest vector of a lattice that is represented by a Lovász-reduced basis $\boldsymbol{b}_1, \ldots, \boldsymbol{b}_n$ in $2^{O(n^3)}$ steps via enumerating the candidates $\sum_{j=1}^n \lambda_j \boldsymbol{b}_j$, where $|\lambda_j| \leq 2^{n(n-1)/4}$ and choosing the shortest nonzero vector from this set.

Kannan [64, 66] provided an algorithm for the shortest vector problem, whose dependence on the dimension is $2^{O(n \log n)}$. Helfrich [57] improved Kannan's algorithm. Recently, Ajtai, Kumar and Sivakumar [8] presented a randomized algorithm for the shortest vector problem, with an expected dependence of $2^{O(n)}$. In the following, we briefly review the main idea of Kannan's algorithm and the improvement by Helfrich, see also [65]. Recall the Gram-Schmidt orthogonalization $\boldsymbol{b}_1^*, \ldots, \boldsymbol{b}_n^*$ of a lattice basis $\boldsymbol{b}_1, \ldots, \boldsymbol{b}_n$ from Section 2.2.

A lattice basis $\boldsymbol{b}_1, \ldots, \boldsymbol{b}_n$ is *Korkine-Zolotareff reduced*, or *K-Z reduced* for short, if the following conditions hold.

1. The vector $\boldsymbol{b}_1$ is a shortest vector of the lattice generated by $\boldsymbol{b}_1, \ldots, \boldsymbol{b}_n$.
2. The numbers μ_{jk} in the Gram-Schmidt orthogonalization of $\boldsymbol{b}_1, \ldots, \boldsymbol{b}_n$ satisfy $|\mu_{jk}| \leq 1/2$, cf. Section 3.2, Expression (8).
3. If $\boldsymbol{b}_2', \ldots, \boldsymbol{b}_n'$ denotes the projection of $\boldsymbol{b}_2, \ldots, \boldsymbol{b}_n$ onto the orthogonal complement of the space generated by $\boldsymbol{b}_1$, then $\boldsymbol{b}_2', \ldots, \boldsymbol{b}_n'$ is Korkine-Zolotareff reduced.

A two-dimensional lattice basis that is K-Z reduced is also called *Gauß* reduced, see [49]. The algorithm of Kannan computes a Korkine-Zolotareff reduced basis in dimension n by first computing a partially Korkine-Zolotareff reduced lattice basis, from which a shortest vector is among $2^{O(n \log n)}$ candidates. The basis is partially Korkine-Zolotareff reduced with the help of an algorithm for Korkine-Zolotareff reduction in dimension $n-1$.

With a shortest vector at hand, one can then compute a fully K-Z reduced basis by K-Z reducing the projection along the orthogonal complement of this shortest vector. A lattice basis $\boldsymbol{b}_1, \ldots, \boldsymbol{b}_n$ is *partially Korkine-Zolotareff reduced* or *partially K-Z reduced* for short, if it satisfies the following properties.

1. If $\boldsymbol{b}_2', \ldots, \boldsymbol{b}_n'$ denotes the projection of $\boldsymbol{b}_2, \ldots, \boldsymbol{b}_n$ onto the orthogonal complement of the space generated by $\boldsymbol{b}_1$, then $\boldsymbol{b}_2', \ldots, \boldsymbol{b}_n'$ is Korkine-Zolotareff reduced.
2. The numbers μ_{jk} in the Gram-Schmidt orthogonalization of $\boldsymbol{b}_1, \ldots, \boldsymbol{b}_n$ satisfy $|\mu_{jk}| \leq 1/2$.
3. $\|\boldsymbol{b}_2'\| \geq 1/2 \, \|\boldsymbol{b}_1\|$.

Notice that, once Conditions 1 and 3 hold, Condition 2 can be satisfied, as explained in Section 3.2, via a size reduction step. Size reduction does not destroy Conditions 1 and 3. Condition 1 can be satisfied by applying Kannan's algorithm for full K-Z reduction to $\boldsymbol{b}_2', \ldots, \boldsymbol{b}_n'$, and applying the transformation to the original vectors $\boldsymbol{b}_2, \ldots, \boldsymbol{b}_n$. If then Condition 3 is not satisfied, then Helfrich [57] has proposed to replace $\boldsymbol{b}_1$ and $\boldsymbol{b}_2$ with the *Gauß-reduction* of this pair, or equivalently its K-Z reduction. Clearly, if $\boldsymbol{b}_1$, $\boldsymbol{b}_2$ is Gauß-reduced, which means that $\|\boldsymbol{b}_1\| \leq \|\boldsymbol{b}_2\|$ and the angle enclosed by $\boldsymbol{b}_1$ and $\boldsymbol{b}_2$ is at least 60° and at most 120°, then Condition 3 holds.

The following algorithm computes a partially K-Z reduced basis from a given input basis $\boldsymbol{b}_1, \ldots, \boldsymbol{b}_n$. It uses as a subroutine an algorithm to K-Z reduce the lattice basis $\boldsymbol{b}_2', \ldots, \boldsymbol{b}_n'$.

Algorithm 2 (Partial K-Z reduction).

1. Apply Lovász' basis reduction algorithm to $\boldsymbol{b}_1, \dots, \boldsymbol{b}_n$.
2. K-Z reduce $\boldsymbol{b}_2', \dots, \boldsymbol{b}_n'$ and apply the corresponding transformation to $\boldsymbol{b}_2, \dots, \boldsymbol{b}_n$.
3. Perform size reduction on $\boldsymbol{b}_1, \dots, \boldsymbol{b}_n$.
4. If $\|\boldsymbol{b}_2'\| < 1/2\,\|\boldsymbol{b}_1\|$, then replace $\boldsymbol{b}_1$, $\boldsymbol{b}_2$ by its Gauß reduction and go to Step 2.

We show in a moment that we can extract a shortest vector from a partially K-Z reduced basis in $2^{O(n \log n)}$ steps, but before, we analyze the running time of the algorithm.

Theorem 3 ([57]). *Step 4 of Algorithm 2 is executed at most* $\log n + 6$ *times.*

Proof. Let $\boldsymbol{v}$ be a shortest vector and let $\boldsymbol{b}_1, \dots, \boldsymbol{b}_n$ be the lattice basis immediately before Step 4 of Algorithm 2 and let $\boldsymbol{b}_2', \dots, \boldsymbol{b}_n'$ denote the projection of $\boldsymbol{b}_2, \dots, \boldsymbol{b}_n$ onto the orthogonal complement of $\boldsymbol{b}_1$.

If Step 4 is executed, then $\boldsymbol{v}$ is not equal to $\boldsymbol{b}_1$. Then clearly, the projection of $\boldsymbol{v}$ onto the orthogonal complement of $\boldsymbol{b}_1$ is nonzero. Since $\boldsymbol{b}_2', \dots, \boldsymbol{b}_n'$ is K-Z reduced it follows that $\|\boldsymbol{v}\| \geq \|\boldsymbol{b}_2'\|$ holds. Denote the Gauß reduction of $\boldsymbol{b}_1, \boldsymbol{b}_2$ by $\tilde{\boldsymbol{b}}_1, \tilde{\boldsymbol{b}}_2$. The determinant of $L(\boldsymbol{b}_1, \boldsymbol{b}_2)$ is equal to $\|\boldsymbol{b}_1\|\,\|\boldsymbol{b}_2'\|$. After the Gauß reduction in Step 4, we have therefore

$$\|\tilde{\boldsymbol{b}}_1\| \leq 2\sqrt{\|\boldsymbol{b}_1\|\,\|\boldsymbol{b}_2'\|} \tag{16}$$

$$\leq 2\sqrt{\|\boldsymbol{b}_1\|\,\|\boldsymbol{v}\|}. \tag{17}$$

Dividing this inequality by $\|\boldsymbol{v}\|$ gives

$$\frac{\|\tilde{\boldsymbol{b}}_1\|}{\|\boldsymbol{v}\|} \leq 2\sqrt{\frac{\|\boldsymbol{b}_1\|}{\|\boldsymbol{v}\|}}.$$

Thus, if $\boldsymbol{b}_1^{(i)}$ denotes the first basis vector after the i-th execution of Step 4, one has

$$\frac{\|\boldsymbol{b}_1^{(i)}\|}{\|\boldsymbol{v}\|} \leq 4\left(\frac{\|\boldsymbol{b}_1^{(0)}\|}{\|\boldsymbol{v}\|}\right)^{(1/2)^i}. \tag{18}$$

Since we start with a Lovász reduced basis, we know that $\|\boldsymbol{b}_1^{(0)}\|/\|\boldsymbol{v}\| \leq 2^{(n-1)/2}$ holds, and consequently that $\|\boldsymbol{b}_1^{(\log n)}\|/\|\boldsymbol{v}\| \leq 8$. Each further Gauß reduction decreases the length of the first basis vector by at least 3/4. Therefore the number of runs through Step 4 is bounded by $\log n + 6$. □

We now argue that with such a partially K-Z reduced basis $\boldsymbol{b}_1, \ldots, \boldsymbol{b}_n$ at hand, one only needs to check $\mathrm{O}(n)^n$ candidates for the shortest vector. Let $\boldsymbol{v} = \sum_{j=1}^{n} \lambda_j \boldsymbol{b}_j$ be a shortest vector. After rewriting each $\boldsymbol{b}_j$ in terms of the Gram-Schmidt orthogonalization one obtains

$$\begin{aligned}\boldsymbol{v} &= \sum_{j=1}^{n} \sum_{k=1}^{j} (\lambda_j \mu_{jk} \boldsymbol{b}_k^*) \\ &= \sum_{k=1}^{n} \left(\sum_{j=k}^{n} \lambda_j \mu_{jk} \right) \boldsymbol{b}_k^*.\end{aligned}$$

The length of $\boldsymbol{v}$ satisfies

$$\|\boldsymbol{v}\| = \sum_{k=1}^{n} \left| \sum_{j=k}^{n} (\lambda_j \mu_{jk}) \right| \|\boldsymbol{b}_k^*\|. \tag{19}$$

Consider the coefficient $c_n = |\lambda_n \mu_{nn}| = |\lambda_n|$ of $\|\boldsymbol{b}_n^*\|$ in (19). We can bound this absolute value by $|\lambda_n| \leq \|\boldsymbol{v}\|/\|\boldsymbol{b}_n^*\| \leq \|\boldsymbol{b}_1\|/\|\boldsymbol{b}_n^*\|$. This leaves us $1 + 2\|\boldsymbol{b}_1\|/\|\boldsymbol{b}_n^*\|$ possibilities for λ_n. Suppose now that we picked $\lambda_n, \ldots, \lambda_{j+1}$ and inspect the coefficient c_j of $\|\boldsymbol{b}_j^*\|$ in (19), which is

$$\begin{aligned}c_j &= \left| \sum_{k=j}^{n} (\lambda_k \mu_{kj}) \right| \\ &= \left| \lambda_j + \sum_{k=j+1}^{n} (\lambda_k \mu_{kj}) \right|.\end{aligned}$$

Since the inequality $c_j \leq \|\boldsymbol{b}_1\|/\|\boldsymbol{b}_j^*\|$ must hold, this leaves only $1 + 2\,\|\boldsymbol{b}_1\|/\|\boldsymbol{b}_j^*\|$ possibilities to pick λ_j. Thus by choosing the coefficients $\lambda_n, \ldots, \lambda_1$ in this order, one has at most $\prod_{j=1}^{n} (1 + 2\,\|\boldsymbol{b}_1\|/\|\boldsymbol{b}_j^*\|)$ candidates.

Suppose $\|\boldsymbol{b}_j^*\| > \|\boldsymbol{b}_1\|$ for some j. Then $\boldsymbol{b}_j$ can never have a nonzero coefficient λ_j in a shortest vector representation $\boldsymbol{v} = \sum_{j=1}^{n} \lambda_j \boldsymbol{b}_j$. Because in that case, $\boldsymbol{v}$ has a nonzero component in its projection to the orthogonal complement of $\boldsymbol{b}_1\mathbb{R} + \cdots + \boldsymbol{b}_{i-1}\mathbb{R}$ and since $\boldsymbol{b}_2', \ldots, \boldsymbol{b}_n'$ is K-Z reduced, this implies that $\|\boldsymbol{v}\| \geq \|\boldsymbol{b}_j^*\| > \|\boldsymbol{b}_1\|$, which is impossible. Thus we can assume that $\|\boldsymbol{b}_j^*\| \leq \|\boldsymbol{b}_1\|$ holds for all $j = 1, \ldots, n$. Otherwise, $\boldsymbol{b}_j$ can be discarded. Therefore the number of candidates N for the tuples $(\lambda_1, \ldots, \lambda_n)$ satisfies

$$\begin{aligned}N &\leq \prod_{j=1}^{n} (1 + 2\,\|\boldsymbol{b}_1\|/\|\boldsymbol{b}_j^*\|) \\ &\leq \prod_{j=1}^{n} (3\,\|\boldsymbol{b}_1\|/\|\boldsymbol{b}_j^*\|) \\ &= 3^n\,\|\boldsymbol{b}_1\|^n/d(L).\end{aligned}$$

Next we give an upper bound for $\|\boldsymbol{b}_1\|$. If $\boldsymbol{b}_1$ is a shortest vector, then Minkowski's theorem, (Theorem 1 in Section 2.2) guarantees that $\|\boldsymbol{b}_1\| \leq \sqrt{n}\, d(L)^{1/n}$ holds. If $\boldsymbol{b}_1$ is not a shortest vector, then the shortest vector $\boldsymbol{v}$ has a nonzero projection onto the orthogonal complement of $\boldsymbol{b}_1 \mathbb{R}$. Since $\boldsymbol{b}'_2, \ldots, \boldsymbol{b}'_n$ is K-Z reduced, this implies that $\|\boldsymbol{v}\| \geq \|\boldsymbol{b}'_2\| \geq 1/2\, \|\boldsymbol{b}_1\|$, since the basis is partially K-Z reduced. In any case we have $\|\boldsymbol{b}_1\| \leq 2\sqrt{n}\, d(L)^{1/n}$ and thus that $N \leq 6^n n^{n/2}$.

Now it is clear how to compute a K-Z reduced basis and thus a shortest vector. With an algorithm for K-Z reduction in dimension $n-1$, one uses Algorithm 2 to partially K-Z reduce the basis and then one checks all possible candidates for a shortest vector. Then one performs K-Z reduction on the basis for the projection onto the orthogonal complement of the shortest vector. Kannan [66] has shown that this procedure for K-Z reduction requires $\mathrm{O}(n)^n \varphi$ operations, where φ is the binary encoding length of the initial basis and where the operands during the execution of the algorithm have at most $\mathrm{O}(n^2\varphi)$ bits.

Theorem 4 ([66]). *Let $\boldsymbol{b}_1, \ldots, \boldsymbol{b}_n$ be a lattice basis of binary encoding length φ. There exists an algorithm which computes a K-Z reduced basis of $L(\boldsymbol{b}_1, \ldots, \boldsymbol{b}_n)$ with $\mathrm{O}(n)^n \varphi$ arithmetic operations on rationals of size $\mathrm{O}(n^2\varphi)$.*

Further notes. Van Emde Boas [45] proved that the shortest vector problem with respect to the l_∞ norm is NP-hard, and he conjectured that it is NP-hard with respect to the Euclidean norm. In the same paper he proved that the closest vector problem is NP-hard for any norm. Recently substantial progress has been made in gaining more information about the complexity status of the two problems. Ajtai [7] proved that the shortest vector problem is NP-hard for randomized problem reductions. This means that the reduction makes use of results of a probabilistic algorithm. These results are true with probability arbitrarily close to one. Ajtai also showed that approximating the length of a shortest vector in a given lattice within a factor $1 + 1/2^{n^c}$ is NP-hard for some constant c. The non-approximability factor was improved to $(1 + 1/n^{\epsilon})$ by Cai and Nerurkar [21]. Micciancio [81] improved this factor substantially by showing that it is NP-hard to approximate the shortest vector in a given lattice within any constant factor less that $\sqrt{2}$ for randomized problem reductions, and that the same result holds for deterministic problem reductions (the "normal" type of reductions used in an NP-hardness proof) under the condition that a certain number theoretic conjecture holds. Micciancio's results hold for any l_p norm. Goldreich and Goldwasser [51] proved that it is not NP-hard to approximate the shortest vector, or the closest vector, within a factor $\sqrt{n}$ unless the polynomial-time hierarchy collapses. Goldreich et al. [52] show that, given oracle access to a subroutine that returns approximate closest vectors in a given lattice, one can find in polynomial time approximate shortest vectors in the same lattice with the same approximation factor. This implies that the shortest vector problem is not harder than the closest vector problem. From the other side, Kannan [65] showed that any algorithm producing an approximate shortest vector with approximation factor $f(n)$, where $f(n)$ is a nondecreasing function, can be used to produce an approximate closest vector to within $n^{3/2}f(n)^2$. For a recent overview on complexity results related to lattice problems, see for instance Cai [20], and Nguyen and Stern [87].

Kannan [66] also developed an exact algorithm for the closest vector problem, see also Helfrich [57] and Blömer [14].

3.4 The generalized basis reduction algorithm

In the generalized basis reduction algorithm a norm related to a full-dimensional compact convex set C is used, instead of the Euclidean norm as in Lovász' algorithm. A compact convex set $C \subseteq \mathbb{R}^n$ that is symmetric about the origin gives rise to a norm $F(\boldsymbol{c}) = \inf\{t \geq 0 \mid \boldsymbol{c}/t \in C\}$. Lovász and Scarf [79] call the function F the *distance function* with respect to C. As in Lovász' basis reduction algorithm, the generalized basis reduction algorithm finds short basis vectors with respect to the chosen norm. Moreover, the first basis vector is an approximation of the shortest nonzero lattice vector.

Given the convex set C we define a dual set $C^* = \{\boldsymbol{y} \mid \boldsymbol{y}^T\boldsymbol{c} \leq 1 \text{ for all } \boldsymbol{c} \in C\}$. We also define a distance function associated with a projection of C. Let $\boldsymbol{b}_1, \ldots, \boldsymbol{b}_n$ be a basis for $\mathbb{Z}^n$, and let C_j be the projection of C onto the orthogonal complement of $\boldsymbol{b}_1, \ldots, \boldsymbol{b}_{j-1}$. We have that $\boldsymbol{c} = \beta_j\boldsymbol{b}_j + \cdots + \beta_n\boldsymbol{b}_n \in C_j$ if and only if there exist $\alpha_1, \ldots, \alpha_{j-1}$ such that $\boldsymbol{c} + \alpha_1\boldsymbol{b}_1 + \cdots + \alpha_{j-1}\boldsymbol{b}_{j-1} \in C$. The distance function associated with C_j is defined as:

$$F_j(\boldsymbol{c}) = \min_{\alpha_1,\ldots,\alpha_{j-1}} F(\boldsymbol{c} + \alpha_1\boldsymbol{b}_1 + \cdots + \alpha_{j-1}\boldsymbol{b}_{j-1}). \tag{20}$$

Using duality, one can show that $F_j(\boldsymbol{c})$ is also the optimal value of the maximization problem:

$$F_j(\boldsymbol{c}) = \max\{\boldsymbol{c}^T\boldsymbol{z} \mid \boldsymbol{z} \in C^*,\ \boldsymbol{b}_1^T\boldsymbol{z} = 0, \ldots, \boldsymbol{b}_{j-1}^T\boldsymbol{z} = 0\}. \tag{21}$$

In Expression (21), note that only vectors $\boldsymbol{z}$ that are orthogonal to the basis vectors $\boldsymbol{b}_1, \ldots, \boldsymbol{b}_{j-1}$ are considered. This is similar to the role played by the Gram-Schmidt basis in Lovász' basis reduction algorithm. Also, notice that if C is a polytope, then (21) is a linear program. The distance function F has the following properties:

- F can be computed in polynomial time,
- F is convex,
- $F(-\boldsymbol{x}) = F(\boldsymbol{x})$,
- $F(t\boldsymbol{x}) = tF(\boldsymbol{x})$ for $t > 0$.

Lovász and Scarf use the following definition of a reduced basis. A basis $\boldsymbol{b}_1, \ldots, \boldsymbol{b}_n$ is called *reduced* in the sense of Lovász and Scarf if

$$F_j(\boldsymbol{b}_{j+1} + \mu \boldsymbol{b}_j) \geq F_j(\boldsymbol{b}_{j+1}) \quad \text{for } 1 \leq j \leq n-1 \text{ and all integers } \mu, \tag{22}$$

$$F_j(\boldsymbol{b}_{j+1}) \geq (1-\epsilon) F_j(\boldsymbol{b}_j) \quad \text{for } 1 \leq j \leq n-1, \tag{23}$$

where ϵ satisfies $0 < \epsilon < \frac{1}{2}$. A basis $\boldsymbol{b}_1, \ldots, \boldsymbol{b}_n$, not necessarily reduced, is called *proper* if

$$F_k(\boldsymbol{b}_j + \mu \boldsymbol{b}_k) \geq F_k(\boldsymbol{b}_j) \quad \text{for } 1 \leq k < j \leq n. \tag{24}$$

The algorithm is called *generalized* basis reduction since it generalizes Lovász' basis reduction algorithm in the following sense. If the convex set C is an ellipsoid, then a proper reduced basis is precisely a Lovász-reduced basis.

An important question is how to check whether Condition (22) is satisfied for all integers μ. Here we make use of the dual relationship between Formulations (20) and (21). We have the following equality: $\min_{\alpha \in \mathbb{R}} F_j(\boldsymbol{b}_{j+1} + \alpha \boldsymbol{b}_j) = F_{j+1}(\boldsymbol{b}_{j+1})$. Let α^* denote the optimal α in the minimization. The function F_j is convex, and hence the integer μ that minimizes $F_j(\boldsymbol{b}_{j+1} + \mu \boldsymbol{b}_j)$ is either $\lfloor \alpha^* \rfloor$ or $\lceil \alpha^* \rceil$. If the convex set C is a rational polytope, then $\alpha^* \in \mathbb{Q}$ is the optimal dual variable corresponding to the constraint $\boldsymbol{b}_j^T \boldsymbol{z} = 0$ in the optimization problem $F_{j+1}(\boldsymbol{b}_{j+1})$, cf. (21), which implies that the integer μ that minimizes $F_j(\boldsymbol{b}_{j+1} + \mu \boldsymbol{b}_j)$ can be determined by solving two additional linear programs, unless α^* is integral.

Condition (24) is analogous to Condition (8) of Lovász' basis reduction algorithm, and is violated if adding an integer multiple of $\boldsymbol{b}_k$ to $\boldsymbol{b}_j$ yields a distance function value $F_k(\boldsymbol{b}_j + \mu \boldsymbol{b}_k)$ that is smaller than $F_k(\boldsymbol{b}_j)$. In the generalized basis reduction algorithm we only check whether the condition is satisfied for $k = j-1$ (cf. Condition (22)), and we use the value of μ that minimizes $F_j(\boldsymbol{b}_{j+1} + \mu \boldsymbol{b}_j)$ as mentioned above. If Condition (22) is violated, we do a *size reduction*, i.e., we replace $\boldsymbol{b}_{j+1}$ by $\boldsymbol{b}_{j+1} + \mu \boldsymbol{b}_j$.

Condition (23) corresponds to Condition (9) in Lovász' algorithm, and ensures that the basis vectors are in the order of increasing distance function value, aside from the factor $(1-\epsilon)$. Recall that we want the first basis vector to be an approximation of the shortest lattice vector. If Condition (23) is violated, we interchange vectors $\boldsymbol{b}_j$ and $\boldsymbol{b}_{j+1}$.

The algorithm works as follows. Let C be a compact convex set, and let $\boldsymbol{b}_1, \ldots, \boldsymbol{b}_n$ be an initial basis for $\mathbb{Z}^n$. Typically $\boldsymbol{b}_j = \boldsymbol{e}_j$, where $\boldsymbol{e}_j$ is the j-th unit vector in $\mathbb{R}^n$. Let j be the first index for which Conditions (22) or (23) are not satisfied. If (22) is violated, we replace $\boldsymbol{b}_{j+1}$ by $\boldsymbol{b}_{j+1} + \mu \boldsymbol{b}_j$ with the appropriate value of μ. If Condition (23) is satisfied after the replacement, we let $j := j+1$. If Condition (23) is violated, we interchange $\boldsymbol{b}_j$ and $\boldsymbol{b}_{j+1}$, and let $j := j-1$ if $j \geq 2$. If $j = 1$, we remain at this level. The operations that the algorithm performs on the basis vectors are elementary column operations as in Lovász' algorithm. The vectors that we obtain as output from the generalized basis reduction algorithm can therefore be written as the product of the initial basis matrix and a unimodular matrix, which implies that the output vectors form a basis for the lattice $\mathbb{Z}^n$. The question is how efficient the algorithm is.

Theorem 5 ([79]). *Let ϵ be chosen as in (23), let $\gamma = 2 + 1/\log(1/(1-\epsilon))$, and let $B(R)$ be a ball with radius R containing C. Moreover, let $U = \max_{1 \le j \le n}\{F_j(\boldsymbol{b}_j)\}$, where $\boldsymbol{b}_1, \ldots, \boldsymbol{b}_n$ is the initial basis, and let $V = 1/(R(nRU)^{n-1})$.*

The generalized basis reduction algorithm runs in polynomial time for fixed n. The maximum number of interchanges performed during the execution of the algorithm is

$$\left(\frac{\gamma^n - 1}{\gamma - 1}\right)\left(\frac{\log(U/V)}{\log(1/(1-\epsilon))}\right).$$

It is important to notice that, so far, the generalized basis reduction algorithm has been proved to run in polynomial time for *fixed n* only, whereas Lovász' basis reduction algorithm runs in polynomial time for arbitrary n (cf. Theorem 2).

We now give a few properties of a Lovász-Scarf reduced basis. If one can obtain a basis $\boldsymbol{b}_1, \ldots, \boldsymbol{b}_n$ such that $F_1(\boldsymbol{b}_1) \le F_2(\boldsymbol{b}_2) \le \cdots \le F_n(\boldsymbol{b}_n)$, then one can prove that $\boldsymbol{b}_1$ is the shortest integer vector with respect to the distance function. The generalized basis reduction algorithm does not produce a basis with the above property, but it gives a basis that satisfies the following weaker condition.

Theorem 6 ([79]). *Let $0 < \epsilon < \frac{1}{2}$, and let $\boldsymbol{b}_1, \ldots, \boldsymbol{b}_n$ be a Lovász-Scarf reduced basis. Then*

$$F_{j+1}(\boldsymbol{b}_{j+1}) \ge \left(\frac{1}{2} - \epsilon\right) F_j(\boldsymbol{b}_j) \quad \textit{for } 1 \le j \le n-1.$$

We can use this theorem to obtain a result analogous to (14) of Proposition 2.

Proposition 4 ([79]). *Let $0 < \epsilon < \frac{1}{2}$, and let $\boldsymbol{b}_1, \ldots, \boldsymbol{b}_n$ be a Lovász-Scarf reduced basis. Then*

$$F(\boldsymbol{b}_1) \le \left(\frac{1}{2} - \epsilon\right)^{1-n} F(\boldsymbol{x}) \quad \textit{for all } \boldsymbol{x} \in \mathbb{Z}^n, \boldsymbol{x} \ne \boldsymbol{0}.$$

We can also relate the distance function $F_j(\boldsymbol{b}_j)$ to the *j-th successive minimum* of F on the lattice $\mathbb{Z}^n$ (cf. Proposition 3). $\nu_1, \ldots, \nu_n$ are the *successive minima* of F on $\mathbb{Z}^n$ if there are vectors $\boldsymbol{x}_1, \ldots, \boldsymbol{x}_n \in \mathbb{Z}^n$ with $\nu_j = F(\boldsymbol{x}_j)$, such that for each

$1 \le j \le n$, $\boldsymbol{x}_j$ is the shortest lattice vector (with respect to F) that is linearly independent of $\boldsymbol{x}_1, \ldots, \boldsymbol{x}_{j-1}$.

Proposition 5 ([79]). *Let $\nu_1, \ldots, \nu_n$ denote the successive minima of F on the lattice $\mathbb{Z}^n$, let $0 < \epsilon < \frac{1}{2}$, and let $\boldsymbol{b}_1, \ldots, \boldsymbol{b}_n$ be a Lovász-Scarf reduced basis. Then*

$$\left(\frac{1}{2} - \epsilon\right)^{j-1} \nu_j \le F_j(\boldsymbol{b}_j) \le \left(\frac{1}{2} - \epsilon\right)^{j-n} \nu_j \quad \text{for } 1 \le j \le n.$$

The first reduced basis vector is an approximation of the shortest lattice vector (Proposition 4). In fact the generalized basis reduction algorithm can be used to find the shortest vector in the lattice in polynomial time for fixed n. This algorithm is used as a subroutine of Lovász and Scarf's algorithm for solving the integer programming problem "Is $X \cap \mathbb{Z}^n \neq \emptyset$?" described in Section 4.3. To find the shortest lattice vector we proceed as follows. If the basis $\boldsymbol{b}_1, \ldots, \boldsymbol{b}_n$ is Lovász-Scarf reduced, we can obtain a bound on the coordinates of lattice vectors $\boldsymbol{c}$ that satisfy $F_1(\boldsymbol{c}) \le F_1(\boldsymbol{b}_1)$. We express the vector $\boldsymbol{c}$ as an integer linear combination of the basis vectors, i.e., $\boldsymbol{c} = \lambda_1 \boldsymbol{b}_1 + \cdots + \lambda_n \boldsymbol{b}_n$, where $\lambda_j \in \mathbb{Z}$. We have

$$F_1(\boldsymbol{b}_1) \ge F_1(\boldsymbol{c}) \ge F_n(\boldsymbol{c}) = F_n(\lambda_n \boldsymbol{b}_n) = |\lambda_n| F_n(\boldsymbol{b}_n), \tag{25}$$

where the second inequality holds since $F_n(\boldsymbol{c})$ is more constrained than $F_1(\boldsymbol{c})$ (cf. (21)), the first equality holds due to the constraints $\boldsymbol{b}_i^T \boldsymbol{z} = 0, 1 \le i \le n-1$, and the second equality holds as $F(t\boldsymbol{x}) = tF(\boldsymbol{x})$ for $t > 0$. We can now use (25) to obtain the following bound on $|\lambda_n|$:

$$|\lambda_n| \le \frac{F_1(\boldsymbol{b}_1)}{F_n(\boldsymbol{b}_n)} \le \frac{1}{(\frac{1}{2} - \epsilon)^{n-1}},$$

where the last inequality is obtained by applying Theorem 6 iteratively. Notice that the bound on λ_n is a constant for fixed n. In a similar fashion we can obtain a bound on λ_j for $n-1 \ge j \ge 1$. Suppose that we have chosen multipliers $\lambda_n, \ldots, \lambda_{j+1}$ and that we want to determine a bound on λ_j. Let γ^* be the value of γ that minimizes $F_j(\lambda_n \boldsymbol{b}_n + \cdots + \lambda_{j+1} \boldsymbol{b}_{j+1} + \gamma \boldsymbol{b}_j)$. If this minimum is greater than $F_1(\boldsymbol{b}_1)$, then there does not exist a vector $\boldsymbol{c}$, with $\lambda_n, \ldots, \lambda_{j+1}$ fixed such that $F_1(\boldsymbol{c}) \le F_1(\boldsymbol{b}_1)$, since in that case $F_1(\boldsymbol{b}_1) < F_j(\lambda_n \boldsymbol{b}_n + \cdots + \lambda_{j+1} \boldsymbol{b}_{j+1} + \gamma^* \boldsymbol{b}_j) \le F_j(\lambda_n \boldsymbol{b}_n + \cdots + \lambda_j \boldsymbol{b}_j) = F_j(\boldsymbol{c}) \le F_1(\boldsymbol{c})$, which yields a contradiction. If the minimum is less than or equal to $F_1(\boldsymbol{b}_1)$, then we can obtain the bound:

$$|\lambda_j - \gamma^*| \le 2 \frac{F_1(\boldsymbol{b}_1)}{F_j(\boldsymbol{b}_j)} \le \frac{2}{(\frac{1}{2} - \epsilon)^{j-1}}.$$

Hence, we obtain a search tree that has at most n levels, and, given the bounds on the multipliers λ_j, each level consists of a constant number of nodes if n is fixed.

The generalized basis reduction algorithm was implemented by Cook, Rutherford, Scarf, and Shallcross [29] and by Wang [104]. Cook et al. used generalized basis reduction to derive a heuristic version of the integer programming algorithm by Lovász and Scarf (see Section 4.3) to solve difficult integer network design instances. Wang [104] solved both linear and nonlinear integer programming problems using the generalized basis reduction algorithm as a subroutine. For a small example on how to use the generalized basis reduction algorithm, we refer to Section 4.3, Example 2.

3.5 Fast algorithms in the bit model when the dimension is fixed

The running times of the algorithms for lattice basis reduction depend on the number of bits that are necessary to represent the numbers of the input basis. The complexity model that reflects the fact that arithmetic operations on large numbers do not come for free is the *bit-complexity* model. Addition and subtraction of φ-bit integers takes $O(\varphi)$ time. The current state of the art method for multiplication [97] shows that the bit complexity $M(\varphi)$ of multiplication and division is $O(\varphi \log \varphi \log\log \varphi)$, see [6, p. 279].

The use of this complexity model is best illustrated with algorithms to compute the *greatest common divisor* of two integers. The *Euclidean algorithm* for computing the greatest common divisor $\gcd(a_0, a_1)$ of two integers a_0, $a_1 > 0$ computes the remainder sequence $a_0, a_1, \ldots, a_{k-1}, a_k \in \mathbb{N}_{>0}$, where a_i, $i \geq 2$ is given by $a_{i-2} = a_{i-1}q_{i-1} + a_i$, with $q_i \in \mathbb{N}$, $0 < a_i < a_{i-1}$, and where a_k divides a_{k-1} exactly. If $a_0 = F_n$ and $a_1 = F_{n-1}$, were F_i denotes the i-th Fibonacci number, then the remainder sequence, generated by the Euclidean algorithm, is the sequence of Fibonacci numbers $F_n, F_{n-1}, \ldots, F_0$. Since the size of the n-th Fibonacci number is $\Theta(n)$, it follows that the Euclidean algorithm requires $\Omega(\varphi^2)$ bit-operations on an input of size φ. It can be shown, that the Euclidean algorithm runs in time $\Theta(\varphi^2)$ even if one uses the naive algorithms for basic arithmetic operations, see [71]. However, a gcd can be computed in $O(M(\varphi) \log \varphi)$ bit operations with the algorithm of Schönhage [95].

The greatest common divisor of two integers a and b is the absolute value of the shortest vector of the 1-dimensional lattice $a\mathbb{Z} + b\mathbb{Z}$. Thus shortest vector computation and lattice basis reduction form a natural generalization of greatest common divisor computation. In this section, we treat the dimension n as a constant and consider the bit-complexity of the shortest vector problem and lattice basis reduction in fixed dimension.

Schönhage [96]) and Yap [105] proved that a 2-dimensional lattice basis can be K-Z reduced (or Gauß reduced) with $O(M(\varphi) \log \varphi)$ bit-operations. In fact, 2-dimensional K-Z reduction can be solely based on Schönhage's [95] classical algorithm on the fast computation of continued fractions and the original reduction algorithm of Gauß [49], see [39].

Theorem 7 ([96, 105]). *Let $B \in \mathbb{Z}^{2\times 2}$ be a two dimensional lattice basis with* size$(B) = \varphi$. *Then B can be K-Z reduced with $O(M(\varphi)\log\varphi)$ bit-operations.*

Eisenbrand and Rote [43] showed that a lattice basis $B = (b_1, \ldots, b_n) \in \mathbb{Z}^{n\times n}$ of binary encoding length φ can be reduced in $O(M(\varphi)\log^{n-1}\varphi)$ bit-operations when n is fixed. In this section we describe how this result can be obtained with the algorithm for partial K-Z reduction, presented in Section 3.3. For the three-dimensional case, van Sprang [103] and Semaev [100] provided an algorithm which requires $O(\varphi^2)$ bit-operations, using the naive quadratic algorithms for multiplication and division.

Theorem 8. *Let $B \in \mathbb{Z}^{n\times n}$ be a lattice basis with* size$(B) = \varphi$. *Then B can be K-Z reduced with $O(M(\varphi)(\log\varphi)^{n-1})$ bit operations when n is fixed.*

To prove this theorem, recall Algorithm 2 for a partial K-Z reduction. We modify this algorithm as follows.

- Instead of computing a Lovász reduced basis in Step 1, compute the Hermite normal form of B
- The stopping condition in Step 4 is modified, such that we go to Step 2 as long as $\|b_1\| > 8\sqrt{n}\, d(L)^{1/n}$.

We assume that a $(n-1)$-dimensional rational lattice basis $B' \in \mathbb{Z}^{(n-1)\times(n-1)}$ of size φ can be K-Z reduced with $O(M(\varphi)(\log\varphi)^{n-2})$ bit operations.

We now analyze this modified algorithm. Recall that the HNF can be computed with a constant number of extended-gcd computations and a constant number of arithmetic operations, thus with $O(M(\varphi)\log\varphi)$ bit-operations. If $b_1, \ldots, b_n$ is in Hermite normal form, then b_1 is a vector which has zeroes in its $n-1$ first components, and a factor of the determinant in its last component. Thus, by swapping b_1 and b_n one has a basis, whose first vector b_1 satisfies $\|b_1\| \leq d(L)$.

Minkowski's theorem (Theorem 1 in Section 2.2) implies that the length of the shortest vector v of L is bounded by $\|v\| \leq \sqrt{n}\, d(L)^{1/n}$. Thus in the proof of Theorem 3 we can replace inequality (17) by the inequality

$$\|\tilde{b}_1\| \leq 2\sqrt{\|b_1\|\sqrt{n}\, d(L)^{1/n}}.$$

Following the proof, we replace inequality (18) by

$$\frac{\|b_1^{(i)}\|}{\sqrt{n}\, d(L)^{1/n}} \leq 4\left(\frac{\|b_1^{(0)}\|}{\sqrt{n}\, d(L)^{1/n}}\right)^{(1/2)^i}. \tag{26}$$

This means that after $O(\log\log(d(L))$ iterations of the outer loop of the modified Algorithm 2, one has $\|\boldsymbol{b}_1\| \leq 8\sqrt{n}\, d(L)^{1/n}$. It follows that the number of runs through the outer loop is bounded by $O(\log\varphi)$. Thus using the assumption that an $(n-1)$-dimensional lattice basis can be K-Z reduced in $O(M(\varphi)(\log\varphi)^{n-2})$, we see that the modified Algorithm 2 runs with $O(M(\varphi)(\log\varphi)^{n-1})$ bit-operations.

How quickly can the shortest vector be determined from the returned basis? Following the discussion preceding Theorem 4 we obtain the upper bound $N \leq 3^n(8\cdot 8\cdot\sqrt{n}\cdot d(L)^{1/n})^n/d(L) = 24^n n^{n/2}$, which is a constant in fixed dimension. This proves Theorem 8.

It is currently not known whether a shortest vector can be computed in $O(M(\varphi)\log\varphi)$ bit-operations.

4 Algorithms for the integer feasibility problem in fixed dimension

Let A be a rational $m\times n$-matrix and let $\boldsymbol{d}$ be a rational m-vector. Let $X = \{\boldsymbol{x}\in\mathbb{R}^n \mid A\boldsymbol{x} \leq \boldsymbol{d}\}$. We consider the integer feasibility problem in the following form:

$$\text{Does there exist an integer vector } \boldsymbol{x}\in X? \tag{27}$$

Karp [69] showed that the zero-one integer feasibility problem is NP-complete, and Borosh and Treybig [17] proved that the integer feasibility problem (27) belongs to NP. Combining these results implies that (27) is NP-complete. The NP-completeness of the zero-one version is a fairly straightforward consequence of the proof by Cook [26] that the satisfiability problem is NP-complete. An important open question was still: Can the integer feasibility problem be solved in polynomial time in bounded dimension? If the dimension $n=1$, the affirmative answer is trivial. Some special cases of $n=2$ were proven to be polynomially solvable by Hirschberg and Wong [60], and by Kannan [63]. Scarf [90] showed that (27), for the general case $n=2$, is polynomially solvable. Both Hirschberg and Wong, and Scarf conjectured that the integer feasibility problem could be solved in polynomial time if the dimension is fixed. The proof of this conjecture was given by H. W. Lenstra, Jr. [76].

Let K be a full-dimensional closed convex set in $\mathbb{R}^n$ given by integer input. The *width of K along the nonzero integer vector* $\boldsymbol{v}$ is defined as

$$w_{\boldsymbol{v}}(K) = \max\{\boldsymbol{v}^T\boldsymbol{x} : \boldsymbol{x}\in K\} - \min\{\boldsymbol{v}^T\boldsymbol{x} : \boldsymbol{x}\in K\}. \tag{28}$$

The *width* of K, $w(K)$, is the minimum of its widths along nonzero integer vectors $\boldsymbol{v}\in\mathbb{Z}^n\backslash\{\boldsymbol{0}\}$. Notice that this is different from the definition of the geometric width of a polytope (see p 6 in [54]). Khinchine [70] proved that if K does not contain a lattice point, then there exists a nonzero integer vector $\boldsymbol{c}$ such that $w_{\boldsymbol{c}}(K)$ is bounded from above by a constant depending only on the dimension.

Theorem 9 (Khinchine's flatness theorem [70]). *There exists a constant $f(n)$ depending only on the dimension n, such that each convex body $K \subseteq \mathbb{R}^n$ containing no integer points has width at most $f(n)$.*

Currently the best asymptotic bounds on $f(n)$ are given in [9]. Tight bounds seem to be unknown already in dimension 3.

To appreciate Khinchine's results, we first have to interpret what the width of K in direction $\boldsymbol{v}$ means. To do that it is easier to look at the *integer width of K in the nonzero integer direction* $\boldsymbol{v}$, $w_{\boldsymbol{v}}^I(K) = \lfloor \max\{\boldsymbol{v}^T\boldsymbol{x} : \boldsymbol{x} \in K\} \rfloor - \lceil \min\{\boldsymbol{v}^T\boldsymbol{x} : \boldsymbol{x} \in K\} \rceil + 1$. The integer width of K in the direction $\boldsymbol{v}$ is the number of lattice hyperplanes intersecting K in direction $\boldsymbol{v}$. The width $w_{\boldsymbol{v}}(K)$ is an approximation of the integer width, so Khinchine's results says that if K is lattice point free, then there exists an integer vector $\boldsymbol{c}$ such that the number of lattice hyperplanes intersecting K in direction $\boldsymbol{c}$ is small. The direction $\boldsymbol{c}$ is often referred to as a "thin" direction, and we say that K is "thin" or "flat" in direction $\boldsymbol{c}$.

The algorithms we are going to describe in this section do not directly use Khinchine's flatness theorem, but they do use ideas that are related. First, we are going to find a point $\boldsymbol{x}$, not necessarily integer, that lies approximately in the center of the polytope X. Given the point $\boldsymbol{x}$ we can quickly find a lattice point $\boldsymbol{y}$ reasonably close to $\boldsymbol{x}$. Either $\boldsymbol{y}$ is also in X, in which case our feasibility problem is solved, or it is outside of X. If $\boldsymbol{y} \notin X$, then we know X cannot be too big since $\boldsymbol{x}$ and $\boldsymbol{y}$ are close. In particular, we can show that if we use a reduced basis and branch in the direction of the longest basis vector, then the number of lattice hyperplanes intersecting X is going to be bounded by a constant depending only on n. Then, for each of these hyperplanes we consider the polytope formed by the intersection of X with that polytope. This is a polytope in dimension less than or equal to $n-1$. For the new polytope we repeat the process. We can illustrate the algorithm by a search tree that has at most n levels, and a number of nodes at each level that is bounded by a constant depending only on the dimension on that level.

In the following three subsections we describe algorithms, based on the above idea, for solving the integer feasibility problem (27) in polynomial time for fixed dimension. Lenstra's algorithm is presented in Section 4.1. In Section 4.2 we present a version of Lenstra's algorithm that follows from Lovász' theorem on thin directions. Both of these algorithms use Lovász' basis reduction algorithm. In Section 4.3 we describe the algorithm of Lovász and Scarf [79], which is based on the generalized basis reduction algorithm. Finally, in Section 4.4 we give an outline of Barvinok's algorithm to count integer points in integer polytopes. This algorithm does not use "width" as the main concept, but exponential sums and decomposition of cones. Barvinok's algorithm runs in polynomial time if the dimension is fixed, so his result generalizes Lenstra's result.

4.1 Lenstra's algorithm

If one uses branch-and-bound for solving problem (27) it is possible, even in dimension (2), to create an arbitrarily deep search tree for certain thin polytopes, see e.g. [5]. Lenstra [76] suggested to transform the polytope using a linear transformation τ such that the polytope τX becomes "round" according to a certain measure. Assume, without loss of generality, that the polytope X is full-dimensional and bounded, and let $B(\boldsymbol{p}, z) = \{\boldsymbol{x} \in \mathbb{R}^n : \|\boldsymbol{x} - \boldsymbol{p}\| \leq z\}$ be the closed ball with center $\boldsymbol{p}$ and radius z. The transformation τ that we apply to the polytope is constructed such that $B(\boldsymbol{p}, r) \subset \tau X \subset B(\boldsymbol{p}, R)$ for some $\boldsymbol{p} \in \tau X$, with r, R satisfying

$$\frac{R}{r} \leq c_2, \tag{29}$$

where c_2 is a constant that depends only on the dimension n. Relation (29) is the measure of "roundness" that Lenstra uses. For an illustration, see Figure 4. Once we have transformed the polytope, we need to apply the same transformation to the lattice, which gives us the following feasibility problem that is equivalent to problem (27):

$$\text{Is } \tau\mathbb{Z}^n \cap \tau X \neq \emptyset? \tag{30}$$

The vectors $\tau\boldsymbol{e}_j$, $1 \leq j \leq n$, where $\boldsymbol{e}_j$ is the j-th unit vector in $\mathbb{R}^n$, form a basis for the lattice $\tau\mathbb{Z}^n$. If the polytope X is thin, then this will translate to the

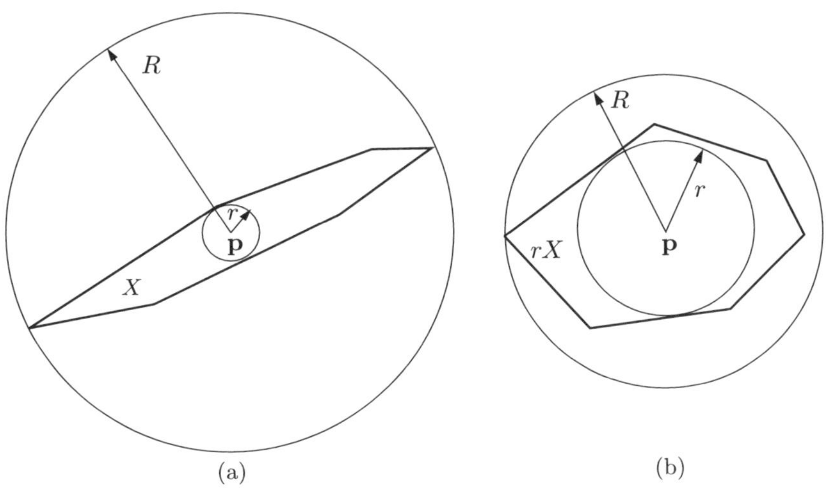

Figure 4. (a) The original polytope X is thin, and the ratio R/r is large. (b) The transformed polytope τX is "round", and R/r is relatively small.

lattice basis vectors τe_j, $1 \leq j \leq n$ in the sense that these vectors are long and non-orthogonal. This is where lattice basis reduction becomes useful. Once we have the transformed polytope τX, Lenstra uses the following lemma to find a lattice point quickly.

Lemma 1 ([76]). *Let $\boldsymbol{b}_1, \ldots, \boldsymbol{b}_n$ be any basis for L. Then for all $\boldsymbol{x} \in \mathbb{R}^n$ there exists a vector $\boldsymbol{y} \in L$ such that*

$$\|\boldsymbol{x} - \boldsymbol{y}\|^2 \leq \frac{1}{4}(\|\boldsymbol{b}_1\|^2 + \cdots + \|\boldsymbol{b}_n\|^2).$$

The proof of this lemma suggests a fast construction of the vector $\boldsymbol{y} \in L$ given the vector $\boldsymbol{x}$.

Next, let $L = \tau\mathbb{Z}^n$, and let $\boldsymbol{b}_1, \ldots, \boldsymbol{b}_n$ be a basis for L such that (10) holds. Notice that (10) holds if the basis is reduced. Also, reorder the vectors such that $\|\boldsymbol{b}_n\| = \max_{1 \leq j \leq n}\{\|\boldsymbol{b}_j\|\}$. Let $\boldsymbol{x} = \boldsymbol{p}$ where $\boldsymbol{p}$ is the center of the closed balls $B(\boldsymbol{p}, r)$ and $B(\boldsymbol{p}, R)$. Apply Lemma 1 to the given $\boldsymbol{x}$. This gives a lattice vector $\boldsymbol{y} \in \tau\mathbb{Z}^n$ such that

$$\|\boldsymbol{p} - \boldsymbol{y}\|^2 \leq \frac{1}{4}(\|\boldsymbol{b}_1\|^2 + \cdots + \|\boldsymbol{b}_n\|^2) \leq \frac{1}{4} \cdot n \cdot \|\boldsymbol{b}_n\|^2 \tag{31}$$

in polynomial time. We now distinguish two cases. Either $\boldsymbol{y} \in \tau X$ or $\boldsymbol{y} \notin \tau X$. In the first case we are done, so assume we are in the second case. Since $\boldsymbol{y} \notin \tau X$ we know that $\boldsymbol{y}$ is not inside the ball $B(\boldsymbol{p}, r)$ as $B(\boldsymbol{p}, r)$ is completely contained in τX. Hence we know that $\|\boldsymbol{p} - \boldsymbol{y}\| > r$, or using (31), that

$$r < \frac{1}{2} \cdot \sqrt{n} \cdot \|\boldsymbol{b}_n\|. \tag{32}$$

Below we will describe the tree search algorithm and argue why it is polynomial for fixed n. The distance between any two consecutive lattice hyperplanes, as defined in Corollary 1, is equal to h. We now create t subproblems by considering intersections between the polytope τX with t of these parallel hyperplanes. Each of the subproblems has dimension at least one lower than the parent problem and they are solved recursively. The procedure of splitting the problem into subproblems of lower dimension is called "branching", and each subproblem is represented by a node in the enumeration tree. In each node we repeat the whole process of transformation, basis reduction and, if necessary, branching. The enumeration tree created by this recursive process is of depth at most n, and the number of nodes at each level is bounded by a constant that depends only on the dimension. The value of t will be computed below.

Let H, h and L' be defined as in Corollary 1 of Section 3.2, and its proof. We can write L as

$$L = L' + \mathbb{Z}\boldsymbol{b}_n \subset H + \mathbb{Z}\boldsymbol{b}_n = \cup_{k \in \mathbb{Z}}(H + k\boldsymbol{b}_n). \tag{33}$$

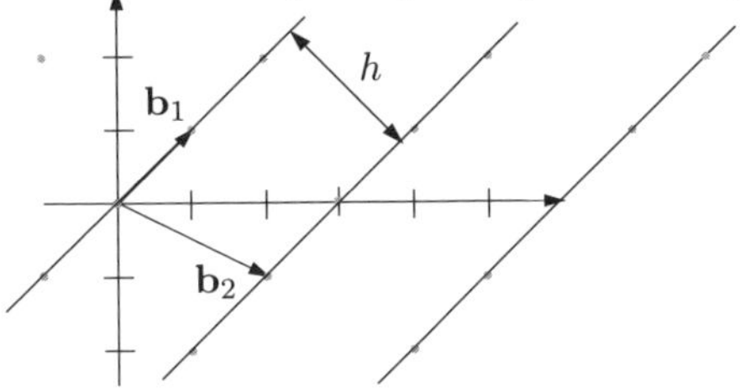

Figure 5.

So the lattice L is contained in countably many parallel hyperplanes. For an example we refer to Figure 5. The distance between the two consecutive hyperplanes is h, and Corollary 1 says that h is bounded from below by $c_1^{-1}\|\boldsymbol{b}_n\|$, which implies that not too many hyperplanes intersect τX. To determine precisely how many hyperplanes intersect τX, we approximate τX by the ball $B(\boldsymbol{p}, R)$. If t is the number of hyperplanes intersecting $B(\boldsymbol{p}, R)$ we have

$$t - 1 \leq \frac{2R}{h}.$$

Using the relationship (29) between the radii R and r we have $2R \leq 2rc_2 < c_2\sqrt{n}\|\boldsymbol{b}_n\|$, where the last inequality follows from (32). Since $h \geq c_1^{-1}\|\boldsymbol{b}_n\|$, we get the following bound on the number of hyperplanes that we need to consider:

$$t - 1 \leq \frac{2R}{h} < c_1 c_2 \sqrt{n},$$

which depends on the dimension only. The values of the constants c_1 and c_2 that are used by Lenstra are: $c_1 = 2^{n(n-1)/4}$ and $c_2 = 2n^{3/2}$. Lenstra discusses ways of improving these values. To determine the values of k in expression (33), we express $\boldsymbol{p}$ as a linear combination of the basis vectors $\boldsymbol{b}_1, \ldots, \boldsymbol{b}_n$. Recall that $\boldsymbol{p}$ is the center of the ball $B(\boldsymbol{p}, R)$ that was used to approximate τX.

So far we have not mentioned how to determine the transformation τ and hence the balls $B(\boldsymbol{p}, r)$ and $B(\boldsymbol{p}, R)$. We give the general idea here without going into detail. First, determine an n-simplex contained in X. This can be done in polynomial time by repeated calls to the ellipsoid algorithm. The resulting simplex is described by its extreme points $\boldsymbol{v}_0, \ldots, \boldsymbol{v}_n$. By again applying the ellipsoid algorithm repeatedly we can decide whether there exists an extreme point $\boldsymbol{x}$ of X such that if we replace $\boldsymbol{v}_j$ by $\boldsymbol{x}$ we obtain a new simplex whose volume is at least a factor of $\frac{3}{2}$ larger than the current simplex. We stop the procedure if we cannot find such a new simplex. The factor $\frac{3}{2}$ can be modified, but the choice will affect the value of the constant c_2, see [76] for further details. We now map the extreme points of the simplex to the unit vectors of $\mathbb{R}^{n+1}$ so as to obtain a regular n-simplex, and we denote this transformation by τ. Lenstra [76] shows that τ has the property that if we let $\boldsymbol{p} = 1/(n+1)\sum_{j=0}^{n} \boldsymbol{e}_j$, where $\boldsymbol{e}_j$ is the j-th unit vector of $\mathbb{R}^{n+1}$ (i.e., $\boldsymbol{p}$ is the center of the regular simplex), then there exist closed balls $B(\boldsymbol{p}, r)$ and $B(\boldsymbol{p}, R)$ such that $B(\boldsymbol{p}, r) \subset \tau X \subset B(\boldsymbol{p}, R)$ for some $\boldsymbol{p} \in \tau X$, with r, R satisfying $R/r \leq c_2$.

Kannan [66] developed a variant of Lenstra's algorithm. The algorithm follows Lenstra's algorithm up to the point where he has applied a linear transformation to the polytope X and obtained a polytope τX such that $B(\boldsymbol{p}, r) \subset \tau X \subset B(\boldsymbol{p}, R)$ for some $\boldsymbol{p} \in \tau X$. Here Kannan applies K-Z basis reduction to a basis of the lattice $\tau \mathbb{Z}^n$. As in Lenstra's algorithm two cases are considered. Either τX is relatively large which implies that τX contains a lattice vector, or τX is small, which means that not too many lattice hyperplanes can intersect τX. Each such intersection gives rise to a subproblem of at least one dimension lower. Kannan's reduced basis makes it possible to improve the bound on the number of hyperplanes that has to be considered to $O(n^{5/2})$. Lenstra's algorithm has been implemented by Gao and Zhang [47], and a heuristic version of the algorithm has been developed and implemented by Aardal et al. [1], and Aardal and Lenstra [4].

4.2 Lovász' theorem on thin directions

Let $E(\boldsymbol{z}, \boldsymbol{D}) = \{\boldsymbol{x} \in \mathbb{R}^n \mid (\boldsymbol{x} - \boldsymbol{z})^T \boldsymbol{D}^{-1} (\boldsymbol{x} - \boldsymbol{z}) \leq 1\}$. $E(\boldsymbol{z}, \boldsymbol{D})$ is the *ellipsoid* in $\mathbb{R}^n$ associated with the vector $\boldsymbol{z} \in \mathbb{R}^n$ and the positive definite $n \times n$ matrix $\boldsymbol{D}$. The vector $\boldsymbol{z}$ is the *center* of the ellipsoid. Goffin [50] showed that for any full-dimensional rational polytope X it is possible, in polynomial time, to find a vector $\boldsymbol{p} \in \mathbb{Q}^n$ and a positive definite $n \times n$ matrix $\boldsymbol{D}$ such that

$$E\left(\boldsymbol{p}, \frac{1}{(n+1)^2} \boldsymbol{D}\right) \subseteq X \subseteq E(\boldsymbol{p}, \boldsymbol{D}). \tag{34}$$

Grötschel, Lovász and Schrijver [54] showed a similar result for the case where the polytope is not given explicitly, but by a separation algorithm.

The norm // // defined by the matrix $\boldsymbol{D}^{-1}$ is given by $//\boldsymbol{x}// = \sqrt{\boldsymbol{x} \boldsymbol{D}^{-1} \boldsymbol{x}}$. Lovász used basis reduction with the norm // //, and the result by Goffin to obtain the following theorem.

Theorem 10 (see [99]). *Let $\boldsymbol{Ax} \leq \boldsymbol{d}$ be a system of m rational inequalities in n variables, let $X = \{\boldsymbol{x} \in \mathbb{R}^n \mid \boldsymbol{Ax} \leq \boldsymbol{d}\}$, and let $w_{\boldsymbol{c}}(X)$ be defined as in Expression (28). There exists a polynomial algorithm that finds either an integer vector $\boldsymbol{y} \in X$, or a vector $\boldsymbol{c} \in \mathbb{Z}^n \backslash \{\boldsymbol{0}\}$ such that*

$$w_{\boldsymbol{c}}(X) \leq n(n+1)2^{n(n-1)/4}$$

We will sketch the proof of the theorem for the case that X is full-dimensional and bounded. For the not full-dimensional case, and the case where P is unbounded we refer to the presentation by Schrijver [99]. Notice that the algorithm of Theorem 10 is polynomial for arbitrary n.

Proof of the full-dimensional bounded case: Assume that $dim(X) = n$. Here we will not make a transformation to a lattice $\tau\mathbb{Z}^n$, but remain in the lattice $\mathbb{Z}^n$. First, find two ellipsoids $E(\boldsymbol{p}, \frac{1}{(n+1)^2}\boldsymbol{D})$ and $E(\boldsymbol{p}, \boldsymbol{D})$, such that (34) holds, by the algorithm of Goffin. Next, we apply basis reduction, using the norm // // defined by $\boldsymbol{D}^{-1}$, to the unit vectors $\boldsymbol{e}_1, \ldots, \boldsymbol{e}_n$ to obtain a reduced basis $\boldsymbol{b}_1, \ldots, \boldsymbol{b}_n$ for the lattice $\mathbb{Z}^n$ that satisfies (cf. the second inequality of (10))

$$\Pi_{j=1}^n //\boldsymbol{b}_j// \le 2^{n(n-1)/4}\sqrt{\det(\boldsymbol{D}^{-1})}. \tag{35}$$

Next, reorder the basis vectors such that $//\boldsymbol{b}_n// = \max_{1\le j\le n}\{//\boldsymbol{b}_j//\}$. After reordering, inequality (35) still holds. Write $\boldsymbol{p} = \sum_{j=1}^n \alpha_j \boldsymbol{b}_j$, and let $\boldsymbol{y} = \sum_{j=1}^n \lceil\alpha_j\rfloor \boldsymbol{b}_j$. Notice that $\boldsymbol{y} \in \mathbb{Z}^n$. If $\boldsymbol{y} \in$ X we are done, and if not we know that $\boldsymbol{y} \notin E(\boldsymbol{p}, (1/(n+1)^2)\,\boldsymbol{D})$, so

$$\frac{1}{(n+1)^2} < (\boldsymbol{y}-\boldsymbol{p})^T\boldsymbol{D}^{-1}(\boldsymbol{y}-\boldsymbol{p}) = //\boldsymbol{y}-\boldsymbol{p}//^2 = \Big/\!\!\Big/ \sum_{j=1}^n (\alpha_j - \lceil\alpha_j\rfloor)\boldsymbol{b}_j \Big/\!\!\Big/^2.$$

From this expression we obtain

$$\frac{1}{(n+1)} < \sum_{j=1}^n (\alpha_j - \lceil\alpha_j\rfloor)//\boldsymbol{b}_j// \le \frac{n}{2}//\boldsymbol{b}_n//,$$

so

$$//\boldsymbol{b}_n// > \frac{2}{n(n+1)}. \tag{36}$$

Choose a direction $\boldsymbol{c}$ such that the components of $\boldsymbol{c}$ are relatively prime integers, and such that $\boldsymbol{c}$ is orthogonal to the subspace generated by the basis vectors $\boldsymbol{b}_1, \ldots, \boldsymbol{b}_{n-1}$. One can show, see Schrijver [99], pp 257–258, that if we consider a vector $\boldsymbol{x}$ such that $\boldsymbol{x}^T\boldsymbol{D}^{-1}\boldsymbol{x} \le 1$, then

$$\begin{aligned}|\boldsymbol{c}^T\boldsymbol{x}| &\le \sqrt{\det(\boldsymbol{D})}//\boldsymbol{b}_1// \cdot \cdots \cdot //\boldsymbol{b}_{n-1}// \\ &\le 2^{n(n-1)/4}//\boldsymbol{b}_n//^{-1} < \frac{n(n+1)}{2}2^{n(n-1)/4},\end{aligned} \tag{37}$$

where the second inequality follows from inequality (35), and the last inequality follows from (36). If $z \in E(\boldsymbol{p}, \boldsymbol{D})$, then

$$|c^T(z - p)| \leq \frac{n(n+1)}{2} 2^{n(n-1)/4},$$

which implies

$$\begin{aligned} w_c(X) &= \max\{c^T x \mid x \in X\} - \min\{c^T x \mid x \in X\} \\ &\leq \max\{c^T x \mid x \in E(p, D)\} - \min\{c^T x \mid x \in E(p, D)\} \\ &\leq n(n+1)2^{n(n-1)/4}, \end{aligned} \tag{38}$$

which gives the desired result. □

Lenstra's result that the integer feasibility problem can be solved in polynomial time for fixed n follows from Theorem 10. If we apply the algorithm implied by Theorem 10, we either find an integer point $y \in X$ or a thin direction c, i.e., a direction c such that equation (38) holds. Assume that the direction c is the outcome of the algorithm. Let $\mu = \lceil \min\{c^T x \mid x \in X\} \rceil$. All points in $X \cap \mathbb{Z}^n$ are contained in the parallel hyperplanes $c^T x = t$ where $t = \mu, \ldots, \mu + n(n+1)2^{n(n-1)/4}$, so if n is fixed, then the number of hyperplanes is constant, and each of them gives rise to a subproblem of dimension less than or equal to $n - 1$. For each of these lower-dimensional problems we repeat the algorithm of Theorem 10. The search tree has at most n levels and the number of nodes at each level is bounded by a constant depending only on the dimension.

Remark. The ingredients of Theorem 10 are actually present in Lenstra's paper [76]. In the preprinted version, however, the two auxiliary algorithms used by Lenstra; the algorithm to make the set X appear round, and the basis reduction algorithm, were polynomial for fixed n only, which was enough to prove his result that the integer programming feasibility problem can be solved in polynomial time in fixed dimension. Later, Lovász' basis reduction algorithm [75] was developed, and Lovász also pointed out that the "rounding" of X can be done in polynomial time for varying n due to the ellipsoid algorithm. Lenstra uses both these algorithms in the published version of the paper.

4.3 The Lovász-Scarf algorithm

The integer feasibility algorithm of Lovász and Scarf [79] determines, in polynomial time for fixed n, either a certificate for feasibility, or a thin direction of X. If a thin direction is found, then one needs to branch, i.e., divide the problem into lower-dimensional subproblems, in order to determine whether or not a feasible vector exists, but then the number of branches is bounded by a constant for fixed n. If the algorithm indicates that X contains an integer vector, then one needs to determine a so-called Korkine-Zolotareff basis in order to construct a feasible vector. The Lovász-Scarf algorithm avoids the approximations by balls as in Lenstra's algorithm, or by ellipsoids as in the algorithm implied by Lovász' result. Again, we assume that $X = \{x \in \mathbb{R}^n \mid Ax \leq d\}$ is bounded, rational, and full-dimensional.

Let $(X - X) = \{(x - y) \mid x \in X,\ y \in X)\}$ be the difference set corresponding to X. Recall that $(X - X)^*$ denotes the dual set corresponding to $(X - X)$, and notice that $(X - X)^*$ is symmetric about the origin. The distance functions associated with $(X - X)^*$ are:

$$\begin{aligned} F_j(c) &= \min_{\alpha_1,\ldots,\alpha_{j-1} \in \mathbb{Q}} F(c + \alpha_1 b_1 + \cdots + \alpha_{j-1} b_{j-1}) \\ &= \max\{c^T(x - y) \mid x \in X,\ y \in X,\ b_1^T(x - y) = 0, \ldots, b_{j-1}^T(x - y) = 0\}, \end{aligned}$$

(cf. Expressions (20) and (21)). Here, we notice that $F(c) = F_1(c)$ is the width of X in the direction c, $w_c(X)$ (see Expression (28) in the introduction to Section 4). From the above we see that a lattice vector c that minimizes the width of the polytope X is a *shortest lattice vector* for the polytope $(X - X)^*$.

To outline the algorithm by Lovász and Scarf we need the results given in Theorem 11 and 12 below, and the definition of a *generalized Korkine-Zolotareff basis*. Let b_j, $1 \leq j \leq n$ be defined recursively as follows. Given $b_1, \ldots, b_{j-1}$, the vector b_j minimizes $F_j(x)$ over all lattice vectors that are linearly independent of $b_1, \ldots, b_{j-1}$. A generalized Korkine-Zolotareff (KZ) basis is defined to be any proper basis $b'_1, \ldots, b'_n$ associated with b_j, $1 \leq j \leq n$ (see Expression (24) for the definition of a proper basis). The notion of a generalized KZ basis was introduced by Kannan and Lovász [67], [68]. Kannan and Lovász [67] gave an algorithm for computing a generalized KZ basis in polynomial time for fixed n. Notice that b'_1 in a generalized KZ basis is the shortest non-zero lattice vector.

Theorem 11 ([68]). *Let $F(c)$ be the length of the shortest non-zero lattice vector c with respect to the set $(X - X)^*$, and let $\rho_{\mathrm{KZ}} = \sum_{j=1}^n F_j(b'_j)$, where b'_j, $1 \leq j \leq n$ is a generalized Korkine-Zolotareff basis. There exists a universal constant c_0 such that*

$$F(c)\rho_{\mathrm{KZ}} \leq c_0 \cdot n \cdot (n + 1)/2.$$

To derive their result, Kannan and Lovász used a lower bound on the product of the volume of a convex set $C \subset \mathbb{R}^n$ that is symmetric about the origin, and the volume of its dual C^*. The bound, due to Bourgain and Milman [18], is equal to $\frac{c_{\mathrm{BM}}^n}{n^n}$, where c_{BM} is a constant depending only on n. In Theorem 11 we have $c_0 = \frac{4}{c_{\mathrm{BM}}}$, see also the remark below.

Theorem 12 ([68]). *Let $b_1, \ldots, b_n$ be any basis for $\mathbb{Z}^n$, and let X be a bounded convex set that is symmetric about the origin. If $\rho = \sum_{j=1}^n F_j(b_j) \leq 1$, then X contains an integer vector.*

The first step of the Lovász-Scarf algorithm is to compute the shortest vector $\boldsymbol{c}$ with respect to $(X-X)^*$ using the algorithm described in Section 3.4. If $F(\boldsymbol{c}) \geq c_0 \cdot n \cdot (n+1)/2$, then $\rho_{\mathrm{KZ}} \leq 1$, which by Theorem 12 implies that X contains an integer vector. If $F(\boldsymbol{c}) < c_0 \cdot n \cdot (n+1)/2$, then we need to branch. Due to the definition of $F(\boldsymbol{c})$ we know in this case that $w_{\boldsymbol{c}}(X) < c_0 \cdot n \cdot (n+1)/2$, which implies that the polytope X in the direction $\boldsymbol{c}$ is "thin". As in the previous subsection we create one subproblem for every hyperplane $\boldsymbol{c}^T\boldsymbol{x} = \mu, \ldots, \boldsymbol{c}^T\boldsymbol{x} = \mu + c_0 \cdot n \cdot (n+1)/2$, where $\mu = \lceil \min\{\boldsymbol{c}^T\boldsymbol{x} \mid \boldsymbol{x} \in X\} \rceil$. Once we have fixed a hyperplane $\boldsymbol{c}^T\boldsymbol{x} = t$, we have obtained a problem in dimension less than or equal to $n-1$, and we repeat the process. This procedure creates a search tree that is at most n deep, and that has a constant number of branches at each level when n is fixed. The algorithm called in each branch is, however, polynomial for fixed dimension only. First, the generalized basis reduction algorithm runs in polynomial time for fixed dimension, and second, computing the shortest vector $\boldsymbol{c}$ is done in polynomial time for fixed dimension. An alternative would be to use the first reduced basis vector with respect to $(X-X)^*$, instead of the shortest vector $\boldsymbol{c}$. According to Proposition 4, $F(\boldsymbol{b}_1) \leq (\frac{1}{2}-\epsilon)^{1-n}F(\boldsymbol{c})$. In this version of the algorithm we would first check whether $F(\boldsymbol{b}_1) \geq c_0 \cdot n \cdot (n+1)/(2(\frac{1}{2}-\epsilon)^{1-n})$. If yes, then X contains an integer vector, and if no, we need to branch, and we create at most $c_0 \cdot n \cdot (n+1)/(2(\frac{1}{2}-\epsilon)^{n-1})$ hyperplanes.

If the algorithm terminates with the result that X contains an integer vector, then Lovász and Scarf describe how such a vector can be constructed by using the Korkine-Zolotareff basis (see [79], proof of Theorem 10).

Lagarias, Lenstra, and Schnorr [73] derive bounds on the Euclidean length of Korkine-Zolotareff reduced basis vectors of a lattice and its dual lattice. The bounds are given in terms of the successive minima of L and the dual lattice L^*. Later, Kannan and Lovász [67], [68] introduced the generalized Korkine-Zolotareff basis, as defined above, and derived bounds of the same type as in the paper by Lagarias et al. These bounds were used to study covering minima of a convex set with respect to a lattice, such as the covering radius, and the lattice width. An important result by Kannan and Lovász is that the product of the first successive minima of the lattices L and L^* is bounded from above by $c_0 \cdot n$. This improves on a similar result of Lagarias et al. and implies Theorem 11 above. There are many interesting results on properties of various lattice constants. Many of them are described in the survey by Kannan [65], and will not be discussed further here.

Example 2. The following example demonstrates a few iterations with the generalized basis reduction algorithm. Consider the polytope $X = \{\boldsymbol{x} \in \mathbb{R}^2_{\geq 0} \mid x_1 + 7x_2 \geq 7, 2x_1 + 7x_2 \leq 14, -5x_1 + 4x_2 \leq 4\}$. Let $j = 1$ and $\epsilon = \frac{1}{4}$. Assume we want to use the generalized basis reduction algorithm to find a direction in which the width of X is small. Recall that a lattice vector $\boldsymbol{c}$ that minimizes the width of X is a shortest lattice vector with respect to the set $(X-X)^*$. The first reduced basis vector is an approximation of the shortest vector for $(X-X)^*$ and hence an approximation of the thinnest direction for X. The distance functions associated with $(X-X)^*$ are

$$F_j(\boldsymbol{c}) = \max\{\boldsymbol{c}^T(\boldsymbol{x}-\boldsymbol{y}) \mid \boldsymbol{x} \in X, \boldsymbol{y} \in X, \boldsymbol{b}_i^T(\boldsymbol{x}-\boldsymbol{y}) = 0, 1 \le i \le j-1\}.$$

The initial basis is

$$\boldsymbol{b}_1 = \begin{pmatrix} 1 \\ 0 \end{pmatrix} \quad \boldsymbol{b}_2 = \begin{pmatrix} 0 \\ 1 \end{pmatrix}.$$

We obtain $F_1(\boldsymbol{b}_1)=7.0$, $F_1(\boldsymbol{b}_2)=1.8$, $\mu=0$, and $F_1(\boldsymbol{b}_2+0\boldsymbol{b}_1)=1.8$, see Figure 6. Here we see that the number of lattice hyperplanes intersecting X in direction $\boldsymbol{b}_1$ is 8. The hyperplane are $x_1=0$, $x_1=1,\ldots,x_1=7$. The number of hyperplanes intersecting X in direction $\boldsymbol{b}_2$ is 2: $x_2=0$, $x_2=1$.

Checking Conditions (22) and (23) shows that Condition (22) is satisfied as $F_1(\boldsymbol{b}_2+0\boldsymbol{b}_1) \ge F_1(\boldsymbol{b}_2)$, but that Condition (23) is violated as $F_1(\boldsymbol{b}_2) \ngeq (3/4)F_1(\boldsymbol{b}_1)$, so we interchange $\boldsymbol{b}_1$ and $\boldsymbol{b}_2$ and remain at $j=1$.

Now we have $j=1$ and

$$\boldsymbol{b}_1 = \begin{pmatrix} 0 \\ 1 \end{pmatrix} \quad \boldsymbol{b}_2 = \begin{pmatrix} 1 \\ 0 \end{pmatrix}.$$

$F_1(\boldsymbol{b}_1)=1.8$, $F_1(\boldsymbol{b}_2)=7.0$, $\mu=4$, and $F_1(\boldsymbol{b}_2+4\boldsymbol{b}_1)=3.9$.
Condition (22) is violated as $F_1(\boldsymbol{b}_2+4\boldsymbol{b}_1) \ngeq F_1(\boldsymbol{b}_2)$, so we replace $\boldsymbol{b}_2$ by $\boldsymbol{b}_2+4\boldsymbol{b}_1=(1,4)^T$. Given the new basis vector $\boldsymbol{b}_2$ we check Condition (23) and we conclude that this condition is satisfied. Hence the basis

$$\boldsymbol{b}_1 = \begin{pmatrix} 0 \\ 1 \end{pmatrix} \quad \boldsymbol{b}_2 = \begin{pmatrix} 1 \\ 4 \end{pmatrix}$$

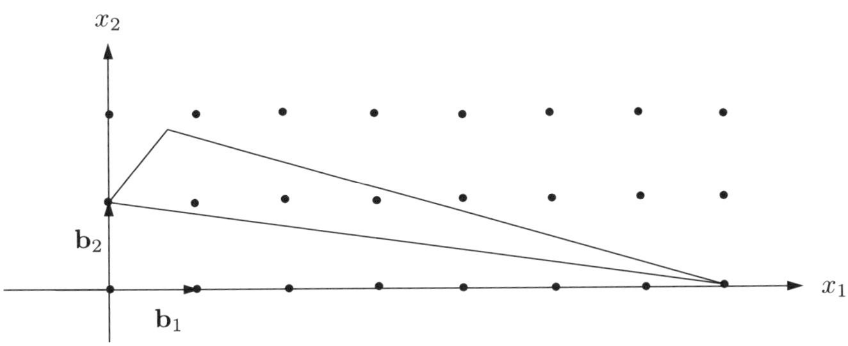

Figure 6. The unit vectors form the initial basis.

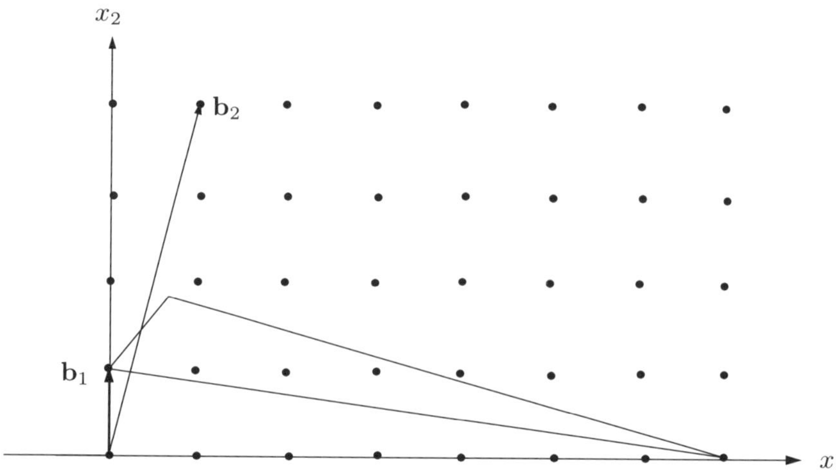

Figure 7. The reduced basis yields thin directions for the polytope.

is Lovász-Scarf reduced, see Figure 7. In the root node of our search tree we would create two branches corresponding to the lattice hyperplanes $x_2 = 0$ and $x_2 = 1$. □

4.4 *Counting integer points in polytopes*

Barvinok [11] showed that there exists a polynomial time algorithm for *counting* the number of integer points in a polytope if the dimension is fixed. Barvinok's result therefore generalizes the result of Lenstra [76]. Before Barvinok developed his counting algorithm, polynomial algorithms were only known for dimensions $n = 1, 2, 3, 4$. The cases $n = 1, 2$ are relatively simple, and for the challenging cases $n = 3, 4$, algorithms were developed by Dyer [37]. On the approximation side, Cook, Hartmann, Kannan, and McDiarmid [28] developed an algorithm that for a given rational number $\epsilon > 0$ counts the number of points in a polytope with a relative error less than ϵ in time polynomial in the input size and $1/\epsilon$.

Barvinok based his algorithm on an identity by Brion for exponential sums over polytopes. Later, Dyer and Kannan [38] developed a simplification of Barvinok's algorithm in which the step of the algorithm that uses the property that the exponential sum can be continued to define a meromorphic function over $\mathbb{C}^n$ (cf. Proposition 1) is unnecessary. In addition, Dyer and Kannan observed that Lenstra's algorithm is no longer needed as a subroutine of Barvinok's algorithm. See also the paper by Barvinok and Pommersheim [12] for a more elementary description of the algorithm. De Loera et al. [36] introduced further practical improvements over Dyer and Kannan's version, and implemented their version of the algorithm, which uses Lovász' basis reduction algorithm. De Loera et al. report on the first computational results from using an algorithm to count the number of lattice points in a polytope. These results are encouraging.

To describe Barvinok's algorithm in detail would require the introduction of quite a lot of new material, which would take us outside the scope of this chapter. The results is so important though that we still want to give a high-level presentation here.

Barvinok's algorithm counts integer points in an integer simplex; given $k+1$ integer vectors such that their convex hull is a k-dimensional simplex Δ, compute the number of integer points in Δ. Dyer [37] had previously shown that the problem of counting integer points in a polytope can be reduced to counting integer points in polynomially many integer simplices. See also Cook et al. [28], who proved that if P_I is the integer hull of the rational polyhedron $P \subset \mathbb{R}^n$ given by m inequalities whose size is at most φ, then for fixed n an upper bound on the number of vertices of P_I is $O(m^n \varphi^{n-1})$.

The main tools of Barvinok's algorithm are *decompositions of rational cones in so-called primitive cones*, and *exponential sums over polytopes*. The decomposition of cones will be treated very briefly. For details we refer to Section 5 of Barvinok's paper. For an exponential sum over a polytope P we write

$$\sum_{\boldsymbol{x} \in (P \cap \mathbb{Z}^n)} \exp\{\boldsymbol{c}^T \boldsymbol{x}\}, \tag{39}$$

where P is a polytope in $\mathbb{R}^n$, and $\boldsymbol{c}$ is an n-dimensional real vector.

Before giving an outline of the algorithm we need to introduce new notation. A convex cone $K \in \mathbb{R}^n$ is *rational* if it is the conic hull of finitely many integer generators, i.e., $K = \text{cone}\{\boldsymbol{u}_1, \ldots, \boldsymbol{u}_k\}$, $\boldsymbol{u}_i \in \mathbb{Z}^n$, $1 \leq i \leq k$. A cone K is *simple* if it can be generated by linearly independent vectors. A simple rational cone K is *primitive* if $K = \text{cone}\{\boldsymbol{u}_1, \ldots, \boldsymbol{u}_k\}$, where $\boldsymbol{u}_1, \ldots, \boldsymbol{u}_k$ form a basis of the lattice $\mathbb{Z}^n \cap \text{lin}(K)$, where $\text{lin}(K)$ is the linear hull of K. A *meromorphic function* $f(z)$ is a single-valued function that can be expressed as $f(z) = g(z)/h(z)$, where $g(z)$ and $h(z)$ are functions that are analytic at all finite points of the complex plane $\mathbb{C}$. We can associate a meromorphic function with each rational cone.

Proposition 6. *Let K be a simple rational cone. Let $\boldsymbol{c} \in \mathbb{R}^n$ be a vector such that the inner product $(\boldsymbol{c}^T \cdot)$ decreases along the extreme rays of K. Then the series*

$$\sum_{\boldsymbol{x} \in (K \cap \mathbb{Z}^n)} \exp\{\boldsymbol{c}^T \boldsymbol{x}\}$$

converges and defines a meromorphic function in $\boldsymbol{c} \in \mathbb{C}^n$. This function is denoted by $\sigma(K; \boldsymbol{c})$. If $\boldsymbol{u}_1, \ldots, \boldsymbol{u}_k \in \mathbb{Z}^n$ are linearly independent generators of K, then for *all* $c \in \mathbb{C}^n$ *the following holds,*

$$\sigma(K;\, \boldsymbol{c}) = p_K(\exp\{c_1\}, \ldots, \exp\{c_n\}) \cdot \Pi_{i=1}^{k} \frac{1}{1 - \exp\{\boldsymbol{c}^T \boldsymbol{u}_i\}}, \tag{40}$$

where p_K *is a Laurent polynomial in n variables.*

We observe that the set of singular points of $\sigma(K;\, c)$ is the set of hyperplanes $H_i = \{\boldsymbol{c} \in \mathbb{R}^n \mid \boldsymbol{c}^T \boldsymbol{u}_i = 0\}$, $1 \le i \le k$. The question now is how we can obtain an explicit expression for the number of points in a polytope from the result above. The key of such an expression is the following theorem by Brion.

Theorem 13 ([19]). *Let* $P \subset \mathbb{R}^n$ *be a rational polytope, and let V be the set of vertices of P. For each vertex* $\boldsymbol{v} \in V$*, the supporting cone* $K_{\boldsymbol{v}}$ *of P at* $\boldsymbol{v}$ *is defined as* $K_{\boldsymbol{v}} = \{\boldsymbol{u} \in \mathbb{R}^n \mid \boldsymbol{v} + \delta \boldsymbol{u} \in P$ *for all sufficiently small* $\delta > 0\}$*. Then*

$$\sum_{\boldsymbol{x} \in (P \cap \mathbb{Z}^n)} \exp\{\boldsymbol{c}^T \boldsymbol{x}\} = \sum_{\boldsymbol{v} \in V} \exp\{\boldsymbol{c}^T \boldsymbol{v}\} \cdot \sigma(K_{\boldsymbol{v}};\, \boldsymbol{c}) \tag{41}$$

for all $\boldsymbol{c} \in \mathbb{R}^n$ *that are not singular points for any of the functions* $\sigma(K_{\boldsymbol{v}};\, \boldsymbol{c})$.

Considering the left-hand side of expression (41), it seems tempting to use $\boldsymbol{c} = \boldsymbol{0}$ in Expression (41) for $P = \Delta$, since this will contribute 1 to the sum from every integer point, but this is not possible since $\boldsymbol{0}$ is a singular point for the functions $\sigma(K_{\boldsymbol{v}};\, \boldsymbol{c})$. Instead we take a vector $\boldsymbol{c}$ that is regular for all of the functions $\sigma(K_{\boldsymbol{v}};\, \boldsymbol{c})$, $\boldsymbol{v} \in V$, and a parameter t, and we compute the constant term of the Taylor expansion of the function $\sum_{\boldsymbol{x} \in \Delta \cap \mathbb{Z}^n} \exp\{t \cdot (\boldsymbol{c}^T \boldsymbol{x})\}$ in the neighborhood of the point $t = 0$. Equivalently, due to Theorem 13, we can instead compute the constant terms of the Laurent expansions of the functions $\exp\{t \cdot (\boldsymbol{c}^T \boldsymbol{v})\} \cdot \sigma(K_{\boldsymbol{v}};\, t \cdot \boldsymbol{c})$ for all vertices $\boldsymbol{v}$ of Δ. These constant terms are denoted by $R(K_{\boldsymbol{v}}, \boldsymbol{v}, \boldsymbol{c})$. In general there does not exist an explicit formula for $R(K_{\boldsymbol{v}}, \boldsymbol{v}, \boldsymbol{c})$, but if $K_{\boldsymbol{v}}$ is primitive, then such an explicit expression does exist, and is based on the fact that the function $\sigma(K;\, \boldsymbol{c})$ in Expression (40) looks particularly simple if K is a primitive cone, namely, the polynomial p_K is equal to one.

Proposition 7. *Assume that* $K \subset \mathbb{R}^n$ *is a primitive cone with primitive generators* $\{\boldsymbol{u}_1, \ldots, \boldsymbol{u}_k\}$*. Then*

$$\sigma(K;\, \boldsymbol{c}) = \Pi_{i=1}^{k} \frac{1}{1 - \exp\{\boldsymbol{c}^T \boldsymbol{u}_i\}}.$$

A simple rational cone can be expressed as an integer linear combination of primitive cones in polynomial time if the dimension n is fixed (see also Section 5 in [11]) as is stated in the following important theorem by Barvinok.

Theorem 14 ([11]). *Let us fix $n \in \mathbb{N}$. Then there exists a polynomial algorithm that for any given rational cone K constructs a family $K_i \subset \mathbb{R}^n$, $i \in I$ of rational primitive cones and computes integer numbers ϵ_i, $i \in I$ such that*

$$K = \sum_{i \in I} \epsilon_i K_i \quad \text{and} \quad \sigma(K; \mathbf{c}) = \sum_{i \in I} \epsilon_i \sigma(K_i; \mathbf{c}) \tag{42}$$

for all $c \in \mathbb{R}^n$ that are regular points for the functions $\sigma(K; \mathbf{c})$, $\sigma(K_i; \mathbf{c})$, $i \in I$.

Notice that the numbers ϵ_i, $i \in I$, in Expression (42) are either equal to $+1$ or -1.

Barvinok's decomposition of rational cones leads to a polynomial algorithm for fixed n for computing the constant term $R(K, \mathbf{v}, \mathbf{c})$ for an arbitrary rational cone K and an arbitrary vector $\mathbf{v}$. Lenstra's algorithm is used as a subroutine in the decomposition. As mentioned earlier, Lenstra's algorithm is not necessary in the algorithm presented by Dyer and Kannan.

The only component of the overall algorithm that we are missing is how to construct a generic vector $\mathbf{c}$ that is not a singular point for $\sigma(K_{\mathbf{v}}; \mathbf{c})$. This can be done in polynomial time as is stated in the following lemma.

Lemma 2 ([11]). *There exists a polynomial time algorithm that for any given $n \in \mathbb{N}$, for any given $m \in \mathbb{N}$, and for any rational vectors $\mathbf{u}_1, \ldots, \mathbf{u}_m \in \mathbb{Q}^n$ constructs a rational vector $\mathbf{c}$ such that $\mathbf{c}^T \mathbf{u}_i \neq 0$ for $1 \leq i \leq m$.*

To summarize, a sketch of Barvinok's algorithm is as follows. First, for each vertex $\mathbf{v}$ of the simplex Δ, compute the integer generators of the supporting cone $K_{\mathbf{v}}$. Each cone $K_{\mathbf{v}}$ is then expressed as an integer linear combination of primitive cones K_i, i.e., $K_{\mathbf{v}} = \sum_{i \in I_{\mathbf{v}}} \lambda_i K_i$ for integer λ_i. By using Lemma 2 we can now construct a vector $\mathbf{c}$ that is not orthogonal to any of the generators of the cones K_i, $i \in \cup_{\mathbf{v}} I_{\mathbf{v}}$, which means that $\mathbf{c}$ is not a singular point for the functions $\sigma(K_i; \mathbf{c})$. Next, for all $\mathbf{v}$ and $I_{\mathbf{v}}$ compute the constant term $R(K_i, \mathbf{v}, \mathbf{c})$ of the function $\exp\{t \cdot (\mathbf{c}^T \mathbf{v})\} \cdot \sigma(K_i; t \cdot \mathbf{c})$ as $t \to 0$. Let $\#(\Delta \cap \mathbb{Z}^n)$ denote the number of integer points in the simplex Δ. Through Brion's expression (41) we have now obtained

$$\#(\Delta \cap \mathbb{Z}^n) = \sum_{\mathbf{v} \in V} \sum_{i \in I_{\mathbf{v}}} \lambda_i \cdot R(K_i, \mathbf{v}, \mathbf{c}).$$

5 Algorithms for the integer optimization problem in fixed dimension

So far we have only dealt with the integer feasibility problem in fixed dimension n. We now come to algorithms that solve the *integer optimization problem* in fixed dimension. Here one is given an integer matrix $\mathbf{A} \in \mathbb{Z}^{m \times n}$ and integer vectors $\mathbf{d} \in \mathbb{Z}^m$ and $\mathbf{c} \in \mathbb{Z}^n$, where the dimension n is fixed. The task is to find an integer vector $x^* \in \mathbb{Z}^n$ that satisfies $Ax \leq d$, and that maximizes $c^T x$. Thus the integer feasibility problem is a subproblem of the integer optimization problem. Let φ be the maximum size of c and a constraint $a_i x \leq d_i$ of $Ax \leq d$. The running time of the methods described here will be estimated in terms of the number of constraints m and the number φ.

The integer optimization problem can be reduced to the integer feasibility problem (27) via binary search, see e.g. [54, 99]. This approach yields a running time of $O(m\varphi + \varphi^2)$, and is described in Section 5.1.

There have been many efficient algorithms for the 2-dimensional integer optimization problem. Feit [46], and Zamanskij and Cherkasskij [106] provided an algorithm for the 2-dimensional integer optimization problem that runs in $O(m \log m + m\varphi)$ steps. Other algorithms are by Kanamaru et al. [62] ($O(m \log m + \varphi)$), and by Eisenbrand and Rote [42] ($O(m + \log(m)\varphi)$). Eisenbrand and Laue [41] recently provided a linear time algorithm ($O(m + \varphi)$).

A randomized algorithm for arbitrary fixed dimension was proposed by Clarkson [25], which we present in Section 5.3. His result can be stated in the more general framework of the *LP-type problems*. Applied to integer programming, the result is as follows. An integer optimization problem that is defined by m constraints can be solved with an expected number of $O(m)$ basic operations and $O(\log m)$ calls to another algorithm that solves an integer optimization problem with a fixed number of constraints, see also [48]. In the description of Clarkson's algorithm here, we ignore the dependence of the running time on the dimension. Clarkson's algorithm has played an important role in the search for faster algorithms in varying dimension for *linear programming* in the ram-model of complexity. For more on this fascinating topic, see [80] and [48].

We also sketch a recent result of Eisenbrand [40] in Section 5.2, which shows that an integer optimization problem of binary encoding size φ with a fixed number of constraints can be solved with $O(\varphi)$ arithmetic operations on rationals of size $O(\varphi)$. Thus with Clarkson's result one obtains an expected running time of $O(m + (\log m)\varphi)$ arithmetic operations for the integer optimization problem.

First we will transform the integer optimization problem into a more convenient form. If $U \in \mathbb{Z}^{n \times n}$ is a unimodular matrix, then by substituting $y = U^{-1}x$, the integer optimization problem above is the problem to find a vector $y^* \in \mathbb{Z}^n$ that satisfies $AUy^* \leq d$ and maximizes $c^T Uy$. With a sequence of the extended-greatest common divisor operations, one can compute a unimodular $U \in \mathbb{Z}^{n \times n}$ of binary encoding length $O(\varphi)$ (n is fixed) such that $c^T U = (\gcd(c_1, \ldots, c_n), 0 \ldots, 0)$. Therefore we can assume that the objective vector c is the first unit vector.

The algorithms for the integer feasibility problem (27), which we discussed in Section 4, require $O(m + \varphi)$ arithmetic operations to be solved. This is linear in the input encoding. Therefore we can assume that the system $Ax \leq d$ is integer feasible.

Now, there exists an optimal $\boldsymbol{x}^* \in \mathbb{Z}^n$ whose binary encoding length is $O(\varphi)$, see, e.g. Schrijver [99, p. 239]. This means that we can assume that the constraints $\boldsymbol{Ax} \leq \boldsymbol{d}$ describe a polytope. This polytope can be translated with an integer vector into the positive orthant.

Notice that the above described transformation can be carried out with $O(m+\varphi)$ basic operations. Furthermore the number of constraints of the transformed system is $O(m)$ and the binary encoding length of each constraint remains $O(\varphi)$. Thus given $\boldsymbol{A}, \boldsymbol{d}$, and $\boldsymbol{c}$, we can in $O(m+\varphi)$ steps check whether the system $\boldsymbol{Ax} \leq \boldsymbol{d}$ is integer feasible and carry out the above described transformation. We therefore define the integer optimization problem as being the following:

Given an integer matrix $\boldsymbol{A} \in \mathbb{Z}^{m\times n}$ and an integer vector $\boldsymbol{d} \in \mathbb{Z}^m$ defining a polytope $P = \{\boldsymbol{x} \in \mathbb{R}^n \mid \boldsymbol{Ax} \leq \boldsymbol{d}\}$ such that $P \subseteq \mathbb{R}^n_{\geq 0}$ and $P \cap \mathbb{Z}^n \neq \emptyset$:

$$\text{Find an integer vector } \boldsymbol{x}^* \in \mathbb{Z}^n, \text{ with maximal first component, satisfying } \boldsymbol{Ax}^* \leq \boldsymbol{d}. \tag{43}$$

5.1 *Binary search*

We first describe and analyze the binary search technique for the integer optimization problem. As we argued, we can assume that $P \subseteq [0, M]^n$, where $M \in \mathbb{N}$, and that M is part of the input. In the course of binary search, one keeps two integers $l, u \in \mathbb{N}$ such that $l \leq x_1^* \leq u$. We start with $l = 0$ and $u = M$. In the i-th iteration, one checks whether the system $\boldsymbol{Ax} \leq \boldsymbol{d}$, $x_1 \geq \lfloor (l+u)/2 \rfloor$ is integer feasible. If it is feasible, then one sets $l = \lfloor (l+u)/2 \rfloor$. If the system is integer infeasible, one sets $u = \lfloor (l+u)/2 \rfloor$. After $O(\text{size}(M))$ steps one has either $l = u$ or $l+1 = u$ and the optimum can be found with at most two more calls to an integer feasibility algorithm.

The binary encoding length of M is at most $O(\varphi)$, see, e.g. [99, p. 120]. Therefore the integer optimization problem can be solved with $O(\varphi)$ queries to an integer feasibility algorithm.

Theorem 15. *An integer optimization problem* (43) *in fixed dimension defined by m constraints, each of binary encoding length at most φ, can be solved with* $O(m\varphi + \varphi^2)$ *basic operations on rational numbers of size* $O(\varphi)$.

5.2 *A linear algorithm*

In this section, we outline a recent algorithm by Eisenbrand [40] that solves an integer optimization problem with a fixed number of constraints in linear time. Thus, the complexity of integer feasibility with a fixed number of variables and a fixed number of constraints can be matched with the complexity of the Euclidean algorithm in the arithmetic model.

As in the algorithms in Sections 4.2 and 4.3 one makes use of the lattice width concept, see Expression (28) and Theorem 9 in the introduction of Section 4.

The first step of the algorithm is to reduce the integer optimization problem over a full-dimensional polytope to a disjunction of integer optimization problems over *two-layer simplices*. A two layer simplex is a full-dimensional simplex, whose vertices can be partitioned into two sets V and W, such that the first components of the elements in each of the sets V and W agree, i.e., for all $\boldsymbol{v}_1, \boldsymbol{v}_2 \in V$ one has $v_1^1 = v_2^1$ and for all $\boldsymbol{w}_1, \boldsymbol{w}_2 \in W$ one has $w_1^1 = w_2^1$.

How can one reduce the integer optimization problem over a polytope P to a sequence of integer optimization problems over two-layer simplices? Simply consider the hyperplanes $x_1 = v_1$ for each vertex $\boldsymbol{v}$ of P. If the number of constraints defining P is fixed, then these hyperplanes partition P into a constant number of polytopes, whose vertices can be grouped into two groups, according to the value of their first component. Thus we can assume that the vertices of P itself can be partitioned into two sets V and W, such that the first components of the elements in each of the sets V and W agree. Carathéodory's theorem, see Schrijver [99, p. 94], implies that P is covered by the simplices that are spanned by the vertices of P. These simplices are two-layer simplices. Therefore, the integer optimization problem in fixed dimension with a fixed number of constraints can be reduced in constant time to a constant number of integer optimization problems over a two-layer simplex.

The key idea is then to let the objective function slide into the two-layer simplex, until the width of the truncated simplex exceeds the flatness bound. In this way, one can be sure that the optimum of the integer optimization problem lies in the truncation, which is still flat. Thereby one has reduced the *integer optimization problem* in dimension n to a constant number of *integer optimization problems* in dimension $n-1$ and binary search can be avoided.

How do we determine a parameter π such that truncated two-layer simplex $\Sigma \cap (x_1 \geq \pi)$ just exceeds the flatness bound? We explain the idea with the help of the 3-dimensional example in Figure 8. Here we have a two-layer simplex Σ in 3-space. The set V consists of the points $\mathbf{0}$ and $\boldsymbol{v}_1$ and W consists of $\boldsymbol{w}_1$ and $\boldsymbol{w}_2$. The picture on the left describes a particular point in time, where the objective function slid into Σ. So we consider the truncation $\Sigma \cap (x_1 \geq \pi)$ for some $\pi \geq w_1^1$. This truncation is the convex hull of the points

$$\mathbf{0}, \boldsymbol{v}_1, \mu \boldsymbol{w}_1, \mu \boldsymbol{w}_2, (1-\mu)\boldsymbol{v}_1 + \mu \boldsymbol{w}_1, (1-\mu)\boldsymbol{v}_1 + \mu \boldsymbol{w}_2, \tag{44}$$

where $\mu = \pi / w_1^1$. Now consider the simplex $\Sigma_{V,\mu W}$, which is spanned by the points $\mathbf{0}, \boldsymbol{v}_1, \mu \boldsymbol{w}_1, \mu \boldsymbol{w}_2$. This simplex is depicted on the right in Figure 8. If this simplex is scaled by 2, then it contains the truncation $\Sigma \cap (x_1 \geq \pi)$. This is easy to see, since the scaled simplex contains the points $2(1-\mu)\boldsymbol{v}_1, 2\mu \boldsymbol{w}_1$ and $2\mu \boldsymbol{w}_2$. So we have the condition $\Sigma_{V,\mu W} \subseteq \Sigma \cap (x_1 \geq \pi) \subseteq 2\Sigma_{V,\mu W}$. From this we can infer the important observation

$$w(\Sigma_{V,\mu W}) \leq w(\Sigma \cap (x_1 \geq \pi)) \leq 2w(\Sigma_{V,\mu W}). \tag{45}$$

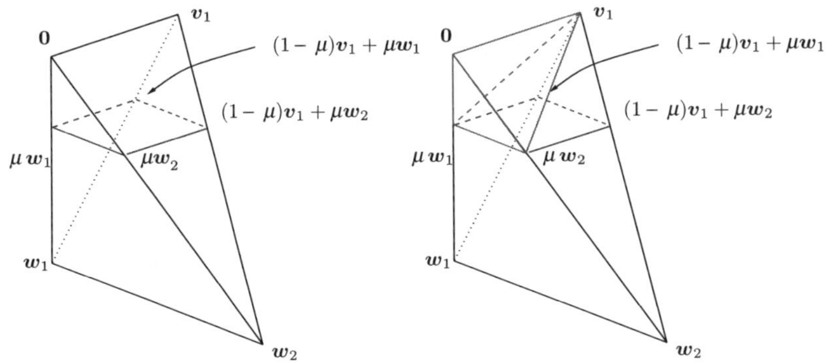

Figure 8. Solving the parametric lattice width problem.

This means that we essentially determine the correct π by determining a $\mu \geq 0$, such that the width of the simplex $\Sigma_{V,\mu W}$ just exceeds the flatness bound. The width of $\Sigma_{V,\mu W}$ is roughly (up to a constant factor) the length of the shortest vector of the lattice $L_\mu = L(A)$, where A is the matrix

$$A = \begin{pmatrix} \mu w_1^T \\ \mu w_2^T \\ v_1 \end{pmatrix}.$$

Thus we have to find a parameter μ, such that the shortest vector of L_μ is sandwiched between $f(n)+1$ and $\gamma \cdot (f(n)+1)$ for some constant γ. This problem can be understood as a *parametric shortest vector* problem.

To describe this problem, let us introduce some notation. We define for an $n \times n$-matrix $A = (a_{ij})_{\forall i,j}$, the matrix $A^{\mu,k} = (a_{ij})^{\mu,k}_{\forall i,j}$, as

$$a_{ij}^{\mu,k} = \begin{cases} \mu \cdot a_{ij}, & \text{if } i \leq k, \\ a_{ij}, & \text{otherwise.} \end{cases} \tag{46}$$

In other words, the matrix $A^{\mu,k}$ results form A by scaling the first k rows with μ. The parametric shortest vector problem is now defined as follows.

> Given a nonsingular matrix $A \in \mathbb{Z}^{n\times n}$ and some $U \in \mathbb{N}$, find a parameter $p \in \mathbb{N}$ such that $U \leq \mathrm{SV}(L(A^{p,k})) \leq 2^{n+1/2} \cdot U$, or assert that $\mathrm{SV}(L) > U$.

It turns out that the parametric shortest vector problem can be solved in linear time when the dimension in fixed. From this, it follows that the integer optimization problem in fixed dimension with a fixed number of constraints can be solved in linear time.

Theorem 16 ([40]). *An integer program of binary encoding length φ in fixed dimension, which is defined by a fixed number of constraints, can be solved with $O(\varphi)$ arithmetic operations on rational numbers of binary encoding length $O(\varphi)$.*

5.3 Clarkson's random sampling algorithm

Clarkson [25] presented a *randomized algorithm* for problems of *linear programming type*. This algorithm solves an integer optimization problem in fixed dimension that is defined by m constraints with an expected number of $O(m)$ basic arithmetic operations and $O(\log m)$ calls to an algorithm that solves an integer optimization problem defined by a fixed-size subset of the constraints. The expected running time of this method for an integer optimization problem defined by m constraints, each of size at most φ, can thus be bounded by $O(m+(\log m)\varphi)$ arithmetic operation on rationals of size $O(\varphi)$.

Let P be the polytope defined by $P=\{\boldsymbol{x}\in\mathbb{R}^n \mid A\boldsymbol{x}\leq \boldsymbol{d},\ 0\leq x_j\leq M,\ 1\leq j\leq n\}$. The integer vectors $\tilde{\boldsymbol{x}}\in\mathbb{Z}^n\cap P$ satisfy $0\leq\tilde{\boldsymbol{x}}_j\leq M$ for $1\leq j\leq n$, where M is an integer of binary encoding length $O(\varphi)$. A feasible integer point $\tilde{\boldsymbol{x}}$ is optimal with respect to the objective vector $\boldsymbol{c}=((M+1)^{n-1},(M+1)^{n-2},\ldots,(M+1)^0)^T$ if and only if it has maximal first component. Observe that the binary encoding length of this perturbed objective function vector $\boldsymbol{c}$ is $O(\varphi)$. Moreover, for each pair of distinct points $\tilde{\boldsymbol{x}}_1,\tilde{\boldsymbol{x}}_2\in[0,M]^n\cap\mathbb{Z}^n$, $\tilde{\boldsymbol{x}}_1\neq\tilde{\boldsymbol{x}}_2$, we have $\boldsymbol{c}^T\tilde{\boldsymbol{x}}_1\neq\boldsymbol{c}^T\tilde{\boldsymbol{x}}_2$.

In the sequel we use the following notation. If H is a set of linear integer constraints, then the *integer optimum* defined by H is the *unique* integer point $\boldsymbol{x}^*(H)\in\mathbb{Z}^n\cap[0,M]^n$ which satisfies all constraints $h\in H$ and maximizes $\boldsymbol{c}^T\boldsymbol{x}$. Observe that, due to the perturbed objective function $\boldsymbol{c}^T\boldsymbol{x}$, the point $\boldsymbol{x}^*(H)$ is uniquely defined for any set of constraints H. The integer optimization problem now reads as follows:

$$\text{Given a set } H \text{ of integer constraints in fixed dimension, find } \boldsymbol{x}^*(H). \tag{47}$$

A *basis* of a set of constraints H, is a minimal subset B of H such that $\boldsymbol{x}^*(B)=\boldsymbol{x}^*(H)$. The following is a consequence of a theorem of Bell [13] and Scarf [89], see Schrijver [99, p. 234].

Theorem 17. *Any set H of constraints in dimension n has a basis B of cardinality $|B|\leq 2^n-1$.*

In the following, we use the letter D for the number 2^n-1. Clarksons algorithm works for many *LP-type* problems, see Gärtner and Welzl [48] for more examples. The maximal cardinality of a basis is generally referred to as the *combinatorial dimension* of the LP-type problem.

Now we are ready to describe the algorithm. It comes in two layers that we call *Clarkson 1* and *Clarkson 2* respectively. The input of both algorithms is a set of constraints H and the output $\boldsymbol{x}^*(H)$. The algorithm *Clarkson 1* keeps a constraint set G, which is initially empty and grows in the course of the algorithm. In one iteration, one draws a subset $R\subseteq H$ of cardinality $|R|=\lceil D\sqrt{m}\rceil$ at random and computes the optimum $\boldsymbol{x}^*(G\cup R)$ with the algorithm *Clarkson 2* described later. Now one identifies the constraints $V\subseteq H$ that are violated by $\boldsymbol{x}^*(G \cup R)$. We will prove below that the expected cardinality of V is $\sqrt{m}$. In Step (2c), the constraints V are added to the set G, if the cardinality of V does not exceed twice its expected cardinality. In this case, i.e., if $|V| \leq 2\sqrt{m}$, then an iteration of the REPEAT-loop is called *successful*.

Algorithm 3 (Clarkson 1).

1. $r \leftarrow \lceil D\sqrt{m}\rceil$, $G \leftarrow \emptyset$
2. REPEAT
   - (a) Choose random $R \in \binom{H}{r}$
   - (b) Compute $\boldsymbol{x}^* = \boldsymbol{x}^*(G \cup R)$ *with Clarkson 2*
   - (c) $V \leftarrow \{h \in H \mid \boldsymbol{x}^* \text{ violates } h\}$
   - (*d*) IF $|V| \leq 2\sqrt{m}$, THEN $G \leftarrow G \cup V$
3. UNTIL $V = \emptyset$
4. RETURN $\boldsymbol{x}^*$

How many expected iterations will *Clarkson 1* perform? To analyze this, let $B \subseteq H$ be a basis of H. Observe that, if the set V, which is computed in Step (2c), is nonempty, then there must be a constraint $b \in B$ that also belongs to V. Because, if no constraint in B is violated by $\boldsymbol{x}^*(G \cup R)$, then one has $\boldsymbol{x}^*(G \cup R) = \boldsymbol{x}^*(G \cup R \cup B) = \boldsymbol{x}^*(H)$ and V must be empty. Thus at each successful iteration, at least one new element of B enters the set G. We conclude that the number of successful iterations is bounded by D. The *Markov inequality*, see, e.g. Motwani and Raghavan [84] says that the probability that a random variable exceeds k-times its expected value is bounded by $1/k$. Therefore the expected number of iterations of the REPEAT-loop is bounded by $2D$. The additional arithmetic operations of each iteration is $O(m)$ if n is fixed, and each iteration requires the solution of an integer optimization problem in fixed dimension with $O(\sqrt{m})$ constraints.

Theorem 18 ([25]). *Given a set H of m integer linear constraints in fixed dimension, the algorithm* Clarkson 1 *computes $\boldsymbol{x}^*(H)$ with a constant number of expected calls to an algorithm which solves the integer optimization problem for a subset of* $O(\sqrt{m})$ *constraints and an expected number of* $O(m)$ *basic operations.*

We still need to prove that the expected cardinality of V in Step (2c) is at most $\sqrt{m}$. Following the exposition of Gärtner and Welzl [48], we do this in the slightly more general setting where H can be a multiset of constraints.

Lemma 3 ([25, 48]). *Let G be a set of integer linear constraints and let H be a multiset of m integer constraints in dimension n. Let $R \in \binom{H}{r}$ be a random subset of H of cardinality r. The expected cardinality of the set $V_R = \{h \in H \mid \boldsymbol{x}^*(G \cup R)$ violates $h\}$ is at most $D(m-r)/(r+1)$.*

This lemma establishes our desired bound on the cardinality of V in Step 2c, because there we have $r = \lceil D\sqrt{m} \rceil$ and thus

$$D(m-r)/(r+1) \leq Dm/r \leq \sqrt{m}. \tag{48}$$

Proof of Lemma 3. The expected cardinality of V_R is equal to the sum of all the cardinalities of V_R, where R is an r-element subset of H, divided by the number of ways that r elements can be drawn from H,

$$E(|V_R|) = \left(\sum_{R \in \binom{H}{r}} |V_R| \right) \Big/ \binom{m}{r}. \tag{49}$$

Let $\chi_G(Q, h)$, $Q \subseteq H$, $h \in H$ be the characteristic function for the event that $\boldsymbol{x}^*(G \cup Q)$ violates h. Thus

$$\chi_G(Q,h) = \begin{cases} 1 & \text{if } \boldsymbol{x}^*(G \cup Q) \text{ violates } h, \\ 0 & \text{otherwise.} \end{cases} \tag{50}$$

With this we can write

$$\binom{m}{r} E(|V_R|) = \sum_{R \in \binom{H}{r}} \sum_{h \in H \setminus R} \chi_G(R, h) \tag{51}$$

$$= \sum_{Q \in \binom{H}{r+1}} \sum_{h \in Q} \chi_G(Q - h, h) \tag{52}$$

$$\leq \sum_{Q \in \binom{H}{r+1}} D \tag{53}$$

$$= \binom{m}{r+1} D. \tag{54}$$

From (51) to (52) we used the fact that the ways in which we can choose a set R of cardinality r from H and then a constraint h from $H \setminus R$ are exactly the ways in which we can choose a set Q of cardinality $r+1$ from H and then one constraint h from Q. To justify the step from (52) to (53), consider a basis B_Q of $Q \cup G$. If h is not from the basis B_Q, then $\boldsymbol{x}^*(G \cup Q) = \boldsymbol{x}^*(G \cup (Q \setminus \{h\}))$. Therefore $\sum_{h \in Q} \chi_G(Q - h, h) \leq D$. □

The algorithm *Clarkson 2* proceeds from another direction. Instead of randomly sampling large sets of constraints and augmenting a set of constraints G, one at the time, a set R of cardinality $6D^2$ is drawn and the optimum $\boldsymbol{x}^*(R)$ is determined in each iteration with the algorithm outlined in Section 5.2. As in *Clarkson 1* one determines the constraints $V = \{h \in H \mid \boldsymbol{x}^*(R) \text{ violates } h\}$. If this set is nonempty, then there must be constraints of a basis B of H that are in V. One then doubles the probability of each constraint $h \in V$ to be drawn in the next round. This procedure is repeated until $V = \emptyset$. Instead of explicitly speaking about the probabilities of a constraint $h \in H$, we follow again the exposition of Gärtner and Welzl [48], who assign a *multiplicity* $\mu(h) \in \mathbb{N}$ to each constraint of H. In this way, one can think of H as being a multiset and apply Lemma 3 in the analysis. Let $Q \subseteq H$ be a subset of the constraints, then $\mu(Q)$ denotes the sum of the multiplicities of Q, $\mu(Q) = \sum_{h \in Q} \mu(h)$. In the beginning $\mu(h) = 1$ for each $h \in H$.

Algorithm 4 (Clarkson 2).

1. $r \leftarrow 6D^2$
2. REPEAT:
   (a) Choose random $R \in \binom{H}{r}$
   (b) Compute $\boldsymbol{x}^* = \boldsymbol{x}^*(R)$
   (c) $V \leftarrow \{h \in H \mid \boldsymbol{x}^* \text{ violates } h\}$
   (d) IF $\mu(V) \leq \mu(H)/(3D)$ THEN for all $h \in V$ do $\mu_h \leftarrow 2\mu_h$
3. UNTIL $V = \emptyset$
4. RETURN $\boldsymbol{x}^*$

An iteration through the REPEAT-loop is called a *successful iteration*, if the condition in the IF-statement in Step (2d) is true. Using Lemma 3 the expected cardinality of V (as a multiset) is at most $\mu(H)/(6D)$. Again with the Markov inequality, the expected number of total iterations is at most twice the number of the successful iterations of the algorithm.

Let $B \subseteq H$ be a basis of H. In each successful iteration, the multiplicity of at least one element of B is doubled. Since $|B| \leq D$, the multiplicity of at least one element of B will be at least 2^k after kD successful iterations. Therefore one has $2^k \leq \mu(B)$ after kD successful iterations.

The number $\mu(B)$ is bounded by $\mu(H)$. In the beginning $\mu(H) = m$. After Step (2d) one has $\mu(H) := \mu(H) + \mu(V) \leq \mu(H)(1 + 1/(3D))$. Thus after kD successful iterations one has $\mu(B) \leq m(1 + 1/(3D))^{kD}$. Using the inequality $e^t \geq (1 + t)$ for $t \geq 0$, we obtain the following lemma on the number of successful iterations.

Lemma 4. *Let B be a basis of H and suppose that H has at least $6D^2$ elements. After kD successful iterations of* Clarkson 2 *one has*

$$2^k \leq \mu(B) \leq me^{k/3}.$$

This implies that the number of successful iterations is bounded by $O(\log m)$. The expected number of iterations is therefore also $O(\log m)$. In each iteration, one computes one integer optimization problem with a fixed number of constraints. If φ is the maximal binary encoding length of a constraint in H, this costs $O(\varphi)$ basic operations with the linear algorithm of Section 5.2. Then one has to check each constraint in H, whether it is violated by $\boldsymbol{x}^*(R)$. This costs $O(m)$ arithmetic operations. Altogether we obtain the following running time.

Lemma 5 ([25]). *Let H be a set of integer linear constraints in fixed dimension and let φ be the maximal binary encoding length of a constraint $h \in H$. Then the integer optimization problem* (47) *can be solved with the randomized algorithm* Clarkson 2 *with an expected number of* $O(m \log m + (\log m)\,\varphi)$ *basic operations.*

Now we estimate the running time of *Clarkson 1* where we plug in the running time bound for Step (2b). We obtain an expected constant number of calls to *Clarkson 2* on $O(\sqrt{m})$ constraints and an additional cost of $O(m)$ basic operations for the other steps. Thus we have a total amount of $O(m + \sqrt{m} \log \sqrt{m} + (\log \sqrt{m})\varphi) = O(m + (\log m)\varphi)$ basic operations.

Theorem 19 ([25]). *Let H be a set of integer linear constraints in fixed dimension and let φ be the maximal binary encoding length of a constraint $h \in H$. Then the integer optimization problem* (43) *can be solved with a randomized algorithm with an expected number of* $O(m + (\log m)\varphi)$ *basic operations.*

6 Using lattices to reformulate the problem

Here we will study some special types of integer feasibility problems that have been successfully solved by the following approach. Create a lattice L such that we can say that feasible solutions to our problem are short vectors in L. Once we have L, we write down an initial basis $\boldsymbol{B}$ for L, we then apply basis reduction to $\boldsymbol{B}$, which produces $\boldsymbol{B}'$. The columns of $\boldsymbol{B}'$ are relatively short and some might be feasible for our problem. If not, do a search for a feasible solution, or prove than none exists.

In Section 6.1 we present results for subset sum problems arising in the knapsack cryptosystems. In cryptography, researchers have made extensive use of lattices and basis reduction algorithms to *break* cryptosystems; their computational experiments were among the first to establish the practical effectiveness of basis reduction algorithms. On the "constructive side" recent complexity results on lattice problems have also inspired researchers to *develop* cryptographic schemes based on the hardness of certain lattice problems. Even though cryptography is not within the central scope of this chapter, and even though the knapsack cryptosystems have long been broken, we still wish to present the main result by Lagarias and Odlyzko [74], since it illustrates a nice application of lattice basis reduction, and since it has inspired the work on integer programming presented in Section 6.2. There, we will see how systems of linear diophantine equations with lower and upper bounds on the variables can be solved by similar techniques.

For comprehensive surveys on the topic of lattices in cryptography we refer to the surveys of Joux and Stern [61], and of Nguyen and Stern [86, 87].

6.1 Cryptosystems – solving subset sum problems

A sender wants to transmit a message to a receiver. The plaintext message of the sender consists of a 0–1 vector $\boldsymbol{x} = (x_1, \ldots, x_n)$, and this message is encrypted by using integer weights $a_1, \ldots, a_n$ leading to an encrypted message $a_0 = \sum_{j=1}^{n} a_j x_j$. The coefficients a_j, $1 \le j \le n$, are known to the public, but there is a hidden structure in the relation between these coefficients, called a trapdoor, which only the receiver knows. If the trapdoor is known, then the subset sum problem:

$$\text{Determine a 0-1 vector } \boldsymbol{x} \text{ such that } \sum_{j=1}^{n} a_j x_j = a_0 \tag{55}$$

can be solved easily. For an eavesdropper who does not know the trapdoor, however, the subset sum problem should be hard to solve in order to obtain a secure transmission.

The *density* of a set of coefficients a_j, $1 \le j \le n$ is defined as

$$\delta(\boldsymbol{a}) = d(\{a_1, \ldots, a_n\}) = \frac{n}{\log_2(\max_{1 \le j \le n}\{a_j\})}.$$

The density, as defined above, is an approximation of the information rate at which bits are transmitted. The interesting case is $\delta(\boldsymbol{a}) \le 1$, since for $\delta(\boldsymbol{a}) > 1$ the subset sum problem (55) will in general have several solutions, which makes it unsuitable for generating encrypted messages. Lagarias and Odlyzko [74] proposed an algorithm based on basis reduction that often finds a solution to the subset sum problem (55) for instances having relatively low density. Earlier research had found methods based on recovering trapdoor information. If the information rate is high, i.e., $\delta(\boldsymbol{a})$ is high, then the trapdoor information is relatively hard to conceal. The result of Lagarias and Odlyzko therefore complements the earlier results by providing a method that is successful for low-density instances. In their algorithm Lagarias and Odlyzko consider a lattice $\mathbb{Z}^{n+1}$ consisting of vectors of the following form:

$$L_{\boldsymbol{a},a_0} = \{(x_1, \ldots, x_n, (\boldsymbol{ax} - a_0\xi))^T\} \tag{56}$$

where ξ is a variable associated with the right-hand side of $\boldsymbol{ax} = a_0$. Notice that the lattice vectors that are interesting for the subset sum problem all have $\xi = 1$ and $\boldsymbol{ax} - a_0\xi = 0$. It is easy to write down an initial basis $\boldsymbol{B}$ for $L_{\boldsymbol{a},a_0}$:

$$B = \begin{pmatrix} \boldsymbol{I}^{(n)} & \boldsymbol{0}^{(n\times 1)} \\ \boldsymbol{a} & -a_0 \end{pmatrix}. \tag{57}$$

To see that $\boldsymbol{B}$ is a basis for $L_{\boldsymbol{a},a_0}$, we note that taking integer linear combinations of the column vectors of $\boldsymbol{B}$ generates vectors of type (56). Let $\boldsymbol{x} \in \mathbb{Z}^n$ and $\xi \in \mathbb{Z}$. We obtain

$$\begin{pmatrix} \boldsymbol{x} \\ \boldsymbol{ax} - a_0\xi \end{pmatrix} = \boldsymbol{B}\begin{pmatrix} \boldsymbol{x} \\ \xi \end{pmatrix}.$$

The algorithm SV (Short Vector) by Lagarias and Odlyzko consists of the following steps.

1. Apply Lovász' basis reduction algorithm to the basis $\boldsymbol{B}$ (57), which yields a reduced basis $\tilde{\boldsymbol{B}}$.
2. Check if any of the columns $\tilde{\boldsymbol{b}}_k = (\tilde{b}_k^1, \ldots, \tilde{b}_k^{n+1})$ has all $\tilde{b}_k^j = 0$ or γ for some fixed constant γ, for $1 \leq j \leq n$. If such a reduced basis vector is found, check if the vector $x_j = \tilde{b}_k^j/\gamma$, $1 \leq j \leq n$, is a solution to $\sum_{j=1}^n a_j x_j = a_0$, and if yes, stop. Otherwise go to Step 3.
3. Repeat Steps 1 and 2 for the basis $\boldsymbol{B}$ with $a_0 = \sum_{j=1}^n a_j - a_0$, which corresponds to complementing all x_j-variables, i.e., considering $1 - x_j$ instead of x_j.

Algorithm SV runs in polynomial time as Lovász' basis reduction algorithm runs in polynomial time. It is not certain, however, that algorithm SV actually produces a solution to the subset sum problem. As Theorem 20 below shows, however, we can expect algorithm SV to work well on instances of (55) having low density. Consider a 0-1 vector $\boldsymbol{x}$, which we will consider as fixed. We assume that $\sum_{j=1}^n x_j \leq \frac{n}{2}$. The reason for this assumption is that either $\sum_{j=1}^n x_j \leq \frac{n}{2}$, or $\sum_{j=1}^n x'_j \leq \frac{n}{2}$, where $x'_j = (1 - x_j)$, and since algorithm SV is run for both cases, one can perform the analysis for the vector that does satisfy the assumption. Let $\bar{\boldsymbol{x}} = (x_1, \ldots, x_n, 0)$. Let the sample space $\Lambda(A, \bar{\boldsymbol{x}})$ of lattices be defined to consist of all lattices $\boldsymbol{L}_{\boldsymbol{a},a_0}$ generated by the basis (57) such that

$$1 \leq a_j \leq A, \quad \text{for } 1 \leq j \leq n, \tag{58}$$

and

$$a_0 = \sum_{j=1}^n a_j \bar{x}_j.$$

There is precisely one lattice in the sample space for each vector $\boldsymbol{a}$ satisfying (58). Therefore the sample space consists of A^n lattices.

Theorem 20 ([74]). *Let $\bar{\boldsymbol{x}}$ be a 0-1 vector for which $\sum_{j=1}^{n} \bar{x}_j \leq \frac{n}{2}$. If $A = 2^{\beta n}$ for any constant $\beta > 1.54725$, then the number of lattices $L_{\boldsymbol{a}, a_0}$ in $\Lambda(A, \bar{\boldsymbol{x}})$ that contain a vector $\boldsymbol{v}$ such that $\boldsymbol{v} \neq k\bar{\boldsymbol{x}}$ for all $k \in \mathbb{Z}$, and such that $\|\boldsymbol{v}\|^2 \leq \frac{n}{2}$ is*

$$O(A^{n-c_1(\beta)}(\log A)^2), \tag{59}$$

where $c_1(\beta) = 1 - \frac{1.54725}{\beta} > 0$.

For $A = 2^{\beta n}$, the density of the subset sum problems associated with the lattices in the sample space can be proved to be equal to β^{-1}. This implies that Theorem 20 applies to lattices having density $\delta(\boldsymbol{a}) < (1.54725)^{-1} \approx 0.6464$. Expression (59) gives a bound on the number of lattices we need to subtract from the total number of lattices in the sample space, A^n, in order to obtain the number of lattices in $\Lambda(A, \bar{\boldsymbol{x}})$ for which $\bar{\boldsymbol{x}}$ is the *shortest* non-zero vector. Here we notice that the term (59) grows slower than the term A^n as n goes to infinity, and hence we can conclude that "almost all" lattices in the sample space $\Lambda(A, \bar{\boldsymbol{x}})$ have $\bar{\boldsymbol{x}}$ as the shortest vector. So, the subset sum problems (55) with density $\delta(\boldsymbol{a}) < 0.6464$ could be solved in polynomial time if we had an oracle that could compute the shortest vector in the lattice $L_{\boldsymbol{a},a_0}$. Lagarias and Odlyzko also prove that the algorithm SV actually finds a solution to "almost all" feasible subset sum problems (55) having density $\delta(\boldsymbol{a}) < (2-\epsilon)(\log(\frac{4}{3}))^{-1} n^{-1}$ for any fixed $\epsilon > 0$.

Coster, Joux, LaMacchia, Odlyzko, Schnorr, and Stern [34] proposed two ways of improving Theorem 20. They showed that "almost all" subset sum problems (55) having density $\delta(\boldsymbol{a}) < 0.9408$ can be solved in polynomial time in presence of an oracle that finds the shortest vector in certain lattices. Both ways of improving the bound on the density involve some changes in the lattice considered by Lagarias and Odlyzko. The first lattice $L'_{\boldsymbol{a},a_0} \in \mathbb{Q}^{n+1}$ considered by Coster et al. is defined as

$$L'_{\boldsymbol{a},a_0} = \left\{ \left(x_1 - \frac{1}{2}\xi, \ldots, x_n - \frac{1}{2}\xi, N(\boldsymbol{a}\boldsymbol{x} - a_0\xi) \right)^T \right\},$$

where N is a natural number. The following basis $\bar{\boldsymbol{B}}$ spans $L'_{\boldsymbol{a},a_0}$:

$$\bar{\boldsymbol{B}} = \begin{pmatrix} \boldsymbol{I}^{(n)} & \left(-\frac{1}{2}\right)^{(n\times 1)} \\ N\boldsymbol{a} & -Na_0 \end{pmatrix}. \tag{60}$$

As in the analysis by Lagarias and Odlyzko, we consider a fixed vector $\boldsymbol{x} \in \{0, 1\}^n$, and we let $\bar{\boldsymbol{x}} = (x_1, \ldots, x_n, 0)$. The vector $\bar{\boldsymbol{x}}$ does not belong to the lattice $L'_{\boldsymbol{a},a_0}$, but the vector $\boldsymbol{w} = (w_1, \ldots, w_n, 0)$, where $w_j = x_j - \frac{1}{2}$, $1 \le j \le n$ does. So, if Lovász' basis reduction algorithm is applied to $\bar{\boldsymbol{B}}$ and if the reduced basis $\bar{\boldsymbol{B}}'$ contains a vector $(w_1, \ldots, w_n, 0)$ with $w_j = \{-\frac{1}{2}, \frac{1}{2}\}$, $1 \le j \le n$, then the vector $(w_j + \frac{1}{2})$, $1 \le j \le n$ solves the subset sum problem (55). By shifting the feasible region to be symmetric about the origin we now look for vectors of shorter Euclidean length. Coster et al. prove the following theorem that is analogous to Theorem 20.

Theorem 21 ([34]). *Let A be a natural number, and let $a_1, \ldots, a_n$ be random integers such that $1 \le a_j \le A$, for $1 \le j \le n$. Let $\boldsymbol{x} = (x_1, \ldots, x_n)$, $x_j \in \{0, 1\}$, be fixed, and let $a_0 = \sum_{j=1}^n a_j x_j$. If the density $\delta(\boldsymbol{a}) < 0.9408$, then the subset sum problem (55) defined by $a_1, \ldots, a_n$ can "almost always" be solved in polynomial time by a single call to an oracle that finds the shortest vector in the lattice $L'_{\boldsymbol{a},a_0}$.*

Coster et al. prove Theorem 21 by showing that the probability that the lattice $L'_{\boldsymbol{a},a_0}$ contains a vector $\boldsymbol{v} = (v_1, \ldots, v_{n+1})$ satisfying

$$\boldsymbol{v} \neq k\boldsymbol{w} \quad \text{for all } k \in \mathbb{Z}, \text{ and } \|\boldsymbol{v}\|^2 \le \|\boldsymbol{w}\|^2$$

is bounded by

$$n\big(4n\sqrt{n} + 1\big)\frac{2^{c_0 n}}{A} \tag{61}$$

for $c_0 = 1.0628$. Using the lattice $L'_{\boldsymbol{a},a_0}$, note that $\|\boldsymbol{w}\|^2 \le \frac{n}{4}$. The number N in basis (60) is used in the following sense. Any vector in the lattice L' is an integer linear combination of the basis vectors. Hence, the $(n+1)$-st element of a such a lattice vector is an integer multiple of N. If N is chosen large enough, then a lattice vector can be "short" only if the $(n+1)$-st element is equal to zero. Since it is known that the length of $\boldsymbol{w}$ is bounded by $\frac{1}{2}\sqrt{n}$, then it suffices to choose $N > \frac{1}{2}\sqrt{n}$ in order to conclude that for a vector $\boldsymbol{v}$ to be shorter than $\boldsymbol{w}$ it should satisfy $v_{n+1} = 0$. Hence, Coster et al. only need to consider lattice vectors $\boldsymbol{v}$ in their proof that satisfy $v_{n+1} = 0$. In the theorem we assume that the density $\delta(\boldsymbol{a})$ of the subset sum problems is less than 0.9408. Using the definition of $\delta(\boldsymbol{a})$ we obtain $\delta(\boldsymbol{a}) = n/\log_2(\max_{1\le j\le n}\{a_j\}) < 0.9408$, which implies that $\max_{1\le j\le n}\{a_j\} > 2^{n/0.9408}$, giving $A > 2^{c_0 n}$. For $A > 2^{c_0 n}$, the bound (61) goes to zero as n goes to infinity, which shows that "almost all" subset sum problems having density $\delta(\boldsymbol{a}) < 0.9408$ can be solved in polynomial time given the existence of a shortest vector oracle. Coster et al. also gave another lattice $L''_{\boldsymbol{a},a_0} \in \mathbb{Z}^{n+2}$ that could be used to obtain the result given in Theorem 21. The lattice $L''_{\boldsymbol{a},a_0}$

consists of vectors

$$L''_{\boldsymbol{a},a_0} = \begin{pmatrix} (n+1)x_1 - \sum_{\substack{k=1\\k\neq 1}}^{n} x_k - \xi \\ \vdots \\ (n+1)x_n - \sum_{\substack{k=1\\k\neq n}}^{n} x_k - \xi \\ (n+1)\xi - \sum_{j=1}^{n} x_j \\ N(\boldsymbol{ax} - a_0\xi) \end{pmatrix}$$

and is spanned by the basis

$$\begin{pmatrix} (n+1) & -1 & -1 & \cdots & -1 \\ -1 & (n+1) & -1 & \cdots & -1 \\ \cdots & & \ddots & & \vdots \\ -1 & \cdots & -1 & (n+1) & -1 \\ -1 & \cdots & \cdots & -1 & (n+1) \\ Na_1 & Na_2 & \cdots & Na_n & -Na_0 \end{pmatrix}. \tag{62}$$

Note that the lattice $L''_{\boldsymbol{a},a_0}$ is not full dimensional as the basis consists of $n+1$ vectors. Given a reduced basis vector $\boldsymbol{b} = (b_1, \ldots, b_{n+1}, 0)$, we solve the system of equations

$$b_j = (n+1)x_j - \sum_{\substack{k=1\\k\neq j}}^{n} x_k - \xi, \ 1 \le j \le n,$$

$$b_{n+1} = (n+1)\xi - \sum_{j=1}^{n} x_j$$

and check whether $\xi = 1$, and the vector $\boldsymbol{x} \in \{0,1\}^n$. If so, $\boldsymbol{x}$ solves the subset sum problem (55). Coster et al. show that for $\boldsymbol{x} \in \{0,1\}^n$, $\xi = 1$, we obtain $\|\boldsymbol{b}\|^2 \le \frac{n^3}{4}$, and they indicate how to show that most of the time there will be no shorter vectors in $L''_{\boldsymbol{a},a_0}$.

6.2 Solving systems of linear Diophantine equations

Aardal, Hurkens, and Lenstra [2], [3] considered the following integer feasibility problem:

$$\text{Does there exist a vector } \boldsymbol{x} \in \mathbb{Z}^n \text{ such that } \boldsymbol{A}\boldsymbol{x} = \boldsymbol{d},\ \boldsymbol{l} \leq \boldsymbol{x} \leq \boldsymbol{u}? \tag{63}$$

Here $\boldsymbol{A}$ is an integer $m \times n$-matrix, with $m \leq n$, and the integer vectors $\boldsymbol{d}$, $\boldsymbol{l}$, and $\boldsymbol{u}$ are of compatible dimensions. Problem (63) is NP-complete, but if we remove the bound constraints $\boldsymbol{l} \leq \boldsymbol{x} \leq \boldsymbol{u}$, it is polynomially solvable. A standard way of tackling problem (63) is by branch-and-bound, but for the applications considered by Aardal et al. this method did not work well. Let $X = \{x \in \mathbb{Z}^n \mid \boldsymbol{A}\boldsymbol{x} = \boldsymbol{d},\ \boldsymbol{l} \leq \boldsymbol{x} \leq \boldsymbol{u}\}$. Instead of using a method based on the linear relaxation of the problem, they considered the following integer relaxation of X, $X_{\mathrm{IR}} = \{\boldsymbol{x} \in \mathbb{Z}^n \mid \boldsymbol{A}\boldsymbol{x} = \boldsymbol{d}\}$. Determining whether X_{IR} is empty can be carried out in polynomial time for instance by generating the Hermite normal form of the matrix $\boldsymbol{A}$. Assume that X_{IR} is nonempty. Let $\boldsymbol{x}_f$ be an integer vector satisfying $\boldsymbol{A}\boldsymbol{x}_f = \boldsymbol{d}$, and let $\boldsymbol{B}^{\mathbf{0}}$ be an $n \times (n-m)$-matrix consisting of integer, linearly independent column vectors $\boldsymbol{b}_j^{\mathbf{0}}$, $1 \leq j \leq n-m$, such that $\boldsymbol{A}\boldsymbol{b}_j^{\mathbf{0}} = \mathbf{0}$ for $1 \leq j \leq n-m$. Notice that the matrix $\boldsymbol{B}^{\mathbf{0}}$ is a basis for the lattice $L_{\mathbf{0}} = \{\boldsymbol{x} \in \mathbb{Z}^n \mid \boldsymbol{A}\boldsymbol{x} = \mathbf{0}\}$. We can now rewrite X_{IR} as

$$X_{\mathrm{IR}} = \{\boldsymbol{x} \in \mathbb{Z}^n \mid \boldsymbol{x} = \boldsymbol{x}_f + \mathbf{B}^{\mathbf{0}}\boldsymbol{\lambda},\ \boldsymbol{\lambda} \in \mathbb{Z}^{n-m}\}. \tag{64}$$

Since a lattice has infinitely many bases if the dimension is greater than 1, reformulation (64) is not unique if $n - m > 1$.

The intuition behind the approach of Aardal et al. is as follows. Suppose it is possible to obtain a vector $\boldsymbol{x}_f$ that is short with respect to the bounds. Then, we may hope that $\boldsymbol{x}_f$ satisfies $\boldsymbol{l} \leq \boldsymbol{x}_f \leq \boldsymbol{u}$, in which case we are done. If $\boldsymbol{x}_f$ does not satisfy the bounds, then one can observe that $\boldsymbol{A}(\boldsymbol{x}_f + \lambda \boldsymbol{y}) = \boldsymbol{d}$ for any integer multiplier λ and any vector $\boldsymbol{y}$ satisfying $\boldsymbol{A}\boldsymbol{y} = \mathbf{0}$. Hence, it is possible to derive an enumeration scheme in which we branch on integer linear combinations of vectors $\boldsymbol{b}_j^{\mathbf{0}}$ satisfying $\boldsymbol{A}\boldsymbol{b}_j^{\mathbf{0}} = \mathbf{0}$, which explains the reformulation (64) of X_{IR}. Similar to Lagarias and Odlyzko, Aardal et al. choose a lattice, different from the standard lattice $\mathbb{Z}^n$, and then apply basis reduction to the initial basis of the chosen lattice. Since they obtain both $\boldsymbol{x}_f$ and the basis $\boldsymbol{B}^{\mathbf{0}}$ by basis reduction, $\boldsymbol{x}_f$ is relatively short and the columns of $\boldsymbol{B}^{\mathbf{0}}$ are near-orthogonal.

Aardal et al. [3] suggested a lattice $L_{\boldsymbol{A},\boldsymbol{d}} \in \mathbb{Z}^{n+m+1}$ that contains vectors of the following form:

$$(\boldsymbol{x}^T, N_1\xi, N_2(\boldsymbol{a}^1\boldsymbol{x} - d_1\xi), \ldots, N_2(\boldsymbol{a}^m\boldsymbol{x} - d_m\xi))^T, \tag{65}$$

where $\boldsymbol{a}^i$ is the i-th row of the matrix $\boldsymbol{A}$, where N_1 and N_2 are natural numbers, and where ξ, as in Section 6.1, is a variable associated with the right-hand side vector $\boldsymbol{d}$. The basis $\boldsymbol{B}$ given below spans the lattice $L_{\boldsymbol{A},\boldsymbol{d}}$:

$$\boldsymbol{B} = \begin{pmatrix} I^{(n)} & \mathbf{0}^{(n\times 1)} \\ \mathbf{0}^{(1\times n)} & N_1 \\ N_2\boldsymbol{A} & -N_2\boldsymbol{d} \end{pmatrix}. \tag{66}$$

The lattice $L_{\boldsymbol{A},\boldsymbol{d}} \subset \mathbb{Z}^{m+n+1}$ is not full-dimensional as $\boldsymbol{B}$ only contains $n+1$ columns. The numbers N_1 and N_2 are chosen so as to guarantee that certain elements of the reduced basis are equal to zero (cf. the similar role of the number N used in the bases (60) and (62)). The following proposition states precisely which type of vectors one wishes to obtain.

Proposition 8 ([3]). *The integer vector $\boldsymbol{x}_f$ satisfies $\boldsymbol{A}\boldsymbol{x}_f = \boldsymbol{d}$ if and only if the vector*

$$((\boldsymbol{x}_f)^T, N_1, \mathbf{0}^{(1\times m)})^T = \mathbf{B}\begin{pmatrix} \boldsymbol{x}_f \\ 1 \end{pmatrix} \tag{67}$$

belongs to the lattice L, and the integer vector $\boldsymbol{y}$ satisfies $\boldsymbol{A}\boldsymbol{y} = \mathbf{0}$ if and only if the vector

$$(\boldsymbol{y}^T, 0, \mathbf{0}^{(1\times m)})^T = \mathbf{B}\begin{pmatrix} \boldsymbol{y} \\ 0 \end{pmatrix} \tag{68}$$

belongs to the lattice L.

Let $\hat{\boldsymbol{B}}$ be the basis obtained by applying Lovász' basis reduction algorithm to the basis $\boldsymbol{B}$, and let $\hat{\boldsymbol{b}}_j = (\hat{b}_j^1, \ldots, \hat{b}_j^{n+m+1})$ be the j-th column vector of $\hat{\boldsymbol{B}}$. Aardal et al. [3] prove that if the numbers N_1 and N_2 are chosen appropriately, then the $(n-m+1)$-st column of $\hat{\boldsymbol{B}}$ is of type (67), and the first $n-m$ columns of $\hat{\boldsymbol{B}}$ are of type (68), i.e., the first $n-m+1$ columns of $\hat{\boldsymbol{B}}$ are of the following form:

$$\begin{pmatrix} \mathbf{B}^0 & \boldsymbol{x}_f \\ \mathbf{0}^{(1\times(n-m))} & N_1 \\ \mathbf{0}^{(m\times(n-m))} & \mathbf{0}^{(m\times 1)} \end{pmatrix}. \tag{69}$$

This result is stated in the following theorem.

Theorem 22 ([3]). *Assume that there exists an integer vector $\boldsymbol{x}$ satisfying the system $\boldsymbol{A}\boldsymbol{x} = \boldsymbol{d}$. There exist numbers N_{01} and N_{02} such that if $N_1 > N_{01}$, and* *if* $N_2 > 2^{n+m}N_1^2 + N_{02}$, *then the vectors* $\hat{\boldsymbol{b}}_j \in \mathbb{Z}^{n+m+1}$ *of the reduced basis* $\hat{\boldsymbol{B}}$ *have the following properties:*

1. $\hat{b}_j^{n+1} = 0$ *for* $1 \leq j \leq n-m$,
2. $\hat{b}_j^i = 0$ *for* $n+2 \leq i \leq n+m+1$ *and* $1 \leq j \leq n-m+1$,
3. $|\hat{b}_{n-m+1}^{n+1}| = N_1$.

Moreover, the sizes of N_{01} *and* N_{02} *are polynomially bounded in the sizes of* $\boldsymbol{A}$ *and* $\boldsymbol{d}$.

In the proof of Properties 1 and 2 of Theorem 22, Aardal et al. make use of inequality (15) of Proposition 2.

Once we have obtained the matrix $\boldsymbol{B}^{\mathbf{0}}$ and the vector $\boldsymbol{x}_f$, we can derive the following equivalent formulation of problem (63):

$$\text{Does there exist a vector } \boldsymbol{\lambda} \in \mathbb{Z}^{n-m} \text{ such that } \boldsymbol{l} \leq \boldsymbol{x}_f + \boldsymbol{B}^{\mathbf{0}}\boldsymbol{\lambda} \leq \boldsymbol{u}? \qquad (70)$$

Aardal, Hurkens, and Lenstra [3], and Aardal, Bixby, Hurkens, Lenstra, and Smeltink [1] investigated the effect of the reformulation on the number of nodes of a linear programming based branch-and-bound algorithm. They considered three sets of instances: instances obtained from Philips Research Labs, the Frobenius instances of Cornuéjols, Urbaniak, Weismantel, and Wolsey [33], and the market split instances of Cornuéjols and Dawande [31]. The results were encouraging. For instance, after transforming problem (63) to problem (70), the size of the market split instances that could be solved doubled.

Aardal et al. [1] also investigated the performance of *integer branching*. They implemented a branching-on-hyperplanes search algorithm, such as the algorithms in Section 4. Instead of finding provably good directions they branched on hyperplanes in the directions of the unit vectors $\boldsymbol{e}_j$, $1 \leq j \leq n-m$ in the space of the $\boldsymbol{\lambda}$-variables.

Their computational study indicated that integer branching on the unit vectors taken in the order $j = n-m, \ldots, 1$, was quite effective, and in general much better than the order $1, \ldots, n-m$. This can be explained as follows. Due to Lovász' algorithm, the vectors of $\boldsymbol{B}^{\mathbf{0}}$ are more or less in order of increasing length, so typically, the $(n-m)$-th vector of $\boldsymbol{B}^{\mathbf{0}}$ is the longest one. Branching on this vector first should generate relatively few hyperplanes intersecting the linear relaxation of X, if this set has a regular shape, or equivalently, the polytope $P = \{\boldsymbol{\lambda} \in \mathbb{R}^{n-m} \mid \boldsymbol{l} \leq \boldsymbol{x}_f + \boldsymbol{B}^{\mathbf{0}}\boldsymbol{\lambda} \leq \boldsymbol{u}\}$ is relatively thin in the unit direction $\boldsymbol{e}_{n-m}$ compared to direction $\boldsymbol{e}_1$. In this context Aardal and Lenstra [4] studied infeasible instances of the knapsack problem

$$\text{Does there exist a vector } \boldsymbol{x} \in \mathbb{Z}^n_{\geq 0} \text{ such that } \boldsymbol{ax} = a_0?$$

Write a_j as $a_j = p_j M + r_j$ with p_j, $M \in \mathbb{N}_{>0}$, and $r_j \in \mathbb{Z}$. Aardal and Lenstra showed the following:

Theorem 23 ([4]). *Let $\boldsymbol{b}^0_{n-1}$ be the last vector of the basis matrix $\boldsymbol{B}^0$ as obtained in (69). The following holds:*

- $d(L_0) = \|\boldsymbol{a}^T\|$,
- $\|\boldsymbol{b}^0_{n-1}\| \geq \frac{|\boldsymbol{a}^T|}{\sqrt{|\boldsymbol{p}|^2 \cdot |\boldsymbol{r}|^2 - (\boldsymbol{p}\boldsymbol{r}^T)^2}}$.

If M is large, then $d(L_0) = \|\boldsymbol{a}^T\|$ will be large, and if $\boldsymbol{p}$ and $\boldsymbol{r}$ are short compared to $\boldsymbol{a}$ the vector $\boldsymbol{b}^0_{n-1}$ is going to be long, so in this case the value of $d(L_0)$ essentially comes from the length of the last basis vector. In their computational study it was clear that branching in the direction of the last basis vector first gave rise to extremely small search trees.

Example 3. Let $\boldsymbol{a} = (12223, 12224, 36671)$. We can decompose $\boldsymbol{a}$ as

$a_1 = M + 0$
$a_2 = M + 1$
$a_3 = 3M + 2$

with $M = 12223$. For this example we obtain

$$\boldsymbol{x}_f = \begin{pmatrix} -4075 \\ 4074 \\ 4074 \end{pmatrix} \qquad \boldsymbol{B}^0 = \begin{pmatrix} -1 & 14261 \\ -2 & -8149 \\ 1 & -2037 \end{pmatrix}.$$

The polytope P is:

$$P = \{\boldsymbol{y} \in \mathbb{R}^2 \mid -\lambda_1 + 14261\lambda_2 \geq 4075,$$
$$-2\lambda_1 - 8149\lambda_2 \geq -4074, \lambda_1 - 2037\lambda_2 \geq -4074\}.$$

The constraints imply that $0 < \lambda_2 < 1$, so branching first in the direction of $\boldsymbol{e}_2$ immediately yields a certificate of infeasibility. Searching in direction $\boldsymbol{e}_1$ first yields 4752 search nodes at the first level of our search tree. Solving the instance using the original formulation in $\boldsymbol{x}$-variables requires 1,262,532 search nodes using CPLEX 6.5 with default settings. □

Recently, Louveaux and Wolsey [78] considered the problem: "Does there exist a matrix $\boldsymbol{X} \in \mathbb{Z}^{m \times n}_{\geq 0}$ such that $\boldsymbol{XA} = \boldsymbol{C}$, and $\boldsymbol{BX} = \boldsymbol{D}$?", where $\boldsymbol{A} \in \mathbb{Z}^{n \times p}$ and $\boldsymbol{B} \in \mathbb{Z}^{q \times m}$. Their study was motivated by a portfolio planning problem, where variable x_{ij} denotes the number of shares of type j included in portfolio i. This problem can be written in the same form as problem (63), so in principle the approach discussed in this section could be applied. For reasonable problem sizes Louveaux and Wolsey observed that the basis reduction step became too time consuming. Instead they determined reduced bases for the lattices $L_0^A = \{y \in \mathbb{Z}^n \mid y^T A = \mathbf{0}\}$, and $L_0^B = \{z \in \mathbb{Z}^m \mid Bz = \mathbf{0}\}$. Let B_A be a basis for the lattice L_0^A, and let B_B be a basis for the lattice L_0^B. They showed that taking the so-called *Kronecker product* of the matrices B_A^T and B_B yields a basis for the lattice $L_0 = \{X \in \mathbb{Z}^{m \times n} \mid XA = \mathbf{0}, BX = \mathbf{0}\}$. The Kronecker product of two matrices $M \in \mathbb{R}^{m \times n}$, and $N \in \mathbb{R}^{p \times q}$ is defined as:

$$M \otimes N = \begin{pmatrix} m_{11}N & \cdots & m_{1n}N \\ \cdots & \ddots & \cdots \\ m_{m1}N & \cdots & m_{mn}N \end{pmatrix}.$$

Moreover, they showed that the basis of L_0 obtained by taking the Kronecker product between B_A^T and B_B is reduced, up to a reordering of the basis vectors, if the bases B_A and B_B are reduced. Computational experience is reported.

7 Integer hulls and cutting plane closures in fixed dimension

An integer optimization problem $\max\{c^T x \mid Ax \le b, x \in \mathbb{Z}^n\}$, for integral A and b, can be interpreted as the linear programming problem $\max\{c^T x \mid A'x \le b', x \in \mathbb{R}^n\}$, where $A'x \le b'$ is an inequality description of the integer hull of the polyhedron $\{x \in \mathbb{R}^n \mid Ax \le b\}$. We have seen that the integer optimization problem in fixed dimension can be solved in polynomial time. The question now is, how large can the integer hull of a polyhedron be if the dimension if fixed? Can the integer hull be described with a polynomial number of inequalities and if the answer is "yes", can these inequalities be computed in polynomial time? It turns out that the answer to both the questions is "yes", as we will see in the following section.

One of the most successful methods to attack an integer optimization problem in practice is branch-and-bound combined with the addition of cutting planes. Cutting planes are valid inequalities for the integer hull, which are not necessarily valid for the linear relaxation of the problem. A famous family of cutting planes, also historically the first ones, are Gomory-Chvátal cutting planes [53]. In the second part of this section, we consider the question, whether the polyhedron that results from the application of all possible Gomory-Chvátal cutting planes, the so-called elementary closure, has a polynomial representation in fixed dimension. Furthermore we address the problem of constructing the elementary closure in fixed dimension.

7.1 *The integer hull*

In this section we describe a result of Hayes and Larman [56] and its generalization by Schrijver [99] which states that P_I can be described with a polynomial number of inequalities in fixed dimension, provided that P is rational.

We start by proving a polynomial upper bound on the number of vertices of the integer hull of a full-dimensional simplex $\Sigma = \text{conv}\{\mathbf{0}, \mathbf{v}_1, \ldots, \mathbf{v}_n\}$. Let φ denote the maximum binary encoding length of a vertex $\varphi = \max_{i=1,\ldots,n} \text{size}(\mathbf{v}_i)$. A full dimensional simplex in $\mathbb{R}^n$ is defined by $n+1$ inequalities. Each choice of n inequalities in such a definition has linearly independent normal vectors, defining one of the vertices of Σ. Since $\mathbf{0}$ is one of the vertices, Σ is the set of all $\mathbf{x} \in \mathbb{R}^n$ satisfying $\mathbf{B}\mathbf{x} \geq \mathbf{0}$, $\mathbf{c}^T\mathbf{x} \leq \beta$, where $\mathbf{B} \in \mathbb{Z}^{n \times n}$ is a nonsingular matrix, and $\mathbf{c}^T\mathbf{x} \leq \beta$ is an inequality. It follows from the Hadamard bound that we can choose $\mathbf{B}$ such that $\text{size}(\mathbf{B}) = O(\varphi)$. The inequality $\mathbf{c}^T\mathbf{x} \leq \beta$ can be rewritten as $\mathbf{a}^T\mathbf{B}\mathbf{x} \leq \beta$, with $\mathbf{a}^T = \mathbf{c}^T\mathbf{B}^{-1} \in \mathbb{Q}^n$. Let K be the knapsack polytope $K = \{\mathbf{x} \in \mathbb{R}^n \mid \mathbf{x} \geq \mathbf{0}, \mathbf{a}^T\mathbf{x} \leq \beta\}$. The vertices of Σ_I correspond exactly to the vertices of $\text{conv}(K \cap L(\mathbf{B}))$.

Proposition 9. *Let $K \subseteq \mathbb{R}^n$ be a knapsack polytope given by the inequalities $\mathbf{x} \geq \mathbf{0}$ and $\mathbf{a}^T\mathbf{x} \leq \beta$. Let $L(\mathbf{B})$ be a lattice with integer and nonsingular $\mathbf{B} \subseteq \mathbb{Z}^{n \times n}$, then:*

1. *A vector $\mathbf{B}\hat{\mathbf{x}} \in L(\mathbf{B})$ is a vertex of $\text{conv}(K \cap L(\mathbf{B}))$ if and only if $\hat{\mathbf{x}}$ is a vertex of the integer hull of the simplex Σ defined by $\mathbf{B}\mathbf{x} \geq \mathbf{0}$ and $\mathbf{a}^T\mathbf{B}\mathbf{x} \leq \beta$;*
2. *if $\mathbf{v}_1$ and $\mathbf{v}_2$ are distinct vertices of $\text{conv}(K \cap L(\mathbf{B}))$, then there exists an index $i \in \{1, \ldots, n\}$ such that $\text{size}(\mathbf{v}_1^i) \neq \text{size}(\mathbf{v}_2^i)$.*

Proof. The convex hull of $K \cap L(\mathbf{B})$ can be written as

$$\begin{aligned}\text{conv}(K \cap L(\mathbf{B})) &= \text{conv}(\{\mathbf{x} \mid \mathbf{x} \geq \mathbf{0}, \mathbf{a}^T\mathbf{x} \leq \beta, \mathbf{x} = \mathbf{B}\mathbf{y}, \mathbf{y} \in \mathbb{Z}^n\}) \\ &= \text{conv}(\{\mathbf{B}\mathbf{y} \mid \mathbf{B}\mathbf{y} \geq \mathbf{0}, \mathbf{a}^T\mathbf{B}\mathbf{y} \leq \beta, \mathbf{y} \in \mathbb{Z}^n\}).\end{aligned}$$

If one transforms this set with $\mathbf{B}^{-1}$, one is faced with the integer hull of the described simplex Σ. Thus Point (1) in the proposition follows.

For Point (2) assume that $\mathbf{v}_1$ and $\mathbf{v}_2$ are vertices of $\text{conv}(K \cap L(\mathbf{B}))$, with $\text{size}(\mathbf{v}_1^i) = \text{size}(\mathbf{v}_2^i)$ for all $i \in \{1, \ldots, n\}$. Then clearly $2\mathbf{v}_1 - \mathbf{v}_2 \geq 0$ and $2\mathbf{v}_2 - \mathbf{v}_1 \geq 0$. Also

$$\mathbf{a}^T(2\mathbf{v}_1 - \mathbf{v}_2 + 2\mathbf{v}_2 - \mathbf{v}_1) = \mathbf{a}^T(\mathbf{v}_1 + \mathbf{v}_2) \leq 2\beta,$$

therefore one of the two lattice points lies in K. Assume without loss of generality that $2\mathbf{v}_1 - \mathbf{v}_2 \in K \cap L(\mathbf{B})$. Then $\mathbf{v}_1$ cannot be a vertex since

$$\mathbf{v}_1 = 1/2(2\mathbf{v}_1 - \mathbf{v}_2) + 1/2\mathbf{v}_2. \qquad \square$$

If $K = \{\mathbf{x} \in \mathbb{R}^n \mid \mathbf{x} \geq \mathbf{0}, \mathbf{a}^T\mathbf{x} \leq \beta\}$ is the corresponding knapsack polytope to the simplex Σ, then any component $\hat{x}_j$, $j = 1, \ldots, n$ of an arbitrary point $\hat{\mathbf{x}}$ in K satisfies $0 \leq \hat{x}_j \leq \beta/a_j$. Thus the size of a vertex $\hat{\mathbf{x}}$ of $\text{conv}(K \cap L(\mathbf{B}))$ is of $O(\text{size}(K)) = O(\text{size}(\Sigma))$ in fixed dimension. This is because $\text{size}(\mathbf{B}^{-1}) = O(\text{size}(\mathbf{B}))$ in fixed dimension. It follows from Proposition 9 that Σ_I can have at most $O(\text{size}(\Sigma)^n)$ vertices.

By translation with the vertex $-\boldsymbol{v}_0$, we can assume that $\Sigma = \text{conv}(\boldsymbol{v}_0, \ldots, \boldsymbol{v}_n)$ is a simplex whose first vertex $\boldsymbol{v}_0$ is *integral.*

Lemma 6 ([56, 99]). *Let* $\Sigma = \text{conv}(\boldsymbol{v}_0, \ldots, \boldsymbol{v}_n)$ *be a rational simplex with* $\boldsymbol{v}_0 \in \mathbb{Z}^n, \boldsymbol{v}_i \in \mathbb{Q}^n, i = 1, \ldots, n$. *The number of vertices of the integer hull* Σ_I *is bounded by* $O(\varphi^n)$, *where* $\varphi = \max_{i=0,\ldots,n} \text{size}(\boldsymbol{v}_i)$.

A polynomial bound for general polyhedra can then be found by triangulation.

Theorem 24 ([56, 99]). *Let* $P = \{\boldsymbol{x} \in \mathbb{R}^n \mid \boldsymbol{A}\boldsymbol{x} \le \boldsymbol{d}\}$, *where* $\boldsymbol{A} \in \mathbb{Z}^{m \times n}$ and $\boldsymbol{d} \in \mathbb{Z}^m$, *be a rational polyhedron where each inequality in* $\boldsymbol{A}\boldsymbol{x} \le \boldsymbol{d}$ *has size at most* φ. *The integer hull* P_I *of* P *has at most* $O(m^{n-1}\varphi^n)$ *vertices.*

The following upper bound on the number of vertices of P_I was proved by Cook et al. [28]. Barany et al. [10] showed that this bound is tight if P is a simplex.

Theorem 25. *If* $P \subseteq \mathbb{R}^n$ *is a rational polyhedron that is the solution set of a system of at most* m *linear inequalities whose size is at most* φ, *then the number of vertices of* P_I *is at most* $2m^d(6n^2\varphi)^{d-1}$, *where* $d = \dim(P_I)$ *is the dimension of the integer hull of* P.

Tight bounds for varying number of inequalities m seem to be unknown.

7.2 Cutting planes

Rather than computing the integer hull P_I of P, the objective pursued by the *cutting plane method* is a better approximation of P_I. Here the idea is to intersect P with the integer hull of halfspaces containing P. These will still include P_I but not necessarily P.

In the following we will study the theoretical framework of Gomory's cutting plane method [53] as given by Chvátal [23] and Schrijver [98] and derive a polynomiality result on the number of facets of the polyhedron that results from the application of all possible cutting planes.

If the halfspace $(\boldsymbol{c}^T\boldsymbol{x} \le \delta)$, $\boldsymbol{c} \in \mathbb{Z}^n$, with $\gcd(c_1, \ldots, c_n) = 1$, contains the polyhedron P, i.e. if $\boldsymbol{c}^T\boldsymbol{x} \le \delta$ is valid for P, then $\boldsymbol{c}^T\boldsymbol{x} \le \lfloor \delta \rfloor$ is valid for the integer hull P_I of P. The inequality $\boldsymbol{c}^T\boldsymbol{x} \le \lfloor \delta \rfloor$ is called a *cutting plane* or *Gomory-Chvátal cut* of P. The geometric interpretation behind this process is that $(\boldsymbol{c}^T\boldsymbol{x} \le \delta)$ is "shifted inwards" until an integer point of the lattice is in the boundary of the halfspace.

The idea, pioneered by Gomory [53], is to apply these cutting planes to the integer optimization problem. Cutting planes tighten the linear relaxation of an integer program and Gomory showed how to apply cutting planes successively until the resulting relaxation has an integer optimal solution.

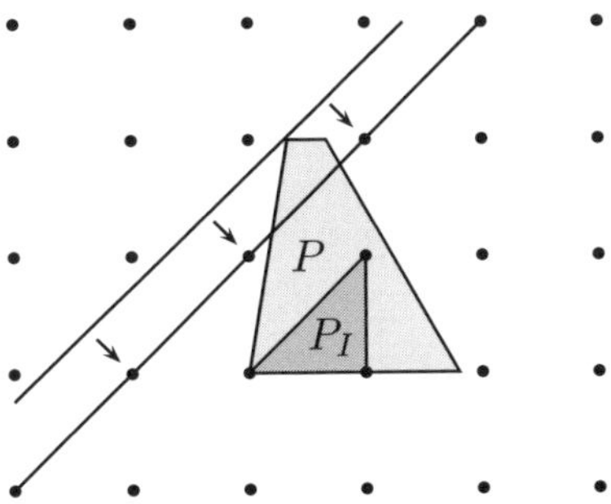

Figure 9. The halfspace $(-x_1+x_2\leq\delta)$ containing P is replaced by its integer hull $(-x_1+x_2\leq\lfloor\delta\rfloor)$. The darker region is the integer hull P_I of P.

7.2.1 *The elementary closure*

Cutting planes $\boldsymbol{c}^T\boldsymbol{x}\leq\lfloor\delta\rfloor$ of $P(\boldsymbol{A},\boldsymbol{d})$, $\boldsymbol{A}\in\mathbb{R}^{m\times n}$ obey a simple *inference rule*. Clearly $\max\{\boldsymbol{c}^T\boldsymbol{x}\mid\boldsymbol{A}\boldsymbol{x}\leq\boldsymbol{d}\}\leq\delta$ and it follows from duality and Carathéodory's theorem that there exists a weight vector $\boldsymbol{\lambda}\in\mathbb{Q}^m_{\geq 0}$ with at most n positive entries such that $\boldsymbol{\lambda}^T\boldsymbol{A}=\boldsymbol{c}^T$ and $\boldsymbol{\lambda}^T\boldsymbol{d}\leq\delta$. Thus $\boldsymbol{c}^T\boldsymbol{x}\leq\lfloor\delta\rfloor$ follows from the following inequalities by weakening the right-hand side if necessary:

$$\boldsymbol{\lambda}^T\boldsymbol{A}\boldsymbol{x}\leq\lfloor\boldsymbol{\lambda}^T\boldsymbol{d}\rfloor,\ \boldsymbol{\lambda}\in\mathbb{Q}^m_{\geq 0},\ \boldsymbol{\lambda}^T\boldsymbol{A}\in\mathbb{Z}^n. \tag{71}$$

Instead of applying cutting planes successively, one can apply all possible cutting planes at once. P intersected with all Gomory-Chvátal cutting planes

$$P' = \bigcap_{\substack{(\boldsymbol{c}^T\boldsymbol{x}\leq\delta)\supseteq P\\ \boldsymbol{c}\in\mathbb{Z}^n}} (\boldsymbol{c}^T\boldsymbol{x}\leq\lfloor\delta\rfloor) \tag{72}$$

is called the *elementary closure* of P.

The set of inequalities in (71) that describe P' is infinite. However, as observed by Schrijver [98], a finite number of inequalities in (71) imply the rest.

Lemma 7. *Let P be the polyhedron $P=\{\boldsymbol{x}\in\mathbb{R}^n\mid\boldsymbol{A}\boldsymbol{x}\leq\boldsymbol{d}\}$ with $\boldsymbol{A}\in\mathbb{Z}^{m\times n}$ and $\boldsymbol{d}\in\mathbb{Z}^m$. The elementary closure P' is the polyhedron defined by $\boldsymbol{A}\boldsymbol{x}\leq\boldsymbol{d}$ and the set of all inequalities $\boldsymbol{\lambda}^T\boldsymbol{A}\boldsymbol{x}\leq\lfloor\boldsymbol{\lambda}^T\boldsymbol{d}\rfloor$, where $\boldsymbol{\lambda}\in[0,1)^m$ and $\boldsymbol{\lambda}^T\boldsymbol{A}\in\mathbb{Z}^n$.*

Proof. An inequality $\boldsymbol{\lambda}^T\boldsymbol{A}\boldsymbol{x}\leq\lfloor\boldsymbol{\lambda}^T\boldsymbol{d}\rfloor$, with $\boldsymbol{\lambda}\in\mathbb{Q}^m_{\geq 0}$ and $\boldsymbol{\lambda}^T\boldsymbol{A}\in\mathbb{Z}^n$ is implied by $\boldsymbol{A}\boldsymbol{x}\leq\boldsymbol{d}$ and $(\boldsymbol{\lambda}-\lfloor\boldsymbol{\lambda}\rfloor)^T\boldsymbol{A}\boldsymbol{x}\leq\lfloor(\boldsymbol{\lambda}-\lfloor\boldsymbol{\lambda}\rfloor)^T\boldsymbol{d}\rfloor$, since

$$\boldsymbol{\lambda}^T\boldsymbol{A}\boldsymbol{x}=(\boldsymbol{\lambda}-\lfloor\boldsymbol{\lambda}\rfloor)^T\boldsymbol{A}\boldsymbol{x}+\lfloor\boldsymbol{\lambda}\rfloor^T\boldsymbol{A}\boldsymbol{x}\leq\lfloor(\boldsymbol{\lambda}-\lfloor\boldsymbol{\lambda}\rfloor)^T\boldsymbol{d}\rfloor+\lfloor\boldsymbol{\lambda}\rfloor^T\boldsymbol{d}=\lfloor\boldsymbol{\lambda}^T\boldsymbol{d}\rfloor. \tag{73}$$

□

Corollary 2 ([98]). *If P is a rational polyhedron, then P' is a rational polyhedron.*

Proof. P can be described as $P(\boldsymbol{A}, \boldsymbol{d})$ with integral $\boldsymbol{A}$ and $\boldsymbol{d}$. There is only a finite number of vectors $\boldsymbol{\lambda}^T\boldsymbol{A} \in \mathbb{Z}^n$ with $\boldsymbol{\lambda} \in [0, 1)^m$. □

This yields an exponential upper bound on the number of facets of the elementary closure of a polyhedron. The infinity norm $\|\boldsymbol{c}\|_\infty$ of a possible candidate $\boldsymbol{c}^T\boldsymbol{x} \leq \lfloor \delta \rfloor$ is bounded by $\|\boldsymbol{A}^T\|_\infty$, where the matrix norm $\|\cdot\|_\infty$ is the row sum norm. Therefore we have an upper bound of $O(\|\boldsymbol{A}^T\|_\infty^n)$ for the number of facets of the elementary closure of a polyhedron. We will later prove a polynomial upper bound of the size of P' in fixed dimension.

7.2.2 *The Chvátal-Gomory procedure*

The elementary closure operation can be iterated, so that successively tighter relaxations of the integer hull P_I of P are obtained. We define $P^{(0)} = P$ and $P^{(i+1)} = (P^{(i)})'$, for $i \geq 0$. This iteration of the elementary closure operation is called the *Chvátal-Gomory procedure*. The *Chvátal rank* of a polyhedron P is the smallest $t \in \mathbb{N}_0$ such that $P^{(t)} = P_I$. In analogy, the *depth* of an inequality $\boldsymbol{c}^T\boldsymbol{x} \leq \delta$ which is valid for P_I is the smallest $t \in \mathbb{N}_0$ such that $(\boldsymbol{c}^T\boldsymbol{x} \leq \delta) \supseteq P^{(t)}$.

Chvátal [23] showed that every bounded polyhedron $P \subseteq \mathbb{R}^n$ has a finite rank. Schrijver [98] extended this result to rational polyhedra. The main ingredient of his result is the following result.

Lemma 8 ([98]). *Let F be a face of a rational polyhedron P. If $\boldsymbol{c}_F^T\boldsymbol{x} \leq \lfloor \delta_F \rfloor$ is a cutting plane for F, then there exists a cutting plane $\boldsymbol{c}_P^T\boldsymbol{x} \leq \lfloor \delta_P \rfloor$ for P with*

$$F \cap \left(\boldsymbol{c}_P^T\boldsymbol{x} \leq \lfloor \delta_P \rfloor\right) = F \cap \left(\boldsymbol{c}_F^T\boldsymbol{x} \leq \lfloor \delta_F \rfloor\right).$$

Intuitively, this result means that a cutting plane of a face F of a polyhedron P can be "rotated" so that it becomes a cutting plane of P and has the same effect on F. This implies that a face F of P behaves under its closure F' as it behaves under the closure P' of P.

Corollary 3. *Let F be a face of a rational polyhedron P. Then*

$$F' = P' \cap F.$$

From this, one can derive that the Chvátal rank of rational polyhedra is finite.

Theorem 26 ([98]). *If P is a rational polyhedron, then there exists some $t \in \mathbb{N}$ such that $P^{(t)} = P_I$.*

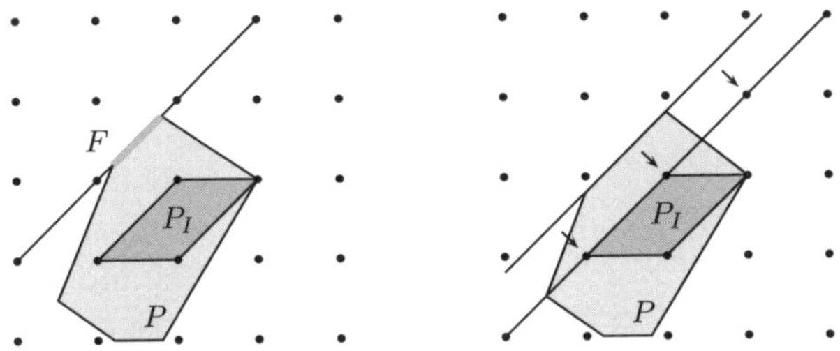

Figure 10. After a finite number of iterations F is empty. Then the halfspace defining F can be pushed further down. This is basically the argument why every inequality, valid for P_I, eventually becomes valid for the outcome of the successive application of the elementary closure operation.

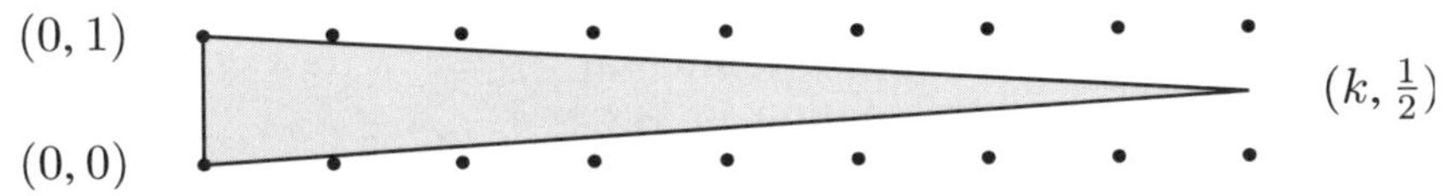

Figure 11. The polytope P_k.

Already in dimension 2, there exist rational polyhedra of arbitrarily large Chvátal rank [23]. To see this, consider the class of polytopes

$$P_k = \mathrm{conv}\left\{(0, 0), (0, 1), \left(k, \frac{1}{2}\right)\right\}, k \in \mathbb{N}. \tag{74}$$

One can show that $P_{(k-1)} \subseteq P'_k$. For this, let $\boldsymbol{c}^T\boldsymbol{x} \leq \delta$ be valid for P_k with $\delta = \max\{\boldsymbol{c}^T\boldsymbol{x} \mid \boldsymbol{x} \in P_k\}$. If $c_1 \leq 0$, then the point $(0,0)$ or $(0,1)$ maximizes $\boldsymbol{c}^T\boldsymbol{x}$, thus $(\boldsymbol{c}^T\boldsymbol{x} = \delta)$ contains integer points. If $c_1 > 0$, then $\boldsymbol{c}^T(k, \frac{1}{2}) \geq \boldsymbol{c}^T(k-1, \frac{1}{2}) + 1$. Therefore the point $(k-1, \frac{1}{2})$ is in the halfspace $(\boldsymbol{c}^T\boldsymbol{x} \leq \delta - 1) \subseteq (\boldsymbol{c}^T\boldsymbol{x} \leq \lfloor \delta \rfloor)$. Unfortunately, this lower bound on the Chvátal rank of P_k is exponential in the encoding length of P_k which is $O(\log(k))$.

Bockmayr et al. [16] have shown that the Chvátal rank of polytopes in the 0/1 cube is polynomial. The current best bound [44] on the Chvátal rank of polytopes in the 0/1 cube is $O(n^2 \log n)$. Lower bounds on the Chvátal rank for polytopes stemming from combinatorial optimization problems have been provided by Chvátal, Cook and Hartmann [24]. Cook and Dash [30] provided lower bounds on the matrix-cut rank of polytopes in the 0/1 cube. In particular they provide examples with rank n and so do Cornuéjols and Li [32] for the split closure in the 0/1 cube.

7.2.3 Cutting plane proofs

An important property of polyhedra is the following rule to derive valid inequalities, which is a consequence of linear programming duality. If P is defined by the inequalities $\boldsymbol{Ax} \leq \boldsymbol{d}$, then the inequality $\boldsymbol{c}^T\boldsymbol{x} \leq \delta$ is valid for P if and only if there exists some $\boldsymbol{\lambda} \in \mathbb{R}^m_{\geq 0}$ with

$$\boldsymbol{c} = \boldsymbol{\lambda}^T\boldsymbol{A} \text{ and } \delta \geq \boldsymbol{\lambda}^T\boldsymbol{d}. \tag{75}$$

This implies that linear programming (in its decision version) belongs to the class NP $\cap$ co $-$ NP, because $\max\{\boldsymbol{c}^T\boldsymbol{x} \mid \boldsymbol{Ax} \leq \boldsymbol{d}\} \leq \delta$ if and only if $\boldsymbol{c}^T\boldsymbol{x} \leq \delta$ is valid for $P(\boldsymbol{A}, \boldsymbol{d})$. A "No" certificate would be some vertex of P which violates $\boldsymbol{c}^T\boldsymbol{x} \leq \delta$.

In integer programming there is an analogy to this rule. A sequence of inequalities

$$\boldsymbol{c}_1^T\boldsymbol{x} \leq \delta_1,\ \boldsymbol{c}_2^T\boldsymbol{x} \leq \delta_2, \ldots, \boldsymbol{c}_m^T\boldsymbol{x} \leq \delta_m \tag{76}$$

is called a *cutting-plane proof* of $\boldsymbol{c}^T\boldsymbol{x} \leq \delta$ from a given system of linear inequalities $\boldsymbol{Ax} \leq \boldsymbol{d}$, if $\boldsymbol{c}_1, \ldots, \boldsymbol{c}_m$ are integral, $\boldsymbol{c}_m = \boldsymbol{c}$, $\delta_m = \delta$, and if $\boldsymbol{c}_i^T\boldsymbol{x} \leq \delta_i'$ is a nonnegative linear combination of $\boldsymbol{Ax} \leq \boldsymbol{d}$, $\boldsymbol{c}_1^T\boldsymbol{x} \leq \delta_1, \ldots, \boldsymbol{c}_{i-1}^T\boldsymbol{x} \leq \delta_{i-1}$ for some δ_i' with $\lfloor \delta_i' \rfloor \leq \delta_i$. In other words, if $\boldsymbol{c}_i^T\boldsymbol{x} \leq \delta_i$ can be obtained from $\boldsymbol{Ax} \leq \boldsymbol{d}$ and the previous inequalities as a Gomory-Chvátal cut, by weakening the right-hand-side if necessary. Obviously, if there is a cutting-plane proof of $\boldsymbol{c}^T\boldsymbol{x} \leq \delta$ from $\boldsymbol{Ax} \leq \boldsymbol{d}$ then every integer solution to $\boldsymbol{Ax} \leq \boldsymbol{d}$ must satisfy $\boldsymbol{c}^T\boldsymbol{x} \leq \delta$. The number m here, is the *length* of the cutting plane proof.

The following proposition shows a relation between the length of cutting plane proofs and the depth of inequalities (see also [24]). It comes in two flavors, one for the case $P_I \neq \emptyset$ and one for $P_I = \emptyset$. The latter can then be viewed as an analogy to Farkas' lemma.

Proposition 10 ([24]). *Let $P(\boldsymbol{A}, \boldsymbol{d}) \subseteq \mathbb{R}^n$, $n \geq 2$ be a rational polyhedron.*

1. *If $P_I \neq \emptyset$ and $\mathbf{c}^T\mathbf{x} \leq \delta$ with integer $\mathbf{c}$ has depth t, then $\mathbf{c}^T\mathbf{x} \leq \delta$ has a cutting plane proof of length at most $(n^{t+1} - 1)/(n-1)$.*
2. *If $P_I = \emptyset$ and rank$(P) = t$, then there exists a cutting plane proof of $\mathbf{0}^T\boldsymbol{x} \leq -1$ of length at most $(n+1)(n^t - 1)/(n-1) + 1$.*

We have seen for the class of polytopes P_k (74) that, even in fixed dimension, a cutting plane proof of minimal length can be exponential in the binary encoding length of the given polyhedron.

Yet, if $P_I = \emptyset$ and $P \subseteq \mathbb{R}^n$, Cook, Coullard and Turán [27] showed that there exists a number $t(n)$, such that $P^{(t(n))} = \emptyset$.

Theorem 27 ([27]). *There exists a function $t(d)$, such that if $P \subseteq \mathbb{R}^n$ is a d-dimensional rational polyhedron with empty integer hull, then $P^{t(d)} = \emptyset$.*

Proof. If P is not full dimensional, then there exists a rational hyperplane $(\boldsymbol{c}^T\boldsymbol{x} = \delta)$ with $\boldsymbol{c} \in \mathbb{Z}^n$ and $\gcd(c_1, \ldots, c_n) = 1$ such that $P \subseteq (\boldsymbol{c}^T\boldsymbol{x} = \delta)$. If $\delta \notin \mathbb{Z}$, then $P' = \emptyset$. If $\delta \in \mathbb{Z}$, then there exists a unimodular matrix, transforming $\boldsymbol{c}$ into the first unit vector $\boldsymbol{e}_1$. Thus P can be transformed via a unimodular transformation into a polyhedron where the first variable is fixed to an integer.

Thus we can assume that P is full-dimensional. The function $t(d)$ is inductively defined. Let $t(0)=1$. For $d>0$, let $\boldsymbol{c} \in \mathbb{Z}^n$, $\boldsymbol{c} \neq \boldsymbol{0}$ be a direction in which P is flat (c.f. Theorem 9), i.e., $\max\{\boldsymbol{c}^T\boldsymbol{x} \mid x \in P\} - \min\{\boldsymbol{c}^T\boldsymbol{x} \mid \boldsymbol{x} \in P\} \leq f(d)$. We "slice off" in this direction using Corollary 3. If $\boldsymbol{c}^T\boldsymbol{x} \leq \delta$, $\delta \in \mathbb{Z}$ is valid for P, then $\boldsymbol{c}^T\boldsymbol{x} \leq \delta - 1$ is valid for $P^{(t(d-1)+1)}$, since the face $F = P \cap (\boldsymbol{c}^T\boldsymbol{x} = \delta)$ has at most dimension $d-1$. Thus $\boldsymbol{c}^T\boldsymbol{x} \leq \delta - k$ is valid for $P^{(k(t(d-1)+1))}$. Since the integer vector $\boldsymbol{c}$ is chosen such that $\max\{\boldsymbol{c}^T\boldsymbol{x} \mid \boldsymbol{x} \in P\} - \min\{\boldsymbol{c}^T\boldsymbol{x} \mid \boldsymbol{x} \in P\} \leq f(d)$, $t(d) = (f(d)+2)(t(d-1)+1)$ satisfies our needs. □

The validity of an inequality $\boldsymbol{c}^T\boldsymbol{x} \leq \delta$ for P_I can be established by showing that $P \cap (\boldsymbol{c}^T\boldsymbol{x} \geq \delta + 1)$ is integer infeasible. A cutting plane proof for the integer infeasibility of $P \cap (\boldsymbol{c}^T\boldsymbol{x} \geq \delta + 1)$ is called an *indirect cutting plane proof* of $\boldsymbol{c}^T\boldsymbol{x} \leq \delta$. Combining Proposition 10 and Theorem 27 one obtains the following result.

Theorem 28 ([27]). *Let P be a rational polyhedron in fixed dimension n and let* $\boldsymbol{c}^T\boldsymbol{x} \leq \delta$ *be a valid inequality for P, then* $\boldsymbol{c}^T\boldsymbol{x} \leq \delta$ *has an indirect cutting plane proof of constant length.*

In varying dimension, the length of a cutting plane proof of infeasibility of 0/1 systems can be exponential. This was shown by Pudlák [88]. Exponential lower bounds for other types of cutting-plane proofs provided by lift-and-project or Lovász–Schrijver cuts were derived by Dash [35].

7.3 The elementary closure in fixed dimension

In this section we will show that the elementary closure of rational polyhedra in fixed dimension can be described with a polynomial number of inequalities.

7.3.1 Simplicial cones

Consider a *rational simplicial cone*, i.e., a polyhedron $P = \{x \in \mathbb{R}^n \mid \boldsymbol{A}\boldsymbol{x} \leq \boldsymbol{d}\}$, where $\boldsymbol{A} \in \mathbb{Z}^{m \times n}$, $\boldsymbol{d} \in \mathbb{Z}^m$ and $\boldsymbol{A}$ has full row rank. If $\boldsymbol{A}$ is a square matrix, then P is called *pointed.*

Observe that P, P' and P_I are all full-dimensional. The elementary closure P' is given by the inequalities

$$(\boldsymbol{\lambda}^T\boldsymbol{A})\boldsymbol{x} \leq \lfloor \boldsymbol{\lambda}^T\boldsymbol{d} \rfloor, \text{ where } \boldsymbol{\lambda} \in [0,1]^m, \text{ and } \boldsymbol{\lambda}^T\boldsymbol{A} \in \mathbb{Z}^n. \tag{77}$$

Since P' is full-dimensional, there exists a unique (up to scalar multiplication) minimal subset of the inequalities in (77) that suffices to describe P'.

These inequalities are the facets of P'. We will derive a polynomial upper bound on their number in fixed dimension.

The vectors $\boldsymbol{\lambda}$ in (77) belong to the dual lattice $L^*(\boldsymbol{A})$ of the lattice $L(\boldsymbol{A})$. Recall that each element in $L^*(\boldsymbol{A})$ is of the form $\boldsymbol{\mu}/d_L$, where $d_L = d(L(\boldsymbol{A}))$ is the lattice determinant. It follows from the Hadamard inequality that size(d_L) is polynomial in size($\boldsymbol{A}$), even for varying n. Now (77) can be rewritten as

$$\frac{\boldsymbol{\mu}^T \boldsymbol{A}}{d_L} \boldsymbol{x} \leq \left\lfloor \frac{\boldsymbol{\mu}^T \boldsymbol{d}}{d_L} \right\rfloor, \text{ where } \boldsymbol{\mu} \in [0, \ldots, d]^m, \text{ and } \boldsymbol{\mu}^T \boldsymbol{A} \in (d_L \cdot \mathbb{Z})^n. \tag{78}$$

Notice here that $\boldsymbol{\mu}^T \boldsymbol{d}/d_L$ is a rational number with denominator d_L. There are two cases: either $\boldsymbol{\mu}^T \boldsymbol{d}/d_L$ is an integer, or $\boldsymbol{\mu}^T \boldsymbol{d}/d_L$ misses the nearest integer by at least $1/d_L$. Therefore $\lfloor \boldsymbol{\mu}^T \boldsymbol{d}/d_L \rfloor$ is the only integer in the interval

$$\left[\frac{\boldsymbol{\mu}^T \boldsymbol{d} - d_L + 1}{d_L}, \frac{\boldsymbol{\mu}^T \boldsymbol{d}}{d_L} \right].$$

These observations enable us to construct a polytope Q, whose integer points will correspond to the inequalities (78). Let Q be the set of all $(\boldsymbol{\mu}, \boldsymbol{y}, z)$ in $\mathbb{R}^{2n+1}$ satisfying the inequalities

$$\begin{aligned} \boldsymbol{\mu} &\geq \boldsymbol{0} \\ \mu_i &\leq d_L, \quad i = 1, \ldots, n \\ \boldsymbol{\mu}^T \boldsymbol{A} &= d_L \boldsymbol{y}^T \\ (\boldsymbol{\mu}^T \boldsymbol{d}) - d_L + 1 &\leq d_L z \\ (\boldsymbol{\mu}^T \boldsymbol{d}) &\geq d_L z. \end{aligned} \tag{79}$$

If $(\boldsymbol{\mu}, \boldsymbol{y}, z)$ is integral, the $\boldsymbol{\mu} \in [0, \ldots, d]^m$, $\boldsymbol{y} \in \mathbb{Z}^n$ enforces $\boldsymbol{\mu}^T \boldsymbol{A} \in (d_L \cdot \mathbb{Z})^n$ and z is the only integer in the interval $[(\boldsymbol{\mu}^T \boldsymbol{d} + 1 - d_L)/d_L, \boldsymbol{\mu}^T \boldsymbol{d}/d_L]$. It is not hard to see that Q is indeed a polytope. We call Q the *cutting plane polytope* of the simplicial cone $P(\boldsymbol{A}, \boldsymbol{d})$.

The correspondence between inequalities (their syntactic representation) in (78) and integer points in the cutting plane polytope Q is obvious. We now show that the facets of P' are among the vertices of Q_I.

Proposition 11 ([15]). *Each facet of P' is represented by an integer vertex of Q_I.*

Proof. Consider a facet $\boldsymbol{c}^T \boldsymbol{x} \leq \delta$ of P'. If we remove this inequality (possibly several times, because of scalar multiples) from the set of inequalities in (78),

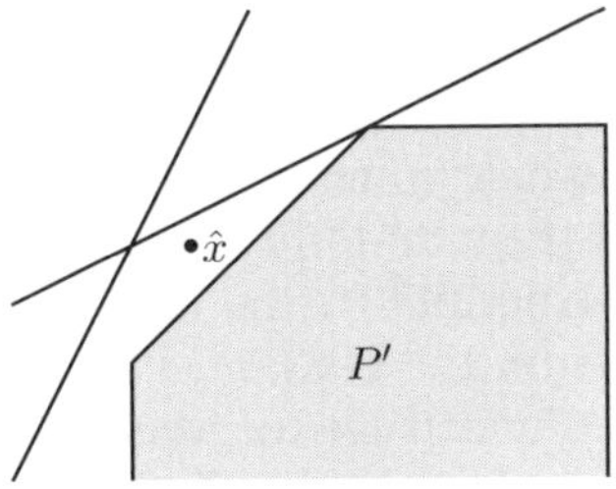

Figure 12. The point $\hat{\boldsymbol{x}}$ lies "above" the facet $\boldsymbol{c}^T\boldsymbol{x} \le \delta$ and "below" each other inequality in (78).

then the polyhedron defined by the resulting set of inequalities differs from P', since P' is full-dimensional. Thus there exists a point $\hat{\boldsymbol{x}} \in \mathbb{Q}^n$ that is violated by $\boldsymbol{c}^T\boldsymbol{x} \le \delta$, but satisfies any other inequality in (78) (see Figure 12). Consider the following integer program:

$$\max\{(\boldsymbol{\mu}^T\boldsymbol{A}/d_L)\hat{\boldsymbol{x}} - z \mid (\boldsymbol{\mu}, \boldsymbol{y}, z) \in Q_I\}. \tag{80}$$

Since $\hat{\boldsymbol{x}} \notin P'$ there exists an inequality $(\boldsymbol{\mu}^T\boldsymbol{A}/d_L)\boldsymbol{x} \le \lfloor \boldsymbol{\mu}^T\boldsymbol{d}/d_L \rfloor$ in (78) with

$$(\boldsymbol{\mu}^T\boldsymbol{A}/d_L)\hat{\boldsymbol{x}} - \lfloor \boldsymbol{\mu}^T\boldsymbol{d}/d_L \rfloor > 0.$$

Therefore, the optimal value will be strictly positive, and an integer optimal solution $(\boldsymbol{\mu}, \boldsymbol{y}, z)$ must correspond to the facet $\boldsymbol{c}^T\boldsymbol{x} \le \delta$ of P'. Since the optimum of the integer linear program (80) is attained at a vertex of Q_I, the assertion follows. □

Not each vertex of Q_I represents a facet of P'. In particular, if P is defined by nonnegative inequalities only, then $\boldsymbol{0}$ is a vertex of Q_I but not a facet of P'.

Lemma 9 ([15]). *The elementary closure of rational simplicial cone* $P = \{\boldsymbol{x} \in \mathbb{R}^n \mid \boldsymbol{A}\boldsymbol{x} \le \boldsymbol{d}\}$, *where* $\boldsymbol{A}$ *and* $\boldsymbol{d}$ *are integral and* $\boldsymbol{A}$ *has full row rank, is polynomially bounded in* size(P) *when the dimension is fixed.*

Proof. Each facet of P' corresponds to a vertex of Q_I by Proposition 11. Recall from the Hadamard bound that $d_L \le \|\boldsymbol{a}_1\| \cdots \|\boldsymbol{a}_n\|$, where $\boldsymbol{a}_i$ are the columns of $\boldsymbol{A}$. Thus the number of bits needed to encode d_L is in O(n size(P)). Therefore the size of Q is in O(n size(P)). It follows from Theorem 25 that the number of vertices of Q_I is in O(size$(P)^n$) for fixed n, since the dimension of Q is $n+1$. □

It is possible to explicitly construct, in polynomial time, a minimal inequality system defining P' when the dimension is fixed.

Observe first that the lattice determinant d_L in (79) can be computed with some polynomial Hermite normal form algorithm. If H is the HNF of A, then $L(A) = L(H)$ and the determinant of H is simply the product of its diagonal elements. Notice then that the system (79) can be written down. In particular its size is polynomial in the size of A and d, even in varying dimension, which follows from the Hadamard bound.

As noted in [28], one can construct the vertices of Q_I in polynomial time. This works as follows. Suppose one has a list of vertices $v_1, \ldots, v_k$ of Q_I. Let Q_k denote the convex hull of these vertices. Find an inequality description of Q_k, $Cx \leq d$. For each row-vector c_i of C, find with Lenstra's algorithm a vertex of Q_I maximizing $\{c^T x \mid x \in Q_I\}$. If new vertices are found, add them to the list and repeat the preceding steps, otherwise the list of vertices is complete. The list of vertices of Q_I yields a list of inequalities defining P'. With the ellipsoid method or your favorite linear programming algorithm in fixed dimension, one can decide for each individual inequality, whether it is necessary. If not, remove it. What remains are the facets of P'.

Proposition 12. *There exists an algorithm which, given a matrix $A \in \mathbb{Z}^{m \times n}$ of full row rank and a vector $d \in \mathbb{Z}^m$, constructs the elementary closure P' of $P(A, d)$ in polynomial time when the dimension n is fixed.*

7.3.2 Rational polyhedra

Let $P = \{x \in \mathbb{R}^n \mid Ax \leq d\}$, with integer A and d, be a rational polyhedron. Any Gomory-Chvátal cut can be derived from a set of rank(A) inequalities out of $Ax \leq d$ where the corresponding rows of A are linear independent. Such a choice represents a simplicial cone C and it follows from Theorem 9 that the number of inequalities of C' is polynomially bounded by size(C) $\leq$ size(P).

Theorem 29 ([15]). *The number of inequalities needed to describe the elementary closure of a rational polyhedron $P = P(A, d)$* with $A \in \mathbb{Z}^{m \times n}$ *and $d \in \mathbb{Z}^m$, is polynomial in* size(P) *in fixed dimension.*

Following the discussion at the end of Section 7.3.1 and using again Lenstra's algorithm, it is now easy to come up with a polynomial algorithm for constructing the elementary closure of a rational polyhedron $P(A, d)$ in fixed dimension. For each choice of rank(A) rows of A defining a simplicial cone C, compute the elementary closure C' and put the corresponding inequalities in the partial list of inequalities describing P'. At the end, redundant inequalities can be deleted.

Theorem 30. *There exists a polynomial algorithm that, given a matrix $A \in \mathbb{Z}^{m \times n}$* and a vector *$d \in \mathbb{Z}^m$, constructs an inequality description of the elementary closure of $P(A, d)$.*

References

[1] K. Aardal, R. E. Bixby, C. A. J. Hurkens, A. K. Lenstra, and J. W. Smeltink. Market split and basis reduction: Towards a solution of the Cornuéjols-Dawande instances. *INFORMS Journal on Computing*, 12(3):192–202, 2000.

[2] K. Aardal, C. Hurkens, and A. K. Lenstra. Solving a linear diophantine equation with lower and upper bounds on the variables. In R. E. Bixby, E. A. Boyd, and R. Z. Ríos-Mercado, editors, *Integer Programming and Combinatorial Optimization, 6th International IPCO Conference*, volume 1412 of *Lecture Notes in Computer Science*, pages 229–242, Berlin, 1998. Springer-Verlag.

[3] K. Aardal, C. A. J. Hurkens, and A. K. Lenstra. Solving a system of liner Diophantine equations with lower and upper bounds on the variables. *Mathematics of Operations Research*, 25(3):427–442, 2000.

[4] K. Aardal and A. K. Lenstra. Hard equality constrained integer knapsacks. *Mathematics of Operations Research*, 29(3):724–738, 2004.

[5] K. Aardal, R. Weismantel, and L. A. Wolsey. Non-standard approaches to integer programming. *Discrete Applied Mathematics*, 123(1-3):5–74, 2002.

[6] A. V. Aho, J. E. Hopcroft, and J. D. Ullman. *The Design and Analysis of Computer Algorithms*. Addison-Wesley, Reading, 1974.

[7] M. Ajtai. The shortest vector problem in *L2* is NP-hard for randomized reductions. In *Proceedings of the 30th Annual ACM Symposium on Theory of Computing*, pages 10–19, New York, 1998. ACM Press.

[8] M. Ajtai, R. Kumar, and D. Sivakumar. A sieve algorithm for the shortest lattice vector problem. In *Proceedings of the 33rd Annual ACM symposium on Theory of Computing*, pages 601–610, New York, 2001. ACM Press.

[9] W. Banaszczyk, A. E. Litvak, A. Pajor, and S. J. Szarek. The flatness theorem for nonsymmetric convex bodies via the local theory of Banach spaces. *Mathematics of Operations Research*, 24(3):728–750, 1999.

[10] I. Bárány, R. Howe, and L. Lovász. On integer points in polyhedra: A lower bound. *Combinatorica*, 12(2):135–142, 1992.

[11] A. I. Barvinok. A polynomial time algorithm for counting integral points in polyhedra when the dimension is fixed. *Mathematics of Operations Research*, 19(4):769–779, 1994.

[12] A. Barvinok and J. E. Pommersheim. An algorithmic theory of lattice points in polyhedra. *New Perspectives in Algebraic Combinatorics, MSRI Publications*, 38:91–147, 1999.

[13] D. E. Bell. A theorem concerning the integer lattice. *Studies in Applied Mathematics*, 56(2): 187–188, 1976/77.

[14] J. Blömer. Closest vectors, successive minima, and dual HKZ-bases of lattices. In *Proceedings of the 17th ICALP*, volume 1853 of *Lecture Notes in Computer Science*, pages 248–259, Berlin, 2000. Springer-Verlag.

[15] A. Bockmayr and F. Eisenbrand. Cutting planes and the elementary closure in fixed dimension. *Mathematics of Operations Research*, 26(2):304–312, 2001.

[16] A. Bockmayr, F. Eisenbrand, M. E. Hartmann, and A. S. Schulz. On the Chvátal rank of polytopes in the 0/1 cube. *Discrete Applied Mathematics*, 98:21–27, 1999.

[17] I. Borosh and L. B. Treybig. Bounds on positive integral solutions of linear diophantine equations. *Proceedings of the American Mathematical Society*, 55:299–304, 1976.

[18] J. Bourgain and V. D. Milman. Sections Euclidiennes et volume des corps symétriques convexes dans $\mathbf{R}^n$. *Comptes Rendus de l'Académie des Sciences. Série I. Mathématique*, 300(13):435–438, 1985.

[19] M. Brion. Points entiers dans polyèdres convexes. *Annales Scientifiques de l'École Normale Supérieure*, 21(4):653–663, 1988.

[20] J.-Y. Cai. Some recent progress on the complexity of lattice problems. *Electronic Colloquium on Computational Complexity*, (6), 1999. ECCC is available at: http://www.eccc.uni-trier.de/eccc/.

[21] J.-Y. Cai and A. P. Nerurkar. Approximating the svp to within a factor $(1 + 1/dim^{\varepsilon})$ is NP-hard under randomized reductions. In *Proceedings of the 38th IEEE Conference on Computational Complexity*, pages 46–55, Pittsburgh, 1998. IEEE Computer Society Press.

[22] J. W. S. Cassels. *An Introduction to the Geometry of Numbers*. Classics in Mathematics. Springer-Verlag, Berlin, 1997. Second Printing, Corrected, Reprint of the 1971 ed.

[23] V. Chvátal. Edmonds polytopes and a hierarchy of combinatorial problems. *Discrete Mathematics*, 4:305–337, 1973.

[24] V. Chvátal, W. Cook, and M. Hartmann. On cutting-plane proofs in combinatorial optimization. *Linear Algebra and its Applications*, 114/115:455–499, 1989.

[25] K. L. Clarkson. Las Vegas algorithms for linear and integer programming when the dimension is small. *Journal of the Association for Computing Machinery*, 42:488–499, 1995.

[26] S. A. Cook. The complexity of theorem-proving procedures. In *Proceedings of the 3rd Annual ACM Symposium on Theory of Computing*, pages 151–158, New York, 1971. ACM Press.

[27] W. Cook, C. R. Coullard, and G. Turán. On the complexity of the cutting plane proofs. *Discrete Applied Mathematics*, 18:25–38, 1987.

[28] W. Cook, M. E. Hartmann, R. Kannan, and C. McDiarmid. On integer points in polyhedra. *Combinatorica*, 12(1):27–37, 1992.

[29] W. Cook, T. Rutherford, H. E. Scarf, and D. Shallcross. An implementation of the generalized basis reduction algorithm for integer programming. *ORSA Journal on Computing*, 5(2):206–212, 1993.

[30] W. J. Cook and S. Dash. On the matrix-cut rank of polyhedra. *Mathematics of Operations Research*, 26(1):19–30, 2001.

[31] G. Cornuéjols and M. Dawande. A class of hard small 0-1 programs. In R. E. Bixby, E. A. Boyd, and R. Z. Ríos-Mercado, editors, *Integer Programming and Combinatorial Optimization, 6th International IPCO Conference*, volume 1412 of *Lecture Notes in Computer Science*, pages 284–293, Berlin, 1998. Springer-Verlag.

[32] G. Cornuéjols and Y. Li. On the rank of mixed 0,1 polyhedra. *Mathematical Programming*, 91(2):391–397, 2002.

[33] G. Cornuéjols, R. Urbaniak, R. Weismantel, and L. Wolsey. Decomposition of integer programs and of generating sets. In R. Burkard and G. Woeginger, editors, *Algorithms—ESA '97*, volume 1284 of *Lecture Notes in Computer Science*, pages 92–103, Springer-Verlag, Berlin, 1997.

[34] M. J. Coster, A. Joux, B. A. LaMacchia, A. M. Odlyzko, C.-P. Schnorr, and J. Stern. Improved low-density subset sum algorithms. *Computational Complexity*, 2(2):111–128, 1992.

[35] S. Dash. An exponential lower bound on the length of some classes of branch-and-cut proofs. In W. J. Cook and A. S. Shulz, editors, *Integer Programming and Combinatorial Optimization, 9th International IPCO Conference*, volume 2337 of *Lecture Notes in Computer Science*, pages 145–160, Berlin, 2002. Springer-Verlag.

[36] J. A. De Loera, R. Hemmecke, J. Tauzer, and R. Yoshida. Effective lattice point counting in rational polytopes. *Journal of Symbolic Computation*. To appear. Available at: http://www.math.ucdavis.edu/~deloera.

[37] M. E. Dyer. On integer points in polyhedra. *SIAM Journal on Computing*, 20:695–707, 1991.

[38] M. E. Dyer and R. Kannan. On Barvinok's algorithm for counting lattice points in fixed dimension. *Mathematics of Operations Research*, 22(3):545–549, 1997.

[39] F. Eisenbrand. Short vectors of planar lattice via continued fractions. *Information Processing Letters*, 79(3):121–126, 2001.

[40] F. Eisenbrand. Fast integer programming in fixed dimension. In G. D. Battista and U. Zwick, editors, *Algorithms – ESA 2003*, volume 2832 of *Lecture Notes in Computer Science*, pages 196–207, Berlin, 2003. Springer-Verlag.

[41] F. Eisenbrand and S. Laue. A linear algorithm for integer programming in the plane. *Mathematical Programming*, 2004. To appear.

[42] F. Eisenbrand and G. Rote. Fast 2-variable integer programming. In K. Aardal and B. Gerards, editors, *Integer Programming and Combinatorial Optimization, 8th International*

IPCO Conference, volume 2081 of *Lecture Notes in Computer Science*, pages 78–89, Berlin, 2001. Springer-Verlag.

[43] F. Eisenbrand and G. Rote. Fast reduction of ternary quadratic forms. In J. Silverman, editor, *Cryptography and Lattices, International Conference, CaLC 2001*, volume 2146 of *Lecture Notes in Computer Science*, pages 32–44, Berlin, 2001. Springer-Verlag.

[44] F. Eisenbrand and A. S. Schulz. Bounds on the Chvátal rank of polytopes in the 0/1 cube. In G. Cornuéjols, R. E. Burkard, and G. J. Woeginger, editors, *Integer Programming and Combinatorial Optimization, 7th International IPCO Conference*, volume 1610 of *Lecture Notes in Computer Science*, pages 137–150. Springer-Verlag, 1999.

[45] P. van Emde Boas. Another NP-complete partition problem and the complexity of computing short vectors in a lattice. Technical Report MI-UvA-81-04, Mathematical Institute, University of Amsterdam, Amsterdam, 1981.

[46] S. D. Feit. A fast algorithm for the two-variable integer programming problem. *Journal of the Association for Computing Machinery*, 31(1):99–113, 1984.

[47] L. Gao and Y. Zhang. Computational experience with Lenstra's algorithm. Technical Report TR02-12, Department of Computational and Applied Mathematics, Rice University, Houston, TX, 2002.

[48] B. Gärtner and E. Welzl. Linear programming—randomization and abstract frameworks. In *STACS 96*, volume 1046 of *Lecture Notes in Computer Science*, pages 669–687, Berlin, 1996. Springer-Verlag.

[49] C. F. Gauß. *Disquisitions arithmeticae*. Gerh. Fleischer Iun., 1801.

[50] J.-L. Goffin. Variable metric relaxation methods. II. The ellipsoid method. *Mathematical Programming*, 30(2):147–162, 1984.

[51] O. Goldreich and S. Goldwasser. On the limits of non-approximability of lattice problems. In *Proceedings of the 30th Annual ACM Symposium on Theory of Computing*, pages 1–9, New York, 1998. ACM Press.

[52] O. Goldreich, D. Micciancio, S. Safra, and J.-P. Seifert. Approximating shortest lattice vectors is not harder than approximating closest lattice vectors. *Information Processing Letters*, 71(2):55–61, 1999.

[53] R. E. Gomory. Outline of an algorithm for integer solutions to linear programs. *Bulletin of the American Mathematical Society*, 64:275–278, 1958.

[54] M. Grötschel, L. Lovász, and A. Schrijver. *Geometric Algorithms and Combinatorial Optimization*. Springer-Verlag, Berlin, 1988.

[55] M. Grötschel, L. Lovász, and A. Schrijver. Geometric methods in combinatorial optimization. In W. R. Pulleyblank, editors, *Progress in Combinatorial Optimization*, pages 167–183. Academic Press, Toronto, 1984.

[56] A. C. Hayes and D. G. Larman. The vertices of the knapsack polytope. *Discrete Applied Mathematics*, 6:135–138, 1983.

[57] B. Helfrich. Algorithms to construct Minkowski reduced and Hermite reduced lattice basis. *Theoretical Computer Science*, 41:125–139, 1985.

[58] C. Hermite. Extraits de lettres de M. Ch. Hermite à M. Jacobi sur différents objects de la théorie des nombres. *Journal für die reine und angewandte Mathematik*, 40, 1850.

[59] C. Hermite. Deuxième lettre à Jacobi. In *Oevres de Hermite I*, pages 122–135, Gauthier-Villary, Paris, 1905.

[60] D. S. Hirschberg and C. K. Wong. A polynomial algorithm for the knapsack problem in two variables. *Journal of the Association for Computing Machinery*, 23(1):147–154, 1976.

[61] A. Joux and J. Stern. Lattice reduction: a toll box for the cryptanalyst. *Journal of Cryptology*, 11(3):161–185, 1998.

[62] N. Kanamaru, T. Nishizeki, and T. Asano. Efficient enumeration of grid points in a convex polygon and its application to integer programming. *International Journal of Computational Geometry & Applications*, 4(1):69–85, 1994.

[63] R. Kannan. A polynomial algorithm for the two-variable integer programming problem. *Journal of the Association for Computing Machinery*, 27(1):118–122, 1980.

[64] R. Kannan. Improved algorithms for integer programming and related problems. In *Proceedings of the 15th Annual ACM Symposium on Theory of Computing*, pages 193–206, New York, 1983. ACM Press.

[65] R. Kannan. Algorithmic geometry of numbers. *Annual Review of Computer Science*, 2:231–267, 1987.

[66] R. Kannan. Minkowski's convex body theorem and integer programming. *Mathematics of Operations Research*, 12(3):415–440, 1987.

[67] R. Kannan and L. Lovász. Covering minima and lattice point free convex bodies. In *Foundations of Software Technology and Theoretical Computer Science*, volume 241 of *Lecture Notes in Computer Science*, pages 193–213. Springer-Verlag, Berlin, 1986.

[68] R. Kannan and L. Lovász. Covering minimal and lattice-point-free convex bodies. *Annals of Mathematics*, 128:577–602, 1988.

[69] R. M. Karp. Reducibility among combinatorial problems. In *Complexity of Computer Computations (Proc. Sympos., IBM Thomas J. Watson Res. Center, Yorktown Heights, N.Y., 1972)*, pages 85–103, Plenum Press, New York, 1972.

[70] A. Khinchine. A quantitative formulation of Kronecker's theory of approximation (in russian). *Izvestiya Akademii Nauk SSR Seriya Matematika*, 12:113–122, 1948.

[71] D. Knuth. *The Art of Computer Programming*, volume 2. Addison-Wesley, Reading 1969.

[72] A. Korkine and G. Zolotareff. Sur les formes quadratiques. *Mathematische Annalen*, 6:366–389, 1873.

[73] J. C. Lagarias, H. W. Lenstra, Jr., and C. P. Schnorr. Korkin-Zolotarev bases and successive minima of a lattice and its reciprocal lattice. *Combinatorica*, 10(4):333–348, 1990.

[74] J. C. Lagarias and A. M. Odlyzko. Solving low-density subset sum problems. *Journal of the Association for Computing Machinery*, 32(1):229–246, 1985.

[75] A. K. Lenstra, H. W. Lenstra, Jr., and L. Lovász. Factoring polynomials with rational coefficients. *Mathematische Annalen*, 261:515–534, 1982.

[76] H. W. Lenstra, Jr. Integer programming with a fixed number of variables. *Mathematics of Operations Research*, 8(4):538–548, 1983.

[77] LiDIA – A Library for Computational Number Theory. TH Darmstadt/Universität des Saarlandes, Fachbereich Informatik, Institut für Theoretische Informatik. http://www.informatik.th-darmstadt.de/pub/TI/LiDIA.

[78] Q. Louveaux and L. A. Wolsey. Combining problem structure with basis reduction to solve a class of hard integer programs. *Mathematics of Operations Research*, 27(3):470–484, 2002.

[79] L. Lovász and H. E. Scarf. The generalized basis reduction algorithm. *Mathematics of Operations Research*, 17(3):751–764, 1992.

[80] J. Matoušek, M. Sharir, and E. Welzl. A subexponential bound for linear programming. *Algorithmica*, 16(4-5):498–516, 1996.

[81] D. Micciancio. The shortest vector in a lattice is hard to approximate to within some constant. In *Proceedings of the 39th Annual Symposium on Foundations of Computer Science*, pages 92–98, Los Alamitos, CA, 1998. IEEE Computer Society.

[82] H. Minkowski. Über die positiven quadratischen Formen und über kettenbruchänliche Algorithmen. *Journal für die reine und angewandte Mathematik*, 107:278–297, 1891.

[83] H. Minkowski. *Geometrie der Zahlen* Teubner, Leipzig, 1896.

[84] R. Motwani and P. Raghavan. *Randomized Algorithms*. Cambridge University Press, Cambridge, 1995.

[85] G. L. Nemhauser and L. A. Wolsey. *Integer and Combinatorial Optimization*. John Wiley & Sons, New York, 1988.

[86] P. Q. Nguyen and J. Stern. Lattice reduction in cryptology: An update. In W. Bosma, editor, *Algorithmic Number Theory, 4th International Symposium, ANTS-IV*, volume 1838 of *Lecture Notes in Computer Science*, pages 85–112, Berlin, 2000. Springer-Verlag.

[87] P. Q. Nguyen and J. Stern. The two faces of lattices in cryptology. In J. H. Silverman, editor, *Cryptography and Lattices, International Conference, CaLC 2001*, volume 2146 of *Lecture Notes in Computer Science*, pages 146–180, Berlin, 2001. Springer-Verlag.

[88] P. Pudlák. Lower bounds for resolution and cutting plane proofs and monotone computations. *Journal of Symbolic Logic*, 62(3):981–988, 1997.
[89] H. E. Scarf. An observation on the structure of production sets with indivisibilities. *Proceedings of the National Academy of Sciences, U.S.A.*, 74(9):3637–3641, 1977.
[90] H. E. Scarf. Production sets with indivisibilities. Part I: generalities. *Econometrica*, 49:1–32, 1981.
[91] C.-P. Schnorr. A hierarchy of polynomial time lattice basis reduction algorithms. *Theoretical Computer Science*, 53(2-3):201–224, 1987.
[92] C.-P. Schnorr. Block reduced lattice bases and successive minima. *Combinatorics Probability and Computing*, 3(4):507–522, 1994.
[93] C.-P. Schnorr and M. Euchner. Lattice basis reduction: improved practical algorithms and solving subset sum problems. *Mathematical Programming*, 66(2):181–199, 1994.
[94] C. P. Schnorr and H. H. Hörner. Attacking the Chor-Rivest cryptosystem by improved lattice reduction. In *Advances in Cryptology—EUROCRYPT '95*, volume 921 of *Lecture Notes in Computer Science*, pages 1–12, Springer-Verlag, Berlin, 1995.
[95] A. Schönhage. Schnelle Berechung von Kettenbruchentwicklungen. (Speedy computation of expansions of continued fractions). *Acta Informatica*, 1:139–144, 1971.
[96] A. Schönhage. Fast reduction and composition of binary quadratic forms. In *International Symposium on Symbolic and Algebraic Computation, ISSAC 91*, pages 128–133, New York, 1991. ACM Press.
[97] A. Schönhage and V. Strassen. Schnelle Multiplikation grosser Zahlen (Fast multiplication of large numbers). *Computing*, 7:281–292, 1971.
[98] A. Schrijver. On cutting planes. *Annals of Discrete Mathematics*, 9:291–296, 1980.
[99] A. Schrijver. *Theory of Linear and Integer Programming*. John Wiley & Sons, Chichester, 1986.
[100] I. Semaev. A 3-dimensional lattice reduction, algorithm. In J. H. Silverman, editor, *Cryptography and Lattices, International Conference, CaLC 2001*, volume 2146 of *Lecture Notes in Computer Science*, pages 181–193, Berlin, 2001. Springer-Verlag.
[101] M. Seysen. Simultaneous reduction of a lattice basis and its reciprocal basis. *Combinatorica*, 13(3):363–376, 1993.
[102] V. Shoup. NTL: A Library for doing Number Theory. Courant Institute, New York. http://www.shoup.net/.
[103] O. van Sprang. *Basisreduktionsalogirthmen für Gitter kleiner Dimension*. PhD thesis, Fachbereich Informatik, Universität des Saarlandes, Saarbrücken, Germany, 1994. In German.
[104] X. Wang. *A New Implementation of the Generalized Basis Reduction Algorithm for Convex Integer Programming*. PhD thesis, Yale University, 1997.
[105] C. K. Yap. Fast unimodular reduction: Planar integer lattices. In *Proceedings of the 33rd Annual Symposium on Foundations of Computer Science*, pages 437–446, Pittsburgh, 1992. IEEE Computer Society Press.
[106] L. Y. Zamanskij and V. D. Cherkasskij. A formula for determining the number of integral points on a straight line and its applications. *Ehkon. Mat. Metody*, 20:1132–1138, 1984.

K. Aardal et al., Eds., *Handbooks in OR & MS, Vol. 12*

Chapter 5

Primal Integer Programming

Bianca Spille and Robert Weismantel

University of Magdeburg, Universitätsplatz 2, D-39106 Magdeburg, Germany
E-mail: [spille,weismantel]@imo.math.uni-magdeburg.de

Abstract

Primal Integer Programming is concerned with the design of algorithms for linear integer programs that move from a feasible solution to a better feasible solution until optimality is proved. We refer to such a method as a *primal* (or *augmentation*) *algorithm*. We study such algorithms and address the questions related to making such an approach theoretically efficient and practically work. In particular, we address the question of computational complexity with respect to the number of augmentation steps. From a theoretical point of view, the study of the augmentation problem leads to the theory of irreducible lattice points and integral generating sets. We present the algorithmic approaches to attack *general integer programs*; the first approach is based on the use of cutting planes, the Integral Basis Method is a second approach. For specific *combinatorial optimization problems* such a min-cost flow, matching, matroid intersection and the problem of minimizing a submodular function, we discuss the basics of the related combinatorial algorithms.

1 Introduction

Enumerative methods in combination with primal or dual algorithms form the basic algorithmic building blocks for tackling linear integer programs today.

Dual type algorithms start solving a linear programming relaxation of the underlying problem, typically with the dual simplex method. In the course of the algorithm one maintains as an invariant, both primal and dual feasibility of the solution of the relaxation. While the optimal solution to the relaxation is not integral, one continues adding cutting planes to the problem formulation and reoptimizes.

In contrast to the dual methods, primal type algorithms work with integral solutions, usually with primal feasible integer solutions, hence the name. More precisely, given a feasible solution for a specified discrete set of points $\mathcal{F} \subseteq \mathbb{Z}^n$, one applies an augmentation strategy: starting with the feasible solution one iteratively tries to detect an improving direction that is applicable at the current solution for as long as possible. We will study such augmentation algorithms or primal algorithms in the following and address the questions related to making such an approach theoretically efficient and practically work. Throughout this chapter we investigate optimization problems over discrete sets of points,

$$\max c^T x : x \in \mathcal{F} := \{x \in \mathbb{Z}^n : Ax = b, 0 \leq x \leq u\}, \tag{1}$$

with data $A \in \mathbb{Z}^{m \times n}$, $b \in \mathbb{Z}^m$, $u \in (\mathbb{Z}_+ \cup \{\infty\})^n$, and $c \in \mathbb{Z}^n$, i.e., linear integer programming problems with or without upper bounds on the variables. The object of our investigation is a solution of the following optimization problem.

> **The Optimization Problem (OPT)**
>
> Given a vector $c \in \mathbb{Z}^n$ and a point $x \in \mathcal{F}$, find a vector $x^* \in \mathcal{F}$ that maximizes c over $\mathcal{F}$, if it exists.

The generic form of an algorithm that we will apply to solve (OPT) is a *primal algorithm* or an *augmentation algorithm* that works as follows.

Algorithm 1.1. (*Augmentation algorithm for a maximization problem*)
Input. $x^0 \in \mathcal{F}$, $c \in \mathbb{Z}^n$.
Output. *An optimal solution* $x^* \in \mathcal{F}$ *or a direction* $z \in \mathbb{Z}^n$ *and a feasible point* $x \in \mathcal{F}$, *such that* $c^T z > 0$ *and* $x + \lambda z \in \mathcal{F}$ *for all* $\lambda \in \mathbb{Z}_+$.

(1) *Set* $x := x^0$.
(2) *While x is not optimal,*
   (a) *Determine an augmenting direction, i.e., an integral vector z such that* $c^T z > 0$ *and* $x + z \in \mathcal{F}$ *and*
   (b) *Determine a step length, i.e., a maximal number* $\lambda \in \mathbb{Z}_+$ *such that* $x + \lambda z \in \mathcal{F}$. *If this number does not exist, return x and z. Stop.*
   (c) *Set* $x := x + \lambda z$.
(3) *Return* $x^* := x$.

The augmentation algorithms have been designed for and applied to a range of linear integer programming problems: the augmenting path methods for solving maximum flow problems or algorithms for solving the min-cost flow problem via augmentation along negative cycles are of this type. Other examples include the greedy algorithm for solving the matroid optimization problem, alternating path algorithms for solving the maximum (weight) matching problem, or methods for optimizing over the intersection of two matroids.

There are three elementary questions that arise in the analysis of an augmentation algorithm for a linear integer program:

(i) How can one solve the subproblem of detecting an augmenting direction?
(ii) How can one verify that a given point is optimal?
(iii) What is a bound on the number of augmentation steps one has to apply in order to reach an optimal point?

We begin with the question (iii) in Section 2. The subproblem (i) of detecting an augmenting direction establishes a natural link to the theory of irreducible lattice points. This issue is discussed in Section 3. It provides at least conceptually an answer to question (ii). Whereas algorithmic approaches to attack general integer programs are discussed in Section 4, primal algorithms for specific combinatorial optimizations problems are the central topic of Section 5.

2 Efficient primal algorithms

One may certainly doubt in the beginning whether an augmentation algorithm can be made effective in terms of the number of augmentations that one needs to find an optimal solution. It however turns out that one can reach an optimal solution by solving a *directed augmentation subproblem* a polynomial number of times. We will make precise below what we mean by this. In case of a 0/1-program, the directed augmentation subproblem is in fact identical with an *augmentation subproblem* that we introduce next.

The Augmentation Problem (AUG)

Given a vector $c \in \mathbb{Z}^n$ and a point $x \in \mathcal{F}$, find a point $y \in \mathcal{F}$ such that $c^T y > c^T x$, or assert that no such y exists.

A classical example of an augmentation algorithm is the cycle canceling algorithm for the min-cost flow problem:

Let $D = (V, A)$ be a digraph with specified nodes $r, s \in V$, $u \in \mathbb{Z}_+^A$ a capacity function on the arcs, $c \in \mathbb{Z}^A$ a cost function on the arcs, and $f \in \mathbb{Z}_+$. A vector $x \in \mathbb{R}^A$ is a *flow* if

$$\begin{aligned} x(\delta^+(r)) - x(\delta^-(r)) &= f, & \\ x(\delta^+(v)) - x(\delta^-(v)) &= 0 & \text{for all } v \in V \setminus \{r, s\}, \\ x(\delta^+(s)) - x(\delta^-(s)) &= -f, & \\ 0 \le x_a &\le u_a & \text{for all } a \in A, \\ x_a &\in \mathbb{Z} & \text{for all } a \in A. \end{aligned}$$

The *min-cost flow problem* is to find a flow of minimum cost $\sum_{a \in A} c_a x_a$. For any flow x, define an augmentation digraph $D(x)$ with node set V and arcs

$$\begin{array}{llllll} (v,w) & \text{with cost } c_{vw} & \text{for} & vw \in A & \text{with} & x_{vw} < u_{vw}, \\ (w,v) & \text{with cost } -c_{vw} & \text{for} & vw \in A & \text{with} & x_{vw} > 0. \end{array}$$

The first kind of arcs are called *forward arcs*, the latter *backward arcs*. A flow x is minimal if and only if there is no negative dicycle in $D(x)$. The cycle canceling algorithm works as follows: beginning with a flow x, repeatedly find a negative dicycle C in $D(x)$ and augment x along it, i.e., raise x'_{vw} by 1 on each forward arc (v, w) of C and lower x_{vw} by 1 on each backward arc (w, v).

A generalization of this augmentation strategy to integer programs requires an investigation of a directed version of the augmentation problem.

The Directed Augmentation Problem (DIR-AUG)

Given vectors $c, d \in \mathbb{Z}^n$ and a point $x \in \mathcal{F}$, find vectors $z^1, z^2 \in \mathbb{Z}^n_+$ such that $\text{supp}(z^1) \cap \text{supp}(z^2) = \emptyset$,

$$c^T z^1 - d^T z^2 > 0, \quad \text{and} \quad x + z^1 - z^2 \quad \text{is feasible},$$

or assert that no such vectors z^1, z^2 exist.

For the min-cost flow problem, the directed augmentation problem can be solved as follows. Let $c, d \in \mathbb{Z}^A$ and let x be a flow. Define the augmentation digraph $D(x)$ as above but with modified cost: assign a cost c_{vw} to each forward arc (v, w) and a cost $-d_{vw}$ to each backward arc (w, v). Let C be a dicycle in $D(x)$ that is negative w.r.t. the new costs. Let z be the vector associated with the set C, i.e., $z_{vw} = +1$ if (v, w) is a forward arc in C, $z_{vw} = -1$ if (w, v) is a backward arc in C, and $z_{vw} = 0$, otherwise. We denote by z^1 the positive part of z and by z^2 the negative part of z. Then $z^1, z^2 \in \mathbb{Z}^A_+$ satisfy the following conditions: $\text{supp}(z^1) \cap \text{supp}(z^2) = \emptyset$, $c^T z^1 - d^T z^2 < 0$, and $x + z^1 - z^2 = x + z$ is a flow. Therefore, z^1 and z^2 constitute a solution to the directed augmentation problem.

In case of the min-cost flow problem, it is also well known that a cycle cancelling algorithm does not necessarily converge to an optimal solution in polynomial time in the encoding length of the input data. Indeed, a more sophisticated strategy for augmenting is required. In the min-cost flow application it is for instance the augmentation of flow along the maximum mean ratio cycles that makes the primal algorithm work efficiently. The maximum mean ratio cycles are very special objects and there is no obvious counterpart in the case of general integer programs.

A generalization of this strategy to the integer programming problems with bounded feasible region is our plan for the remainder of this section.

Our approach follows Schulz and Weismantel (2002), see also Wallacher (1992) and McCormick and Shioura (1996). The analysis of this augmentation algorithm is based on a lemma about geometric improvement (Ahuja Magananti and Orlin, 1993) that characterizes the improvement for each augmentation step.

Algorithm 2.1. (*An efficient augmentation procedure*)
Input. *$\mathcal{F}$ bounded, $x^0 \in \mathcal{F}$, $c \in \mathbb{Z}^n$*
Output. *An optimal solution $x^* \in \mathcal{F}$.*

(1) *Set $x := x^0$.*
(2) *If x is an optimal solution, return $x^* := x$. Stop.*
(3) *Otherwise, solve the following problem:*

$$\begin{aligned} \max \quad & |c^T(z^1 - z^2)|/(p(x)^T z^1 + n(x)^T z^2) \\ s.t. \quad & x + z^1 - z^2 \in \mathcal{F}, \\ & c^T(z^1 - z^2) > 0, \\ & z^1, z^2 \in \mathbb{Z}^n_+, \end{aligned}$$

where, for $j \in \{1, \ldots, n\}$

$$p(x)_j = \begin{cases} \dfrac{1}{u_j - x_j} & \text{if } x_j < u_j, \\ 0 & \text{otherwise,} \end{cases}$$

and

$$n(x)_j = \begin{cases} \dfrac{1}{x_j} & \text{if } x_j > 0, \\ 0 & \text{otherwise.} \end{cases}$$

(4) *Determine $\lambda \in \mathbb{Z}_+$ such that*

$$x + \lambda(z^1 - z^2) \in \mathcal{F}, \quad x + (\lambda + 1)(z^1 - z^2) \notin \mathcal{F}.$$

(5) *Set $x := x + \lambda(z^1 - z^2)$ and return to Step* (2).

Resorting to the technique of reducing a fractional programming problem to a series of linear optimization problems and using binary search, one may implement Step (3) of Algorithm 2.1 by solving the subproblem (DIR-AUG) a polynomial number of times in the encoding length of the input data. We use the following two symbols,

$$K := \max\{|c_i|: i = 1, \ldots, n\}, \qquad U := \max\{|u_i|: i = 1, \ldots, n\}.$$

Lemma 2.2. *Let $U < \infty$. Then there is a number α that is polynomial in n and* $\log(nKU)$ *such that Step* (3) *of Algorithm* 2.1 *can be implemented by solving a subproblem of the form* (*DIR-AUG*) *at most α times.*

Proof. Let $x \in \mathcal{F}$ be not optimal. We have to detect an optimal solution of Problem (2),

$$\begin{aligned} \max \quad & |c^T(z^1 - z^2)|/(p(x)^T z^1 + n(x)^T z^2) \\ \text{s.t.} \quad & x + z^1 - z^2 \in \mathcal{F}, \\ & c^T(z^1 - z^2) > 0, \\ & z^1, z^2 \in \mathbb{Z}^n_+. \end{aligned}$$

Let μ^* be the (unknown) optimal value of this program. Inspecting the objective function we notice that the numerator $c^T(z^1 - z^2)$ is an integer value that is bounded by nKU. The denominator $p(x)^T z^1 + n(x)^T z^2$ is a fractional value that is in the interval $[(1/U), n]$. For any estimate μ of μ^*, we define two rational vectors,

$$\begin{aligned} c' &= c - \mu p(x), \\ d &= c + \mu n(x). \end{aligned}$$

With input c', d and x we solve the subproblem (DIR-AUG). Since $(c')^T z^1 - d^T z^2 > 0$ if and only if $|c^T(z^1 - z^2)|/(p(x)^T z^1 + n(x)^T z^2) > \mu$, it follows that (DIR-AUG) returns a solution if and only if $\mu < \mu^*$. Hence, depending on the output, μ is either an upper bound or a lower bound for μ^*. We use binary search to find μ^* and the corresponding vectors z^1, z^2 with which we can augment the current solution x. □

We are now ready for analyzing Algorithm 2.1.

Theorem 2.3. [Schulz and Weismantel (2002)] *Let $U < \infty$. For any $x \in \mathcal{F}$ and $c \in \mathbb{Z}^n$, Algorithm* 2.1 *detects an optimal solution with α applications of the subproblem* (*DIR-AUG*), *where α is a polynomial in n and* $\log(nKU)$.

Proof. Let $x^0 \in \mathcal{F}$, $c \in \mathbb{Z}^n$ be the input of Algorithm 2.1. By x^* we denote an optimal solution. We assume that Algorithm 2.1 produces a sequence of points $x^0, x^1, \ldots \in \mathcal{F}$.

Assuming that x^k is not optimal, let z^1, z^2 be the output of Step (3) of Algorithm 2.1. Apply Step (4), i.e., choose $\lambda \in \mathbb{Z}_+$ such that

$$x^k + \lambda(z^1 - z^2) \in \mathcal{F}, \quad x^k + (\lambda + 1)(z^1 - z^2) \notin \mathcal{F}.$$

Define $z := \lambda(z^1 - z^2)$. Then $x^{k+1} = x^k + z$ and there exists $j \in \{1, \ldots, n\}$ such that $x_j^k + 2z_j > u_j$ or $x_j^k + 2z_j < 0$. Therefore, $z_j^+ > (u_j - x_j^k)/2$ or $z_j^- > x_j^k/2$ and hence, $p(x^k)^T z^+ + n(x^k)^T z^- \geq (1/2)$. Let $z^* := x^* - x^k$. It is $p(x^k)^T(z^*)^+ + n(x^k)^T(z^*)^- \leq n$. On account of the condition

$$|c^T z|/(p(x^k)^T z^+ + n(x^k)^T z^-) \geq |c^T z^*|/(p(x^k)^T(z^*)^+ + n(x^k)^T(z^*)^-)$$

we obtain that

$$|c^T(x^{k+1} - x^k)| = |c^T z| \geq \frac{|c^T z^*|}{2n} = \frac{|c^T(x^* - x^k)|}{2n}$$

Consider a consecutive sequence of $4n$ iterations starting with iteration k. If each of these iterations improves the objective function value by at least $(|c^T(x^* - x^k)|/4n)$, then x^{k+4n} is an optimal solution. Otherwise, there exists an index l such that

$$\frac{|c^T(x^* - x^l)|}{2n} \leq |c^T(x^{l+1} - x^l)| \leq \frac{|c^T(x^* - x^k)|}{4n}.$$

It follows that

$$|c^T(x^* - x^l)| \leq \frac{1}{2}|c^T(x^* - x^k)|,$$

i.e., after $4n$ iterations we have halved the gap between $c^T x^*$ and $c^T x^k$. Since the objective function value of any feasible solution is integral and bounded by nKU, the result follows. □

Consequently, the study of directed augmentation problems is a reasonable attempt to attack on optimization problem. This may be viewed as a sort of "primal counterpart" of the fact that a polynomial number of calls of a separation oracle suffices to solve an optimization problem with a cutting plane algorithm.

Note that one can also use the method of *bit-scaling* [see Edmonds and Karp (1972)] in order to show that an optimal solution of a 0/1-integer programming can be found by solving a polynomial number of augmentation subproblems. This is discussed in Grötschel and Lovász (1995) and Schulz, Weismantel, Ziegler (1995).

3 Irreducibility and integral generating sets

Realizing that optimization problems in 0/1 variables can be solved with not too many calls of a subroutine that returns a solution to the augmentation subproblem, a natural question is to study the latter in more detail. In the case of a min-cost flow problem in digraphs it is clear that every augmentation vector in the augmentation digraph associated with a feasible solution corresponds to a zero-flow of negative cost. Any such zero-flow can be decomposed into directed cycles. A generalization of this decomposition property is possible with the notion of irreducibility of integer points.

Definition 3.1. *Let* $S \subseteq \mathbb{Z}^n$.

(1) *A vector* $z \in S$ *is reducible if* $z = 0$ *or there exist* $k \geq 2$ *vectors* $z^1, \ldots, z^k \in S \setminus \{0\}$ *and integral multipliers* $\lambda_1, \ldots, \lambda_k \geq 1$ *such that* $z = \sum_{i=1}^{k} \lambda_i z^i$. *Otherwise,* z *is irreducible.*
(2) *An integral generating set of* S *is a subset* $S_0 \subseteq S$ *such that every vector* $z \in S$ *is a nonnegative integral linear combination of vectors of* S_0. *It is called an integral basis if it is minimal w.r.t. inclusion.*

Using integral generating sets, we can define a set that allows to verify whether a given feasible point of an integer program is optimal. Also, with the help of such a set one can solve the irreducible augmentation problem, at least conceptually.

> **The Irreducible Augmentation Problem (IRR-AUG)**
>
> Given a vector $c \in \mathbb{Z}^n$ and a point $x \in \mathcal{F}$, find an irreducible vector $z \in S := \{y - x \colon y \in \mathcal{F}\}$ such that $c^T z > 0$ and $x + z \in \mathcal{F}$, or assert that no such z exists.

The approach as we introduce it now is however not yet algorithmically tractable because the size of such a set for an integer program is usually exponential in the dimension of the integer program.

Here we deal with general families of integer programs of the form

$$\max c^T x : Ax = b, 0 \leq x \leq u, x \in \mathbb{Z}^n, \tag{3}$$

with fixed matrix $A \in \mathbb{Z}^{m \times n}$ and varying data $c \in \mathbb{R}^n$, $b \in \mathbb{Z}^m$, and $u \in \mathbb{Z}^n$.

Definition 3.2. *Consider the family of integer programs* (3). *Let O_j be the j-th orthant in $\mathbb{R}^n$, let $C_j := \{x \in O_j: Ax = 0\}$* and H_j be an integral basis of $C_j \cap \mathbb{Z}^n$. *The set*

$$\mathcal{H} := \bigcup_j H_j \setminus \{0\}$$

is called the Graver set *for this family.*

Note that we have so far not established that $\mathcal{H}$ is a finite set. This however will follow from our analysis of the integral generating sets. Next we show that $\mathcal{H}$ can be used to solve the irreducible augmentation problem for the family of integer programs of the above form.

Theorem 3.3. *Let x^0 be a feasible point for an integer program of the form* (3). *If x^0 is not optimal there exists an irreducible vector $h \in \mathcal{H}$ that solves* (*IRR-AUG*).

Proof. Let $b \in \mathbb{Z}^m$, $u \in \mathbb{Z}^n$, and $c \in \mathbb{R}^n$ and consider the corresponding integer program $\max c^T x : Ax = b, 0 \leq x \leq u, x \in \mathbb{Z}^n$. Let x^0 be a feasible solution for this program, that is not optimal and let y be an optimal solution. It follows that $A(y - x^0) = 0$, $y - x^0 \in \mathbb{Z}^n$ and $c^T(y - x^0) > 0$. Let O_j denote an orthant that contains $y - x^0$. As $y - x^0$ is an integral point in C_j, there exist multipliers $\lambda_h \in \mathbb{Z}_+$ for all $h \in H_j$ such that

$$y - x^0 = \sum_{h \in H_j} \lambda_h h.$$

As $c^T(y - x^0) > 0$ and $\lambda_h \geq 0$ for all $h \in H_j$, there exists a vector $h^* \in H_j$ such that $c^T h^* > 0$ and $\lambda_{h^*} > 0$. Since h^* lies in the same orthant as $y - x^0$, we have that $x^0 + h^*$ is feasible. Hence, $h^* \in \mathcal{H}$ is an irreducible vector that solves (IRR-AUG). □

If one can solve (IRR-AUG), then one can also solve (AUG). However, the other direction is difficult even in the case of 0/1-programs, see Schulz et al. (1995). This fact is not surprising, because it is ($\mathcal{NP}$-complete to decide whether an integral vector in some set $S \subseteq \mathbb{Z}^n$ is reducible.

The Integer Decomposition Problem (INT-DEC)

Given a set $S \subseteq \mathbb{Z}^n \setminus \{0\}$ by a membership-oracle and a point $x \in S$. Decide, whether x is reducible.

Theorem 3.4. [Sebö (1990)] *The integer decomposition problem is $\mathcal{NP}$-complete.*

Theorem 3.4 asserts the difficulty of deciding whether an integral vector is reducible. On the other hand, every such vector can be decomposed into a finite number of irreducible ones. In fact, we can write every integral vector in a pointed cone in $\mathbb{R}^n$ as the nonnegative integer combination of at most $2n-2$ irreducible vectors, see Sebö (1990).

Next we deal with the question on how to compute the irreducible members of a set of integral points. This topic will become important for the remaining sections when we deal with primal integer programming algorithms. In order to make sure that an algorithm for the computing irreducible solutions is finite, it is important to establish the finiteness of the set of irreducible solutions. We will analyze this property for systems of the form

$$S := \{z \in \mathbb{Z}^n_+ \colon Az \leq b\} \quad with \quad A \in \mathbb{Z}^{m \times n}, b \in \mathbb{Z}^m_+. \tag{4}$$

Note that when $b=0$, an integral basis is also known as a (minimal) *Hilbert basis* of the pointed cone $C = \{z \in \mathbb{R}^n_+ \colon Az \leq 0\}$. In the case of cones the set S is closed under addition, i.e., if $z, z' \in S$. However, this property does not apply to the inhomogeneous case.

Example 3.5. Consider the integral system

$$\begin{aligned} -z_1 + z_2 + z_3 &\leq 1, \\ z_1 - z_2 + z_3 &\leq 1, \\ z_1 + z_2 - z_3 &\leq 1, \\ z_1, z_2, z_3 &\in \mathbb{Z}_+. \end{aligned} \tag{5}$$

The unit vectors $(1,0,0)$, $(0,1,0)$ and $(0,0,1)$ are solutions to (5). The vector $(1,1,1)$ is a solution to (5) that is generated by the unit vectors but it is not the sum of two other solutions.

As a consequence of Theorem 3.7 to be stated below we obtain that integral generating sets are finite. In fact, an integral basis of a set S as in (4) is uniquely determined. This result follows essentially from the Gordan Lemma.

Theorem 3.6. (*Gordan lemma*)
Let $\emptyset \neq S \subseteq \mathbb{Z}^n_+$. There exists a unique minimal and finite subset $\{s^1, \ldots, s^m\}$ of S such that $s \in S$ implies that $s^j \leq s$ for at least one index $j \in \{1, \ldots, m\}$.

Theorem 3.7. *Let $S = \{x \in \mathbb{Z}^n_+ \colon Ax \leq b\}$ where $A \in \mathbb{Z}^{m \times n}$ and $b \in \mathbb{Z}^m_+$. There exists a unique integral basis of S.*

Proof. We split the proof into two parts. Part (a) shows the existence of a finite integral generating set of S. In Part (b) we establish uniqueness of an integral basis for S.

(a) We define a subset $P \subseteq \mathbb{Z}_+^{n+2m}$ as follows.

$$P := \{(x, (Ax)^+, (Ax)^-) \colon x \in S \setminus \{0\}\}.$$

The Gordan Lemma tells us that there exists a unique minimal and finite set

$$P_0 = \{(x[1], (Ax[1])^+, (Ax[1])^-), \ldots, (x[t], (Ax[t])^+, (Ax[t])^-)\}$$

of elements in P such that for every $p \in P$ there exists a vector in P_0 dominated by p, i.e., there is an index $j \in \{1, \ldots, t\}$ with $p \geq (x[j], (Ax[j])^+, (Ax[j])^-)$. We claim that the set $\{x[1], \ldots, x[t]\}$ is an integral generating set of S.

By definition, $\{x[1], \ldots, x[t]) \subseteq S$. Let $y \in S \setminus \{0\}$. Then there exists an index $j \in \{1, \ldots, t\}$ such that $(y, (Ay)^+, (Ay)^-) \geq (x[j], (Ax[j])^+, (Ax[j])^-)$. Therefore, $y - x[j] \in \mathbb{Z}_+^n$ and

$$A(y - x[j]) \leq (A(y - x[j]))^+ = (Ay)^+ - (Ax[j])^+ \leq b.$$

Hence, $y' := y - x[j] \in S$. If $y' \neq 0$, apply the previous arguments iteratively to y' instead of y. Due to strictly decreasing l_1-norms, this procedure terminates, showing that y is a nonnegative integral combination of the vectors $x[1], \ldots, x[t]$.

(b) Let $\mathcal{H}(S)$ be the set of all irreducible vectors of S. By definition, every integral generating set of S must contain $\mathcal{H}(S)$. On account of (a), $\mathcal{H}(S)$ is finite. We claim that $\mathcal{H}(S)$ is already an integral generating set of S. Suppose the contrary. Let $y \in S \setminus \{0\}$ be a point of minimal l_1-norm that cannot be represented as a nonnegative integer combination of the elements in $\mathcal{H}(S)$. By definition of S, we have $y = \sum_{i=1}^{k} \lambda_i v^i$ with $k \geq 2$ vectors $v^1, \ldots, v^k \in S \setminus \{0\}$ and integral multipliers $\lambda_1, \ldots, \lambda_k \geq 1$. We obtain

$$\sum_{i=1}^{k} \lambda_i \|v^i\|_1 = \|y\|_1, \qquad \|v^i\|_1 > 0 \text{ for } i = 1, \ldots, k.$$

Since $\|v^i\|_1 < \|y\|_1$ for $i = 1, \ldots, k$, all summands v^i can be written as a nonnegative integral combination of the elements in $\mathcal{H}(S)$, and hence, y too. □

Having realized that integral generating sets for sets S of the form (4) are finite, it is a natural question to ask how to compute them. There is a finite algorithm for performing this task that may be viewed as a combinatorial variant of the Buchberger algorithm (Buchberger, 1985) for computing Gröbner bases of polynomial ideals. We refer to Urbaniak et al. (1997) and Cornuejols et al. (1997) for earlier versions of this algorithm as well as other proofs of their correctness. For other algorithms along these lines were refer to Hemmecke (2002).

Starting with input $T := \{e^i\colon i = 1, \ldots, n\}$ one takes repeatedly all the sums of two vectors in T, reduces each of these vectors as long as possible by the elements of T and adds all the reduced vectors that are different from the origin to the set T. When we terminate with this step, the set T will contain the set of all irreducible vectors w.r.t. the set S. Note that the set T is usually a strict superset of the set of all irreducible vectors w.r.t. S.

Algorithm 3.8. (*A combinatorial Buchberger algorithm*)
Input. $A \in \mathbb{Z}^{m\times n}$, $b \in \mathbb{Z}^m_+$.
Output. *A finite set T containing all the irreducible vectors of the set*

$$S = \{x \in \mathbb{Z}^n_+\colon Ax \le b\}.$$

(1) *Set $T_{\text{old}} := \emptyset$ and $T := \{e^i\colon i = 1, \ldots, n\}$.*
(2) *While $T_{\text{old}} \ne T$ repeat the following steps:*
  (a) *Set $T_{\text{old}} := T$.*
  (b) *For all pairs of vectors $v, w \in T_{\text{old}}$, set $z := v + w$.*
    (i) *While there exists $y \in T$ such that*

$$y \le z, (Ay)^+ \le (Az)^+, \text{ and } (Ay)^- \le (Az)^-,$$

    set $z := z - y$.
    (ii) *If $z \ne 0$, update $T := T \cup \{z\}$.*
(3) *Set $T_{\text{old}} := \emptyset$ and $T := T \cap S$.*
(4) *While $T_{\text{old}} \ne T$ repeat the following steps:*
  (a) *Set $T_{\text{old}} := T$.*
  (b) *For every $z \in T$, perform the following steps:*
    (i) *$T := T \setminus \{z\}$.*
    (ii) *While there exists $y \in T$ such that $y \le z$ and $(z - y) \in S$, set $z := z - y$.*
    (iii) *If $z \ne 0$, update $T := T \cup \{z\}$.*
(5) *Return T.*

Theorem 3.9. *Algorithm* 3.8 *is finite. The set T that is returned by the algorithm contains the set of all irreducible vectors w.r.t. the set S.*

Proof. Let $\mathcal{H}(S)$ denote the set of all irreducible elements w.r.t. S. Let T^u denote the current set T of Algorithm 3.8 before the u-th performance of Step 2. We define a function

$$f: \mathbb{Z}_+^n \to \mathbb{Z},\ f(t) := \|t\|_1 + \|(At)^+\|_1 + \|(At)^-\|_1. \tag{6}$$

Note that for t^1, $t^2 \in \mathbb{Z}_+^n$ we have that $f(t^1) + f(t^2) \geq f(t^1 + t^2)$. Moreover, $f(t^1) + f(t^2) = f(t^1 + t^2)$ if and only if the vectors (t^1, At^1) and (t^2, At^2) lie in the same orthant of $\mathbb{R}^{n+m}$.

Let $t \in \mathcal{H}(S)$. Since $\{e^i: i = 1, \ldots, n\} \subseteq T^u$ there exists a multiset (repetition of vectors is allowed) $\{t^1, \ldots, t^k\} \subseteq T^u$ such that

$$t = t^1 + \cdots + t^k.$$

For every multiset $M = \{t^1, \ldots, t^k\} \subseteq T^u$ with $t = \sum_{i=1}^k t^i$, let

$$\phi(M) := \sum_{i=1}^{k} f(t^i)$$

Let $M(t, u)$ denote a multiset $\{t^1, \ldots, t^k\} \subseteq T^u$ such that $t = \sum_{i=1}^k t^i$ and $\phi(M(t, u))$ is minimal. From the definition of $M(t, u)$ and the irreducibility of t we have that $\phi(M(t, u) > f(t)$ if and only if $t \notin T^u$. W.l.o.g. $t \notin T^u$. Then there exist indices $i, j \in \{1, \ldots, k\}$ such that the vectors (t^i, At^i) and (t^j, At^j) lie in different orthants of $\mathbb{R}^{n+m}$. This implies that $f(t^i) + f(t^j) > f(t^i + t^j)$. On account of the minimality of $\phi(M(t, u))$, $g = t^i + t^j$ is not in T^u. Moreover, there do not exist $g^1, \ldots, g^l \in T^u$ with $g = \sum_{i=1}^l g^i$ and $f(g) = \sum_{i=1}^l f(g^i)$. However, g will be considered in the u-th performance of Step (2). Then $g = t^i + t^j$ will be added to T^{u+1} or there exist $g^1, \ldots, g^l \in T^{u+1}$ with $g = \sum_{i=1}^l g^i$ and $f(g) = \sum_{i=1}^l f(g^i)$. In any case, the value $\phi(M(t, u + 1))$ will be strictly smaller than the value $\phi(M(t, u))$. Since $\phi(M(t, u)) > f(t)$ for all iterations of Step (2) in which $t \notin T^u$, the algorithm will detect t in a finite number of steps. These arguments apply to any irreducible vector. There is only a finite number of irreducible vectors, and hence, the algorithm is finite. We remark that Steps (3) and (4) just eliminate reducible vectors in S or vectors that do not belong to S. □

We illustrate the performance of Algorithm 3.8 on a small example.

Example 3.10. Consider the three-dimensional problem

$$\{x \in \mathbb{Z}_+^3: x_1 + 3x_2 - 2x_3 \leq 0\}.$$

Algorithm 3.8 starts with $T=\{e^1, e^2, e^3\}$. Taking all the sums of vectors of T and performing Step (2) results in an updated set

$$T = \{e^1, e^2, e^3, (e^1 + e^3), (e^2 + e^3)\}.$$

We again perform Step (2). The following sums of vectors of T become interesting:

$$e^1 + (e^1 + e^3), (e^1 + e^3) + (e^2 + e^3), e^3 + (e^2 + e^3).$$

Note that, for instance, $f(e^1 + (e^2 + e^3)) = f(e^1) + f(e^2 + e^3)$ where f is defined according to (6). Therefore, the vector $e^1 + e^2 + e^3$ will not be included in T. We obtain an updated set

$$T = \{e^1, e^2, e^3, (e^1 + e^3), (e^2 + e^3), (2e^1 + e^3),$$
$$(e^1 + e^2 + 2e^3), (e^2 + 2e^3)\}.$$

Again performing Step (2) yields one additional vector $(e^2 + e^3) + (e^2 + 2e^3)$ that is irreducible and added to T. Algorithm 3.8 terminates before Step (3) with the following set

$$T = \{e^1, e^2, e^3, (e^1 + e^3), (e^2 + e^3), (2e^1 + e^3), (e^1 + e^2 + 2e^3),$$
$$(e^2 + 2e^3), (2e^2 + 3e^3)\}.$$

It remains to analyze Steps (3) to (5). We first eliminate from T all the vectors that are not in S. This gives a new set

$$T = \{e^3, (e^1 + e^3), (2e^1 + e^3), (e^1 + e^2 + 2e^3), (e^2 + 2e^3), (2e^2 + 3e^3)\}.$$

Performing Step (4) we realize that this set is the set of all irreducible vectors w.r.t. the set S.

4 General integer programming algorithms

This section is devoted to the design of augmentation algorithms for a general integer program when no a priori knowledge about the structure of the side constraints is available. More precisely, we deal with integer programs of the form

$$\max c^T x : x \in \mathcal{F} := \{x \in \mathbb{Z}^n : Ax = b, x \geq 0\}, \tag{7}$$

with integral data $A \in \mathbb{Z}^{m \times n}$, $b \in \mathbb{Z}^m$, and $c \in \mathbb{Z}^n$.

There are two different algorithmic ways to design an augmentation method for this problem. Both methods resort to the power of linear programming duality. Starting with an integral feasible solution x^0 one wants to detect an augmenting direction that is applicable at x^0 or provide a proof that x^0 is optimal. To achieve this we derive in the first step a system of linear inequalities such that x^0 becomes a basic feasible solution of this system. There is a canonical way to achieve this, if $\mathcal{F}$ is contained in the unit cube. In the general integer case we might have to add additional columns to the original system to turn x^0 into a basic feasible solution. A general procedure can be found in Haus, Köppe, and Weismantel (2001b). Once x^0 has been made a basic feasible solution of a system describing $\mathcal{F}$, we make use of the simplex method for performing our task. Clearly, if the reduced cost of all the nonbasic variables are nonpositive, we have a certificate that x^0 is optimal. Suppose this is not the case then there exist nonbasic variables in the current tableau with positive reduced cost. The usual simplex method would then perform a pivot operation on a column with positive reduced cost. This is of course not feasible in an integer setting because in general after the execution of a simplex pivot the new basic feasible solution is no longer integral. We present two different ways to overcome this difficulty: The first approach, described in Subsection 4.1, is based on the use of *cutting planes* in a way that the cut generates a feasible pivot element attaining the value one in the cut inequality and that the cut itself becomes a legitimate pivot row. The *Integral Basis Method* that we introduce in Subsection 4.2 is a second approach. It refrains from adding cuts, but replaces the cutting step by a step in which the columns of the given system are manipulated.

In the following we will assume that a basic feasible integer solution $x^0 \in \mathcal{F}$ is known with basic variables B and nonbasic variables N and that the following tableau is a reformulation of (7):

$$\begin{aligned} \max \quad & \gamma + \bar{c}^T x_N \\ \text{s.t.} \quad & x_B = \bar{b} - \bar{A}_N x_N \geq 0, x_N \geq 0, \\ & x \in \mathbb{Z}^n, \end{aligned} \tag{8}$$

where $\bar{b} \in \mathbb{Z}^m_+$, $B \cup N = \{1, \ldots, n\}$, $B \cap N = \emptyset$. Associated with this tableau is the integral feasible solution $x^0 = (\bar{b}, 0) \in \mathbb{Z}^n$ attaining an objective function value of γ.

Definition 4.1. The tableau (8) is called integral if $\bar{A}_N$ is integral.

4.1 Augmenting with cuts

A general way of designing a primal integer programming algorithm is based on the use of cutting planes. One starts with an integral basic feasible solution that is not optimal and generates a Chvátal-Gomory cut from the corresponding simplex tableau in a way which ensures that pivoting on this cut guarantees the integrality of the new improved solution. This approach is based on the primal simplex algorithm and was first proposed by Ben-Israel and Charnes (1962). Simplified variants were given by Young (1965, 1968) and Glover (1968), see also Garfinkel and Nemhauser (1972) and Hu (1969) for further information. We remark that a variant of this method has been proposed by Padberg and Hong (1980) for the traveling salesman problem. These two authors resort to combinatorial cuts for the TSP instead of making use of the Gomory cutting planes.

Algorithm 4.2. (*Algorithm of Gomory–Young*)
Input. *An integral tableau* (8) *and feasible solution* $x^0 = (\bar{b}, 0) \in \mathbb{Z}^n$.
Output. "*Optimal*", *if* x^0 *maximizes* c;
otherwise, $t \in \mathbb{Z}^n$ *such that* $c^T t > 0$ *and* $x^0 + t \in \mathcal{F}$.

(1) *Set* $N_+ := \{i \in N: \bar{c}_i > 0\}$.
(2) *While* $N_+ \neq \emptyset$ *perform the following steps:*
   (a) *Select* $j \in N_+$.
   (b) *If* $\{i \in \{1, \ldots, m\}: \bar{A}_{ij} > \bar{b}_i\} = \emptyset$, *return the augmenting vector* $t \in \mathbb{Z}^n$ *that corresponds to the nonbasic column* $\bar{A}_{\cdot j}$. *Stop.*
   (c) *Choose a pivot row* r *such that*

$$\bar{b}_r/\bar{a}_{rj} = \min_{1 \le i \le m} \{\bar{b}_i/\bar{a}_{ij}: \bar{a}_{ij} \ge 1\}.$$

   (d) *If* $\bar{a}_{rj} = 1$, *then perform a primal simplex pivot step with pivot element* $\bar{a}_{rj}$. *Go to Step* (*f*).
   (e) *If* $\bar{a}_{rj} > 1$, *then derive a Chvátal–Gomory cut from the source row* r,

$$x_j + \sum_{k \in N \setminus \{j\}} \left\lfloor \frac{\bar{a}_{rk}}{\bar{a}_{rj}} \right\rfloor x_k \le \left\lfloor \frac{\bar{b}_r}{\bar{a}_{rj}} \right\rfloor. \tag{9}$$

   Add a new slack variable s *and adjoin this cut as the bottom row to the initial simplex tableau. Modify the tableau. Perform a primal simplex pivot step on the new tableau with pivot column* j. *Choose as the pivot row the one corresponding to the cut. Update the basis,* N, $\bar{A}_N$, $\bar{c}$, *and* N_+.
(3) *Return* "*Optimal.*"

One reason why this approach can work in principle is that for an integral tableau pivoting on a pivot element of value one leads to an integral basis and an integral tableau. If for a given column j, the pivot element $\bar{a}_{rj}$ of Step 2(c) does not attain the value one, then the coefficient of j in the cut (9) derived in Step 2(e) is equal to one and since

$$\left\lfloor \frac{\bar{b}_r}{\bar{a}_{rj}} \right\rfloor \Big/ \left\lfloor \frac{\bar{a}_{rj}}{\bar{a}_{rj}} \right\rfloor = \left\lfloor \frac{\bar{b}_r}{\bar{a}_{rj}} \right\rfloor \leq \frac{\bar{b}_r}{\bar{a}_{rj}},$$

the cut (9) yields indeed a valid source row for performing the pivot operation.

Let $(x^1, 0)$ denote the new basic integer solution after applying this pivot operation. The difference vector of the feasible solutions $x^1 - x^0$, if different from 0, is called a *Gomory–Young augmentation vector*. Geometrically, a Gomory–Young augmentation vector is the difference vector of adjacent extreme points of the convex hull defined by the feasible integral solutions of the given problem.

Unfortunately, Algorithm 4.2 does not automatically support a proof of finiteness because the right hand side of the cut may be zero. In this case the value of all variables remain unchanged and we do not move away from the old basic feasible solution but represent it by a new basis only. This problem is related to degeneracy that can occur in linear programming. To make the algorithm finite, it requires careful selection rules for the pivot columns and source rows. The first finitely convergent algorithm based on the cuts (9) was given by Young (1965). It uses however, complicated rules for the selection of pivot columns and rows. Simplified versions including finiteness proofs were given by Glover (1968) and Young (1968) [see also Garfinkel and Nemhauser (1972)].

We demonstrate the performance of Algorithm 4.2 on a small example.

Example 4.3. Consider the integer program in equation form,

$$\begin{aligned} \max \quad & x_1 \\ \text{s.t.} \quad & x_3 - 3x_1 + 5x_2 = 1, \\ & x_4 + x_1 - 4x_2 = 1, \\ & x_5 + 5x_1 - 4x_2 = 2, \\ & x \in \mathbb{Z}^5_+. \end{aligned}$$

Associated with this program is a primal feasible solution $x^0 = (0, 0, 1, 1, 2)$. Thus, $B = \{3, 4, 5\}$ and $N = \{1, 2\}$. The reduced cost of variable x_1 is positive. Hence, we select the column corresponding to variable x_1 as a pivot column. Determining ratios shows that the x_5-row is a valid pivot row. Since the value of the pivot element is different from 1, we perform Step 2(e) of Algorithm 4.2. The cut reads.

$$x_1 - x_2 \leq 0.$$

We denote x_6 as the slack variable associated with this cut and perform on the extended system a pivot operation. The following system is obtained

$$\begin{aligned} \max \quad & -x_6 + x_2 \\ \text{s.t.} \quad & x_3 + 3x_6 + 2x_2 = 1, \\ & x_4 - \ \ x_6 - 3x_2 = 1, \\ & x_5 - \ \ 5x_6 + x_2 = 2, \\ & x_1 + x_6 - x_2 = 0, \\ & x \in \mathbb{Z}^6_+ \end{aligned}$$

The solution $(x^0, 0) \in \mathbb{Z}^6$ is a primal feasible solution for this new system. Thus, $B = \{1, 3, 4, 5\}$ and $N = \{2, 6\}$. We again perform Step (2) of Algorithm 4.2. We select the x_2-column as the (unique) pivot column and the x_3-row as the source row. Since the pivot element has a coefficient bigger than 1, we enter Step 2(e). We generate the Chvátal-Gomory cut as defined in (9), adjoin it as the bottom row to the system and add a new slack variable x_7 to the current basis. The cut reads

$$x_6 + x_2 \leq 0.$$

We now perform a pivot operation using the cut. This leads to a new system

$$\begin{aligned} \max \quad & -2x_6 - x_7 \\ \text{s.t.} \quad & x_3 + x_6 - 2x_7 = 1 \\ & x_4 + 2x_6 + 3x_7 = 1, \\ & x_5 - 6x_6 - x_7 = 2, \\ & x_1 \ + 2x_6 + x_7 = 0, \\ & x_2 + x_6 + x_7 = 0, \\ & x \in \mathbb{Z}^7_+. \end{aligned}$$

The final tableau is dual feasible. Hence, the corresponding basic solution $(x^0, 0, 0) \in \mathbb{Z}^7$ is optimal for the problem. Therefore, x^0 is an optimal solution to the initial system.

As we have mentioned before, in order to make Algorithm 4.2 always finite, it requires a careful selection of rules for the pivot columns and source rows that we do not present here. In fact, if we start with an optimal integer solution x^* then we can never move away from this particular solution. As a consequence, the algorithm then requires the addition of cuts that are all tight at x^*. More generally, we are then interested in solving the following variant of the separation problem:

> **The primal separation problem**
>
> Let $S \subseteq \mathbb{Z}^n$ be a feasible region. Given a point $x \in S$ and a vector $x^* \in \mathbb{Z}^n$, find a hyperplane $c^T y = \gamma$ such that
>
> $$S \subseteq \{y \in \mathbb{R}^n | c^T y \leq \gamma\}, c^T x = \gamma, \text{ and } c^T x^* > \gamma$$
>
> or assert that such a hyperplane does not exist.

When one investigates this augmentation approach via cutting for a specific integer programming problem, then there is no need to generate the Chvátal-Gomory cut as defined in (9). Instead, any family of valid inequalities can be used. In fact, it turns out that often the solution to the primal separation problem is substantially easier than a solution to the general separation problem. We illustrate this on an example.

Example 4.4. [Eisenbrand, Rinaldi, and Ventura (2002)] Given a graph $G = (V, E)$ with weights c_e on the edges $e \in E$. *A* perfect matching in G is a set of edges no two of which have a common end such that every node is covered. The maximum weighted perfect matching problem is to find a perfect matching of maximum weight. An integer programming formulation reads

$$\begin{aligned} \max \quad & \sum_{e \in E} c_e x_e \\ \text{s.t} \quad & \sum_{e \in \delta(u)} x_e = 1 \quad \text{for all } u \in V, \\ & x_e \in \{0, 1\} \quad \text{for all } e \in E. \end{aligned} \tag{10}$$

Edmonds (1965) showed that the family of odd cutset inequalities

$$\sum_{e \in \delta(U)} x_e \geq 1 \quad \text{for all } U \subseteq V, |U| \text{ odd}$$

is satisfied by the incidence vector of any perfect matching of G.

Interestingly, the primal separation problem for the odd cutset inequalities can be solved substantially easier than without the requirement that $M \cap \delta(U) = 1$ for a specific perfect matching M. Given a perfect matching M

and a point $x^* \geq 0$ satisfying $x^*(\delta(u)) = 1$ for all $u \in V$, we want to detect whether there exists an odd cutset induced by $U \subseteq V$, $|U|$ odd such that

$$M \cap \delta(U) = 1 \quad \text{and} \quad x^*(\delta(U)) > 1.$$

For $ij \in M$, let $G_{ij} = (V_{ij}, E_{ij})$ be the graph obtained by contracting the two end nodes of e for every edge $e \in M \backslash \{ij\}$. Let $\delta(U_{ij})$ be a minimum (i,j)-cut in G_{ij} with respect to the edge weights given by x^*. Then U_{ij} consists of the node i and some new nodes in V_{ij}, each of these new nodes corresponds to the two nodes in G that are paired via M. Since M is a perfect matching in G, the extension of U_{ij} in G corresponds to a set of nodes $U \subseteq V$ of odd cardinality such that $M \cap \delta(U) = 1$. Therefore, determining such a minimal cut U_{ij} in G_{ij} for every $ij \in M$, solves the primal separation problem for the family of odd cutset inequalities in polynomial time.

For recent developments on primal separation and primal cutting plane algorithms we refered the papers Eisenbrand et al. (2002) and Letchford and Lodi (2002, 2003).

4.2 The integral basis method

We now discuss a second possibility of manipulating a tableau so as to obtain a primal integer programming algorithm. We will again perform operations that enable us to either detect an augmenting direction or prove that a given basic feasible solution $x^0 = (\bar{b}, 0) \in \mathbb{Z}^n$ is optimal. The idea is to eliminate step by step a nonbasic column of positive reduced cost in a simplex tableau and substitute it by a couple of other columns in a way that the nonbasic part of every feasible direction with respect to $(\bar{b}, 0)$ is a nonnegative integral combination of the new nonbasic columns. This is what we call a *proper reformulation* of a tableau.

Theorem 4.5. [Haus et al. (2001b)] *For a tableau* (8), *let*

$$S_N = \{x_N \in \mathbb{Z}_+^{n-m}\colon \bar{A}_N x_N \leq \bar{b}\}.$$

Let $j \in N$, and let $\{t^1, \ldots, t^r\} \subseteq \mathbb{Z}_+^{n-m}$ be all the elements in an integral generating set of S_N with $t_j^i > 0$ for $i = 1, \ldots, r$. Then $x_N \in S_N$ if and only if there exist $z \in \mathbb{Z}_+^{n-m-1}$, $y \in \mathbb{Z}_+^r$ such that

$$\bar{A}_{N \backslash \{j\}} z + \sum_{i=1}^{r} (\bar{A} t^i) y_i \leq \bar{b} \tag{11}$$

and

$$\begin{aligned} x_j &= \textstyle\sum_{i=1}^{r} t_j^i y_i, \\ x_k &= z_k + \textstyle\sum_{i=1}^{r} t_k^i y_i, \quad \textit{for all } k \in N \setminus \{j\}. \end{aligned} \tag{12}$$

Proof. Let $x_N \in S_N$. If $x_j = 0$, then $z := x_{N\setminus\{j\}}$ and $y := 0$ satisfy (11) and (12). Otherwise, $x_j > 0$. Let H be an integral generating set of S_N. Then $\{t^1, \ldots, t^r\} \subseteq H$. We can write H in the following form,

$$H = \{h^1, \ldots, h^l\} \cup \{t^1, \ldots, t^r\},$$

where $h_j^i = 0$ for all $i = 1, \ldots, l$. We conclude that

$$x_N = \sum_{i=1}^{l} \lambda_i h^i + \sum_{i=1}^{r} y_i t^i$$

with $\lambda_i \in \mathbb{Z}_+$ for all $i = 1, \ldots, l$ and $y_i \in \mathbb{Z}_+$ for all $i = 1, \ldots, r$. Let $z_N = \sum_{i=1}^{l} \lambda_i h^i$. Then $z_{N\setminus\{j\}}$ and y satisfy (11) and (12). For the converse direction, assume that there exist $z \in \mathbb{Z}^{n-m-1}$, $y \in \mathbb{Z}_+^r$ satisfying (11). Then define x as in (12). It follows that $x \in S_N$. □

Theorem 4.5 suggests an algorithm for manipulating our initial tableau: if all the reduced costs are negative, we have a proof that our given basic feasible integer solution is optimal. Otherwise, we select a nonbasic variable x_j with positive reduced cost. We eliminate column j and introduce r new nonbasic columns $\bar{A}t^i$ that correspond to all the elements $\{t^1, \ldots, t^r\}$ in an integral generating set of S_N such that $t_j^i > 0$ for all $i = 1, \ldots, r$. According to Theorem 4.5 this step corresponds to a proper reformulation of the original tableau. We obtain the rudimentary form of a general integer programming algorithm that we call Integral Basis Method, because the core of the procedure is to replace columns by new columns corresponding to the elements in an integral basis or an integral generating set. A predecessor of the method for the special case of set partitioning problems has been invented by Balas and Padberg (1975).

Algorithm 4.6. (*Integral Basis Method*) [Haus, Köppe, and Weismantel (2001a)]
Input. *A tableau* (8) *and a feasible solution* $x^0 = (\bar{b}, 0) \in \mathbb{Z}^n$. *Let*

$$S_N = \{x_N \in \mathbb{Z}_+^{n-m} \colon \bar{A}_N x_N \le \bar{b}\}.$$

Output. "*Optimal,*" *if* x^0 *maximizes* c;
otherwise, $t \in \mathbb{Z}^n$ *such that* $c^T t > 0$ and $x^0 + t \in \mathcal{F}$.

(1) *Set* $N_+ := \{i \in N\colon \bar{c}_i > 0\}$.

(2) *While* $N_+ \neq \emptyset$ *perform the following steps:*

   (a) *Select* $j \in N_+$.

   (b) *If* $\{i \in \{1, \ldots, m\}\colon \bar{A}_{ij} > \bar{b}_i\} = \emptyset$, *return the augmenting vector* $t \in \mathbb{Z}^n$ *that corresponds to the nonbasic column* $\bar{A}_j$. *Stop.*

   (c) *Determine the subset* $\{t^1, \ldots, t^r\}$ *of an integral generating set of* S_N *such that* $t^i_j > 0$ *for all* $i = 1, \ldots, r$.

   (d) *Delete column j from the current tableau and define a new tableau as*

$$\begin{array}{ll} \max & \gamma + \bar{c}^T_{N\setminus\{j\}} z + \bar{g}^T y \\ s.t. & x_B + \bar{A}_{N\setminus\{j\}} z + \bar{D} y = \bar{b}, \\ & x_B \in \mathbb{Z}^m_+,\ z \in \mathbb{Z}^{n-m-1}_+,\ y \in \mathbb{Z}^r_+, \end{array}$$

   where $\bar{g}_i = \bar{c}^T t^i$, $\bar{D}_i = \bar{A} t^i$, $i = 1, \ldots, r$. *Update* N, N_+, S_N, $\bar{A}$, $\bar{c}$, $\bar{b}$.

(3) *Return "Optimal."*

As a direct consequence of Theorem 3.7 we obtain that in each performance of Step 2 the number of columns that we add to the system is finite. The analysis carried out in Haus et al. (2001b) shows that the number of times we perform the while-loop in Algorithm 4.6 is finite.

Theorem 4.7. [Haus et al. (2001b)] *The Integral basis method is finite. It either returns an augmenting direction that is applicable at* x^0, *or asserts that* x^0 *is optimal.*

Next we demonstrate on two pathological examples the possible advantages of the Integral basis method.

Example 4.8. [Haus et al. (2001b)] For $k \in \mathbb{Z}_+$ consider the 0/1 integer program

$$\begin{array}{lll} \max & \sum_{i=1}^k (x_i - 2y_i) & \\ \text{s.t.} & 2x_i - y_i \le 1 & \text{for } i = 1, \ldots, k, \\ & x_i, y_i \in \{0, 1\} & \text{for } i = 1, \ldots, k. \end{array} \tag{13}$$

The origin 0 is a feasible integral solution that is optimal to (13). The linear-programming relaxation will yield $x_i = 1/2$, $y_i = 0$ for all variables. Branching on one of these fractional x_i-variables will lead to two subproblems of the same kind with index $k - 1$. Therefore, an exponential number of branching nodes will be required to solve (13) via branch and bound.

The Integral basis method, applied at the basic feasible solution 0, identifies the nonbasic variables x_i as integrally nonapplicable improving columns and eliminates them sequentially. For $i=1,\ldots,k$, the variable x_i is replaced by some variable x_i', say, which corresponds to x_i+y_i. This yields the reformulated problem

$$
\begin{array}{lll}
\max & \sum_{i=1}^{k}(-x_i' - 2y_i) & \\
\text{s.t.} & x_i' - y_i \leq 1 & \text{for} \quad i=1,\ldots,k, \\
 & x_i' + y_i \leq 1 & \text{for} \quad i=1,\ldots,k, \\
 & x_i', y_i \in \{0,1\} & \text{for} \quad i=1,\ldots,k.
\end{array}
\tag{13'}
$$

providing a linear-programming certificate for optimality.

One can also compare the strength of an operation of the Integral basis method to that of a pure Gomory cutting plane algorithm.

Example 4.9. [Haus et al. (2001b)] For $k \in \mathbb{Z}_+$ consider

$$
\begin{array}{ll}
\max & x_2 \\
\text{s.t.} & kx_1 + x_2 \leq k, \\
 & -kx_1 + x_2 \leq 0, \\
 & x_1, x_2 \geq 0, \\
 & x_1, x_2 \in \mathbb{Z}.
\end{array}
\tag{CG$_k$}
$$

There are only two integer solutions to (CG$_k$), namely $(0,0)$ and $(1,0)$, which are both optimal. The LP solution, however, is $((1/2),(1/2)k)$. Note that the Chvátal rank 1 closure of (CG$_k$) is (CG$_{k-1}$). Therefore the inequality $x_2 \leq 0$, which describes a facet of the integer polytope, has a Chvátal rank of k.

The Integral basis method analyzes the second row of (CG$_k$), in order to handle the integrally nonapplicable column x_2. This yields that column x_2 can be replaced by columns corresponding to $x_1 + 1x_2, \ldots, x_1 + kx_2$. Each of these columns however violates the generalized upper-bound constraint in the first row of (CG$_k$), so the replacement columns can simply be dropped. The resulting tableau only has a column for x_1. This proves optimality.

The core of Algorithm 4.6 is to perform column substitutions. For this we need to compute all the elements of an integral generating set that involve a particular variable j. In Section 3 we have introduced a method to accomplish this task. The method is however computationally intractable, even for very small instances. This fact requires a reformulation technique that is based upon systems that partially describe the underlying problem but for which integral generating sets can be easily computed.

Definition 4.10. For a tableau (8) let

$$S_N = \{x_N \in \mathbb{Z}_+^{n-m}\colon \bar{A}_N x_N \leq \bar{b}\}.$$

For $A' \in \mathbb{Q}^{m' \times (n-m)}$ and $b' \in \mathbb{Q}_+^{n-m}$ we call a set

$$\tilde{S}_N = \{x_N \in \mathbb{Z}_+^{n-m}\colon A' x_N \leq b'\}$$

a discrete relaxation of S_N if $S_N \subseteq \tilde{S}_N$.

It can be shown that resorting to an integral generating set of a discrete relaxation of S_N still allows to properly reformulate a tableau. There are numerous possibilities to derive interesting discrete relaxations that we refrain from discussing here in detail. We refer to Haus et al. (2001b) for further details regarding the Integral basis method and its variants.

5 Combinatorial optimization

Besides the min-cost flow problem there are many other combinatorial optimization problems for which there exist primal combinatorial algorithms that run in polynomial time, e.g., the maximum flow problem, the matching problem, the matroid optimization problem, the matroid intersection problem, the independent path-matching problem, the problem of minimizing a submodular function, and the stable set problem in claw-free graphs. We will present the basics of these algorithms and give answers to the two questions that we posed in the beginning of this chapter:

(i) How can one solve the subproblem of detecting an augmenting direction?
(ii) How can one verify that a given point is optimal?

Given a digraph $D=(V,A), r,s \in V$, and $u \in \mathbb{Z}_+^A$. The *maximum flow problem* is the following linear programming problem:

$$\begin{aligned} &\max x(\delta^+(r)) - x(\delta^-(r)) && \\ &x(\delta^+(v)) - x(\delta^-(v)) = 0 && \text{for all} \quad v \in V \setminus \{r, s\} \\ &0 \leq x_a \leq u_a && \text{for all} \quad a \in A. \end{aligned}$$

A feasible vector $x \in \mathbb{R}^A$ is an (r, s)*-flow*, its *flow value* is $x(\delta^+(r) - x(\delta^-(r)$.

The Maximum Flow Problem

Find an (r, s)-flow of maximum flow value.

Theorem 5.1. [Ford and Fulkerson (1956)] *If there is a maximum* (r, s)*-flow, then*

$$\max\{x(\delta^+(r)) - x(\delta^-(r)) \colon x \ (r, s)\text{-}flow\} = \min\{u(X) \colon X \ (r, s)\text{-}cut\},$$

where an (r, s)*-cut is a set* $\delta^+(R)$ *for some* $R \subseteq V$ *with* $r \in R$ *and* $s \notin R$.

An *x-incrementing path* is a path in D such that every forward are a of the path satisfies $x_a < u_a$ and every backward arc a satisfies $x_a > 0$. An *x-augmenting path* is an (r, s)-path that is x-incrementing. Given an x-augmenting path P, we can raise x_a by some positive α on each forward arc of P and lower x_a by α on each backward arc of P; this yields an (r, s)-flow of larger flow value. If there is no x-augmenting path in D, let R be the set of nodes reachable by an x-incrementing path from r. Then R determines an (r, s)-cut $X := \delta^+(R)$ with $x(\delta^+(r)) - x(\delta^-(r)) = u(X)$. By the min–max theorem x is a maximum (r, s)-flow.

The classical maximum flow algorithm of Ford and Fulkerson (1956) proceeds as follows: beginning with an (r, s)-flow x (e.g., $x = 0$), repeatedly find an x-augmenting path P in D and augment x by the maximum value permitted, which is the minimum of $\min\{u_a - x_a \colon a \text{ forward arc in } P\}$ and $\min\{x_a \colon a \text{ backward arc in } P\}$. If this minimum is ∞, no maximum flow exists and the algorithm terminates. If there is no x-augmenting path in D, x is maximum and the algorithm terminates. For more details we refer to Ahuja et al. (1993).

We next consider the *matching problem*. Given a graph $G = (V, E)$. A *matching* in G is a set of edges no two of which have a common end.

> **The Matching Problem**
>
> Find a matching in G of maximum cardinality.

Theorem 5.1. [König (1931)] *For a bipartite graph* $G = (V, E)$,

$$\max\{|M| \colon M \subseteq E \ matching\} = \min\{|C| \colon C \subseteq V \ cover\},$$

where a cover C *is a set of nodes such that every edge of* G *has at least one end in* C.

Theorem 5.2. [Berge (1958), Tutte (1947)] *For a graph* $G = (V, E)$

$$\begin{aligned}&\max\{|M| \colon M \subseteq E \ matching\}\\ &\quad = \min\{(|V| - odd(G \setminus X) + |X|)/2 \colon X \subseteq V\},\end{aligned}$$

where $odd\ (G \backslash X)$ *denotes the number of connected components of* $G \backslash X$ *which have an odd number of nodes.*

Let M be a matching in G. An *M-alternating path* is a path in G whose edges are alternately in and not in M. An *M-augmenting path* is an M-alternating path whose both end nodes are M-exposed. If P is an M-augmenting path, then $M\Delta P$ is a larger matching than M. Berge (1957) showed that a matching M in G is maximum if and only if there is no M-augmenting path in G. This suggests a possible approach to construct a maximum matching: repeatedly find an augmenting path and obtain a new matching using the path, until we discover a maximum matching.

The basic idea to find an M-augmenting path is to grow a forest of alternating paths rooted at M-exposed nodes. Then if a leaf of the tree is also M-exposed, an M-augmenting path has been found.

For a bipartite graph G with bipartition (V_1, V_2), each M-exposed node in V_1 is made the root of an M-alternating tree. If an M-exposed node in V_2 is added to one of the trees, the matching M is augmented and the tree-building procedure is repeated with respect to the new matching. If it is not possible to add more nodes and arcs to any of the trees and no M-exposed node in V_2 is added to one of the trees, let C be the union of all out-of-tree nodes in V_1 and all in-tree nodes in V_2. Then C is a cover of cardinality $|M|$ and by Theorem 5.1, M is a maximum matching. The approach used in this algorithm is called the *Hungarian Method* since it seems to have first appeared in the work of König (1916) and of Egerváry (1931). For more details we refer to Lawler (1976) and Lovász and Plummer (1986).

The algorithm may fail to find an M-augmenting path if the graph is not bipartite. Edmonds (1965) invented the idea of "shrinking" certain odd cycles, called *blossoms*. We detect them during the construction of an M-alternating forest by finding two nodes in the same tree that are adjacent via an edge that is not part of the tree. Shrinking the blossom leads to a shrinked matching in a shrinked graph. It turns out that a maximum matching in the shrinked graph and a corresponding minimizer X, see Theorem 5.2, has a straightforward corresponding maximum matching in G with the same minimizer X. Thus, we apply the same ideas recursively to the shrinked matching in the shrinked graph. If the constructed alternating forest is complete, i.e., it is not possible to add further edges or the shrink blossoms, let X be the set of nodes in the forest that has an odd distance to its root. The algorithm is called *Edmonds' matching algorithm*. For more details we refer to Edmonds (1965), Lovász and Plummer (1986), Cook, Cunningham, Pulleybank, and Schrijver (1998), Korte and Vygen (2000), Schrijver (2003).

One of the fundamental structures in combinatorial optimization are matroids. Let S be a finite set. An *independence system* $\mathcal{I}$ on S is a family of subsets of S such that

$$\emptyset \in \mathcal{I} \quad \text{and} \quad \text{if} \quad J' \subseteq J \quad \text{and} \quad J \in \mathcal{I} \quad \text{then} \quad J' \in \mathcal{I}.$$

The subsets of S belonging to $\mathcal{I}$ are *independent*. A maximal independent subset of a set $A \subseteq S$ is a *basis* of A. The *rank* of A, denoted $r(A)$, is the cardinality of a maximal basis of A. A *matroid* $\mathcal{M}$ on S is an independence system on S such that, for every $A \subseteq S$ every basis of A has the same cardinality. We assume that a matroid $\mathcal{M}$ is given by an *independence oracle*, i.e., an oracle which, when given a set $J \subseteq S$, decides whether $J \in \mathcal{M}$ or not.

> **The Matroid Optimization Problem**
>
> Given a matroid $\mathcal{M}$ on S and a weight vector $c \in \mathbb{R}^S$. Find an independent set J of maximum weight $c(J) := \sum_{i \in J} c_i$.

The matroid optimization problem can be solved by a simple *greedy algorithm* that is in fact a primal algorithm.

Algorithm 5.3. [Rado (1957)] (*Greedy algorithm*)
Input. *A matroid $\mathcal{M}$ on S and $c \in \mathbb{R}^S$.*
Output. *An independent set of maximum weight.*

(1) *Set $J := \emptyset$.*
(2) *While there exists $i \notin J$ with $c_i > 0$ and $J \cup \{i\} \in \mathcal{M}$*
  (a) *Choose such i with c_i maximum;*
  (b) *Replace J by $J \cup \{i\}$.*
(3) *Return J.*

We next consider a generalization of both the bipartite matching problem and the matroid optimization problem.

> **The Matroid Intersection Problem**
>
> Given matroids $\mathcal{M}_1$ and $\mathcal{M}_2$ on S. Find a common independent set $J \in \mathcal{M}_1 \cap \mathcal{M}_2$ of maximum cardinality.

Theorem 5.4. [Edmonds (1970)] *For matroids $\mathcal{M}_1$, $\mathcal{M}_2$ on S,*

$$\max\{|J|: J \in \mathcal{M}_1 \cap \mathcal{M}_2\} = \min\{r_1(A) + r_2(S \setminus A): A \subseteq S\},$$

where r_i denotes the rank function of the matroid $\mathcal{M}_i$, $i = 1, 2$.

For $J \in \mathcal{M}_1 \cap \mathcal{M}_2$, we define a digraph $D(J)$ with node set S and arcs

$$\begin{array}{llll} (b,a) & \text{for} & a \in J, b \notin J & \text{with} \quad J \cup \{b\} \notin \mathcal{M}_1, J \cup \{b\} \setminus \{a\} \in \mathcal{M}_1; \\ (a,b) & \text{for} & a \in J, b \notin J & \text{with} \quad J \cup \{b\} \notin \mathcal{M}_2, J \cup \{b\} \setminus \{a\} \in \mathcal{M}_2. \end{array}$$

A *J-augmenting path* is a dipath in $D(J)$ that starts in a node $b \notin J$ with $J \cup \{b\} \in \mathcal{M}_2$ and ends in a node $b' \notin J$ with $J \cup \{b'\} \in \mathcal{M}_1$. Note that the nodes of the path are alternately in and not in J and that the arcs alternately fulfill conditions with respect to $\mathcal{M}_1$ and $\mathcal{M}_2$.

Lemma 5.5 [Lawler (1976)] *Any chordless J-augmenting path P leads to an augmentation, i.e., $J\Delta P$ is a common independent set of larger size.*

If there exists no J-augmenting path, let $A \subseteq S$ be the set of end nodes of dipaths in $D(J)$ that start in nodes $b \notin J$ with $J \cup \{b\} \in \mathcal{M}_2$. Then $|J| = r_1(A) + r_2(S \backslash A)$ and Theorem 5.4 implies that J is maximum.

The primal algorithm for the matroid intersection problem now works as follows: starting with a common independent set J (e.g., $J = \emptyset$), repeatedly find a cordless J-augmenting path P and replace J by $J\Delta P$ until there is no J-augmenting path.

In the remainder of this section, we state three further problems that can be solved by a combinatorial primal approach, namely the independent path matching problem, the problem of minimizing a submodular function and the stable set problem in claw-free graphs. The combinatorial algorithms for these problems are fairly involved and require many technical definitions that we refrain from giving here.

Cunningham and Geelen (1997) proposed a common generalization of the matching problem and the matroid intersection problem: the *independent path-matching problem.* Let $G = (V, E)$ be a graph, T_1, T_2 disjoint stable sets of G, and $R := V \backslash (T_1 \cup T_2)$. Moreover, for $i = 1, 2$, let $\mathcal{M}_i$ be a matroid on T_i. An *independent path-matching* K in G is a set of edges such that every component of $G(V, K)$ having at least one edge is a path from $T_1 \cup R$ to $T_2 \cup R$ all of whose internal nodes are in R, and such that the set of nodes of T_i in any of these paths is independent in $\mathcal{M}_i$, for $i = 1, 2$. An edge e of K is a *matching-edge* of K if e is an edge of a one-edge component of $G(V, K)$ having both ends in R, otherwise e is a *path-edge* of K. The *size* of K is the number of path-edges K plus twice the number of matching-edges of K.

The Independent Path-Matching Problem
Find an independent path-matching in G of maximum size.

Cunningham and Geelen (1997) solved the independent path-matching problem via the ellipsoid method. They and also Frank and Szegö (2002) presented min–max theorems for this problem.

Theorem 5.2. [Frank and Szegö (2002)]

$$\max\{\text{size of } K\text{: } K \text{ path-matching in } G\} = |R| + \min_{X \text{ cut}}(|X| - \text{odd}_G(X)),$$

where a cut is a subset $X \subseteq V$ such that there is no path between $T_1 \backslash X$ and $T_2 \backslash X$ in $G \backslash X$ and $\text{odd}_G(X)$ denotes the number of connected components of $G \backslash X$ which are disjoint from $T_1 \cup T_2$ and have an odd number of nodes.

Combining the augmenting path methods for the matching problem and the matroid intersection problem, Spille and Weismantel (2001, 2002b) gave a polynomial-time combinatorial primal algorithm for the independent path-matching problem.

We next turn to *submodular function minimization*. A function $f: 2^V \to \mathbb{R}$ is called *submodular* if

$$f(X) + f(Y) \geq f(X \cup Y) + f(X \cap Y) \quad \text{for all } X, Y \subseteq V.$$

We assume that f is given by a *value-giving oracle* and that the numbers $f(X)$ $(X \subseteq V)$ are rational.

The Problem of Minimizing a Submodular Function

Find $\min\{f(X): X \subseteq V\}$ for a submodular function f on V.

The task of finding a minimum for f is a very general combinatorial optimization problem which includes for example the matroid intersection problem.

Associated with a submoduar function f on V is the so-called *base polytope*

$$B_f := \{x \in \mathbb{R}^V: x(X) \leq f(X) \quad \text{for all } X \subseteq V,\ x(V) = f(V)\}.$$

Theorem 5.3 [Edmonds (1970)] *For a submodular function f on V,*

$$\max\{x^-(V): x \in B_f\} = \min\{f(X): X \subseteq V\}.$$

Grötschel, Lovász, and Schrijver (1981, 1988) solved the submodular function minimization problem in strongly polynomial-time with the help of the ellipsoid method. Cunningham (1985) gave a pseudopolynomial-time combinatorial primal algorithm for minimizing a submodular function. Schrijver (2000) and Iwata, Fleischer, and Fujishige (2000) developed strongly polynomial-time combinatorial primal algorithms for minimizing the submodular functions, both extending Cunningham's approach. These combinatorial primal algorithms use an augmenting path approach with reference to a convex combination x of vertices of B_f. They seek to increase $x^-(V)$ by performing exchange operations along a certain path.

The *stable set problem* generalizes the matching problem. Given a graph G. A *stable set* in G is a set of nodes not two of which are adjacent.

The Stable Set Problem

Find a stable set in G of maximum cardinality.

Karp (1972) showed that the stable set problem is NP-hard in general and hence, one cannot expect to derive a "compact" combinatorial min–max formula. In the case of claw-free graphs the situation is simplified. A graph is *claw-free* if whenever three distinct nodes u, v, w are adjacent to a single node, the set $\{u, v, w\}$ is not stable. The stable set problem for claw-free graphs is a generalization of the matching problem. Minty (1980) and Sbini (1980) solved the stable set problem for claw-free graphs in polynomial time via a primal approach that extends Edmonds' matching algorithm.

Acknowledgment

The authors were supported by the European Union, contract ADONET 504438.

References

Ahuja, R. K., Magnanti, T., Orlin, J. B. (1993), *Network Flows*, Prentice Hall, New Jersey.

Balas, E., M. Padberg (1975). On the set covering problem II. An algorithm for set partitioning. *Operations Research* 23, 74–90.

Berge, C. (1957). Two theorems in graph theory. *Proc. of the National Academy of Sciences (U.S.A.)* 43, 842–844.

Berge, C. (1958). Sur le couplage maximum d'un graphe. *Comptes Rendus de l' Académie des Sciences Paris,* series 1, *Mathématique* 247, 258–259.

Ben-Israel, A., Charnes, A. (1962). On some problems of diophantine programming. *Cahiers du Centre d'Etudes de Recherche Operationelle* 4, 215–280.

Buchberger, B. Gröbner bases: an algorithmic method in polynomial ideal theory, in: N. K. Bose (ed.), Multidimensional Systems Theory, 184–232D. Reidel Publications.

Cook, W. J., W. H. Cunningham, W. R. Pulleyblank, A. Schrijver (1998). *Combinatorial Optimization*, Wiley-Interscience, New York.

Cornuejols, G., R. Urbaniak, R. Weismantel, L. A. Wolsey (1997). Decomposition of integer programs and of generating sets, Algorithms-ESA97. in: R. Burkard, G. Woeginger (eds.), *Lecture Notes in Computer Science 1284*, Springer, Berlin, 92–103.

Cunningham, W. H., J. F. Geelen (1997). The optimal path-matching problem. *Combinatorica* 17, 315–337.

Cunningham, W. H. (1995). On submodular function minimization. *Combinatorica* 5, 185–192.

Edmonds, J. (1965). Paths, trees, and flowers. *Canadian Journal of Mathematics* 17, 449–467.

Edmonds, J. (1970). Submodular functions, matroids, and certain polyhedra. in: R. K. Guy, H. Hanai, N. Sauer, J. Schönheim (eds.), *Combinatorial Structures and their Applications*, Gordon and Brach, New York, 69–87.

Edmonds, J., R. M. Karp (1972). Theoretical improvement in algorithmic efficiency for network flow problems. *J. ACM* 19, 248–264.

Egerváry, E. (1931). Matrixok kombinatorius tulajdonságairól (On combinatorial properties of matrices). *Matematikai és Fizikai Lapok* 38, 16–28.

Eisenbrand, F., G. Rinaldi, P. Ventura (2002). 0/1 optimizations and 0/1 primal separation are equivalent. *Proceedings of SODA* 02, 920–926.

Ford, L. R. Jr, D. R. Fulkerson (1956). Maximal flow through a network. *Canadian Journal of Mathematics* 8, 399–404.

Frank, A., L. Szegö (2002). Note on the path-matching formula. *Journal of Graph Theory* 41, 110–119.

Garfinkel, R. S., G. L. Nemhauser (1972). *Integer Programming*, Wiley, New York.

Glover, F. (1968). A new foundation for a simplified primal integer programming algorithm. *Operations Research* 16, 727–740.

Grötschel, M., L. Lovász (1995). Combinatorial optimization. *Handbook of Combinatorics*. in: M. Graham, R. Grötschel, L. Lovász, North-Holland, Amsterdam.

Grötschel, M., L. Lovász, A. Schrijver (1981). The ellipsoid method and its consequences in combinatorial optimization. *Combinatorica* 1, 169–197.

Grötschel, M., L. Lovász, A. Chrijver (1988). *Geometric Algorithms and Combinatorial Optimization*, Springer Verlag.

Haus, U., M. Köppe, R. Weismantel (2001a). The integral basis method for integer programming. *Math. Methods of Operations Research* 53, 353–361.

Haus, U., Köppe, M., Weismantel, R. (2001b). A primal all-integer algorithm based on irreducible solutions, Manuscript. To appear in *Math. Programming Series B* (*Algebraic Methods in Discrete Optimization*).

Hemmecke, R. (2002), On the computation of Hilbert bases and extreme rays of cones, eprint arXiv:math.CO/0203105.

Hu, T. C. (1969). *Integer Programming and Network Flows*, Addison-Wesley Publishing Company, Inc., Reading, Massachusetts.

Iwata, S., Fleischer, L., Fujishige, S. (2000). A combinatorial, strongly polynomial-time algorithm for minimizing submodular functions, *Proceedings of the 32nd ACM Symposium on Theory of Computing*, Submitted to *J. ACM*.

Karp, R. M. (1972). Reducibility among combinatorial problems. in: R. E. Miller, J. W. Thatcher (eds.), *Complexity of Computer Computations*, Plenum Press, New York, 85–103.

König, D. (1961). Über graphen und ihre anwendung auf determinantentheorie und mengenlehre. *Mathematische Annalen* 77, 453–465.

König, D. (1931). Graphok és matrixok (Graphs and matrices). *Matematikai és Fizikai Lapok* 38, 116–119.

Korte, B., J. Vygen (2000). *Combinatorial Optimization: Theory and Algorithms*, Springer.

Lawler, E. L. (1976). Combinatorial optimization: networks and matroids, Holt, Rinehart and Winston, New York etc.

Letchford, A. N., A. Lodi (2002). Primal cutting plane algorithms. revisited. *Math. Methods of Operations Research* 56, 67–81.

Letchford, A. N., Lodi, A. (2003). An augment-and-branch-and-cut framework for mixed 0-1 programming, *Combinatorial Optimization: Eureka, you Shrink! Lecture Notes in Computer Science 2570*, M. Jünger, G. Reinelt, G. Rinaldi (eds.), Springer, pp. 119–133.

Lovász, L., M. Plummer (1986). *Matching Theory*, North-Holland, Amsterdam.

McCormick, T., Shioura, A. (1996), A minimum ratio cycle canceling algorithm for linear programming problems with applications to network optimization, Manuscript.

Minty, G. J. (1980). On maximal independent sets of vertices in claw-free graphs. *Journal of Combinatorial Theory B* 28, 284–304.

Padberg, M., S. Hong (1980). On the symmetric traveling salesman problem: a computational study. *Mathematical Programming Study* 12, 78–107.

Rado, R. (1957). Note on independence functions. *Proceedings of the London Mathematical Society* 7, 300–320.

Sbihi, N. (1980). Algorithme de recherché d'un stable de cardinalité maximum dans un graphe sand étoile. *Discrete Mathematics* 29, 53–76.

Schrijver, A. (2000). A combinatorial algorithm minimizing submodular functions in strongly polynomial time. *Journal of Combinatorial Theory B* 80, 346–355.

Schrijver, A. (2003). *Combinatorial Optimization: Polyhedra and Efficiency*, Springer.

Schulz, A., R. Weismantel (2002). The complexity of generic primal algorithms for solving general integer programs. *Mathematics of Operations Research* 27, 681–692.

Schulz, A. S., R. Weismantel, G. M. Ziegler (1995). 0/1 integer programming: optimization and augmentation are equivalent, Algorithms ESA95. in: P. Spirakis. (eds.), *Lecture Notes in Computer Science 979* Springer, Berlin, 473–483.

Sebö, A. (1990), *Hilbert bases, Carathéodory's theorem and combinatorial optimization*, Integer programming and combinatorial optimization, R. Kannan, W. P. Pulleyblank (eds.), Proceedings of the IPCO Conference, Waterloo, Canada, pp. 431–455.

Spille, B., Weismantel, R. (2001), A combinatorial algorithm for the independent path-matching problem, Manuscript.

Spille, B., Weismantel, R. (2002), A generalization of Edmonds' Matching and matroid intersection algorithms. *Proceedings of the Ninth International Conference on Integer Programming and Combinatorial Optimization, Lecture Notes in Computer Science* 2337, Springer, 9–20.

Tutte, W. T. (1947). The factorization of linear graphs. *Journal of the London Mathematical Society* 22, 107–111.

Urbaniak, R., R. Weismantel, G. M. Ziegler (1997). A variant of Buchberger's algorithm for integer programming. *SIAM Journal on Discrete Mathematics* 1, 96–108.

Wallacher, C. (1992). *Kombinatorische Algorithmen für Flubprobleme und submodulare Flubprobleme*, PhD thesis, Technische Universität zu Braunschweig.

Young, R. D. (1965). A primal (all integer) integer programming algorithm. *Journal of Research of the National Bureau of Standard* 69B, 213–250.

Young, R. D. (1968). A simplified primal (all integer) integer programming algorithm. *Operation Research* 16, 750–782.

K. Aardal et al., Eds., *Handbooks in OR & MS, Vol. 12*

Chapter 6

Balanced Matrices[#]

Michele Conforti

Dipartimento di Matematica Pura ed Applicata, Università di Padova,
Via Belzoni 7, 35131 Padova, Italy
E-mail: conforti@math.unipd.it

Gérard Cornuéjols

Carnegie Mellon University, Schenley Park, Pittsburgh, PA 15213, USA and Laboratoire d'Informatique Fondamentale, Faculté des Sciences de Luminy, 13288 Marseilles, France
E-mail: gc0v@andrew.cmu.edu

Abstract

A $0, \pm 1$ matrix A is *balanced* if, in every submatrix with two nonzero entries per row and column, the sum of the entries is a multiple of four. This definition was introduced by Truemper and generalizes the notion of balanced 0, 1 matrix introduced by Berge. In this tutorial, we survey what is currently known about these matrices, including polyhedral results, structural theorems and recognition algorithms.

1 Introduction

A $0, \pm 1$ matrix H is a *hole matrix* if H contains two nonzero entries per row and per column and no proper submatrix of H has this property. A hole matrix H is square, say of order n, and its rows and columns can be permuted so that its nonzero entries are $h_{i,i}$, $1 \leq i \leq n$, $h_{i,i+1}$, $1 \leq i \leq n-1$, $h_{n,1}$ and no other. Note that $n \geq 2$ and the sum of the entries of H is even.

A hole matrix is *odd* if the sum of its entries is congruent to 2 mod 4 and *even* if the sum of its entries is congruent to 0 mod 4.

A $0, \pm 1$ matrix A is *balanced* if no submatrix of A is an odd hole matrix. This notion is due to Truemper (1982) and it extends the definition of balanced 0, 1 matrices introduced by Berge (1970). The class of balanced $0, \pm 1$ matrices includes balanced 0, 1 matrices and totally unimodular $0, \pm 1$ matrices. (A matrix is *totally unimodular* if every square submatrix has determinant equal to $0, \pm 1$. The fact that total unimodularity implies balancedness follows, for example, from Camion's theorem (1963) which

[#] Dedicated to the memory of Claude Berge.

states that a $0, \pm 1$ matrix A is totally unimodular if and only if A does not contain a square submatrix with an *even number* of nonzero entries per row and per column whose sum of the entries is congruent to 2 mod 4). In this tutorial, we survey what is currently known about balanced matrices, including polyhedral results, structural theorems and recognition algorithms. A previous survey on this topic appears in Conforti, Cornuéjols, Kapoor, Vušković, and Rao (1994).

2 Integral polytopes

A polytope is *integral* if all its vertices have only integer-valued components. Given an $n \times m$ 0, 1 matrix A, the *set packing polytope* is

$$P(A) = \{x \in R^n \colon Ax \leq \mathbf{1},\ 0 \leq x \leq \mathbf{1}\},$$

where $\mathbf{1}$ denotes a column vector of appropriate dimension whose entries are all equal to 1.

The next theorem characterizes a balanced 0, 1 matrix A in terms of the set packing polytope $P(A)$ as well as the set covering polytope $Q(A)$ and the set partitioning polytope $R(A)$:

$$Q(A) = \{x \colon Ax \geq \mathbf{1},\ 0 \leq x \leq \mathbf{1}\},$$
$$R(A) = \{x \colon Ax = \mathbf{1},\ 0 \leq x \leq \mathbf{1}\}.$$

Theorem 2.1. [Berge (1972), Fulkerson, Hoffman, and Oppenheim (1974)] *Let M be a* 0, 1 *matrix. Then the following statements are equivalent:*

(i) *M is balanced.*
(ii) *For each submatrix A of M, the set covering polytope $Q(A)$ is integral.*
(iii) *For each submatrix A of M, the set packing polytope $P(A)$ is integral.*
(iv) *For each submatrix A of M, the set partitioning polytope $R(A)$ is integral.*

Given a $0, \pm 1$ matrix A, let $p(A)$, $n(A)$ denote respectively the column vectors whose ith components $p_i(A)$, $n_i(A)$ are the number of $+1$'s and the number of -1's in the ith row of matrix A. Theorem 2.1 extends to $0, \pm 1$ matrices as follows.

Theorem 2.2. [Conforti and Cornuéjols (1995)] *Let M be a $0, \pm 1$ matrix. Then the following statements are equivalent:*

(i) *M is balanced.*
(ii) *For each submatrix A of M, the generalized set covering polytope $Q(A) = \{x \colon Ax \geq \mathbf{1} - n(A),\ 0 \leq x \leq \mathbf{1}\}$ is integral.*
(iii) *For each submatrix A of M, the generalized set packing polytope $P(A) = \{x \colon Ax \leq \mathbf{1} - n(A),\ 0 \leq x \leq \mathbf{1}\}$ is integral.*

(iv) *For each submatrix A of M, the generalized set partitioning polytope* $R(A) = \{x\colon Ax = \mathbf{1} - n(A),\ 0 \leq x \leq \mathbf{1}\}$ *is integral.*

To prove this theorem, we need the following two results. The first one is an easy application of the computation of determinants by cofactor expansion.

Remark 2.3. Let H be a $0, \pm 1$ hole matrix. If H is an even hole matrix, H is singular and if H is an odd hole matrix, $\det(H) = \pm 2$.

Lemma 2.4. *If A is a balanced* $0, \pm 1$ *matrix, then the generalized set partitioning polytope R(A) is integral.*

Proof. Assume that A contradicts the theorem and has the smallest size (number of rows plus number of columns). Then $R(A)$ is nonempty. Let $\bar{x}$ be a fractional vertex of $R(A)$. By the minimality of A, $0 < \bar{x}_j < 1$ for all j and it follows that A is square and nonsingular. So $\bar{x}$ is the unique vector in $R(A)$.

Let $a^1, \ldots, a^n$ denote the row vectors of A and let A_i be the $(n-1) \times n$ submatrix of A obtained by removing row a^i. By the minimality of A, the set partitioning polytope $R(A_i) = \{x \in R^n\colon A_i x = \mathbf{1} - n(A_i),\ 0 \leq x \leq \mathbf{1}\}$ is an integral polytope. Since A is square and nonsingular, the polytope $R(A_i)$ has exactly two vertices, say x^S, x^T. Since $\bar{x}$ is in $R(A_i)$, then $\bar{x} = \lambda x^S + (1 - \lambda) x^T$. Since $0 < \bar{x}_j < 1$ for all j and x^S, x^T have $0, 1$ components, it follows that $x^S + x^T = \mathbf{1}$. Let k be any row of A_i. Since both x^S and x^T satisfy $a^k x = 1 - n(a^k)$, this implies that $a^k \mathbf{1} = 2(1 - n(a^k))$, i.e., row k contains exactly two nonzero entries. Applying this argument to two different matrices A_i, it follows that every row of A contains exactly two nonzero entries.

If A has a column j with only one nonzero entry a_{kj}, remove column j and row k. Since A is nonsingular, the resulting matrix is also nonsingular and the absolute value of the determinant is unchanged. Repeating this process, we get a square nonsingular matrix B of order at least 2, with exactly two nonzero entries in each row and column (possibly $B = A$). Now B can be put in block-diagonal form, where all the submatrices are hole matrices. Since B is nonsingular, all these submatrices are also nonsingular and by Remark 2.3 they are odd hole matrices. Hence A is not balanced. □

Theorem 2.5. *Let A be a balanced* $0, \pm 1$ *matrix with rows* a^i, $i \in S$, *and let* S_1, S_2, S_3 *be a partition of S. Then*

$$\begin{aligned} T(A) = \{x \in R^n\colon\quad & a^i x \geq 1 - n(a^i) \quad \textit{for} \quad i \in S_1, \\ & a^i x = 1 - n(a^i) \quad \textit{for} \quad i \in S_2, \\ & a^i x \leq 1 - n(a^i) \quad \textit{for} \quad i \in S_3, \\ & 0 \leq x \leq \mathbf{1}\} \end{aligned}$$

is an integral polytope.

Proof. If $\bar{x}$ is a vertex of $T(A)$, it is a vertex of the polytope obtained from $T(A)$ by deleting the inequalities that are not satisfied with equality by $\bar{x}$. By Theorem 2.4, every vertex of this polytope has 0, 1 components. □

Proof of Theorem 2.2. Since balanced matrices are closed under taking submatrices, Theorem 2.5 shows that (i) implies (ii), (iii) and (iv).

Assume that A contains an odd hole submatrix H. By Remark 2.3, the vector $x=((1/2),\ldots,(1/2))$ is the unique solution of the system $Hx=\mathbf{1}$. This proves all three reverse implications. □

3 Bicoloring

Berge (1970) introduced the following notion. A 0, 1 matrix is *bicolorable* if its columns can be partitioned into blue and red columns in such a way that every row with two or more 1's contains a 1 in a blue column and a 1 in a red column. This notion provides the following characterization of balanced 0, 1 matrices.

Theorem 3.1. [Berge (1970)] *A* 0, 1 *matrix A is balanced if and only if every submatrix of A is bicolorable.*

Ghouila-Houri (1962) introduced the notion of *equitable bicoloring* for a $0, \pm 1$ matrix A as follows. The columns of A are partitioned into blue columns and red columns in such a way that, for every row of A, the sum of the entries in the blue columns differs from the sum of the entries in the red columns by at most one.

Theorem 3.2. [Ghouila-Houri (1962)] *A* $0, \pm 1$ *matrix A is totally unimodular if and only if every submatrix of A has an equitable bicoloring.*

This theorem generalizes a result of Heller and Tompkins (1956) for matrices with at most two nonzero entries per row.

A $0, \pm 1$ matrix A is *bicolorable* if its columns can be partitioned into blue columns and red columns in such a way that every row with two or more nonzero entries either contains two entries of opposite sign in columns of the same color, or contains two entries of the same sign in columns of different colors. For a 0, 1 matrix, this definition coincides with Berge's notion of bicoloring. Clearly, if a $0, \pm 1$ matrix has an equitable bicoloring as defined by Ghouila-Houri, then it is bicolorable. So the theorem below implies that every totally unimodular matrix is balanced.

Theorem 3.3. [Conforti and Cornuéjols (1995)] *A* $0, \pm 1$ *matrix A is balanced if and only if every submatrix of A is bicolorable.*

Proof. Assume first that A is balanced and let B be any submatrix of A. Remove from B any row with fewer than two nonzero entries. Since B is balanced, so is the matrix $(B, -B)$. It follows from Theorem 2.5 that the inequalities

$$\begin{aligned} Bx &\geq \mathbf{1} - n(B) \\ -Bx &\geq \mathbf{1} - n(-B) \\ 0 \leq x &\leq \mathbf{1} \end{aligned} \tag{1}$$

define an integral polytope. Since it is nonempty (the vector $((1/2), \ldots, (1/2))$ is a solution), it contains a 0, 1 vector $\bar{x}$. Color a column j of B red if $\bar{x}_j = 1$ and blue otherwise. By (1), this is a valid bicoloring of B.

Conversely, assume that A contains an odd hole matrix H. We claim that H is not bicolorable. Suppose otherwise. Since H contains exactly 2 nonzero entries per row, the bicoloring condition shows that the vector of all zeroes can be obtained by adding the blue columns and subtracting the red columns. So H is singular, a contradiction to Remark 2.3. □

In Section 5 we prove a bicoloring theorem that extend all the above results.

Cameron and Edmonds (1990) observed that the following simple algorithm finds a bicoloring of a balanced matrix.

Algorithm [Cameron and Edmonds (1990)]
Input. A $0, \pm 1$ *matrix* A.
Output. *A bicoloring of* A *or a proof that the matrix* A *is not balanced.*

Stop if all columns are colored or if some row is incorrectly colored. Otherwise, color a new column red or blue as follows.
If some row of A *forces the color of a column, color this column accordingly.*
If no row of A *forces the color of a column, arbitrarily color one of the uncolored columns.*

In the above algorithm, a row a^i *forces* the color of a column when all the columns corresponding to the nonzero entries of a^i have been colored except one, say column k, and row a^i, restricted to the colored columns, violates the bicoloring condition. In this case, the bicoloring rule dictates the color of column k.

When the algorithm fails to find a bicoloring, the sequence of forcings that results in an incorrectly colored row identifies an odd hole submatrix of A.

Note that a matrix A may be bicolorable even if A is not balanced. In fact, the algorithm may find a bicoloring of A even if A is not balanced. For example, if

$$A = \begin{pmatrix} 1 & 1 & 1 & 0 \\ 1 & 1 & 0 & 1 \\ 1 & 0 & 1 & 1 \end{pmatrix},$$

the algorithm may color the first two columns blue and the last two red, which is a bicoloring of A. For this reason, the algorithm cannot be used as a recognition of balancedness.

4 Total dual integrality

A system of linear constraints is *totally dual integral* (TDI) if, for each integral objective function vector c, the dual linear program has an integral optimal solution (if an optimal solution exists). Edmonds and Giles (1977) proved that, if a linear system $Ax \le b$ is TDI and b is integral, then $\{x\colon Ax \le b\}$ is an integral polyhedron.

Theorem 4.1. [Fulkerson et al. (1974)] *Let*

$$A = \begin{pmatrix} A_1 \\ A_2 \\ A_3 \end{pmatrix}$$

be a balanced 0, 1 *matrix. Then the linear system*

$$\begin{cases} A_1 x \ge \mathbf{1} \\ A_2 x \le \mathbf{1} \\ A_3 x = \mathbf{1} \\ \quad x \ge 0 \end{cases}$$

is TDI.

Theorem 4.1 and the Edmonds–Giles theorem imply Theorem 2.1. In this section, we prove the following, more general result.

Theorem 4.2. [Conforti and Cornuéjols (1995b)] *Let*

$$A = \begin{pmatrix} A_1 \\ A_2 \\ A_3 \end{pmatrix}$$

be a balanced $0, \pm 1$ *matrix. Then the linear system*

$$\begin{cases} A_1 x \geq \mathbf{1} - n(A_1) \\ A_2 x \leq \mathbf{1} - n(A_2) \\ A_3 x = \mathbf{1} - n(A_3) \\ \quad 0 \leq x \leq \mathbf{1} \end{cases}$$

is TDI.

The following transformation of a $0, \pm 1$ matrix A into a $0, 1$ matrix B is often seen in the literature: to every column a_j of $A, j = 1, \ldots, p$, associate two columns of B, say b_j^P and b_j^N, where $b_{ij}^P = 1$ if $a_{ij} = 1, 0$ otherwise, and $b_{ij}^N = 1$ if $a_{ij} = -1, 0$ otherwise. Let D be the $0, 1$ matrix with p rows and $2p$ columns d_j^P and d_j^N such that $d_{jj}^P = d_{jj}^N = 1$ and $d_{ij}^P = d_{ij}^N = 0$ for $i \neq j$.

Given a $0, \pm 1$ matrix

$$A = \begin{pmatrix} A_1 \\ A_2 \\ A_3 \end{pmatrix}$$

and the associated 0,1 matrix

$$B = \begin{pmatrix} B_1 \\ B_2 \\ B_3 \end{pmatrix},$$

define the following linear systems:

$$\begin{cases} A_1 x \geq \mathbf{1} - n(A_1) \\ A_2 x \leq \mathbf{1} - n(A_2) \\ A_3 x = \mathbf{1} - n(A_3) \\ \quad 0 \leq x \leq \mathbf{1}, \end{cases} \tag{2}$$

and

$$\begin{cases} B_1 y \geq \mathbf{1} \\ B_2 y \leq \mathbf{1} \\ B_3 y = \mathbf{1} \\ D y = \mathbf{1} \\ \quad y \geq 0. \end{cases} \tag{3}$$

The vector $x \in R^p$ satisfies (2) if and only if the vector $(y^P, y^N) = (x, \mathbf{1} - x)$ satisfies (3). Hence the polytope defined by (2) is integral if and only if the polytope defined by (3) is integral. We show that, if A is a balanced $0, \pm 1$ matrix, then both (2) and (3) are TDI.

Lemma 4.3. *If*

$$A = \begin{pmatrix} A_1 \\ A_2 \\ A_3 \end{pmatrix}$$

is a balanced $0, \pm 1$ *matrix, the corresponding system* (3) *is TDI.*

Proof. The proof is by induction on the number m of rows of B. Let $c = (c^P, c^N) \in Z^{2p}$ denote an integral vector and R_1, R_2, R_3 the index sets of the rows of B_1, B_2, B_3 respectively. The dual of min $\{cy: y$ satisfies (3)$\}$ is the linear program

$$\begin{aligned} \max \quad & \sum_{i=1}^{m} u_i + \sum_{j=1}^{p} v_j \\ & uB + vD \leq c \\ & u_i \geq 0, i \in R_1 \\ & u_i \leq 0, i \in R_2. \end{aligned} \tag{4}$$

Since v_j only appears in two of the constraints $uB + vD \leq c$ and no constraint contains v_j and v_k, it follows that any optimal solution to (4) satisfies

$$v_j = \min\left(c_j^P - \sum_{i=1}^{m} b_{ij}^P u_i, c_j^N - \sum_{i=1}^{m} b_{ij}^N u_i \right). \tag{5}$$

Let $(\bar{u}, \bar{v})$ be an optimal solution of (4). If $\bar{u}$ is integral, then so is $\bar{v}$ by (5) and we are done. So assume that $\bar{u}_\ell$ is fractional. Let b^ℓ be the corresponding row of B and let B_ℓ be the matrix obtained from B by removing row b^ℓ. By induction on the number of rows of B, the system (3) associated with B_ℓ is TDI. Hence the system

$$\begin{aligned} \max \quad & \sum_{i \neq \ell} u_i + \sum_{j=1}^{p} v_j \\ & u_\ell B_\ell + vD \leq c - \lfloor \bar{u}_\ell \rfloor b^\ell \\ & u_i \geq 0, i \in R_1 \setminus \{\ell\} \\ & u_i \leq 0, i \in R_2 \setminus \{\ell\} \end{aligned} \tag{6}$$

has an integral optimal solution $(\tilde{u}, \tilde{v})$. Since $(\bar{u}_1, \ldots, \bar{u}_{\ell-1}, \bar{u}_{\ell+1}, \ldots, \bar{u}_m, \bar{v}_1, \ldots, \bar{v}_p)$ is a feasible solution to (6) and Theorem 2.5 shows that $\sum_{i=1}^{m} \bar{u}_i + \sum_{j=1}^{p} \bar{v}_j$ is an integer,

$$\sum_{i \neq \ell} \tilde{u}_i + \sum_{j=1}^{p} \tilde{v}_j \geq \left\lceil \sum_{i \neq \ell} \bar{u}_i + \sum_{j=1}^{p} \bar{v}_j \right\rceil = \sum_{i=1}^{m} \bar{u}_i + \sum_{j=1}^{p} \bar{v}_j - \lfloor \bar{u}_\ell \rfloor.$$

Therefore the vector $(u^*, v^*) = (\tilde{u}_1, \ldots, \tilde{u}_{\ell-1}, \lfloor \bar{u}_\ell \rfloor, \tilde{u}_{\ell+1}, \ldots, \tilde{u}_m, \tilde{v}_1, \ldots, \tilde{v}_p)$ is integral, is feasible to (4) and has an objective function value not smaller than $(\bar{u}, \bar{v})$, proving that the system (3) is TDI. □

Proof of Theorem 4.2. Let R_1, R_2, R_3 be the index sets of the rows of A_1, A_2, A_3. By Lemma 4.3, the linear system (3) associated with (2) is TDI. Let $d \in R^p$ be any integral vector. The dual of $\min\{dx\colon x \text{ satisfies (2)}\}$ is the linear program

$$\begin{aligned} \max\quad & w(1 - n(A)) - t\mathbf{1} \\ & wA - t \leq d \\ & w_i \geq 0, i \in R_1 \\ & w_i \leq 0, i \in R_2 \\ & t \geq 0. \end{aligned} \tag{7}$$

For every feasible solution $(\bar{u}, \bar{v})$ of (4) with $c = (c^P, c^N) = (d, 0)$, we construct a feasible solution $(\bar{w}, \bar{t})$ of (7) with the same objective function value as follows:

$$\begin{aligned} &\bar{w} = \bar{u} \\ &\bar{t}_j = \begin{cases} 0 & \text{if } \bar{v}_j = -\sum_i b_{ij}^N \bar{u}_i \\ \sum_i b_{ij}^P \bar{u}_i - \sum_i b_{ij}^N \bar{u}_i - d_j & \text{if } \bar{v}_j = d_j - \sum_i b_{ij}^P \bar{u}_i. \end{cases} \end{aligned} \tag{8}$$

When the vector $(\bar{u}, \bar{v})$ is integral, the above transformation yields an integral vector $(\bar{w}, \bar{t})$. Therefore (7) has an integral optimal solution and the linear system (2) is TDI. □

It may be worth noting that this theorem does not hold when the upper bound $x \leq \mathbf{1}$ is dropped from the linear system. In fact, the resulting polyhedron may not even be integral [see Conforti and Cornuéjols (1995) for an example].

5 k-Balanced matrices

We introduce a hierarchy of balanced $0, \pm 1$ matrices that contains as its two extreme cases the balanced and totally unimodular matrices. The following well known result of Camion will be used.

A $0, \pm 1$ matrix which is not totally unimodular but whose proper submatrices are all totally unimodular is said to be *almost totally unimodular*. Camion (1965) proved the following:

Theorem 5.1. [Camion (1965) and Gomory [cited in Camion (1965)]] *Let A be an almost totally unimodular $0, \pm 1$ matrix. Then A is square,* $\det A = \pm 2$ *and A^{-1} has only $\pm(1/2)$ entries. Furthermore, each row and each column of A has an even number of nonzero entries and the sum of all entries in A equals 2 modulo 4.*

Proof. Clearly A is square, say $n \times n$. If $n=2$, then indeed, $\det A = \pm 2$. Now assume $n>2$. Since A is nonsingular, it contains an $(n-2)\times(n-2)$ nonsingular submatrix B. Let

$$A = \begin{pmatrix} B & C \\ D & E \end{pmatrix} \quad \text{and} \quad U = \begin{pmatrix} B^{-1} & 0 \\ -DB^{-1} & I \end{pmatrix}.$$

Then $\det U = \pm 1$ and

$$UA = \begin{pmatrix} I & B^{-1}C \\ 0 & E - DB^{-1}C \end{pmatrix}.$$

We claim that the 2×2 matrix $E - DB^{-1}C$ has all entries equal to $0, \pm 1$. Suppose to the contrary that $E - DB^{-1}C$ has an entry different from $0, \pm 1$ in row i and column j. Denoting the corresponding entry of E by e_{ij}, the corresponding column of C by c^j and row of D by d^i,

$$\begin{pmatrix} B^{-1} & 0 \\ -d^iB^{-1} & 1 \end{pmatrix}\begin{pmatrix} B & c^j \\ d^i & e_{ij} \end{pmatrix} = \begin{pmatrix} I & B^{-1}c^j \\ 0 & e_{ij} - d^iB^{-1}c^j \end{pmatrix}$$

and consequently A has an $(n-1)\times(n-1)$ submatrix with a determinant different from $0, \pm 1$, a contradiction.

Consequently, $\det A = \pm\det UA = \pm\det(E - DB^{-1}C) = \pm 2$.

So, every entry of A^{-1} is equal to $0, \pm(1/2)$. Suppose A^{-1} has an entry equal to 0, say in row i and column j. Let $\bar{A}$ be the matrix obtained from A by removing column i and let h^j be the jth column of A^{-1} with row i removed. Then $\bar{A}h^j = u^j$, where u^j denotes the jth unit vector. Since $\bar{A}$ has rank $n-1$, this linear system of equations has a unique solution h^j. Since $\bar{A}$ is totally unimodular and u^j is integral, this solution h^j is integral. Since $h^j \neq 0$, this contradicts the fact that every entry of h^j is equal to $0, \pm(1/2)$. So A^{-1} has only $\pm(1/2)$ entries.

This property and the fact that AA^{-1} and $A^{-1}A$ are integral, imply that A has an even number of nonzero entries in each row and column.

Finally, let α denote a column of A^{-1} and $S = \{i\colon \alpha_i = +(1/2)\}$ and $\bar{S} = \{i\colon \alpha_i = -(1/2)\}$. Let k denote the sum of all entries in the columns of A indexed by $\bar{S}$. Since $A\alpha$ is a unit vector, the sum of all entries in the columns of A indexed by S equals $k+2$. Since every column of A has an even number of nonzero entries, k is even, say $k = 2p$ for some integer p. Therefore, the sum of all entries in A equals $4p+2$. □

For any positive integer k, we say that a $0, \pm 1$ matrix A is *k-balanced* if A does not contain any almost totally unimodular submatrix with at most $2k$ nonzero entries in each row.

Note that every almost totally unimodular matrix contains at least 2 nonzero entries per row and per column. So the odd hole matrices are the almost totally unimodular matrices with at most 2 nonzero entries per row. Therefore the balanced matrices are the 1-balanced matrices and the totally unimodular matrices with n columns are the k-balanced matrices for $k \geq \lfloor n/2 \rfloor$. The class of k-balanced matrices was introduced by Truemper and Chandrasekaran (1978) for 0, 1 matrices and by Conforti et al. (1994) for $0, \pm 1$ matrices. Let $\mathbf{k}$ denote a column vector whose entries are all equal to k.

Theorem 5.2. [Conforti et al. (1994)] *Let A be an $m \times n$ k-balanced $0, \pm 1$ matrix with rows a^i, $i \in [m]$, b be a vector with entries b_i, $i \in [m]$, and let S_1, S_2, S_3 be a partition of $[m]$. Then*

$$P(A, b) = \{x \in R^n\colon \quad \begin{aligned} & a^i x \leq b_i \ \textit{for}\ i \in S_1 \\ & a^i x = b_i \ \textit{for}\ i \in S_2 \\ & a^i x \geq b_i \ \textit{for}\ i \in S_3 \\ & 0 \leq x \leq \mathbf{1}\} \end{aligned}$$

is an integral polytope for all integral vectors b such that $-n(A) \leq b \leq$ $\mathbf{k} - n(A)$.

Proof. Assume the contrary and let A be a k-balanced matrix of the smallest order such that $P(A, b)$ has a fractional vertex $\bar{x}$ for some vector b such that $-n(A) \leq b \leq \mathbf{k} - n(A)$ and some partition S_1, S_2, S_3 of $[m]$. Then by the minimality of A, $\bar{x}$ satisfies all the constraints in $S_1 \cup S_2 \cup S_3$ at equality. So we may assume $S_1 = S_3 = \emptyset$. Furthermore all the components of x are fractional, otherwise let A^f be the column submatrix of A corresponding to the fractional components of $\bar{x}$ and A^p be the column submatrix of A corresponding to the components of $\bar{x}$ that are equal to 1. Let $b^f = b - p(A^p) + n(A^p)$. Then $-n(A^f) \leq b^f \leq \mathbf{k} - n(A^f)$ since $b^f = b - p(A^p) + n(A^p) = A^f \bar{x} \geq n(A^f)$ and because $b^f = b - p(A^p) + n(A^p) \leq b + n(A^p) \leq k - n(A) + n(A^p) \leq k - n(A^f)$.

Since the restriction of $\bar{x}$ to its fractional components is a vertex of $P(A^f, b^f)$ with $S_1 = S_3 = \emptyset$, the minimality of A is contradicted. So A is a square non-singular matrix which is not totally unimodular. Let G be an almost totally unimodular submatrix of A. Since A is not k-balanced, G contains a row i such that $p_i(G) + n_i(G) > 2k$. Let A^i be the submatrix of A obtained by removing row i and let b^i be the corresponding subvector of b. By the minimality of A, $P(A^i, b^i)$ with $S_1 = S_3 = \emptyset$ is an integer polytope and since A is nonsingular,

$P(A^i, b^i)$ has exactly two vertices, say z^1 and z^2. Since $\bar{x}$ is a vector whose components are all fractional and $\bar{x}$ can be written as the convex combination of the 0,1 vectors z^1 and z^2, then $z^1 + z^2 = \mathbf{1}$. For $\ell = 1, 2$, define

$$L(\ell) = \{j;\ \text{either } g_{ij} = 1 \text{ and } z_i^\ell = 1 \text{ or } g_{ij} = -1 \text{ and } z_i^\ell = 0\}.$$

Since $z^1 + z^2 = \mathbf{1}$, it follows that $|L(1)| + |L(2)| = p_i(G) + n_i(G) > 2k$. Assume w.l.o.g. that $|L(1)| > k$. Now this contradicts

$$|L(1)| = \sum_j g_{ij} z_j^1 + n_i(G) \leq b_i + n_i(A) \leq k$$

where the first inequality follows from $A^i z^1 = b^i$. □

This theorem generalizes the previous results by Hoffman and Kruskal (1956) for totally unimodular matrices, Berge (1972) for 0,1 balanced matrices. Conforti and Cornuéjols (1995b) for $0, \pm 1$ balanced matrices, and Truemper and Chandrasekaran (1978) for k-balanced 0, 1 matrices.

A $0, \pm 1$ matrix A has a *k-equitable bicoloring* if its columns can be partitioned into blue columns and red columns so that:

- The bicoloring is equitable for the row submatrix A' determined by the rows of A with at most $2k$ nonzero entries,
- Every row with more than $2k$ nonzero entries contains k pairwise disjoint pairs of nonzero entries such that each pair contains either entries of opposite sign in columns of the same color or entries of the same sign in columns of different colors.

Obviously, an $m \times n$ $0, \pm 1$ matrix A is bicolorable if and only if A has a 1-equitable bicoloring, while A has an equitable bicoloring if and only if A has a k-equitable bicoloring for $k \geq \lfloor n/2 \rfloor$. The following theorem provides a new characterization of the class of k-balanced matrices, which generalizes the bicoloring results of Section 3 for balanced and totally unimodular matrices.

Theorem 5.3. [Conforti, Cornuéjols, and Zambelli (2004)] *A* $0, \pm 1$ *matrix A is k-balanced if and only if every submatrix of A has a k-equitable bicoloring.*

Proof. Assume first that A is k-balanced and let B be any submatrix of A. Assume, up to row permutation, that

$$B = \begin{pmatrix} B' \\ B'' \end{pmatrix}$$

where B' is the row submatrix of B determined by the rows of B with $2k$ or fewer nonzero entries. Consider the system

$$\begin{aligned} B'x &\geq \left\lfloor \frac{B'\mathbf{1}}{2} \right\rfloor \\ -B'x &\geq -\left\lfloor \frac{B'\mathbf{1}}{2} \right\rfloor \\ B''x &\geq \mathbf{k} - n(B'') \\ -B''x &\geq \mathbf{k} - n(-B'') \\ 0 \leq\ & x \leq \mathbf{1} \end{aligned} \tag{9}$$

Since B is k-balanced, $\binom{B}{-B}$ also is k-balanced. Therefore the constraint matrix of system (9) above is k-balanced. One can readily verify that $-n(B') \leq \lfloor (B'\mathbf{1}/2) \rfloor \leq \mathbf{k} - n(B')$ and $-n(-B') \leq -\lceil (B'\mathbf{1}/2) \rceil \leq \mathbf{k} - n(-B')$. Therefore, by Theorem 5.2 applied with $S_1 = S_2 = \emptyset$, system (9) defines an integral polytope. Since the vector $((1/2), \ldots, (1/2))$ is a solution for (9), the polytope is nonempty and contains a 0,1 point $\bar{x}$. Color a column i of B blue if $\bar{x}_i = 1$, red otherwise. It can be easily verified that such a bicoloring is, in fact, k-equitable.

Conversely, assume that A is not k-balanced. Then A contains an almost totally unimodular matrix B with at most $2k$ nonzero elements per row. Suppose that B has a k-equitable bicoloring, then such a bicoloring must be equitable since each row has, at most, $2k$ nonzero elements. By Theorem 5.1, B has an even number of nonzero elements in each row. Therefore the sum of the columns colored blue equals the sum of the columns colored red, therefore B is a singular matrix, a contradiction. □

Given a $0, \pm 1$ matrix A and a positive integer k, one can find in polynomial time a k-equitable bicoloring of A or a certificate that A is not k-balanced as follows:

Find a basic feasible solution of (9). If the solution is not integral, A is not k-balanced by Theorem 5.2. If the solution is a $0, 1$ vector, it yields a k-equitable bicoloring as in the proof of Theorem 5.3.

Note that, as with the algorithm of Cameron and Edmonds (1990) discussed in Section 3, a $0, 1$ vector may be found even when the matrix A is not k-balanced.

Using the fact that the vector $((1/2), \ldots, (1/2))$ is a feasible solution of (9), a basic feasible solution of (9) can actually be derived in strongly polynomial time using an algorithm of Megiddo (1991).

6 Perfection and idealness

A 0,1 matrix A is said to be *perfect* if the set packing polytope $P(A)$ is integral. A 0,1 matrix A is *ideal* if the set covering polytope $Q(A)$ is integral.

The study of perfect and ideal 0,1 matrices is a central topic in polyhedral combinatorics. Theorem 2.1 shows that every balanced 0, 1 matrix is both perfect and ideal.

The integrality of the set packing polytope associated with a (0, 1) matrix A is related to the notion of the perfect graph. A graph G is *perfect* if, for every induced subgraph H of G, the chromatic number of H equals the size of its largest clique. The fundamental connection between the theory of perfect graphs and integer programming was established by Fulkerson (1972), Lovász (1972) and Chvátal (1975). The *clique-node matrix* of a graph G is a 0, 1 matrix whose columns are indexed by the nodes of G and whose rows are the incidence vectors of the maximal cliques of G.

Theorem 6.1. [Lovász (1972), Fulkerson (1972), Chvátal (1975)] *Let A be a* 0,1 *matrix. The set packing polytope $P(A)$ is integral if and only if the rows of A of maximal support form the clique-node matrix of a perfect graph.*

Now we extend the definition of perfect and ideal 0, 1 matrices to $0, \pm 1$ matrices. A $0, \pm 1$ matrix A is *ideal* if the generalized set covering polytope $Q(A) = \{x\colon\ Ax > \mathbf{1} - n(A),\ 0 \leq x \leq \mathbf{1}\}$ is integral. A $0, \pm 1$ matrix A is *perfect* if the generalized set packing polytope $P(A) = \{x\colon\ Ax \leq \mathbf{1} - (A),\ 0 \leq x \leq \mathbf{1}\}$ is integral.

Hooker (1996) was the first to relate idealness of a $0, \pm 1$ matrix to that of a family of 0, 1 matrices. A similar result for perfection was obtained in Conforti, Cornuéjols, and De Francesco (1997). These results were strengthened by Guenin (1998) and by Boros and Čepek (1997) for perfection, and by Nobili and Sassano (1998) for idealness. The key tool for these results is the following:

Given a $0, \pm 1$ matrix A, let P and R be 0, 1 matrices of the same dimension as A, with entries $p_{ij} = 1$ if and only if $a_{ij} = 1$, and $r_{ij} = 1$ if and only if $a_{ij} = -1$. The matrix

$$D_A = \begin{pmatrix} P & R \\ I & I \end{pmatrix}$$

is the 0, 1 *extension* of A. Note that the transformation $x^+ = x$ and $x^- = \mathbf{1} - x$ maps every vector x in $P(A)$ into a vector in $\{(x^+, x^-) \geq 0\colon\ Px^+ + Rx^- \leq \mathbf{1},\ x^+ + x^- = \mathbf{1}\}$ and every vector x in $Q(A)$ into a vector in $\{(x^+, x^-) \geq 0\colon\ Px^+ + Rx^- \geq \mathbf{1},\ x^+ + x^- = 1\}$. So $P(A)$ and $Q(A)$ are respectively the faces of $P(D_A)$ and $Q(D_A)$, obtained by setting the inequalities $x^+ + x^- \leq \mathbf{1}$ and $x^+ + x^- \geq \mathbf{1}$ at equality.

Given a $0, \pm 1$ matrix A, let a^1 and a^2 be two rows of A, such that there is one index k such that $a^1_k a^2_k = -1$ and, for all $j \neq k$, $a^1_j a^2_j = 0$. A *disjoint implication* of A is the $0, \pm 1$ vector $a^1 + a^2$. The matrix A^+ obtained by recursively adding all disjoint implications and removing all dominated rows (those whose support is not maximal in the packing case; those whose support is not minimal in the covering case) is called the *disjoint completion* of A.

Theorem 6.2. [Nobili and Sassano (1998)] *Let A be a* $0, \pm 1$ *matrix. Then A is ideal if and only if the* 0,1 *matrix* D_{A+} *is ideal.*

Furthermore A is ideal if and only if $\min\{cx\colon x \in Q(A)\}$ has an integer optimum for every vector $c \in \{0, \pm 1, \pm\infty\}^n$.

Theorem 6.3. [Guenin (1998)] *Let A a* $0, \pm 1$ *matrix such that* $P(A)$ *is not contained in any of the hyperplanes* $\{x\colon x_j = 0\}$ *or* $\{x\colon x_j = 1\}$. *Then A is perfect if and only if the* 0, 1 *matrix* D_{A+} *is perfect.*

Theorem 6.4. [Guenin (1998)] *Let A is a* $0, \pm 1$ *matrix such that* $P(A)$ *is not contained in any of the hyperplanes* $\{x\colon x_j = 0\}$ *or* $\{x\colon x_j = 1\}$. *Then A is perfect if and only if* $\max\{cx\colon x \in P(A)\}$ *admits an integral optimal solution for every* $c \in \{0, \pm 1\}^n$. *Moreover, if A is perfect, the linear system* $Ax \leq \mathbf{1} - n(A)$, $0 \leq x \leq \mathbf{1}$ *is TDI.*

This is the natural extension of Lovász's theorem for perfect 0, 1 matrices. The next theorem characterizes perfect $0, \pm 1$ matrices in terms of excluded submatrices. A row of a $0, \pm 1$ matrix A is *trivial* if it contains at most one nonzero entry. Note that trivial rows can be removed without changing $P(A)$.

Theorem 6.5. [Guenin (1998)] *Let A is a* $0, \pm 1$ *matrix such that* $P(A)$ *is not contained in any of the hyperplanes* $\{x\colon x_j{=}0\}$ *or* $\{x\colon x_j{=}1\}$. *Then A is perfect if and only if* A^+ *does not contain.*

(1)

$$\begin{pmatrix} 1 & 1 \\ -1 & 1 \end{pmatrix} \quad \text{or} \quad \begin{pmatrix} 1 & -1 \\ -1 & -1 \end{pmatrix}$$

as a submatrix, or

(2) *a column submatrix which, without its trivial rows, is obtained from a minimally imperfect* 0, 1 *matrix B by switching signs of all entries in a subset to the columns of B.*

For ideal $0, \pm 1$ matrices, a similar characterization was obtained in terms of excluded "weak minors" by Nobili and Sassano (1998).

7 Propositional logic

In propositional logic, *atomic propositions* $x_1, \ldots, x_j, \ldots, x_n$ can be either *true* or *false*. A *truth assignment* is an assignment of "true" or "false" to every atomic proposition. A *literal* is an atomic proposition x_j or its negation $\neg x_j$.

A *clause* is a disjunction of literals and is *satisfied* by a given truth assignment if at least one of its literals is true.

A survey of the connections between propositional logic and integer programming can be found in Hooker (1988).

A truth assignment satisfies the set S of clauses

$$\bigvee_{j\in P_i} x_j \vee \left(\bigvee_{j\in N_i} \neg x_j\right) \quad \text{for all} \quad i \in S$$

if and only if the corresponding 0, 1 vector satisfies the system of inequalities

$$\sum_{j\in P_i} x_j - \sum_{j\in N_i} x_j \geq 1 - |N_i| \quad \text{for all} \quad i \in S.$$

The above system of inequalities is of the form

$$Ax \geq \mathbf{1} - n(A). \tag{10}$$

We consider three classical problems in logic. Given a set S of clauses, the *satisfiability problem* (SAT) consists of finding a truth assignment that satisfies all the clauses in S or showing that none exists. Equivalently, SAT consists of finding a 0, 1 solution x to (10) or showing that none exists.

Given a set S of clauses and a weight vector w whose components are indexed by the clauses in S, the *weighted maximum satisifiabilty problem* (MAXSAT) consists of finding a truth assignment that maximizes the total weight of the satisfied clauses. MAXSAT can be formulated as the integer program

$$\begin{aligned} &\text{Min} \sum_{i=1}^{m} w_i s_i \\ &\quad Ax + s \geq \mathbf{1} - n(A) \\ &\quad x \in \{0,1\}^n,\, s \in \{0,1\}^m. \end{aligned}$$

Given a set S of clauses (the premises) and a clause C (the conclusion), *logical inference* in propositional logic consists of deciding whether every truth assignment that satisfies all the clauses in S also satisfies the conclusion C.

To the clause C, using transformation (10), we associate an inequality

$$cx \geq 1 - n(c),$$

where c is a $0, \pm 1$ vector. Therefore C cannot be deduced from S if and only if the integer program

$$\text{Min}\{cx\colon Ax \geq \mathbf{1} - n(A),\, x \in \{0,1\}^n\} \tag{11}$$

has a solution with values $-n(c)$.

These three problems are NP-hard in general but SAT and logical inference can be solved efficiently for Horn clauses, clauses with at most two literals and several related classes Boros, Crama, and Hammer (1990), Chandru and Hooker (1991), Truemper (1990). MAXSAT remains NP-hard for Horn clauses with at most two literals Georgakopoulos, Kavvasdias, and Papdimitriou (1988). A set S of clauses is *balanced* if the corresponding $0, \pm 1$ matrix A defined in (10) is balanced. Similarly, a set of clauses *ideal* if A is ideal. If S is ideal, SAT, MAXSAT, and logical inference can be solved by linear programming. The following theorem is an immediate consequence of Theorem 2.2.

Theorem 7.1. *Let S be a balanced set of clauses. Then the SAT, MAXSAT, and logical inference problems can be solved in polynomial time by linear programming.*

This has consequences for probabilistic logic as defined by Nilsson (1986). Being able to solve MAXSAT in polynomial time provides a polynomial time separation algorithm for probabilistic logic via the ellipsoid method, as observed by Georgakopoulos et al. (1988). Hence probabilistic logic is solvable in polynomial time for ideal sets of clauses.

Remark 7.2. Let S be an ideal set of clauses. If every clause of S contains more than one literal then, for every atomic proposition x_j, there exist at least two truth assignments satisfying S, one in which x_j is true and one in which x_j is false. □

Proof. Since the point $x_j = 1/2$, $j = 1, \ldots, n$ belongs to the polytope $Q(A) = \{x\colon Ax \geq \mathbf{1} - n(A),\ 0 \leq x \leq \mathbf{1}\}$ and $Q(A)$ is an integral polytope, then the above point can be expressed as a convex combination of 0, 1 vectors in $Q(A)$. Clearly, for every index j, there exists in the convex combination a 0, 1 vector with $x_j = 0$ and another with $x_j = 1$.

A consequence of Remark 7.2 is that, for an ideal set of clauses, SAT can be solved more efficiently than by general linear programming.

Theorem 7.3. [Conforti and Cornuéjols (1995a)] *Let S be an ideal set of clauses. Then S is satisfiable if and only if a recursive application of the following procedure stops with an empty set of clauses.*

7.1 Recursive step

If $S = \emptyset$ then S is satisfiable.

If S contains a clause C with a single literal (unit clause), set the corresponding atomic proposition x_j so that C is satisfied. Eliminate from S all *clauses that become satisfied and remove x_j from all the other clauses. If a clause becomes empty, then S is not satisfiable (unit resolution).*

If every clause in S contains at least two literals, choose any atomic proposition x_j appearing in a clause of S and add to S an arbitrary clause x_j or $\neg x_j$.

The above algorithm for SAT can also be used to solve the logical inference problem when S is an ideal set of clauses, see Conforti and Cornuéjols (1995a). For balanced (or ideal) sets of clauses, it is an open problem to solve MAXSAT in polynomial time by a direct method, without appearing to polynomial time algorithms for general linear programming.

8 Nonlinear 0, 1 optimization

Consider the nonlinear 0, 1 maximization problem

$$\max_{x\in\{0,1\}^n} \sum_k a_k \prod_{j\in T_k} x_j \prod_{j\in R_k} (1 - x_j),$$

where, w.l.o.g., all ordered pairs (T_k, R_k) are distinct and $T_k \cap R_k = \emptyset$. This is an NP-hard problem. A standard linearization of this problem was proposed by Fortet (1976):

$$\begin{array}{lll}
\max & \sum a_k y_k & \\
& y_k - x_j \leq 0 & \text{for all } k \text{ s.t. } a_k > 0, \text{ for all } j \in T_k \\
& y_k + x_j \leq 1 & \text{for all } k \text{ s.t. } a_k > 0, \text{ for all } j \in R_k \\
& y_k - \sum_{j\in T_k} x_j + \sum_{j\in R_k} x_j \geq 1 - |T_k| & \text{for all } k \text{ s.t. } a_k < 0 \\
& y_k, x_j \in \{0, 1\} & \text{for all } k \text{ and } j.
\end{array}$$

When the constraint matrix is balanced, this integer program can be solved as a linear program, as a consequence of Theorem 4.2. Therefore, in this case, the nonlinear 0, 1 maximization problem can be solved in polynomial time. The relevance of balancedness in this context was pointed out by Crama (1993).

9 Balanced hypergraphs

A 0, 1 matrix A can be represented by a hypergraph (the columns of A represent nodes and the rows represent edges). Then the definition of balancedness for 0, 1 matrices is a natural extension of the property of not containing odd cycles for graphs. In fact, this is the motivation that led Berge (1970) to introduce the notion of balancedness: A hypergraph H is *balanced* if every odd cycle C of H has an edge containing at least three nodes of C. We refer to Berge (1989) for an introduction to the theory of hypergraphs. Several results on bipartite graphs generalize to balanced hypergraphs, such as König's bipartite matching theorem, as stated in the next theorem. In a hypergraphs, a *matching* is a set of pairwise nonintersecting edges and a *transversal* is a node set intersecting all the edges.

Theorem 9.1. [Berge and Las Vergnas (1970)] *In a balanced hypergraph, the maximum cardinality of a matching equals the minimum cardinality of a transversal.*

Proof. Follows form Theorem 4.1 applied with $A_1 = A_3 = \emptyset$ and the primal objective function $\max \sum_j x_j$. □

The next result generalizes a theorem of Gupta (1978) on bipartite multigraphs.

Theorem 9.2. [Berge (1980)] *In a balanced hypergraph, the minimum number of nodes in an edge equals the maximum cardinality of a family of disjoint transversals.*

One of the first results on matchings in graphs is the following celebrated theorem of Hall.

Theorem 9.3. [Hall (1935)] *A bipartite graph has no perfect matching if and only if there exist disjoint node sets R and B such that $|B| > |R|$ and every edge having one endnode in B has the other in R.*

The following result generalizes Hall's theorem to balanced hypergraphs.

Theorem 9.4. [Conforti, Cornuéjols, Kapoor, and Vušković (1996)] *A balanced hypergraphs has no perfect matching if and only if there exist disjoint node sets R and B such that $|B| > |R|$ and every edge contains at least as many nodes in R as in B.*

The proof of Theorem 9.4 uses integrality properties of some polyhedra associated with balanced 0, 1 $m \times n$ matrix A. Let a^i denote the ith row of A, I the identity matrix.

Lemma 9.5. *The polyhedron $P = \{x, s, t \mid Ax + Is - It = \mathbf{1},\ x, s, t \geq 0\}$ is integral when A is a balanced* 0,1 *matrix.*

Proof. Let $\bar{x}, \bar{s}, \bar{t}$ be a vertex of P. Then $\bar{s}_i \bar{t}_i = 0$ for $i = 1, \ldots, m$ since the corresponding columns are linearly dependent. Let $Q = \{x | a^i x \geq 1$, if $\bar{t}_i > 0$, $a^i x \leq 1$, if $\bar{s}_i > 0$, $a^i x = 1$, otherwise, $x \geq 0\}$. By Theorem 4.1, Q is an integer polyhedron. Since $\bar{x}$ is a vertex of Q, then $\bar{x}$ is an integral vector and so are $\bar{s}$ and $\bar{t}$. □

Lemma 9.6. *The linear system* $Ax + Is - It = \mathbf{1}, x, s, t, \geq 0$ *is TDI when* A *is a balanced* 0, 1 *matrix.*

Proof. Consider the linear program:

$$\begin{aligned} \max \quad & bx + cs + dt \\ & Ax + Is - It = \mathbf{1} \\ & x, s, t \geq 0 \end{aligned} \tag{12}$$

and its dual:

$$\begin{aligned} \min \quad & y\mathbf{1} \\ & yA \geq b \\ & \;\; y \geq c \\ & -y \geq d. \end{aligned} \tag{13}$$

Let A be a 0, 1 balanced matrix with smallest number of rows such that the lemma does not hold. Then there exist integral vectors b, c, d, such that an optimal solution of (13), say $\bar{y}$, has a fractional component $\bar{y}_i$. Consider the following linear program:

$$\begin{aligned} \min \quad & y\mathbf{1} \\ & yA^i \geq b - \lceil \bar{y}_i \rceil a^i \\ & \;\; y \geq c^i \\ & -y \geq d^i \end{aligned} \tag{14}$$

where A^i denotes the matrix obtained from A by removing row a^i, and where c^i and d^i denote the vectors obtained from c and d respectively by removing the ith component. Let $\tilde{y} = (\tilde{y}_1, \ldots, \tilde{y}_{i-1}, \tilde{y}_{i+1}, \ldots, \tilde{y}_m)$ be an optimal integral solution of (14). Define $y^* = (\tilde{y}_1, \ldots, \tilde{y}_{i-1}, \lceil \bar{y}_i \rceil, \tilde{y}_{i+1}, \ldots, \tilde{y}_m)$. Then y^* is integral and feasible to (13). We claim that y^* is in fact optimal to (13). To prove this claim, note that $(\bar{y}_1, \ldots, \bar{y}_{i-1}, \bar{y}_{i+1}, \ldots, \bar{y}_m)$ is feasible to (14). Therefore

$$\sum_{k \neq i} \tilde{y}_k \leq \sum_{k \neq i} \bar{y}_k.$$

In fact,

$$\sum_{k\neq i}\bar{y}_k - \sum_{k\neq i}\tilde{y}_k \geq \lceil \bar{y}_i \rceil - \bar{y}_i$$

because $\sum_{k\neq i}\bar{y}_k + \bar{y}_i$ is an integer by Lemma 9.5 and $\bar{y}_i$ is fractional. So

$$\sum_{k\neq i}\tilde{y}_k + \lceil \bar{y}_i \rceil \leq \sum_{k=1}^{m}\bar{y}_k,$$

i.e., y^* is an optimal integral solution to (13), and so the lemma must hold.□

Proof of Theorem 9.4. Let A be the node-edge incidence matrix of a balanced hypergraphs H. Then by Lemma 9.5, H has no perfect matching if and only if the objective value of the linear program

$$\begin{aligned} \max \quad & 0x - \mathbf{1}s - \mathbf{1}t \\ & Ax + Is - It = \mathbf{1} \\ & x, s, t \geq 0 \end{aligned} \tag{15}$$

is strictly negative. By Lemma 9.6, this occurs if and only if there exists an integral vector y such that

$$\begin{aligned} & y\mathbf{1} < 0 \\ & yA \geq 0 \\ & -\mathbf{1} \leq y \leq \mathbf{1}. \end{aligned} \tag{16}$$

Let B denote the set of nodes i such that $y_i = -1$ and R the set of nodes such that $y_i = 1$.

Then $yA \geq 0$ implies that each edge of H contains at least as many nodes in R as in B, and $y\mathbf{1} < 0$ implies $|B| > |R|$. □

It is well known that a bipartite graph with maximum degree Δ contains Δ edge-disjoint matchings. The same property holds for balanced hypergraphs. This result can be proved using Theorem 9.4.

Corollary 9.7. *The edges of a balanced hypergraph H with maximum degree Δ can be partitioned into Δ matchings.*

Proof. By adding edges containing a unique node, we can assume that H is Δ-regular. (This operation does not destroy the property of being balanced). We now show that H has a perfect matching. Assume not. By Theorem 9.4, there exist disjoint node sets R and B such that $|B| > |R|$ and $|R \cap E| \geq |B \cap E|$ for every edge E of H. Adding these inequalities over all the edges, we get $|R| \geq |B|$ since H is Δ-regular, a contradiction. So H contains a perfect matching M. Removing the edges of M, the result now follows by induction. □

10 Bipartite representation

In an undirected graph G, a cycle is *balanced* if its length is a multiple of 4. The graph G is *balanced* if all its chordless cycles are balanced. Clearly, a balanced graph is simple and bipartite.

Given a 0, 1 matrix A, the *bipartite representation of A* is the bipartite graph $G(A) = (V^r \cup V^c, E)$ having a node in V^r for every row of A, a node in V^c for every column of A and an edge ij joining nodes $i \in V^r$ and $j \in V^c$ if and only if the entry a_{ij} of A equals 1.

Note that a 0, 1 matrix is balanced if and only if its bipartite representation is a balanced graph.

Given a $0, \pm 1$ matrix A, the *bipartite representation of A* is the weighted bipartite graph $G(A) = (V^r \cup V^c, E)$ having a node in V^r for every row of A, a node in V^c for every column of A and an edge ij joining nodes $i \in V^r$ and $j \in V^c$ if and only if the entry a_{ij} is nonzero. Furthermore a_{ij} is the weight of the edge ij. This concept extends the one introduced above. Conversely, given a bipartite graph $G = (V^r \cup V^c, E)$, with weights ± 1 on its edges, there is a unique matrix A for which $G = G(A)$ (up to transposition of the matrix, permutations of rows and columns).

11 Totally balanced 0,1 matrices

In this section, statements about a 0, 1 matrix A are formulated in terms of its bipartite representation $G(A)$, whenever this is more convenient. A bipartite graph is *totally balanced* if every hole has length 4. Totally balanced bipartite graphs arise in location theory and were the first balanced graphs to be the object of an extensive study. Several authors (Golumbic and Goss, 1978; Anstee and Farber, 1984; Hoffman, Kolen, and Sakarovitch, 1985 among others) have given properties of these graphs.

A *biclique* is a complete bipartite graph with at least one node from each side of the bipartition. For a node u, let $N(u)$ denote the set of all neighbors of u. An edge uv is *bisimplicial* if the node set $N(u) \cup N(v)$ induces a biclique. The following theorem of Golumbic and Goss (1978) characterizes totally balanced bipartite graphs.

Theorem 11.1. [Golumbic and Goss (1978)] *A totally balanced bipartite graph has a bisimplicial edge.*

A 0, 1 matrix A is in *standard greedy form* if it contains no 2×2 submatrix of the form

$$\begin{pmatrix} 1 & 1 \\ 1 & 0 \end{pmatrix},$$

where the order of the rows and columns in the submatrix is the same as in the matrix A. This name comes from the fact that the linear program

$$\begin{aligned} \max \quad & \sum y_i \\ & yA \leq c \\ & 0 \leq y \leq p \end{aligned} \tag{17}$$

can be solved by a greedy algorithm. Namely, given $y_1, \ldots, y_{k-1}$ such that $\sum_{i=1}^{k-1} a_{ij} y_i \leq c_j$, $j = 1, \ldots, n$ and $0 \leq y_i \leq p_i$, $i = 1, \ldots, k-1$, set y_k to the largest value such that $\sum_{i=1}^{k} a_{ij} y_i \leq c_j$, $j = 1, \ldots, n$ and $0 \leq y_k \leq p_k$. The resulting greedy solution is an optimum solution to this linear program. What does this have to do with totally balanced matrices? The answer is in the next theorem.

Theorem 11.2. [Hoffman et al. (1985)] *A* 0, 1 *matrix is totally balanced if and only if its rows and columns can be permuted in standard greedy form.*

This transformation can be performed in time $O(nm^2)$ (Hoffman et al., 1985).

Totally balanced 0, 1 matrices come up in various ways in the context of facility location problems on trees. For example, the *covering problem*

$$\begin{aligned} \min \quad & \sum_1^n c_j x_j + \sum_1^m p_i z_i \\ & \sum_j a_{ij} x_j + z_i \geq 1, \quad i = 1, \ldots, m \\ & x_j, z_i \in \{0, 1\} \end{aligned} \tag{18}$$

can be interpreted as follows: c_j is the setup cost of establishing a facility at site j, p_i is the penalty if client i is not served by any facility, and $a_{ij} = 1$ if a facility at site j can serve client i, 0 otherwise.

When the underlying network is a tree and the facilities and clients are located at nodes of the tree, it is customary to assume that a facility at site j can serve all the clients in a *neighborhood subtree* of j, namely, all the clients within distance r_j from node j.

An *intersection matrix* of the set $\{S_1, \ldots, S_m\}$ vs. $\{R_1, \ldots, R_n\}$, where S_i, $i = 1, \ldots, m$, and R_j, $j = 1, \ldots, n$, are subsets of a given set, is defined to be the $m \times n$ 0, 1 matrix $A = (a_{ij})$ where $a_{ij} = 1$ if and only if $S_i \cap R_j \neq \emptyset$.

Theorem 11.3. [Giles (1978)] *The intersection matrix of neighborhood subtrees versus nodes of a tree is totally balanced.*

It follows that the above location problem on trees (18) can be solved as a linear program (by Theorem 2.1 and the fact that totally balanced matrices are balanced). In fact, by using the standard greedy form of the neighborhood subtrees versus nodes matrix, and by noting that (18) is the dual of (17), the greedy solution described earlier for (17) can be used, in conjunction with complementary slackness, to obtain an elegant solution of the covering problem. The above theorem of Giles has been generalized as follows.

Theorem 11.4. [Tamir (1983)] *The intersection matrix of neighborhood subtrees versus neighborhood subtrees of a tree is totally balanced.*

Other classes of totally balanced 0, 1 matrices arising from location problems on trees can be found in (Tamir, 1987).

12 Signing 0, 1 matrices

A 0, 1 matrix is *balanceable* if its nonzero entries can be signed $+1$ or -1 so that the resulting $0, \pm 1$ matrix is balanced. A bipartite G graph is *balanceable* if $G = G(A)$ and A is a balanceable matrix.

Camion (1965) observed that the signing of a balanceable matrix into a balanced matrix is unique up to multiplying rows or columns by -1, and he gave a simple algorithm to obtain this signing. We present Camion's result next.

Let A be a $0, \pm 1$ matrix and let A' be obtained from A by multiplying set S of rows and columns by -1. A is balanced if and only if A' is. Note that, in the bipartite representation of A, this corresponds to switching signs on all edges of the cut $(S, \bar{S})$. Now let R be a 0, 1 matrix and $G(R)$ is its bipartite representation. Since every edge of a maximal forest F of $G(R)$ is contained in a cut that does not contain any other edge of F, it follows that if R is balanceable, there exists a balanced signing of R in which the edges of F have any specified (arbitrary) signing.

This implies that, if a 0, 1 matrix A is balanceable, one can find a balanced signing of A as follows.

12.1 Camion's signing algorithm

Input. *A balanceable* 0, 1 *matrix* A *and its bipartite representation* $G(A)$, *a maximal forest* F *of* $G(A)$ *and an arbitrary signing of the edges of* F.
Output. *The unique balanced signing of* $G(A)$ *such that the edges of* F *are signed as specified in the input.*

Index the edges of G $e_1, \ldots, e_n$, so that the edges of F come first, and every edge $e_j, j \geq |F| + 1$, together with edges having smaller indices, closes a chordless cycle H_j of G. For $j = |F| + 1, \ldots, n$, sign e_j so that the sum of the weights of H_j is congruent to 0 *mod* 4.

Note that the rows and columns corresponding to the nodes of H_j define a hole submatrix of A.

The fact that there exists an indexing of the edges of G as required in the signing algorithm follows from the following observation. For $j \geq |F| + 1$, we can select e_j so that the path connecting the endnodes of e_j in the subgraph $(V(G), \{e_1, \ldots, e_{j-1}\})$ is the shortest possible one. The chordless cycle H_j identified this way is also a chordless cycle in G. This forces the signing of e_j, since all the other edges of H_j are signed already. So, once the (arbitrary) signing of F has been chosen, the signing of G is unique. Therefore we have the following results.

Theorem 12.1. *If the input matrix A is a balanceable* $0, 1$ *matrix, Camion's signing algorithm produces a balanced* $0, \pm 1$ *matrix B. Furthermore every balanced* $0, \pm 1$ *matrix that arises from A by signing its nonzero entire either* $+1$ *or* -1, *can be obtained by switching signs on rows and columns of B.*

One can easily check (using Camion's algorithm, for example) that the following matrix is not balanceable.

$$\begin{pmatrix} 1 & 1 & 0 & 1 \\ 1 & 0 & 1 & 1 \\ 0 & 1 & 1 & 1 \end{pmatrix}$$

Assume that we have an algorithm to check if a bipartite graph is balanceable. Then, we can check whether a weighted bipartite graph G is balanced as follows. Let G' be an unweighted copy of G. Test whether G' is balanceable. If it is not, then G is not balanced. Otherwise, let F be a maximal forest of G'. Run the signing algorithm on G' with the edges of F signed as they are in G. Then G is balanced if and only if the signing of G' coincides with the signing of G.

13 Truemper's theorem

In a bipartite graph, a *wheel* (H, v) consists of a hole H and a node v having at least three neighbors in H. The wheel (H, v) is *odd* if v has an odd number of neighbors in H. A *3-path configuration* is an induced subgraph consisting of three internally node-disjoint paths connecting two nonadjacent nodes u and v and containing no edge other than those of the paths. If u and v are in

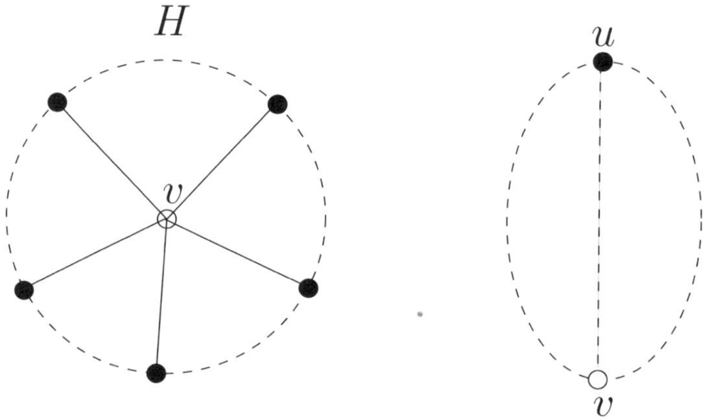

Fig. 1. An odd wheel and a 3-odd-path configuration.

opposite sides of the bipartition, i.e., the three paths have an odd number of edges, the 3-path configuration is called a *3-odd-path configuration.* In Fig. 1, solid lines represent edges and dotted lines represent paths with at least one edge.

Both a 3-odd-path configuration and an odd wheel have the following properties: each edge belongs to exactly two holes and the total number of edges is odd. Therefore in any signing, the sum of the labels of all holes is equal to 2 *mod* 4. This implies that at least one of the holes is not balanced, showing that neither 3-odd-path configurations nor odd wheels are balanceable. These are in fact the only minimal bipartite graphs that are not balanceable, as a consequence of a theorem of Truemper (1992).

Theorem 13.1. [Truemper (1992)] *A bipartite graph is balanceable if and only if it does not contain an odd wheel or a 3-odd-path configuration as an induced subgraph.*

We prove Theorem 13.1 following Conforti, Gerards, and Kapoor (2000).

For a connected bipartite graph G that contains a clique cutset K_t with t nodes, let $G'_1, \ldots, G'_n$ be the connected components of $G \backslash K_t$. The *blocks* of G are the subgraphs G_i induced by $V(G'_i) \cup K_t$ for $i = 1, \ldots, n$.

Lemma 13.2. *If a connected bipartite graph G contains a K_1 or K_2 cutset, then G is balanceable if and only if each block is balanceable.*

Proof. If G is balanceable, then so are the blocks. Therefore we only have to prove the converse. Assume that all the blocks are balanceable. Give each block a balanced signing. If the cutset is a K_1 cutset, this yields a balanced signing of G. If the cutset is a K_2 cutset, resign each block so that the edge of that K_2 has the sign $+1$. Now take the union of these signings. This yields a balanced signing of G again. □

Thus, in the remainder of the proof, we can assume that G is a connected bipartite graph with no K_1 or K_2 cutset.

Lemma 13.3. *Let H be a hole of G. If $G \neq H$, then H is contained in a 3-path configuration or a wheel of G.*

Proof. Choose two nonadjacent nodes u and w in H and a uw-path $P=u, x, \ldots, z, w$ whose intermediate nodes are in $G\backslash H$ such that P is as short as possible. Such a pair of nodes u, w exists since $G \neq H$ and G has no K_1 or K_2 cutset. If $x=z$, then H is contained in a 3-path configuration or a wheel. So assume $x \neq z$. By our choice of P, u is the only neighbor of x in H and w is the only neighbor of z in H.

Let Y be the set of nodes in $V(H) - \{u,w\}$ that have a neighbor in P. If Y is empty, H is contained in a 3-path configuration. So assume Y is nonempty. By the minimality of P, the nodes of Y are pairwise adjacent and they are adjacent to u and w. This implies that Y contains a single node y and the y is adjacent to u and w. But then $V(H) \cup V(P)$ induces a wheel with center y.

For $e \in E(G)$, let G^e denote the graph with a node v_H for each hole H of G containing e and an edge $v_{H_i}v_{H_j}$ if and only if there exists a wheel or a 3-path configuration containing both holes H_i and H_j.

Lemma 13.4. *G^e is a connected graph.*

Proof. Suppose not. Let $e=uw$. Choose two holes H_1 and H_2 of G with v_{H_1} and v_{H_2} in different connected components of G^e, with the minimum distance $d(H_1, H_2)$ in $G\backslash\{u, v\}$ between $V(H_1) - \{u, w\}$ and $V(H_2) - \{u, w\}$ and, subject to this, with the smallest $|V(H_1) \cup V(H_2)|$.

Let T be a shortest path from $V(H_1) - \{u, v\}$ to $V(H_2) - \{u, v\}$ in $G\backslash\{u, v\}$. Note that T is just a node of $V(H_1) \cap V(H_2)\backslash\{u, v\}$ when this set is nonempty. The graph G' induced by the nodes in H_1, H_2, and T has no K_1 or K_2 cutset. By Lemma 13.3, H_1 is contained in a 3-path configuration or a wheel of G'. Since each edge of a 3-path configuration or a wheel belongs to two holes, there exists a hole $H_3 \neq H_1$ containing edge e in G'. Since v_{H_1} and v_{H_3} are adjacent in G^e, it follows that v_{H_2} and v_{H_3} are in different components of G^e. Since H_1 and H_3 are distinct holes, H_3 contains a node in $V(H_2) \cup V(T)\backslash V(H_1)$. If H_3 contains a node in $V(T)\backslash(V(H_1) \cup V(H_2))$, then $V(H_1) \cap V(H_2) = \{u, v\}$ and $d(H_3, H_2) < d(H_1, H_2)$ a contradiction to the choice of H_1, H_2.

Therefore H_3 contains a node x in $V(H_2)\backslash V(H_1)$. By our choice of H_1, H_2, we have that $V(H_1) \cap V(H_2)\backslash\{u, v\}$ is nonempty. Let $P_1 = H_1\backslash e$ and $P_2 = H_2\backslash e$ and let s, t be the nodes in $V(H_1) \cap V(H_2)$ such that the st-subpath P_2^{st} of P_2 contains x and is shortest. Let P_1^{st} be the st-subpath of P_1. Since H_2 is hole, P_1^{st} contains an intermediate node $z \in V(H_1)\backslash V(H_2)$. Now $V(H_3) \cup V(H_2)$ is contained in $V(H_1) \cup V(H_2)\backslash z$, a contradiction to our choice of H_1, H_2. □

Proof of Theorem 13.1. We showed already that odd wheels and 3-odd-path configurations are not balanceable. It remains to show that, conversely, if G contains no odd wheel or 3-odd-path configuration, then G is balanceable. Suppose G is a counterexample with the smallest number of nodes. By Lemma 13.2, G is connected and has no K_1 or K_2 cutset. Let $e = uv$ be an edge of G. Since $G \backslash \{u, v\}$ is connected, there exists a spanning tree F of G where u and v are leaves. Arbitrarily sign F and use Camion's signing algorithm in $G \backslash \{u\}$ and $G \backslash \{v\}$. By the minimality of G, these two graphs are balanceable and therefore Camion's algorithm yields a unique signing of all the edges except e. Furthermore, all holes not going through edge e are balanced. Since G is not balanceable, any signing of e yields some holes going through e that are balanced and some that are not. By Lemma 13.4, there exists a wheel or a 3-path configuration C containing an unbalanced hole H_1 and a balanced hole H_2 both going through edge e. Now we use the fact that each edge of C belongs to exactly two holes of C. Since the holes of C, distinct form H_1 and H_2 do not go through e, they are balanced. Furthermore, applying the above fact to all edges of C, the sum of all labels in C is 1 mod 2, which implies that C has an odd number of edges. Thus C is an odd wheel or a 3-odd-path configuration, a contradiction. □

14 Decomposition theorem

In this section, we present a decomposition theorem for balanced $0, \pm 1$ matrices due to Conforti, Cornuéjols, and Rao (1999) and Conforti et al. (2001), and we give an outline of its proof.

By the result of Section 12, it suffices to decompose balanceable 0, 1 matrices. We state the decomposition theorem in terms of the bipartite representation, as defined in Section 10.

14.1 Cutsets

A set S of nodes (edges) of a connected graph G is a *node* (*edge*) *cutset* if the subgraph of G obtained by removing the nodes (edges) in S, is disconnected.

For a node x, let $N(x)$ denote the set of all neighbors of x. In a bipartite graph, an *extended star* is defined by disjoint subsets T, A, N of $V(G)$ and a node $x \in T$ such that

(i) $A \neq \emptyset$ and $A \cup N \subseteq N(x)$,
(ii) Every node of A is adjacent to every node of T,
(iii) If $|T| \geq 2$, then $|A| \geq 2$.

This concept was introduced by Conforti et al. (1999) and is illustrated in Fig. 2. An *extended star cutset* is one where $T \cup A \cup N$ is a node cutset. An extended star cutset with $N = \emptyset$ is called a *biclique cutset*. An extended star

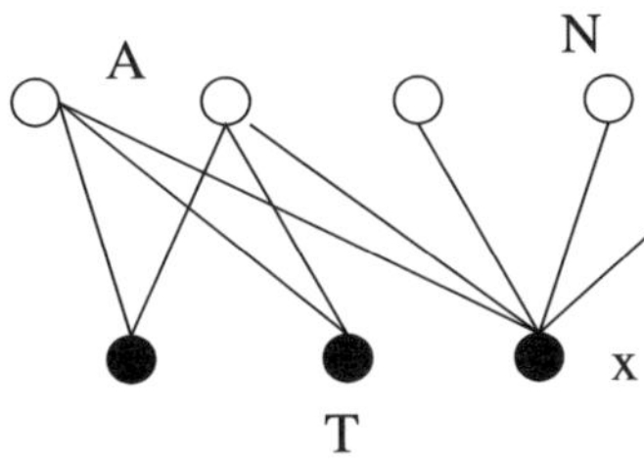

Fig. 2. Extended star.

cutset having $T=\{x\}$ is called a *star cutset*. Note that a star cutset is a special case of a biclique cutset.

A graph G has a 1-*join* if its nodes can be partitioned into sets H_1 and H_2 with $|H_1|\geq 2$ and $|H_2|\geq 2$, so that $A_1\subseteq H_1$, $A_2\subseteq H_2$ are nonempty, all nodes of A_1 are adjacent to all nodes of A_2 and these are the only adjacencies between H_1 and H_2. This concept was introduced by Cunningham and Edmonds (1980).

A graph G has 2-*join* if its nodes can be partitioned into sets H_1 and H_2 so that $A_1, B_1 \subseteq H_1, A_2, B_2 \subseteq H_2$ where A_1, B_2, A_2, B_2 are nonempty and disjoint, all nodes of A_1 are adjacent to all nodes of A_2, all nodes of B_1 are adjacent to all nodes of B_2 and these are the only adjacencies between H_1 and H_2. Also, for $i=1,2$, H_i has at least one path from A_i to B_i and if A_i and B_i are both of cardinality 1, then the graph induced by H_i is not a chordless path. We also say that $E(K_{A_1A_2})\cup E(K_{B_1B_2})$ is a 2-join of G. This concept was introduced by Cornuéjols and Cunningham (1985).

In a connected bipartite graph G, let $A_i, i=1,\ldots,6$, be disjoint nonempty node sets such that, for each i, every node in A_i is adjacent to every node in $A_{i-1}\cup A_{i+1}$ (indices are taken modulo 6), and these are the only edges in the subgraph A induced by the node set $\cup_{i=1}^{6} A_i$. Assume that $E(A)$ is an edge cutset but that no subset of its edges forms a 1-join or a 2-join. Furthermore assume that no connected component of $G\backslash E(A)$ contains a node in $A_1\cup A_3\cup A_5$ and a node in $A_2\cup A_4\cup A_6$. Let G_{135} be the union of the components of $G\backslash E(A)$ containing a node in $A_1\cup A_3\cup A_5$ and G_{246} be the union of components containing a node in $A_2\cup A_4\cup A_6$. The set $E(A)$ constitutes a 6-*join* if the graphs G_{135} and G_{246} contain at least four nodes each (Fig. 3). This concept was introduced by Conforti et al. (2001).

14.2 Main theorem

A graph is *strongly balanceable* if it is balanceable and contains no cycle with exactly one chord. This class of bipartite graphs is well studied in the literature, see Conforti and Rao (1987). We discuss it in a later section. R_{10} is the bipartite graph on ten nodes defined by the cycle $x_1,\ldots,x_{10},x_1$ of length ten with chords x_ix_{i+5}, $1\leq i\leq 5$, see Fig. 4.

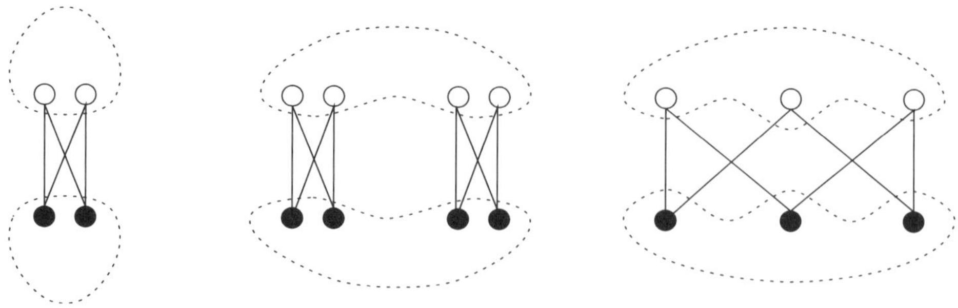

Fig. 3. A 1-join, a 2-join, and a 6-join.

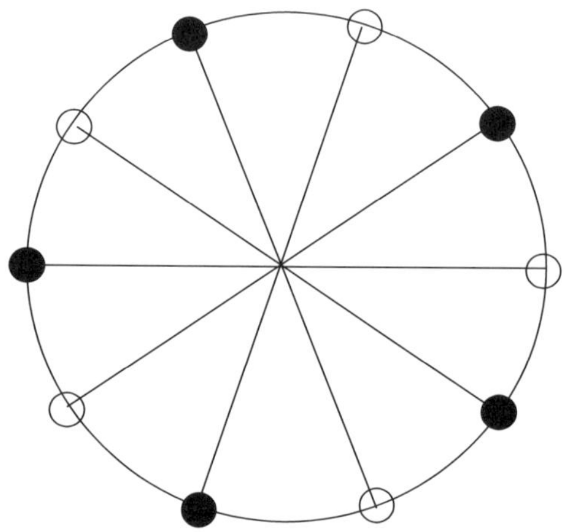

Fig. 4. R_{10}.

Theorem 14.1. [Conforti et al. (2001)] *A balanceable bipartite graph that is not strongly balanceable is either* R_{10} *or contains a* 2*-join, a* 6*-join or an extended star cutset.*

14.3 Outline of the proof

The key idea in the proof of Theorem 14.1 is that if a balanceable bipartite graph G is not strongly balanceable or R_{10}, then G contains one of several induced subrgraphs, which force a decomposition of G with one of the cutsets described in Section 14.1.

14.3.1 Parachutes

A *parachute* is defined by four chordless paths of positive lengths, $T = v_1, \ldots, v_2$; $P_1 = v_1, \ldots, z$; $P_2 = v_2, \ldots, z$; $M = v, \ldots, z$, where v_1, v_2, v, z

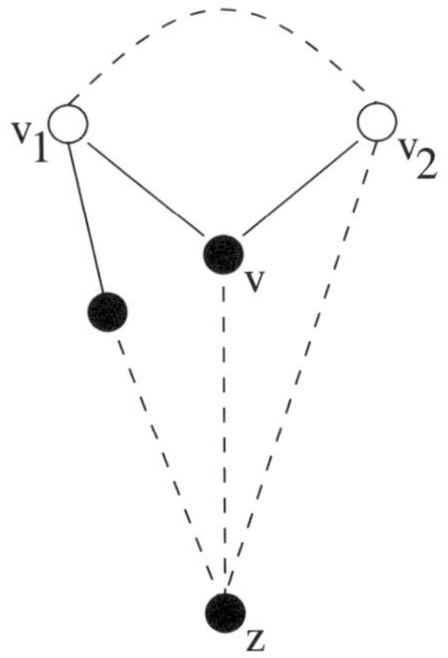

Fig. 5. Parachute.

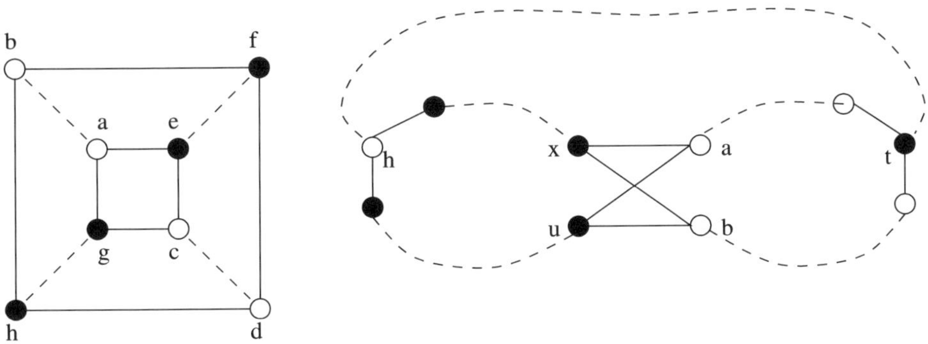

Fig. 6. Connected squares and goggles.

are distinct nodes, and two edges vv_1 and vv_2. No other edges exist in the parachute, except the ones mentioned above. Furthermore $|E(P_1)| + |E(P_2)| \geq 3$ (See Fig. 5).

Note that if G is balanceable then nodes v, z belong to the same side of the bipartition, else the parachute contains a 3-path configuration connecting v and z or an odd wheel (H, v) with three spokes.

14.3.2 Connected squares and goggles

Connected squares are defined by four chordless paths $P_1 = a, \ldots, b$; $P_2 = c, \ldots, d$; $P_3 = e, \ldots, f$; $P_4 = g, \ldots, h$, where nodes a and c are adjacent to both e and g, and b and d are adjacent to both f and h, as in Fig. 6. No other adjacency exists in the connected squares. Note that nodes a and b belong to the same side of the bipartition, else the connected squares contain a 3-path configuration connecting a and b or, if $|E(P_1)| = 1$, an odd wheel with center a. Therefore the nodes a, b, c, d are on one side of the bipartition and e, f, g, h are on the other.

Goggles are defined by a cycle $C = h, P, x, a, Q, t, R, b, u, S, h$, with two chords ua and xb, and chordless paths P, Q, R, S of length greater that one, and a chordless path $T = h, \ldots, t$ of length at least one, such that no intermediate node of T belongs to C. Not other edge exists, connecting nodes of the goggles, see Fig. 6.

14.3.3 Connected 6-holes

A triad consists of three internally node-disjoint paths $t, \ldots, u$; $t, \ldots, v$ and $t, \ldots, w$, where t, u, v, w are distinct nodes and u, v, w belong to the same side of the bipartition. Furthermore, the graph induced by the nodes of the triad contains no other edges than those of the three paths. Nodes u, v and w are called the *attachments* of the triad.

A *fan* consists of a chordless path $x, \ldots, y$ together with a node z adjacent to at least one node of the path, where x, y, and z are distinct nodes all belonging to the same side of the bipartition. Nodes x, y, and z are called the *attachments* of the fan.

A *connected* 6*-hole* Σ is a graph induced by two disjoint node sets $T(\Sigma)$ and $B(\Sigma)$ such that each induces either a triad or a fan, the attachments of $T(\Sigma)$ and $B(\Sigma)$ induce a 6-hole and there are no other adjacencies between the nodes of $T(\Sigma)$ and $B(\Sigma)$. Figure 7 depicts the four types of connected 6-holes.

The following theorem concerns the class of balanceable bipartite graphs that do not contain a connected 6-hole or R_{10} as induced subgraph.

Theorem 14.2. [Conforti et al. (1999)] *A balanceable bipartite graph not containing R_{10} or a connected 6-hole as induced subgraph either is strongly balanceable or contains a 2-join or an extended star cutset.*

The proof of this theorem involves the following intermediate results.

Theorem 14.3. *Let G be a balanceable bipartite graph that is not strongly balanceable. If G contains no wheel or parachute as induced subgraph, then G has a 2-join.*

Theorem 14.4. *Let G be a balanceable bipartite graph. If G contains a wheel but no connected 6-hole as induced subgraph, then G has an extended star cutset.*

Theorem 14.5. *Let G be a balanceable bipartite graph that is not strongly balanceable. If G contains a parachute but no wheel, no R_{10} and no connected 6-hole as induced subgraph, then G has an extended star cutset or G contains connected squares or goggles as induced subgraph.*

Theorem 14.6. *Let G be a balanceable bipartite graph. If G contains connected squares but no wheel as induced subgraph, then G has a biclique cutset or a 2-join.*

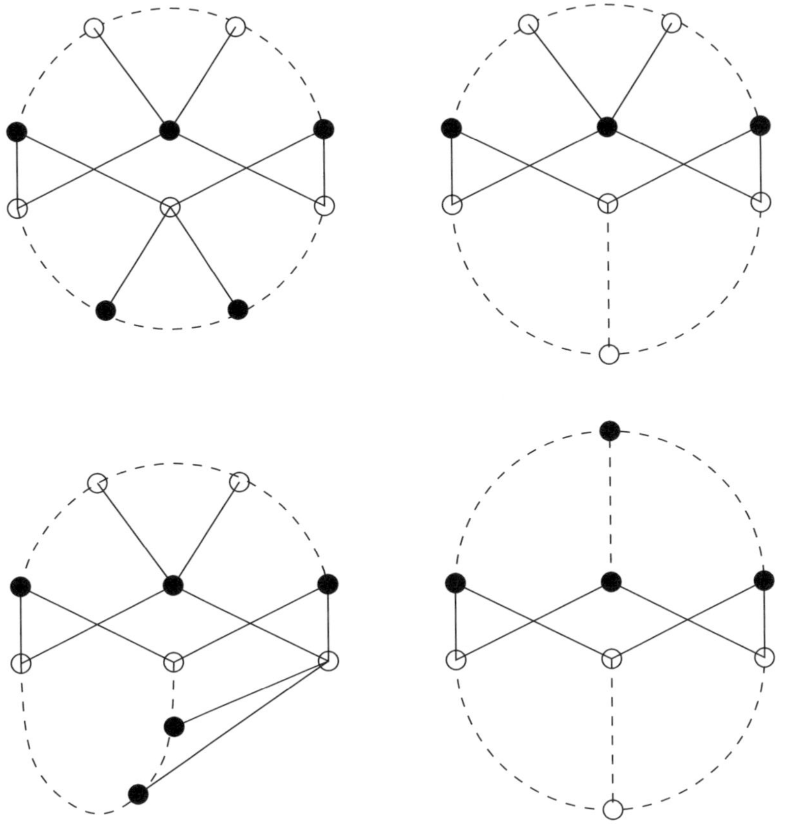

Fig. 7. The four types of connected 6-holes.

Theorem 14.7. *Let G be a balanceable bipartite graph. If G contains goggles but no wheel, no* R_{10} *and no connected* 6*-hole as induced subgraph, then G has an extended star cutset or a* 2*-join.*

Together, these results prove Theorem 14.2. So it remains to find a decomposition of balanceable bipartite graphs that contain R_{10} or connected 6-holes as induced subgraph. This is accomplished as follows.

Theorem 14.8. [Conforti et al. (2001)] *A balanceable bipartite graph containing* R_{10} *as a proper induced subgraph has a biclique cutset.*

Theorem 14.9. [Conforti et al. (2001)] *A balanceable bipartite graph that contains a connected* 6*-hole as induced subgraph, has an extended star cutset or a* 6*-join.*

Now Theorem 14.1 follows from Theorems 14.2, 14.8 and 14.9.

15 Recognition algorithm

Conforti et al. (2001) give a polynomial time algorithm to check whether a $0, \pm 1$ matrix A is balanced. We describe the recognition algorithm using the bipartite representation $G(A)$ introduced in Section 10. Since each edge of $G(A)$ is signed $+1$ or -1 according to the corresponding entry in the matrix A, we call G a *signed bipartite graph.*

15.1 Balancedness preserving decomposition

Let G be a connected signed bipartite graph. The removal of a node or edge cutset disconnects G into two or more connected components. From these components we construct *blocks* by adding some new nodes and signed edges. We say that a decomposition is *balancedness preserving* when it has the following property: all the blocks are balanced if and only if G itself is balanced. The central idea in the algorithm is to decompose G using balancedness preserving decompositions into a polynomial number of basic bocks that can be checked for balancedness in polynomial time.

For the 2-join and 6-join, the blocks can be defined so that the decompositions are balancedness preserving. For the extended star cutset this is not immediately possible.

15.1.1 2-Join decomposition

Let $E(K_{A_1A_2}) \cup E(K_{B_1B_2})$ be a 2-join of G and let H_1 and H_2 be the sets defined in Section 14.1. We construct the block G_1 from H_1 as follows:

- Add two nodes a_2 and b_2, connected respectively to all nodes in A_1 and to all nodes in B_1.
- Let P be a shortest path in the graph induced by H_2 connecting a node in A_2 to a node in B_2. If the weight of P is 0 or 2 mod 4, nodes d and f are connected by a path of length 2 in G_1. If the weight of P is 0 mod 4, one edge of Q is signed $+1$ and the other -1, and if the weight of P is 2 mod 4, both edges of Q are signed $+1$. Similarly if the weight of P is 1 or 3 mod 4, nodes a_2 and b_2 are connected by a path of length 3 with edges signed so that Q and P have the same weight modulo 4. Let α and β be the endnodes of P in A_2 and B_2 respectively. Sign the edges between node a_2 and the nodes in A_1 exactly the same as the corresponding edges between α and the nodes of A_1 in G. Similarly, sign the edges between b_2 and B_1 exactly the same as the corresponding edges between β and the nodes in B_1.

The block G_2 is defined similarly.

Theorem 15.1. *Let G be a signed bipartite graph with a 2-join $E(K_{A_1A_2}) \cup E(K_{B_1B_2})$ where $K_{A_1A_2}$ and $K_{B_1B_2}$ are balanced and neither $A_1 \cup B_1$ nor $A_2 \cup B_2$*

induces a biclique. Then G is balanced if and only if both blocks G_1 and G_2 are balanced.

15.1.2 6-Join decomposition

Let G be a signed bipartite graph and let $A_1, \ldots, A_6$ be disjoint nonempty node sets such that the edges of the graph A induced by $\cup_{i=1}^{6} A_i$ form a 6-join. Let G_{135} and G_{246} be the graphs defined earlier, in the definition of 6-join. We construct the block G_1 from G_{135} as follows:

- Add a node a_2 adjacent to all the nodes in A_1 and A_3, a node a_4 adjacent to all the nodes in A_3 and A_5, and a node a_6 adjacent to all the nodes in A_5 and A_1.
- Pick any three nodes $a'_2 \in A_2$, $a'_4 \in A_4$ and $a'_6 \in A_6$ and, in G_1, sign the edges incident with a_2, a_4, and a_6 according to the signs of the corresponding edges of G incident with a'_2, a'_4, and a'_6.

The block G_2 is defined similarly.

Theorem 15.2. *Let G be a signed bipartite graph with a 6-join $E(A)$ such that A is balanced. Then G is balanced if and only if both blocks G_1 and G_2 are balanced.*

15.2 Extended star cutset decomposition

Consider the following way of defining the blocks for the extended star decomposition of a connected signed bipartite graph G. Let S be an extended star cutset of G and $G'_1, \ldots, G'_k$ the connected components of $G \backslash S$. Define the blocks to be $G_1, \ldots, G_k$ where G_i is the subgraph of G induced by $V(G'_i) \cup S$ with all edges keeping the same sign as in G.

The extended star decomposition defined in this way is not balancedness preserving. Consider, for example, a signed odd wheel (H, x) where H is an *unbalanced hole* (a hole of weight congruent to 2 mod 4). If we decompose (H, x) by the extended star cutset $\{x\} \cup N(x)$, then it is possible that all of the blocks are balanced, whereas (H, x) itself is not since H is an unbalanced hole. Two other classes of bipartite graphs that can present a similar problem when decomposing with an extended star cutset are tents and short 3-odd-path configurations, see Fig. 8. A *tent*, denoted by $\tau(H, u, v)$, is a bipartite graph induced by a hole H and two adjacent nodes $u, v \notin V(H)$ each having two neighbors on H, say u_1, u_2 and v_1, v_2 respectively, with the property that u_1, u_2, v_2, v_1 appear in this order on H. A *short 3-odd-path configuration* is a 3-odd-path configuration in which one of the paths contains three edges.

To overcome the fact that our extended star decomposition is not balancedness preserving, we proceed in the following way. We transform the

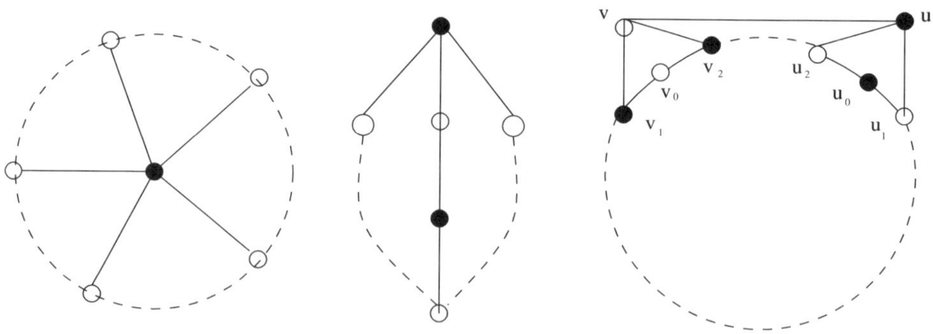

(a) Odd wheel (b) Short 3PC (c) Tent

Fig. 8. Odd wheel, short-3-odd-path configuration and tent.

input graph G into a graph G' that contains a polynomial number of connected components, each of which is an induced subgraph of G, and which has the property that if G is not balanced, then G' contains an unbalanced hole that will either never be broken by any of the decompositions we use, or else be detected while performing the decomposition. We call this process a *cleaning procedure*. To do this, we have to study the structure of signed bipartite graphs that are not balanced, in particular the structure of a smallest (in the number of edges) unbalanced hole. For such a hole we prove the following theorem.

Theorem 15.3. *In a nonbalanced signed bipartite graph, a smallest unbalanced hole H^* contains two edges x_1x_2 and y_1y_2 such that*:

- *The set $N(x_1) \cup N(x_2) \cup N(y_1) \cup N(y_2)$ contains all nodes with an odd number (greater than* 1) *of neighbors in H.*
- *For every tent $\tau(H^*, u, v)$, u or v is contained in $N(x_1) \cup N(x_2) \cup N(y_1) \cup N(y_2)$.*

Let x_0, x_1, x_2, x_3 and y_0, y_1, y_2, y_3 be subpaths of H^*. The above theorem shows that if we remove from G the nodes $N(x_1) \cup N(x_2) \cup N(y_1) \cup N(y_2) \backslash \{x_0, x_1, x_2, x_3, y_0, y_2, y_3, y_4\}$, then H^* will be *clean* (i.e., it will not be contained in any odd wheel or tent). If H^* is contained in a short 3-odd-path configuration, this can be detected during the decomposition (before it is broken). It turns out that, by this process, all the problems are eliminated. So the cleaning procedure consists of enumerating all possible pairs of chordless paths of length 3, and in each case, generating the subgraph of G as described above. The number of subgraphs thus generated is polynomial and, if G is not balanced, then at least one of these subgraphs contains a clean unbalanced hole.

15.3 Algorithm outline

The recognition algorithm takes a signed bipartite graphs as input and recognizes whether or not it is balanced. The algorithm consists of four phases:

- **Preprocessing:** *The cleaning procedure is applied to the input graph.*
- **Extended stars:** *Extended star decompositions are performed, until no block contains an extended star cutset.*
- **6-joins:** 6-*join decompositions are performed until no block contains at* 6-*join.*
- **2-joins:** *Finally,* 2-*join decompositions are performed until no block contains a* 2-*join.*

The 2-join and 6-join decompositions cannot create any new extended star cutset, except in one case which can be dealt with easily. Also a 2-join decomposition does not create any new 6-joins. So, when the algorithm terminates, none of the blocks have an extended star cutset, a 2-join or a 6-join. By the decomposition theorem (Theorem 14.1), if the original signed bipartite graph is balanced, the blocks must be copies of R_{10} or strongly balanced (i.e., a balanced signed bipartite graph where no cycle has exactly one chord). R_{10} is a graph with only ten nodes and so it can be checked in constant time. Checking whether a signed bipartite graph is strongly balanced can be done in polynomial time (Conforti and Rao, 1987).

The preprocessing phase and the decomposition phases using 2-joins and 6-joins are easily shown to be polynomial. For the extended star decomposition phase, it is shown that each bipartite graph which is decomposed has a path of length three which is not present in any of the blocks. This bounds the number of such decompositions by a polynomial in the size of the graph. Thus the entire algorithm is polynomial. See Conforti et al. (2001) for details. Very recently, Zambelli (2003) has obtained a polynomial recognition algorithm for balancedness that does not use decomposition.

The algorithm outlined in this section recognizes in polynomial time whether a signed bipartite graph contains an unbalanced hole. Interestingly Kapoor (1993) has shown that it is NP-complete to recognize whether a signed bipartite graph contains an unbalanced hole going through a prespecified node.

16 More decomposition theorems

A signed bipartite graph is *restricted balanced* if the weight of every cycle is a multiple of four. A signed bipartite graph is *strongly balanced* if every cycle of weight 2 mod 4 has at least two chords. Restricted (strongly, resp.) balanced $0, \pm 1$ matrices are defined accordingly. It follows from the definition that restricted balanced $0, \pm 1$ matrices are strongly balanced, and it can be shown that strongly balanced $0, \pm 1$ matrices are totally unimodular, see Conforti and Rao (1987). Restricted (strongly, resp.) balanceable 0, 1 matrices are those where the nonzero entries can be signed $+1$ or -1 so that the resulting $0, \pm 1$ matrix is restricted (strongly, resp.) balanced. Restricted (strongly, resp.) balanceable 0, 1 matrices can be signed to be restricted (strongly, resp.) balanced using Camion's signing algorithm described in Section 12. Conforti and Rao (1987) have shown that a strongly balanceable 0, 1 matrix that is not restricted balanceable has a 2-separation (the bipartite graph representation has a 1-join).

Theorem 16.1. [Conforti and Rao (1987)] *A strongly balanceable bipartite graph either is restricted balanceable or contains a* 1*-join.*

Crama, Hammer, and Ibaraki (1986) say that a $0, \pm 1$ matrix A is *strongly unimodular* if every basis of (A, I) can be put in a triangular form by permutation of rows and columns.

Theorem 16.2. [Crama et al. (1986)] *A* $0, \pm 1$ *matrix is strongly unimodular if and only if it is strongly balanced.*

Yannakakis (1985) has shown that a restricted balanceable 0, 1 matrix having both a row and a column with more than two nonzero entries has a very special 3-separation: the bipartite graph representation has a 2-join consisting of two single edges. A bipartite graph is *2-bipartite* if all the nodes in one side of the bipartition have degree at most 2.

Theorem 16.3. [Yannakakis (1985)] *A restricted balanceable bipartite graph either is 2-bipartite or contains a cutnode or contains a 2-join consisting of two edges.*

Based on this theorem, Yannakakis designed a linear time algorithm for checking whether a $0, \pm 1$ matrix is restricted balanced. A different algorithm for this recognition problem was given by Conforti and Rao (1987):

Construct a spanning forest in the bipartite graph and check if there exists a cycle of weight 2 mod 4 which is either fundamental or is the symmetric difference of fundamental cycles. If no such cycle exists, the signed bipartite graph is restricted balanced.

A bipartite graph is *linear* if it does not contain a cycle of length 4. Note than an extended star cutset in a linear bipartite graph is always a star cutset, due to Condition (iii) in the definition of extended star cutsets. Conforti and Rao (1992) proved the following theorem for linear balanced bipartite graphs.

Theorem 16.4. [Conforti and Rao (1992)] *A linear balanced bipartite graph either is restricted balanced or contains a star cutset.*

17 Some conjectures and open questions

17.1 Eliminating edges

Conjecture 17.1. [Conforti et al. (2001)] *In a balanced signed bipartite graph G, either every edge belongs to some R_{10}, or some edge can be removed from G so that the resulting signed bipartite graph is still balanced.*

The condition on R_{10} is necessary since removing any edge from R_{10} yields a wheel with three spokes or a 3-odd-path configuration as induced subgraph. This conjecture implies that given a $0, \pm 1$ balanced matrix we can sequentially turn the nonzero entries to zero until every nonzero belongs to some R_{10} matrix, while maintaining balanced $0, \pm 1$ matrices at each step. For 0, 1 matrices, the above conjecture reduces to the following.

Conjecture 17.2. [Conforti and Rao (1992)] *Every balanced bipartite graph contains an edge which is not the unique chord of a cycle.*

It follows from the definition that restricted balanced signed bipartite graphs are exactly the ones for which the removal of *any* subset of edges leaves a restricted balanced signed bipartite graph.

Conjecture 17.1 holds for signed bipartite graphs that are strongly balanced since, by definition, the removal of any edge leaves a chord in every unbalanced cycle.

Theorem 11.1 shows that the graph obtained by eliminating a bisimplicial edge in a totally balanced bipartite graph is totally balanced. Hence Conjecture 17.2 holds for totally balanced bipartite graphs.

17.2 Strengthening the decomposition theorems

The extended star decomposition is not balancedness preserving. This heavily affects the running time of the recognition algorithm for balancedness. Therefore it would be desirable to find strengthenings of Theorem 14.1 that only use operations that preserve balancedness. We have been unable to obtain these results even for linear balanced bipartite graphs (Conforti and Rao, 1993).

Another direction in which the main theorem might be strengthened is as follows.

Conjecture 17.3. [Conforti et al. (2001)] *Every balanceable bipartite graph G which is not signable to be totally unimodular has an extended star cutset.*

This conjecture was shown to hold when G is the bipartite representation of a balanced 0, 1 matrix (Conforti et al., 1999).

17.3 Holes in graphs

17.3.1 α-Balanced graphs

Let G be a signed graph (not necessarily bipartite), i.e., each edge of G has weight $+1$ or -1. let α be a vector whose components are in one-to-one correspondence with the chordless cycles of G and take values in $\{0, 1, 2, 3\}$. The signed graphs G is said to be *α-balanced* if the sum of the weights on each chordless cycle H of G is congruent to α_H mod 4. In the special case where G is bipartite and $\alpha = 0$, this definition coincides with the notion of balanced signed bipartite graph, introduced earlier in this survey.

A graph is *α-balanceable* if there is a signing of its edges such that the resulting signed graph is α-balanced. A *3-path configuration* is one of the three graphs represented in Fig. 9(a), (b), or (c). A *wheel* consists of a chordless cycle H and a node $v \notin V(H)$ with at least three neighbors on H, see Fig. 9(d).

Theorem 17.4. [Truemper (1982)] *A graph G is α-balanceable if and only if*

- $\alpha_H \equiv |H|$ *mod 2 for every chordless cycle H of G,*
- *Every 3-path configuration and wheel of G is α-balanceable.*

Theorem 13.1 is the special case of this theorem where G is bipartite and $\alpha = 0$. A difficult open problem is to extend the decomposition Theorem 14.1 to α-balanceable graphs.

Acknowledgment

The work was supported in part by NSF grant DMII-0352885 and ONR grant N00014-97-1-0196.

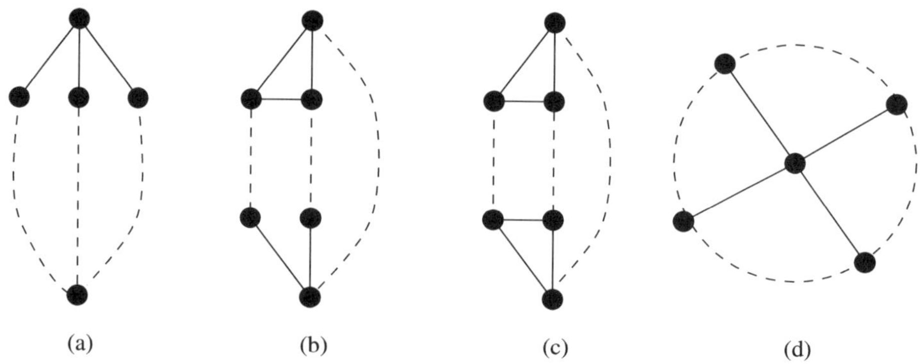

Fig. 9. 3-path configurations and wheel.

References

Anstee, R., M. Farber (1984). Characterizations of totally balanced matrices. *Journal of Algorithms* 5, 215–230.

Berge, C. (1970). Sur certains hypergraphes généralisant les graphes bipartites, in: P. Erdös, A. Rényi, V. Sós (eds.), *Combinatorial Theory and its Applications I. Colloq. Math. Soc. János Bolyai 4*, North Holland, Amsterdam, pp. 119–133.

Berge, C. (1972). Balanced matrices. *Mathematical Programming* 2, 19–31.

Berge, C. (1980). Balanced matrices and the property G. *Mathematical Programming Study* 12, 163–175.

Berge, C. (1989). *Hypergraphs*, North Holland.

Berge, C., M. Las Vergnas (1970). Sur un théorème du type König pour hypergraphes, *International Conference on Combinatorial Mathematics, Annals of the New York Academy of Sciences* 175, 32–40.

Boros, E. O., O. Čepek (1997). On perfect $0, \pm 1$ matrices. *Discrete Mathematics* 165, 81–100.

Boros, E., Y. Crama, P. L. Hammer (1990). Polynomial-time inference of all valid implications for Horn and related formulae. *Annals of Mathematics and Artificial Intelligence* 1, 21–32.

Cameron, K., J. Edmonds (1990). Existentially polytime theorems. *DIMACS Series in Discrete Mathematics and Theoretical Computer Science 1*, American Mathematical Society, Providence, R.I, 83–100.

Camion, P. (1963). Caractérisation des matrices unimodulaires. *Cahiers du Centre d' Études de Recherche Opérationelle* 5, 181–190.

Camion, P. (1965). Characterization of totally unimodular matrices. *Proceedings of the American Mathematical Society* 16, 1068–1073.

Chandru, V., J. N. Hooker (1991). Extended Horn set in propositional logic. *Journal of the ACM* 38, 205–221.

Chvátal, V. (1975). On certain polytopes associated with graphs. *Journal of Combinatorial Theory B* 18, 138–154.

Conforti, M., G. Cornuéjols (1995a). A class of logic problems solvable by linear programming. *Journal of the ACM* 42, 1107–1113.

Conforti, M., G. Cornuéjols (1995b). Balanced $0, \pm 1$ matrices, bicoloring and total dual integrality. *Mathematical Programming* 71, 249–258.

Conforti, M., G. Cornuéjols, C. De Francesco (1997). Perfect $0, \pm 1$ matrices. *Linear Algebra and its Applications* 43, 299–309.

Conforti, M., G. Cornuéjols, A. Kapoor, K. Vušković (1996). Perfect matching in balanced hypergraphs. *Combinatorica* 16, 325–329.

Conforti, M., G. Cornuéjols, A. Kapoor, K. Vušković (2001). Balanced $0, \pm 1$ matrices, Parts I–II. *Journal of Combinatorial Theory B* 81, 243–306.

Conforti, M., G. Cornuéjols, A. Kapoor, K. Vušković, M. R. Rao (1994). Balanced matrices, in: J. R. Birge, K. G. Murty (eds.), *Mathematical Programming, State of the Art 1994*, University of Michigan Press, 1–33.

Conforti, M., G. Cornuéjols, M. R. Rao (1999). Decomposition of balanced matrices. *Journal of Combinatorial Theory B* 77, 292–406.

Conforti, M., G. Cornuéjols, K. Truemper (1994). From totally unimodular to balanced $0, \pm 1$ matrices: a family of integer polytopes. *Mathematics of Operations Research* 19, 21–23.

Conforti, M., G. Cornuéjols, G. Zambelli (2004). Bicolorings and Equitable Bicolorings of matrices, in M. Grötschel (ed.), *The Sharpest Cut: The Impact of Manfred Padberg and his Work*, NPS-SIAM Series on Optimization, 33–37.

Conforti, M., A. M. H. Gerards, A. Kapoor (2000). A theorem of Truemper. *Combinatorica* 20, 15–26.

Conforti, M., M. R. Rao (1987). Structural properties and recognition of restricted and strongly unimodular matrices. *Mathematical Programming* 38, 17–27.

Conforti, M., M. R. Rao (1992). Structural properties and decomposition of linear balanced matrices. *Mathematical Programming* 55, 129–168.

Conforti, M., M. R. Rao (1993). Testing balancedness and perfection of linear matrices. *Mathematical Programming* 61, 1–18.

Cornuéjols, G., W. H. Cunningham (1985). Compositions for perfect graphs. *Discrete Mathematics* 55, 245–254.

Crama, Y. (1993). Concave extensions for nonlinear 0–1 maximization problems. *Mathematical Programming* 61, 53–60.

Crama, Y., P. L. Hammer, T. Ibaraki (1986). Strong unimodularity for matrices and hypergraphs. *Discrete Applied Mathematics* 15, 221–239.

Cunningham, W. H., J. Edmonds (1980). A combinatorial decomposition theory. *Canadian Journal of Mathematics* 32, 734–765.

Edmonds, J., R. Giles (1977). A min–max relation for submodular functions on graphs. *Annals of Discrete Mathematics* 1, 185–204.

Fortet, R. (1976). Applications de l'algèbre de Boole en recherchè opérationelle. *Revue Française de Recherche Opérationelle* 4, 251–259.

Fulkerson, D. R. (1972). Anti-blocking polyhedra. *Journal of Combinatorial Theory B* 12, 50–71.

Fulkerson, D. R., A. Hoffman, R. Oppenheim (1974). On balanced matrices. *Mathematical Programming Study* 1, 120–132.

Georgakopoulos, G., D. Kavvadias, C. H. Papadimitriou (1988). Probabilistic satisfiability. *Journal of Complexity* 4, 1–11.

Ghouila-Houri, A. (1962). Charactérisations des matrices totalement unimodulaires. *C.R. Acad. Sc. Paris* 254, 1192–1193.

Giles, R. (1978). A balanced hypergraph defined by subtrees of a tree. *ARS Combinatorica* 6, 179–183.

Golumbic, M. C., C. F. Goss (1978). Perfect elimination and chordal bipartite graphs. *Journal of Graph Theory* 2, 155–163.

Guenin, B. (1998). Perfect and ideal $0, \pm 1$ matrices. *Mathematics of Operations Research* 23, 322–338.

Gupta, R. P. (1978). An edge-coloration theorem for bipartite graphs of paths in trees. *Discrete Mathematics* 23, 229–233.

Hall, P. (1935). On representatives of subsets. *J. London Math. Soc.* 26–30.

Heller, I., C. B. Tompkins (1956). An extension of a theorem of Dantzig's, in: H. W. Kuhn, A. W. Tucker (eds.), *Linear Inequalities and Related Systems*, Princeton University Press, 247–254.

Hoffman, A. J., J. K. Kruskal (1956). Integral boundary points of convex polyhedra, in: H. W. Kuhn, A. W. Tucker (eds.), *Linear Inequalities and Related Systems*, Princeton University Press, 223–246.

Hoffman, A. J., A. Kolen, M. Sakarovitch (1985). Characterizations of totally balanced and greedy matrices. *SIAM Journal of Algebraic and Discrete Methods* 6, 721–730.

Hooker, J. N. (1988). A quantitative approach to logical inference. *Decision Support Systems* 4, 45–69.

Hooker, J. N. (1996). Resolution and the integrality of satisfiability polytopes. *Mathematical Programming* 74, 1–10.

Kapoor, A. (1993). On the complexity of finding holes in bipartite graphs, preprint, Carnegie Mellon University.

Lovász, L. (1972). Normal hypergraphs and the perfect graph conjecture. *Discrete Mathematics* 2, 253–267.

Megiddo, N. (1991). On finding primal- and dual-optimal bases. *Journal of Computing* 3, 63–65.

Nilsson, N. J. (1986). Probabilistic logic. *Artificial Intelligence* 28, 71–87.

Nobili, P., A. Sassano (1998). $(0, \pm 1)$ Ideal matrices. *Mathematical Programming* 80, 265–281.

Tamir, A. (1983). A class of balanced matrices arising from locations problems. *SIAM Journal on Algebraic and Discrete Methods* 4, 363–370.

Tamir, A. (1987). Totally balanced and totally unimodular matrices defined by center location problems. *Discrete Applied Mathematics* 16, 245–263.

Truemper, K. (1982). Alpha-balanced graphs and matrices and GF(3)-representability of matroids. *Journal of Combinatorial Theory B* 32, 112–139.

Truemper, K. (1990). Polynomial theorem proving I. Central matrices. *Technical Report UTDCS* 34–90.

Truemper, K. (1992). A decomposition theory for matroids. VII. Analysis of minimal violation matrices. *Journal of Combinatorial Theory B* 55, 302–335.

Truemper, K., R. Chandrasekaran (1978). Local unimodularity of matrix-vector pairs. *Linear Algebra and its Applications* 22, 65–78.

Yannakakis, M. (1985). On a class of totally unimodular matrices. *Mathematics of Operations Research* 10, 280–304.

Zambelli, G. (2003). A polynomial recognition algorithm for balanced matrices, preprint.

K. Aardal et al., Eds., *Handbooks in OR & MS, Vol. 12*

Chapter 7

Submodular Function Minimization

S. Thomas McCormick

Sauder School of Business, University of British Columbia, Vancouver, BC V6T 1Z2 Canada

Abstract

This chapter describes the submodular function minimization problem (SFM); why it is important; techniques for solving it; algorithms by Cunningham, by Schrijver as modified by Fleischer and Iwata, by Iwata, Fleischer and Fujishige, and by Iwata for solving it; and extensions of SFM to more general families of subsets.

1 Introduction

We start with a guide for the reader. If you don't know about submodularity, you should start here. If you are already familiar with submodular functions but don't know the algorithms, start with Section 2. If you just want to learn about recent algorithms, start with Section 3. This chapter assumes some familiarity with network flow concepts, particularly those of Max Flow; see, e.g., Ahuja, Magnanti, and Orlin (1993) for coverage of these.

1.1 What is submodularity?

Suppose that our factory has the capability to make any subset of a given set E of potential products. If we decide to produce subset $S \subseteq E$ of products, then we must pay a setup cost $c(S)$ to make the factory ready to produce S. This setup cost is a particular instance of a set function: Given a finite set E (the *ground set*), the notation 2^E stands for the family of all subsets of E. Then a scalar-valued function $f: 2^E \to \mathbb{R}$ is called a *set function*. We write $f(S)$ for the value of f on subset $S \subseteq E$, and use n for $|E|$.

Suppose that we have tentatively decided to produce subset S in our factory, and that we are considering whether to add product $e \notin S$ to our product mix. Then the incremental setup cost that we would have to pay is $c(S \cup \{e\}) - c(S)$. We deal with a lot of singleton sets, so to unclutter things we use the standard notation that $S+e$ means $S \cup \{e\}$, $S-e$ means $S-\{e\}$, and $f(e)$ means $f(\{e\})$. In this notation the incremental cost of adding e is $c(S+e)-c(S)$. We use $S \subset T$ to mean that $S \subseteq T$ but $S \neq T$.

Now economics suggests that in most real-world situations, this incremental cost is a nonincreasing function of S. That is, adding new product e to a larger set should produce an incremental cost no more than adding e to a smaller set. In symbols, for a general function f we should have

$$\text{for all} \quad S \subset T \subset T+e, \quad f(S+e) - f(S) \geq f(T+e) - f(T). \tag{1}$$

When any set function f satisfies (1), we say that f is *submodular*. The connection between submodularity and economics suggested here is very deep; many more details about this are available in Topkis' book (Topkis, 1998).

We say that f is *supermodular* if $-f$ is submodular, and *modular* if it is both sub- and supermodular. It is easy to see that f is supermodular iff it satisfies (1) with the inequality reversed, and modular iff it satisfies (1) with equality. The canonical (and essentially only) example of a modular functions is derived from a vector $v \in \mathbb{R}^E$: For $S \subseteq E$, define $v(S) = \sum_{e \in S} v_e$ (so that $v(\emptyset) = 0$), and then $v(S)$ is modular. For example, if π_e is the net present value (NPV) of profits expected from producing product e (the value of the future stream of profits from producing e discounted back to the present), then $\pi(S)$ is the total NPV expected from producing subset S, and $\pi(S) - c(S)$ is the present value of net profits expected from producing S. Note that, because $\pi(S)$ is modular and $c(S)$ is submodular, $\pi(S) - c(S)$ is supermodular.

There is an alternate and more standard definition of submodularity that is sometimes more useful for proofs:

$$\text{for all} \quad X, Y \subseteq E, \quad f(X) + f(Y) \geq f(X \cup Y) + f(X \cap Y). \tag{2}$$

We now show that these definitions are equivalent:

Lemma 1.1. *Set function f satisfies* (1) *if and only if it satisfies* (2).

Proof. To show that (2) implies (1), apply (2) to the sets $X=S+e$ and $Y=T$ to get $f(S+e)+f(T)\geq f((S+e)\cup T)+f((S+e)\cap T)=f(T+e)+f(S)$, which is equivalent to (1).

To show that (1) implies (2), first rewrite (1) as $f(S+e)-f(T+e)\geq f(S)-f(t)$ for $S\subset T\subset T+e$. Now, enumerate the elements of $Y-X$ as $e_1,e_2,\ldots,e_k$ and note that, for $i<k$, $[(X\cap Y)\cup\{e_1,e_2,\ldots,e_i\}]\subset[X\cup\{e_1,e_2,\ldots,e_i\}]\subset[X\cup\{e_1,e_2,\ldots,e_i\}]+e_{i+1}$, so the rewritten (1) implies that

$$\begin{aligned} f(X\cap Y)-f(X) &\leq f((X\cap Y)+e_1)-f(X+e_1)\\ &\leq f((X\cap Y)\cup\{e_1,e_2\})-f(X\cup\{e_1,e_2\})\\ &\vdots\\ &\leq f((X\cap Y)\cup\{e_1,e_2,\ldots,e_k\})-f(X\cup\{e_1,e_2,\ldots,e_k\})\\ &=f(Y)-f(X\cup Y), \end{aligned}$$

and this is equivalent to (2). □

Here are some examples of submodular functions that arise often in practice:

Example 1.2. Suppose that $G=(N,A)$ is a directed graph with nodes N and arcs A. For $S\subseteq N$ define $\delta^+(S)$ to be the set of arcs $i\to j$ with $i\in S$ but $j\notin S$; similarly, $\delta^-(S)$ is the set of $i\to j$ with $i\notin S$ and $j\in S$, and $\delta(S)=\delta^+(S)\cup\delta^-(S)$ (for an undirected graph, $\delta(S)$ is the set of edges with exactly one end in S). Recall that for $w\in\mathbb{R}^A$, notation $w(\delta^+(S))$ means $\sum_{e\in\delta^+(S)}w_e$. Then if $w\geq 0$, $w(\delta^+(S))$ (or $w(\delta^-(S))$, or $w(\delta(S))$), is a submodular function on ground set N.

Example 1.3. Suppose that $\mathcal{M}=(E,\ r)$ is a matroid (see Welsh (1976) for further details) on ground set E with rank function r. Then r is a submodular function on ground set E. More generally, if r is a set function on E, we call r a polymatroid rank function if (i) $r(\emptyset)=0$, (ii) $S\subseteq T\subseteq E$ implies $r(S)\leq r(T)$ (r is increasing), and (iii) r is submodular. Then the polyhedron $\{x\in\mathbb{R}^E|x\geq 0$ and $x(S)\leq r(S)$ for all $S\subseteq E\}$ is the associated polymatroid. For example, let $G=(N,A)$ be a Max Flow network with source s, sink t, and capacities $u\in\mathbb{R}^A$. Define $E=\{i\to j\in A|i=s\}=\delta^+(s)$, the subset of arcs with tail s. Then $\{x_{sj}\,|\,x$ is a feasible flow in $G\}$ (i.e., the projection of the set of feasible flows onto E) is a polymatroid on E. If S is a subset of the arcs with tail s, then $r(S)$ is the max flow value when we set the capacities of the arcs in $E-S$ to zero.

Example 1.4. Suppose that we have a set L of potential locations for warehouses. These warehouses are intended to serve the set R of retail stores. There is a fixed cost φ_l for opening a warehouse at $l \in L$, and the benefit to us of serving retail store $r \in R$ from $l \in L$ is b_{rl} (where $b_{rl} = -\infty$ if location l is too far away to serve store r). Thus if we choose to open warehouses $S \subseteq L$, our net benefit would be $f(S) = \sum_{r \in R} \max_{l \in S} b_{rl} - \sum_{l \in S} \varphi_l$. This is a submodular function.

Example 1.5. Suppose that we have a system of queues (waiting lines) $E = \{1, 2, \ldots, n\}$. For queue i, let x_i denote its throughput (the amount of work it processes) under some control policy (allocation of resources to the queues). Then the set of feasible throughputs is some set X in $\mathbb{R}^n$. We say that the system satisfies conservation laws if the maximum amount of work possible from the set of queues S, namely $f(S) = \max_{x \in X} \sum_{i \in S} x_i$, depends only on whether the queues in S have priority over other queues, and not on the priority order within S. Shanthikumar and Yao (1992) show that if the system satisfies conservation laws, then $f(S)$ is submodular. Since any feasible x is nonnegative, and this f is clearly increasing, then X is the polymatroid associated with f.

For some applications f is not defined on all subsets of E. Suppose that $\mathcal{F} \subseteq 2^E$ is a family of subsets of E. If $\mathcal{F}$ is closed under unions and intersections, then we say that $\mathcal{F}$ is a *ring family*, or a *distributive lattice*, or a *lattice family*. If we require (2) to hold only for members of $\mathcal{F}$, then we say that f is *ring submodular*. If instead we require that $S \cap T$ and $S \cup T$ are also in $\mathcal{F}$ only for all S, $T \in \mathcal{F}$ with $S \cap T \neq \emptyset$, then we call $\mathcal{F}$ an *interesecting family*. If we require (2) to hold only for members of $\mathcal{F}$ with nonempty intersection, then we say that f is *intersecting sumodular*. Finally, if we require that $S \cap T$ and $S \cup T$ are also in $\mathcal{F}$ only for all S, $T \in \mathcal{F}$ with $S \cap T \neq \emptyset$ and $S \cup T \neq E$, then we call $\mathcal{F}$ a *crossing family*. If we require (2) to hold only for members of $\mathcal{F}$ with nonempty intersection and whose union is not E, then we say that f is *crossing submodular*. We consider more general families in Section 5.2.

Here are two examples of these specialized submodular functions:

Example 1.6. Continuing with our introductory factory example, suppose we have some precedences among products expressed by a directed graph $G = (E, A)$ on node set E, where arc $i \rightarrow j \in A$ means that any set containing product i must also contain product j. Then feasible sets are those $S \subseteq E$ such that $\delta^+(S) = \emptyset$, called closed sets. It is easy to see that these sets form a ring family, and would be reasonable to assume that the cost of function $c(S)$ should be ring submodular on this family. (Birkhoff's Representation Theorem (Birkhoff, 1967) says that all ring families arise in this way, as the family of closed sets of a directed graph.)

Example 1.7. Suppose that we have a connected directed graph $G=(N, A)$ with node $r \in N$ designated as the root, and weights $w \in \mathbb{R}^A$. We want to find a minimum weight arborescence rooted at r (spanning tree such that exactly one arc enters every node besides r, so that the unique path from r to any other node is a directed path). It can be shown (see [(Schrijver, 2003), Section 5.2.4]) that one way to formulate this as an integer program is as follows: Make a decision variable x_a for each $a \in A$ with the intended interpretation that $x_a=1$ is a is included in the arborescence, and 0 otherwise. Let $\mathcal{T}$ be the family of nonempty subsets of N not containing r. Then the family of constraints $x(\delta^-(S)) \geq 1$ for all $S \in \mathcal{T}$ expresses that each such subset should have at least one arc entering it. The family $\mathcal{T}$ is intersecting, and the right-hand side $f(S)=1$ for all $S \in \mathcal{T}$ is intersecting supermodular. Note that this is a very common way for submodular functions to arise, as right-hand sides in integer programming formulations (and their linear relaxations) of combinatorial problems.

It is useful to have a mental model of submodularity to better understand it. Definition (1) tends to suggest that submodularity is related to concavity. Indeed, suppose that $g: \mathbb{R} \to \mathbb{R}$ is a scalar function, and set function f is defined by $f(S)=g(|S|)$. Then it is easy to show that f is submodular iff g is concave.

A deeper result by Lovász (1983) suggests instead that submodularity is related to convexity. For $S \subseteq E$ define the *incidence vector* $\chi(S)$ of S as $\chi(S)_e$ equals 1 if $e \in S$, and 0 otherwise (we use χ_u to stand for $\chi(\{u\})$. This is a 1–1 map between 2^E and the vertices of the n-cube $C_n=[0,1]^n$. If $v=\chi(S)$ is such a vertex, then f gives the value $f(S)$ to v. It is well-known that C_n can be dissected into $n!$ simplices, where the simplex $\sigma(\pi)$ corresponding to permutation π contains all $x \in C_n$ with $0 \leq x_{\pi(1)} \leq x_{\pi(2)} \leq \cdots \leq x_{\pi(n)} \leq 1$. Since f gives values to the vertices of $\sigma(\pi)$, there is a unique way to extend f to the interior of $\sigma(\pi)$ in a linear way. Let $\hat{f}: C_n \to \mathbb{R}$ denote the piecewise linear function which is these $n!$ linear extensions pasted together. This particular piecewise linear extension of f is called the *Lovász extension*.

Theorem 1.8. [Lovász (1983)] *Set function f is submodular iff its Lovász extension $\hat{f}$ is convex.* □

It turns out that this "convex" view of submodularity is much more fruitful than the "concave" view. In particular, Section 2.3 shows that, similar to convexity, minimizing a submodular function is "easy," whereas maximizing one is "hard." In fact, Murota (1998, 2003) has developed a theory of discrete convexity based on submodularity, in which many of the classic theorems of convexity find analogues.

For a more extensive look at submodular functions and their applications, consult Fujishige's book (Fujishige, 1991), Lovász's article (Lovász, 1983), or Nemhauser and Wolsey (1988) [Section III.3].

1.2 What is submodular function minimization?

Returning to our factory example, which subset should we choose? Clearly we should choose a subset that maximizes our future NPV minus our costs. That is, among the 2^n subsets of E, we want to find one that maximizes the supermodular function $\pi(S) - c(S)$. This is formally equivalent to minimizing the submodular function $c(S) - \pi(S)$, so we consider the core problem of this chapter.

Submodular Function Minimization (SFM)
$\min_{S \subseteq E} f(S)$, where f is submodular.

Here are some applications of SFM:

Example 1.9. Let's change Example 1.2 a bit. Now we are given a directed graph $G = (N, A)$ with source $s \in N$ and sink $t \in N$ $(t \neq s)$ and with nonnegative weights $w \in \mathbb{R}^A$. Let $E = N - \{s,t\}$, and for $S \subseteq E$ define $f(S) = w(\delta^+(S + s))$. This f is again submodular, and SFM with this f is just the familiar s–t Min Cut problem. This also works if G is undirected, by redefining $f(S) = w(\delta(S + s))$.

Example 1.10. Continuing with Example 1.3, let $\mathcal{M}_1 = (E, r_1)$ and $\mathcal{M}_2 = (E, r_2)$ be two matroids on the same ground set. Then Edmonds' Matroid Intersection Theorem (Edmonds, 1970) says that the size of the largest common independent set equals $\min_{s \subseteq E} r_1(S) + r_2(E - S)$. The set function $f(S) = r_1(S) + r_2(E - S)$ is submodular, so this is again SFM. This also works for the intersection of polymatroids.

Example 1.11. As a different continuation of Example 1.3, suppose we have a polymatroid P with rank function r, and that we are given some point $\bar{x} \in \mathbb{R}^E$ that satisfies $\bar{x} \geq 0$. The question is to determine whether $\bar{x} \in P$. To do this we need to verify the exponential number of inequalities $x(S) \leq r(S)$ for all $S \subseteq E$. We could do this by computing $g = \min_{S \subseteq E} r(S) - \bar{x}(S)$ via SFM (note that $r(S) - \bar{x}(S)$ is submodular), because if $g \geq 0$ then $\bar{x} \in P$, and if $g < 0$ then $\bar{x} \notin P$ (and the minimizing S gives a violated constraint). This *separation problem* (see Section 2.3) is a common application of SFM.

Three recent models in supply chain management use SFM to compute solutions. Shen, Coullard, and Daskin (2003) model a facility location-inventory problem related to Example 1.4, which they solve using a linear programming column generation algorithm. The column generation subproblem needs to find optimal subsets of demand points to be served by a facility, and this is a SFM problem. Lu and Song (2002) model inventory of components in a assemble-to-order system where demand is for final products assembled from subsets of components. Then the problem of minimizing expected long-run cost is a discretely convex problem, which uses SFM in its solution. Huh and Roundy (2002) model capacity expansion sequencing decisions in the semiconductor industry, where we trade off the declining cost of buying fabrication tools with the cost of lost sales from buying tools too late. The problem of determining an optimal sequence with general costs uses a (parametric) SFM subroutine.

1.3 Computational models for SFM

A naive algorithm for SFM is to use brute force to look at the 2^n values of $f(S)$ and select the smallest, but this would take 2^n time, which is exponential, and hence impractical for all but the smallest instances. We would very much prefer to have an algorithm that is polynomial in n. The running time of an algorithm might also depend on the "size" of f as measured by, e.g., some upper bound M on $\max_S|f(S)|$. Since we could scale f to make M arbitrarily small, this makes sense only when we assume that f is integer-valued, and hence we implicitly so assume whenever we use M. An SFM algorithm that is polynomial in n and M is called *pseudo-polynomial*. To be truly polynomial, the running time must be a polynomial in n and $\log M$, leading to a *weakly polynomial* algorithm. If f is real-valued, or if M is very large, then it would be better to have an algorithm whose running time is independent of M, i.e., a polynomial function of n only, which is then called a *strongly polynomial* algorithm.

The first polynomial algorithms for SFM used the Ellipsoid method, see Section 2.3. Algorithms that avoid using Ellipsoid-like methods are called *combinatorial*. There appears to be no intrinsic reason why an SFM algorithm would have to use multiplication or division, so Schrijver (2000) asks whether an SFM algorithm exists that is strongly polynomial, and which uses only additions, subtractions, and comparisons (such an algorithm would have to be combinatorial). Schrijver calls such an algorithm *fully combinatorial*. It is sometimes more convenient to hide logarithmic factors in running times, so we use the common notation that $\tilde{O}(f(n))$ stands for $O(f(n)\cdot(\log n)^k)$ for some positive constant k.

This brings up the problem of how to represent the apparently exponential-sized input f in an algorithm. If we explicitly listed the values of f, then just reading the input would already be super-polynomial. The assumption we make to deal with this is that we have an *evaluation oracle* $\mathcal{E}$ available. We assume that $\mathcal{E}$ is a black box whose input in some set $S \subseteq E$, and whose output is $f(S)$. We use EO to stand for the time needed for one call to $\mathcal{E}$. For Example 1.2 with a reasonable representation for the graph, we would have $\text{EO} = O(|A|)$. Since the input S to EO has size $\Theta(n)$, it is reasonable to assume that $\text{EO} = \Omega(n)$. Section 2.2 shows how to compute a bound M on the size of f in $O(n\text{EO})$ time. Thus our hope is to solve SFM with a polynomial number of calls to $\mathcal{E}$, and a polynomial amount of other work.

1.4 Overview, and short history of SFM

SFM has been recognized as an important problem since the early days of combinatorial optimization, when in the early 1970s Edmonds (1970) established many of the fundamental results that we use, which we cover in Sections 2.1 and 2.2.

When the Ellipsoid Algorithm arrived, in 1981 Grötschel, Lovász, and Schrijver (1981) realized that it is a useful tool for finding polynomial algorithms for problems such as SFM; we cover these developments in Section 2.3. However, this result is ultimately unsatisfactory, since Ellipsoid is not very practical, and does not give much combinatorial insight. The problem shifted from "Is SFM polynomial?" to "Is there a combinatorial (i.e., non-Ellipsoid) polynomial algorithm for SFM?". In 1985 Cunningham (1985) said that:

> It is an outstanding open problem to find a practical combinatorial algorithm to minimize a general submodular function, which also runs in polynomial time.

Cunningham made what turned out to be key contributions to this effort in the mid-80s by using a linear programming duality result of Edmonds (1970) to set up a Max Flow-style algorithmic framework for SFM. We cover the LPs in Section 2.4, the network flow framework in Section 2.6, and Cunningham's applications of it (Bixby et al., 1985; Cunningham, 1984, 1985) that yield a pseudo-polynomial algorithm for SFM in Section 3.1.

Then, nearly simultaneously in 1999, two working papers appeared giving quite different combinatorial strongly polynomial algorithms for SFM. These were by Schrijver (2000) (formally published in 2000) and Iwata, Fleischer, and Fujishige (IFF) (2001) (formally published in 2001). Both of them are based on Cunningham's framework. We describe Schrijver's Algorithm in Section 3.2, and the IFF Algorithm in Section 3.3.

Both of these algorithms use a "Carathéodory" subroutine whose input is a representation of a vector $y \in \mathbb{R}^E$ as a convex combination of vertices: $y = \sum_{i \in I} \lambda_i v^i$, and whose output is a set of at most n of the v^i whose convex hull still contains y, see Section 2.5. This can be done using standard linear algebra techniques, but it is aesthetically unpleasant. This led Schrijver

(2000) to pose the question as to whether there exists a fully combinatorial SFM algorithm. Iwata (2002) found such an algorithm, based on the IFF Algorithm, which we describe in Section 3.3.3. An alternate version of Schrijver's Algorithm using push-relabel ideas from Max Flow is given by Fleischer and Iwata (2001) (which we call Schrijver-PR and incorporate into Section 3.2). A speedup of the IFF Algorithm (which uses ideas from both Schrijver and IFF, and which we call the Hybrid Algorithm) and Iwata's fully combinatorial version of it is given by Iwata (2002), which we describe in Section 3.3.4. We compare and contrast these algorithms in Section 4, where we also give some guidelines on solving SFM in practice. We discuss various solvable extensions of SFM in Section 5, and we speculate about the future of SFM algorithms in Section 6. We note that Fleischer (2000), Fujishige (2002), and Schrijver [(Schrijver, 2003), Chapter 45] wrote other surveys of submodular function minimization.

We cannot cover it here in detail, but we note that there also exists some work on the structure of solutions to *parametric* SFM problems (where we want to solve a parametrized sequence of SFM problems), notably the work of Topkis (1978, 1998). He shows that when a parametric SFM problem satisfies certain properties, then optimal SFM solutions are nested as a function of the parameter. Granot and Veinott (1985) later extended this work. Fleischer and Iwata (2001) extend their Push-Relabel version of Schrijver's Algorithm to solve some parametric SFM problems in the same running time.

The SFM algorithms share a common heritage with algorithms for the Submodular Flow problem, a common generalization of Min Cost Flow and Matroid Intersection developed by Edmonds and Giles (1977): in particular IFF grew out of a Submodular Flow algorithm of Fleischer, Iwata, and McCormick (2002). In return, Fleischer and Iwata were able to show how to solve Submodular Flow in the same time as one call to IFF in (Fleischer and Iwata, 2000). The IFF algorithms have been further extended to minimizing *bisubmodular* functions. These are a directed, or signed, analogue of submodular functions, see Fujishige and Iwata (2001), or McCormick and Fujishige (2003).

2 Building blocks for SFM algorithms

These subsections build up some tools that are common to all the SFM algorithms.

2.1 Greedy optimizes over submodular polyhedra

Generalizing the polymatroids of Example 1.3 somewhat, for a submodular function f it is natural to consider the *submodular polyhedron* $P(f) = \{x \in \mathbb{R}^E \mid x(S) \leq f(S) \text{ for all } S \subseteq E\}$. For our arguments to be consistent for every case we need to worry about the constraint $0 = x(\emptyset) \leq f(\emptyset)$. To ensure that this makes sense, from this point forward we redefine $f(S)$ to be $f(S) - f(\emptyset)$ so that $f(\emptyset) = 0$; note that this change affects neither submodularity nor SFM. It turns out to be quite useful to consider the face of $P(f)$ satisfying $x(E) = f(E)$, the *base polyhedron*: $B(f) = \{x \in P(f) \mid x(E) = f(E)\}$. We prove below that $B(f)$ is never empty.

Given weights $w \in \mathbb{R}^E$, it is natural to wonder about maximizing the linear objective $w^T x$ over $P(f)$ and $B(f)$. Note that $y \leq x \in P(f)$ implies that $y \in P(f)$. Hence if $w_e < 0$ for some $e \in E$, then max $w^T x$ is unbounded on $P(f)$, since we can let $x_e \to -\infty$. If $w \geq 0$, then the results below imply that an optimal x^* must belong to $B(f)$. Hence we can restrict our attention to solving max $\{w^T x \mid x \in B(f)\}$. The dual of this LP has dual variable π_S for each $\emptyset \subset S \subseteq E$ and is min $\{\sum_{S \subseteq E} f(S)\pi_S \mid \sum_{S \ni e} \pi_S = w_e$ for each $e \in E$, $\pi_S \geq 0$ for all $S \subset E\}$.

One remarkable property of submodularity is that the naive *Greedy Algorithm* solves this problem. For a linear order $\prec$ of the elements of E as $e_1 \prec e_2 \prec \cdots \prec e_n$, and any $e \in E$, define $e^{\prec}$ as $\{e' \in E \mid e' \prec e\}$, a subset of E, and define $e^{\prec}_{n+1} = E$. Then Greedy takes $\prec$ as input, and outputs a vector $v^{\prec} \in \mathbb{R}^E$; component e_i of $v^{\prec}$ is then $v^{\prec}_{e_i}$.

The Greedy Algorithm with Linear Order $\prec$

For $i = 1, \ldots, n$

Set $v^{\prec}_{e_i} = f(e^{\prec}_{i+1}) - f(e^{\prec}_{e_i})$ $(= f(e^{\prec}_i + e_i) - f(e^{\prec}_i))$.

Return $v^{\prec}$.

To use this to maximize $w^T x$, let $\prec_w$ denote a linear order of E as $e_1 \prec_w, e_2 \prec_w \ldots \prec_w e_n$ such that $w_{e_1} \geq w_{e_2} \geq \cdots \geq w_{e_n}$, and apply Greedy to $\prec_w$ to get $v^{\prec_w}$. Further define $w_{n+1} = 0$, and dual variables π^w_S as having value $w_{e_{i-1}} - w_{e_i}$ if $S = e^{\prec_w}_i (i = 2, \ldots, n+1)$, and zero otherwise.

Theorem 2.1. *The optimization version of Greedy runs in* $O(n \log n + n\text{EO})$ *time,* $v^{\prec_w}$ *is primal optimal,* π^w *is dual optimal, and* $v^{\prec_w}$ *is a vertex of* $B(f)$.

Proof. Computing $\prec_w$ involves sorting the weights, which takes $O(n \log n)$ time. Otherwise, Greedy takes $O(n\text{EO})$ time.

Now we prove that $v^{\prec_w} \in B(f)$. Note that $v^{\prec_w}(E) = \sum_{i=1}^n [f(e^{\prec_w}_{i+1}) - f(e^{\prec_w}_i)] = f(E) - f(\emptyset) = f(E)$. So we just need to verify that for $\emptyset \subset S \subset E$, $v^{\prec_w}(S) \leq f(S)$. Define k as the largest index such that $e_k \in S$. We proceed by induction on k. For $k = 1$ we must have $S = \{e_1\}$, and $v^{\prec_w}(e_1) = v^{\prec_w}_{e_1} = f(e^{\prec_w}_2) - f(e^{\prec_w}_1) = f(e_1) - 0 = f(e_1)$, so $v^{\prec_w}(e_1) \leq f(e_1)$ is true.

For $1<k<n$, note that $S \cup e_k^{\prec_w} = e_{k+1}^{\prec_w}$ and $S \cap e_k^{\prec_w} = S - e_k$. Hence (2) gives $f(S) \geq f(e_{k+1}^{\prec_w}) + f(S - e_k) - f(e_k^{\prec_w})$. Now $v^{\prec_w}(S) = f(e_{k+1}^{\prec_w}) - f(e_k^{\prec_w}) + v^{\prec_w}(S - e_k)$. By induction $v^{\prec_w}(S - e_k) \leq f(S - e_k)$, so we get $v^{\prec_w}(S) \leq f(e_{k+1}^{\prec_w}) - f(e_k^{\prec_w}) + f(S - e_k) \leq f(S)$, as required.

Now we prove that π^w is dual feasible. Suppose that $e = e_k$. Then $\sum_{S \ni e} \pi_S^w = \sum_{i=k}^{n} (w_{e_i} - w_{e_{i+1}}) = w_{e_k} = w_e$ as desired. By the ordering of E, $\pi_S^w \geq 0$ for all $S \subset E$.

Next, we prove that $v^{\prec_w}$ and π^w are complementary slack. First, $\pi_S^w > 0$ implies that $S = e_k^{\prec_w}$ for some k, and $v^{\prec_w}(e_k^{\prec_w}) = \sum_{i=1}^{k-1} [f(e_{i+1}^{\prec_w}) - f(e_i^{\prec_w}) = f(e_k^{\prec_w})$. Next, if $v^{\prec_w}(S) < f(S)$, then S cannot be one of the $e_k^{\prec_w}$, so $\pi_S = 0$. Hence $v^{\prec_w}$ and π^w are feasible and complementary slack, and thus optimal.

Recall that $v^{\prec_w}$ is a vertex of $B(f)$ if the submatrix of constraints where $\pi_S^w > 0$ is nonsingular. This submatrix has rows which are a subset of $\chi(e_2^{\prec_w})$, $\chi(e_3^{\prec_w}), \ldots, \chi(e_{n+1}^{\prec_w})$, and these vectors are clearly linearly independent. □

Suppose that $x \in P(f)$. We say that $S \subseteq E$ is *tight* for x if $x(S) = f(S)$, and we denote the family of tight sets for x by $\mathcal{T}(x)$. A corollary to this proof is that

$$\text{If } v^{\prec} \text{ is generated by Greedy from } \prec, \text{ then } e^{\prec} \text{ is tight for } v^{\prec} \text{ for all } e \in E. \tag{3}$$

Note that when $w \geq 0$ then we get that $\pi_E^w \geq 0$ also, showing that the given solutions are also optimal over $P(f)$ in this case. We can also conclude from this proof that $B(f) \neq \emptyset$, and that every permutation of E generates a vertex of $B(f)$, and hence that $B(f)$ has a maximum of $n!$ vertices. Our ability to generate vertices of $B(f)$ as desired is a key part of the SFM algorithms that follow.

The strongly polynomial version of IFF in Section 3.3.2 reduces SFM over 2^E to SFM over a ring family $\mathcal{D}$ represented by the closed sets of the directed graph (E, C), so we need to understand how these concepts generalize in that case. (We therefore henceforth refer to $e \in E$ as "nodes" as well as "elements".) In this case $B(f)$ is in general not bounded (we continue to write $B(f)$ for the base polyhedron over a ring family), because some of the constraints $x(S) \leq f(S)$ needed to bound $B(f)$ do not exist when $S \notin \mathcal{D}$. In particular, if (E, C) has a directed cycle Q and $l \neq k$ are nodes of Q, then for any $z \in B(f)$ we have $z + \alpha(\chi_l - \chi_k) \in B(f)$ for any (positive or negative) value of α, and so $B(f)$ cannot have any vertices. Section 3.3.2 deals with this by contracting strong components of (E, C), so we can assume that (E, C) has no directed cycles. Then we say that linear order $\prec$ is *consistent* with (E, C) (a consistent linear order is called a *linear extension* in (Fujishige, 1991; Iwata, 2002a)) if $k \to l \in C$ implies that $l \prec k$, which implies that $e^{\prec} \in \mathcal{D}$ for every $e \in E$. The proof of Theorem 2.1 shows that when $\prec$ is consistent with $\mathcal{D}$, then $v^{\prec}$ is a vertex of $B(f)$.

If x is a flow (not necessarily satisfying conservation) on (E, C), define ∂x: $E \rightarrow \mathbb{R}$ by $\partial x_k = \sum_l x_{kl} - \sum_j x_{jk}$, the net x-flow out of node k, or *boundary* of x. Then it can be shown (see Fujishige, 1991, [Theorem 3.36]) that $w \in B(f)$ iff there is some y which is a convex combination of vertices $v^{\prec}$ for consistent $\prec$, and some flow $x \geq 0$ such that $w = y + \partial x$. Thus the boundaries of nonnegative flows in (E, C) are precisly the directions of unboundedness of $B(f)$.

Section 3.3.2 also needs sharper bounds than M on y_e for $y \in B(f)$. For $e \in E$ define D_e, the *descendants* of e, as the set of nodes reachable from e via directed paths in (E, C). We know from (1) and Greedy that the earlier that e appears in $\prec$, the larger the value of $v_e^{\prec}$ is. Any consistent order must have all elements of $D_e - e$ coming before e. Therefore, an order $\prec^e$ putting D_e before all other nodes should maximum y_e, so we should have that $y_e \leq v_e^{\prec^e} = f(D_e) - f(D_e - e)$. The next lemma formalizes this.

Lemma 2.2. *If $y \in B(f)$ and y is in the convex hull of the vertices of $B(f)$, then $y_e \leq f(D_e) - f(D_e - e)$.*

Proof. It suffices to show that, for any $\prec$ consistent with (E, C), $v_e^{\prec} \leq f(D_e) - f(D_e - e)$. From Greedy, $v^{\prec} = f(e^{\prec} + e) - f(e^{\prec})$. By consistency, $D_e \subseteq e^{\prec} + e$, and so by (1), $f(e^{\prec} + e) - f(e^{\prec}) \leq f(D_e) - f(D_e - e)$. □

Here is a useful observation about how Greedy computes vertices for closely-related linear orders, used in Section 3.3.4. Suppose that we have linear orders $\prec$ and $\prec'$ such that $\prec = (e_1, e_2, \ldots, e_n)$ and $\prec' = (e_1, e_2, \ldots, e_k, e'_{k+1}, e'_{k+2}, \ldots, e'_l, e_{l+1}, \ldots, e_n)$, i.e., $\prec'$ differs from $\prec$ only in that we have permuted the elements $e_{k+1}, e_{k+2}, \ldots, e_l$ of $\prec$ into some other order $e'_{k+1}, e'_{k+2}, \ldots, e'_l$ in $\prec'$. We call this move from $\prec$ to $\prec'$ a *block* modification of the block of size $b = l - k$. Then

If we've already computed $v^{\prec}$, we can compute $v^{\prec'}$ using only $O(b)$ calls to EO instead of $O(n)$ calls. (4)

This is because $e_j^{\prec} = e_j^{\prec'}$, and so $v_j^{\prec} = v_j^{\prec'}$, for $j \leq k$ and $j > l$.

2.2 *Algorithmic tools for submodular polyhedra*

Here is one of the most useful implications of submodularity:

Lemma 2.3. *If S, $T \in \mathcal{T}(x)$, then $S \cap T$, $S \cup T \in \mathcal{T}(x)$, i.e., the union and intersection of tight sets are also tight.*

Proof. Since $x(S)$ is modular, $f(S) - x(S)$ is submodular. Suppose that S, $T \in \mathcal{T}(x)$. Then by (2) and $x \in P(f)$ we get that $0 = (f(S) - x(S)) + (f(T) - x(T)) \geq (f(S \cup T) - x(S \cup T) + (f(S \cap T) - x(S \cap T)) \geq 0$, which implies that we have equality everywhere, so we get that $S \cap T$, $S \cup T \in \mathcal{T}(x)$. □

We use this to prove the useful fact that every vector in $P(f)$ is dominated by a vector in $B(f)$.

Lemma 2.4. *If $z \in P(f)$ and T is tight for z, then there exists some $y \in B(f)$ with $y \geq z$ and $y_e = z_e$ for $e \in T$.*

Proof. Apply the following generalization of the Greedy Algorithm: Start with $y = z$. Then for each $e \notin T$, compute (by brute force) $\alpha = \min\{f(S) - y(s) | e \in S\}$, and set $y \leftarrow y + \alpha\chi_{e_i}$. Since we start with $z \in P(f)$ and maintain feasibility thoughout, we always have that $\alpha \geq 0$, and the final y must still being to $P(f)$. Since only $e \notin T$ are changed, for the final y we have $y_e = z_e$ for $e \in T$.

At iteration e we find some set S_e that achieves the minimum. Thus, after iteration e, S_e is tight for y, and S_e remains tight for y for all iterations until the end. Then Lemma 2.3 says that $E = T \cup \bigcup_{e \notin T} S_e$ is also tight, and hence the final y belongs to $B(f)$. □

The Greedy Algorithm in this proof raises the natural question: Given $y \in P(f)$ and $k \in E$, find the maximum step length we can move in direction χ_k while remaining in $P(f)$. Equivalently, compute $c(k; y) = \max\{\alpha | y + \alpha\chi_k \in P(f)\}$, which is easily seen to be equivalent to $\min\{f(S) - y(S) | k \in S\}$. A similar problem arises for $y \in B(f)$. In order to stay in $B(f)$ we must lower some component l while raising component k to keep $y(E) = f(E)$ satisfied. Equivalently, compute $c(k, l; y) = \max\{\alpha | y + \alpha(\chi_k - \chi_l) \in B(f)\}$, which is easily seen to be equivalent to $\min\{f(S) - y(S) | k \in S,\ l \notin S\}$ (which is closely related to Example 1.11). This $c(k, l; y)$ is called an *exchange capacity*. If we choose a large number K and define the modular weight function $w(S)$ to be $-K$ when k but not l is in S, $+K$ if l but not k is in S, and 0 otherwise, then $f(S) - y(S) + w(S)$ is submodular, and solving SFM on this function computes $c(k, l; y)$. The same trick works for $c(k; y)$.

In fact it can be shown that the converse is also true: Given an algorithm to compute $c(k, l; y)$ or $c(k, y)$, we can use it solve general SFM. This is unfortunate, as the algorithmic framework we'll see later would like to be able to compute $c(k, l; y)$ and/or $c(k, y)$, but this is as hard as the problem we started out with. However, there is one case where computing $c(k, l; y)$ is easy. We say that (l, k) is *consecutive* in $\prec$ if $l \prec k$ and there is no j with $l \prec j \prec k$. It can be shown (Bixby et al., 1985) that the following result corresponds to a move along an edge of $B(f)$.

Lemma 2.5. *Suppose that $y = v^{\prec}$ is an extreme point of $B(f)$ arising from the Greedy Algorithm using linear order $\prec$. If (l, k) is consecutive in $\prec$, then*

$$c(k, l; y) = [f(l^{\prec} + k) - f(l^{\prec})] - [f(k^{\prec} + k) - f(k^{\prec})] = [f(l^{\prec} + k) - f(l^{\prec})] - v_k^{\prec},$$

which is nonnegative.

Proof. Since (l,k) is consecutive in $\prec$, we have $k^{\prec}=l^{\prec}+l$, and so the expression is nonnegative by (1).

Let y' be the result of the Greedy Algorithm with the linear order $\prec'$ that matches $\prec$ except that $k\prec' l$ (the same order with l and k switched). Note that y and y' match in every component except that $y_l=f(k^{\prec})-f(l^{\prec})$ whereas $y'_l=f(k^{\prec}+k)-f(l^{\prec}+k)$, and $y_k=f(k^{\prec}+k)-f(k^{\prec})$, whereas $y'_k=f(l^{\prec}+k)-f(l^{\prec})$. Thus $y'=y+(\chi_k-\chi_l)\cdot([f(l^{\prec}+k)-f(l^{\prec})]-[f(k^{\prec}+k)-f(k^{\prec})])$. Since the line segment defined by y and y' clearly belongs to $B(f)$, we get that $c(k,l;y)\geq[f(l^{\prec}+k)-f(l^{\prec})]-[f(k^{\prec}+k)-f(k^{\prec})]$. But if $f(k,l;y)$ was strictly larger, then y' would not be an extreme point, so we get the desired result. □

There is a similar result for $c(k;y)$.

For vector v, define v^- so that $v^-_e=\min(0,v_e)\leq 0$, and $v^+_e=\max(0,v_e)\geq 0$. Computing the exact value $\max_{S\subseteq E}|f(S)|$ is hard (see Section 2.3.1), but we can easily compute a good enough bound M such that $|f(S)|\leq M$ for all $S\subseteq E$: Pick any linear order $\prec$ and use Greedy to compute $v=v^{\prec}$. Then for any $S\subseteq E$, by (2) $v^-(E)\leq v(S)\leq f(S)\leq\sum_{e\in E}f(e)^+$. Thus $M=\max(|v^-(E)|,\sum_{e\in E}f(e)^+)$ works as a bound, and takes O(nEO) time to compute.

2.3 *Optimization, separation and complexity*

Suppose that we have a class $\mathcal{L}$ of linear programs that we want to solve. We say that OPT($\mathcal{L}$) is the problem of computing an optimal solution for any LP in $\mathcal{L}$. The Ellipsoid Algorithm gives a generic way to solve OPT($\mathcal{L}$) as long as we have a subroutine to solve the associated *separation problem* SEP($\mathcal{L}$): Given an LP $L\in\mathcal{L}$ and a point $\bar{x}$, either prove that $\bar{x}$ is feasible for L, or find a constraint $a^Tx\leq b$ that is satisfied by all feasible points of L, but violated by $\bar{x}$. Then Ellipsoid says that if SEP($\mathcal{L}$) is polynomial, then OPT($\mathcal{L}$) is also polynomial. In fact, Grötschel et al. (1981) were able to use polarity of polyhedra (which interchanges OPT and SEP) to also show the converse (modulo certain technicalities that we skip here):

Theorem 2.6. *OPT($\mathcal{L}$) is solvable in polynomial time iff SEP($\mathcal{L}$) is solvable in polynomial time.* □

For ordinary LPs, SEP($\mathcal{L}$) is trivially polynomial: just look through all the constraints of L and plug $\bar{x}$ into each one. Either $\bar{x}$ satisfies each one, or we find some constraint violated by $\bar{x}$, and we output that. Thus the Ellipsoid Algorithm is polynomial for ordinary LPs.

However, consider "combinatorial" LPs where the number of constraints is exponential in the number of variables, as is the case for polymatroids in Example 1.3. Here the trivial separation algorithm is no longer polynomial in the number of variables, although Theorem 2.6 is still valid.

This is important for SFM since we can use an idea from Cunningham (1983) to reduce SFM to a separation problem over a polymatroid. For $e \in E$ define $\gamma_e = f(E-e) - f(E)$. If $\gamma_e < 0$, then by (1) for any $S \subseteq E$ containing e we have $f(S-e) - f(S) \leq f(E-e) - f(E) = \gamma_e < 0$, or $f(S) > f(S-e)$. Hence e cannot belong to any solution to SFM, and without loss of optimality we can delete e from E and solve SFM on the reduced problem. Thus we can assume that $\gamma \geq 0$. Define $\tilde{f}(S) = f(S) + \gamma(S)$. Clearly $\tilde{f}$ is submodular, and for any $S \subset S+e \subseteq E$, $\tilde{f}(S+e) = \tilde{f}(S) + (f(E-e) - f(E)) + (f(S+e) - f(S)) \geq \tilde{f}(S)$ by (1), so $\tilde{f}$ is increasing. Thus $\tilde{f}$ is a polymatroid rank function.

Now consider the separation problem over $P(\tilde{f})$ with $\bar{x} = \gamma$. The optimization $\max_S \gamma(S) - \tilde{f}(S)$ yields the set S with maximum violation. But $\gamma(S) - \tilde{f}(S) = -f(S)$, so this also would solve SFM for f. So, if we could solve SEP for $P(\tilde{f})$, we could then use binary search to find a maximum violation, and hence solve SFM for f. But by Theorem 2.6 we can solve SEP for $P(\tilde{f})$ in polynomial time iff we can solve OPT for $P(\tilde{f})$ in polynomial time. But Theorem 2.1 showed that we can in fact solve OPT over $P(\tilde{f})$ in polynomial time. We have proved that the Ellipsoid Algorithm leads to a weakly polynomial algorithm for SFM (recently, Fujishige and Iwata (2002) showed that there is a direct algorithm that needs only $O(n^2)$ calls to a separation routine to solve SFM). In fact, later Grötschel, Lovász, and Schrijver were able to extend this result to show how to use Ellipsoid to get a strongly polynomial algorithm for SFM:

Theorem 2.7. [Grötschel, Lovász and Schrijver (1988)] *The Ellipsoid Algorithm can be used to construct a strongly polynomial algorithm for SFM that runs in* $\tilde{O}(n^5 \text{EO} + n^7)$ *time.* □

(The running time of this algorithm is quoted as $O(n^4\text{EO})$ in Queyranne (1998), but Lovász (2002) relates that the previous computation was "too optimistic," and that the running time above is correct.)

This theorem establishes that SFM is technically "easy", but it is unsatisfactory in atleast two ways:

- The Ellipsoid Algorithm has proven to be very slow in practice.
- This algorithm gives us very little insight into the combinatorial structure of SFM.

2.3.1 Submodular function maximization is hard

Note that in Example 1.4 we are interested in *maximizing* the submodular function, i.e., solving $\max_S f(S)$. However, this example of submodular function maximization is known to be NP Hard (even when all $\varphi_l = 1$ and all b_{rl} are 1 or $-\infty$, since it is a special case of Min Dominating Set in a graph, see Garey and Johnson (1979), Problem GT2), so the general problem is also NP Hard. (However, Shen et al. (2003) propose a related problem where we do want to solve SFM.) There are also applications where we want to maximize the submodular function in Example 1.2, leading to the Max Cut problem [see Laurent (1997)], and this is also NP Hard [see (Garey and Johnson, 1979), Problem ND16]. Nemhauser and Wolsey (1988), Section II.3.9] survey other results about maximizing submodular functions.

2.4 A useful LP formulation of SFM

Edmonds developed many of the basic concepts and results that led to SFM algorithms. In particular, all combinatorial SFM algorithms to date derive from the following idea from (Edmonds, 1970) (which considered only polymatroids, but the extension to general submodular functions is easy): Let $\mathbb{1}$ denote the vector of all ones, so that if $z \in \mathbb{R}^E$, then, $\mathbb{1}^T z = z(E)$. Suppose that we are given an upper bound vector $x \in \mathbb{R}^E$ (data, not a variable), and we want to find a maximal vector (i.e., a vector $z \in \mathbb{R}^E$ whose sum of components $\mathbb{1}^T z$ is as large as possible) in $P(f)$ subject to this upper bound. This naturally formulates as the following linear program and its dual:

$$
\begin{array}{ll}
\max \mathbb{1}^T z & \min \sum_e x_e \sigma_e + \sum_{S \subseteq E} f(S) \pi_S \\
z_e \le x_e \quad \text{for all } e \in E, & \sigma_e + \sum_{S \ni e} \pi_S = 1 \quad \text{for all } e \in E \\
z(S) \le f(S) \quad \text{for all } S \subseteq E & \sigma_e \ge 0 \quad \text{for all } e \in E \\
z_e \quad \text{free for all } e \in E & \pi_S \ge 0 \quad \text{for all } S \subseteq E.
\end{array}
$$

One consequence of submodularity is that LPs like these often have integral optimal solutions when the data is integral. Edmonds saw that these LPs not only have integral optimal solutions, but also have the special property that there is a 0–1 dual solution with exactly one π_S having value 1. Assuming that this is true, let S^* be the subset of E such that $\pi_{S^*} = 1$. Then an optimal solution must have that $\sigma = \chi(E - S^*)$ to satisfy the dual constraint, and the dual objective becomes $x(E - S^*) + f(S^*)$. We now prove this:

Theorem 2.8. *The dual LP has a* 0–1 *optimal solution with exactly one* $\pi_S = 1$. *This implies that*

$$\max\{\mathbb{1}^T z | z \in P(f), z \le x\} = \min_{S \subseteq E}\{f(S) + x(E - S)\}. \tag{5}$$

If f and x are integer-valued, then the primal LP also has an integral optimal solution.

Proof. Note that (weak duality) $z(E) = z(S) + z(E - S) \le f(S) + x(E - S)$. Hence we just need to show that an optimal solution satisfies this with equality.

Recall that $\mathcal{T}(z)$ is the family of tight sets for z. By Lemma 2.3 we have that $S^* = \cup_{T \in \mathcal{T}(z)} T$ is also tight. If z is optimal and $z_e < x_e$, then there must be some

$T \in \mathcal{T}(z)$ containing e, else we could feasibly increase z_e. Hence $z_e = x_e$ for all $e \notin S^*$. Thus we have $z(S^*) + z(E-S^*) = f(S^*) + x(E-S^*)$, and so the 0–1 π with only $\pi_{S^*} = 1$ is optimal.

If f and x are integer-valued, define $M' = \min(-M, \min_e x_e)$, so that $z = M'\mathbb{1}$ satisfies $z \in P(f)$ and $z \leq x$. Now apply Greedy starting from this z and ensuring that $z \leq x$ is preserved. By induction, z is integral at the current iteration, so that the exchange capacity used to determine the next step is also integral, so the next z is also integral. Hence the final, optimal z is also integral. □

One way we could apply this LP to SFM, which we call the *polymatroid approach*, is to recall from Section 2.3 Cunningham's reduction of SFM to a separation problem for the derived polymatroid function $\tilde{f}$ w.r.t. the point γ. Since $\tilde{f}(S) + \gamma(E-S) = f(S) + \gamma(E)$ (and since $\gamma(E)$ is a constant), minimizing $f(S)$ is equivalent to minimizing $\tilde{f}(S) + \gamma(E-S)$. As noted in Section 2.3 we can assume that $\gamma \geq 0$. Since $\tilde{f}$ is a polymatroid function we can use the detailed knowledge about polymatroids developed in (Bixby et al., 1985). Since $\tilde{f}(S) + \gamma(E-S)$ matches the RHS of (5), we can use Theorem 2.8 and its proof for help. Since we can assume that $\gamma \geq 0$, we can in fact replace the condition $z \in P(f)$ in the LHS of (5) with $z \in \tilde{P}(\tilde{f}) = \{z \in P(\tilde{f}) | z \geq 0\}$, i.e., the polymatroid itself. We can recognize optimality when we have a point $z \in \tilde{P}(\tilde{f})$ and a set $S \subseteq E$ with $z(E) = \tilde{f}(S) + \gamma(E-S)$.

Alternatively, we could use the *base polyhedron approach*, which is to use Theorem 2.8 directly without modifying f, by choosing $x = 0$. Then (5) simplifies to

$$\max\{\mathbb{1}^T z | z \in P(f), z \leq 0\} = \min_{S \subseteq E} f(S). \tag{6}$$

The RHS of this is just SFM. In this approach, it is more convenient to enforce that $z \in B(f)$ instead of $z \in P(f)$. When we switch from $z \in P(f)$ to $y \in B(f)$, to faithfully represent $z \leq 0$ we change the objective function from $\max \mathbb{1}^T z$ to $\max \sum_e \min(0, y_e)$. The proof of Theorem 2.8 shows that for optimal z and S we have $z(S) = f(S)$, and Lemma 2.4 shows that this z is dominated by some $y \in B(f)$ with $y_e = z_e$ for $e \in S$, so this change does not harm the objective value. Recall that we defined y_e^- to be $\min(0, y_e)$. Then (6) becomes

$$\max\{y^-(E) | y \in B(f)\} = \min_{S \subseteq E} f(S). \tag{7}$$

(This result could also be derived directly from LP duality and an argument similar to Theorem 2.8.) For any $y \in B(f)$ and $S \subseteq E$, $y^-(E) \leq y^-(S) \leq y(S) \leq f(S)$, which is weak duality for (7). Complementary slackness is equivalent to these inequalities becoming equalities, which is equivalent to

$y_e < 0 \Rightarrow e \in S$ (first inequality), $e \in S \Rightarrow y_e \leq 0$ (second), and $y(S) = f(S)$ (third). Thus joint optimality is equivalent to $y^-(E) = f(S)$. Note that $y(E) = f(E) = y^+(E) + y^-(E)$, or $y^-(E) = f(E) - y^+(E)$, so we can think of the LHS of (7) as $\min y^+(E)$ if we prefer.

2.5 *How do we know that our current point is feasible?*

In either approach we face a difficult problem: How can the algorithm ensure that either $z \in \tilde{P}(\tilde{f})$ or $y \in B(f)$? Since both are described by an exponential number of constraints, there is no straightforward way to verify these.

A way around this comes from the following facts: (a) Since $B(f)$ and $\tilde{P}(\tilde{f})$ are bounded, a point belongs to them iff it is a convex combination of extreme points; (b) The extreme points $v^{\prec}$ of $B(f)$ and $\tilde{P}(\tilde{f})$ are available to us from the Greedy Algorithm (or a simple modification of it, in the case of $\tilde{P}(\tilde{f})$); (c) By Carathéodory's Theorem, it suffices to use at most n extreme points for $B(f)$ (since $y \in B(f)$ satisfies the linear constraint $y(E) = f(E)$ the dimension of $B(f)$ is at most $n-1$), or $n+1$ extreme points of $\tilde{P}(\tilde{f})$. We concentrate on the $B(f)$ case here, as the $\tilde{P}(\tilde{f})$ case is similar. Therefore, to prove that $y \in B(f)$ it suffices to keep linear orders $\prec_i$ with associated extreme points $v^{\prec_i}$ and multipliers $\lambda_i \geq 0$ for i in index set $\mathcal{I}$, such that

$$\sum_{i \in \mathcal{I}} \lambda_i = 1, \quad y = \sum_{i \in \mathcal{I}} \lambda_i v^{\prec_i}, \tag{8}$$

and $|\mathcal{I}| \leq n$. To reduce clutter, we'll usually write $v^{\prec_i}$ as v^i, as we'll abuse notation by considering $i \in \mathcal{I}$ to be both $\prec_i$ and v^i. Since the Greedy Algorithm is a strongly polynomial algorithm for checking if $\prec_i$ truly does generate v^i, we can use this to prove that y really does belong to $B(f)$ in strongly polynomial time.

Most of our algorithms after this use such a representation of the current point, and they dynamically change the set $\mathcal{I}$ by adding one or more new vertices v^j to $\mathcal{I}$ to allow a move away from the current point. To keep $|\mathcal{I}|$ small, such algorithms need to reduce the set of v^i to the Carathéodory minimum from time to time. This is a simple matter, handled by subroutine ReduceV. Its input is a representation of y in terms of $\mathcal{I}$ and λ as in (8) with $|\mathcal{I}| \leq 2n$, and the output is a new representation with $|\mathcal{I}| \leq n$. It could happen that a v^j we want to add to $\mathcal{I}$ already belongs to $\mathcal{I}$. We could search $\mathcal{I}$ to detect such duplicates, but this would add an overhead of $O(n^2)$ per addition. The simpler, more efficient method that we use is to allow $\mathcal{I}$ to contain duplicates, which get removed by a later ReduceV.

Let V be the matrix whose columns are the current (too large set of) v^i's, and V' be V with a row of ones added at the top. When we reduce $\mathcal{I}$ (remove columns from V') we must compute and maintain the invariant that there are nonnegative multipliers λ_i satisfying (8), which is equivalent to

$$V'\lambda = \begin{pmatrix} 1 \\ y \end{pmatrix}.$$

By standard linear algebra manipulations (essentially converting a feasible solution to a basic feasible solution), REDUCEV finds a linearly independent set of columns of V' with corresponding new λ. Since V' has at most $2n$ columns, the initial reduction of V' to $(I\ N)$ takes $O(n^3)$ time. Each of the at most n columns subsequently deleted requires reducing at most one column to a unit vector, which can be done in $O(n^2)$ time. Thus REDUCEV takes $O(n^3)$ total time.

Carathéodory Subroutine REDUCEV

Let V be the matrix whose columns are the v^i.

Let V' be $\begin{pmatrix} \mathbb{1} \\ V \end{pmatrix}$, i.e., V with a row of ones added.

While $|\mathcal{I}| > n$ do
 Use linear algebra to reduce V' to $(I\ N)$, where i is an identity matrix. [$(I\ N)$ might have fewer rows than V'; if $|\mathcal{I}| > n$, N has at least one column]
 Let $\mathcal{B}$ index the columns of I.
 Select a column j of N, call it N^j.
 Compute the vector μ with entries N^j in positions $\mathcal{B}$, and $-\chi_j$ otherwise. [thus $(I\ N)\mu = 0 \Rightarrow V'\mu = 0 \Rightarrow V'(\lambda + \alpha\mu) = \begin{pmatrix} 1 \\ y \end{pmatrix}$ for any α]
 Compute $\alpha = \min\{-\lambda_i/\mu_i | \mu_i < 0\}$, with the min achieved at indices in $\mathcal{M}$.
 Set $\lambda \leftarrow \lambda + \alpha\mu$. [this makes $\lambda_k = 0$ for $k \in \mathcal{M}$ and keeps $\lambda \geq 0$]
 Set $\mathcal{I} \leftarrow \mathcal{I} - \mathcal{M}$, and delete columns in $\mathcal{M}$ from V'.

2.6 *From LPs to network flow-like problems*

Our descriptions of the network-like formulations of SFM are somewhat vague, since each algorithm makes different choices about the details of implementation. The two approaches outlined in Section 2.4 lead to two slightly different networks.

2.6.1 The base polyhedron approach

This approach suggests the following generic algorithm: Pick an arbitrary linear order $\prec$ and use it to generate extreme point $y = v^{\prec} \in B(f)$. Define $S^-(y) = \{e \in E | y_e < 0\}$, $S^+(y) = \{e \in E | y_e > 0\}$, and $S^0(y) = \{e \in E | y_e = 0\}$. Then if we could find $k, l \in E$ with $k \in S^-(y)$, $l \in S^+(y)$, and $c(k, l; y) > 0$, then we could update $y \leftarrow y + (\chi_k - \chi_l)c(k, l; y)$ and increase $y^-(E)$ by $c(k, l; y)$. The difficulty with this is that it would require knowing the exchange capacities $c(k, l; y)$, and this is already as hard as SFM, as discussed in Section 2.2.

However, we can at least use Lemma 2.5, which says that $c(k, l; y)$ is easily computable when (l, k) is consecutive in $\prec$. Suppose that (l, k) is consecutive in $\prec$, and let $\prec'$ be $\prec$ with k and l reversed (so that (k, l) is consecutive in $\prec'$). Then if we move step length α along the $\chi_k - \chi_l$ direction, we would have new point $y' = (1 - \alpha)y + \alpha v^{\prec'}$, and we need to add $v^{\prec'}$ to $\mathcal{I}$ to keep y in the convex hull of the v^i. Note that $y' = y + \alpha(\chi_k - \chi_l)$, as desired. This is the mechanism by which new vertices are added to $\mathcal{I}$.

More generally (see Fig. 1), suppose that (k_2, k_1) is consecutive in $\prec$, $k_1 \in S^-(y)$, $k_2 \in S^0(y)$, and $c(k_1, k_2; y) > 0$; (k_3, k_2) is consecutive in $\prec$, $k_3 \in S^0(y)$ and $c(k_2, k_3; y) > 0$ and (k_4, k_3) is consecutive in $\prec$, $k_4 \in S^+(y)$ and $c(k_3, k_4; y) > 0$ (thus $\prec$ contains the block $\cdots k_4 k_3 k_2 k_1 \cdots$). Define v^{21} to be generated by $\prec$ with k_1 and k_2 exchanged, v^{32} to be generated by $\prec$ with k_2 and k_3 exchanged, and v^{43} to be generated by $\prec$ with k_3 and k_4 exchanged. Choose $\alpha = \min(1/3, |y_{k_1}|, y_{k_4}, c(k_1, k_2; y), c(k_2, k_3; y), c(k_3, k_4; y)) > 0$, and $y' = (1 - 3\alpha)y + \alpha(v^{21} + v^{32} + v^{43})$. Then, despite the fact that none of these three changes by itself improves $y^-(E)$, doing all three changes simultaneously has the net effect of $y' = y + (\chi_{k_1} - \chi_{k_4})\alpha$, which does improve $y^-(E)$ by α, at the expense of adding three new vertices to $\mathcal{I}$.

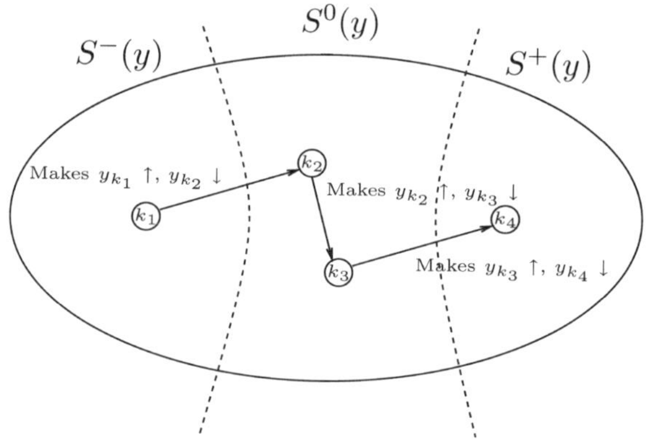

Fig. 1. Example showing why we need to consider paths of arcs in the network. None of these three changes improves $y^-(E)$ by itself, but their union does improve $y^-(E)$.

This suggests that we define a network with node set E, and arc $k \to l$ with capacity $c(k, l; v^i)$ whenever there is an $i \in \mathcal{I}$ with (l, k) consecutive in $\prec_i$. (This definition has our arcs in the reverse direction of most of the literature. We choose this convention to get the natural sense of augmenting from $S^-(y)$ towards $S^+(y)$, but somewhat nonintuitively, it means that arc $k \to l$ corresponds to $l \prec k$.) Then we look for paths from $S^-(y)$ to $S^+(y)$. If we find a path, then we "augment" by making changes as above, and call REDUCEV to keep $|\mathcal{I}|$ small.

Schrijver's Algorithm and the Hybrid Algorithm both consider changes to the v^i more general than swaps of consecutive elements. Hence both use this more liberal definition of arcs: $k \to l$ exists whenever there is an $i \in \mathcal{I}$ with $l \prec_i k$.

Lemma 2.9. *For either definition of arcs, if no augmenting path exists, then the node subset S defined as $\{e \in E|$ there is a partial augmenting path from some node $e' \in S^-(y)$ to node $e\}$ solves SFM.*

Proof. Since no augmenting path exists, $S^-(y) \subseteq S \subseteq S^-(y) \cup S^0(y)$, implying that $y^-(E) = y(S)$. Since no arcs exit S we must have that for each $i \in \mathcal{I}$, there is some $e_i \in E$ such that $S = e_i^{\prec_i}$, hence by (3) $f(S) = v^i(S)$. But then for any $T \subseteq E$, since $y \in B(f)$, $f(S) = \sum_{i \in \mathcal{I}} \lambda_i f(S) = \sum_{i \in \mathcal{I}} \lambda_i v^i(S) = y(S) = y^-(E) \leq y(T) \leq f(T)$, proving that S is an optimal solution to SFM. □

Here is another way to think about this. For some v^i in $\mathcal{I}$, consider the pattern of signs of the y_e when ordered by $\prec_i$. If $\oplus$ is a nonnegative entry and $\ominus$ is a nonpositive entry, we are trying to find an $S \subseteq E$ such that this sign pattern looks like this for every $i \in \mathcal{I}$:

$$\Big(\overbrace{\ominus \ominus \cdots \ominus \ominus}^{S} \oplus \oplus \cdots \oplus \oplus\Big).$$

If we find such an S, then (3) says that S is tight for v^i, and then by (8) S is tight also for y. Then we must have that $y^-(E) = y(S) = f(S)$, and by (7) y and S must be optimal. Thus to move closer to optimality we try to move positive components of the v^i to the right, and negative components to the left.

2.6.2 The polymatroid approach

This approach suggests a similar generic algorithm: start with $z = 0$ and try to increase $\mathbb{1}^T z$ while maintaining $z \leq \gamma$ and $z \in P(f)$. In theory, we could do this via the sort of modified Greedy Algorithm used in the proof of Theorem 2.8. The difficulty with this is that it would require knowing the exchange capacities $c(k; z)$, and this is already as hard as SFM, as discussed in Section 2.2.

We define a similar network. This time we add a source s and a sink t to E to get the node set. The arcs not incident to s and t are as above. We make arc $s \to e$ if $z_i < \gamma_e$ for some $i \in \mathcal{I}$. We make arc $e \to t$ if there is some $i \in \mathcal{I}$ such that e belongs to no tight set of v^i. Now an s–t augmenting path in this network allows us to bring z closer to γ, and $z(E)$ closer to $\tilde{f}(E)$. When there is no augmenting path, define S as the elements of E reachable from s by augmenting paths. As above, S is tight. Since $e \notin S$ is not reachable, it must have $z_e = \gamma_e$, so we have $z(E) = z(S) + z(E - S) = \tilde{f}(S) + \gamma(S)$, proving that S is optimal for SFM.

2.7 *Strategies for getting polynomial bounds*

In both cases we end up with generic algorithms that greatly resemble Max Flow/Min Cut: We have a network, we look for augmenting paths, we have a theorem that says that an absence of augmenting paths implies optimality, we have general capacities on the arcs, but we have 0–1 objective coefficients. In keeping with this analogy, we consider the flow problems to be the *primal* problems, and the "min cut" problems to be the *dual* problems, despite the fact that our original problem of SFM then turns out to be a dual problem.

This analogy helps us think about ways in which we might make these generic algorithms have polynomial bounds. There are two broad strategies that have been successful for Max Flow/Min Cut:

(1) Give a *distance-based* argument that some measure bounded by a polynomial function of n is monotone nondecreasing, and strictly increases in a polynomial number of iterations. The canonical instance of this for Max Flow is Edmonds and Karp's Shortest Augmenting Path (Edmonds and Karp, 1972) bound. They show that the length of the shortest augmenting path from s to each node is monotone nondecreasing, and that each new time an arc is the bottleneck arc on an augmenting path, this shortest distance must strictly increase by 2 at one of its nodes. With $m = |A|$, this leads to their $O(nm^2)$ bound on Max Flow. The same sort of argument is used in Goldberg and Tarjan's Push-Relabel Max Flow Algorithm (Goldberg and Tarjan, 1988) to get an $O(mn \log (n^2/m))$ bound.

This strategy is attractive since it typically yields a strongly polynomial bound without extra work, and it implies that we don't have to worry about how large the change in objective value is at each iteration. It also doesn't require precomputing the bound M on the size of f. For Max Flow, these algorithms also seem to work well in practice [see, e.g., Cherkassky and Goldberg (1997)].

(2) Give a *sufficient decrease* argument that when one iteration changes y to y', the difference in objective value between y and y' is a sufficiently large fraction of the gap between the objective value of y and the optimal objective value that we can get a polynomial bound. The

canonical instance of this for Max Flow also comes from Edmonds and Karp (1972), the Maximum Capacity Path bound. Here we augment on an augmenting path with maximum capacity at each iteration. This can be shown to reduce the gap between the current solution and an optimal solution by a factor of $(1 - 1/m)$, leading to an overall $O(m(m + n \log n) \log (nU))$ bound, where U is the maximum capacity. Capacity scaling algorithms (scaling algorithms were first suggested also by Edmonds and Karp (1972), and capacity scaling for Max Flow was suggested by Gabow (1985)) can also be seen as a way of achieving sufficient decrease.

This strategy leads to quite simple proofs of polynomiality. However, it does require starting off with the assumption that all data are integral (so that an optimality gap of less than one implies optimality), and precomputing the bound M on the size of f. Therefore it leads to algorithms which are naturally only weakly polynomial, not strongly polynomial (in fact, Queyranne (1980) showed that Maximum Capacity Path for Max Flow is *not* strongly polynomial). However, it is usually possible to modify these algorithms so they become strongly polynomial, and so can deal with nonintegral data. It is generally believed that these algorithms do not perform well in practice, partly because their average-case behavior tends to be close to their worst-case behavior, unlike the distance-based algorithms.

There are two aspects of these network-based SFM algorithms that are significantly more difficult than Max Flow. In Max Flow, if we augment flow on s–t path P, then this does not change the residual capacity of any arc not on P. In SFM, augmenting from y to y' along a path P not containing $k \to l$ can cause $c(k, l; y')$ to be positive despite $c(k, l; y) = 0$. A technique that has been developed to handle this is called *lexicographic* augmenting paths (also called *consistent breadth-first search* in Cunningham (1984), which was discovered independently by Lawler and Martel (1982) and Schönsleben (1980). It is an extension of the shortest augmenting path idea. We choose some fixed linear order on the nodes, and we select augmenting paths which are lexicographically minimum, i.e., among shortest paths, choose those whose first node is as small as possible, and among these choose those whose second node is as small as possible, etc. Then, despite the exchange arcs changing dynamically, one can mimic a Max Flow-type distance label-based convergence proof.

Second, the coefficients λ_i in the representation (8) can be arbitrarily small even with integral data. Consider this example due to Iwata: Let L be a large integer. Then f defined by $f(S) = 1$ if $1 \in S$, $n \notin S$, $f(S) = L$ if $n \in S$, $1 \notin S$, and $f(S) = 0$ otherwise is a submodular function. The base polyhedron $B(f)$ is the line segment between the vertices $v^1 = (1, 0, \ldots, 0, -1)$ and $v^2 = (-L, 0, \ldots, 0, L)$. Then the zero vector, i.e., the unique primal optimal solution, has a unique representation as in (8) with $\lambda_1 = 1 - 1/(L+1)$ and $\lambda_2 = 1/(L+1)$. This phenomenon means that it is difficult to carry through a sufficient decrease argument, since we may be forced to take very small steps to keep the λ_i nonnegative.

Another choice is whether an algorithm augments along paths as in the classic Edmonds and Karp (1972) or Dinic (1970) Max Flow Algorithms, or augments arc by arc, as in the Goldberg and Tarjan (1988) Push-Relabel Max Flow Algorithm. Augmenting along a path here is tricky since several arcs of the path might correspond to the same v^i, so that tracking the changes to $\mathcal{I}$ is difficult. In terms of worst-case running time, the Dinic (1970) layered network approach speeds up the standard Edmonds and Karp shortest augmenting path approach and has been extended to situations such as SFM by Tardos, Tovey, and Trick (1986), but the Goldberg and Tarjan approach is even faster. In terms of running time in practice, the evidence shows [see, e.g., Cherkassky and Goldberg (1997)] that for Max Flow, the arc by arc approach seems to work better in practice than the path approach. Schrijver's Algorithm uses the arc by arc method. The IFF Algorithm and its variants blend the two methods: A relaxed current point is augmented arc by arc, but the flow mediating the difference between the relaxed point and the feasible point is augmented on paths.

The algorithms have the generic outline of keeping a current point y and moving in some direction to improve $y^-(E)$. This movement is achieved by modifying the $\prec_i$ from (8) into better orders. A natural choice for a set of directions is *unit differences* $\chi_k - \chi_l$ for $k, l \in E$, since these are simple and are the edge directions of $B(f)$ (Bixby et al., 1985). Alternatively, we could choose directions based on *vertex differences*, i.e., $v^j - v^h$. When we choose unit differences, computing a step length that keeps the point inside $B(f)$ involves computing $c(k, l; y)$, which is as difficult as SFM unless l and k are a consecutive pair in $\prec$, in which case we can use Lemma 2.5. This has the virtue of having an easy-to-compute exchange capacity, but the vice of being a slow way to make big changes in the linear orders. Alternatively, we could modify larger *blocks* of elements. This has the vice that exchange capacities are hard to compute (but at least we can use (4) to quickly compute new vertices), but the virtue that big changes in the linear orders are faster. Cunningham's Algorithm uses unit differences and consecutive pairs. Schrijver's Algorithm uses unit differences, but blocks; modifying by blocks means that it is complicated to synthesize a unit difference, but it does give a good enough bound on $c(k, l; y)$. Basic IFF uses unit differences and consecutive pairs, but the Hybrid Algorithm changes to vertex differences and blocks; blocks represent vertex differences easily, and staying within $B(f)$ is easy since we are effectively just replacing v^h by v^j in (8).

Cunningham's Algorithm for General SFM (Cunningham, 1985) uses the polymatroid approach, augmenting on paths, unit differences, modifying consecutive pairs, and the sufficient decrease strategy. However, he is able to prove only a pseudo-polynomial bound. Schrijver's Algorithm (Schrijver,

2000) and Schrijver-PR use the base polyhedron approach, augmenting arc by arc, unit differences, modifying blocks, and the distance-based strategy, and so they easily get a strongly polynomial bound. Iwata, Fleischer, and Fujishige's Algorithm (IFF) (Iwata et al., 2001) uses the base polyhedron approach, augmenting both on paths and arc by arc, unit differences, modifying consecutive pairs, and the sufficient decrease strategy. IFF are able to modify their algorithm to make it strongly polynomial. Iwata's Algorithm (Iwata, 2002a) is a fully combinatorial extension of IFF. Iwata's Hybrid Algorithm (Iwata, 2002b) largely follows IFF, but adds some distance-based ideas that lead to vertex differences and modifying blocks instead of unit differences and consecutive pairs.

There is some basis to believe that the distance-based strategy is more "natural" than scaling for Max Flow-like problems such as SFM. Despite this, the running time for the IFF Algorithm is in most cases faster than the running time for Schrijver's Algorithm. However, Iwata's Hybrid Algorithm, which adds some distance-based ideas to IFF, is even faster than IFF, see Section 4.

3 The SFM algorithms

We describe Cunningham's Algorithms in Section 3.1, Schrijver's Algorithm in Section 3.2, and the IFF algorithms in Section 3.3.

3.1 Cunningham's SFM algorithms

We skip most of the details of these algorithms, as more recent algorithms appear to be better in both theory and practice.

In a series of three papers in the mid-1980s (Bixby et al., 1985; Cunningham, 1984, 1985), Cunningham developed the ideas of the polymatroid approach and gave three SFM algorithms. The first (Cunningham, 1984) is for Example 1.11, for separating point $\bar{x}$ from the matroid polytope defined by rank function r, which is the special case of SFM where $f(S) = r(S) - \bar{x}(S)$. Here Cunningham takes advantage of the special structure of f and carefully analyzes how augmentations happen in a lexicographic shortest augmenting path framework. This allows him to prove that the algorithm needs $O(n^3)$ total augmenting paths; each path adds $O(n)$ new v^i (which are the incidence vectors of independent sets in this case) to $\mathcal{I}$, so when it doesn't call REDUCEV the algorithm must manage $O(n^4)$ vertices in $\mathcal{I}$. To construct the graph of augmenting paths, for each of the $O(n^4)$ $i \in \mathcal{I}$ and each of the $O(n^2)$ pairs $k, l \in E$, we must consider whether i implies an arc $k \to l$, for a total of $O(n^6\text{EO})$ time per augmenting path. This yields a total time of $O(n^9\text{EO})$, and a fully combinatorial algorithm for this case (without calling REDUCEV). If we do use REDUCEV, then the size of $\mathcal{I}$ stays $O(n)$, so the time per augmentation is now only $O(n^3\text{EO})$, for a total of $O(n^6\text{EO})$

(although the resulting algorithm is no longer fully combinatorial, but only strongly polynomial).

In the second paper, Bixby et al. (1985) extend some of these ideas to the general case. It uses the polymatroid approach and augmenting on paths. Because of degeneracy, there might be several different linear orders that generate the same vertex v of $\tilde{P}(\tilde{f})$. A given pair (l, k) might be consecutive in some of these orders but not others. They show that, for each vertex v, there is a partial order $\prec_v$ (note that $\prec_v$ is in general *not* a linear order) such that $c(k, l; v) > 0$ iff k *covers* l in $\prec_v$, i.e., if $l \prec_v k$ but there is no $j \in E$ with $l \prec_v j \prec_v k$ (if $\prec_v$ is linear, then k covers l in $\prec_v$ iff (l, k) is consecutive). Furthermore, they gave an $O(n^2 \text{EO})$ algorithm for computing $\prec_v$. Finally, they note that if k covers l in $\prec_v$, then $c(k, l; v)$ (and also $c(k; v)$) can be computed in O(EO) time, similar to Lemma 2.5. They define the arcs to include $k \rightarrow l$ if there is some $i \in \mathcal{I}$ such that k covers l in v^i, and thus they know that the capacity of every arc is positive. When this is put into the polymatroid approach using ReduceV, it is easy to argue that no set of vertices $\mathcal{I}$ can repeat, leading to a finite algorithm.

In the third paper, Cunningham (1985) modified this second algorithm into what we call Cunningham's Algorithm for General SFM. It adds a weak version of the sufficient decrease strategy to the second algorithm. The fact that the λ_i can be arbitrarily small (discussed in Section 2.7) prevents Cunningham from using a stronger sufficient decrease argument. Suppose that we restrict our search for augmenting paths only to arcs $s \rightarrow e$ with $\gamma_e - z_e \geq 1/Mn(n+1)^2$ and arcs $k \rightarrow l$ with $\lambda_i c(k, l; z) \geq 1/M(n+1)^2$. If we find an augmenting path P of such arcs, then it can be seen that augmenting along P increase $\mathbb{1}^T z$ by atleast $1/M(n+1)^2$. Then the key to Cunningham's argument is the following lemma:

Lemma 3.1. [(Cunningham, 1985), Theorem 3.1] *If no such path exists, then there is some* $S \subseteq E$ *with* $z(E) > f(S) + \gamma(E - S) - 1$, *and because all data are integral, we conclude that* S *solves SFM.* □

Cunningham suggests some speedups, which are essentially variants of implicit capacity scaling (look for augmenting paths of capacity at least K until none are left, then set $K \leftarrow K/2$ until $K < 1/M(n+1)^2$ and maximum capacity augmenting path. These lead to the overall time bound of $O(Mn^6 \log(Mn) \cdot \text{EO})$, which is pseudo-polynomial.

3.2 *Schrijver's SFM algorithm*

Schrijver's Algorithm (Schrijver, 2000) uses the base polyhedron approach, augmenting arc by arc, modifying blocks, and the distance-based strategy. Schrijver's big innovation is to avoid being constrained to consecutive pairs, but to allow arcs $k \rightarrow l$ if $l \prec_i k$ for some $i \in \mathcal{I}$, even if l and k are not consecutive in $\prec_i$. This implies that Schrijver has a looser definition of arcs than some other algorithms. Of course, the problem that computing $c(k, l; v)$ is equivalent to SFM still remains; Schrijver's solution is to compute a lower bound on $c(k, l; v)$.

Let's focus on a particular arc $k \to l$, associated with $\prec_h$, which we'd like to include in an augmentation. For simplicity call $\prec_h$, just $\prec$ and v^h just v. Define $(l, k]_\prec = \{e \in E | \ l \prec e \preceq k\}$ (and similarly $[l, k]_\prec$ and $[l, k)_\prec$), so that $[l, k]_\prec = \emptyset$ if $k \preceq l$. Then Lemma 2.5 says that $c(k, l; v)$ is easy to compute if $|(l, k]_\prec| = 1$. In order to get combinatorial progress, we would like to represent the direction we want to move in, $v + \alpha(\chi_k - \chi_l)$, as a combination of new vertices w^j with linear orders $\prec'_j$ with $(l, k]_{\prec'_j} \subset (l, k]_\prec$ for each j. That is, we would like to drive arcs which are not consecutive more and more towards being consecutive.

Schrijver gives a subroutine for achieving this, which we call ExchBd$(k, l; \prec)$ (and describe in Section 3.2.1). It chooses the following linear orders to generate its w^j: For each j with $l \prec j$ define $\prec^{l,j}$ as the linear order with j moved just before l. That is, if $\prec$'s order is

$$\ldots s_{a-1} s_1 l t_1 t_2 \ldots t_b j u_1 u_2 \ldots,$$

then $\prec^{t,j}$'s order is

$$\ldots s_{a-1} s_a j l t_1 t_2 \ldots t_b u_1 u_2 \ldots.$$

Note that if $l \prec j \preceq k$, then $(l, k]_{\prec^{l,j}} \subset (l, k]_\prec$, as desired.

ExchBd $(k,l; \prec)$ has the following properties. The input is linear order $\prec$ and $k,l \in E$ with $l \prec k$. The output is a step length $\alpha \geq 0$, and the collection of vertices $w^j = v^{\prec^{l,j}}$ with coefficients $\mu_j \geq 0$ for $j \in \mathcal{J} = (l, k]_\prec$. This implies that $|\mathcal{J}| \leq |(l,k]_\prec| \leq n$. The μ_j satisfy $\sum_{j \in \mathcal{J}} \mu_j = 1$, and

$$v^\prec + \alpha(\chi_k - \chi_l) = \sum_{j \in \mathcal{J}} \mu_j w^j. \tag{9}$$

That is, $v^\prec + \alpha(\chi_k - \chi_l)$ is a convex combination of the w^j. Also, this implies that $v^\prec + \alpha(\chi_k - \chi_l) \in B(f)$, and hence that $\alpha \leq c(k, l; v)$. We show below that ExchBd takes $O(n^2\text{EO})$ time.

We now describe Schrijver's Algorithm, assuming ExchBd as a given. We actually present a Push-Relabel variant due to Fleischer and Iwata (2001) that we call Schrijver-PR, because it is simpler to describe, and seems to run faster in practice than Schrijver's original algorithm (see Section 4). Schrijver-PR originally also had a faster time bound than Schrijver, but Vygen (2003) recently showed that in fact the time bound for Schrijver's Algorithm is the same as for Schrijver-PR. Roughly speaking, Schrijver's original algorithm is similar to Dinic's Max Flow Algorithm (Dinic, 1970), in that it uses exact distance labels to define a layered network, whereas Schrijver-PR is similar to Goldberg and Tarjan's Push-Relabel Max Flow Algorithm (Goldberg and Tarjan, 1988), in that it uses approximate distance labels to achieve the same thing.

Similar to Goldberg and Tarjan (1988), we put nonnegative, integer *distance labels d* on the nodes. We call labels *d valid* if $d_e = 0$ for all $e \in \mathcal{S}^-(y)$, and we have $d_l \leq d_k + 1$ for every arc $k \to l$ (i.e., whenever $l \prec_i k$ for some $i \in \mathcal{I}$). This implies that d_e is a lower bound on the number of arcs in a shortest path from $S^-(y)$ to e, so that $d_e < n$; we use $d_e = n$ to signify that no path from $S^-(y)$ to e exists. We choose $d_e = 0$ for all $e \in E$ as an initial valid labeling.

The algorithm defines the set of *active* nodes as $\mathcal{A} = \{e \in S^+(y) | \ d_e < n\}$, i.e., the set of positive nodes which still have a hope of being decreased. The basic idea of the algorithm is to choose a node $l \in \mathcal{A}$ with maximum d_l, some node k such that $d_k = d_l - 1$ and arc $k \to l$ exists due to $l \prec_h k$ for some $h \in \mathcal{I}$, and call PUSH, which applies EXCHBD repeatedly to $k \to l$. Each EXCHBD decreases y_l, and makes the $(l, k]_{\prec_i}$ smaller. We apply EXCHBD until either (1) y_l drops to 0, called *nonsaturating*, or (2) arc $k \to l$ disappears because $k \prec_i l$ for all $i \in \mathcal{I}$ (i.e., $|(l,k]_{\prec_i}| = 0$ for all $i \in \mathcal{I}$), called *saturating*. To keep combinatorial monotonicity, we always choose an associated $\prec_h$ achieving $\max_{i \in \mathcal{I}} |(l,k]_{\prec_i}|$. To be lexicographic, we scan through the possible nodes k in a fixed linear order.

When we work on arc $k \to l$, we are increasing y_k and decreasing y_l. We enforce that y_l stays nonnegative (since $d_l > 0$, if we allowed y_l to become negative, this would violate that $d_e = 0$ for $e \in S^-(y)$), but if y_k is negative, we allow it to become positive. To see the algebraic details of this, note that (8) and (9) imply that

$$y + \alpha\lambda_h(\chi_k - \chi_l) = \sum_{i \neq h} \lambda_i v^i + \lambda_h \sum_j \mu_j w^j. \tag{10}$$

If $\alpha\lambda_h > y_l$, then this would make $y_l < 0$, which we don't allow. So we set $\beta = \min(y_l, \alpha\lambda_h)$, and we want to take the step $y + \beta(\chi_k - \chi_l)$. Note that $\beta = y_l$ means that the new $y_l = 0$, leading to a nonsaturating PUSH; and $\beta = \alpha\lambda_h$ means that h leaves $\mathcal{I}$, so there is one less index in $\mathcal{I}$ with a maximum value of $|(l,k]_{\prec_i}|$, so we are closer to being saturating. To get this effect we add $(1 - \beta/(\alpha\lambda_h))$ times (8) to $\beta/(\alpha\lambda_h)$ times (10) to get:

$$y + \beta(\chi_k - \chi_l) = \sum_{i \neq h} \lambda_i v^i + (\lambda_h - \beta/\alpha)v^h + \sum_j (\beta\mu_j/\alpha)w^j.$$

We put these pieces together into the subroutine PUSH(k, l).

Push (k, l) Subroutine for the Schrijver-PR Algorithm

While $y_l > 0$ and arc $k \to l$ exists,
  Select h that solves $\max_{i \in \mathcal{I}} |(l,k]_{\prec i}|$.
  Call EXCHBD $(k, l; v^h)$ to get $\alpha, \mathcal{J}, \mu_j, w^j$.
  Set $\beta = \min(y_l, \alpha\lambda_h)$.
  Update $y \leftarrow y + \beta(\chi_k - \chi_l)$, $\mathcal{I} - \mathcal{I} \cup \mathcal{J}$, and $\lambda_h \leftarrow \lambda_h - \beta/\alpha$.
  For $j \in \mathcal{J}$, set $\lambda_j \leftarrow \beta\mu_j/\alpha$.
  Call REDUCEV.

If we have selected l but every arc $k \to l$ has $d_k \geq d_l$ (i.e., no arc $k \to l$ satisfies the distance criterion for applying PUSH(k, l) that $d_k = d_l - 1$), then we apply RELABEL(l).

RELABEL(l) Subroutine for the Schrijver-PR Algorithm

Set $d_l \leftarrow d_l + 1$.
If $d_l = n$, then $\mathcal{A} \leftarrow \mathcal{A} - l$.

Now we are ready to describe the whole algorithm. For simplicity, assume that $E = \{1, 2, \ldots, n\}$. To get our running time bound, we need to ensure that for each fixed node l, we do at most n saturating PUSHes before RELABELing l. To accomplish this, we do PUSHes to l from nodes k for each k in order from 1 to n; to ensure that we restart where we left off if PUSHes to l are interrupted by a nonsaturating PUSH, we keep a pointer p_l for each node l that keeps track of the next k where we want to do a PUSH(k, l).

The Schrijver-PR Algorithm for SFM

Initialize by choosing $\prec_1$ to be any linear order, $y = v^1$, and $\mathcal{I} = \{1\}$.
Set $d = 0$ and $p = \mathbb{1}$.
Compute $S^-(y)$ and $S^+(y)$ and set $\mathcal{A} = S^+(y)$.
While $\mathcal{A} \neq \emptyset$ and $S^-(y) \neq \emptyset$,
  Find l solving $\max_{e \in \mathcal{A}} d_e$.
  [try to push to max distance node l]
  While $p_l \leq n$ do
  [scan through possible nodes that could push to l]
    If $d_{pl} = d_l - 1$ then
      PUSH(p_l, l)
      IF $y_l = 0$ set $\mathcal{A} \leftarrow \mathcal{A} - l$, and break out of the "While p_l" loop.
    SET $p_l \leftarrow p_l + 1$.
  IF $p_l > n$, set $p_l = 1$ and RELABEL (l).
Compute S as the set of nodes reachable from $S^-(y)$, and return S.

We now prove that this works, and give its running time. We give one big proof, but we pick out the key claims along the way in boldface.

Theorem 3.2. *Schrijver-PR correctly solves SFM, and runs in* $O(n^7 \text{EO} + n^8)$ *time.*

Proof.
Distance labels d stay valid. We use induction on the iterations of the algorithm; d starts out being valid. Only PUSH and RELABEL could make d invalid.

PUSH preserves validity of *d*. Suppose that a call to EXCHBD($k, l; v^h$) in PUSH(k, l) introduces a new arc $u \to t$. Since $u \to t$ didn't exist before we must have had $u \prec_h t$, and since it does exist now we must have that $t \prec_h^{l,j} u$ for some $j \in (l, k]_{\prec_h}$. The only way for this to happen is if $j = t$ and we had $l \preceq_h u \prec_h t \preceq_h k$ and now have $t \prec_h^{l,t} l \preceq_h^{l,t} u \prec_h^{l,t} k$. Doing PUSH($k, l$) means that $d_k + 1 = d_l$. Since d was valid before the PUSH(k, l), we have $d_t \leq d_k + 1 = d_l \leq d_u + 1$, so d is still valid.

RELABEL preserves validity of *d*. We must show that when the algorithm calls RELABEL(t), every arc $u \to t$ has $d_u \geq d_t$. Since RELABEL(t) gets called when $p_t = n + 1$, if we can show that $u < p_t$ and $u \to t$ an arc imply that $d_u \geq d_t$, then we are done. We prove this by induction; it is trivially true when $p_t = 1$, and so also true just after RELABEL(t). A RELABEL(u) for $u \neq t$ also only improves things, so we need worry only about PUSHes. The algorithm increases p_l only when all $p_l \to l$ arcs have been made to disappear in PUSH, so the only problem that could arise is when a call to PUSH(k, l) (with $k = p_l$) creates a new arc $u \to t$. Suppose that the claim remains true until this point. The previous paragraph showed that in this case we had $l \preceq_h u \prec_h t \preceq_h k$, implying that $d_t \leq d_k + 1 = d_l \leq d_u + 1$. If $t = k$ then $d_k = d_t$ which gives that $d_u \geq d_t$; similarly if $u = l$. If $k < p_t$, then $t \prec_h k$ implies that $k \to t$ was an arc, an induction gives that $d_k \geq d_t$, implying $d_u \geq d_t$. Otherwise, we have $p_l = k \geq p_t$, and we are assuming that $u < p_t$, so we get $u < p_l$. Then $l \prec_h u$ implies that $u \to l$ was an arc, so induction gives $d_u \geq d_l$, again implying that $d_u \geq d_t$.

The algorithm performs at most n^2 total RELABELS. Each RELABEL(l) increases d_l by 1, and $d_l \leq n$, so we call RELABEL(l) at most n times, so that the total is at most n^2.

The algorithm performs at most n^3 total saturating PUSHes. Because of the p_l, for each l we do at most n saturating PUSHes to l before doing a RELABEL(l). Since there are at most n RELABEL(l)s, there are at most n^2 saturating PUSHes to l, or n^3 total saturating PUSHes.

The algorithm performs at most n^3 total nonsaturating PUSHes. We have a nonsaturating PUSH(k, l) because y_l drops to 0. For the next nonsaturating PUSH to happen at l, some other PUSH(l, u) must make $y_t > 0$ first. Since we always PUSH from the highest label, and since distance labels are monotone nondecreasing, we must have that d_u at the time of the PUSH(l,u) is at last one larger than d_l at the time of the nonsaturating PUSH, so a RELABEL(u) must have happened in between. Since there are at most n^2 RELABELS, and each RELABEL can reactivate at most n such l's, there are at most n^3 nonsaturating PUSHes.

Each call to PUSH (k, l) iterates at most n^2 times. An iteration of the while loop of PUSH(k, l) might cause $y_l = 0$ (a nonsaturating PUSH), in which case we exit.

Each iteration that does not cause $y_l = 0$ has $\beta = \alpha\lambda_h$, meaning that the new coefficient of v^h is 0, so that h drops out of $\mathcal{I}$. This either reduces $\max_{i \in \mathcal{I}} |(l, k]_{\prec_i}|$, or reduces the number of $i \in \mathcal{I}$ achieving this maximum (calling REDUCEV can only help here). Since $|(l, k]_{\prec_i}| < n$, this implies the claim.

The running time is $\mathbf{O(n^7 EO + n^8)}$. There are $O(n^3)$ calls to PUSH, each of which iterates at most n^2 times, and each iteration calls EXCHBD and REDUCEV once each, for a total of $O(n^5)$ calls to EXCHBD and REDUCEV. Each call to EXCHBD costs $O(n^2 EO)$ time, and each call to REDUCEV costs $O(n^3)$ time.

The algorithm terminates with an optimal solution. By Lemma 2.9. □

3.2.1 The exchange capacity bound subroutine

Recall that for each $j \in (l, k]_\prec$ we define $\prec^{l,j}$ as the linear order with j moved just before l. The task of EXCHBD $(k, l; \prec)$ is to find a step length $\alpha \geq 0$ and a representation of $v^\prec + \alpha(\chi_k - \chi_l)$ as a convex combination of vertices $v^{l,j}$ corresponding to the linear orders $\prec^{l,j}$.

Let $q = |(l, k]_\prec|$, enumerate $\prec$ as $\ldots l u_1 u_2 \ldots u_{q-1} k \ldots$, and define $u_q = k$. Define $V^{l,k}$ to be the matrix whose columns are the $v^{l,j}$ for $j \in (l, k]_\prec$, so that $V^{l,k}$ has n rows and q columns, and $V^\prec$ to be the matrix of the same dimension with every column equal to $v^\prec$. Since $\prec^{l,j}$ is the same order as $\prec$ except for $j \in [l, k]_\prec$, by (4) the only places where two columns of $V^{l,k}$ might differ is in the $q+1$ rows $[l, k]_\prec$. Again using $\oplus$ for nonnegative and $\ominus$ for nonpositive, the next lemma proves that the sign pattern of this submatrix of $V^{l,k} - V^\prec$ is:

$$
\begin{array}{c} \\ l \\ u_1 \\ u_2 \\ u_3 \\ \vdots \\ k = u_q \end{array}
\begin{array}{c} \begin{array}{ccccc} v^{l,u_1} & v^{l,u_2} & v^{l,u_3} & \cdots & v^{l,u_q} \end{array} \\
\begin{pmatrix}
\ominus & \ominus & \ominus & \cdots & \ominus \\
\oplus & \ominus & \ominus & \cdots & \ominus \\
0 & \oplus & \ominus & \cdots & \ominus \\
0 & 0 & \oplus & \cdots & \ominus \\
\vdots & \vdots & \vdots & \ddots & \vdots \\
0 & 0 & 0 & \cdots & \ominus
\end{pmatrix} \end{array}
\qquad (11)
$$

Lemma 3.3. *If $h \in [l, u)_\prec$, then $v_h^{l,u} \geq v_h^\prec$. If $h = u$, then $v_h^{l,u} \geq v_h^\prec$. Otherwise $(h \notin [l, u]_\prec)$, $v_h^{l,u} = v_h^\prec$.*

Proof. If $h \in [l, u)_\prec$, then $h^{\prec^{l,u}} = h^\prec + u$. Thus, by Greedy and (1), $v_h^{l,u} = f(h^{\prec^{l,u}} + h) - f(h^{\prec^{l,u}}) = f(h^\prec + h + u) - f(h^\prec + u) \leq f(h^\prec + h) - f(h^\prec) = v_h^\prec$. Also, $u^{\prec^{l,u}} = u^\prec - [l, u)_\prec$, so again $f(u^\prec + u) - f(u^\prec) = v_u^\prec$. Finally, for $h \notin [l, u]_\prec$ we have that $h^{\prec^{l,u}} = h^\prec$, so by Greedy $v_h^{l,u} = f(h^{\prec^{l,u}} + h) - f(h^{\prec^{l,u}}) = f(h^\prec + h) - f(h^\prec) = v_h^\prec$. □

Suppose that diagonal element $v_u^{l,u} - v_u^{\prec}$ of (11) equals zero. Then, since $v^{l,u}(E) = v^{\prec}(E) = f(E)$, from (11) we would get $v^{l,u} = v^{\prec}$. In this case we choose $\alpha = 0$ and represent $v^{\prec} + \alpha(\chi_k - \chi_l) = v^{\prec}$ as $1 \cdot v^{l,u}$ for our convex combination.

Suppose instead that all diagonal elements of (11) are positive. Consider the following equation in unknowns η:

$$(V^{l,k} - V^{\prec})\eta = \chi_k - \chi_l. \tag{12}$$

Since (11) is triangular with positive diagonal, (12) has a unique solution with $\eta > 0$. We than set $\alpha = 1/\eta(E)$ and $\mu = \alpha\eta$, which then satisfy $(V^{l,k} - V^{\prec})\mu = \alpha(\chi_k - \chi_l)$. Since $\mu(E) = 1$, this is equivalent to (9), as desired. Suppose that $q = 1$, i.e., (l, k) is consecutive in $\prec$. Then $\prec^{l,k}$ is just $\prec$ with l and k interchanged. In this case Lemma 2.5 tells us that $v^{l,k} = v^{\prec} + c(k, l; v^{\prec})(\chi_k - \chi_l)$. This implies that when $c(k, l; v^{\prec}) > 0$, the solution of (12) in this case $\eta = 1/c(k, l; v^{\prec})$, which means that we would compute $\alpha = c(k, l; v^{\prec})$. Thus in this case, as we would expect, ExchBd computes the exact exchange capacity.

Now we consider the running time of ExchBd. Computing the $v^{l,u}$ requires at most n calls to Greedy, which takes $O(n^2\text{EO})$ time (we can save time in practice by using (4), but this doesn't seem to improve the overall bound). Setting up and solving (12) takes only $O(n^2)$ time (because it is triangular), for a total of $O(n^2\text{EO})$ time.

3.3 *Iwata, Fleischer and Fujishige's SFM algorithms*

We describe the weakly polynomial version of the IFF algorithm in Section 3.3.1, a strongly polynomial version in Section 3.3.2, Iwata's fully combinatorial version in Section 3.3.3, and Iwata's faster Hybrid Algorithm in Section 3.3.4.

3.3.1 *The basic weakly polynomial IFF algorithm*

Iwata, Fleischer, and Fujishige's Algorithm (IFF) (Iwata et al., 2001) uses the base polyhedron approach, augmenting both on paths and arc by arc, modifying consecutive pairs, and the sufficient decrease strategy. IFF are able to modify their algorithm to make it strongly polynomial. The IFF Algorithm would like to use capacity scaling. A difficulty is that here the "capacities" are derived from the values of f, and scaling a submodular function typically destroys its submodularity. One way to deal with this is suggested by Iwata (1997) in the context of algorithms for Submodular Flow: Add a sufficiently large perturbation to f and the scaled function is submodular. However this proved to be slow, yielding a run time of $\tilde{O}(n^7\text{EO})$ compared to $\tilde{O}(n^4\text{EO})$ for the current fastest algorithm for Submodular Flow (Fleischer et al., 2002).

A different approach is suggested by Goldberg and Tarjan's Successive Approximation Algorithm for Min Cost Flow (Goldberg and Tarjan, 1990), using an idea first proposed by Bertsekas (1986): Instead of scaling the data, relax the data by a parameter δ and scale δ instead. As δ is scaled closer to zero, the scaled problem more closely resembles the original problem, and when the scale factor is small enough and the data are integral, it can be shown that the scaled problem gives a solution to the original problem. Tardos-type (Tardos, 1985) proximity theorems can then be applied to turn this weakly polynomial algorithm into a strongly polynomial algorithm.

The idea here is to relax the capacities of arcs by δ. This idea was first used for Min Cost Flow by Ervolina and McCormick (1993). For SFM, every pair of nodes could potentially form an arc, so we introduce a complete directed network on nodes E with *relaxation arcs* $R=\{k\to l \mid k\neq l\in E\}$. We maintain $y\in B(f)$ as before, but we also maintain a flow x in (E,R). We say that x is *δ-feasible* if $0\leq x_{kl}\leq +\delta$ for all $k\neq l\in E$. We enforce that x is δ-feasible, and that for every $k\neq l\in E$, $x_{kl}\cdot x_{lk}=0$, i.e., at least one of x_{kl} and x_{lk} is zero. (Some versions of IFF instead enforce that for all $k\neq l\in E$, $x_{kl}=-x_{lk}$, i.e., that x is skew-symmetric, which leads to a simpler description. However, we later sometimes have infinite bounds on some arcs of R which are incompatible with skew-symmetry, so we choose to use this more general representation from the start.) Recall that ∂x: $E\to\mathbb{R}$ is defined as $\partial x_k=\sum_l x_{kl}-\sum_j x_{jk}$. We perturb $y\in B(f)$ by ∂x to get $z=y+\partial x$. If we define $\kappa(S)=|S|\cdot|E-S|$ (which is $|\delta(S)|$ in (E,R), and hence submodular), we could also think of this as relaxing the condition $y\in B(f)$ to $z\in B(f+\delta\kappa)$ (this is the relaxation originated by (Iwata, 1997)). The perturbed vector z has enough flexibility that we are able to augment z on paths even though we augment the original vector y arc by arc. The flow x buffers the difference between these two augmentation methods.

The idea of scaling δ instead of $f+\delta\kappa$ is developed for use in Submodular Flow algorithms by Iwata, McCormick, and Shigeno (1999), and in an improved version by Fleischer, Iwata, and McCormick (2002). Indeed, some parts of the IFF SFM Algorithm (notably the Swap subroutine below) were inspired by the Submodular Flow algorithm from (Fleischer et al., 2002). It is formally similar to an excess scaling Min Cost Flow algorithm of Goldfarb and Jin (1999), with the flow x playing the role of arc excesses.

As $\delta\to 0$, Lemma 3.4 below shows that $\mathbb{1}^T z^-$ converges towards $\mathbb{1}^T y^-$, so we concentrate on maximizing $\mathbb{1}^T z^-$ instead of $\mathbb{1}^T y^-$. We do this by looking for augmenting paths from S^- to S^+ with capacity at least δ (called *δ-augmenting paths*). We modify y arc by arc as needed to try to create further such augmenting paths for z. Roughly speaking, we call z *δ-optimal* if there is no further way to construct a δ-augmenting path. Augmenting on δ-augmenting paths turns out to imply that we make enough progress at each iteration that the number of iterations in a δ-scaling phase is strongly polynomial (only the number of scaling phases is weakly polynomial).

The outline of the outer scaling framework is now clear: We start with $y = v^1$ for an arbitrary order $\prec_1$, and a sufficiently large value of δ (it turns out that $\delta = |y^-(E)|n^2 \leq 2M/n^2$ suffices). We then cut the value of δ in half, and apply a REFINE procedure to make the current values δ-optimal. We continue until the value of δ is small enough that we know that we have an optimal SFM solution (it turns out that $\delta = 1/n^2$ suffices). Thus the number of outer iterations is $1 + \lfloor \log_2(2M/n^2)/(1/n^2) \rfloor = O(\log M)$.

> **IFF Outer Scaling Framework**
>
> Initialize by choosing $\prec_1$ to be any linear order, $y = v^1$, and $\mathcal{I} = \{1\}$.
> Initialize $\delta = |y^-(E)|/n^2$, $x = 0$, and $z = y$.
>   [$z = y + \partial x$ is δ-optimal]
> While $\delta \geq 1/n^2$, [when $\delta < 1/n^2$ we are optimal]
>   Set $\delta \leftarrow \delta/2$.
>   Call REFINE. [converts 2δ-optimality to δ-optimality]
> Return last approximate solution from REFINE as optimal SFM solution.

Since the outer scaling framework cuts δ in half, REFINE starts by halving the 2δ-feasible flow x to make it a δ-feasible flow.

To find δ-augmenting paths, we must restrict the starting and ending nodes to have sufficiently large and small values of z_l, so we define $S^{-\delta}(z) = \{l \in E | z_l \leq -\delta\}$, and $S^{+\delta}(z) = \{l \in E | z_l \geq +\delta\}$. Further define the subset of arcs of R with residual capacity δ as $R(\delta) = \{k \to l \mid x_{kl} = 0\}$. We look for a directed augmenting path P from some $k \in S^{-\delta}(z)$ to some $l \in S^{+\delta}(z)$ using only arcs of $R(\delta)$. Since P contains only relaxation arcs (no exchange arcs), somewhat surprisingly we do *not* need to ensure that P is a lexicographic shortest path, or even a shortest path at all. Define the set $S = \{l \in E \mid$ there is a path in $(E, R(\delta))$ from $S^{-\delta}(z)$ to $l\}$. If we find such a P (if $S \cap S^{+\delta}(z) \neq \emptyset$), we call AUGMENT($P$) to increase x on arcs in P by δ. If $t \to u \in P$, then $x_{tu} = 0$ and the old contribution of $t \to u$ and $u \to t$ to ∂x_t is $-x_{ut}$. AUGMENT(P) updates $x_{tu} = \delta - x_{ut}$ and $x_{ut} = 0$, so that the new contribution of $t \to u$ and $u \to t$ to ∂x_t is $\delta - x_{ut}$, which is δ larger than before as desired (and their contribution to ∂x_u decreases by δ). Over all arcs of P, this has the effect of increasing ∂x_k by δ, decreasing ∂x_l by δ, and leaving ∂x_h the same for $h \neq k,l$. The corresponding update to $z = y + \partial x$ increases z_k by δ, decreases z_1 by δ, and leaves z_h the same for $h \neq k, l$, thereby increasing $\mathbb{1}^T z^-$ by δ. The running time of AUGMENT is dominated by recomputing S, which takes $O(n^2)$ time (since $|R| = O(n^2)$).

IFF Subroutine Augment (P) for P from $k \in S^{-\delta}(z)$ to $l \in S^{+\delta}(z)$

For all $t \to u \in P$ do [augment each arc of P, update $R(\delta)$]
  Set $x_{tu} \leftarrow \delta - x_{ut}, x_{ut} \leftarrow 0$.
  If $x_{tu} > 0$ set $R(\delta) \leftarrow R(\delta) - (t \to u)$, and set $R(\delta) \leftarrow R(\delta) \cup (u \to t)$.

Set $\left\{ \begin{matrix} z_k \leftarrow z_k + \delta \\ z_l \leftarrow z_l - \delta \end{matrix} \right\}$.

[update z, $S^{-\delta}(z)$, $S^{+\delta}(z)$, and S.]

If $\left\{ \begin{matrix} z_k > -\delta & \text{set} & S^{-\delta}(z) \leftarrow S^{-\delta}(z) - k \\ z_l < +\delta & \text{set} & S^{+\delta}(z) \leftarrow S^{+\delta}(z) - l \end{matrix} \right\}$.

Set $S = \{l \in E \mid \exists$ a path in $(E, R(\delta))$ from $S^{-\delta}(z)$ to $l\}$.

What do we do if no augmenting path from $S^{-\delta}(z)$ to $S^{+\delta}(z)$ using only arcs of $R(\delta)$ exists? Suppose that there is some $i \in \mathcal{I}$ such that (l, k) is consecutive in $\prec_i$, $k \in S$ and $l \notin S$. We call such a $(k, l; v^i)$ a *boundary triple*, and let B denote the current set of boundary triples. Note that if $\prec_i$ has no boundary triple, then all $s \in S$ must occur first in $\prec_i$, implying by (3) that $v^i(S) = f(S)$. Thus

$$\begin{aligned} &\text{If } B = \emptyset, \text{ then } v^i(S) = f(S) \ (S \text{ is tight for } v^i) \text{ for all } i \in \mathcal{I}, \\ &\text{so that } y(S) = \textstyle\sum_{i \in \mathcal{I}} \lambda_i v^i(S) = \sum_{i \in \mathcal{I}} \lambda_i f(S) = f(S), \\ &\text{and so } S \text{ is also tight for } y. \end{aligned} \tag{13}$$

We develop a Swap$(k, l; v^i)$ procedure below (called *double-exchange* in Fleischer et al. (2002); Iwata et al. (2001)) to deal with boundary triples.

Note that two different networks are being used here to change two different sets of variables that are augmented in different ways: Augmentations happen on paths, affect variables z, and are defined by and implemented on the network of relaxation arcs. Swaps happen arc by arc, affect variables y, and are defined by and implemented on the network of arcs of potential boundary triples (where $k \to l$ is an arc iff (l, k) is consecutive in some $\prec_i$). The flow variables x are used to mediate between these different changes.

Let $\prec_j$ be $\prec_i$ with k and l interchanged. Then Lemma 2.5 says that

$$v^j = v^i + c(k, l; v^i)(\chi_k - \chi_l). \tag{14}$$

Then (14) together with (8) implies that

$$y + \lambda_i c(k, l; v^i)(\chi_k - \chi_l) = \lambda_i v^j + \sum_{h \neq i} \lambda_h v^h, \tag{15}$$

so we could take a step of $\lambda_i c(k, l; v^i)$ in direction $\chi_k - \chi_l$ from y. The plan is to choose a step length $\alpha \leq \lambda_i c(k, l; v^i)$ and then update $y \leftarrow y + \alpha(\chi_k - \chi_l)$. Then we are sure that the new y also belongs to $B(f)$. This increases y_k and decreases y_l by α. To keep $z = y + \partial x$ invariant, we also modify x_{kl} by α so as to decrease ∂x_k and increase ∂x_l by α.

Recall that x_{kl} was positive (else $k \to l \in R(\delta)$, implying that $l \in S$). As long as $\alpha \leq x_{kl}$, updating $x_{kl} \leftarrow x_{kl} - \alpha$ (and keeping $x_{lk} = 0$) modifies ∂x as desired, and keeps x δ-feasible. But there is no reason to use $\alpha > x_{kl}$, since we could instead use $\alpha = x_{kl}$ so that the updated $x_{kl} = 0$, meaning that l would join S, and we would make progress. Thus we choose $\alpha = \min(\lambda_i c(k, l; v^i), x_{kl})$. If $\alpha = x_{kl}$ so that l joins S, we call the SWAP *partial* (since we take only part of the full step from v^i to v^j; *nonsaturating* in (Iwata et al., 2001)), else we call it *full* (*saturating* in (Iwata et al., 2001)). Every full SWAP has $\alpha = \lambda_i c(k, l; v^i)$, which implies that $|\mathcal{I}|$ does not change; a partial SWAP increases $|\mathcal{I}|$ by at most one. Since there are clearly at most n partial SWAPS before calling AUGMENT, $|\mathcal{I}|$ can be at most $2n$ before calling REDUCEV.

IFF Subroutine SWAP $(k, l; v^i)$

Set $\alpha \leftarrow \min(x_{kl}, \lambda_i c(k, l; v^i))$.
[compute step length and new linear order]
Define $\prec_j$ as $\prec_i$ with k and l interchanged and compute v^j.
If $\alpha = x_{kl}$ then
[a partial SWAP, so $k \to l$ joins $R(\delta)$ and at least l joins S]
  Set $\lambda_j \leftarrow x_{kl}/c(k, l; v^i)$, $\mathcal{I} \leftarrow \mathcal{I} + j$
  If $\alpha < \lambda_i c(k, l; v^i)$ set $\lambda_i \leftarrow \lambda_i - x_{kl}/c(k, l; v^i)$, else set $\mathcal{I} \leftarrow \mathcal{I} - i$.
Else ($\alpha = \lambda_i c(k, l; v^i)$) [a full SWAP]
  Set $\lambda_j \leftarrow \lambda_i$, $\mathcal{I} \leftarrow \mathcal{I} + j - i$.

Set $x_{kl} \leftarrow x_{kl} - \alpha$, $\begin{Bmatrix} y_k \leftarrow y_k + \alpha \\ y_l \leftarrow y_l - \alpha \end{Bmatrix}$, and update $R(\delta)$ and S.

For each new member h of S do
  Delete any boundary triples $(u, h; v^h)$ from B.
  Add any new boundary triples $(h, u; v^h)$ to B.

If $\alpha = x_{kl} < \lambda_i c(k, l; v^i)$ then we'll want to take a step of only x_{kl}. To achieve this, take $x_{kl}/(\lambda_i c(k, l; v^i))$ times (15) plus $(1 - x_{kl}/(\lambda_i c(k, l; v^i)))$ times (8) to get

$$y + x_{kl}(\chi_k - \chi_l) = (x_{kl}/(c(k, l; v^i))v^j + (\lambda_i - x_{kl}/c(k, l; v^i))v^i + \sum_{h \neq i} \lambda_h v^h, \tag{16}$$

which shows how to update the λs in SWAP. The running time of SWAP is O(EO) plus the time for updating B. Thus a full SWAP is O(EO). For a partial SWAP, for each h added to S we can update B in $O(n^2)$ time. Thus a partial swap costs O(EO) plus $O(n^2)$ per element added to S. Note that if $x_{kl} = \lambda_i c(k, l; v^i)$ then we have a "degenerate" SWAP that is both partial and full. Although it is partial, $|\mathcal{I}|$ does not change, and although it is full we need to update B anyway. In the complexity analysis we double-count such a SWAP as being both partial and full. The key idea here is trading off (hard to manage) exchange capacity for (easy to manage) flow on the relaxation arcs, and this idea comes from Fleischer et al. (2002).

REFINE stops and concludes that the current point is δ-optimal when it can no longer find any augmenting paths and $B = \emptyset$. We show later that the running time of REFINE is $O(n^5\text{EO})$.

IFF Subroutine REFINE

Set $x \leftarrow x/2$. [make x δ-feasible]
For all $l \in E$ do [update z]
  Set $z_l \leftarrow y_l + \partial x_l$.
Compute $S^{-\delta}(z)$, $S^{+\delta}(z)$, $R(\delta)$, S, and B.
While augmenting paths exist ($S \cap S^{+\delta}(z) \neq \emptyset$), or $B \neq \emptyset$ do
  While $\exists$ path P from $S^{-\delta}(z)$ to $S^{+\delta}(z)$ using arcs from $R(\delta)$, do
    AUGMENT(P) and set B to be boundary triples w.r.t. new S.
  While $\nexists$ path P from $S^{-\delta}(z)$ to $S^{+\delta}(z)$ using arcs from $R(\delta)$ and $B \neq \emptyset$, do
    Find a boundary triple $(k, l; v^i)$ and SWAP$(k, l; v^i)$.
  Call REDUCEV
Return S as an approximate optimum solution.

Recall from Section 2.6.1 that our optimality condition for S solving SFM is that $y^-(E) = f(S)$. The following lemma (which is a relaxed version of Lemma 2.9) shows for both y and z how close these approximate solutions are to exactly satisfying $y^-(E) = f(S)$ and $z^-(E) = f(S)$, as a function of δ.

Lemma 3.4. *When a δ-scaling phase ends, S is tight for y, and we have $y^-(E) \geq f(S) - n^2\delta$ and $z^-(E) \geq f(S) - n\delta$.*

Proof. Note that for any $l \in E$ and any δ-feasible x, $-(n-1)\delta \leq \partial x_l \leq (n-1)\delta$.

Because the δ-scaling phase ended, we have $S^{-\delta}(z) \subseteq S \subseteq E - S^{+\delta}(z)$. This implies that for every $l \in S$, $z_1 < +\delta$, equivalent to $y_l < \delta - \partial x_l \leq +n\delta$; and for every $l \in E - S$, $z_l > -\delta$, equivalent to $y_l > -\delta - \partial x_l \geq -n\delta$. This implies that $y^-(S) = \sum_{l \in S: y_l \leq 0} y_l + \sum_{l \in S: y_l > 0} 0 \geq \sum_{l \in S: y_l \leq 0} y_l + \sum_{l \in S: y_l > 0} (y_l - n\delta) \geq y(S) - n\delta|S|$. Thus we get $y^-(E) = y^-(S) + y^-(E - S) \geq (y(S) - n\delta|S|) - n\delta|E - S| = f(S) - n^2\delta$. For $l \in S$, $z_l = y_l + \partial x_l < +\delta$ implies that $z_l^- > y_l + \partial x_l - \delta$. When REFINE ends, $B = \emptyset$, and then (13) says that S is tight for y. Note that $\partial x(S) = \sum_{k \in S, l \notin S} x_{kl} > 0$, since every $k \to l$ with $k \in S$ and $l \notin S$ must have $x_{kl} > 0$. Thus we get $z^-(E) = z^-(S) + z^-(E - S) \geq [(y(S) + \partial x(S)) - \delta|S|] - \delta|E - S| \geq f(S) - n\delta$. □

We now use this to prove correctness and running time. We now formally define z to be *δ-optimal* (for set T) if there is some $T \subseteq E$ such that $z^-(E) \geq f(T) - n\delta$. Lemma 3.4 shows that the z at the end of each δ-scaling phase is δ-optimal for the current approximate solution S. As before, we pick out the main points in boldface.

Theorem 3.5. *The IFF SFM Algorithm is correct for integral data and runs in* $O(n^5 \log M \cdot \text{EO})$ *time.*

Proof.
The current approximate solution T at the end of a δ-scaling phase with $\delta < 1/n^2$ solves SFM. Lemma 3.4 shows that $y^-(E) \geq f(T) - n^2\delta > f(T) - 1$. But for any $U \subseteq E$, $f(U) \geq y(U) \geq y^-(E) > f(T) - 1$. Since f is integer-valued, T solves SFM.

The first δ-scaling phase calls AUGMENT $O(n^2)$ times. Denote initial values with hats. Recall that $\hat{\delta} = |\hat{y} - (E)|/n^2$. Now $\hat{x} = 0$ implies that $\hat{z} = \hat{y}$, so that $\hat{z}^-(E) = \hat{y}^-(E)$. Since $z^-(E)$ monotonically increases during REFINE and is always nonpositive, the total increase in $z^-(E)$ is no greater than $|\hat{y}^-(E)| = n^2\hat{\delta}$. Since each AUGMENT increases $z^-(E)$ by δ, there are only $O(n^2)$ calls to AUGMENT.

Subsequent δ-scaling phases call AUGMENT $O(n^2)$ times. After halving δ, for the data at the end of the previous scaling phase we had $z^-(E) \geq f(T) - 2n\delta$. Making x δ-feasible at the beginning of REFINE changes each x_{kl} by at most δ, and so degrades this to at worst $z^-(E) \geq f(T) - (2n + n^2)\delta$. Each call to AUGMENT increases $z^-(E)$ by δ, and $z^-(E)$ can't get bigger than $f(T)$, so AUGMENT gets called at most $2n + n^2 = O(n^2)$ times.

There are $O(n^3)$ full SWAPs before each call to AUGMENT. Each full SWAP$(k, l; v^i)$ replaces v^i by v^j where l is one position higher in v^j than in v^i. Consider one v^i and the sequence of v^j's generated from v^i by full SWAPs. Since each such SWAP moves an element l of $E - S$ one position higher in its linear order, and no operations before AUGMENT allow elements of $E-S$ to become lower, no pair k, l occurs more than once in a boundary triple. There are $O(n^2)$ such pairs for each v^i, and $O(n)$ v^is, for a total of $O(n^3)$ full SWAPS before calling AUGMENT.

The total amount of work in all calls to SWAP before a call to AUGMENT is $O(n^3\text{EO})$. There are $O(n^3)$ full SWAPS before the AUGMENT, and each costs O(EO). Each node added to S by a partial SWAP costs $O(n^2)$ time to update B, and this happens at most n times before we must include a node of $S^{+\delta}(z)$, at which point we call AUGMENT. Each partial SWAP adds at least one node to S and costs O(EO) other than updating B. Hence the total SWAP-cost before the AUGMENT is $O(n^3\text{EO})$.

The time for one call to REFINE is $O(n^5\text{EO})$. Each call to REFINE calls AUGMENT $O(n^2)$ times. The call to AUGMENT costs $O(n^2)$ time, the work in calling SWAP before the AUGMENT is $O(n^3\text{EO})$, and the work in calling REDUCEV after the AUGMENT is $O(n^3)$, so we charge $O(n^3\text{EO})$ to each AUGMENT.

There are $O(\log M)$ calls to REFINE. For the initial $\hat{y}$, $\hat{y}(E) = f(E) \geq -M$. Let T be the set of elements where $\hat{y}$ is positive. Then $\hat{y}^+(E) = \hat{y}(T) \leq f(T) \leq M$. Thus $\hat{y}^-(E) = \hat{y}(E) - \hat{y}^+(E) \geq -2M$, so $\hat{\delta} = |y^-(E)|/n^2 \leq 2M/n^2$. Since δ's initial value is at most $2M/n^2$, it ends at $1/n^2$, and is halved at each REFINE, there are $O(\log M)$ calls to REFINE.

The total running time of the algorithm is $O(n^5 \log M \cdot \text{EO})$. Multiplying together the factors from the last two paragraphs gives the claimed total time. □

3.3.2 *Making the IFF algorithm strongly polynomial*

We now develop a strongly polynomial version of the IFF algorithm that we call IFF-SP. The challenge in making a weakly polynomial scaling algorithm like the IFF Algorithm strongly polynomial is to avoid having to call REFINE for each scaled value of δ, since the weakly polynomial factor $O(\log M)$ is really $\Theta(\log M)$. The rough idea is to find a way for the current data of the problem to reveal a good starting value of δ, and then to apply $O(\log n)$ calls to REFINE to get close enough to optimality that we can "fix a variable," which can happen only a strongly polynomial number of times. Letting the current data determine the value of δ can also be seen as a way to allow the algorithm to make much larger decreases in δ than would be available in the usual scaling framework.

The general mechanism for fixing a variable is to prove a "proximity lemma" as in Tardos (1985) that says that if the value of a variable gets too far from a bound, then we can remove that bound, and then reduce the size of the problem. In this case, the proximity lemma below says that if we have some $y \in B(f)$ such that y_l is negative enough w.r.t. δ, then we know that l belongs to every minimizer of f. This is a sort of approximate complementary slackness for LP (7): Complementary slackness for exact optimal solutions y^* and S^* says that $y^*_e < 0$ implies that $e \in S^*$, and the lemma says that for δ-optimal y, $y_e < -n^2\delta$ implies that $e \in S^*$.

Lemma 3.6. *At the end of a δ-scaling phase, if there is some $l \in E$ such that the current y satisfies $y_l < -n^2\delta$, then l belongs to every minimizer of f.*

Proof. By Lemma 3.4, at the end of a δ-scaling phase, for the current approximate solution S, we have $y^-(E) \geq f(S) - n^2\delta$. If S^* solves SFM, we have $f(S) \geq f(S^*) \geq y(S^*) \geq y^-(S^*)$. These imply that $y^-(E) \geq y^-(S) - n^2\delta$, or $y^-(E - S^*) \geq -n^2\delta$. Then if $l \in E - S^*$, we could add $-y_l > n^2\delta$ to this get $y^-(E - S^* - l) > 0$, a contradiction, so we must have $l \in S^*$. □

There are two differences between how we use this lemma and how IFF (Iwata et al., 2001) use it. First, we apply the lemma in a more relaxed way than IFF proposed that is shorter and simpler to describe, and which extends to the bisubmodular case (McCormick and Fujishige, 2003), whereas the IFF approach seems not to extend (Fujishige and Iwata, 2001). Second, we choose to implement the algorithm taking the structure it builds on the optimal solution explicitly into account (as is done in Iwata (2002a)) instead of implicitly into account (as is done in Iwata et al. (2001)), which requires us to slightly generalize Lemma 3.6 into Lemma 3.7 below.

We compute and maintain a set OUT of elements proven to be *out of* every optimal solution, effectively leading to a reduced problem on $E - \text{OUT}$. Previously we used M to estimate the "size" of f. The algorithm deletes "big" elements, so that the reduced problem consists of "smaller" elements, and we need a sharper initial estimate δ^0 of the size of the reduced problem. At first we choose $f(u) = \max_{l \in E} f(l)$ and $\delta^0 = f(u)^+$. Let $\hat{y} \in B(f)$ be an initial point coming from Greedy. Then $\hat{y}^+(E) = \sum_e \hat{y}^+_e \leq n\delta^0$, so that $\hat{y}^-(E) = \hat{y}(E) - \hat{y}^+(E) \geq f(E) - n\delta^0$. Thus, if we choose $x = 0$, then $\hat{z} = \hat{y} + \partial\hat{x} = \hat{y}$, so that E proves that $\hat{z}$ is δ^0-optimal. Thus we could start calling REFINE with $y = \hat{y}$ and $\delta = \delta^0$.

Suppose we have some set T such that $f(T) \leq -\delta^0$; we call such a set *highly negative*. Then $\lceil \log_2(2n^3) \rceil = O(\log n)$ (a strongly polynomial number) calls to REFINE produces some δ-optimal y with $\delta < \delta^0/n^3$. Subroutine FIX makes these $O(\log n)$ calls to REFINE. But $y(T) \leq f(T) \leq -\delta^0 < -n^3\delta$ implies that there is at least one $t \in T$ with $y_t < -n^2\delta$, and Lemma 3.6 then shows that such t belongs to every minimizer of f. We call such a t a *highly negative element*. This would be great, but IFF must go through some trouble to manufacture such a highly negative T.

Instead we adapt a more relaxed version of the IFF idea of considering the set function on $E - u$ defined by $f_u(S) = f(S + u) - f(u) = f(S + u) - \delta^0$. Clearly

f_u is submodular on $E-u$ with $f_u(\emptyset)=0$. Now apply FIX to f_u. Suppose that FIX does not find any highly negative element for f_u. This implies that there cannot be a highly negative set T for f_u. Then we know that for every T not containing u, $-\delta^0 < f_u(T) = f(T+u) - f(u) = f(T+u) - \delta^0$, or $f(T+u) > 0 = f(\emptyset)$. This proves that u cannot belong to any minimizer of f, and so we add u to OUT. On the other hand, suppose that FIX identifies at least one highly negative element t (which is guaranteed if there exists a highly negative set T for f_u). Then t belongs to every minimizer of f_u. Note that any minimizer of f_u actually solves the problem of minimizing $f(S)$ over subsets of E containing u. Therefore we would get the *condition* that every minimizer of f that contains u must also contain t. Note that it is possible that there is no highly negative set for f_u but that FIX identifies some highly negative element t anyway. This is not a problem, since Lemma 3.6 still implies the condition that any minimizer containing u must also contain t. Each new condition arc $u \to t$ means that we no longer need to consider sets containing u but not t as possible SFM solutions, thereby reducing the problem. Only $O(n^2)$ condition arcs can be added before the reduced problem becomes trivial, so this real progress.

As the algorithm proceeds we need some way of tracking such conditions. We do this by maintaining a set of arcs C on node set E, where arc $k \to l$ in C means that every minimizer of f containing k must also contain l. We start with $C=\emptyset$, and add arcs to C as we go along. If adding an arc creates a directed cycle Q in (E, C), then the nodes in Q either all belong to every minimizer of f, or none belong to every minimizer of f.

Dealing with (E, C) adds a new layer of complexity to the algorithm. For $u \in E$ define the *descendants* of u as $D_u = \{l \in E \mid$ there is a directed path from u to l in $(E, C)\}$, and the *ancestors* of u as $A_u = \{l \in E\}$ there is a directed path from l to u in $(E, C)\}$. If FIX finds a highly negative l (so that l belongs to every minimizer of f_u), then we know that D_l must also belong to every minimizer of f_u. Similarly, if we add u to OUT, we must also add all of A_u to OUT. Doing this ensures that whenever we call FIX, the arcs we find for C are indeed new, and so that we make real progress.

Let $\mathcal{C}$ be the set of strongly connected components of $(E - \text{OUT}, C)$. By the above comments, for every $\sigma \in \mathcal{C}$, every solution to SFM either includes all or no nodes of σ. Thus C is better though of as being a set of arcs on the node subset $\mathcal{C}$. Thus we should redefine descendants (resp. ancestors) from D_u (A_u) for $u \in E - \text{OUT}$ to D_σ (A_σ) for $\sigma \in \mathcal{C}$, again as the set of nodes of $\mathcal{C}$ reachable from σ (that σ can reach) via arcs of C. If $\mathcal{S} \subseteq \mathcal{C}$, define $E(\mathcal{S}) = \cup_{\sigma \in \mathcal{S}} \sigma$, the set of original elements contained in the union of strong components in $\mathcal{S}$. Therefore our general situation is that we have $\text{OUT} \subset E$ as the set of nodes out of an optimal solution, and we are essentially solving a reduced SFM problem on the contracted set of elements $\mathcal{C}$, which partitions $E - \text{OUT}$.

Subset $\mathcal{S} \subseteq \mathcal{C}$ can be part of an SFM solution only if no arc of C exits $\mathcal{S}$, i.e., if $\delta^+(\mathcal{S}) = \emptyset$. In this case we call $\mathcal{S}$ *closed* (or an *ideal*). Note that the family $\mathcal{D}$ of closed sets is closed under unions and intersections (it is a ring family), and we say that $(\mathcal{C}, C)$ *represents* $\mathcal{D}$ (in the sense of Birkhoff's Theorem

(Birkhoff, 1967)). Thus a solution to SFM for f has the form $E(\mathcal{S})$ for some $\mathcal{S}\in\mathcal{D}$. For $\mathcal{S}\in\mathcal{D}$, define $\hat{f}(\mathcal{S})=f(E(\mathcal{S}))$, so that $\hat{f}(\emptyset)=0$ and $\hat{f}$ is submodular on $\mathcal{D}$. Essentially $\hat{f}$ is just f restricted to $E-$ OUT, and then with each of the components of $\mathcal{C}$ contracted to a single new element. With good data structures for representing $\mathcal{C}$ we can evaluate $\hat{f}$ using just one call to the evaluation oracle $\mathcal{E}$ for f, so we use EO to also count evaluations of $\hat{f}$. We also need to redefine f_u for $u\in E$ to be a set function $\hat{f}_\tau$ for $\tau\in\mathcal{C}$. Since D_τ is closed, $D_\tau\in\mathcal{D}$. Define $\mathcal{D}_\tau$ to be the subsets $\mathcal{S}\subseteq\mathcal{C}-D_\tau$ such that $\mathcal{S}\cup D_\tau$ is closed (again a ring family). The graph representing $\mathcal{D}_\tau$ is $(\mathcal{C}-D_\tau, C)$ which is $(\mathcal{C}, C)$ with the nodes of D_τ (and any incident arcs) deleted. For $\mathcal{S}\in\mathcal{D}_\tau$ define $\hat{f}_\tau(\mathcal{S})=\hat{f}(\mathcal{S}\cup D_\tau)-\hat{f}(D_\tau)$. Then $\hat{f}_\tau$ is submodular, has $\hat{f}_\tau(\emptyset)=0$, and can be evaluated using only two calls to the evaluation oracle for $\hat{f}$. Thus we also use EO for $\hat{f}_\tau$.

Instead of restricting $\hat{f}$ to the closed subsets of $\mathcal{C}$, we could define it on all subsets of $\mathcal{C}$ via $\hat{f}(\mathcal{S})=f(E(\mathcal{S}))$ for any $\mathcal{S}\subseteq\mathcal{C}$ (and similarly for $\hat{f}_\tau$). Since we call FIX on the set of contracted elements $\mathcal{C}-D_\tau$, we would still be sure that any condition arcs found by FIX are new (do not already belong to C), and we could use Lemma 3.6 as it stands. This *implicit* method of handling $\mathcal{D}_\tau$ is used by IFF (Iwata et al., 2001). Here we use choose to use the slightly more complicated *explicit* method (developed for Iwata's fully combinatorial version of IFF (Iwata, 2002a)) that does restrict $\hat{f}_\tau$ to $\mathcal{D}_\tau$ because it yields better insight into the structure of the problem, and it is needed for Lemma 3.9 (which is crucial for making the fully combinatorial version work). It also allows us to demonstrate how to modify REFINE to work over a ring family, which is needed in Section 5. (The published version of (Iwata, 2002a) contains an error pointed out by Matthias Kriesell: It handles flow x as needed for the explicit method, but uses the implicit method Lemma 3.6 instead of the explicit method Lemma 3.7; a corrected version is available at `http://www.sr3.t.u-tokyo.ac.jp/~iwata/`)

We call the extended version of REFINE (that can deal with optimizing over a ring family such as $\mathcal{D}_\tau$ instead of 2^E) REFINER. There are only two changes that we need to make to REFINE. First, we must ensure that our initial $y=v^{\prec}$ comes from an order $\prec$ that is consistent with $\mathcal{D}_\tau$ (recall that this means that $\sigma\to\rho\in C$ implies that $\rho\prec\sigma$; this change is needed for both the implicit and explicit methods). This is easy to achieve, since we can take any order coming from an acyclic labeling of $(\mathcal{C}-D_\tau, C)$. Second, we must ensure that all $v^i\in\mathcal{I}$ that arise in the algorithm also have $\prec_i$ consistent with $\mathcal{D}_\tau$. We do this by setting the capacity of each $\sigma\to\rho\in R$ equal to $+\infty$ when $\sigma\to\rho\in C$ (this change occurs only in the explicit method, and is the big difference between the implicit and explicit methods). Then such arcs always belong to $R(\delta)$, so that $(\sigma,\rho;v^i)$ can never be a boundary triple (since $\sigma\in S$ and $\sigma\to\rho\in R(\delta)$ imply that $\rho\in S$), so an inconsistent $\prec_j$ is never created. This also implies that S always belongs to $\mathcal{D}_\tau$, so the optimal solution belongs to $\mathcal{D}_\tau$.

We also now need to revisit Lemma 3.6, since its proof assumed that all x were bounded by δ, and if $\sigma\to\rho\in C$ then $x_{\sigma\rho}$ could be much larger than δ.

This implies that we need to handle the boundary of arcs in C separately, so we define $\partial_C x_\sigma = \sum_{\sigma\to\rho\in C} x_{\sigma\rho} - \sum_{\psi\to\sigma\in C} x_{\psi\sigma}$, and $w = y + \partial_C x$. Note that every constraint $y(S) \le \hat{f}_\tau(S)$ defining $B(\hat{f}_\tau)$ comes from some closed $S \in \mathcal{D}_\tau$, and each such S has no arcs of C exiting it. Hence for any $S \in \mathcal{D}_\tau$ (since $x \ge 0$) $\partial_C x(S) \le 0$, and so $y \in B(\hat{f}_\tau)$ implies that $w \in B(\hat{f}_\tau)$ (recall that $w = y + \partial_C x$ is how all points in the (now unbounded) $B(\hat{f}_\tau)$ arise).

Lemma 3.7. *At the end of a δ-scaling phase, if there is some $\sigma \in \mathcal{C} - D_\tau$ such that the current w satisfies $w_\sigma < -n^2\delta$, then σ belongs to every minimizer of $\hat{f}_\tau$.*

Proof. By Lemma 3.4, at the end of δ-scaling phase, for the current approximate solution S, we have $z^-(\mathcal{C} - D_\tau) \ge \hat{f}_\tau(S) - n\delta$. Since $x_{\sigma\rho} \le \delta$ for each $\sigma \to \rho \notin C$, for each σ we have $z_\sigma - w_\sigma = \partial x_\sigma - \partial_C x_\sigma \le (n-1)\delta$. Hence $w^-(\mathcal{C} - D_\tau) \ge z^-(\mathcal{C} - D_\tau) - n(n-1)\delta \ge \hat{f}_\tau(S) - n^2\delta$.

If S^* solves SFM, we have $\hat{f}_\tau(S) \ge \hat{f}_\tau(S^*) \ge w(S^*) \ge w^-(S^*)$. These imply that $w^-(\mathcal{C} - D_\tau) \ge w^-(S^*) - n^2\delta$, or $w^-((\mathcal{C} - D_\tau) - S^*) \ge -n^2\delta$. Then if $\sigma \in (\mathcal{C} - D_\tau) - S^*$, we could add $-w_\sigma > n^2\delta$ to this to get $w^-((\mathcal{C} - D_\tau) - S^* - \sigma) > 0$, a contradiction, so we must have $\sigma \in S^*$. □

Define $\delta^0 = \max_{\sigma\in\mathcal{C}} \hat{f}(D_\sigma) - \hat{f}(D_\sigma - \sigma)$. Lemma 2.2 shows that δ^0 is an upper bound on the components of any y in the convex hull of the vertices of $B(\hat{f})$, and we show below that if $\delta^0 \le 0$, then $E - $ OUT solves SFM for f (it is not hard to show that δ^0 is monotone nonincreasing during the algorithm). So we can assume that $\delta^0 > 0$, and we take this as the "size" of the current solution. Suppose that τ achieves the max for δ^0, i.e., that $\delta^0 = \hat{f}(D_\tau) - \hat{f}(D_\tau - \tau)$. We then apply FIX to $\hat{f}_\tau$. If FIX finds a highly negative σ then we add $\tau \to \sigma$ to C; if it finds no highly negative elements, then we add $E(A_\tau)$ to OUT.

IFF-SP Subroutine FIX ($\hat{f}_\tau, (\mathcal{C} - D_\tau, C), \delta^0$)

Applies to $\hat{f}_\tau$ defined on closed sets of $\mathcal{C}(-D_\tau, C)$, and $y_\sigma \le \delta^0$ for all $y \in B(\hat{f}_\tau)$.

> Initialize $\prec$ as any linear order consistent with C, $y \leftarrow v^\prec$, $\delta \leftarrow \delta^0$, and $\mathcal{N} = \emptyset$.
> Initialize $x = 0$ and $z = y + \partial x (= y)$.
> While $\delta \ge \delta^0/n^3$ do
>   Set $\delta \leftarrow \delta/2$.
>   Call REFINER.
> For $\sigma \in \mathcal{C} - D_\tau$ do [add descendants of highly negative nodes to $\mathcal{N}$]
>   If $w_\sigma = y_\sigma + \partial_C x_\sigma < -n^2\delta$ set $\mathcal{N} \leftarrow \mathcal{N} \cup \mathcal{D}_\sigma$.
> Return $\mathcal{N}$.

IFF Strongly Polynomial Algorithm (IFF-SP)

Initialize OUT $\leftarrow \emptyset$, $C \leftarrow \emptyset$, $\mathcal{C} \leftarrow E$.
While $|\mathcal{C}| > 1$ do
  Compute $\delta^0 = \max_{\sigma \in \mathcal{C}} \hat{f}(D_\sigma) - \hat{f}(D_\sigma - \sigma)$ and let $\tau \in \mathcal{C}$ attain the maximum.
  If $\delta^2 \leq 0$ then return $E -$ OUT as an optimal SFM solution.
  Else ($\delta^0 > 0$)
    Set $\mathcal{N} \leftarrow$ Fix($\hat{f}_\tau$, ($\mathcal{C} - D_\tau$, C), δ^0).
    If $\mathcal{N} \neq \emptyset$, for all $\sigma \in \mathcal{N}$ and $\tau \to \sigma$ to C, update $\mathcal{C}$, and all D_ρ's, A_ρ's.
    Else ($\mathcal{N} \neq \emptyset$) set OUT $\leftarrow$ OUT $\cup E(A_\tau)$.
Return whichever of $\emptyset$ and $E -$ OUT has a smaller function value.

Theorem 3.8. *IFF-SP is correct, and runs in* $O(n^7 \log n \cdot \text{EO})$ *time.*

Proof.
If $\delta \leq 0$ then $E -$ OUT solves SFM for f. Lemma 2.2 shows that for the current y and $\sigma \in \mathcal{C}$, $y_\sigma \leq 0$. Thus $y^-(\mathcal{C}) = y(\mathcal{C}) = \hat{f}(\mathcal{C})$, proving that $\mathcal{C}$ solves SFM for $\hat{f}$. We know that any solution T of SFM for f must be of the form $E(\mathcal{T})$ for $\mathcal{T} \in \mathcal{D}$. By optimality of $\mathcal{C}$ for $\hat{f}$, $\hat{f}(\mathcal{C}) \leq \hat{f}(\mathcal{T})$, or $f(E -$ OUT$) = f(E(\mathcal{C})) \leq f(E(\mathcal{T})) = f(T)$, so $E -$ OUT is optimal for f.

In Fix ($\hat{f}_\tau$, ($\mathcal{C}$, C), δ^0) with $\delta^0 > 0$, the first call to RefineR calls Augment O(n) times. Lemma 2.2 shows that for the current y and any $\sigma \in \mathcal{C}$, $y_\sigma \leq \delta^0$. In the first call to RefineR we start with $z = y$, so that $z^+(\mathcal{C}) = y^+(\mathcal{C})$. Since $y_\sigma \leq \delta^0$ for each $\sigma \in \mathcal{C}$, we get $z^+(\mathcal{C}) = y^+(\mathcal{C}) \leq n\delta^0$. Each call to Augment reduces $z^+(\mathcal{C})$ by $\delta^0/2$. Thus there are at most $2n$ calls to Augment during the first call to RefineR.

When a highly negative $\mathcal{T} \in \mathcal{D}_\tau$ exists, a call to Fix ($\hat{f}_\tau$, ($\mathcal{C} - D_\tau$, C), δ^0) results in at least one element added to $\mathcal{N}$. The call to Fix reduces δ from δ^0 to below δ^0/n^3. Then $\mathcal{T}$ highly negative and $\mathcal{T} \in \mathcal{D}_\tau$ imply that $w(\mathcal{T}) \leq y(\mathcal{T}) \leq \hat{f}(\mathcal{T}) \leq -\delta^0 < -n^3\delta$. This implies that there is at least one $\rho \in \mathcal{C}$ with $w_\rho < -n^3\delta$, so at least one element gets added to $\mathcal{N}$.

If Fix($\hat{f}_\tau$, ($\mathcal{C} - D_\tau$, C), δ^0) finds no highly negative element, then $E(A_\tau)$ belongs to no minimizer of f. As above, if there were a highly negative set $\mathcal{T}$ for $\hat{f}_\tau$, then the call to Fix would find a highly negative element. Thus for all $\mathcal{T} \in \mathcal{D}_\tau$ we have $-\delta^0 < \hat{f}_\tau(\mathcal{T})$, or $-\hat{f}(D_\tau) + \hat{f}(D_\tau - \tau) < \hat{f}(\mathcal{T} \cup D_\tau) - \hat{f}(D_\tau)$, or $f(E(D_\tau - \tau)) < f(E(\mathcal{T} \cup D_\tau))$. Since $E(\mathcal{T} \cup D_\tau)$ is a generic feasible set containing τ and $E(D_\tau - \tau)$ is a specific set not containing τ, no set containing τ can be optimal. Thus adding $E(A_\tau)$ to OUT is correct.

The algorithm returns a solution to SFM. If some $\delta^0 \leq 0$, then we showed above that the returned $E - \text{Out}$ is optimal. Otherwise the algorithm terminates because $|\mathcal{C}| = 1$. In this case the only two choices left for solving SFM are $E(\mathcal{C}) = \text{E} - \text{Out}$ and $\emptyset$, and the algorithm returns the better of these.

Fix calls RefineR O(log *n*) times. Parameter δ starts at δ^0, ends at its first value below δ^0/n^3, and is halved at each iteration. Thus there are $\lceil \log_2(2n^3) \rceil = O(\log n)$ calls to RefineR.

The algorithm calls Fix O(n^2) times. Each call to Fix either (i) adds at least one element to Out, or (ii) adds at least one arc to C. Case (i) happens at most n times. Since there are only $n(n-1)$ possible arcs for C, case (ii) happens $O(n^2)$ times.

The algorithm runs in O(n^7log *n* · EO) time. From the proof of Theorem 3.5, one call to RefineR costs $O(n^5\text{EO})$ time. Each call to Fix calls RefineR O(log n) times, so the time of one call to Fix is $O(n^5 \log n \cdot \text{EO})$. The algorithm calls Fix $O(n^2)$ times, for a total time of $O(n^7 \log n \cdot \text{EO})$. □

3.3.3 Iwata's fully combinatorial SFM algorithm

Iwata's algorithm (Iwata, 2002a) is a fully combinatorial extension of IFF-SP, and so we call it IFF-FC. Recall that a fully combinatorial algorithm cannot use multiplication or division, and must also be strongly polynomial. This implies that it cannot call ReduceV, since the linear algebra in ReduceV apparently needs to use multiplication and division in a way that cannot be simulated with addition and subtraction. This suggests that we adapt an existing algorithm by avoiding the calls to ReduceV; this would probably degrade the running time since $|\mathcal{I}|$ would be allowed to get much larger than n, but as long as we could show that $|\mathcal{I}|$ remained polynomially-bounded, we should still be ok.

Let's try to imagine a fully combinatorial version of (either version of) Schrijver's Algorithm. A key part of the running time proof of Theorem 3.2 is that Push has $O(n^2)$ iterations since each saturating Push either reduces $\max_i |(l,k]_{\prec_i}|$, or the number of $i \in \mathcal{I}$ attaining this max. Without ReduceV, the first saturating Push could have $|(l,k]_{\prec_i}| = n-1$ and could create $n-2$ v^j's with $|(l,k]_{\prec_i}| = n-2$; these could each cause $n-2$ saturating Pushes, each of which creates $n-3$ v^j's with $|(l,k]| = n-3$; these $(n-2)(n-3)$ v_j's could each cause $n-3$ saturating Pushes, each of which creates $n-4$ v^j's with $|(l,k]_{\prec_i}| = n-4$; these $(n-2)(n-3)(n-4)$ v_j's could.... Thus $|\mathcal{I}|$ could become super-polynomial. Also, Schrijver's ExchBd subroutine needs to solve the system (12), and this seems to require using multiplication and division. For either of these reasons, a fully combinatorial version of Schrijver's Algorithm appears to be unattainable.

IFF-SP adds new v^j's only at partial SWAPS, and only one new v^j at a time. Since there are at most n partial SWAPS per AUGMENT, this means that each AUGMENT creates at most n new v^j's. In the strongly polynomial version of the algorithm, each call to FIX calls REFINER $O(\log n)$ times. Each call to REFINER does $O(n^2)$ AUGMENTS, for a total of $O(n^2 \log n)$ AUGMENTS for each call to FIX, for a total of $O(n^3 \log n)$ v^j's added in each call to FIX. Each call to FIX starts out with $|\mathcal{I}| = 1$, so $|\mathcal{I}|$ stays bounded by $O(n^3 \log n)$ when we don't use REDUCEV.

When we do use REDUCEV, the running time for REFINER comes from ($O(n^2)$ calls to AUGMENT) times ($O(n^3\text{EO})$ work from full SWAPS between each AUGMENT). This last term comes from ($O(n^2)$ possible boundary triples per vertex) times ($O(n)$ vertices in $\mathcal{I}$) times (O(EO) work per boundary triple).

When we don't use REDUCEV, we instead have $O(n^3 \log n)$ vertices in $\mathcal{I}$. Each one again has $O(n^2)$ possible boundary triples, so now the work from full SWAPS between each AUGMENT is $O(n^5 \log n \cdot \text{EO})$. Multiplied times the $O(n^2)$ AUGMENTS, this gives $O(n^7 \log n \cdot \text{EO})$ as the time for REFINER. Multiplied times the $O(\log n)$ calls to REFINER per cal to FIX, and times the $O(n^2)$ calls to FIX overall, we would get a total of $O(n^9 \log^2 n \cdot \text{EO})$ time for the algorithm without calling REDUCEV. Thus there is some real hope for making a fully combinatorial version of IFF-SP.

However, getting rid of REDUCEV is not sufficient to make IFF-SP fully combinatorial. There is also the matter of the various other multiplications and divisions in IFF-SP. The only nontrivial remaining multiplication in IFF-SP is the term $\lambda_i c(k, l; v^i)$ that arises in SWAP. Below we modify the representation (8) by implicity multiplying through by a common denominator so that each λ_i is an integer bounded by a polynomial in n. Then this product can be dealt with using repeated addition.

IFF-SP has two nontrivial divisions. One is the computation of δ^0/n^3 in FIX. We change from having δ at each iteration to doubling a scaling parameter, and we need another factor of n for technical reasons, so we need to compute instead n^4. This can again be done via $O(n)$ repeated additions. The second is the division $x_{kl}/c(k, l; v^i)$ in (16). We would like to simulate this division via repeated subtractions. To do this we need to know that the quotient $x_{kl}/c(k, l; v^i)$ has strongly polynomial size in terms of a scale factor. Here we take advantage of some flexibility in the choice of the step length α. Recall that when the full step length $\lambda_i c(k, l; v^i)$ is "big", we chose to set $\alpha = x_{kl}$. But (with appropriate modification of the update to x) the analysis of the algorithm remains the same for any α satisfying $x_{kl} \le \alpha \le \min(x_{kl} + \delta, \lambda_i c(k, l; v^i))$, since for any such value of α x remains δ-feasible and we can still add l to S. Our freedom to choose α in this range gives us enough flexibility to discretize the quotient. The setup of IFF-SP facilities making such arguments, since it has the explicit bound δ^0 on the components of y available at all times. Indeed, this is essentially what Iwata (Iwata, 2002a) does.

IFF-FC adapts IFF-SP as follows: We denote corresponding variables in IFF-FC by tildes, so where IFF-SP has x, y, z, λ, δ, etc., IFF-FC has $\tilde{x}, \tilde{y}, \tilde{z}, \tilde{\lambda}, \tilde{\delta}$, etc. Since FIX is always working with $\hat{f}_\tau$ defined on $(\mathcal{C} - D_\tau, \mathcal{C})$, we use σ and ρ in place of k and l. Recall from (8) that IFF-SP keeps $y \in B(\hat{f}_\tau)$ as a convex combination of vertices $y = \sum_{i \in \mathcal{I}} \lambda_i v^i$. The λ_i satisfy $\lambda_i \geq 0$ and $\sum_{i \in \mathcal{I}} \lambda_i = 1$, but are otherwise arbitrary. To make the arithmetic discrete in IFF-FC, we keep a scale factor $\mathrm{SF} = 2^a$ (for a a nonnegative integer). We now insist that each λ_i be a fraction with integer numerator, and denominator SF. To clear the fractions we represent $\tilde{y}$ as $\mathrm{SF}y \in B(\mathrm{SF}\hat{f})$ and $\tilde{\lambda}_i = \mathrm{SF}\lambda_i$, so that $\tilde{y} = \sum_{i \in \mathcal{I}} \tilde{\lambda}_i v^i$ with each $\tilde{\lambda}_i$ a positive integer, and $\sum_{i \in \mathcal{I}} \tilde{\lambda}_i = \mathrm{SF}$. At the beginning of each call to FIX, as before we choose an arbitrary $\prec_1$ consistent with $\mathcal{D}$ and set $\tilde{y} = v^1$. Thus we choose $a = 0$, $\mathrm{SF} = 2^0 = 1$, and $\tilde{\lambda}_1 = 1$ to satisfy this initially.

IFF-SP starts each call to FIX with $\delta = \delta^0$ and halves it before each call to REFINER. IFF-FC starts with $\tilde{\delta} = (n+1)\delta^0$, and instead of halving it, IFF-FC doubles SF (increases a by 1). This extra factor of $n+1$ is needed to make Lemma 3.9 work, which in turn is needed to make the fully combinatorial discrete approximation of $\tilde{x}_{\sigma\rho}/c(\sigma, \rho; v^i)$ lead to a $\tilde{\delta}$-feasible update to $\tilde{x}$. The proof of Lemma 3.9 also obliges using the explicit method of handling $\mathcal{D}_\tau$, since it needs to know that all vertices generated during REFINER are consistent with $\mathcal{D}_\tau$, and this may not be true with the implicit method.

Lemma 3.9 also needs that $\hat{f}(\mathcal{C})$ is not too negative, which necessitates changing IFF-SP: If $\hat{f}(\mathcal{C}) \leq -\delta^0$ then it is highly negative, and we can call FIX directly on $\hat{f}$ (instead of $\hat{f}_\tau$) to find some $\sigma \in \mathcal{C}$ that is contained in all SFM solutions via Lemma 3.6, and then we add $E(D_\sigma)$ to a set IN of elements *in* all SFM solutions. We than delete D_σ from $\mathcal{C}$ and reset $\hat{f} \leftarrow \hat{f}_\sigma$. This change clearly does not impair the running time of the algorithm. This also means that we need the same sort of bound for $B(\hat{f})$.

Lemma 3.9. *If $\hat{f}(\mathcal{C}) > -\delta^0$, then for any two vertices v^i and v^j of $B(\hat{f}_\tau)$ and $\sigma \in \mathcal{C} - D_\tau$, $|v^i_\sigma - v^j_\sigma| \leq \tilde{\delta}$. In particular $c(\sigma, \rho; v^i) \leq \tilde{\delta}$ in $B(\hat{f}_\tau)$ (and also $B(\hat{f})$).*

Proof. Note that $c(\sigma, \rho; v^i)$ equals $|v^i_\sigma - v^j_\sigma|$ for the vertex v^j coming from $\prec_i$ with σ and ρ interchanged, so it suffices to prove the first statement. Lemma 2.2 shows that for any y in $B(\hat{f}_\tau)$, in particular $y = v^\prec$, and any $\rho \in \mathcal{C} - D_\tau$, we have $y_\rho \leq \delta^0$. We have that $y(\mathcal{C} - D_\tau) = \hat{f}_\tau(\mathcal{C} - D_\tau) = \hat{f}_\tau(\mathcal{C}) - \hat{f}(D_\tau)$. Then $\hat{f}(\mathcal{C}) > -\delta^0$ and $\hat{f}(D_\tau) \leq \sum_{\sigma \in D_\tau} (\hat{f}(D_\sigma) - \hat{f}(D_\sigma - \sigma)) \leq |D_\tau|\delta^0$ imply that $y(\mathcal{C} - D_\tau) \geq -(|D_\tau| + 1)\delta^0$. Adding $-y_\sigma \geq -\delta^0$ to this for all $\sigma \in \mathcal{C} - D_\tau$ other than ρ implies that $-n\delta^0 \leq y_\rho \leq \delta^0$ for any $\rho \in \mathcal{C} - D_\tau$. Thus any exchange capacity is at most $(n+1)\delta^0 = \tilde{\delta}$. A simpler version of the same proof works for $B(\hat{f})$. □

IFF Fully Combinatorial Algorithm (IFF-FC)

Initialize IN$\leftarrow \emptyset$, OUT$\leftarrow \emptyset$, $C \leftarrow \emptyset$, $\mathcal{C} \leftarrow E$.
While $|\mathcal{C}| > 1$ do
  Compute $\delta^0 = \max_{\sigma\in\mathcal{C}} \hat{f}(D_\sigma) - \hat{f}(D_\sigma - \sigma)$ and let $\tau \in \mathcal{C}$ attain the maximum.
  If $\delta^0 \le 0$ then return $E -$ OUT as an optimal SFM solution.
  If $\hat{f}(\mathcal{C}) \le -\delta^0$
    Set $\mathcal{N} \leftarrow$ FIX($\hat{f}$, ($\mathcal{C}$, C), δ^0).
    For each $\sigma \in \mathcal{N}$ add $E(D_\sigma)$ to IN, and reset $\mathcal{C} \leftarrow \mathcal{C} - D_\sigma, \hat{f} \leftarrow \hat{f}_\sigma$.
  Else ($\delta^0 > 0$ and $\hat{f}(\mathcal{C}) > -\delta^0$)
    Set $\mathcal{N} \leftarrow$ FIX($\hat{f}_\tau$, ($\mathcal{C} - D_\tau$, C), δ^0).
    If $\mathcal{N} \neq \emptyset$, for each $\sigma \in \mathcal{N}$ add $\tau \rightarrow \sigma$ to C, update $\mathcal{C}$, and all D_ρ's, A_ρ's.
    Else ($\mathcal{N} \neq \emptyset$) set OUT $\leftarrow$ OUT $\cup\, E(A_\tau)$.
Return whichever of IN $E -$ OUT has a smaller function value.

Thus, where IFF-SP kept δ, IFF-FC keeps the pair $\tilde{\delta}$ and SF, which we could translate into IFF-SP terms via $\delta = \tilde{\delta}/\text{SF}$. Also, in IFF-SP δ dynamically changes during FIX, whereas in IFF-FC $\tilde{\delta}$ keeps its initial value and only SF changes. Since $\tilde{y} = \text{SF}y$, we get the effect of scaling by keeping $\tilde{x} = x$ (so that doubling SF makes x half as large relative to y, implying that we do not need to halve the flow $\tilde{x}$ at each call to REFINER), and continue to keep the invariant that $\tilde{z} = \tilde{y} + \partial\tilde{x}$. However, to keep $\tilde{y} = \text{SF}y$ we do need to double y and each $\tilde{\lambda}_i$ when SF doubles.

When IFF-SP chose the step length α, if $x_{\sigma\rho} \geq \lambda_i c(\sigma, \rho; v^i)$, then we chose $\alpha = \lambda_i c(\sigma, \rho; v^i)$ and took a full step. Since this implied replacing v^i by v^j in $\mathcal{I}$ with the same coefficient, we can translate it directly to IFF-FC without harming discreteness. Because both $\tilde{x}$ and $\tilde{\lambda}$ are multiplied by SF, this translates to saying that if $\tilde{x}_{\sigma\rho} \geq \tilde{\lambda}_i c(\sigma, \rho; v^i)$, then we choose $\tilde{\alpha} = \tilde{\lambda}_i c(\sigma, \rho; v^i)$ and take a full step.

In IFF-SP, if $x_{\sigma\rho} < \lambda_i c(\sigma, \rho; v^i)$, then we chose $\alpha = x_{\sigma\rho}$ and took a partial step. This update required computing $x_{\sigma\rho}/c(\sigma, \rho; v^i)$ in (16), which is not allowed in a fully combinatorial algorithm. To keep the translated $\tilde{\lambda}_i$ and $\tilde{\lambda}_j$ integral, we need to compute an integral approximation to $\tilde{x}_{\sigma\rho}/c(\sigma, p; v^i)$. To ensure that $\tilde{x}_{\sigma\rho} - \alpha$ hits zero (so that ρ joins S), we need this approximation to be at least as large as $\tilde{x}_{\sigma\rho}/c(\sigma, \rho; v^i)$.

The natural thing to do is to compute $\tilde{\beta} = \lceil x_{\sigma\rho}/c(\sigma, \rho; v^i)\rceil$ and update λ_i and λ_j to $\lambda_i - \tilde{\beta}$ and $\tilde{\beta}$ respectively, which are integers as required. This implies choosing $\tilde{\alpha} = \tilde{\beta}c(\sigma, \rho; v^i)$. Because $\lceil\tilde{x}_{\sigma\rho}/c(\sigma, \rho; v^i)\rceil < (\tilde{x}_{\sigma\rho}/c(\sigma, \rho; v^i)) + 1$, $\tilde{\alpha}$ is less than $c(\sigma, \rho; v^i)$ larger than α. Hence the increase we make to $\tilde{x}_{\rho\sigma}$ to keep the invariant $\tilde{z} = \tilde{y} + \partial\tilde{x}$ is at most $c(\sigma, \rho; v^i)$. By Lemma 3.9, $c(\sigma, \rho; v^i) \leq \tilde{\delta}$, so we would have that the updated $\tilde{x}_{\rho\sigma} \leq \tilde{\delta}$, so it remains $\tilde{\delta}$-feasible, as desired.

Furthermore, we could compute $\tilde{\beta}$ by repeatedly subtracting $c(\sigma, \rho; v^i)$ from $\tilde{x}_{\sigma\rho}$ until we get a nonpositive answer. We started from the assumption that $\tilde{x}_{\sigma\rho} < \tilde{\lambda}_i c(\sigma, \rho; v^i)$, or $\tilde{x}_{\sigma\rho} < c(\sigma, \rho; v^i) < \tilde{\lambda}_i$, implying that $\tilde{\beta} \leq \tilde{\lambda}_i \leq \text{SF}$. Thus the number of subtractions needed is at most SF, which we show below remains small. In fact, we can do better by using repeated doubling: Initialize $q = c(\sigma, \rho; v^i)$ and set $q \leftarrow 2q$ until $q \geq x_{\sigma\rho}$. The number d of doublings is $O(\log \text{SF}) = O(a)$. Along the way we save $q_i = 2^i q$ for $i = 0, 1, \ldots, d$. Then set $q \leftarrow q_{d-1}$, and for $i = d-2, d-3, \ldots, 0$, if $q + q_i \leq x_{\sigma\rho}$ set $q \leftarrow q + q_i$. If the final $q > x_{\sigma\rho}$, set $q \leftarrow q + 1$. Thus the final q is of the form $pc(\sigma, \rho; v^i)$ for some integer p, we have $q \geq x_{\sigma\rho}$, and $(p-1)\ c(\sigma, \rho; v^i) < x_{\sigma\rho}$. Thus $q = \tilde{\beta}$, and we have computed this in $O(\log \text{SF})$ time.

IFF-FC Subroutine SWAP $(\sigma, \rho; v^i)$

Define $\prec_j$ as $\prec_i$ with σ and ρ interchanged and compute v^j.
If $\tilde{x}_{\sigma\rho} \geq \tilde{\lambda}_i c(\sigma, \rho; v^i)$ [a full SWAP]
  Set $\tilde{\alpha} = \tilde{\lambda}_i c(\sigma, \rho; v^i)$, and $\tilde{x}_{\sigma\rho} \leftarrow \tilde{x}_{\sigma\rho} - \tilde{\alpha}$.
  Set $\mathcal{I} \leftarrow \mathcal{I} + j - i$ and $\tilde{\lambda}_j \leftarrow \tilde{\lambda}_i$.
Else ($\tilde{x}_{\sigma\rho} < \tilde{\lambda}_i c(\sigma, \rho; v^i)$) [a partial SWAP, so at least ρ joins S]
  Compute $\tilde{\beta} = \lceil \tilde{x}_{\sigma\rho} / c(\sigma, \rho; v^i) \rceil$ and $\tilde{\alpha} = \tilde{\beta} c(\sigma, \rho; v^i)$.
  Set $\tilde{x}_{\rho\sigma} \leftarrow \tilde{\alpha} - \tilde{x}_{\sigma\rho}$ and $\tilde{x}_{\sigma\rho} \leftarrow 0$. [makes ∂x_σ drop by $\tilde{\alpha}$ as required]
  Set $\tilde{\lambda}_j \leftarrow \tilde{\beta}$ and $\mathcal{I} \leftarrow \mathcal{I} + j$.
  If $\tilde{\beta} < \tilde{\lambda}_i$ set $\tilde{\lambda}_i \leftarrow \tilde{\lambda}_i - \tilde{\beta}$, else ($\tilde{\beta} = \tilde{\lambda}_i$) set $\mathcal{I} \leftarrow \mathcal{I} - i$.

Set $\left\{ \begin{array}{l} \tilde{y}_\sigma \leftarrow \tilde{y}_\sigma + \tilde{\alpha} \\ \tilde{y}_\rho \leftarrow \tilde{y}_\rho - \tilde{\alpha} \end{array} \right\}$, and update $R(\delta)$ and S.

For each new member η of S do
  Delete any boundary triples $(\mu, \eta; v^h)$ from B.
  Add any new boundary triples $(\eta, \mu; v^h)$ to B.

Due to choosing the initial value of $\tilde{\delta} = (n+1)\delta^0$ instead of δ^0, we now need to run FIX for $\lceil \log_2((n+1)2n^3) \rceil$ iterations instead of $\lceil \log_2(2n^3) \rceil$, but this is still $O(\log n)$. This implies that SF stays bounded by a polynomial in n, so that the computation of $\tilde{\beta}$ and our simulated multiplications are fully combinatorial operations.

From this point the analysis of IFF-FC proceeds just like the analysis of IFF-SP when it doesn't call REDUCEV that we did at the beginning of this section, so we end up with a running time of $O(n^9 \log^2 n \cdot \text{EO})$.

IFF-FC Subroutine Fix $(\tilde{f}_\tau(\mathcal{C}-D_\tau, \mathcal{C}), \tilde{\delta})$

Applies to $\tilde{f}_\tau$ defined on closed sets of $(\mathcal{C}-D_\tau, \mathcal{C})$, and $c(\sigma, \rho; v^i) \le \tilde{\delta}$ for all $y \in B(\tilde{f}_\tau)$

Initialize $\prec$ as any linear order consistent with C, $\tilde{y} \leftarrow v^{\prec}$, SF$\rightarrow$1, and $\mathcal{N} = \emptyset$.

Initialize $\tilde{x} = 0$ and $\tilde{z} = \tilde{y} + \partial\tilde{x}(= \tilde{y})$.

While SF $\le 2n^4$ do

Set SF $\leftarrow$2SF, $y \leftarrow 2y$, and $\tilde{\lambda}_i \leftarrow 2\tilde{\lambda}_i$ for $i \in \mathcal{I}$.

Call RefineR.

For $\sigma \in \mathcal{C} - D_\tau$ do [add descendants of highly negative nodes to $\mathcal{N}$]

If $\tilde{w}_\sigma = \tilde{y}_\sigma + \partial_C x_\sigma < -n^2\tilde{\delta}$ set $\mathcal{N} \leftarrow \mathcal{N} \cup D_\sigma$.

Return $\mathcal{N}$.

3.3.4 Iwata's faster hybrid algorithms

In Iwata (2002b) Iwata shows a way to adopt some of the ideas behind Schrijver's SFM Algorithm, in particular the idea of modifying the linear orders by blocks instead of consecutive pairs, to speed up the IFF Algorithm, including the fully combinatorial version of the previous section. The high-level view of the IFF-based algorithms is that they all depend on the $O(n^5\text{EO})$ running time of Refine: The weakly polynomial version embeds this in $O(\log M)$ iterations of a scaling loop; the strongly polynomial version calls Fix $O(n^2)$ times, and each call to Fix requires $O(\log n)$ calls to Refine (actually RefineR). For the fully combinatorial version we need to look more closely at the running time of RefineR. One term in the bottleneck experession determining the running time of RefineR is $|\mathcal{I}|$. Ordinarily we have $|\mathcal{I}| = O(n)$, but in the fully combinatorial version we don't call ReduceV, so $|\mathcal{I}|$ balloons up to $O(n^3 \log n)$. This makes RefineR run a factor of $O(n^2 \log n)$ slower. Otherwise the analysis is the same as for the strongly polynomial version.

Therefore, if we can make Refine run faster, then all three versions should also run faster. One place to look for an improvement is the action that Refine takes when no augmenting path exists: it finds any boundary triple $(k, l; v^i)$ and does a Swap. Potentially a more constrained choice of boundary triple would lead to a faster running time. The Hybrid Algorithm implements this idea in HRefine by adding distance labels as in Schrijver's Algorithm.

But a problem arises with this: the pair of elements (k, l) picked out by distance labels need not be consecutive in $\prec_i$. Schrijver's Algorithm deals with this by using ExchBd to come up with a representation of $\chi_k - \chi_l$ in terms of vertices with smaller $(l, k]_{\prec_j}$. Indeed, all previous non-Ellipsoid SFM algorithms move in $\chi_k - \chi_l$ directions. The Hybrid Algorithm introduces a new idea (originally suggested by Fujishige as a heuristic speedup for IFF): instead of focusing on $\chi_k - \chi_l$, do a BLOCKSWAP (called *Multiple-Exchange* in Iwata (Iwata, 2002b)) that makes multiple changes to the block $[l, k]_{\prec i}$ of $\prec_i$ to get a new $\prec_j$ that is much closer to our ideal (of having all elements of the current set of reachable elements appear consecutively at the beginning of $\prec_j$), and the move in direction $v^j - v^i$. Using such directions means that at most one new vertex (namely v^j) needs to be added to $\mathcal{I}$ at each iteration, so the fully combinatorial machinery still works.

By (4), when we generate $\prec_j$ from $\prec_i$ by rearranging some block of b elements, Greedy needs O(bEO) time to compute v^j. for a(n ordinary) SWAP, $b = 2$, so it costs only O(EO) time (plus overhead for updating the set of boundary triples). A BLOCKSWAP is more complicated and costs O(bEO) $\leq$ O(nEO) time. However, we still come out ahead because the sum of these times over all calls to BLOCKSWAP in one call to HREFINE is only O(n^4EO), whereas we called SWAP O(n^5) times per REFINE. This leads to the improved running time of O(n^4EO) for HREFINE, exclusive of calls to REDUCEV. As with IFF, the Hybrid Algorithm needs to call REDUCEV once per AUGMENT, for a total of O(n^5) linear algebra work (which dominates other overhead). Thus the running time of HREFINE is O(n^4EO + n^5), compared to O(n^5EO) for REFINE. Since we can safely assume that EO is at least O(n) (because the length of its input is a subset of size O(n)), this is a speedup over all three versions of IFF by a factor of O(n).

The top-level parts of the Hybrid Algorithm look much like the IFF Algorithm: We relax $y \in B(f)$ to $z \in B(f + \delta_k)$ via flows x in the relaxation network and keep the invariant $z = y + \partial x$, and we put this into a loop that scales δ. We again define $S^{-\delta}(z) = \{l \in E | \, z_l \leq -\delta\}$, $S^{+\delta}(z) = \{l \in E | \, z_l \geq +\delta\}$, and $R(\delta) = \{k \to l | \, x_{kl} \leq 0\}$. We look for a directed augmenting path P from $S^{-\delta}(z)$ to $S^{+\delta}(z)$ using only arcs of $R(\delta)$ and then AUGMENT as before.

Hybrid Outer Scaling Framework

Initialize by choosing $\prec_1$ to be any linear order, $y = v^1$, and $\mathcal{I} = \{1\}$.
Initialize $\delta = |y^-(E)|/n^2$, $x = 0$, and $z = y$. [$z = y + \partial x$ is δ-optimal]
While $\delta \geq 1/n^2$, [when $\delta < 1/n^2$ we are optimal]
  Set $\delta \leftarrow \delta/2$.
  Call HREFINER. [converts 2δ-optimality to δ-optimality]
Return last approximate solution from HREFINER as optimal SFM solution.

Since we no longer require consecutive pairs, we now define the set of arcs available for augmenting y to be $A(\mathcal{I}) = \{k \to l \mid \exists i| \in \mathcal{I} \text{ s.t. } l \prec_i k\}$ (the same set of arcs as in Schrijver's Algorithm), which includes many more arcs than in IFF. We use distance labels d w.r.t. $A(\mathcal{I})$ in a similar way as in Schrijver's Algorithm: For now we say that d is valid if $d_s = 0$ for all $s \in S^{-\delta}(z)$, and $d_l \leq d_k + 1$ for all $k \to l \in A(\mathcal{I})$ ($l \prec_i k$). As usual, d_l is a lower bound on the number of arcs in a path in $(E, A(\mathcal{I}))$ from $S^{-\delta}(z)$ to l, so that $d_l = n$ signifies that no such path exists.

With IFF we keep iterating until $B = \emptyset$, i.e., until the set S has no arcs of $A(\mathcal{I})$ exiting it, ensuring via (13) that S is tight for y. Allowing Hybrid to iterate until S has no arcs of $A(\mathcal{I})$ exiting it would take too much time, so instead Hybrid iterates only until $d_t \geq n$ for all $t \notin S$, and then defines S' to be the set of nodes reachable from $S^{-\delta}(z)$ via arcs of $A(\mathcal{I})$. Since no node t with $d_t \geq n$ is reachable via such arcs we have $S' \subseteq S$. Also, S' clearly has no arcs of $A(\mathcal{I})$ exiting it, so we could use S' in place of S in the proof of Lemma 3.4.

However, there is a problem with this strategy when we try to put infinite bounds on arcs of C for an explicit strongly polynomial version of Hybrid, which is needed for a fully combinatorial version of Hybrid: There is nothing to prevent having an arc $t \to s$ of C entering S' with $x_{ts} \gg \delta$ (note that this could not happen with $t \to s$ entering S in IFF, since such an arc would belong to $R(\delta)$, implying that $t \in S$). Such a rogue arc would then invalidate the proof of Lemma 3.4 since the inequality $-(n-1)\delta \leq \partial x_l$ for $l \in S'$ might no longer be true. This problem causes the argument for the fully combinatorial version of Hybrid in Iwata (2002b) to be incorrect as it stands.

A fix for this problem was suggested by Fujishige: Let's keep a separate flow φ on C. Flows x_{st} have the bounds $0 \leq x_{st} \leq \delta$, and φ_{st} have the bounds $0 \leq \varphi_{st} \leq \infty$. Augmentations will affect only x, and $R(\delta)$ contains only δ-augmentable arcs w.r.t. x. We now keep the invariant that $z = y + \partial x + \partial\varphi$, and (for the SP and FC versions) define $w = y + \partial\varphi$ so that $z = w + \partial x$. We change the definition of validity of d to ensure that no rogue arcs enter S': We now say that d is valid if (i) $d_s = 0$ for all $s \in S^{-\delta}(z)$, and (ii) $d_l \leq d_k + 1$ for all $k \to l \in A(\mathcal{I})$ ($l \prec_i k$), all $l \to k \in C$, and all $k \to l \in C$, with $\varphi_{kl} \geq \delta$. Defining $C(\delta) = C \cup \{l \to k \mid k \to l \in C \text{ and } \varphi_{kl} \geq \delta\}$ (the set of δ-augmentable arcs of C), then d_l is a lower bound on the number of arcs in a path in $(E, A(\mathcal{I}) \cup C(\delta))$ from $S^{-\delta}(z)$ to l, and $d_l = n$ signifies that no such path exists. We use this modified explicit method throughout our discussion of Hybrid.

When no augmenting path exists, we use d to guide the algorithm as follows. HREFINER defines the set of nodes reachable from $S^{-\delta}(z)$ as $S = \{k \in E \mid \text{there is a path in } (E, R(\delta)) \text{ from } S^{-\delta}(z) \text{ to } k\}$. Define the set of nodes not in S with minimum distance label as $D = \{l \notin S \mid d_l = \min_{h \notin S} d_h\}$.

If there is some $s \to t \in C(\delta)$ with $t \in D$ and $d_s = d_t - 1$ (which implies that $s \in S$, and that $x_{st} > 0$ else t would be in S), then we call FLOWSWAP: if $s \to t \in C(\delta)$ corresponds to $s \to t \in C$ then we update $\varphi_{st} \leftarrow \varphi_{st} + x_{st}$; else ($s \to t \in C(\delta)$ corresponds to $t \to s \in C$ with $\varphi_{ts} \geq \delta$) update $\varphi_{ts} \leftarrow \varphi_{ts} - x_{st}$. Finally update $x_{st} \leftarrow 0$. Note that this update leaves $\partial\varphi + \partial x$ invariant, and causes t to join S. Furthermore, since it is applied only when $|d_s - d_t| = 1$, it cannot cause d to become invalid. FLOWSWAP is the only operation that changes φ.

If no FLOWSWAP applies, suppose that there is some arc $p \rightarrow q \in A(\mathcal{I})$ (so there is some $i \in \mathcal{I}$ with $q \prec_i p$) with $p \in S$, $q \in D$, and $d_q = d_p + 1$. Then we choose the left-most such q in $\prec_i$ as l and the right-most such p in $\prec_i$ as k, and call the triple $(i; k, l)$ *active*. Thus $h \prec_i l$ implies that $h \in S$, and $k \prec_i h$ implies that $d_h > d_k$. This definition of active is the only delicate lexicographic choice here.

It is a bit tricky to efficiently find active triples. Define a *re-ordering phase* to be the set of BLOCKSWAPs between consecutive calls to RELABEL or AUGMENT. At each re-ordering phase, we SCAN $\prec_i$ for each $i \in \mathcal{I}$ to find out LEFT_{ir}, the left-most element of $\prec_i$ with distance label r, and RIGHT_{ir}, the right-most such element. Then, when we look for an active triple $(i; k, l)$ with $d_l = m$, we can restrict our SEARCH to $[\text{LEFT}_{im}, \text{RIGHT}_{i,m-1}]$.

Define $S(i; k, l)$ to be the elements in $[l, k]_{\prec i}$ in S, and $T(i; k, l)$ to be the elements in $[l, k]_{\prec_i}$ not in S, i.e., $S(i; k, l) = \{h \in S | l \prec_i h \preceq_i k\}$ and $T(i; k, l) = \{h \notin S | l \preceq_i h \prec_i k\}$. Thus $k \in S(i; k, l)$ and $l \in T(i; k, l)$. Define $\prec_j$ to be $\prec_i$ with all elements of $S(i; k, l)$ moved ahead of the elements of $T(i; k, l)$ (without changing the order within $S(i; k, l)$ and $T(i; k, l)$), i.e., just before l. For example (using t_a to denote elements of $T(i; k, l)$ and s_b to denote elements of $S(i; k, l)$, if $\prec_i$ looks like

$$\ldots u_3 u_4 l t_1 t_2 s_1 t_3 t_4 t_5 s_2 s_3 t_6 s_4 k u_5 u_6 \ldots,$$

then $\prec_j$ looks like

$$\ldots u_3 u_4 s_1 s_2 s_3 s_4 k l t_1 t_2 t_3 t_4 t_5 t_6 u_5 u_6 \ldots.$$

Let v^j be the vertex associated with $\prec_j$ by the Greedy Algorithm. By (4), for $b = |[l, k]_{\prec_i}|$, computing v_j costs O(bEO) time. We ideally want to move y in the direction $v^j - v^i$ by replacing the term $\lambda_i v^i$ in (8) by $\lambda_i v^j$. To do this we need to change x to ensure that $z = y + \partial x$ is preserved, and so we must find a flow q to subtract from x whose boundary is $v^j - v^i$.

First we determine the sign of $v^i_u - v^j_u$ depending on whether u is in $S(i; k, l)$ or $T(i; k, l)$ (for $u \notin [l, k]_{\prec_i}$ we have $v^i_u - v^j_u = 0$ since $u^{\prec_j} = u^{\prec_i}$). For $s \in S(i; k, l)$ we have that $s^{\prec_j} \subset s^{\prec_i}$, so by (1) and Greedy we get that $v^j_s = f(s^{\prec_j} + s) - f(s^{\prec_j}) \geq f(s^{\prec_i} + s) - f(s^{\prec_i}) = v^i_s$. Smilarly for $t \in T(i; k, l)$, we have $t^{\prec_j} \supset t^{\prec_i}$, implying that $v^j_t \leq v^i_t$.

Now set up a transportation problem with left nodes $S(i; k, l)$, right nodes $T(i; k, l)$ and all possible arcs. Make the supply at $s \in S(i; k, l)$ equal to $v^j_s - v^i_s \geq 0$, and the demand at $t \in T(i; k, l)$ equal to $v^i_t - v^j_t \geq 0$. Now use, e.g., the Northwest Corner Rule (see Ahuja, Magnanti, and Orlin (1993)) to find a basic feasible flow $q \geq 0$ in this network. This can be done in $O(|[l, k]_{\prec_i}|) = O(b)$ time, and the number of arcs with $q_{st} > 0$ is also $O(b)$ (Ahuja et al., 1993).

Hence computing q and using it to update x takes only O(b) time. Now reimagining q as a flow in (E, R) we see that $\partial q = v^j - v^i$, as desired.

Hybrid Subroutine BlockSwap $(i; k, l)$.

Applies to active triple $(i; k, l)$

Use l and k to compute $S(i; k, l)$, $T(i; k, l)$, $\prec_j$, and v^j.
  Set up the transportation network and compute q.
  Compute $\eta = \max_{st} q_{st}$ and $\alpha = \min(\lambda_i, \delta/\eta)$. [compute step length, then update]
  Set $y \leftarrow y + \alpha(v^j - v^i)$, $\lambda_j \leftarrow \alpha$, and $\mathcal{I} \leftarrow \mathcal{I} + j$.
  If $\alpha = \delta/\eta$ then [a partial BlockSwap, so at least t with $q_{st} = \eta$ joins S]
    Set $\lambda_i \leftarrow \lambda_i - \alpha$.
  Else $(\alpha = \lambda_i)$ [a full BlockSwap, so i leaves $\mathcal{I}$]
    Set $\mathcal{I} \leftarrow \mathcal{I} + i$.
  For $s \to t$ s.t. $q_{st} > 0$, [update x_{st} and x_{ts}]
    If $\alpha q_{st} \le x_{st}$, set $x_{st} \leftarrow x_{st} - \alpha q_{st}$;
    Else $(\alpha q_{st} > x_{st})$ set $x_{ts} \leftarrow \alpha q_{st} - x_{st}$, and $x_{st} \leftarrow 0$.
  Update $R(\delta)$, S, and D.

As with IFF, the capacities of δ on the xs might prevent us from taking the full step from $\lambda_i v^i$ to $\lambda_i v^j$, and modifying x_{st} and x_{ts} by $\lambda_i q_{st}$. So we choose a step length $\alpha \le \lambda_i$ and investigate constraints on α. If $\alpha q_{st} \le x_{st}$ then our update is $x_{st} \leftarrow x_{st} - \alpha q_{st}$, which is no problem. If $\alpha q_{st} > x_{st}$ then our update is $x_{ts} \leftarrow \alpha q_{st} - x_{st}$ and $x_{st} \leftarrow 0$, which requires that $\alpha q_{st} - x_{st} \le \delta$, or $\alpha \le (\delta + x_{st})/q_{st}$. Since $x_{st} \ge 0$, if we choose $\eta = \max_{st} q_{st}$ and $\alpha = \min(\lambda_i, \delta/\eta)$, then this suffices to keep x feasible.

Since x is changed only on arcs from S to $E - S$, S can only get bigger after BlockSwap (since z doesn't change, neither $S^{+\delta}(z)$ nor $S^{-\delta}(z)$ changes). If $\alpha = \delta/\eta < \lambda_i$, then $\alpha q_{st} = \delta \ge x_{st}$, so the updated x_{st} is zero. Hence $s \to t$ joins $R(\delta)$ and so t joins S, and we call such a step *partial* (*nonsaturating* in (Iwata, 2002)). In this case we need to keep both v^j (with coefficient α) and v^i (if $\alpha < \lambda_i$, with coefficient $\lambda_i - \alpha$) in $\mathcal{I}$ so $|\mathcal{I}|$ possibly goes up by one. Otherwise $(\alpha = \lambda_i)$, we call the step *full* (*saturating* in Iwata (2002b)). In this case v^j just replaces v^i (with coefficient $\lambda_i = \alpha$) in $\mathcal{I}$ and $|\mathcal{I}|$ stays the same. Since there are at most n partial BlockSwaps before calling Augment, $|\mathcal{I}| \le 2n$ before calling ReduceV.

If there are no active triples for the current D and $d_l < n$ for $l \in D$, then HRefineR does a Relabel that increases d_t by one for all $l \in D$. HRefineR stops and concludes that the current point is δ-optimal when it can no longer find any augmenting paths and $d_l = n$ for $l \in D$ (and so for all $l \notin S$). Note that HRefineR recomputes $S^{-\delta}(z)$, $S^{+\delta}(z)$, S, and D after every Augment, and S and D during BlockSwap, so that S dynamically changes and is not necessarily monotonic.

Hybrid Subroutine HRefineR

Initialize $d=0$, $x \leftarrow x/2$, $\varphi \leftarrow \varphi/2$, and update z.
Compute $S^{-\delta}(z)$, $S^{+\delta}(z)$, and S.
While augmenting paths exist ($S \cap S^{+\delta}(z) \neq \emptyset$), or $\exists l \notin S$ with $d_l < n$ do
  If $\exists$ path P from $S^{-\delta}(z)$ to $S^{+\delta}(z)$ using arcs from $R(\delta)$, do
    Augment (P), ReduceV, update $S^{-\delta}(z)$, $S^{+\delta}(z)$, S, and D, Scan.
  Else ($\nexists$ such a path, but $l \in D$ have $d_l < n$) if FlowSwap applies, call it.
  Else (FlowSwap does not apply) do
    Search for an active triple.
    If $\exists$ an active triple $(i; k, l)$, BlockSwap $(i; k, l)$.
    Else (no active triple) Relabel: $d_l \leftarrow d_l + 1$ for all $l \in D$, update D, Scan.
[Now no augmenting paths exist and $d_l = n$ for all $l \notin S$]
Return S as an approximate optimum solution.

Recall that $w = y + \partial_C x \in B(f)$, and that our optimality condition for S solving SFM is that $w^-(E) = f(S)$. The following lemma shows for both w and z how close these approximate solutions are to exactly satisfying $w^-(E) = f(S)$ and $z^-(E) = f(S)$ at the end of HRefineR.

Lemma 3.10. *When* HRefineR *ends, S is tight for y, and we have $w^-(E) \geq f(S) - n^2\delta$ and $z^-(E) \geq f(S) - n(n+1)\delta/2$.*

Proof. If $s \in S$ and $t \notin S$ and for some $i \in \mathcal{I}$, $t \prec_i s$, then $d_t = n$ and $d_s < n$ would violate validity of d. Thus for any $s \in S$ and $t \notin S$ we have $s \prec_i t$ for all $i \in \mathcal{I}$. Thus for each $i \in \mathcal{I}$, there is some $e_i \in E$ such that S equals $e_i^{\prec_i}$, and so $v^i(S) = f(S)$ by (3). But then $y(S) = \sum_i \lambda_i v^i(S) = f(S) \sum_i \lambda_i = f(S)$.

Since $S^{-\delta}(z) \subseteq S \subseteq E - S^{+\delta}(z)$ we get $z_s < +\delta$ for all $s \in S$ and $z_t > -\delta$ for all $t \notin S$. For $s \in S$, if $0 \leq z_s < +\delta$, then $z_s^- = 0 > z_s - \delta$. If $z_s < 0$, then $z_s^- = z_s > z_s - \delta$. For $t \notin S$, $z_t > -\delta$ implies that $z_t^- > -\delta$. Thus $z^-(E) = z^-(S) + z^-(E - S) > [z(S) - \delta|S|] - \delta|E - S| = y(S) + \partial x(S) - n\delta = f(S) + \partial x(S) - n\delta$. The upper bound of δ on x_{ts} implies that $\partial x(S) \geq -\delta|S| \cdot |E - S| \geq -n^2\delta/4$. This yields $z^-(E) > f(S) - (n + n^2/4)\delta \geq f(S) - n(n+1)\delta/2$.

Since $w = z + (\partial_C x - \partial x)$, for any $u \in E$ w_u can be at most $\partial_C x - \partial x_u$ lower than z_u. The term involving arcs of C cancel out, leaving only flows between 0 and δ. Thus the at most $n-1$ non-C arcs $t \to u$ can decrcase w_u at most $(n-1)\delta$ below z_u. Furthermore, since $x_{ut} \cdot x_{tu} = 0$, each x_{ut}, x_{tu} pair decreases at most one of w_u and w_t. Thus the total amount by which $w^-(E)$ is smaller than $z^-(E)$ is at most $n(n-1)\delta/2$. Putting this together with the previous paragraph gives $w^-(E) > f(S) - n(n+1)\delta/2 - n(n-1)\delta/2 = f(S) - n^2\delta$. □

We now use this to prove correctness and running time. As before we pick out the main points in boldface.

Theorem 3.11. *The Hybrid SFM Algorithm is correct for integral data and runs in* $O((n^4EO + n^5)\log M)$ time.

Proof.
The current approximate solution S at the end of a δ-scaling phase with $\delta < 1/n^2$ solves SFM. Lemma 3.10 shows that $w^-(E) \geq f(S) - n^2\delta > f(S) - 1$. But for any $U \subseteq E$, $f(U) \geq w(U) \geq w^-(E) > f(T) - 1$. Since f is integer-valued, T solves SFM.

Distance labels remain valid throughout HRefineR. Only Augment changes z, and in such a way that $S^{-\delta}(z)$ only gets smaller. Hence $d_s = 0$ on $S^{-\delta}(z)$ is preserved. BlockSwap $(i; k, l)$ adds new vertex v^j to $\mathcal{I}$. The only new pairs with $s \prec_j t$ but $s \not\prec_i t$ (that might violate validity) are those with $s \in S(i; k, l)$ and $t \in T(i; k, l)$, and for these we need that $d_s \leq d_t + 1$. Validity applied to $s \preceq_i k$ implies that $d_s \leq d_k + 1 = d_l$. By definition of D, $d_l \leq d_t$, so $d_s \leq d_l \leq d_t < d_t + 1$. Finally, since we Relabel only if there are no active triples, Relabeling preserves validity.

Each scaling phase calls Augment $O(n^2)$ times. At the beginning of HRefineR, for X equal to the final S from the previous call to HRefineR, Lemma 3.10 shows that $z^-(E) > f(X) - n(n+1)\delta$. This is also true for the first call to HRefineR for $X = 0$ by the choice of the initial value of $|y^-(E)|/n^2 = |z^-(E)|/n^2$ for δ. At any point during HRefineR, from the upper bound of δ on x_{ts} we have $z^-(E) \leq z(X) = w(X) + \partial x(X) - \partial_C x(X) \leq f(X) + n(n-1)\delta$. Thus the total rise in value for $z^-(E)$ during HRefineR is at most $2n^2\delta$. Each call to Augment increases $z^-(E)$ by δ, so there are $O(n^2)$ calls to Augment.

There are $O(n^2)$ calls to Relabel during HRefineR. Each d_k is between 0 and n and never decreases during HRefineR. Each Relabel increases at least one d_k by one, so there are $O(n^2)$ Relabels.

The previous two paragraphs establish that there are $O(n^2)$ reordering phases.

The common value m of d_l for $l \in D$ is nondecreasing during a reordering phase. During a reordering phase, d does not change but S, D, and $R(\delta)$ do change. However, all arcs where x changes, and hence where $R(\delta)$ can change, are between $S(i; k, l)$ and $T(i; k, l)$. Thus S can only get larger during a reordering phase, and so m is monotone nondecreasing in a phase.

The work done by all BlockSwaps during a reordering phase is $O(n^2EO)$. Suppose that BlockSwap $(i; k, l)$ adds v^j to $\mathcal{I}$. Then, by how k and l were define in an active triple, for any q with $d_q = d_l$, any p with $d_q = d_p + 1$ must have that $p \prec_j q$, and hence there can be no subsequent active triple $(j; p, q)$ in the phase with $d_q = d_l$. Thus m must increase by at least one before the phase uses a subsequent active triple $(j; p, q)$ involving $\prec_j$. But then $d_q > d_l = d_k + 1$, implying that we must have that $l \prec_i k \prec_i q \prec_i p$. Hence if v^j results from v^i via BlockSwap $(i; k, l)$, and $(j; p, q)$ is the next active triple at j in the same reordering phase, it must be that $[l, k]_{\prec_i}$ is disjoint from $[q, p]_{\prec_i}$.

Suppose that $\prec_j$ appears in $\mathcal{I}$ at some point during a reordering phase, having been derived by a sequence of BlockSwaps starting with $\prec_{i_1}$ (which belonged to $\mathcal{I}$ at the beginning of the phase), applying active triple $(i_1; k_1, l_1)$ to $\prec_{i_1}$ to get $\prec_{i_2}$, applying active triple $(i_2; k_2, l_2)$ to $\prec_{i_2}$ to get $\prec_{i_3}, \ldots,$ and applying active triple $(i_a; k_a, l_a)$ to $\prec_{i_a}$ to get $\prec_{i_{a+1}} = \prec_j$. Continuing the argument in the previous paragraph, we must have that $l_1 \prec_{i_1} k_1 \prec_{i_1} l_2 \prec_{i_1} k_2 \prec_{i_1} \cdots \prec_{i_1} l_a \prec_{i_1} k_a$. Thus the sum of the sizes of the intervals $[l_1, k_1]_{\prec_{i_1}}$, $[l_2, k_2]_{\prec_{i_1}}, \ldots, [l_a, k_a]_{\prec_{i_1}}$ is O(n). We count all these BlockSwaps as belonging to $\prec_j$, so the total BlockSwap work attributable to $\prec_j$ is O(nEO). Since $|\mathcal{I}| = $ O(n), the total work during a reordering phase is O(n^2EO).

The time for one call to HRefineR is O(n^4EO + n^5). The bottleneck in calling Augment is the call to ReduceV, which costs O(n^3) time. There are O(n^2) calls to Augment, for a total of O(n^5) ReduceV work during HRefineR. There are O(n^2) reordering phases during HRefineR, so Scan is called O(n^2) times. The BlockSwaps during a phase cost O(n^2EO) time, for a total of O(n^4EO) BlockSwap work in one call to HRefineR. Each call to Scan costs O(n^2) time, for a total of O(n^4) work per HRefineR. As in the previous paragraph, the intervals [Left$_{im}$, Right$_{i,m-1}$] are disjoint in $\prec_i$, so the total Search work for $\prec_i$ is O(n), or a total of O(n^2) per phase, or O(n^4) work over all phases. The updates to S and D cost O(n) work per phase, or O(n^3) overall.

There are O(log M) calls to HRefineR. As in the proof of Theorem 3.5 the initial $\hat{\delta} = |y^-(E)|/n^2 \leq 2M/n^2$. Each call to HRefineR cuts δ in half, and we terminate when $\delta < 1/n^2$, so there are O(log M) calls to HRefineR.

The total running time of the algorithm is O((n^4EO + n^5) log M). Multiplying together the factors from the last two paragraphs gives the claimed total time. □

We already specified HRefineR so that it optimizes over a ring family, and this suffices to embed HRefineR into the strongly polynomial framework of Section 3.3.2, getting a running time of O((n^4EO + n^5)n^2log n).

Making the Hybrid Algorithm fully combinatorial is similar to the ideas in Section 3.3.3. The ratio δ/η in BlockSwap is handled in the same way as the ratio $x_{kl}/c(k, l; v^i)$ in (16) of IFF-FC. If $\tilde{\lambda}_i q_{st} \leq SF\tilde{x}_{st}$ for all $s \to t$ (where SF is the current scale factor), then we can do a full BlockSwap as before. Otherwise we use binary search to compute the minimum integer $\tilde{\beta}$ such that there is some $s \to t$ with $\tilde{\beta} q_{st} \geq \mathrm{SF}\tilde{x}_{st}$. We then update $\tilde{\lambda}_i \prec \tilde{\lambda}_i - \tilde{\beta}$ and $\tilde{\lambda}_j \prec \tilde{\beta}$.

Since $\tilde{\beta} = \lceil \text{SF}\tilde{x}_{st}/q_{st} \rceil$, the increase in $\tilde{\beta}$ over the usual value $\text{SF}\tilde{x}_{st}/q_{st}$ is at most 1, so the change in $\partial\tilde{x}_s$ is at most $q_{st} \leq v_s^j - v_s^i \leq \tilde{\delta}$ by Lemma 3.9, so the update keeps $\tilde{x}$ $\tilde{\delta}$-feasible (this is why we need the explicit method here). We started from the assumption that there is some $s \to t$ with $\tilde{\lambda}_i q_{st} > SF\tilde{x}_{st}$, implying the $\tilde{\beta} \leq \tilde{\lambda}_i \leq SF$, so this binary search is fully combinatorial.

The running time of all versions of the algorithm depends on the $\text{O}(n^4\text{EO} + n^5)$ time for HREFINER, which comes from $\text{O}(n^2)$ reordering phases times $\text{O}(n^2\text{EO})$ BLOCKSWAP work plus $\text{O}(n^3)$ REDUCEV work in each reordering phase. The $\text{O}(n^2\text{EO})$ BLOCKSWAP work in each reordering phase comes from $\text{O}(n\text{EO})$ BLOCKSWAP work attributable to each $\prec_i$ in $\mathcal{I}$ times the $\text{O}(n)$ size of $\mathcal{I}$. Since $|\mathcal{I}|$ is larger by a factor of $\text{O}(n^2 \log n)$ when we don't call REDUCEV (it grows from $\text{O}(n)$ to $\text{O}(n^3 \log n)$), we might expect that the fully combinatorial running time also grows by a factor of $\text{O}(n^2 \log n)$, from $\text{O}((n^6\text{EO} + n^7)\log n)$ to $\text{O}((n^8\text{EO} + n^9)\log^2 n)$. However, the term $\text{O}(n^9)$ comes only from the $\text{O}(n^3)$ REDUCEV work per reordering phase. The SCAN and SEARCH time in a reordering phase is only $\text{O}(n^2)$, which is dominated by the BLOCKSWAP work. Thus, since the fully combinatorial version avoids calling REDUCEV, the total time is only $\text{O}(n^8\text{EO} \log^2 n)$. (The careful implementation of SCAN and SEARCH are needed to avoid the extra term of $\text{O}(n^9 \log^2 n)$, and this is original to this survey).

4 Comparing and contrasting the algorithms

Table 1 summarizes, compares, and contrasts the four main SFM algorithms we have studied, those of Cunningham for General SFM (Cunningham, 1985), the Fleischer and Iwata (2001), Schrijver-PR Push-Relabel variant of Schrijver (2000), Iwata et al. (2001) and Iwata's fully combinatorial version IFF-FC of it (Iwata, 2002a), and Iwata's Hybrid Algorithm (Iwata, 2002b).

Note that all the algorithms besides Schrijver's Algorithm add just one vertex to $\mathcal{I}$ at each exchange (or at most n vertices per augmenting path). Except for the Hybrid Algorithm, they are able to do this because they all consider only consecutive exchanges; the Hybrid Algorithm considers nonconsecutive exchanges, but moves in a $v^i - v^j$ direction instead of a $\chi_k - \chi_l$ direction, thereby allowing it also to add at most one vertex per exchange. By contrast, Schrijver's Algorithm allows nonconsecutive exchanges, and thus must pay the price of needing to add as many as n vertices to $\mathcal{I}$ for each exchange.

Only Schrijver's Algorithm always yields an exact primal solution y. When f is integer-valued, in the base polyhedron approach we apply Theorem 2.8 with $x = 0$, and in the polymatroid approach we apply it with $x = y$. In either case x is also integral, so Theorem 2.8 shows that there always exists an *integral* optimal y. Despite this fact, none of the algorithms always yields an integral optimal y in this case. However, we could get exact integral primal solutions

Table 1.
Summary comparison table of main results. Running times are expressed in terms of n, the number of elements in the ground set E; M, an upper bound on $|f(S)|$ for any $S; \subseteq E$, a measure of the "size" of f; and EO, the time for one call to the evaluation oracle for f. For comparison, the running time of the strongly polynomial version of the Ellipsoid Algorithm is $O(n^5 EO + n^7)$, see Theorem 2.7.

	Cunningham for General SFM (Cunningham, 1985)	Schrijver and Schrijver-PR (Fleischer and Iwata, 2001; Schrijver, 2000)	Iwata, Fleischer, and Fujishige (Iwata et al., 2001; Iwata, 2002a)	Iwata Hybrid (Iwata, 2002b)
Pseudo-polyn. running time	$O(Mn^6 \log(Mn) \cdot EO)$			
Weakly polyn. running time			$O(n^5 \log M \cdot EO)$ (Iwata et al., 2001)	$O((n^4\ EO + n^5) \cdot \log M)$
Strongly polyn. running time		$O(n^7 EO + n^8)$ (Fleischer and Iwata, 2001)	$O(n^7 \log n \cdot EO)$ (Iwata et al., 2001)	$O((n^6\ EO + n^7) \cdot \log n)$
Fully comb. running time			$O(n^9 \log^2 n \cdot EO)$ (Iwata, 2002a)	$O(n^8\ EO \log^2 n)$
Approach	Polymatroid	Base polyhedron	Base polyhedron	Base polyhedron
Convergence strategy	Weak sufficient decrease, pseudo-polynomial	Distance label	Strong sufficient decrease, strongly polynomial	Both distance label and strong sufficient decrease
Exact primal solution?	No	Yes	No	No
Scaling?	No	No	Relaxation parameter	Relaxation parameter
Max Flow analogy	Max Capacity Path	Max Dist. Push-Relabel	Relaxed Max Cap. Path for z, push across cut for y	Relaxed Max Cap. Path for z, Push-Relabel across cut for y
Arc $k \to l$ for aug. y exists if ...	(l, k) consecutive, $c(k, l; y) > 0$ (minimal)	$l \prec_i k$ (loosest)	(l, k) consecutive (medium)	$l \prec_i k$ (loosest)
Movement directions	Unit, simple representation	Unit, complex representation	Unit, simple representation	Vertex, simple representation
Modifies $\prec_i$ by ...	Consecutive pairs	Blocks	Consecutive pairs	Blocks
Augments ...	On paths	Arc by arc	z on paths, y arc by arc via SWAPS	z on paths, y arc by arc via BLOCKSWAPS
Number of vertices added each exchange	0 or 1	$\leq n$	0 or 1	0 or 1

from n calls to SFM as follows. Use the polymatroid approach, compute γ, and discard any $e \in E$ with $\gamma_e < 0$. Run the modified Greedy Algorithm in the proof of Theorem 2.8 starting with $y = 0$, and look for a vector $y \in P(\tilde{f})$ with $y \leq \gamma$. At each of the n steps we can compute the maximum step length we can take and stay inside $P(\tilde{f})$ via one call to SFM.

4.1 *Solving SFM in practice*

There is very little computational experience with any of these algorithms so far, nor is there any generally accepted test bed of instances of SFM. If we reason by analogy with performance of similar Max Flow algorithms [see, e.g., Cherkassky and Goldberg (1997)], then Schrijver-PR should outperform the IFF Algorithm. The reason is that the Push-Relable Max Flow Algorithm (Goldberg and Tarjan, 1988) that is analogous to Schrijver-PR has proven to be more robust and faster in practice than the sort of capacity-scaling algorithms that IFF is based on. However, the superior practical performance of Push-Relabel-based Max Flow algorithms depends on using heuristics to speed up the native algorithm (Cherkassky and Goldberg, 1997), and the relative inflexibility of Schrijver-PR may prevent this.

Iwata (Iwata, 2002a) and Isotani and Fujishige (Isotani and Fujishige, 2003) have done some computational experiments comparing the performance of Schrijver's Algorithm, Schrijver-PR, IFF and Hybrid. The test problems were dense Min Cut problems perturbed by a modular function in such a way that the optimal SFM solution is always $\{1, 2, \ldots, k\}$ for k equaling about $n/3$. All algorithms were started out with the linear order $(n, n-1, \ldots, 1)$ to ensure that they would have to work hard to move the $\{1, 2, \ldots, k\}$ elements before the $\{k+1, k+2, \ldots, n\}$ elements in the linear orders for the optimal solution. Each algorithm was run on instances of sizes from 50 to 1000 elements.

Table 2 shows the empirical estimates of each algorithm's running time and number of evaluation oracle calls. We see that the empirical performance of all four algorithms is much faster than their theoretical time bounds, and that (based on these limited tests) Hybrid is the fastest of the four. Iwata's data also showed that the dominant factor in determining running time is the number of calls to REDUCEV. The big advantage of the IFF-based algorithms is that they ended up calling REDUCEV many fewer times than the Schrijver-based

Table 2.
Empirical results from Iwata (2002). Estimates of running time and number of evaluation oracle calls come from a log–log regression

Algorithm	Total run time	No. oracle calls
Schrijver	$n^{5.8}$	n^4
Schrijver-PR	$n^{5.5}$	n^4
IFF	$n^{4.0}$	$n^{2.5}$
Hybrid	$n^{3.5}$	$n^{2.5}$

algorithms. However, because these results are based on runs on a single class of instances, and because heuristic improvements to these algorithms (such as the "gap" and "exact distance" heuristics that made such a difference for Max Flow algorithms (Cherkassky and Goldberg, 1997)) have not yet been implemented, these results must be viewed as being only suggestive, not definitive.

All of the combinatorial SFM algorithms we consider call the evaluation oracle EO only as part of the Greedy Algorithm. Greedy calls EO for the nested sequence of subsets $e_1^{\prec}, e_2^{\prec}, \ldots, e_i^{\prec}, \ldots, e_n^{\prec}$. In some applications we can take advantage of this and use some incremental algorithm to evaluate $f(e_i^{\prec})$ based on the value of $f(e_{i-1}^{\prec})$ much faster than evaluating $f(e_i^{\prec})$ from scratch. For example, for Min Cut on a graph with n nodes and m arcs, one evaluation of $f(S)$ costs $O(m)$ time, but all n evaluations within Greedy can be done in just $O(m)$ time. This could lead to a heuristic speedup for such applications, although most such applications (such as Min Cut) have specialized algorithms for solving SFM that are much faster than the general algorithms here.

Indeed, it is very rare that true general SFM arises in practice. Nearly all applications of SFM in our experience have some special structure that can be taken advantage of, resulting in much faster special-purpose algorithms than anything covered here. As one example, a naive application of Queyranne's Algorithm (see Section 5.1) to solve undirected Min Cut would take $O(n^3|A|)$ time, since $EO = O(|A|)$ in that case. But in fact Nagamochi and Ibaraki (1992) show how to take advantage of special structure to reduce this to only $O(n|A| + n^2 \log n)$. In the great majority of these cases we end up solving the SFM problem as a sequence of Min Cut problems; see Picard and Queyranne (1982) for a list of problems reducible to Min Cut.

A recent example of this is where f is the rank function of a graph (the rank of edge subset S is the maximum size of an acyclic subgraph of S) modified by a modular function, which has applications in physics [see Anglès d'Auriac et al. (2002)]. Here E is the set of edges of the graph, so we use $n = |E|$, and use $|N|$ for the number of nodes. In this case $EO = O(n)$ so the fastest SFM algorithm here would take $\tilde{O}(n^5)$ time, but Anglés d'Auriac et al. (2002) shows how to solve SFM using Min Cuts in only $O(|N| \cdot MF(|N|, n))$ time, where $MF(|N|, n)$ is the time to solve a Max Flow problem on a graph with $|N|$ nodes and n edges. One of the best bounds for Max Flow is $O(\min\{|N|^{2/3}, \sqrt{n}\} n \log(|N|^2/n) \log M)$ Goldberg and Rao (1998), which would give a running time of $\tilde{O}(\min\{(|N|^{2/3}, \sqrt{n}\} n^2) \le \tilde{O}(n^{5/2})$, much faster than $\tilde{O}(n^5)$.

Therefore if you are faced with solving a practical SFM problem, you should look very carefully to see if there is some way to solve it via Min Cut before using one of these general SFM algorithms. If there is no apparent way to reduce to a Min Cut problem, then another possible direction is to try a column generation method [see, e.g., du Merle, Villeneuve, Desrosiers and Hansen (1999)], which pairs linear programming technology (for solving (6) or (7)) with a column generation subroutine that would (in this context) come from the Greedy Algorithm. Although such algorithms do not have polynomial bounds, they often can be made to work well in practice.

5 Solvable extensions of SFM

We already saw with RefineR in Section 3.3.2 that it is not hard to adapt SFM algorithms to optimize over ring families instead of 2^E. The same trick works for showing that Schrijver's Algorithm also adapts to solving SFM over ring families. But sometimes we are interested in optimizing over other families of subsets which are not ring families. For example, in some applications we would like to optimize over nonempty sets, or sets other than E, or both; or given elements s and t, optimize over sets containing s but not t; or optimize over sets S with $|S|$ odd; etc [see Nemhauser and Wolsey (1988), Section III for typical applications]. Goemans and Ramakrishnan (1995) derive many such algorithms, and give a nice survey of the state of the art.

As we saw in Section 3.3.2, if we want to solve SFM over subsets containing a fixed $l \in E$, then we can consider $E' = E - l$ and $f_l(S) = f(S + l) - f(l)$, a submodular function on E'. If we want to solve SFM over subsets *not* containing a fixed $l \in E$, then we can consider $E' = E - l$ and $\hat{f}(S) = f(S)$, a submodular function on E'.

More generally, Goemans and Ramakrishnan point out that if the family of interest can be expressed as the union of a polynomial number of ring families, then we can run an SFM algorithm on each family and take the minimum answer. For example, suppose we want to minimize over $2^E - \{\emptyset, E\}$. Define $\mathcal{F}_{st}$ to be the family of subsets of E which contain s but not t. Each $\mathcal{F}_{st}$ is a ring family, so we can apply an SFM algorithm to compute an S_{st} solving SFM on $\mathcal{F}_{st}$. Note that for an ordering of E as $s_1, s_2, \ldots, s_n$ (with $s_{n+1} = s_1$), $2^E - \{\emptyset, E\} = \cup_{i=1}^{n} \mathcal{F}_{s_i, s_{i+1}}$ (since the only nonempty set not in this union must contain all s_i, and so must equal E). Thus we can solve SFM over $2^E - \{\emptyset, E\}$ by taking the minimum of the n values $f(S_{s, s_{i+1}})$, so it costs n calls to SFM to solve this problem.

Suppose that $\mathcal{F}$ is an intersecting family. For $e \in E$ define $\mathcal{F}_e$ as the sets in $\mathcal{F}$ containing e. Then each $\mathcal{F}_e$ is a ring family, and $\mathcal{F} = \cup_{e \in E} \mathcal{F}_e$, so we can optimize over an intersecting family with $O(n)$ calls to SFM. If $\mathcal{C}$ is a crossing family, then for each $s \neq t \in E$, $\mathcal{C}_{st}$ is a ring family. Then for any fixed $s \in E$, $\mathcal{C} = \cup_{t \neq s}(\mathcal{C}_{st} \cup \mathcal{C}_{ts})$, so we can solve SFM over a crossing family in $O(n)$ calls to SFM.

5.1 *Symmetric SFM: Queyranne's algorithm*

A special case of this arises when f is *symmetric*, i.e., when $f(S) = f(E - S)$ for all $S \subseteq E$. From (2) we get that for any $S \subseteq E$, $2f(\emptyset) = 2f(E) = f(\emptyset) + f(E) \leq f(S) + f(E - S) = 2f(S)$, or $f(\emptyset) = f(E) \leq f(S)$, so that $\emptyset$ and E trivially solve SFM. But in many cases such as undirected Min Cut we would like to minimize a symmetric function over $2^E - \{\emptyset, E\}$. We could apply the procedure above to solve this in $O(n)$ calls to SFM, but Queyranne (1998) has provided a special-purpose algorithm that is much faster. It is based on Nagamochi and Ibaraki's Algorithm (Nagamochi and Ibaraki, 1992) for finding Min Cuts in undirected graphs.

Queyranne's Algorithm (QA) is *not* based on the LPs from Section 2.4 and so does not have a current primal point y, hence it has no need of $\mathcal{I}$, v^i, and REDUCEV. Somewhat similar to IFF-SP, QA maintains a partition $\mathcal{C}$ of E. As it proceeds, it gathers information that allows it to contract subsets in the partition, until $|\mathcal{C}| = 1$. If $\mathcal{S} \subseteq \mathcal{C}$, then we interpret $f(\mathcal{S})$ to be $f(\cup_{\sigma \in \mathcal{S}} \sigma)$, which is clearly submodular on $\mathcal{C}$. It uses a subroutine LEAFPAIR $(\mathcal{C}, f, \eta)$. LEAFPAIR builds up a set S element by element starting with element η; let S_i denote the S at iteration i. Iteration i adds an element σ of $Q = \mathcal{C} - S$ having a minimum value of $\text{key}_\sigma = f(S_{i-1} + \sigma) - f(\sigma)$ as the next element of S. The running time of LEAFPAIR is clearly $O(n^2\text{EO})$.

We say that $\mathcal{S} \subset \mathcal{C}$ *separates* $\sigma, \tau \in \mathcal{C}$ if $\sigma \in \mathcal{S}$ and $\tau \notin \mathcal{S}$ or $\sigma \notin \mathcal{S}$ and $\tau \in \mathcal{S}$. Note that $\mathcal{S}$ separates σ, τ iff $\mathcal{C} - \mathcal{S}$ separates them. The name of LEAFPAIR comes from the *cut equivalent tree* of Gomory and Hu (1961), which is a compact way of representing a family of minimum cuts separating any two nodes i and j in a capacitated undirected graph. They give an algorithm that constructs a capacitated tree T on the nodes such that we can construct a Min Cut separating i from j as follows. Find a min-capacity edge e on the unique path from i to j in T. Then $T - e$ has two connected components, which form the two sides of a Min Cut separating i from j, and this cut has value the capacity of e. Goemans and Ramakrishnan (1995) point out that cut trees extend to any symmetric submodular function. Suppose that i is a leaf of T with neighbor j in T. This implies that $\{i\}$ is a Min Cut separating i from j. We would call such a pair (j, i) a *leaf pair*. The following lemma shows that LEAFPAIR computes a leaf pair in the more general context of SFM.

LEAFPAIR $(\mathcal{C}, f, \eta)$ Subroutine for Queyranne's Algorithm

Initialize $\tau_1 \leftarrow \eta$, $S_1 \leftarrow \{\tau_1\}$, $Q \leftarrow \mathcal{C} - \tau_1$, $k \leftarrow |\mathcal{C}|$.
For $i = 2, \ldots, k$ do
  For $\sigma \in Q$ set $\mathbf{key}_\sigma = f(S_{i-1} + \sigma) - f(\sigma)$.
  Find τ_i in Q with minimum **key** value.
  Set $S_i \leftarrow S_{i-1} + \tau_i$, and $Q \leftarrow Q - \tau_i$.
Return (τ_{k-1}, τ_k).

Lemma 5.1. *If* LEAFPAIR $(\mathcal{C}, f, \eta)$ *outputs* (τ_{k-1}, τ_k), *then* $f(\tau_k) = \min\{f(\mathcal{S}) \mid \mathcal{S} \subset \mathcal{C}$ *and* $\mathcal{S}$ *separates* τ_{k-1} *and* $\tau_k\}$.

Proof. Suppose that we could prove that for all i, all $T \subseteq S_{i-1}$, and all $\sigma \in \mathcal{C} - S_i$ that

$$f(S_i) + f(\sigma) \leq f(S_i - T) + f(T + \sigma). \tag{17}$$

If we take $i = k - 1$, then we must have that $\sigma = \tau_k$. Then, since S_{k-1} and $\{\tau_k\}$ are complementary sets, and since $S_{k-1} - T$ and $T + \tau_k$ are complementary sets, (17) would imply that $f(\tau_k) \leq f(T + \tau_k)$. Since $T + \tau_k$ is an arbitrary set separating τ_k from τ_{k-1}, this shows that τ_{k-1} and τ_k are a leaf pair.

So we concentrate on proving (17). We use induction on i; it is trivially true for $i = 1$. Suppose that $j < i$ is the maximum index such that $\tau_j \in T$. If $j = i - 1$, then $f(S_i - T) + f(T + \sigma) = f(S_{i-1} - T + \tau_i) + f(T + \sigma)$. By the inductive assumption at index $i - 1$, element τ_i, and set $S_{i-1} - T$ we get $f(S_{i-1} - T + \tau_i) + f(T + \sigma) \geq f(S_{i-1}) + f(T + \sigma) - f(T) + f(\tau_i)$. Since $[S_{i-1} \cup (T + \sigma)] = S_{i-1} + \sigma$ and $[S_{i-1} \cap (T + \sigma)] = T$, from (2) we get $f(S_{i-1}) + f(T + \sigma) - f(T) + f(\tau_i) \geq f(S_{i-1} + \sigma) + f(\tau_i)$. By the choice of τ_i in LeafPair we get $f(S_{i-1} + \sigma) + f(\tau_i) \geq f(S_{i-1} + \tau_i) + f(\sigma) = f(S_i) + f(\sigma)$, as desired.

Otherwise ($j < i - 1$), by the inductive assumption at index $j + 1$, element σ, and set T we get $f(S_i - T) + f(T + \sigma) \geq f(S_i - T) + f(S_{j+1}) - f(S_{j+1} - T) + f(\sigma)$. Since $[(S_i - T) \cup S_{j+1}] = S_i$ and $[(S_i - T) \cap S_{j+1}] = S_{j+1} - T$, from (2) we get $f(S_i - T) + f(S_{j+1}) - f(S_{j+1} - T) + f(\sigma) \geq f(S_i) + f(\sigma)$, as desired. □

Let S^* solve SFM for f. If S^* separates τ_{k-1} and τ_k, then $E(\tau_k)$ must also solve SFM. If S^* does not separate τ_{k-1} and τ_k, then we can contract τ_{k-1} and τ_k without harming SFM optimality. QA takes advantage of this observation to solve SFM by calling LeafPair $n - 1$ times. The running time of QA is thus $O(n^3\text{EO})$. Note that QA is a fully combinatorial algorithm.

Queyranne's Algorithm for Symmetric SFM over $2^E - \{0, E\}$

Initialize $\mathcal{C} = E$ and η as an arbitrary element of $\mathcal{C}$.
For $i = 1, \ldots, n - 1$ do
  Set $(\sigma, \tau) \leftarrow$ LeafPair$(\mathcal{C}, f, \eta)$.
  Set $T_i \leftarrow E(\tau)$ and $m_i \leftarrow f(T_i)$.
  Contract σ and τ into a new subset of the partition.
Return T_i such that $m_i = \min\{m_j \mid j = 1, \ldots, n - 1\}$.

5.2 *Triple families and parity families*

Let $\mathcal{O} = \{S \subseteq E \mid |S| \text{ is odd}\}$ be the family of *odd sets*, and consider SFM over $\mathcal{O}$. This is not a ring family, as the union of two odd sets might be even. However, it does satisfy the following property: If any three of the four sets

$S, T, S \cap T$, and $S \cup T$ are not in $\mathcal{O}$ (are even), then the fourth set is also not in $\mathcal{O}$ (is even). Families of sets with this property are called *triple* families, and were considered by Grötschel, Lovász, and Schrijver (1988). A general lemma giving examples of triple families is:

Lemma 5.2. [Grötschel, Lovász and Schrijver (1988)] *Let $\mathcal{R} \subseteq 2^E$ be a ring family, and let a_e for $e \in E$ be a given set of integers. Then for any integers p and q, the family $\{S \in \mathcal{R} | a(S) \not\equiv q \,(\mathrm{mod}\, p)\}$ is a triple family.*

Let's consider applications of this where $p=2$. If we take $\mathcal{R}=2^E$, $a = \mathbb{1}$, and $q=0$, then we get that $\mathcal{O}$ is a triple family; taking instead $q=1$ we get that the family of even sets is a triple family. If we take $a=\chi(T)$ and $q=0$, then we get that the family of subsets having odd intersection with T is a triple family. If we have two subsets $T_1, T_2 \subseteq E$ and take $q=0$, $a_e=1$ on $T_1 - T_2$, $a_e=-1$ on $T_2 - T_1$, and $a_e=0$ otherwise, then we get that the family of S such that $|S \cap T_1|$ and $|S \cap T_2|$ have different parity is a triple family.

An even more general class of families is considered by Goemans and Ramakrishnan (1995). For ring family $\mathcal{R} \subseteq 2^E$, they call $\mathcal{P} \subseteq \mathcal{R}$ a *parity family* if $S, T \in \mathcal{R} - \mathcal{P}$ implies that $S \cup T \in \mathcal{P}$ iff $S \cap T \in \mathcal{P}$. An important class of parity families is given by:

Lemma 5.3. [(Goemans and Ramakrishnan (1995)] *Let $\mathcal{R}_1 \subseteq \mathcal{R}_2 \subseteq 2^E$ be ring families. Then $\mathcal{R}_2 - \mathcal{R}_1$ is a parity family.*

Any triple family is clearly a parity family, but the converse is not true. For example, take $E=\{a,b,c\}$, $\mathcal{R}_1=\{\{a\},\{a,b\},\{a,b,c\}\}$, and $\mathcal{R}_2=2^E$. Then $\mathcal{R}_1 \subseteq \mathcal{R}_2$ and both $\mathcal{R}_1$ and $\mathcal{R}_2$ are ring families, so the lemma implies that $\mathcal{R}_2 - \mathcal{R}_1$ is a parity family. Taking $S=\{a,b\}$ and $T=\{a,c\}$, we see that $S \in \mathcal{R}_1$, $S \cap T=\{a\} \in \mathcal{R}_1$, and $S \cup T=\{a,b,c\} \in \mathcal{R}_1$, but $T \notin \mathcal{R}_1$, so $\mathcal{R}_2 - \mathcal{R}_1$ is not a triple family.

As an application of Lemma 5.3, note that (2) implies that the union and intersection of solutions of SFM are also solutions of SFM, so the family $\mathcal{S}$ of solutions of SFM is a ring family. Thus $2^E - \mathcal{S}$ is a parity family. The next theorem shows that we can solve SFM over a parity family with $O(n^2)$ calls to SFM over a ring family, so this gives us a way of finding the second-smallest value of any submodular function.

Theorem 5.4. [Goemans and Ramakrishnan (1995)] *If $\mathcal{R}$ is a ring family and $\mathcal{P} \subseteq \mathcal{R} \subseteq 2^E$ is a parity family, then we can solve SFM over $\mathcal{P}$ using $O(n^2)$ calls to SFM over ring families.* □

Since triple families are a special case of parity families, this give us a tool that can solve many interesting problems: SFM over odd sets, SFM over even sets, SFM over sets having odd intersection with a fixed $T \subseteq E$, second-smallest value of $f(S)$, etc.

5.3 Constrained SFM can be hard

So far we have seen that SFM remains easy when we consider the symmetric case, or when we consider SFM over various well-structured families of sets. However, there are other important cases of SFM with side constraints that are NP Hard to solve. One such case is *cardinality constrained* SFM, where we want to restrict to the family $\mathcal{C}_k$ of sets of size k. The $s-t$ Min Cut problem Example 1.9 with this constraint is NP Hard [(Garey and Johnson, 1979), Problem ND17]. This examples is representative of the fact that SFM often becomes hard when side constraints are added.

6 Future directions for SFM algorithms

The history of SFM has been that expectations have continually grown. SFM was recognized early on as being an important problem, and a big question was whether there existed a finite version of Cunningham's "augmenting path" algorithm. In 1985, Bixby et al. (1985) found such an algorithm. Then the question became whether one could get a good bound on the running time of an SFM algorithm. Also in 1985, Cunningham (1985) found an algorithm with a pseudo-polynomial bound. Then the natural question was whether an algorithm with a (strongly) polynomial bound existed. In 1988, Gröschel et al. (1988) showed that the Ellipsoid Algorithm leads to a strongly polynomial SFM algorithm. However, Ellipsoid is slow, so the question became whether there existed a "combinatorial" (non-Ellipsoid) polynomial algorithm for SFM. Simultaneously in 1999, Schrijver (2000), and Iwata et al. (2001) found quite different strongly polynomial combinatorial SFM algorithms. However, both of these algorithms need to use some multiplication and division, leading Schrijver to pose the question of whether there existed a fully combinatorial SFM algorithm. In 2002 Iwata (2002a) found a way to extend the IFF Algorithm to give a fully combinatorial SFM algorithm. In 2001 Flesicher and Iwata (2001) found Schrijver-PR, an apparent speedup for Schrijver's Algorithm (although Vygen (2003) showed in 2003 that both variants actually have the same running time), and in 2002 Iwata (2002b) used ideas from Schrijver's Aglrithm to speed up the IFF algorithms.

Is this the end of the road for SFM algorithms? I say "no," for two reasons:

(1) The existing SFM algorithms have rather slow running times. Both variants of Schrijver's Algorithm take $O(n^7\text{EO}+n^8)$ time, the strongly polynomial Hybrid Algorithm takes $O(n^6\text{EO}+n^7)\log n)$ time, and the weakly polynomial Hybrid Algorithm takes $O((n^4\text{EO}+n^5)\log M)$ time. The Hybrid Algorithm shows that there may be further opportunities for improvement. There is not yet much practical

experience with any of these algorithms, but experience in other domains suggests that an $O(n^5)$ algorithm is practically useless for large instances. Therefore it is natural to ask whether we can find significantly faster SFM algorithms.

(2) The existing general SFM algorithms use Cunningham's idea of verifying that the current y belongs to $B(f)$ via representing y as $\sum_{i \in \mathcal{I}} \lambda_i v^i$ for vertices v^i coming from Greedy. Naively, this is a rather brute-force way to verify that $y \in B(f)$. However, 30 years of research have not yet produced any better idea.

These two points are closely related. To keep their running times manageable, existing algorithms call REDUCEV from time to time keep $|\mathcal{I}|$ small, and REDUCEV costs $O(n^3)$ per call. Thus the key to finding a faster SFM algorithm might be to avoid representing y as a convex combination of vertices. Hybrid, the fastest SFM algorithm known to this points, runs in $\tilde{O}(n^4\text{EO})$time. No formal lower bound on the complexity of SFM exist, but it is hard to imagine an SFM algorithm computing fewer than n vertices, which takes $O(n^2\text{EO})$ time. It is not unreasonable to hope that an $O(n^3\text{EO})$ SFM algorithm exists.

How far could we go with algorithms based on Push-Relabel technology such as Schrjver's Algorithm and Iwata's Hybrid Algorithm? For networks with $\Theta(n^2)$ arcs (and the networks arising in SFM all can have $\Theta(n^2)$ arcs since each of the $O(n)$ linear orders in $\mathcal{I}$ has $O(n)$ consecutive pairs), the best known running time for a pure Push-Relabel Max Flow algorithm uses $\Theta(n^3)$ pushes [see Ahuja et al. (1993)]. Hence such algorithms could not be faster than $\Theta(n^3\text{EO})$ without a breakthrough in Max Flow algorithms. If each such push potentially adds a new vertex to $\mathcal{I}$, then we need to call REDUCEV $\Theta(n^2)$ times, for an overhead of $\Theta(n^5)$. Note that the Hybrid Algorithm, at $O(n^4\text{EO} + n^5)\log M)$, comes close to this informal lower bound, losing only the $O(\log M)$ factor due to scaling, and inflating $O(n^3\text{EO})$ to $O(n^4\text{EO})$ since each BLOCKSWAP takes $O(b\text{EO})$ time instead of $O(\text{EO})$ time.

Ideally it would be useful to have a formal lower bound stating that at least some number of oracle calls is needed to solve SFM. It is easy to see the trivial lower bound that $\Omega(n)$ calls are necessary, but so far nothing nontrivial is known.

Here are two other reasons to be dissatisfied with the current state of the art. It is hard to be completely happy with the fully combinatorial SFM algorithms, as their use of repeated subtraction or doubling to simulate multiplication and division is aesthetically unpleasant, and probably impractical. Second, we saw in Section 2.4 that the linear programs have integral optimal solutions. All the algorithms find an integral dual solution (an optimal set S solving SFM), but (when f is integer-valued) none of them directly finds an integral optimal primal solution (a $y \in B(f)$ with $y^-(E) = f(S)$ or a $y \in P(f)$ with $y(E) = f(S) + \gamma(E - S)$). We conjecture that a faster SFM algorithm exists that maintains an integral y throughout the algorithm.

One possibility for making faster SFM algorithms without using $\mathcal{I}$ is suggested by Queyranne's Algorithm for symmetric SFM. Notice that Queyranne's Algorithm does not use a $y = \sum_{i \in \mathcal{I}} \lambda_i v^i$ representation at all, which suggests that it might be possible to find a similar algorithm for general SFM. On the other hand, Queyranne's Algorithm also does not use any of the LP machinery used by the general SFM algorithms, and it does not produce anything resembling a primal solution (a $y \in B(f)$ with $y^-(E) = f(S)$). Also, as Queyranne notes in (Queyranne, 1998), general SFM is provably not reducible to symmetric SFM, and even SFM with $f(S) = s(S) - u(S)$ with s symmetric and u modular (u a vector in $\mathbb{R}^E$) is not reducible to the symmetric case.

However, we can still dream. A vague outline of an SFM algorithm not representing y as a convex combination of vertices might go like this: Start with $y = v^{\prec}$ for some linear order $\prec$. Then start doing exchanges that increase $y^-(E)$ in such a way that we are assured that y remains in $B(f)$, until we find some S with $y^-(E) = f(S)$, and we are optimal. There would be some lemma, along the lines of our proof that the α from ExchBd is at most $c(k, l; v^i)$, showing inductively that each step remains inside $B(f)$. Then the proof that the final y is in $B(f)$ would be the sequence of steps taken by the algorithm. Alternatively, one could use the framework outlined by Fujishige and Iwata (2002): Their framework needs only a combinatorial strongly polynomial separation routine that either proves that 0 belongs to a submodular polyhedron $P(\tilde{f})$ (for an $\tilde{f}$ derived from f), or gives a subset $S \subseteq E$ such that $\tilde{f}(S) < 0$ (thereby separating 0 from $P(\tilde{f})$). They show that $O(n^2)$ calls to such a routine would suffice for solving SFM. A third possibility would be to derive a polynomial bound on the number of iterations of the "simplex algorithm" for SFM proposed by Fujishige [(Fujishige, 1991), p. 194], although this seems to involve other unpleasant linear algebra. We leave these questions for future researchers.

Acknowledgments

Supported by an NSERC Operating Grant, and by a visit to LIMOS, Université Blaise Pascal, Clermont-Ferrand. I thank the two anonymous referees, Yves Crama, Bill Cunningham, Lisa Fleischer, Satoru Fujishige, Satoru Iwata, Hervé Kerevin, Laszlo Lovász, Kazuo Murota, Maurice Queyranne, Alexander Schrijver, Bruce Shepherd, and Fabio Tardella for their substantial help with this material.

References

Ahuja, R. K., T. L. Magnanti, J. B. Orlin (1993). *Network Flows: Theory, Algorithms, and Applications*, Prentice-Hall, Englewood Cliffs.

Anglès d'Auriac, J.-C., F. Iglói, M. Preissmann, A. Sebö (2002). Optimal cooperation and submodularity for computing Potts' partition functions with a large number of states. *J. Phys. A: Math. Gen.* 35, 6973–6983.

Bertsekas, D. P. (1986). Distributed asynchronous relaxation methods for linear network flow problems. Working paper, Laboratory for Information and Dccision Systems, MIT, Cambridge, MA.

Birkhoff, G. (1967). Lattice theory. *Amer. Math. Soc.*

Bixby, R. E., W. H. Cunningham, D. M. Topkis (1985). The partial order of a polymatroid extreme point. *Math. of OR* 10, 367–378.

Cherkassky, B. V., Goldberg, A. V. (1997). On implementing push-relabel method for the maximum flow problem. *Algorithmica.* 19, 390–410. The PRF code developed here is available from http://www.star-lab.com/goldberg/soft.html.

Cunningham, W. H. (1983). Decomposition of submodular functions. *Combinatorica* 3, 53–68.

Cunningham, W. H. (1984). Testing membership in matroid polyhedra. *JCT Series B* 36, 161–188.

Cunningham, W. H. (1985). On submodular function minimization. *Combinatorica* 3, 185–192.

Dinic, E. A. (1970). Algorithm for solution of a problem of maximum flow in a network with power estimation. *Soviet Math. Dokl.* 11, 1277–1280.

Edmonds, J. (1970). Submodular functions, matroids, and certain polyhedra. in: R. Guy, H. Hanani, N. Sauer, J. Schönheim (eds.), *Combinatorial Structures and their Applications*, Gordon and Breach, 69–87.

Edmonds, J., R. Giles (1977). A min–max relation for submodular functions on graphs. *Ann. Discrete Math.* 1, 185–204.

Edmonds, J., R. M. Karp (1972). Theoretical improvements in algorithmic efficiency for network flow problems. *Journal of ACM* 19, 248–264.

Ervolina, T. R., S. T. McCormick (1993). Two strongly polynomial cut canceling algorithms for minimum cost network flow. *Discrete Applied Mathematics* 46, 133–165.

Fleischer, L. K. (2000). Recent progress in submodular function minimization. *Optima* September 2000, 1–11.

Fleischer, L. K., Iwata, S. (2000). Improved algorithms for submodular function minimization and submodular flow. *Proceedings of the 32nd Annual ACM Symposium on Theory of Computing*, 107–116.

Fleischer, L. K., Iwata, S. (2001). A push-relabel framework for submodular function minimization and applications to parametric optimization. To appear in "Submodularity" special issue of *Discrete Applied Mathematics*, S. Fujishige (ed).

Fleischer, L. K., S. Iwata, S. T. McCormick (2002). A faster capacity scaling algorithm for minimum cost submodular flow. *Math. Prog.* 92, 119–139.

Fujishige, S. (1991). *Submodular Functions and Optimization*. North-Holland.

Fujishige, S. (2002). Submodular function minimization and related topics. *Discrete Mathematics and Systems Science Research Report 02–04*, Osaka University, Japan.

Fujishige, S., Iwata, S. (2001). Bisubmodular function minimization, in: K. Aardal, B. Gerards (eds.), *Proceedings of the 8th Conference on Integer Programming and Combinatorial Optimization* (*IPCO Utrecht*), Lecture Notes in Computer Science 2081, Springer, Berlin, 160–169.

Fujishige, S., S. Iwata (2002). A descent method for submodular function minimization. *Math. Prog.* 92, 387–390.

Gabow, H. N. (1985). Scaling algorithms for network problems. *J. of Computer and Systems Sciences*, 31, 148–168.

Garey, M. R., D. S. Johnson (1979). *Computers and Intractability, A Guide to the Theory of NP-Completeness*, W.H. Freeman and Company, New York.

Goemans, M. X., V. S. Ramakrishnan (1995). Minimizing submodular functions over families of sets. *Combinatorica* 15, 499–513.

Goldberg, A. V., S. Rao (1998). Beyond the flow decomposition barrier. *Journal of ACM* 45, 753–797.

Goldberg, A. V., R. E. Tarjan (1988). A new approach to the maximum flow problem. *JACM* 35, 921–940.

Goldberg, A. V., R. E. Tarjan (1990). Finding minimum-cost circulations by successive approximation. *Mathematics of Operations Research* 15, 430–466.

Goldfarb, D., Z. Jin (1999). A new scaling algorithm for the minimum cost network flow problem. *Operations Research Letters* 25, 205–211.

Gomory, R. E., T. C. Hu Jr. (1961). Multiterminal network flows. *SIAM J. on Applied Math.* 9, 551–570.

Granot, F., A. F. Veinott (1985). Substitutes, complements, and ripples in network flows. *Math. of OR* 10, 471–497.

Grötschel, M., L. Lovaśz, A. Schrijver (1981). The ellipsoid algorithm and its consequences in combinatorial optimization. *Combinatorica* 1, 499–513.

Grötschel, M., L. Lovaśz, A. Schrijver (1988). *Geometric Algorithms and Combinatorial Optimization*, Springer-Verlag.

Huh, W. T., Roundy, R. O. (2002). A continuous-time strategic capacity planning model. Working paper, SORIE, Cornell University, submitted to *Operations Research*.

Isotani, S., S. Fujishige (2003). Submodular Function Minimization: Computational Experiments Technical Report, RIMS, Kyoto University.

Iwata, S. (1997). A capacity scaling algorithm for convex cost submodular flows. *Math. Programming* 76, 299–308.

Iwata, S. (2002a). A fully combinatorial algorithm for submodular function minimization. *J. Combin. Theory Ser. B* 84, 203–212.

Iwata, S. (2002b). A faster scaling algorithm for minimizing submodular functions. *SIAM J. on Computing*. 32, 833–840; an extended abstract appeared in: W. J. Cook, A. S. Schulz (eds.), *Proceedings of the 9th Conference on Integer Programming and Combinatorial Optimization (IPCO MIT)*, Lecture Notes in Computer Science 2337, Springer, Berlin, 1–8.

Iwata, S. (2002c). Submodular function minimization – theory and practice. Talk given at Workshop in Combinatorial Optimization at Oberwolfach, Germany, November 2002.

Iwata, S., L. Fleischer, S. Fujishige (2001). A combinatorial, strongly polynomial-time algorithm for minimizing submodular functions. *J. ACM* 48, 761–777.

Iwata, S., McCormick, S. T., Shigeno, M. (1999). A strongly polynomial cut canceling algorithm for the submodular flow problem. *Proceedings of the Seventh MPS Conference on Integer Programming and Combinatorial Optimization*, 259–272.

Laurent, M. (1997). The max-cut problem, in: M. Dell'amico, F. Maffioli, S. Martello (eds.), *Annotated Bibliographies in Combinatorial Optimization*, Wiley, Chichester.

Lawler, E. L., C. U Martel (1982). Computing maximal polymatroidal network flows. *Math. Oper. Res.* 7, 334–347.

Lovász, L. (1983). Submodular functions and convexity, in: A. Bachem, M. Grötschel, B. Korte (eds.), *Mathematical Programming – The State of the Art*, Springer, Berlin, 235–257.

Lovász, L. (2002). Email reply to query from S. T. McCormick, 6 August 2002.

Lu, Y., J.-S. Song, (2002). Order-based cost optimization in assemble-to-order systems. Working paper, UC Irvine Graduate School of Management, submitted to *Operations Research*.

McCormick, S. T., Fujishige, S. (2003). Better algorithms for bisubmodular function minimization. Working paper, University of British Columbia Faculty of Commerce, Vancouver, BC.

du Merle, O., D. Villenceuve, J. Desrosiers, P. Hansen (1999). Stabilized column generation. *Discrete Mathematics* 194, 229–237.

Murota, K. (1998). Discrete convex analysis. *Math. Programming* 83, 313–371.

Murota, K. (2003). *Discrete convex analysis. SIAM Monographs on Discrete Mathematics and Applications*, Society for Industrial and Applied Mathematics, Philadelphia.

Nagamochi, H., T. Ibaraki (1992). Computing edge connectivity in multigraphs and capacitated graphs. *SIAM J. on Discrete Math.* 5, 54–66.

Nemhauser, G. L., L.A. Wolsey (1988). *Integer and Combinatorial Optimization*, Wiley, New York.

Picard, J-C., M. N. Queyranne (1982). Selected applications of minimum cuts in networks. *INFOR* 20, 394–422.

Queyranne, M. N. (1980). Theoretical efficiency of the algorithm capacity for the maximum flow problem. *Mathematics of Operations Research* 5, 258–266.

Queyranne, M. N. (1998). Minimizing symmetric submodular functions. *Math. Prog.* 82, 3–12.

Schönsleben, P. (1980). *Ganzzahlige Polymatroid-Intersektions Algorithmen*. PhD dissertation, ETH Zürich.

Schrijver, A. (2000). A combinatorial algorithm minimizing submodular functions in strongly polynomial time. *J. Combin. Theory Ser. B* 80, 346–355.

Schrijver, A. (2003). *Combinatorial Optimization: Polyhedra and Efficiency*, Springer, Berlin.

Shanthikumar, J. G., D. D. Yao (1992). Multiclass Queueing systems: polymatroid structure and optimal scheduling control. *Operations Research* 40, S293–S299.

Shen, Z.-J. M., C. Coullard, M. S. Daskin (2003). A joint location-inventory model transportation. *Science* 37, 40–55.

Tardos, É. (1985). A strongly polynomial minimum cost circulation algorithm. *Combinatorica* 5, 247–256.

Tardos, É., C. A. Tovey, M. A. Trick (1986). Layered augmenting path algorithms. *Math. Oper. Res.* 11, 362–370.

Topkis, D. M. (1978). Minimizing a submodular function on a lattice. *Operations Research* 26, 305–321.

Topkis, D. M. (1998). *Supermodularity and Complementarity*, Princeton University Press, Princeton, NJ.

Vygen, J. A note on Schrijver's submodular function minimization alogorithm. *JCT B* 88.

Welsh, D. J. A. (1976). *Matroid Theory*, Academic Press, London.

K. Aardal et al., Eds., *Handbooks in OR & MS, Vol. 12*

Chapter 8

Semidefinite Programming and Integer Programming

Monique Laurent

CWI, Kruislaan 413, 1098 SJ Amsterdam, The Netherlands
E-mail: monique@cwi.nl

Franz Rendl

Universität Klagenfurt, Institut für Mathematik, Universitätstrasse 65-67, 9020 Klagenfurt, Austria
E-mail: franz.rendl@uni-klu.ac.at

Abstract

This chapter surveys how semidefinite programming can be used for finding good approximative solutions to hard combinatorial optimization problems. The chapter begins with a general presentation of several methods for constructing hierarchies of linear and/or semidefinite relaxations for 0/1 problems. Then it moves to an in-depth study of two prominent combinatorial optimization problems: the maximum stable set problem and the max-cut problem. Details are given about the approximation of the stability number by the Lovász theta number and about the Goemans-Williamson approximation algorithm for max-cut, two results for which semidefinite programming plays an essential role, and we survey some extensions of these approximation results to several other hard combinatorial optimization problems.

1 Introduction

Linear optimization is a relatively young area of applied mathematics. Even though the world is nonlinear, as physicists never stop to point out, it seems that in many practical situations a linearized model describes key features of a problem quite accurately.

The success of linear optimization in many real-world applications has led to the study of integer linear programming, which permits to model optimal decision making under finitely many alternatives. A natural way to approach these types of problems consists in using again linear theory, in this case polyhedral combinatorics, to solve them. Mathematically, one tries to find (at least) a (partial) linear description of the convex hull of all integral solutions. While this approach was successful for many combinatorial optimization problems, it turned out that some graph optimization problems, such as Max-Cut or Max-Clique, cannot be approximated tightly by purely linear methods.

Stronger relaxation methods have therefore attracted the focus of recent research. The extension of linear optimization to semidefinite optimization has turned out to be particularly interesting for the following reasons. First, algorithmic ideas can be extended quite naturally from linear to semidefinite optimization. Secondly, there is theoretical evidence that semidefinite models are sometimes significantly stronger than purely linear ones, justifying the computational overhead to solve them.

It is the purpose of this chapter to explain in detail how semidefinite programming is used to solve integer programming problems. Specifically, we start out in the next section with explaining the relevant mathematical background underlying semidefinite programming by summarizing the necessary duality theory, explaining algorithmic ideas and recalling computational complexity results related to semidefinite programming. In Section 3 we show how semidefinite relaxations arise from integer 0/1 programming by lifting the problem formulated in $\mathbb{R}^n$ to a problem in the space of symmetric matrices.

A detailed study of two prominent special graph optimization problems follows in Section 4, dealing with the stable set problem, and Section 5, devoted to Max-Cut. For both these problems the extension of polyhedral to semidefinite relaxations had led to a significant improvement in the approximation of the original problem. Section 5 also introduces the hyperplane rounding idea of Goemans and Williamson, which opened the way to many other approximation approaches, many of which are discussed in Section 6.

Section 7 discusses possible alternatives to the use of semidefinite models to get stronger relaxations of integer programs.

Finally, we summarize in Section 8 some recent semidefinite and other nonlinear relaxations applied to the Quadratic Assignment Problem, which have led to a computational break-through in Branch and Bound computations for this problem.

2 Semidefinite programming: duality, algorithms, complexity, and geometry

2.1 Duality

To develop a duality theory for semidefinite programming problems, we take a more general point of view, and look at *Linear Programs over Cones.*

Suppose K is a closed convex cone in $\mathbb{R}^n$, $c \in \mathbb{R}^n$, $b \in \mathbb{R}^m$ and A is an $m \times n$ matrix. The problem

$$p^* := \sup\{c^T x \colon Ax = b,\ x \in K\} \tag{1}$$

is called **Cone-LP**, because we optimize a linear function subject to linear equations, and we have the condition that the decision variable x lies in the cone K.

The dual cone K^* is defined as follows:

$$K^* := \{y \in \mathbb{R}^n: y^T x \geq 0 \; \forall x \in K\}.$$

It is a well known fact, not hard to verify, that K^* is also a closed convex cone.

We will derive the dual of (1) by introducing Lagrange multipliers for the equality constraints and by using the Minimax Inequality. Let $y \in \mathbb{R}^m$ denote the Lagrange multipliers for $Ax = b$. Using the Lagrangian $L(x, y) := c^T x + y^T(b - Ax)$ we get

$$\inf_y L(x, y) = \begin{cases} c^T x & \text{if } Ax = b \\ -\infty & \text{otherwise.} \end{cases}$$

Therefore,

$$p^* = \sup_{x \in K} \inf_y L(x, y) \leq \inf_y \sup_{x \in K} L(x, y).$$

The inequality is usually called "Minimax inequality", and holds for any real-valued function $L(x, y)$ where x and y are from some ground sets X and Y, respectively.

We can rewrite L as $L = b^T y - x^T(A^T y - c)$. The definition of K^* implies the following. If $A^T y - c \notin K^*$ then there exists $x \in K$ such that $x^T(A^T y - c) < 0$. Therefore we conclude

$$\sup_{x \in K} L(x, y) = \begin{cases} b^T y & \text{if } A^T y - c \in K^* \\ \infty & \text{otherwise.} \end{cases}$$

This translates into

$$p^* \leq \inf\{b^T y: y \in \mathbb{R}^m, A^T y - c \in K^*\} =: d^*. \tag{2}$$

The problem on the right side of the inequality sign is again a Cone-LP, but this time over the cone K^*. We call this problem the dual to (1). By construction, a pair of dual cone-LP satisfies weak duality.

Lemma 1. (Weak duality) *Let $x \in K$, $y \in \mathbb{R}^m$ be given with $Ax = b$, $A^T y - c \in K^*$. Then, $c^T x \leq b^T y$.*

One crucial issue in the duality theory consists in identifying sufficient conditions that insure equality in (2), also called *Strong Duality*. The following condition insures strong duality. We say that the cone-LP (1) satisfies the *Slater constraint qualification* if there exists $x \in \operatorname{int}(K)$ such that $Ax = b$. (A similar definition holds for the dual problem.) Duffin (1956) shows the following result.

Theorem 2. *If (1) satisfies the Slater constraint qualification and p^* is finite, then $p^* = d^*$, and the dual infimum is attained.*

Returning to the semidefinite programs, we consider the vector space S_n of symmetric $n \times n$ matrices as the ground set for the primal problem. It is equipped with the usual inner product $\langle X, Y\rangle = \mathrm{Tr}(XY)$ for $X, Y \in S_n$. The Frobenius norm of a matrix $X \in S_n$ is defined by $\|X\|_F := \sqrt{\mathrm{Tr}(X^T X)}$. A linear operator A, mapping symmetric matrices into $\mathbb{R}^m$, is most conveniently represented by $A(X)_i := \mathrm{Tr}(A_i X)$ for given symmetric matrices A_i, $i = 1, \ldots, m$. The adjoint in this case has the representation $A^T(y) = \sum y_i A_i$. From Fejer's theorem, which states that

$$A \succeq 0 \text{ if and only if } \mathrm{Tr}(AB) \geq 0 \ \forall B \succeq 0,$$

we see that the cone of positive semidefinite matrices is selfdual. Hence we arrive at the following primal-dual pair of semidefinite programs:

$$\max\{\mathrm{Tr}(CX)\colon A(X) = b, X \succeq 0\}, \tag{3}$$

$$\min\{b^T y\colon A^T(y) - C \succeq 0\}. \tag{4}$$

In our combinatorial applications, we usually have the property that both the primal and the dual problems satisfy the Slater constraint qualification, hence we have strong duality and both optima are attained.

Stronger duals for semidefinite programs have been introduced having the property that there is no duality gap, in particular, by Borwein and Wolkowicz (1981), Ramana (1997); see Ramana, Tunçel, and Wolkowicz (1997) for a comparison. In Section 2.3, we will come back briefly to the implications for the complexity of semidefinite programming.

The semidefiniteness of a matrix X can equivalently be expressed as X having only nonnegative eigenvalues. Thus there is some close connection between semidefinite programs and the spectral theory of matrices. The following simple examples of semidefinite programs throw some more light onto this connection. Throughout, I denotes the identity matrix and I_k the identity matrix of order k.

Example 3. Let C be a symmetric matrix. Consider

$$\max \mathrm{Tr}(CX) \quad \text{such that } \mathrm{Tr}(X) = 1, X \succeq 0.$$

The dual is

$$\min y \quad \text{such that } yI - C \succeq 0.$$

Both problems clearly satisfy the Slater constraint qualification. In fact, dual feasibility implies that $y \geq \lambda_{\max}(C)$, hence at the optimum $y = \lambda_{\max}(C)$. It is, in fact, well known that the primal semidefinite program is equivalent to

$$\max x^T C x \quad \text{such that } x^T x = 1,$$

by taking $X = xx^T$.

Example 4. More generally, the sum $\lambda_1 + \cdots + \lambda_k$ of the k largest eigenvalues of $C \in \mathcal{S}_n$ can be expressed as the optimum value of the following semidefinite program:

$$\max \operatorname{Tr}(CX) \quad \text{such that } I \succeq X \succeq 0, \operatorname{Tr}(X) = k \tag{5}$$

which is equivalent to

$$\max \operatorname{Tr}(CYY^T) \quad \text{such that } Y \text{ is an } n \times k \text{ matrix with } Y^T Y = I_k. \tag{6}$$

The fact that $\lambda_1 + \cdots + \lambda_k$ is equal to the optimum value of (6) is known as Fan's theorem; see Overton and Womersley (1992) for discussion.

Let us sketch the proof. The fact that the optimum values of the two programs (5) and (6) are equal follows from a nice geometric property of the feasible set of (5) (namely, that its extreme points correspond to the feasible solutions of (6); cf. Lemma 7 below). Let $y_1, \ldots, y_k$ be a set of orthonormal eigenvectors of C for its k largest eigenvalues and let Y be the matrix with columns $y_1, \ldots, y_k$. Then Y is feasible for (6) and $\operatorname{Tr}(CYY^T) = \sum_{i=1}^k \operatorname{Tr}(y_i^T C y_i) = \sum_{i=1}^k \lambda_i$, which shows that $\sum_{i=1}^k \lambda_i$ is less than or equal to the maximum of (6). Conversely, let Y be an $n \times k$ matrix such that $Y^T Y = I_k$; we show that $\operatorname{Tr}(CYY^T) \leq \sum_{i=1}^k \lambda_i$. For this, let $C = Q^T D Q$ where $Q \in \mathcal{S}_n$ with $Q^T Q = I_n$ and $D := \operatorname{diag}(\lambda_1, \ldots, \lambda_n)$. Set $Z := QY$ and $X := ZZ^T$. As Z is an $n \times k$ matrix with $Z^T Z = I_k$, it follows that the only nonzero eigenvalue of X is 1 with multiplicity k and thus X is feasible for (5). Hence, $\operatorname{Tr}(CYY^T) = \operatorname{Tr}(DX) = \sum_{i=1}^n \lambda_i x_{ii} \leq \sum_{i=1}^k \lambda_i$ since $0 \leq x_{ii} \leq 1$ for all i.

By taking the dual of the semidefinite program (5), we obtain the following alternative formulation for the sum of the k largest eigenvalues of C:

$$\lambda_1 + \cdots + \lambda_k = \min kz + \operatorname{Tr}(Z) \quad \text{such that } zI + Z \succeq C, Z \succeq 0. \tag{7}$$

This latter formulation permits to derive the following semidefinite programming characterization for minimizing the sum of the k largest eigenvalues of a symmetric matrix satisfying linear constraints (cf. Alizadeh (1995)):

$$\begin{aligned} \min \lambda_1(X) + \cdots + \lambda_k(X) \quad & \text{s.t. } X \in \mathcal{S}_n, \operatorname{Tr}(A_j X) = b_j \quad (j = 1, \ldots, m) \\ = \min kz + \operatorname{Tr}(Z) \quad & \text{s.t. } zI + Z - X \succeq 0, Z \succeq 0, \\ & \operatorname{Tr}(A_j X) = b_j \quad (j = 1, \ldots, m). \end{aligned}$$

More recently, Anstreicher and Wolkowicz (2000) showed a strong connection between a theorem of Hoffman and Wielandt and semidefinite programming.

Theorem 5. (Hoffman and Wielandt (1953)) *Let A and B be symmetric matrices of order n with spectral decomposition $A = PDP^T$, $B = QEQ^T$. We assume that the diagonal matrix D contains the eigenvalues of A in nondecreasing order, and E contains the eigenvalues of B in nonincreasing order. Furthermore, $PP^T = QQ^T = I$. Then*

$$\min\{\operatorname{Tr}(AXBX^T) : X^T X = I\} = \operatorname{Tr}(DE). \tag{8}$$

Moreover, the minimum is attained for $X = PQ^T$.

A proof of this theorem can be found for instance in Hoffman and Wielandt (1953), the result can be traced back to the work of John von Neumann (1962). Anstreicher and Wolkowicz (2000) have recently shown that the nonconvex quadratic minimization problem (8) over the set of orthogonal matrices can equivalently be expressed through semidefinite programming. This connection will be a useful tool to bound the Quadratic Assignment Problem, so we recall how this connection can be established. We have:

$$\operatorname{Tr} DE = \min\{\operatorname{Tr} AYBY^T \colon YY^T = I\} = \min\{\operatorname{Tr} DXEX^T \colon XX^T = I\}.$$

The second equation follows because the mapping $X = P^T YQ$ is a bijection on the set of orthogonal matrices. We next introduce Lagrange multipliers S and T for the equations $XX^T = I$, $X^T X = I$, and we get

$$\begin{aligned}\operatorname{Tr} DE &= \min_X \max_{S,T} \operatorname{Tr}(DXEX^T + S(I - XX^T) + T(I - X^T X))\\ &\geq \max_{S,T} \min_{x=\operatorname{vec}(X)} \operatorname{Tr} S + \operatorname{Tr} T + x^T(E \otimes D - I \otimes S - T \otimes I)x.\end{aligned}$$

If $X = (x_1, \ldots, x_n)$ is a matrix with columns x_i, we define

$$x = \operatorname{vec}(X) = \begin{pmatrix} x_1 \\ \vdots \\ x_n \end{pmatrix}$$

to be the vector obtained from stacking the columns of X. The vec-operator leads to the following identity, see Horn and Johnson (1985):

$$\operatorname{vec}(AXB) = (B^T \otimes A)\operatorname{vec}(X). \tag{9}$$

$A \otimes B$ denotes the Kronecker product of A and B. Formally,

$$A \otimes B = (a_{ij}B).$$

The inner minimization is bounded only if $E \otimes D - I \otimes S - T \otimes I \succeq 0$. Since D and E are diagonal, we may restrict S and T also to be diagonal, $S = \operatorname{diag}(s)$, $T = \operatorname{diag}(t)$. (If s is a vector, $\operatorname{diag}(s)$ denotes the diagonal matrix with s on the main diagonal.) This leads to

$$\operatorname{Tr} DE \geq \max\left\{\sum s_i + \sum t_i \colon d_i e_j - s_i - t_j \geq 0 \;\forall i, j\right\}.$$

The last problem is the dual of the assignment problem. Therefore we get

$$\operatorname{Tr} DE \geq \min\left\{\sum_{ij} d_i e_j z_{ij} \colon Z = (z_{ij}) \text{ doubly stochastic}\right\} = \operatorname{Tr} DE.$$

The first term equals the last, so there must be equality throughout. We summarize this as follows.

Theorem 6. (Anstreicher and Wolkowicz (2000)) *Let A and B be symmetric matrices. Then,*

$$\begin{aligned}&\min\{\operatorname{Tr} AXBX^T \colon XX^T = I\}\\ &= \max\{\operatorname{Tr} S + \operatorname{Tr} T \colon B \otimes A - I \otimes S - T \otimes I \succeq 0\}.\end{aligned}$$

2.2 Algorithms

Semidefinite programs (SDP) are convex minimization problems, hence they can be solved in polynomial time to any fixed prescribed precision using for instance the ellipsoid method (see Grötschel, Lovász, and Schrijver (1988)). More recently, interior point methods have turned out to be the method of choice to solve SDP, since they give faster algorithms than the ellipsoid method whose running time is prohibitively high in practice; see for instance the handbook by Wolkowicz, Saigal, and Vandenberghe (2000).

We will now review the main ideas underlying the interior point approach for SDP. The basis assumption is that both the primal (3) and the dual (4) problems satisfy the Slater constraint qualification, which means we assume that there exists a triple (X, y, Z) such that

$$X \succ 0, Z \succ 0, A(X) = b, Z = A^T(y) - C.$$

To avoid trivialities, it is usually also assumed that the linear equations $A(X) = b$ are linearly independent. In view of Theorem 2, we get the following necessary and sufficient optimality conditions.

A triple (X, y, Z) solves (3) and (4) if and only if

$$A(X) = b, \quad X \succeq 0 \quad \text{(primal feasibility)} \tag{10}$$

$$A^T(y) - Z = C, \quad Z \succeq 0 \quad \text{(dual feasibility)} \tag{11}$$

$$ZX = 0 \quad \text{(complementarity)} \tag{12}$$

To see how (12) follows from Theorem 2, we note that both the primal and the dual optima are attained, and the duality gap is 0. If (X, y, Z) is optimal, we get

$$0 = b^T y - \operatorname{Tr} CX = y^T(A(X)) - \operatorname{Tr} CX = \operatorname{Tr}(A^T(y) - C)X = \operatorname{Tr} ZX.$$

Since $X \succeq 0, Z \succeq 0$, we have also $X = UU^T$, $Z = VV^T$, for U and V of appropriate size. Thus

$$0 = \operatorname{Tr} ZX = \operatorname{Tr} VV^T UU^T = \|V^T U\|_F^2,$$

hence $V^T U = 0$, so that $ZX = VV^T UU^T = 0$.

In the interior point approach, the condition $ZX = 0$ is replaced by $ZX = \mu I$, leading to a parameterized system of equations:

$$F_\mu(X, y, Z) := \begin{pmatrix} A(X) - b \\ Z - A^T(y) + C \\ ZX - \mu I \end{pmatrix} = 0. \tag{13}$$

Under our assumptions, there exists a unique solution (X, y, Z) for every $\mu > 0$; see for instance Wolkowicz, Saigal and Vandenberghe (2000) (Chapter 10). (To get this result, one interprets (13) as the KKT system of a convex problem with strictly convex cost function). Denoting this solution by (X_μ, y_μ, Z_μ), it is not too hard to show that the set

$$\{(X_\mu, y_\mu, Z_\mu) \colon \mu > 0\}$$

defines a smooth curve parameterized by μ, which is usually called the "central path."

The interior point approach, more precisely the "primal-dual interior-point path-following method," consists in applying Newton's method to follow this curve until $\mu \to 0$. This sounds straightforward, and it is, except for the following aspect. The equation (13) has $2\binom{n+1}{2} + m$ variables, but $\binom{n+1}{2} + n^2 + m$ equations. The difference arises from $ZX - \mu I$, which needs not be symmetric, even if X and Z are. Therefore, some sort of symmetrization of the last equation in (13) is necessary to overcome this problem.

The first papers exploiting this approach use some ad-hoc ideas to symmetrize the last equation; see Helmberg, Rendl, Vanderbei and Wolkowicz (1996), Kojima, Shindoh, and Hara (1997). Later, Monteiro (1997) and Zhang (1998) introduced a rather general scheme to deal with the equation $ZX = \mu I$. Let P be invertible. Zhang considers the mapping $H_P(M) := \frac{1}{2}[PMP^{-1} + (PMP^{-1})^T]$ and shows that, for $X \succ 0$, $Z \succ 0$,

$$H_P(ZX) = \mu I \text{ if and only if } ZX = \mu I.$$

Of course, different choices for P produce different search directions after replacing $ZX = \mu I$ by $H_P(ZX) = \mu I$. Various choices for P have been proposed and investigated with respect to the theoretical properties and behavior in practice.

Todd (1999) reviews about 20 different variants for the choice of P and investigates some basic theoretical properties of the resulting search directions. The main message seems to be at present that there is no clear champion among these choices in the sense that it would dominate both with respect to the theoretical convergence properties and the practical efficiency.

The following variant was introduced by Helmberg, Rendl, Vanderbei and Wolkowicz (1996), and independently by Kojima, Shindoh, and Hara (1997). It is simple, and yet computationally quite efficient. To simplify the presentation, we assume that there is some starting triple (X, y, Z) which satisfies $A(X) = b$, $A^T(y) - Z = C$ and $X \succ 0$, $Z \succ 0$. If this triple would lie on the central path, its "path parameter" μ would be $\mu = \frac{1}{n}\operatorname{Tr} ZX$. We do not assume that it lies on the central path, but would like to move from this triple towards the central path, and follow it until $\mu \approx 0$. Therefore we head for a point on the central path, given by the path parameter

$$\mu = \frac{1}{2n}\operatorname{Tr} ZX.$$

Applying a Newton step to $F_\mu(Xy, Z) = 0$ at (X, y, Z), with μ as above, leads to

$$A(\Delta X) = 0 \tag{14}$$

$$\Delta Z = A^T(\Delta y) \tag{15}$$

$$Z(\Delta X) + (\Delta Z)X = \mu I - ZX. \tag{16}$$

The second equation can be used to eliminate ΔZ, the last to eliminate ΔX:

$$\Delta X = \mu Z^{-1} - X - Z^{-1}A^T(\Delta y)X.$$

Substituting this into the first equation gives the following linear system for Δy:

$$A(Z^{-1}A^T(\Delta y)X) = \mu A(Z^{-1}) - b.$$

This system is positive definite and can therefore be solved quite efficiently by standard methods, yielding Δy (see Helmberg, Rendl, Vanderbei and Wolkowicz (1996)). Backsubstitution gives ΔZ, which is symmetric, and

ΔX, which needs not be. Taking the symmetric part of ΔX gives the following new point (X^+, y^+, Z^+):

$$\begin{aligned} X^+ &= X + t\frac{1}{2}(\Delta X + \Delta X^T) \\ y^+ &= y + t\Delta y \\ Z^+ &= Z + t\Delta Z. \end{aligned}$$

The stepsize $t > 0$ is chosen so that $X^+ \succ 0$, $Z^+ \succ 0$. In practice one starts with $t = 1$ (full Newton step), and backtracks by multiplying the current t with a factor smaller than 1, such as 0.8, until positive definiteness of X^+ and Z^+ holds.

A theoretical convergence analysis shows the following. Let a small scalar $\epsilon > 0$ be given. If the path parameter μ to start a new iteration is chosen properly, then the full step ($t = 1$ above) is feasible in each iteration, and a primal feasible solution X and a dual feasible solution y, whose duality gap $b^T y - \mathrm{Tr}(CX)$ is less than ϵ, can be found after $O(\sqrt{n}|\log \epsilon|)$ iterations; see the handbook of Wolkowicz, Saigal and Vandenberghe (2000), Chapter 10.

2.3 *Complexity*

We consider here complexity issues for semidefinite programming. We saw above that for semidefinite programs satisfying the Slater constraint qualification, the primal problem (3) and its dual (4) can be solved in polynomial time to any fixed prescribed precision using interior point methods.

However, even if all input data $A_1, \ldots, A_m, C, b$ are rational valued, no polynomial bound has been established for the bitlengths of the intermediate numbers occurring in interior point algorithms. Therefore, interior point algorithm for semidefinite programming are shown to be a polynomial in the real number model only, not in the bit number model of computation.

As a matter of fact, there are semidefinite programs with no rational optimum solution. For instance, the matrix

$$\begin{pmatrix} 1 & x \\ x & 2 \end{pmatrix} \oplus \begin{pmatrix} 2x & 2 \\ 2 & x \end{pmatrix}$$

is positive semidefinite if and only if $x = \sqrt{2}$. (Given two matrices A, B, $A \oplus B$ denotes the matrix $\left(\begin{smallmatrix} A & 0 \\ 0 & B \end{smallmatrix}\right)$). This contrasts with the situation of linear programming, where every rational linear program has a rational optimal solution whose bitlength is polynomially bounded in terms of the bit lengths of the input data (see Schrijver (1986)).

Another "pathological" situation which may occur in semidefinite programming is that all feasible solutions are doubly exponential. Consider, for instance, the matrix (taken from Ramana (1997)): $Q(x) := Q_1(x) \oplus \cdots \oplus Q_n(x)$, where $Q_1(x) := (x_1 - 2)$ and $Q_i(x) := \begin{pmatrix} 1 & x_{i-1} \\ x_{i-1} & x_i \end{pmatrix}$ for $i = 2, \ldots, n$. Then, $Q(x) \succeq 0$ if and only if $Q_i(x) \succeq 0$ for all $i = 1, \ldots, n$ which implies that $x_i \geq 2^{2^i-1}$ for $i = 1, \ldots, n$. Therefore, every rational feasible solution has an exponential bitlength.

Semidefinite programs can be solved in polynomial time to an arbitrary prescribed precision in the bit model using the ellipsoid (see Grötschel, Lovász and Schrijver (1988)). More precisely, let K denote the set of feasible solutions to (3) and, given $\epsilon > 0$, set $S(K, \epsilon) := \{Y \mid \exists X \in K \text{ with } \|X - Y\| < \epsilon\}$ ("the points that are in the ϵ-neighborhood of K") and $S(K, -\epsilon) := \{X \in K \mid \|X - Y\| > \epsilon \text{ for all } Y \notin K\}$ ("the points in K that are at a distance at least ϵ from the border of K"). Let L denote the maximum bit size of the entries of the matrices $A_1, \ldots, A_m$ and the vector b and assume that there is a constant $R > 0$ such that $\exists X \in K$ with $\|X\| \leq R$ if $K \neq \emptyset$. Then, the ellipsoid based algorithm, given $\epsilon > 0$, either finds $X \in S(K, \epsilon)$ for which $\mathrm{Tr}(CY) \leq \mathrm{Tr}(CX) + \epsilon$ for all $Y \in S(K, -\epsilon)$, or asserts that $S(K, -\epsilon) = \emptyset$. Its running time is polynomial in n, m, L, and $\log \epsilon$.

One of the fundamental open problems in semidefinite programming is the complexity of the following *semidefinite programming feasibility problem*[1] (F): *Given integral $n \times n$ symmetric matrices $Q_0, Q_1, \ldots, Q_m$, decide whether there exist real numbers $x_1, \ldots, x_m$ such that $Q_0 + x_1 Q_1 + \cdots + x_m Q_m \succeq 0$.*

This problem belongs obviously to NP in the real number model (since one can test whether a matrix is positive semidefinite in polynomial time using Gaussian elimination), but it is not known whether it belongs to NP in the bit model of computation. Ramana (1997) shows that problem (F) belongs to co-NP in the real number mode, and that (F) belongs to NP if and only if it belongs to co-NP in the bit model. These two results are based on an extended exact duality theory for semidefinite programming. Namely, given a semidefinite program (P), Ramana defines another semidefinite program (D) whose number of variables and coefficients bitlengths are polynomial in terms of the size of data in (P) and with the property that (P) is feasible if and only if (D) is infeasible.

Porkolab and Khachiyan (1997) show that problem (F) can be solved in polynomial time (in the bit model) for *fixed n* or *m*. (More precisely, problem (F) can be solved in $O(mn^4) + n^{O(\min(m,n^2))}$ arithmetic operations over $Ln^{O(\min(m,n^2))}$-bit numbers, where L is the maximum bitlength of the entries of matrices $Q_0, \ldots, Q_m$.) Moreover, for any *fixed m*, one can decide in polynomial time (in the bit model) whether there exist *rational* numbers $x_1, \ldots, x_m$ such that $Q_0 + x_1 Q_1 + \cdots + x_m Q_m \succeq 0$ (Khachiyan and Porkolab (1997)); this extends the result of Lenstra (1983) about polynomial time

[1] The following is an equivalent form for the feasibility region of a semidefinite program (3). Indeed, a matrix X is the form $Q_0 + \sum_{i=1}^m x_i Q_i$ if and only if it satisfies the system: $\mathrm{Tr}(A_j X) = b_j$ $(j = 1, \ldots, p)$, where $A_1, \ldots, A_p$ span the orthogonal complement of the subspace of $\mathcal{S}_n$ generated by $Q_1, \ldots, Q_m$ and $b_j = \mathrm{Tr}(A_j Q_0)$ for $j = 1, \ldots, p$.

solvability of integer linear programming in fixed dimension to semidefinite programming. More generally, given a convex semi-algebraic set $K \subseteq \mathbb{R}^n$, one can find in polynomial time an integral point in K (if some exists) for any fixed dimension n (Khachiyan and Porkolab (2000)). When all the polynomials defining K are quadratic, this result still holds without the convexity assumption (Barvinok (1993)). Further results have been recently given in Grigoriev, de Klerk, and Pasechnik (2003).

A special instance of the semidefinite programming feasibility problem is the *semidefinite matrix completion problem* (MC), which consists of deciding whether a partially specified matrix can be completed to a positive semidefinite matrix. The complexity of problem (MC) is not known in general, not even for the class of partial matrices whose entries are specified on the main diagonal and on the positions corresponding to the edge set of a circuit. However, for circuits (and, more generally, for graphs with no K_4-minor), problem (MC) is known to be polynomial-time solvable in the real number model (Laurent (2000)). In the bit model, problem (MC) is known to be polynomial time solvable when the graph corresponding to the positions of the specified entries is chordal or can be made chordal by adding a *fixed* number of edges (Laurent (2000)). A crucial tool is a result of Grone, Johnson, Sá, and Wolkowicz (1984) asserting that a partial matrix A whose entries are specified on the edge set of a chordal graph can be completed to a positive semidefinite matrix if and only if every fully specified principal submatrix of A is positive semidefinite.

As mentioned above, one of the difficulties in the complexity analysis of semidefinite programming is the possible nonexistence of rational solutions. However, in the special case of the matrix completion problem, no example is known of a rational partial matrix having only irrational positive semidefinite completions. (Obviously, a rational completion exists if a positive *definite* completion exists.)

Further conditions are known for existence of positive semidefinite matrix completions, involving cut and metric polyhedra (Laurent (1997)); see the surveys Johnson (1990), Laurent (1998b) for more information. In practice, positive semidefinite matrix completions can be computed using, e.g., the interior point algorithm of Johnson, Kroschel, and Wolkowicz (1998). This algorithm solves the problem:

$$\min f(X) \quad \text{subject to } X \succeq 0,$$

where $f(X) := \sum_{i,j=1}^{n} (h_{ij})^2 (x_{ij} - a_{ij})^2$. Here H is a given nonnegative symmetric matrix with a positive diagonal and A is a given symmetric matrix corresponding to the partial matrix to be completed; the condition $h_{ij} = 0$ means that entry x_{ij} is free while $h_{ij} > 0$ puts a weight on forcing entry x_{ij} to be as close as possible to a_{ij}. The optimum value of the above program is equal to 0 precisely when there is a positive semidefinite matrix completion of A, where the entries of A corresponding to $h_{ij} = 0$ are unspecified.

2.4 Geometry

We discuss here some geometric properties of semidefinite programming. We refer to Chapter 3 in Wolkowicz, Saigal and Vandenberghe (2000) for a detailed treatment. Let

$$K := \{X \in \mathrm{PSD}_n \mid \mathrm{Tr}(A_i X) = b_i \quad \text{for } i = 1, \ldots, m\}$$

denote the feasible region of a semidefinite program, where $A_1, \ldots, A_m \in \mathcal{S}_n$ and $b \in \mathbb{R}^m$. The set K is a convex set (called a *spectrahedron* in Ramana and Goldman (1995)) which inherits several of the geometric properties of the positive semidefinite cone PSD_n, in particular, concerning the structure of its faces. Recall that a set $F \subseteq K$ is a *face* of K if X, $Y \in F$ and $Z := \alpha X + (1-\alpha) Y \in K$ for some $0 < \alpha < 1$ implies that $Z \in F$. Given $A \in K$, $F_K(A)$ denotes the smallest face of K containing A. A point $A \in K$ is an *extreme point* if $F_K(A) = \{A\}$. It is well known (see Hill and Waters (1987)) that, given a matrix $A \in \mathrm{PSD}_n$, the smallest face $F_{\mathrm{PSD}}(A)$ of PSD_n that contains A is given by

$$F_{\mathrm{PSD}}(A) = \{X \in \mathrm{PSD}_n \mid \ker A \subseteq \ker X\}.$$

(For a matrix X, $\ker X := \{x \in \mathbb{R}^n \mid Xx = 0\}$.) Hence, if A has rank r, then $F_{\mathrm{PSD}}(A)$ is isomorphic to the cone PSD_r and thus has dimension $\binom{r+1}{2}$. As K is the intersection of PSD_n with the affine space

$$\mathcal{A} := \{X \in \mathcal{S}_n \mid \mathrm{Tr}(A_i X) = b_i \quad \text{for } i = 1, \ldots, m\},$$

the face $F_K(A)$ for $A \in K$ is given by

$$F_K(A) = F_{\mathrm{PSD}}(A) \cap \mathcal{A} = \{X \in K \mid \ker A \subseteq \ker X\}.$$

One can compute the dimension of faces of K in the following manner (see Chapter 31.5 in Deza and Laurent (1997)).

Let r denote the rank of A and let $A = QQ^T$, where Q is a $n \times r$ matrix of rank r. A matrix $B \in \mathcal{S}_n$ is called a *perturbation* of A if $A \pm tB \in K$ for some small $t > 0$. One can verify that B is a perturbation of A if and only if $B = QRQ^T$ for some matrix $R \in S_r$ satisfying $\mathrm{Tr}(RQ^T A_i Q) = 0$ for all $i = 1, \ldots, m$. Then the dimension of $F_K(A)$ is equal to the rank of the set of perturbations of A and, therefore,

$$\dim F_K(A) = \binom{r+1}{2} - \mathrm{rank}\{Q^T A_i Q \mid i = 1, \ldots, m\}.$$

This implies:

$$A \text{ is an extreme point of } K \Leftrightarrow \binom{r+1}{2} = \mathrm{rank}\{Q^T A_i Q \mid i = 1, \ldots, m\}. \tag{17}$$

We will use semidefinite programs as relaxations for 0/1 polytopes associated to combinatorial optimization problems; often the rank one matrices in the feasible region K correspond to the integer solutions of the combinatorial problem at hand. With this in mind, it is desirable to find a matrix $A \in K$ optimizing a given linear objective function over K having the smallest possible rank. The smallest possible ranks are obviously achieved at extremal matrices of K. Some results have been obtained along these lines which we now mention.

As an application of (17), we have that if $K \neq \emptyset$ and $\text{rank}\{A_i \mid i=1,\ldots,m\} < \binom{r+2}{2}$, then there exists a matrix $X \in K$ with rank $X \leq r$ (Barvinok (1995); Pataki (1996)). In fact, every extremal matrix X of K has this property; we will see below how to construct extremal matrices.

Barvinok (2001) shows the following refinement. Suppose that K is a nonempty bounded set and that $\text{rank}\{A_i \mid i=1,\ldots,m\} = \binom{r+2}{2}$ for some $1 \leq r \leq n-2$, then there exists a matrix $X \in K$ with rank $X \leq r$. Barvinok's proof is nonconstructive and it is an open question how to find efficiently such X.

Barvinok (1995) suggests the following approach for finding an extremal matrix in K. Let $C \in \mathcal{S}_n$ be a positive definite matrix and let $A \in K$ minimize $\text{Tr}(CX)$ over K. Barvinok shows that if C is sufficiently generic then A is an extremal point of K.

The following algorithm for constructing an extreme point of K has been suggested by several authors (see Alfakih and Wolkowicz (1998), Pataki (1996)). Suppose we want to minimize the objective function $\text{Tr}(CX)$ over K and assume that the minimum is finite. Given $A \in K$, the algorithm will construct an external matrix $A' \in K$ with objective value $\text{Tr}(CA') \leq \text{Tr}(CA)$. Using (17), one can verify whether A is an extreme point of K. If yes, then stop and return $A' = A$. Otherwise, one can find a nonzero matrix R belonging to the orthogonal complement in $\mathcal{S}_r$ of the space spanned by $Q^T A_i Q$ ($i=1,\ldots,m$); then $B := QRQ^T$ is a perturbation of A. If $\text{Tr}(CB) > 0$ then replace B by $-B$. Let t be the largest possible scalar for which $A + tB \succeq 0$. Then, $A + tB$ belongs to the boundary to the face $F_K(A)$ and thus the face $F_K(A + tB)$ is strictly contained in $F_K(A)$. We iterate with $A + tB$ in place of A. In at most n iterations, the algorithm returns an extreme point of K.

We conclude with some examples.

The max-cut spectrahedron. The following spectrahedron

$$\mathcal{E}_n := \{X \in \text{PSD}_n | X_{ii} = 1 \ \forall i = 1, \ldots, n\}$$

underlies the semidefinite relaxation for Max-Cut and will be treated in detail in Section 5. Its geometric properties have been investigated in Laurent and Poljak (1995, 1996). In particular, it is shown there that the only vertices (that is, the extreme points having a full dimensional normal cone) of $\mathcal{E}_n$ are its rank one matrices (corresponding to the cuts, i.e., the combinatorial objects in which we are interested). The spectrum of possible dimensions for the faces of $\mathcal{E}_n$ is shown to be equal to

$$\left[0, \binom{k_n}{2}\right] \cup \bigcup_{r=k_n+1}^{n} \left[\binom{r+1}{2} - n, \binom{r}{2}\right],$$

where $k_n := \lfloor \frac{n}{2} \rfloor + 1$. Moreover it is shown that the possible dimensions for the polyhedral faces of $\mathcal{E}_n$ are all integers k satisfying $\binom{k+1}{2} \leq n$. Geometric properties of other tighter spectrahedra for max-cut are studied in Anjos and Wolkowicz (2002b) and Laurent (2004).

Sum of largest eigenvalues. We introduced in Example 4 two programs (5) and (6) permitting to express the sum of the k largest eigenvalues of a symmetric matrix. Let K and $\mathcal{Y}$ denote their respective feasible regions; that is,

$$K := \{X \in \mathcal{S}_n | I \succeq X \succeq 0, \ \mathrm{Tr}(X) = k\},$$

$$\mathcal{Y} := \{YY^T | Y \in \mathbb{R}^{n \times k} \quad \text{with } Y^T Y = I_k\}.$$

Lemma 7. *The extreme points of the set K are the matrices of $\mathcal{Y}$. Therefore, K is equal to the convex hull of the set $\mathcal{Y}$.*

Proof. Let X be an extreme point of K. Then all its eigenvalues belong to the segment $[0, 1]$. As $\mathrm{Tr}(X) = k$, it follows that X has at least k nonzero eigenvalues and thus $\mathrm{rank}(X) \geq k$. In fact, $\mathrm{rank}(X) = k$ since X is an extreme point of K. Now this implies that the only nonzero eigenvalue of X is 1 with multiplicity k and thus $X \in \mathcal{Y}$. Conversely, every matrix of $\mathcal{Y}$ is obviously an extreme point of K. □

Note the resemblance of the above result to the Birkhoff-König theorem asserting that the set of stochastic matrices is equal to the convex hull of the set of permutation matrices.

Euclidean distance matrix completions. Let $G = (V, E; d)$ be a weighted graph with $V = \{1, \ldots, n\}$ and nonnegative edge weights $d \in \mathbb{Q}_+^E$. Given an integer r, we say that G is *r-realizable* if there exist points $v_1, \ldots, v_n \in \mathbb{R}^r$ such that $d_{ij} = \|v_i - v_j\|$ for all edges $ij \in E$; G is said to be *realizable* if it is r-realizable for some r. The problem of testing existence of a realization is known as the *Euclidean distance matrix completion problem* (EDM) (see Laurent (1998b) and Chapter 18 in Wolkowicz, Saigal and Vandenberghe (2000) for surveys). It has important applications, e.g., to molecular conformation problems in chemistry and distance geometry (see Crippen and Havel (1988)). As is well known, problem (EDM) can be formulated as a semidefinite programming problem. Namely, G is realizable if and only if the system:

$$X \succeq 0, X_{ii} + X_{jj} - 2X_{ij} = (d_{ij})^2 \quad \text{for } ij \in E \tag{18}$$

is feasible; moreover G is r-realizable if and only if the system (18) has a solution X with rank $X \leq r$. It follows from the above mentioned results about ranks of extremal points that if G is realizable, then G is r-realizable for some r satisfying $\binom{r+1}{2} \leq |E|$. Such a realization can be found using the above mentioned algorithm for finding extremal points (see Alfakih and Wolkowicz (1998), Barvinok (1995)).

It is also well known that the Euclidean distance matrix completion problem can be recast in terms of the positive semidefinite matrix completion problem (MC) treated earlier in Section 2.3 (see Laurent (1998a) for details). As a consequence, the complexity results mentioned earlier for problem (MC) also hold for problem (EDM). Namely, problem (EDM) can be solved in polynomial time in the bit number model when G can be made chordal by adding a fixed number of edges, and (EDM) can be solved in polynomial time in the real number model when G has no K_4-minor (Laurent (2000)).

An interior point algorithm is proposed in Alfakih, Khandani, and Wolkowicz (1999) for computing graph realizations. Alfakih (2000, 2001) studies rigidity properties of graph realizations in terms of geometric properties of certain associated spectrahedra.

When the graph G is not realizable, one can look for the smallest distortion needed to be applied to the edge weights in order to ensure existence of a realization. Namely, define this smallest distortion as the smallest scalar C for which there exist points $v_1, \ldots, v_n \in \mathbb{R}^n$ satisfying

$$\frac{1}{C} d_{ij} \leq \|v_i - v_j\| \leq d_{ij}$$

for all $ij \in E$. The smallest distortion can be computed using semidefinite programming. Bourgain (1985) has shown that $C = O(\log n)$ if $G = K_n$ and d satisfies the triangle inequalities: $d_{ij} \leq d_{ik} + d_{jk}$ for all $i, j, k \in V$ (see also Chapter 10 in Deza and Laurent (1997)). Since then research has been done for evaluating the minimum distortion for several classes of metric spaces including graph metrics (that is, when d is the path metric of a graph G); see in particular Linial, London, and Rabinovich (1995), Linial, Magen, and Naor (2002), Linial and Sachs (2003).

3 Semidefinite programming and integer 0/1 programming

3.1 A general paradigm

Suppose we want to solve a 0/1 linear programming problem:

$$\max c^T x \quad \text{subject to } Ax \leq b, x \in \{0, 1\}^n. \tag{19}$$

The classic polyhedral approach to this problem consists of formulating (19) as a linear programming problem:

$$\max c^T x \quad \text{subject to } x \in P$$

over the polytope

$$P := \text{conv}(\{x \in \{0, 1\}^n \mid Ax \leq b\})$$

and of applying linear programming techniques to it. For this one has to find the linear description of P or, at least, good linear relaxations of P. An initial linear relaxation of P is

$$K := \{x \in \mathbb{R}^n_+ \mid Ax \leq b\}$$

and, if $K \neq P$, one has to find "cutting planes" permitting to strengthen the relaxation K by cutting off its fractional vertices. Extensive research has been done for finding (partial) linear descriptions for many polyhedra arising from specific combinatorial optimization problems by exploiting the combinatorial structure of the problem at hand. Next to that, research has also focused on developing general purpose methods applying to arbitrary 0/1 problems (or, more generally, integer programming problems).

An early such method, developed in the sixties by Gomory and based on integer rounding, permits to generate the so-called Chvátal–Gomory cuts. This class of cutting planes was later extended, in particular, by Balas (1979) who introduced the disjunctive cuts. In the nineties several authors investigated lift-and-project methods for constructing cutting planes, the basic idea being of trying to represent a 0/1 polytope as the projection of a polytope lying in higher dimension. These methods aim at constructing good *linear* relaxations of a given 0/1 polytope, all with the exception of the lift-and-project method of Lovász and Schrijver which permits, moreover, to construct *semidefinite* relaxations. Further constructions for semidefinite relaxations have been recently investigated, based on algebraic results about representations of nonnegative polynomials as sums of squares of polynomials.

This idea of constructing semidefinite relaxations for a combinatorial problem goes back to the seminal work of Lovász (1979) who introduced the semidefinite bound $\vartheta(G)$ for the stability number of a graph G, obtained by optimizing over a semidefinite relaxation TH(G) of the stable set polytope. An important application is the polynomial time solvability of the maximum stable set problem in perfect graphs. This idea was later again used successfully by Goemans and Williamson (1995) who, using a semidefinite relaxation of the cut polytope, could prove an approximation algorithm with a good performance guarantee for the max-cut problem. Since then semidefinite programming has been widely used for approximating a variety of combinatorial optimization problems. This will be discussed in detail in further sections of this chapter.

For now we want to go back to the basic question of how to embed the 0/1 linear problem (19) in a semidefinite framework. A natural way of involving positive semidefiniteness is to introduce the matrix variable

$$Y = \begin{pmatrix} 1 \\ x \end{pmatrix} \begin{pmatrix} 1 & x^T \end{pmatrix}.$$

Then Y can be constrained to satisfy

$$\text{(i) } Y \succeq 0, \quad \text{(ii) } Y_{ii} = Y_{0i} \quad \forall i = 1, \ldots, n.$$

Condition (ii) expresses the fact that $x_i^2 = x_i$ as $x_i \in \{0, 1\}$. One can write (i), (ii) equivalently as

$$Y = \begin{pmatrix} 1 & x^T \\ x & X \end{pmatrix} \succeq 0 \quad \text{where } x := \text{diag}(X). \tag{20}$$

The objective function $c^T x$ can be modeled as $\langle \text{diag}(c), x \rangle$. There are several possibilities for modeling a linear constraint $a^T x \leq \beta$ from the system $Ax \leq b$. The simplest way is to use the *diagonal representation*:

$$\langle \text{diag}(a), X \rangle \leq \beta. \tag{21}$$

One can also replace $a^T x \leq \beta$ by its square $(\beta - a^T x)^2 \geq 0$, giving the inequality $(\beta - a^T) Y \binom{\beta}{-a} \geq 0$ which is however redundant under the assumption $Y \succeq 0$. Instead, when $a, \beta \geq 0$, one can use the *squared representation*: $(a^T x)^2 \leq \beta^2$; that is,

$$\langle aa^T, X \rangle \leq \beta^2 \tag{22}$$

or the *extended square representation*: $(a^T x)^2 \leq \beta(a^T x)$; that is,

$$\langle aa^T - \beta \,\text{diag}(a), X \rangle \leq 0. \tag{23}$$

Another possibility is to exploit the fact that the variable x_i satisfies $0 \leq x_i \leq 1$ and to multiply $a^T x \leq \beta$ by x_i and $1 - x_i$, which yields the system:

$$\sum_{j=1}^{n} a_j X_{ij} \leq \beta X_{ii} \, (i = 1, \ldots, n), \quad \sum_{j=1}^{n} a_j (X_{jj} - X_{ij}) \leq \beta (1 - X_{ii}) \, (i = 1, \ldots, n). \tag{24}$$

One can easily compare the strengths of these various representations of the inequality $a^T x \leq \beta$ and verify that, if (20) holds, then

$$(24) \Rightarrow (23) \Rightarrow (22) \Rightarrow (21).$$

Therefore, the constraints (24) define the strongest relaxation; they are, in fact, at the core of the lift-and-project methods by Lovász and Schrijver and by Sherali and Adams as we will see in Section 3.4. From an algorithmic point of view they are however the most expensive ones, as they involve $2n$ inequalities as opposed to one, for the other relaxations. Helmberg, Rendl, and Weismantel (2000) made an experimental comparison of the various relaxations which seems to indicate that the best trade off between running time and quality is obtained when working with the squared representation.

Instead of treating each inequality of the system $Ax \leq b$ separately, one can also consider pairwise products of inequalities: $(\beta_i - a_i^T x) \cdot (\beta_j - a_j^T x) \geq 0$, yielding the inequalities: $(\beta_i, -a_i^T) Y \binom{\beta_j}{-a_j} \geq 0$. This operation is also central to the lift-and-project methods as we will see later in this section.

3.2 *Introduction on cutting planes and lift-and-project methods*

Given a set $F \subseteq \{0,1\}^n$, we are interested in finding the linear description of the polytope $P := \text{conv}(F)$. At first (easy) step is to find a *linear programming formulation* for P; that is, to find a linear system $Ax \leq b$ for which the polytope $K := \{x \in \mathbb{R}^n \mid Ax \leq b\}$ satisfies $K \cap \{0,1\}^n = F$. If all vertices of K are integral, then $P = K$ and we are done. Otherwise we have to find cutting planes permitting to tighten the relaxation K and possibly find P after a finite number of iterations.

One of the first methods, which applies to general integral polyhedra, is the method of Gomory for constructing cutting planes. Given a linear inequality $\Sigma_i a_i x_i \leq \alpha$ valid for K where all the coefficients a_i are integers, the inequality $\Sigma_i a_i x_i \leq \lfloor \alpha \rfloor$ (known as a *Gomory–Chvátal cut*) is still valid for P but may eliminate some part of K. The *Chvátal closure* K' of K is defined as the solution set of all Chvátal-Gomory cuts; that is,

$$K' := \left\{ x \in \mathbb{R}^n | u^T A x \leq \lfloor u^T b \rfloor \quad \text{for all } u \geq 0 \quad \text{such that } u^T A \text{ integral} \right\}.$$

Then,

$$P \subseteq K' \subseteq K. \tag{25}$$

Set $K^{(1)} := K'$ and define recursively $K^{(t+1)} := (K^{(t)})'$ for $t \geq 1$. Chvátal (1973) proved that K' is a polytope and that $K^{(t)} = \text{conv}(K)$ for some t; the smallest t for which this is true is the *Chvátal rank* of the polytope K. The Chvátal rank may be very large as it depends not only on the dimension n but also on the coefficients of the inequalities involved. However, when K is assumed to be contained in the cube $[0, 1]^n$, its Chvátal rank is bounded by $O(n^2 \log n)$; if, moreover, $K \cap \{0, 1\}^n = \emptyset$, then the Chvátal rank is at most n (Bockmayr, Eisenbrand, Hartmann, and Schulz (1999); Eisenbrand and Schulz (1999)). Even if we can optimize a linear objective function over K in polynomial time, optimizing a linear objective function over the first Chvátal closure K' is a co-NP-hard problem is general (Eisenbrand (1999)).

Further classes of cutting planes have been investigated; in particular, the class of split cuts (Cook, Kannan, and Schrijver (1990)) (they are a special case of the disjunctive cuts studied in Balas (1979)). An inequality $a^T x \leq \alpha$ is a *split cut* for K if it is valid for the polytope $\text{conv}((K \cap \{x | c^T x \leq c_0\}) \cup (K \cap \{x | c^T x \geq c_0 + 1\}))$ for some integral $c \in \mathbb{Z}^n$, $c_0 \in \mathbb{Z}$. Split cuts are known to be equivalent to Gomory's mixed integer cuts (see, e.g., Cornuéjols and Li (2001a)). The *split closure* K' of K, defined as the solution set to all split cuts, is a polytope which satisfies again (25) (Cook, Kannan and Schrijver (1990)). One can iterate this operation of taking the split closure and it follows from results in Balas (1979) that P is found after n steps. However, optimizing over the first split closure is again a hard problem (Caprara and Letchford (2003)). (An alternative proof for NP-hardness of the membership problem in the split closure and in the Chvátal closure, based on a reduction from the closest lattice vector problem, is given in Cornuéjols and Li (2002)). If we consider only the split cuts obtained from the disjunctions $x_j \leq 0$ and $x_j \geq 1$, then we obtain a tractable relaxation of K which coincides with the relaxation obtained in one iteration of the Balas–Ceria–Cornuéjols lift-and-project method (which will be described later in Section 3.4).

Another popular approach is to try to represent P as the projection of another polytope Q lying in a higher (but preferably still polynomial) dimensional space, the idea behind being that the projection of a polytope Q may have more facets than Q itself. Hence it could be that even if P has an exponential number of facets, such Q exists having only a polynomial number of facets and lying in a space whose dimension is a polynomial in the original dimension of P (such Q is then called a *compact representation* of P). If this is the case then we have a proof that any linear optimization problem over P can be solved in polynomial time. At this point let us stress that it is not difficult to find a lift Q of P with a simple structure and lying in a space of *exponential* dimension; indeed, as pointed out in Section 3.3, *any* n-dimensional 0/1 polytope can be realized as the projection of a canonical simplex lying in the $(2^n - 1)$-space.

This idea of finding compact representations has been investigated for several polyhedra arising from combinatorial optimization problems; for instance, Barahona (1993), Barahona and Mahjoub (1986, 1994), Ball, Liu, and Pulleyblank (1989), Maculan (1987), Liu (1988) have provided such representations for certain polyhedra related to Steiner trees, stable sets, metrics, etc. On the negative side, Yannakakis (1988) proved that the matching polytope cannot have a compact representation satisfying a certain symmetry assumption.

Several general purpose methods have been developed for constructing projection representations for general 0/1 polyhedra; in particular, by Balas, Ceria, and Cornuéjols (1993) (the BCC method), by Sherali and Adams (1990) (the SA method), by Lovász and Schrijver (1991) (the LS method) and, recently, by Lasserre (2001b). [These methods are also known under the following names: lift-and-project for BCC, Reformulation-Linearization Technique (RLT) for SA, and matrix-cuts for LS.] A common feature of these methods is the construction of a hierarchy

$$K \supseteq K_1 \supseteq K_2 \supseteq \cdots \supseteq K_n \supseteq P$$

of linear or semidefinite relaxations of P which finds the exact convex hull in n steps; that is, $K_n = P$. The methods also share the following important algorithmic property: If one can optimize a linear objective function over the initial relaxation K in polynomial time, then the same holds for the next relaxations K_t for any *fixed* t, when applying the BCC, SA or LS constructions; for the Lasserre construction, this is true under the more restrictive assumption that the matrix A has a polynomial number of rows.

The first three methods (BCC, SA and LS) provide three hierarchies of *linear* relaxations of P satisfying the following inclusions: the Sherali–Adams relaxation is contained in the Lovász–Schrijver relaxation which in turn is contained in the Balas–Ceria–Cornuéjols relaxation. All three can be described following a common recipe: Multiply each inequality of the system $Ax \leq b$ by certain products of the bound inequalities $x_i \geq 0$ and $1 - x_i \geq 0$, replace each square x_i^2 by x_i, and linearize the products $x_i x_j (i \neq j)$ by introducing a new variable $y_{ij} = x_i x_j$. In this way, we obtain polyhedra in a higher dimensional space whose projection on the subspace $\mathbb{R}^n$ of the original x variable contains P and is contained in K. The three methods differ in the way of chosing the variables employed as multipliers and of iterating the basic step. The Lovász–Schrijver method can be strengthened by requiring positive semidefiniteness of the matrix (y_{ij}), which leads then to a hierarchy of positive semidefinite relaxations of P.

The construction of Lasserre produces a hierarchy of *semidefinite* relaxations of P which refines each of the above three hierarchies (BCC, SA and LS, even its positive semidefinite version). It was originally motivated by results about moment sequences and the dual theory of representation of nonnegative polynomials as sums of squares. It is however closely related to the SA method as both can be described in terms of requiring positive semidefiniteness of certain principal submatrices of the moment matrices of the problem.

We present in Section 3.3 some preliminary results which permit to show the convergence of the Lasserre and SA methods and to prove that every 0/1 polytope can be represented as the projection of a simplex in the $(2^n - 1)$-space. Then we describe in Section 3.4 the four lift-and-project methods and Sections 3.5, 3.6 and 3.7 contain applications of these methods to the stable set polytope, the cut polytope and some related polytopes. Section 3.8 presents extensions to (in general nonconvex) polynomial programming problems.

It will sometimes be convenient to view a polytope in $\mathbb{R}^n$ as being embedded in the hyperplane $x_0 = 1$ of $\mathbb{R}^{n+1}$. The following notation will be used throughout these sections. For a polytope P in $\mathbb{R}^n$, its *homogenization*

$$\tilde{P} := \left\{ \lambda \begin{pmatrix} 1 \\ x \end{pmatrix} \mid x \in P, \lambda \geq 0 \right\}$$

is a cone in $\mathbb{R}^{n+1}$ such that $P = \{x \in \mathbb{R}^n \mid \binom{1}{x} \in \tilde{P}\}$. For a cone C in $\mathbb{R}^n$,

$$C^* := \{y \in \mathbb{R}^n \mid x^T y \geq 0 \ \forall x \in C\}$$

denotes its *dual cone*.

3.3 A canonical lifting construction

Let $\mathcal{P}(V) := 2^V$ denote the collection of all subsets of $V = \{1, \ldots, n\}$ and let Z be the square 0/1 matrix indexed by $\mathcal{P}(V)$ with entries

$$Z(I, J) = 1 \quad \text{if and only if } I \subseteq J. \tag{26}$$

As Z is upper triangular with ones on its main diagonal, it is nonsingular and its inverse Z^{-1} has entries

$$Z^{-1}(I, J) = (-1)^{|J \setminus I|} \quad \text{if } I \subseteq J, Z^{-1}(I, J) = 0 \text{ otherwise.}$$

For $J \subseteq V$, let Z^J denote the J-th column of Z. [The matrix Z is known as the *Zeta matrix* of the lattice $\mathcal{P}(V)$ and the matrix Z^{-1} as its *Möbius matrix*.]

Given a subset $\mathcal{J} \subseteq \mathcal{P}(V)$, let $\mathcal{C}_{\mathcal{J}}$ denote the cone in $\mathbb{R}^{\mathcal{P}(V)}$ generated by the columns Z^J $(J \in \mathcal{J})$ of Z and let $P_{\mathcal{J}}$ be the 0/1 polytope in $\mathbb{R}^n$ defined as the convex hull of the incidence vectors of the sets in $\mathcal{J}$. Then $\mathcal{C}_{\mathcal{J}}$ is a simplicial cone,

$$\mathcal{C}_{\mathcal{J}} = \{y \in \mathbb{R}^{\mathcal{P}(V)} \mid Z^{-1} y \geq 0, (Z^{-1} y)_J = 0 \quad \text{for } J \in \mathcal{P}(V) \setminus \mathcal{J}\},$$

and $P_{\mathcal{J}}$ is the projection on $\mathbb{R}^n$ of the simplex $\mathcal{C}_{\mathcal{J}} \cap \{y \mid y_{\emptyset} = 1\}$. This shows therefore that any 0/1 polytope in $\mathbb{R}^n$ is the projection of a simplex lying in $\mathbb{R}^{2^n - 1}$.

Given $y \in \mathbb{R}^{\mathcal{P}(V)}$, let $M_V(y)$ be the square matrix indexed by $\mathcal{P}(V)$ with entries

$$M_V(y)(I, J) := y(I \cup J) \tag{27}$$

for $I, J \subseteq V$; $M_V(y)$ is known as the *moment matrix* of the sequence y. (See Section 7.1 for motivation and further information.) As noted in Lovász and Schrijver (1991), we have:

$$M_V(y) = Z \operatorname{diag}(Z^{-1}y)Z^T.$$

Therefore, the cone $\mathcal{C}_{\mathcal{P}(V)}$ can be alternatively characterized by any of the following linear and positive semidefinite conditions:

$$y \in \mathcal{C}_{\mathcal{P}(V)} \Leftrightarrow Z^{-1}y \geq 0 \Leftrightarrow M_V(y) \succeq 0. \tag{28}$$

Suppose that $\mathcal{J}$ corresponds to the set of 0/1 solutions of a semi-algebraic system

$$g_\ell(x) \geq 0 \quad \text{for} \quad \ell = 1, \ldots, m$$

where the g_ℓ's are polynomials in x. One can assume without loss of generality that each g_ℓ has degree at most one in every variable x_i and then one can identify g_ℓ with its sequence of coefficients indexed by $\mathcal{P}(V)$. Given $g, y \in \mathbb{R}^{\mathcal{P}(V)}$, define $g * y \in \mathbb{R}^{\mathcal{P}(V)}$ by

$$g * y := M(y)g; \text{ that is, } (g * y)_J := \sum_I g_I y_{I \cup J} \quad \text{for} \quad J \subseteq V. \tag{29}$$

It is noted in Laurent (2003a) that the cone $\mathcal{C}_{\mathcal{J}}$ can be alternatively characterized by the following positive semidefinite conditions:

$$y \in \mathcal{C}_{\mathcal{J}} \Leftrightarrow M_V(y) \succeq 0 \quad \text{and} \quad M_V(g_\ell * y) \succeq 0 \quad \text{for} \quad \ell = 1, \ldots, m. \tag{30}$$

This holds, in particular, when $\mathcal{J}$ corresponds to the set of 0/1 solutions of a linear system $Ax \leq b$, i.e., in the case when each polynomial g_ℓ has degree 1.

3.4 The Balas–Ceria–Cornuéjols, Lovász–Schrijver, Sherali–Adams, and Lasserre methods

Consider the polytope $K = \{x \in [0, 1]^n | Ax \leq b\}$ and let $P = \operatorname{conv}(K \cap \{0, 1\}^n)$ be the 0/1 polytope whose linear description is to be found. It is convenient to assume that the bound constraints $0 \le x_i \le 1 (i=1,\dots,n)$ are explicitly present in the linear description of K; let us rewrite the two systems $Ax \le b$ and $0 \le x_i \le 1$ $(i=1,\dots,n)$ as $\tilde{A}x \le \tilde{b}$ and let m denote the number of rows of A.

The Balas–Ceria–Cornuéjols construction. Fix an index $j \in \{1,\dots,n\}$. Multiply the system $\tilde{A}x \le \tilde{b}$ by x_j and $1-x_j$ to obtain the nonlinear system: $x_j(\tilde{A}x - \tilde{b}) \le 0$, $(1-x_j)(\tilde{A}x - b) \le 0$. Replace x_j^2 by x_j and linearize by introducing new variables $y_i = x_i x_j$ $(i=1,\dots,n)$; thus $y_j = x_j$. This defines a polytope in the (x,y)-space defined by $2(m+2n)$ inequalities: $\tilde{A}y - \tilde{b}x_j \le 0$, $\tilde{A}(x-y) - \tilde{b}(1-x_j) \le 0$. Its projection $P_j(K)$ on the subspace $\mathbb{R}^n$ indexed by the original x-variable satisfies

$$P \subseteq P_j(K) \subseteq K.$$

Iterate by defining $P_{j_1 \dots j_t}(K) := P_{j_t}(P_{j_{t-1}} \dots (P_{j_1}(K)) \dots)$. It is shown in Balas, Ceria and Cornuéjols (1993) that

$$P_{j_1 \dots j_t}(K) = \text{conv}(K \cap \{x | x_{j_1}, \dots, x_{j_t} \in \{0,1\}\}). \tag{31}$$

Therefore,

$$P = P_{j_1 \dots j_n}(K) \subseteq P_{j_1 \dots j_{n-1}}(K) \subseteq \dots \subseteq P_{j_1}(K) \subseteq K.$$

The Sherali–Adams construction. The first step is analogous to the first step of the BCC method except that we now multiply the system $\tilde{A}x \le \tilde{b}$ by x_j and $1-x_j$ for *all* indices $j \in \{1,\dots,n\}$. More generally, for $t=1,\dots,n$, the t-th step goes as follows. Multiply the system $\tilde{A}x \le \tilde{b}$ by each product $f_t(J_1,J_2) := \prod_{j\in J_1} x_j \cdot \prod_{j \in J_2}(1-x_j)$ where J_1 and J_2 are disjoint subsets of V with $|J_1 \cup J_2| = t$. Replace each square x_i^2 by x_i and linearize each product $\prod_{i\in I} x_i$ by a new variable y_I. This defines a polytope $R_t(K)$ in the space of dimension $n + \binom{n}{2} + \dots + \binom{n}{T}$ where $T := \min(t+1, n)$ (defined by $2^t\binom{n}{t}(m+2n)$ inequalities) whose projection $S_t(K)$ on the subspace $\mathbb{R}^n$ of the original x-variable satisfies

$$P \subseteq S_n(K) \subseteq \dots \subseteq S_{t+1}(K) \subseteq S_t(K) \subseteq \dots \subseteq S_1(K) \subseteq K$$

and $P = S_n(K)$. The latter equality follows from facts in Section 3.3 as we now see.

Write the linear system $\tilde{A}x \le \tilde{b}$ as $g_\ell^T\binom{1}{x} \ge 0$ $(\ell = 1,\dots,m+2n)$ where $g_\ell \in \mathbb{R}^{n+1}$. Extend g_ℓ to a vector $\mathbb{R}^{\mathcal{P}(V)}$ by adding zero coordinates. The linearization of the inequality $g_\ell^T\binom{1}{x} \cdot f_t(I,J) \ge 0$ reads:

$$\sum_{I \subseteq H \subseteq I \cup J} (-1)^{|H \setminus I|}(g_\ell * y)(H) \ge 0.$$

Using relation (28), one can verify that the set $R_t(K)$ can be alternatively described by the positive semidefinite conditions:

$$\begin{aligned} &M_U(g_\ell * y) \succeq 0 \quad \text{for } \ell = 1, \ldots, m \quad \text{and} \quad U \subseteq V \quad \text{with} \quad |U| = t, \\ &M_U(y) \succeq 0 \quad \text{for} \quad U \subseteq V \quad \text{with} \quad |U| = t+1 \end{aligned} \tag{32}$$

(where $g_1, \ldots, g_m$ correspond to the system $Ax \leq b$). It then follows from (30) that the projection $S_n(K)$ of $R_n(K)$ is equal to P.

The Lovász–Schrijver construction. Let U be another linear relaxation of P which is also contained in the cube $Q := [0,1]^n$; write U as $\{x \in \mathbb{R}^n \mid u_r^T\binom{1}{x} \geq 0 \ \forall r = 1, \ldots, s\}$. Multiply each inequality $g_\ell^T\binom{1}{x} \geq 0$ by each inequality $u_r^T\binom{1}{x} \geq 0$ to obtain the nonlinear system $u_r^T\binom{1}{x} \cdot g_\ell^T\binom{1}{x} \geq 0$ for all $\ell = 1, \ldots, m+2n$, $r = 1, \ldots, s$. Replace each x_i^2 by x_i and linearize by introducing a new matrix variable $Y = \binom{1}{x}(1 \ x^T)$. This defines the set $M(K, U)$ consisting of the symmetric matrices $Y = (y_{ij})_{i,j=0}^n$ satisfying

$$y_{jj} = y_{0j} \quad \text{for } j = 1, \ldots, n, \tag{33}$$

$$u_r^T Y g_\ell \geq 0 \quad \text{for all } r = 1, \ldots, s, \ \ell = 1, \ldots, m+2n \quad [\text{equivalently, } Y\tilde{U}^* \subseteq \tilde{K}]. \tag{34}$$

The first LS relaxation of P is defined as

$$N(K, U) := \left\{ x \in \mathbb{R}^n \middle| \begin{pmatrix} 1 \\ x \end{pmatrix} = Ye_0 \quad \text{for some } Y \in M(K, U) \right\}.$$

Then, $P \subseteq N(K, U) \subseteq N(K, Q) \subseteq K$ and $N(K, K) \subseteq N(K, U)$ if $K \subseteq U$. One can obtain stronger relaxations by adding positive semidefiniteness. Let $M_+(K, U)$ denote the set of positive semidefinite matrices in $M(K, U)$ and $N_+(K, U) := \{x \in \mathbb{R}^n | \binom{1}{x} = Ye_0$ for some $Y \in M_+(K, U)\}$. Then,

$$P \subseteq N_+(K, U) \subseteq N(K, U) \subseteq K.$$

The most extensively studied choice for U is $U := Q$, leading to the N operator. Set $N(K) := N(K, Q)$ and, for $t \geq 2$, $N^t(K) := N(N^{t-1}(K)) = N(N^{t-1}(K), Q)$. It follows from condition (34) that $N(K) \subseteq \text{conv}(K \cap \{x \mid x_j = 0,1\}) = P_j(K)$, the first BCC relaxation, and thus

$$N(K) \subseteq N_0(K) := \bigcap_{j=1}^{n} P_j(K). \tag{35}$$

[One can verify that $N_0(K)$ consists of the vectors $x \in \mathbb{R}^n$ for which $\binom{1}{x} = Ye_0$ for some matrix Y (not necessarily symmetric) satisfying (33) and (34) (with $U = Q$).] More generally, $N^t(K) \subseteq P_{j_1 \ldots j_t}(K)$ and, therefore, $P = N^n(K)$.

The choice $U := K$ leads to the stronger operator N', where we define $N'(K) := N(K, K)$ and, for $t \geq 2$,

$$(N')^t(K) := N((N')^{t-1}(K), K). \tag{36}$$

This operator is considered in Laurent (2001b) when applied to the cut polytope.

When using the relaxation $U = Q$, the first steps in the SA and LS constructions are identical: that is, $S_1(K) = N(K)$. The next steps are however distinct. A main difference between the two methods is that the LS procedure constructs the successive relaxations by a succession of t lift-and-project steps, each lifting taking place in a space of dimension $O(n^2)$, whereas the SA procedure carries but only one lifting step, occurring now in a space of dimension $O(n^{t+1})$; moreover, the projection step is not mandatory in the SA construction.

The Lasserre construction. We saw in relation (32) that the SA method can be interpreted as requiring positive semidefiniteness of certain principal submatrices of the moment matrices $M_V(y)$ and $M_V(g_\ell * y)$. The Lasserre method consists of requiring positive semidefiniteness of certain other principal matrices of those moment matrices. Namely, given an integer $t = 0, \ldots, n$, let $P_t(K)$ be defined by the conditions

$$M_{t+1}(y) \succeq 0, \quad M_t(g_\ell * y) \succeq 0 \quad \text{for } \ell = 1, \ldots, m \tag{37}$$

and let $Q_t(K)$ denote the projection of $P_t(K) \cap \{y | y_\emptyset = 1\}$ on $\mathbb{R}^n$. (For a vector $z \in \mathbb{R}^{\mathcal{P}(V)}$, $M_t(z)$ denotes the principal submatrix of $M_V(z)$ indexed by all sets $I \subseteq V$ with $|I| \leq t$.) Then,

$$P \subseteq Q_n(K) \subseteq Q_{n-1}(K) \subseteq \cdots \subseteq Q_1(K) \subseteq Q_0(K) \subseteq K$$

and it follows from (30) that $P = Q_n(K)$.

The construction of Lasserre (2000, 2001b) was originally presented in terms of moment matrices indexed by integer sequences (rather than subsets of V) and his proof of convergence used results about moment theory and the representation of nonnegative polynomials as sums of squares. The presentation and the proof of convergence given here are taken from Laurent (2003a).

How do the four hierarchies of relaxations relate? The following inclusions hold among the relaxations $P_{j_1 \ldots j_t}(K)$ (BCC), $S_t(K)$ (SA), $N^t(K)$ and $N_+^t(K)$ (LS), and $Q_t(K)$ (Lasserre):

(i) $Q_1(K) \subseteq N_+(K) \subseteq Q_0(K)$
(ii) (Lovász and Schrijver (1991)) For $t \geq 1$, $S_t(K) \subseteq N^t(K) \subseteq P_{j_1 \ldots j_t}(K)$
(iii) (Laurent (2003a)) For $t \geq 1$, $S_t(K) \subseteq N(S_{t-1}(K))$, $Q_t(K) \subseteq N_+(Q_{t-1}(K))$, and thus $Q_t(K) \subseteq S_t(K) \cap N_+^t(K)$.

Summarizing, the Lasserre relaxation is the strongest among all four types of relaxations.

Algorithmic aspects. Efficient approximations to linear optimization problems over the 0/1 polytope P can be obtained by optimizing over its initial relaxation K or any of the stronger relaxations constructed using the BCC, LS, SA and Lasserre methods. Indeed, if one can optimize in polynomial time any linear objective function over K [equivalently (by the results in Grötschel, Lovász and Schrijver (1988)), one can solve the separation problem for K in polynomial time], then, for any *fixed* t, the same holds for each of the relaxations $P_{j_1 \dots j_t}(K)$, $S_t(K)$, $N^t(K)$, $N^t_+(K)$ in the BCC, SA and LS hierarchies. This holds for the Lasserre relaxation $Q_t(K)$ under the more restrictive assumption that the linear system defining K has polynomial number of rows. Better approximations are obtained for higher values of t, at an increasing cost however. Computational experiments have been carried out using the various methods; see, in particular, Balas, Ceria and Cornuéjols (1993), Ceria (1993), Ceria and Pataki (1998) for results using the BCC method, Sherali and Adams (1997) (and further references there) for results using the SA method, and to Dash (2001) for a computational study of the N_+ operator.

Worst case examples where *n* iterations are needed for finding *P*. Let us define the *rank* of K with respect to a certain lift-and-project method as the smallest number of iterations needed for finding P. Specifically, the N-rank of K is the smallest integer t for which $P = N^t(K)$; define similarly the N_+, N_0, BCC, SA and Lasserre ranks. We saw above that n is a common upper bound for any such rank. We give below two examples of polytopes K whose rank is *equal* to n with respect to all procedures (except maybe with respect to the procedure of Lasserre, since the exact value of the Lasserre rank of these polytopes is not known).

As we will see in Section 3.5, the relaxation of the stable set polytope obtained with the Lovász–Schrijver N operator is much weaker than that obtained with the N_+-operator. For example, the fractional stable set polytope of K_n (defined by nonnegativity and the edge constraints) has N-rank $n-2$ while its N_+-rank is equal to 1! However, in the case of max-cut, no graph is known for which a similar result holds. Thus it is not clear in which situations the N_+-operator is significantly better, especially when applied iteratively. Some geometric results about the comparative strengths of the N, N_+ and N_0 operators are given in Goemans and Tunçel (2001). As a matter of fact, there exist polytopes K having N_+-rank equal to n (thus, for them, adding positive semidefiniteness does not help!).

As a first example, let

$$K := \left\{ x \in [0,1]^n \,|\, \sum_{i=1}^{n} x_i \geq \frac{1}{2} \right\}; \tag{38}$$

then $P = \{x \in [0,1]^n | \sum_{i=1}^{n} x_i \geq 1\}$ and the Chvátal rank of K is therefore equal to 1. The N_+-rank of K is equal to n (Cook and Dash (2001); Dash (2001)) and its SA-rank as well (Laurent (2003a)). As a second example, let

$$K := \left\{ x \in [0,1]^n | \sum_{i \in I} x_i + \sum_{i \notin I} (1 - x_i) \geq \frac{1}{2} \forall I \subseteq \{1, \ldots, n\} \right\}; \tag{39}$$

then $K \cap \{0,1\}^n = \emptyset$ and thus $P = \emptyset$. Then the N_+-rank of K is equal to n (Cook and Dash (2001), Goemans and Tunçel (2001) as well as its SA-rank Laurent (2003a). In fact, the Chvátal rank of K is also equal to n (Chvátal, Cook, and Hartman (1989)). The rank of K remains equal to n for the iterated operator N^* defined by $N^*(K) := N_+(K) \cap K'$, combining the Chvátal closure and the N_+-operator (Cook and Dash (2001); Dash (2001)). The rank is also equal to n if in the definition of N^* we replace the Chvátal closure by the split closure (Cornuéjols and Li (2001b)).

General setting in which the four methods apply. We have described above how the various lift-and-project methods apply to 0/1 linear programs, i.e., to the case when K is a polytope and $P = \text{conv}(K \cap \{0,1\}^n)$. In fact, they apply in a more general context, still retaining the property that P is found after n steps. Namely, the Lovász–Schrijver method applies to the case when K and U are arbitrary convex sets, the condition (34) reading then $Y\tilde{U}^* \subseteq \tilde{K}$. The BCC and SA methods apply to mixed 0/1 linear programs (Balas, Ceria and Cornuéjols (1993), Sherali and Adams (1994)). Finally, the Lasserre and Sherali–Adams methods apply to the case when K is a semi-algebraic set, i.e., when K is the solution set of a system of polynomial inequalities (since relation (30) holds in this context).

Moreover, various strengthenings of the basic SA method have been proposed involving, in particular, products of other inequalities than the bounds $0 \leq x_i \leq 1$ (cf., e.g., Ceria (1993), Sherali and Adams (1997), Sherali and Tuncbilek (1992, 1997)). A comparison between the Lasserre and SA methods for polynomial programming from the algebraic point of view of representations of positive polynomials is made in Lasserre (2002).

3.5 *Application to the stable set problem*

Given a graph $G = (V, E)$, a set $I \subseteq V$ is *stable* if no two nodes of I form an edge and the *stable set polytope* STAB(G) is the convex hull of the incidence vectors χ^S of all stable sets S of G, where $\chi_i^S = 1$ if $i \in S$ and $\chi_i^S = 0$ if $i \in V \backslash S$. As linear programming formulation for STAB(G), we consider the *fractional stable set polytope* FRAC(G) which is defined by the nonnegativity constraints: $x \geq 0$ and the *edge inequalities*:

$$x_i + x_j \leq 1 \quad \text{for } ij \in E. \tag{40}$$

Let us indicate how the various lift-and-project methods apply to the pair $P := \mathrm{STAB}(G)$, $K := \mathrm{FRAC}(G)$.

The LS relaxations $N(\mathrm{FRAC}(G))$ and $N_+(\mathrm{FRAC}(G))$ are studied in detail in Lovász and Schrijver (1991) where the following results are shown. The polytope $N(\mathrm{FRAC}(G))$ is completely described by nonnegativity, the edge constraints (40) and the *odd hole inequalities*:

$$\sum_{i \in V(C)} x_i \leq \frac{|C| - 1}{2} \quad \text{for } C \text{ odd circuit in } G. \tag{41}$$

Moreover, $N(\mathrm{FRAC}(G)) = N_0(\mathrm{FRAC}(G))$. Therefore, this gives a compact representation for the stable set polytope of t-perfect graphs (they are the graphs whose stable set polytope is completely determined by nonnegativity together with edge and odd hole constraints).

Other valid inequalities for $\mathrm{STAB}(G)$ include the *clique inequalities*:

$$\sum_{i \in Q} x_i \leq 1 \quad \text{for } Q \text{ clique in } G. \tag{42}$$

The smallest integer t for which (42) is valid for $N^t(\mathrm{FRAC}(G))$ is $t = |Q| - 2$ while (42) is valid for $N_+(\mathrm{FRAC}(G))$. Hence the N_+ operator yields a stronger relaxation of $\mathrm{STAB}(G)$ and equality $N_+(\mathrm{FRAC}(G)) = \mathrm{STAB}(G)$ holds for perfect graphs (they are the graphs for which $\mathrm{STAB}(G)$ is completely determined by nonnegativity and the clique inequalities; cf. Theorem 9). Odd antihole and odd wheel inequalities are also valid for $N_+(\mathrm{FRAC}(G))$.

Given a graph G on n nodes with stability number $\alpha(G)$ (i.e., the maximum size of a stable set in G), the following bounds hold for the N-rank t of $\mathrm{FRAC}(G)$ and its N_+-rank t_+:

$$\frac{n}{\alpha(G)} - 2 \leq t \leq n - \alpha(G) - 1, \, t_+ \leq \alpha(G).$$

See Lipták and Tunçel (2003) for a detailed study of further properties of the N and N_+ operators applied to $\mathrm{FRAC}(G)$; in particular, they show the bound $t_+ \leq n/3$ for the N_+-rank of $\mathrm{FRAC}(G)$.

The Sherali–Adams method does not seem to give a significant improvement, since the quantity $\frac{n}{\alpha(G)} - 2$ remains a lower bound for the SA-rank (Laurent (2003a)).

The Lasserre hierarchy refines the sequence $N_+^t(\mathrm{FRAC}(G))$. Indeed, it is shown in (Laurent (2003a)) that, for $t \geq 1$, the set $Q_t(\mathrm{FRAC}(G))$ can be alternatively described as the projection of the set

$$M_{t+1}(y) \succeq 0, \, y_{ij} = 0 \quad \text{for all edges } ij \in E, \, y_\emptyset = 1. \tag{43}$$

This implies that $Q_{\alpha(G)-1}(\mathrm{FRAC}(G)) = \mathrm{STAB}(G)$; that is, the Lasserre rank of FRAC(G) is at most $\alpha(G)-1$. The inclusion $Q_{\alpha(G)-1}(\mathrm{FRAC}(G)) \subseteq N_+^{\alpha(G)-1}(\mathrm{FRAC}(G))$ is strict, for instance, when G is the line graph of K_n (n odd) since the N_+-rank of FRAC(G) is then equal to $\alpha(G)$ (Stephen and Tunçel (1999)).

Let us mention a comparison with the basic semidefinite relaxation of STAB(G) by the theta body TH(G), which is defined by

$$\mathrm{TH}(G) := \left\{ x \in \mathbb{R}^n \mid \begin{pmatrix} 1 \\ x \end{pmatrix} = Ye_0 \quad \text{for some } Y \succeq 0 \right. \\ \left. \text{s.t. } Y_{ii} = Y_{0i} (i \in V),\ Y_{ij} = 0 (ij \in E) \right\}. \tag{44}$$

When maximizing $\sum_i x_i$ over TH(G), we obtain the theta number $\vartheta(G)$. Comparing with (43), we see that $Q_t(\mathrm{FRAC}(G))$ $(t \geq 1)$ is a natural generalization of the SDP relaxation TH(G) satisfying the following chain of inclusions:

$$Q_t(\mathrm{FRAC}(G)) \subseteq Q_1(\mathrm{FRAC}(G)) \subseteq N_+(\mathrm{FRAC}(G)) \\ \subseteq \mathrm{TH}(G) \subseteq Q_0(\mathrm{FRAC}(G)).$$

Section 4.2 below contains a detailed treatment of the relaxation TH(G).

Feige and Krauthgamer (2003) study the behavior of the N_+ operator applied to the fractional stable set polytope of $G_{n,1/2}$, a random graph on n nodes in which two nodes are joined by an edge with probability 1/2. It is known that the independence number of $G_{n,1/2}$ is equal, almost surely, to roughly $2\log_2 n$ and that its theta number is, almost surely, $\Theta(\sqrt{n})$. Feige and Krauthgamer show that the maximum value of $\sum_i x_i$ over $N_+^r(\mathrm{FRAC}(G_{n,1/2}))$ is, almost surely, roughly $\sqrt{\frac{n}{2^r}}$ when $r = o(\log n)$. This value can be computed efficiently if $r = O(1)$. Therefore, in that case, the typical value of these relaxations is smaller than that of the theta number by no more than a constant factor. Moreover, it is shown in Feige and Krauthgamer (2003) that the N_+-rank of a random graph $G_{n,1/2}$ is almost surely $\Theta(\log n)$.

3.6 Application to the max-cut problem

We consider here how the various lift-and-project methods can be used for constructing relaxations of the cut polytope. Section 5 will focus on the most basic SDP relaxation of the cut polytope and, in particular, on how it can be used for designing good approximation algorithms for the max-cut problem. As it well known, the max-cut problem can be formulated as an unconstrained quadratic ± 1 problem:

$$\max\ x^T A x \quad \text{subject to } x \in \{\pm 1\}^n \tag{45}$$

for some (suitably defined) symmetric matrix A; see relation (75).

As we are now working with ± 1 variables instead of 0/1 variables, one should appropriately modify some of the definitions given earlier in this section. For instance, the condition (33) in the definition of the LS matrix operator M now reads $y_{ii}=y_{00}$ for all $i\in\{1,\ldots,n\}$ (in place of $y_{ii}=y_{0i}$) and the (I,J)-th entry of the moment matrix $M_V(y)$ is now $y(I\Delta J)$ (instead of $y(I\cup J)$ as in (27)).

There are two possible strategies for constructing relaxations of the max-cut problem (45). The first possible strategy is to linearize the quadratic objective function, to formulate (45) as a linear problem

$$\max \langle A, X\rangle \quad \text{subject to } X \in \mathrm{CUT}_n$$

over the *cut polytope*

$$\mathrm{CUT}_n := \mathrm{conv}(xx^T | x \in \{\pm 1\}^n),$$

and to apply the various lift-and-project methods to some linear relaxation of CUT_n. As linear programming formulation for CUT_n, one can take the *metric polytope* MET_n which is defined as the set of symmetric matrices X with diagonal entries 1 satisfying the triangle inequalities:

$$X_{ij} + X_{ik} + X_{jk} \geq -1,\ X_{ij} - X_{ik} - X_{jk} \geq -1$$

for all distinct $i, j, k\in\{1,\ldots,n\}$.

Given a graph $G=(V,E)$ ($V=\{1,\ldots,n\}$), $\mathrm{CUT}(G)$ and $\mathrm{MET}(G)$ denote, respectively, the projections of CUT_n and MET_n on the subspace $\mathbb{R}^E$ indexed by the edge set of G. Barahona and Mahjoub (1986) show that $\mathrm{CUT}(G)\subseteq\mathrm{MET}(G)$ with equality if and only if G has no K_5-minor. Laurent (2001b) studies how the Lovász–Schrijver construction applies to the pair $P:=\mathrm{CUT}(G)$ and $K:=\mathrm{MET}(G)$. The following results are shown there: Equality $N_0^t(\mathrm{MET}(G)) = \mathrm{CUT}(G)$ holds if G has a set of t edges whose contraction produces a graph with no K_5-minor (recall the definition of N_0 from (35)). In particular, $N^{n-\alpha(G)-3}(\mathrm{MET}(G))=\mathrm{CUT}(G)$ if G has a maximum stable set whose deletion leaves at most three connected components and $N^{n-\alpha(G)-3}(G)=\mathrm{CUT}(G)$. Here, $N^t(G)$ denotes the projection on the subspace indexed by the edge set of G of the set $N^t(\mathrm{MET}(K_n))$. The inclusion $N^t(G)\subseteq N^t(\mathrm{MET}(G))$ holds obviously. Therefore, the N-rank of $\mathrm{MET}(K_n)$ is at most $n-4$, with equality for $n\leq 7$ (equality is conjectured for any n). A stronger relaxation is obtained when using the N' operator (recall the definition of N' from (36)). Indeed, $N'(\mathrm{MET}(K_6))=\mathrm{CUT}(K_6)$ is strictly contained in $N(\mathrm{MET}(K_6))$ and the N'-rank of $\mathrm{MET}(K_n)$ is at most $n-5$ for $n\geq 6$.

Another possible strategy is to apply the lift-and-project constructions to the set $K:=[-1,1]^n$ and to project on the subspace indexed by the set E_n of all pairs ij of points of V (instead of projecting on the space $\mathbb{R}^n$ indexed by the singletons of V). The SA and Lasserre methods converge now in $n-1$ steps (as there is no additional linear constraint beside the constraints expressing membership in the cube).

The t-th relaxation in the SA hierarchy is determined by all the inequalities valid for $\mathrm{CUT}(K_n)$ that are induced by at most $t+1$ points. Thus, the relaxation of order $t=1$ is the cube $[-1,1]^E$ while the relaxation of order $t=2$ is the metric polytope $\mathrm{MET}(K_n)$.

The t-th relaxation in the Lasserre hierarchy, denoted as $Q_t(G)$, is the projection on the subspace $\mathbb{R}^E$ indexed by the edge set of G of the set of vectors y satisfying

$$M_{t+1}(y) = (y_{I\Delta J})_{\substack{I,J\subseteq V \\ |I|,|J|\leq t+1}} \succeq 0,\ y_\emptyset = 1. \tag{46}$$

Equivalently, one can replace in (46) the matrix $M_{t+1}(y)$ by its principal submatrix indexed by the subsets whose cardinality has the same parity as $t+1$. Therefore, for $t=0$, $Q_0(K_n)$ corresponds to the basic semidefinite relaxation

$$\{X = (X_{ij})_{i,j=1}^n | X \succeq 0,\ X_{ii} = 1\ \forall i \in \{1,\ldots,n\}\}$$

of the cut polytope. For $t=1$, $Q_1(K_n)$ consists of the vectors $x \in \mathbb{R}^{E_n}$ for which $\binom{1}{x} = Ye_0$ for some matrix $Y \succeq 0$ indexed by $\{\emptyset\} \cup E_n$ satisfying

$$Y_{ij,ik} = Y_{\emptyset,jk} \tag{47}$$

$$Y_{ij,hk} = Y_{ih,jk} = Y_{ik,jh} \tag{48}$$

for all distinct $i,j,h,k \in \{1,\ldots,n\}$.

Applying Lagrangian duality to some extended formulation of the max-cut problem, Anjos and Wolkowicz (2002a) obtained a relaxation F_n of $\mathrm{CUT}(K_n)$, which can be defined as the set of all $x \in \mathbb{R}^{E_n}$ for which $\binom{1}{x} = Ye_0$ for some $Y \succeq 0$ indexed by $\{\emptyset\} \succeq E_n$ satisfying (47). Thus

$$Q_1(K_n) \subseteq F_n$$

(with strict inclusion if $n \geq 5$). It is interesting to note that the relaxation F_n is stronger than the basic linear relaxation by the metric polytope (Anjos and Wolkowicz (2002a)); that is,

$$F_n \subseteq \mathrm{MET}(K_n).$$

Indeed, let $x \in F_n$ with $\binom{1}{x} = Ye_0$ for some $Y \succeq 0$ satisfying (47). The principal submatrix X of Y indexed by $\{\emptyset, 12, 13, 23\}$ has the form

$$\begin{array}{c} \\ \emptyset \\ 12 \\ 13 \\ 23 \end{array} \begin{array}{c} \begin{array}{cccc} \emptyset & 12 & 13 & 23 \end{array} \\ \begin{pmatrix} 1 & x_{12} & x_{13} & x_{23} \\ x_{12} & 1 & x_{23} & x_{13} \\ x_{13} & x_{23} & 1 & x_{12} \\ x_{23} & x_{13} & x_{12} & 1 \end{pmatrix} \end{array}.$$

Now $e^T X e = 4(1 + x_{12} + x_{13} + x_{23}) \geq 0$ implies one of the triangle inequalities for the triple $(1,2,3)$; the other triangle inequalities follow by suitably flipping signs in X.

Laurent (2004) shows that

$$Q_t(G) \subseteq N_+^{t-1}(G)$$

for any $t \geq 1$. Therefore, the second strategy seems to be the most attractive one. Indeed, the relaxation $Q_t(G)$ is at least as tight as $N_+^{t-1}(G)$ and, moreover, it has a simpler explicit description (given by (46)) while the set $N_+^{t-1}(G)$ has only a recursive definition. We refer to Laurent (2004) for a detailed study of geometric properties of the set of (moment) matrices of the form (46). Laurent (2003b) shows that the smallest integer t for which $Q_t(K_n) = \mathrm{CUT}(K_n)$ satisfies $t \geq \lceil \frac{n}{2} \rceil - 1$; equality holds for $n \leq 7$ and is conjectured to hold for any n.

Anjos (2004) considers higher order semidefinite relaxations for the satisfiability problem involving similar types of constraints as the above relaxations for the cut polytope.

3.7 Further results

Lift-and-project relaxations for the matching and related polytopes. Let $G = (V, E)$ be a graph. A *matching* in G is a set of edges whose incidence vector x satisfies the inequalities

$$x(\delta(v)) = \sum_{e \in \delta(v)} x_e \leq 1 \quad \text{for all } v \in V. \tag{49}$$

(As usual, $\delta(v)$ denotes the set of edges adjacent to v.) Hence, the polytope K consisting of the vectors $x \in [0,1]^E$ satisfying the inequalities (49) is a linear relaxation of the *matching polytope*[2] of G, defined as the convex hull of the

[2] Of course, the matching polytope of G coincides with the stable set polytope of the line graph L_G of G; the linear relaxation K considered here is stronger than the linear relaxation $\mathrm{FRAC}(L_G)$ considered in Section 3.5. This implies, e.g., that $N(K) \subseteq N(\mathrm{FRAC}(L_G))$ and analogously for the other lift-and-project methods.

incidence vectors of all matchings in G. If, in relation (49), we replace the inequality sign "$\le$" by the equality sign "=" (resp., by the reverse inequality sign "$\ge$"), then we obtain the notion of *perfect matching* (resp., of *edge cover*) and the corresponding polytope K is a linear relaxation of the *perfect matching polytope* (resp., of the *edge cover polytope*). Thus, depending on the inequality sign in (49), we obtain three different classes of polytopes.

We now let G be the complete graph on $2n+1$ nodes. Stephen and Tunçel (1999) show that n steps are needed for finding the matching polytope when using the N_+ operator applied to the linear relaxation K. Aguilera, Bianchi, and Nasini (2004) study the rank of the Balas–Ceria–Cornuéjols procedure and of the N and N_+ operators applied to the linear relaxation K for the three (matching, perfect matching, and edge cover) problems. They show the following results, summarized in Fig. 1.

(i) The BCC rank is equal to n^2 for the three problems.
(ii) For the perfect matching problem, the rank is equal to n for both the N and N_+ operators.
(iii) The rank is greater than n for the N operator applied to the matching problem, and for the N and N_+ operators applied to the edge cover problem.

	BCC	N	N_+
Matching polytope	n^2	$>n$	n
Perfect matching polytope	n^2	n	n
Edge cover polytope	n^2	$>n$	$>n$

Fig. 1.

About the rank of the BCC Procedure. Given a graph $G=(V,E)$, the polytope QSTAB(G), consisting of the vectors $x \in \mathbb{R}^V_+$ satisfying the clique inequalities (42), is a linear relaxation of the stable set polytope STAB(G), stronger than the fractional stable set polytope FRAC(G) considered earlier in Section 3.5. Aguilera, Escalante, and Nasini (2002b) show that the rank of the polytope QSTAB(G) with respect to the Balas–Ceria–Cornuéjols procedure is equal to the rank of QSTAB($\bar{G}$), where $\bar{G}$ is the complementary graph of G.

Aguilera, Escalante, and Nasini (2002a) define an extension of the Balas–Ceria–Cornuéjols procedure for up-monotone polyhedra K. Namely, given a subset $F \subseteq \{1, \ldots, n\}$, they define the operator $\bar{P}_F(K)$ by

$$\bar{P}_F(K) = P_F(K \cap [0,1]^n) + \mathbb{R}^n_+,$$

where $P_F(\cdot)$ is the usual BCC operator defined as in (31). Then, the *BCC rank* of K is defined as the smallest $|F|$ for which $\bar{P}_F(K)$ is equal to the convex hull of the integer points in K. It is shown in Aguilera, Bianchi and Nasini (2002a) that, for a clutter $\mathcal{C}$ and its blocker $\mathrm{bl}(\mathcal{C})$, the two polyhedra $P_{\mathcal{C}} = \{x \in \mathbb{R}^n_+ | x(C) \geq 1 \ \forall C \in \mathcal{C}\}$ and $P_{\mathrm{bl}(\mathcal{C})} = \{x \in \mathbb{R}^n_+ \mid x(D) \geq 1 \forall D \in \mathrm{bl}(\mathcal{C})\}$ have the *same* rank with respect to the extended BCC procedure.

An extension of lift operators to subset algebras. As we have seen earlier, the lift-and-project methods are based on the idea of lifting a vector $x \in \{0,1\}^n$ to a higher dimensional vector $y \in \{0,1\}^N$ (where $N > n$) such that $y_i = x_i$ for all $i = 1, \ldots, n$. More precisely, let L denote the lattice of all subsets of $V = \{1, \ldots, n\}$ with the set inclusion as order relation, and let Z_L be its Zeta matrix, defined by (26). Then, the lift of $x \in \{0,1\}^n$ is the vector $y \in \{0,1\}^L$ with components $y_I = \prod_{i \in I} x_i$ for $I \in L$; in other words, y is the column of Z_L indexed by x (after identifying a set with its incidence vector).

Bienstock and Zuckerberg (2004) push this idea further and introduce a lifting to a lattice Σ, larger than L. Namely, let Σ denote the lattice of all subsets of $\{0,1\}^n$, with the reverse set inclusion as order relation; that is, $\alpha \leq \beta$ in Σ if $\beta \subseteq \alpha$. Let Z_Σ denote the Zeta matrix of Σ, with (α, β)-entry 1 if $\alpha \leq \beta$ and 0 otherwise. Then, any vector $x \in \{0,1\}^n$ can be lifted to the vector $z \in \{0,1\}^\Sigma$ with components $z_\alpha = 1$ if and only if $x \in \alpha$ (for $\alpha \in \Sigma$); this is, z is the column of Z_Σ indexed by $\{x\}$.

Note that the lattice L is isomorphic to a sublattice of Σ. Indeed, if we set $H_I = \{x \in \{0,1\}^n | x_i = 1 \ \ \forall i \in I\}$ for $I \subseteq V$, then $I \subseteq J \Leftrightarrow H_I \supseteq H_J \Leftrightarrow H_I \leq H_J$ (in Σ) and, thus, the mapping $I \mapsto H_I$ maps L to a sublattice of Σ. Therefore, given $x \in \{0,1\}^n$ and, as above, y (resp., z) the column of Z_L (resp., of Z_Σ) indexed by x, then $z_{H_I} = y_I$ for all $I \in L$ and $z_{H_i} = x_i$ for all $i \in V$.

Let $F \subseteq \{0,1\}^n$ be the set of $0-1$ points whose convex hull $P := \mathrm{conv}(F)$ has to be found, and let F_L (resp., F_Σ) be the corresponding set of columns of Z_L (resp., of Z_Σ). Then, a vector $x \in \mathbb{R}^n$ belongs to conv(F) if and only if there exists $y \in \mathrm{conv}(F_L)$ such that $y_i = x_i$ $(i \in V)$ or, equivalently, if there exists $z \in \mathrm{conv}(F_\Sigma)$ such that $z_{H_i} = x_i (i \in V)$. The SA, LS and Lasserre methods consist of requiring certain conditions on the lifted vector y (or projections of it); Bienstock and Zuckerberg (2004) present analogous conditions for the vector z.

Bienstock and Zuckerberg work, in fact, with a lifted vector $\tilde{z}$ indexed by a small subset ν of Σ; this set ν is constructed on the fly, depending on the structure of F. Consider, for instance, the set covering problem, where F is the set of 0/1 solutions of a system: $x(A_1) \geq 1, \ldots, x(A_m) \geq 1$ (with $A_1, \ldots, A_m \subseteq \{1, \ldots, n\}$). Then, the most basic lifting procedure presented in Bienstock and Zuckerberg (2004) produces a polyhedron $R^{(2)}$ (whose projection is a linear relaxation of P) in the variable $\tilde{z} \in \mathbb{R}^\nu$, where $\nu \subseteq \Sigma$ consists of F, $Y_i := \{x \in F | x_i = 1\}$, $N_i := F \setminus Y_i (i = 1, \ldots, n)$, and $\cap_{i \in C} N_i$, $Y_{i_0} \cap \cap_{i \in C \setminus i_0} N_i$ $(i_0 \in C)$, and $\cup_{S \subseteq C, |S| \geq 2} \cap_{i \in S} Y_i \cap \cap_{i \in C \setminus S} N_i$, for each of the distinct intersections $C = A_h \cap A_\ell$ $(h \neq \ell = 1, \ldots, m)$ with size ≥ 2. The linear relaxation $R^{(2)}$ has $O(m^4 n^2)$ variables and constraints; hence, one can optimize over $R^{(2)}$ in polynomial time. Moreover, any inequality $a^T x \geq a_0$, valid for P with coefficients in $\{0, 1, 2\}$, is valid for (the projection of) $R^{(2)}$. Note that there exist set covering polytopes having exponentially many facets with coefficients in $\{0, 1, 2\}$. The new lifting procedure is more powerful in some cases. For instance, $R^{(2)} = P$ holds for the polytope K from (38), while the N_+-rank of K is equal to n. As another example, consider the circulant set covering polytope:

$$P = \text{conv}\left(\left\{x \in \{0, 1\}^n | \sum_{i \neq j} x_i \geq 1 \ \forall j = 1, \ldots, n\right\}\right);$$

then the inequality $\sum_{i=1}^n x_i \geq 2$ is valid for P, it is not valid neither for $S_{n-3}(K)$ nor for $N_{n-3}^+(K)$, while it is valid for the relaxation $R^{(2)}$ (Bienstock and Zuckerberg (2004)).

A more sophisticated lifting procedure is proposed in Bienstock and Zuckerberg (2004) yielding stronger relaxations $R^{(k)}$ of P, with the following properties. For fixed $k \geq 2$, one can optimize in polynomial time over $R^{(k)}$; any inequality $a^T x \geq a_0$, valid for P with[3] coefficients in $\{0, 1, \ldots, k\}$, is valid for $R^{(k)}$. For instance, $R^{(3)} = \emptyset$ holds for the polytope K from (39), while n steps of the classic lift-and-project procedures are needed for proving that $P = \emptyset$.

Complexity of cutting plane proofs. Results about the complexity of cutting plane proofs using cuts produced by the various lift-and-project methods can be found, e.g., in Dash (2001, 2002), Grigoriev, Hirsch, and Pasechnik (2002).

3.8 Extensions to polynomial programming

Quadratic programming. Suppose we want to solve the program

$$p^* := \min g_0(x) \quad \text{subject to } g_\ell(x) \geq 0 \, (\ell = 1, \ldots, m) \tag{50}$$

where $g_0, g_1, \ldots, g_m$ are quadratic functions of the form: $g_\ell(x) = x^T Q_\ell x + 2q_\ell^T x + \gamma_\ell$ (Q_ℓ symmetric $n \times n$ matrix, $q_\ell \in \mathbb{R}^n$, $\gamma_\ell \in \mathbb{R}$). For any ℓ, define the matrix $P_\ell := \begin{pmatrix} \gamma_\ell & q_\ell^T \\ q_\ell & Q_\ell \end{pmatrix}$. Then, $g_\ell(x) = \langle P_\ell, \begin{pmatrix} 1 & x^T \\ x & xx^T \end{pmatrix} \rangle$. This suggests the following natural positive semidefinite relaxation of (50):

$$\min \langle P_0, Y \rangle \quad \text{subject to } Y \succeq 0, \ Y_{00} = 1, \ \langle P_\ell, Y \rangle \geq 0 \, (\ell = 1, \ldots, m). \tag{51}$$

[3] Validity holds, more generally, for any inequality $a^T x \geq a_0$ with pitch $\leq k$. If we order the indices in such a way that $0 < a_1 \leq a_2 \leq \cdots \leq a_J, a_{J+1} = \ldots = a_n = 0$, then the *pitch* is the smallest t for which $\sum_{j=1}^t a_j \geq a_0$.

Let $F := \{x \in \mathbb{R}^n | g_\ell(x) \geq 0\,(\ell = 1, \ldots, m)\}$ denote the feasible set of (50) and

$$\hat{F} := \{x \in \mathbb{R}^n | \begin{pmatrix} 1 \\ x \end{pmatrix} = Ye_0 \quad \text{for some } Y \succeq 0 \quad \text{satisfying } \langle P_\ell, Y \rangle \geq 0 \\ \text{for all } \ell = 1, \ldots, m\} \tag{52}$$

its natural semidefinite relaxation. It is shown in Fujie and Kojima (1997) and Kojima and Tunçel (2000) that $\hat{F}$ can be alternatively described by the following quadratic system:

$$\hat{F} := \left\{ x \in \mathbb{R}^n | \sum_{\ell=1}^m t_\ell g_\ell(x) \geq 0 \text{ for all } t_\ell \geq 0 \text{ for which } \sum_{\ell=1}^m t_\ell Q_\ell \preceq 0 \right\}. \tag{53}$$

If, in (52), one omits the condition $Y \succeq 0$ and, in (53), the condition $\sum_\ell t_\ell Q_\ell \preceq 0$ is replaced by $\sum_\ell t_\ell Q_\ell = 0$, then one obtains a linear relaxation $\hat{F}_L$ of F such that $\text{conv}(F) \subseteq \hat{F} \subseteq \hat{F}_L$.

Using this construction of linear/semidefinite relaxations, Kojima and Tunçel (2000) construct a hierarchy of successive relaxations of F that converges asymptotically to conv(F). Lasserre (2001a) also constructs such a hierarchy which applies, more generally, to polynomial programs; we expose it below.

Polynomial programming. Consider now the program (50) where all the g_ℓ's are polynomials in $x = (x_1, \ldots, x_n)$. Let w_ℓ be the degree of g_ℓ, $v_\ell := \lceil \frac{w_\ell}{2} \rceil$ and $v := \max_{\ell=1,\ldots,m} v_\ell$. We need some definitions.

Given a sequence $y = (y_\alpha)_{\alpha \in \mathbb{Z}_+^n}$ indexed by $\mathbb{Z}_+^n$, its *moment matrix* is

$$M^{\mathbb{Z}}(y) := (y_{\alpha+\beta})_{\alpha,\beta \in \mathbb{Z}_+^n} \tag{54}$$

and, given an integer $t \geq 0$, $M_t^{\mathbb{Z}}(y)$ is the principal submatirx of $M^{\mathbb{Z}}(y)$ indexed by the sequences $\alpha \in \mathbb{Z}_+^n$ with $|\alpha| := \sum_i \alpha_i \leq t$. [Note that the moment matrix $M_V(y)$ defined earlier in (27) corresponds to the principal submatrix of $M^{\mathbb{Z}}(y)$ indexed by the sequences $\alpha \in \{0,1\}^n$, after replacing y_α by $y_{\alpha'}$ where $\alpha'_i := \min(\alpha_i, 1)$ for all i.] The operation from (29) extends to sequences indexed by $\mathbb{Z}_+^n$ in the following way:

$$g, y \in \mathbb{R}^{\mathbb{Z}_+^n} \rightsquigarrow g * y := \left(\sum_\beta g_\beta y_{\alpha+\beta} \right)_{\alpha \in \mathbb{Z}_+^n}. \tag{55}$$

Given $x \in \mathbb{R}^n$, define the sequence $y \in \mathbb{R}^{\mathbb{Z}^n_+}$ with α-th entry $y_\alpha := \prod_{i=1}^n x_i^{\alpha_i}$ for $\alpha \in \mathbb{Z}^n_+$. Then, $M_t^{\mathbb{Z}}(y) = yy^T \succeq 0$ (where we use the same symbol y for denoting the truncated vector $(y_\alpha)_{|\alpha| \leq t}$) and $M_t^{\mathbb{Z}}(g_\ell * y) = g_\ell(x) \cdot M_t^{Z}(y) \succeq 0$ if $g_\ell(x) \geq 0$. This observation leads naturally to the following relaxations of the set F, introduced by Lasserre (2001a).

For $t \geq v-1$, let $\mathcal{Q}_t(F)$ be the convex set defined as the projection of the solution set to the system

$$M_{t+1}^{\mathbb{Z}}(y) \succeq 0,\, M_{t-v_\ell+1}^{\mathbb{Z}}(g_\ell * y) \succeq 0 \quad \text{for } \ell = 1, \ldots, m, \ y_0 = 1 \tag{56}$$

on the subspace $\mathbb{R}^n$ indexed by the variables y_α for $\alpha = (1, 0, \ldots, 0), \ldots, (0, \ldots, 0, 1)$ (identified with $x_1, \ldots, x_n$). Then,

$$\text{conv}(F) \subseteq \mathcal{Q}_{t+1}(F) \subseteq \mathcal{Q}_t(F).$$

Lasserre (2001a) shows that

$$\bigcap_{t \geq v-1} \mathcal{Q}_t(F) = \text{conv}(F);$$

that is, the hierarchy $(\mathcal{Q}_t(F))_t$ converges asymptotically to conv(F). This equality holds under some technical assumption on F which holds, for instance, when F is the set of 0/1 solutions of a polynomial system and the constraints $x_i(1-x_i) = 0$ ($i \in \{1, \ldots, n\}$) are present in the description of F, or when the set $\{x \mid g_\ell(x) \geq 0\}$ is compact for at least one of the constraints defining F. Lasserre's result relies on a result about representations of positive polynomials as sums of squares, to which we will come back in Section 7.1.

In the quadratic case, when all g_ℓ are quadratic polynomials, one can verify that the first Lasserre relaxation $\mathcal{Q}_0(F)$ coincides with the basic SDP relaxation $\hat{F}$ defined in (52); that is,

$$\mathcal{Q}_0(F) = \hat{F}.$$

Consider now the 0/1 case when F is the set of 0/1 solutions of a polynomial system; write F as

$$F = \{x \in \mathbb{R}^n \mid g_\ell(x) \geq 0\, (\ell = 1, \ldots, m),\ h_i(x) := x_i - x_i^2 = 0\, (i = 1, \ldots, n)\}.$$

One can assume without loss of generality that each g_ℓ has degree at most 1 in every variable. The set

$$K := \{x \in [0, 1]^n \mid g_\ell(x) \geq 0\, (\ell = 1, \ldots, m)\}$$

is a natural relaxation of F. We have constructed in Section 3.4 the successive relaxations $Q_t(K)$ of conv(F) satisfying conv(F) $= Q_{n+v-1}(K)$; their construction used moment matrices indexed by the subsets of V while the definition of $\mathcal{Q}_t(F)$ involves moment matrices indexed by integer sequences. However, the condition $M_t^{\mathbb{Z}}(h_i * y) = 0$ (present in the definition $\mathcal{Q}_t(F)$) permits to show that the two definitions are equivalent; that is,

$$Q_t(K) = \mathcal{Q}_t(F) \quad \text{for } t \geq v - 1.$$

See Laurent (2003a) for details.

In the quadratic 0/1 case, we find therefore that

$$\hat{F} = \mathcal{Q}_0(F) = Q_0(K).$$

As an example, given a graph $G = (V = \{1, \ldots, n\}, E)$, consider the set

$$F := \{x \in \{0, 1\}^n \mid x_i x_j = 0 \quad \text{for all } ij \in E\};$$

then conv(F) is equal to the stable set polytope of G. It follows from the definitions that $\hat{F}$ coincides with the basic SDP relaxation TH(G) (defined in (44)). Therefore, $\mathcal{Q}_0(F) = \text{TH}(G)$ while the inclusion $\text{TH}(G) \subseteq Q_0(\text{FRAC}(G))$ is strict in general. Hence one obtains stronger relaxations for the stable set polytope STAB(G) when starting from the above quadratic representation F for stable sets rather than from the linear relaxation FRAC(G). Applying the equivalent definition (53) for $\hat{F}$, one finds that

$$\text{TH}(G) = \left\{ x \in \mathbb{R}^n \mid x^T M x - \sum_{i=1}^{n} M_{ii} x_i \leq 0 \quad \text{for } M \succeq 0 \quad \text{with} \right. \\ \left. M_{ij} = 0 \; (i \neq j \in V, ij \notin E) \right\}. \tag{57}$$

(This formulation of TH(G) also follows using the duality between the cone of completable partial positive semidefinite matrices and the cone of positive semidefinite matrices having zeros at the positions of unspecified entries; cf. Laurent (2001a).) See Section 4.2 for further information about the semidefinite relaxation TH(G).

4 Semidefinite relaxation for the maximum stable set problem

Given a graph $G = (V, E)$, its *stability number* $\alpha(G)$ is the maximum cardinality of a stable set in G, and its *clique number* $\omega(G)$ is the maximum cardinality of a clique in G. Given an integer $k \geq 1$, a *k-coloring* of G is an assignment of numbers from $\{1, \ldots, k\}$ (*colors*) to the nodes of G in such a way that adjacent nodes receive distinct colors; in other words, a k-coloring is a partition of V into k stable sets. The *coloring number* (or *chromatic number*) $\chi(G)$ is the smallest integer k for which G has a k-coloring. With $\bar{G} = (V, \bar{E})$ denoting the complementary graph of G, the following holds trivially:

$$\alpha(\bar{G}) = \omega(G) \leq \chi(G).$$

The inequality $\omega(G) \leq \chi(G)$ is strict, for instance, for odd circuits of length ≥ 5 and their complements. Berge (1962) defined a graph G to be *perfect* if $\omega(G') = \chi(G')$ for every induced subgraph G' of G and he conjectured that a graph is perfect if and only if it does not contain a circuit of length ≥ 5 or its complement as an induced subgraph. This is the well known *strong perfect graph conjecture*, which has been recently proved by Chudnovsky, Robertson, Seymour and Thomas (2002). Lovász (1972) proved that the complement of a perfect graph is again perfect, solving another conjecture of Berge. As we will see later in this section, perfect graphs can also be characterized in terms of integrality of certain associated polyhedra.

Computing the stability number or the chromatic number of a graph are hard problems; more precisely, given an integer k, it is an NP-complete problem to decide whether $\alpha(G) \geq k$ or $\chi(G) \leq k$ (Karp (1972)). Deciding whether a graph is 2-colorable can be done in polynomial time (as this happens if and only if the graph is bipartite). On the other hand, while every planar graph is 4-colorable (by the celebrated four color theorem), it is NP-complete to decide whether a planar graph is 3-colorable (Garey, Johnson, and Stockmeyer (1976)). When restricted to the class of perfect graphs, the maximum stable set problem and the coloring problem can be solved in polynomial time. This result relies on the use of the Lovász theta function $\vartheta(G)$ which can be computed (with an arbitrary precision) in polynomial time (as the optimum of a semidefinite program) and satisfies the "sandwich" inequalities:

$$\alpha(G) \leq \vartheta(G) \leq \chi(\bar{G}).$$

The polynomial time solvability of the maximum stable set problem for perfect graphs is one of the first beautiful applications of semidefinite programming to combinatorial optimization and, up to date, no other purely combinatorial method is known for proving this.

4.1 The basic linear relaxation

As before, the *stable set polytope* STAB(G) is the polytope in $\mathbb{R}^V$ defined as the convex hull of the incidence vectors of the stable sets of G, FRAC(G) is its linear relaxation defined by nonnegativity and the edge inequalities (40), and QSTAB(G) denotes the linear relaxation of STAB(G) defined by nonnegativity and the clique inequalities (42). Therefore,

$$\mathrm{STAB}(G) \subseteq \mathrm{QSTAB}(G) \subseteq \mathrm{FRAC}(G)$$

and

$$\alpha(G) = \max(e^T x | x \in \mathrm{STAB}(G))$$

setting $e := (1, \ldots, 1)^T$. One can easily see that equality STAB(G) = FRAC(G) holds if and only if G is a bipartite graph with no isolated nodes; thus the maximum stable set problem for bipartite graphs can be solved in polynomial time as a linear programming problem over FRAC(G). Fulkerson (1972) and Chvátal (1975) show:

Theorem 9. *A graph G is perfect if and only if* STAB(G) = QSTAB(G).

This result does not (yet) help for compute efficiently $\alpha(G)$ for perfect graphs. Indeed, optimizing over the linear relaxation QSTAB(G) is, unfortunately, a hard problem is general (as hard as the original problem, since the membership problem for QSTAB(G) is nothing but a maximum weight clique problem in G.) Proving polynomiality requires the use of the semidefinite relaxation TH(G) as we see later in this section.

4.2 The theta function $\vartheta(G)$ and the basic semidefinite relaxation TH(G)

Lovász (1979) introduced the following parameter $\vartheta(G)$, known as the *theta number*:

$$\begin{aligned} \vartheta(G) := \max \quad & e^T X e \\ \text{s.t.} \quad & \mathrm{Tr}(X) = 1 \\ & X_{ij} = 0 \, (i \neq j, ij \in E) \\ & X \succeq 0. \end{aligned} \tag{58}$$

The theta number has two important properties: it can be computed with an arbitrary precision in polynomial time (as the optimum value of a semidefintie program) and it provides bounds for the stability and chromatic numbers. Namely,

$$\alpha(G) \leq \vartheta(G) \leq \chi(\bar{G}). \tag{59}$$

To see that $\alpha(G) \leq \vartheta(G)$, consider a maximum stable set S; then the matrix $X := \frac{1}{|S|} \chi^S (\chi^S)^T$ is feasible for the program (58) and $\alpha(G) = e^T X e$.

To see that $\vartheta(G) \leq \chi(\bar{G})$, consider a matrix X feasible for (58) and a partition $V = Q_1 \cup \cdots \cup Q_k$ into $k := \chi(\bar{G})$ cliques. Then,

$$0 \leq \sum_{h=1}^{k} (k\chi^{Q_h} - e)^T X (k\chi^{Q_h} - e) = k^2 \operatorname{Tr}(X) - ke^T Xe = k^2 - ke^T Xe,$$

which implies $e^T Xe \leq k$ and thus $\vartheta(G) \leq \chi(\bar{G})$.

Several equivalent definitions are known for $\vartheta(G)$ that we recall below. (See Grötschel, Lovász and Schrijver (1988) or Knuth (1994) for a detailed treatment, and Gruber and Rendl (2003) for an algorithmic comparison.) The dual semidefinite program of (58) reads:

$$\min\left(t | tI + \sum_{ij \in E} \lambda_{ij} E_{ij} - J \succeq 0 \right), \tag{60}$$

where $J := ee^T$ is the all ones matrix and E_{ij} is the elementary matrix with all zero entries except 1 at positions (i,j) and (j,i). As the program (58) has a strictly feasible solution (e.g., $X = \frac{1}{n}I$), there is no duality gap and the optimum value of (60) is equal to the theta number $\vartheta(G)$. Setting $Y := J - \sum_{ij \in E} \lambda_{ij} E_{ij}$, $Z := tI - Y$ and $U := \frac{1}{t-1} Z$ in (60), we obtain the following reformulations for $\vartheta(G)$:

$$\begin{aligned} \vartheta(G) = \min \ & \lambda_{\max}(Y) \\ \text{s.t.} \ & Y_{ij} = 1 \ (i = j \text{ or } ij \in \bar{E}) \\ & Y \ \text{ symmetric matrix,} \end{aligned} \tag{61}$$

$$\begin{aligned} \vartheta(G) = \min \ & t \\ \text{s.t.} \ & Z_{ii} = t - 1 \quad (i \in V) \\ & Z_{ij} = -1 \quad (ij \in \bar{E}) \\ & Z \succeq 0 \\ = \min \ & t \\ \text{s.t.} \ & U_{ii} = 1 \quad (i \in V) \\ & U_{ij} = -\frac{1}{t-1} \quad (ij \in \bar{E}) \\ & U \succeq 0, \ t \geq 2. \end{aligned} \tag{62}$$

The formulation (62) will be used later in Section 6 for the coloring and max k-cut problems. One can also express $\vartheta(G)$ as the optimum value of the linear objective function $e^T x$ maximized over a convex set forming a relaxation of STAB(G). Namely, let $\mathcal{M}_G$ denote the set of positive semidefinite matrices Y indexed by the set $V \cup \{0\}$ satisfying $y_{ii} = y_{0i}$ for $i \in V$ and $y_{ij} = 0$ for $i \neq j \in V$ adjacent in G, and set

$$\operatorname{TH}(G) := \left\{ x \in \mathbb{R}^V | \begin{pmatrix} 1 \\ x \end{pmatrix} = Ye_0 \quad \text{for some } Y \in \mathcal{M}_G \right\}, \tag{63}$$

where $e_0 := (1, 0, \ldots, 0)^T \in \mathbb{R}^{n+1}$. (Same definition as (44).)

Lemma 10. *For any graph G,* $\mathrm{STAB}(G) \subseteq \mathrm{TH}(G) \subseteq \mathrm{QSTAB}(G)$.

Proof. If S is a stable set in G and $x := \chi^S$, then $Y := \binom{1}{x}(1\ x^T) \in \mathcal{M}_G$ and $\binom{1}{x} = Ye_0$; from this follows that $\mathrm{STAB}(G) \subseteq \mathrm{TH}(G)$. Let $x \in \mathrm{TH}(G)$, $Y \in \mathcal{M}_G$ such that $\binom{1}{x} = Ye_0$; and let Q be a clique in G. The principal submatrix Y_Q of Y whose rows and columns are indexed by the set $\{0\} \cup Q$ has the form

$$\begin{pmatrix} 1 & x^T \\ x & \mathrm{diag}(x) \end{pmatrix}.$$

As $Y \succeq 0$, we have $Y_Q \succeq 0$, i.e., $\mathrm{diag}(x) - xx^T \succeq 0$ (taking a Schur complement), which implies that $e^T(\mathrm{diag}(x) - xx^T)e = e^T x(1 - e^T x) \geq 0$ and thus $e^T x = \sum_{i \in Q} x_i \leq 1$. This shows the inclusion $\mathrm{TH}(G) \subseteq \mathrm{QSTAB}(G)$. □

Theorem 11. $\vartheta(G) = \max(e^T x | x \in \mathrm{TH}(G))$.

Proof. We use the formulation of $\vartheta(G)$ from (58). Let μ_G denote the maximum of $e^T x$ over $\mathrm{TH}(G)$. We first show that $\vartheta(G) \leq \mu_G$. For this, let X be an optimum solution to the program (58). Let $v_1, \ldots, v_n \in \mathbb{R}^n$ such that $x_{ij} = v_i^T v_j$ for all $i, j \in V$; thus $\vartheta(G) = \| \sum_{i=1}^n v_i \|^2$, $\sum_{i=1}^n (v_i)^2 = \mathrm{Tr}(X) = 1$, and $v_i^T v_j = 0$ if i, j are adjacent in G. Set $P := \{i \in V | v_i \neq 0\}$, $u_0 := \frac{1}{\sqrt{\vartheta(G)}} \sum_{i=1}^n v_i$, $u_i := \frac{v_i}{\|v_i\|}$ for $i \in P$, and let $u_i (i \in V \setminus P)$ be an orthonormal basis of the orthogonal complement of the space spanned by $\{v_i | i \in P\}$. Let D denote the diagonal matrix indexed by $\{0\} \cup V$ with diagonal entries $u_0^T u_i\ (i = 0, 1, \ldots, n)$, let Z denote the Gram matrix of $u_0, u_1, \ldots, u_n$ and set $Y := DZD$, with entries $y_{ij} = (u_i^T u_j)(u_0^T u_i)(u_0^T u_j)\ (i, j = 0, 1, \ldots, n)$. Then, $Y \in \mathcal{M}_G$ with $y_{00} = 1$. It remains to verify that $\vartheta(G) \leq \sum_{i=1}^n y_{0i}$. By the definition of u_0, we find

$$\begin{aligned} \vartheta(G) &= \left(\sum_{i=1}^n u_0^T v_i \right)^2 = \left(\sum_{i \in P} u_0^T v_i \right)^2 = \left(\sum_{i \in P} u_0^T u_i \|v_i\| \right)^2 \\ &\leq \left(\sum_{i \in P} \|v_i\|^2 \right) \left(\sum_{i \in P} (u_0^T u_i)^2 \right) = \sum_{i=1}^n y_{0i}, \end{aligned}$$

where the inequality follows using the Cauchy–Schwartz inequality. We now show the converse inequality $\mu_G \leq \vartheta(G)$. For this, let $x \in \mathrm{TH}(G)$ be optimum for the program defining μ_G, let $Y \in \mathcal{M}_G$ such that $\binom{1}{x} = Ye_0$, and $v_0, v_1, \ldots, v_n \in \mathbb{R}^{n+1}$ such that $y_{ij} = v_i^T v_j$ for all $i, j = 0, 1, \ldots, n$. It suffices to construct X feasible for (58) satisfying $\sum_{i,j=1}^n x_{ij} \geq \mu_G$. Define the $n \times n$ matrix X with entries $x_{ij} := \frac{1}{\mu_G} v_i^T v_j\ (i, j = 1, \ldots, n)$; then X is feasible for (58). Moreover, $\mu_G = \sum_{i=1}^n y_{0i} = \sum_{i=1}^n v_0^T v_i = v_0^T (\sum_{i=1}^n v_i)$ is less than or equal to $\| \sum_{i=1}^n v_i \|$ (by the Cauchy–Schwartz inequality, since $\|v_0\| = 1$).

As $\sum_{i,j=1}^n x_{ij} = \frac{1}{\mu_G} (\sum_{i=1}^n v_i)^2$, we find that $\mu_G \leq \sum_{i,j=1}^n x_{ij}$. □

An *orthonormal representation* of G *is a set of unit vectors* $u_1, \dots, u_n \in \mathbb{R}^N$ $(N \geq 1)$ *satisfying* $u_i^T u_j = 0$ *for all* $ij \in \bar{E}$.

Theorem 12. $\vartheta(G) = \max_{d,v_i} \sum_{i \in V} (d^T v_i)^2$, *where the maximum is taken over all unit vectors* $d \in \mathbb{R}^N$ *and all orthonormal representations* $v_1, \dots, v_n \in \mathbb{R}^N$ of $\bar{G}$.

Proof. Let $\vartheta(G) = e^T X e$, where X is an optimum solution to the program (58) and let $b_1, \dots, b_n$ be vectors such that $X_{ij} = b_i^T b_j$ for $i, j \in V$. Set $d := (\sum_{i \in V} b_i)/\| \sum_{i \in V} b_i \|$, $P := \{i \in V | b_i \neq 0\}$ and $v_i := \frac{b_i}{\|b_i\|}$ for $i \in P$. Let v_i $(i \in V \backslash P)$ be an orthonormal basis of the orthogonal complement of the space spanned by v_i $(i \in P)$. Then, $v_1, \dots, v_n$ is an orthonormal representation of $\bar{G}$. We have:

$$\sqrt{\vartheta(G)} = \left\| \sum_{i \in P} b_i \right\| = d^T \left(\sum_{i \in P} b_i \right) = \sum_{i \in P} \|b_i\| v_i^T d$$
$$\leq \sqrt{\sum_{i \in P} \|b_i\|^2} \cdot \sqrt{\sum_{i \in P} (v_i^T d)^2} \leq \sqrt{\sum_{i \in V} (v_i^T d)^2}$$

(using the Cauchy–Schwartz inequality and $\mathrm{Tr}(X) = 1$). This implies that $\vartheta(G) \leq \sum_{i \in V} (d^T v_i)^2$.

Conversely, let d be a unit vector and let $v_1, \dots, v_n$ be an orthonormal representation of $\bar{G}$. Let Y denote the Gram matrix of the vectors d, $(d^T v_1) v_1, \dots, (d^T v_n) v_n$. Then, $Y \in \mathcal{M}_G$. Therefore, $((d^T v_1)^2, \dots, (d^T v_n)^2)^T \in \mathrm{TH}(G)$ which implies that $\sum_{i \in V} (d^T v_i)^2 \leq \vartheta(G)$. □

Let A_G denote the convex hull of all vectors $((d^T v_1)^2, \dots, (d^T v_n)^2)^T$ where d is a unit vector and $v_1, \dots, v_n$ is an orthonormal representation of $\bar{G}$, let B_G denote the set of $x \in \mathbb{R}_+^V$ satisfying the *orthonormal representation constraints*:

$$\sum_{i \in V} (c^T u_i)^2 x_i \leq 1 \tag{64}$$

for all unit vectors c and all orthonormal representations $u_1, \dots, u_n$ of G, and let C_G denote the set of $x \in \mathbb{R}_+^V$ satisfying

$$\sum_{i \in V} x_i \leq \min_{c, u_i} \max_{i \in V} \frac{1}{(c^T u_i)^2}$$

where the minimum is taken over all unit vectors c and all orthonormal representations $u_1, \dots, u_n$ of G.

Lemma 13. $A_G \subseteq \mathrm{TH}(G) \subseteq B_G \subseteq C_G$.

Proof. The inclusion $A_G \subseteq \mathrm{TH}(G)$ follows from the second part of the proof of Theorem 12 and the inclusion $B_G \subseteq C_G$ is easy to verify. Let $x \in \mathrm{TH}(G)$ and let $z := ((c^T u_1)^2, \dots, (c^T u_n)^2)^T$ where c is a unit vector and $u_1, \dots, u_n$ is an orthonormal representation of G; we show that $x^T z \le 1$. By the above, $z \in A_{\bar{G}} \subseteq \mathrm{TH}(\bar{G})$. Let $Y \in \mathcal{M}_G$ and $Z \in \mathcal{M}_{\bar{G}}$ such that $\binom{1}{x} = Ye_0$ and $\binom{1}{z} = Ze_0$. Denote by Y' the matrix obtained from Y by changing the signs on its first row and column. Then, $\langle Y', Z\rangle = 1 - 2\sum_{i \in V} y_{0i} z_{0i} + \sum_{i \in V} y_{ii} z_{ii} = 1 - \sum_{i \in V} x_i z_i \ge 0$ (since $Y', Z \succeq 0$) and thus $x^T z \le 1$. This shows the inclusion $\mathrm{TH}(G) \subseteq B_G$. □

Theorem 14. $\vartheta(G) = \min_{c,u_i} \max_{i \in V} \frac{1}{(c^T u_i)^2}$, *where the minimum is taken over all unit vectors* c *and all orthonormal representations* $u_1, \ldots, u_n$ *of* G.

Proof. The inequality $\vartheta(G) \le \min \ldots$ follows from the inclusion $\mathrm{TH}(G) \subseteq C_G$ and Theorem 11. For the reverse inequality, we use the definition of $\vartheta(G)$ from (61). Let Y be a symmetric matrix with $Y_{ii} = 1$ $(i \in V)$ and $Y_{ij} = 1$ $(ij \in \bar{E})$ and $\vartheta(G) = \lambda_{\max}(Y)$. As $\vartheta(G)I - Y \succeq 0$, there exist vectors $b_1, \ldots, b_n$ such that $b_i^2 = \vartheta(G) - 1$ $(i \in V)$ and $b_i^T b_j = -1$ $(ij \in \bar{E})$. Let c be a unit vector orthogonal to all b_i (which exists since $\vartheta(G)I - Y$ is singular) and set $u_i := (c + b_i)/\sqrt{\vartheta(G)}$ $(i \in V)$. Then, $u_1, \ldots, u_n$ is an orthonormal representation of G and $\vartheta(G) = \frac{1}{(c^T u_i)^2}$ for all i. □

Theorems 12 and 14 and Lemma 13 show that one obtains the same optimum value when optimizing the linear objective function $e^T x$ over $\mathrm{TH}(G)$ or over any of the sets A_G, B_G, or C_G. In fact, the same remains true for an arbitrary linear objective function $w^T x$ where $w \in \mathbb{R}_+^V$, as the above extends easily to the weighted case. Therefore,

$$\mathrm{TH}(G) = A_G = B_G = C_G$$

Moreover, $\mathrm{TH}(\bar{G})$ is the *antiblocker* of $\mathrm{TH}(G)$; that is, $\mathrm{TH}(\bar{G}) = \{z \in \mathbb{R}_+^V \mid x^T z \le 1 \ \forall x \in \mathrm{TH}(G)\}$. One can show that the only orthonormal representation inequalities (64) defining facets of $\mathrm{TH}(G)$ are the clique inequalities. From this follows:

$$\mathrm{TH}(G) \text{ is a polytope} \Longleftrightarrow G \text{ is perfect} \Longleftrightarrow \mathrm{TH}(G) = \mathrm{QSTAB}(G)$$
$$\Longleftrightarrow \mathrm{TH}(G) = \mathrm{STAB}(G).$$

We refer to Chapter 12 in Reed and Ramirez (2001) for a detailed exposition on the theta body $\mathrm{TH}(G)$.

4.3 Coloring and finding maximum stable sets in perfect graphs

The stability number $\alpha(G)$ and the chromatic number $\chi(G)$ of a perfect graph G can be computed in polynomial time. (Indeed, it suffices to compute an approximated value of $\vartheta(G)$ with precision $< 1/2$ in order to determine $\alpha(G) = \chi(\bar{G}) = \vartheta(G)$.) We now mention how to find in polynomial time a stable set of size $\alpha(G)$ and a $\chi(G)$-coloring in a perfect graph. The weighted versions of these problems can also be solved in polynomial time (cf. Grötschel, Lovász and Schrijver (1988) for details).

Finding a maximum cardinality stable set in a perfect graph. Let $G=(V,E)$ be a perfect graph and let $v_1,\ldots,v_n$ be an ordering of its nodes. We construct a sequence of graphs $G_0:=G\supseteq G_1\supset\cdots\supseteq G_i\supseteq G_{i+1}\supseteq\cdots\supseteq G_n$ in the following manner: For each $i\geq 1$, compute $\alpha(G_{i-1}\backslash v_i)$; if $\alpha(G_{i-1}\backslash v_i)=\alpha(G)$, then set $G_i:=G_{i-1}\backslash v_i$, otherwise set $G_i:=G_{i-1}$. Then, $\alpha(G_i)=\alpha(G)$ for all i and G_n is a stable set, thus providing a maximum stable set in G. Therefore, a maximum stable set in a perfect graph G can be found by applying n times an algorithm for computing the theta function.

Finding a minimum coloring in a perfect graph. We follow the presentation of Schrijver (2003). Let $G=(V,E)$ be a perfect graph. A crucial observation is that it suffices to find a stable set S which intersects all the maximum cardinality cliques of G. Indeed, if such S is found, then one can recursively color $G\backslash S$ with $\omega(G\backslash S)=\omega(S)-1$ colors and thus G with $\omega(G)=\chi(G)$ colors. For $t\geq 1$, we grow iteratively a list $Q_1,\ldots,Q_t$ of maximum cardinality cliques. Suppose $Q_1,\ldots,Q_t$ have been found. We begin with finding a stable set S meeting each of $Q_1,\ldots,Q_t$. For this, setting $w:=\sum_{i=1}^{t}\chi^{Q_i}$, it suffices to find a maximum weight stable set S. (This can be done by applying the above maximum cardinality stable set algorithm to the graph G' obtained from G by replacing every node i by a set W_i of w_i nonadjacent nodes, making two nodes $u\in W_i$, $v\in W_j$ adjacent in G' if the nodes i,j are adjacent in G.) Then S has weight t which means that S meets each of $Q_1,\ldots,Q_t$. Now, if $\omega(G\backslash S)<\omega(G)$, then S meets *all* the maximum cardinality cliques in G and we are done. Otherwise, we find a clique Q_{t+1} in $G\backslash S$ of size $\omega(G)$ and add it to our list.

The algorithm has a polynomial running time since the number of iterations is bounded by $|V|$. To see it, consider the affine space $L_t:=\{x\in\mathbb{R}^V\,|\,x(Q_i)=1\ \forall i=1,\ldots,t\}$. Then, $L_1\supseteq L_2\supseteq\cdots\supseteq L_t\supseteq L_{t+1}\supseteq\cdots$. The dimension of the spaces L_t decreases at each step since $\chi^S\in L_t\backslash L_{t+1}$, where S is the stable set constructed at the t-th iteration as above.

4.4 *Sharpening the theta function*

The number $\vartheta'(G)$. McEliece, Rodemich, and Rumsey (1978) and Schrijver (1979) introduce the parameter $\vartheta'(G)$ as

$$\begin{aligned}\vartheta'(G):=\max\quad & e^TXe\\ \text{s.t.}\quad & \mathrm{Tr}(X)=1\\ & X_{ij}=0\ (i\neq j,\ ij\in E)\\ & X\succeq 0,\ X\geq 0.\end{aligned}\tag{65}$$

Comparing with (58), it follows that

$$\alpha(G) \le \vartheta'(G) \le \vartheta(G).$$

As was done for $\vartheta(G)$ one can prove the following equivalent formulations for $\vartheta'(G)$:

$$\begin{aligned} \vartheta'(G) = \min \quad & \lambda_{\max}(Y) \\ \text{s.t.} \quad & Y_{ij} \ge 1 \ (i = j \text{ or } ij \in \bar{E}) \\ & Y \text{ symmetric matrix;} \end{aligned} \tag{66}$$

$$\begin{aligned} \vartheta'(G) = \min \quad & t \\ \text{s.t.} \quad & Z_{ii} = t-1 \quad (i \in V) \\ & Z_{ij} \le -1 \quad (ij \in \bar{E}) \\ & Z \succeq 0 \\ = \min \quad & t \\ \text{s.t.} \quad & U_{ii} = 1 \quad (i \in V) \\ & U_{ij} \le -\frac{1}{t-1} \quad (ij \in \bar{E}) \\ & U \succeq 0, \ t \ge 2; \end{aligned} \tag{67}$$

and $\vartheta'(G) = \max(e^T x | \binom{1}{x} = Ye_0$ for some nonnegative matrix $Y \in \mathcal{M}_G)$. The inequality $\vartheta'(G) \le \vartheta(G)$ is strict, for instance, for the graph with node set $\{0,1\}^6$ where two nodes are adjacent if their Hamming distance (i.e., the number of positions where their coordinates are distinct) is at most 3 (then, $\vartheta(G) = \frac{16}{3}$ and $\vartheta'(G) = \alpha(G) = 4$).

The number $\vartheta^+(G)$. In a similar vein, Szegedy (1994) introduced the following parameter $\vartheta^+(G)$ which provides a sharper lower bound for the chromatic number of $\bar{G}$:

$$\begin{aligned} \vartheta^+(G) := \max \quad & e^T X e \\ \text{s.t.} \quad & \text{Tr}(X) = 1 \\ & X_{ij} \le 0 \ (i \ne j, ij \in E) \\ & X \succeq 0. \end{aligned} \tag{68}$$

We have $\vartheta(G) \le \vartheta^+(G) \le \chi(\bar{G})$. The first inequality is obvious and the second one can be proved in the same way as the inequality $\vartheta(G) \le \chi(\bar{G})$ in Section 4.2. Therefore, the following chain of inequalities holds:

$$\alpha(G) \le \vartheta'(G) \le \vartheta(G) \le \vartheta^+(G) \le \chi(\bar{G}). \tag{69}$$

The parameters of $\vartheta'(G)$, $\vartheta(G)$, and $\vartheta^+(G)$ are known, respectively, as the *vector chromatic number*, the *strict vector chromatic number*, and the *strong vector chromatic number* of $\bar{G}$; see Section 6.4. As was done for $\vartheta(G)$, one can prove the following equivalent formulations for $\vartheta^+(G)$:

$$\begin{aligned} \vartheta^+(G) = \min \quad & \lambda_{\max}(Y) \\ \text{s.t.} \quad & Y_{ij} = 1 \ (i = j \text{ or } ij \in \bar{E}) \\ & Y_{ij} \leq 1 \ (ij \in E) \\ & Y \text{ symmetric matrix;} \end{aligned} \tag{70}$$

$$\begin{aligned} \vartheta^+(G) = \min \quad & t \\ \text{s.t.} \quad & Z_{ii} = t - 1 \quad (i \in V) \\ & Z_{ij} = -1 \quad (ij \in \bar{E}) \\ & Z_{ij} \geq -1 \quad (ij \in E) \\ & Z \succeq 0 \\ = \min \quad & t \\ \text{s.t.} \quad & U_{ii} = 1 \quad (i \in V) \\ & U_{ij} = -\frac{1}{t-1} \quad (ij \in \bar{E}) \\ & U_{ij} \geq -\frac{1}{t-1} \quad (ij \in E) \\ & U \succeq 0, \ t \geq 2. \end{aligned} \tag{71}$$

The parameter $\vartheta^+(G)$ (in the formulation (71)) was introduced independently by Meurdesoif (2000) who gives a graph G for which inequality $\vartheta(G) \leq \vartheta^+(G)$ is strict. See Szegedy (1994) for more about this parameter.

Bounding the Shannon capacity. The theta number $\vartheta(G)$ was introduced by Lovász (1979) in connection with a problem of Shannon in coding theory. The strong product $G \cdot H$ of two graphs G and H has node set $V(G) \times V(H)$ with two distinct nodes (u, v) and (u', v') being adjacent if u, u' are equal or adjacent in G and v, v' are equal or adjacent in H. Then G^k is the strong product of k copies of G. The Shannon capacity of G is defined by

$$\Theta(G) := \sup_{k \geq 1} \sqrt[k]{\alpha(G^k)}.$$

As $\alpha(G^k) \geq (\alpha(G))^k$ and $\vartheta(G^k) \leq (\vartheta(G))^k$, one finds

$$\alpha(G) \leq \Theta(G) \leq \vartheta(G).$$

Using these inequalities, Lovász (1979) could show that the Shannon capacity of C_5 is $\sqrt{5}$ (as $\alpha(C_5^2) = 5$ and $\vartheta(C_5) = \sqrt{5}$). For $n \geq 7$ odd,

$$\vartheta(C_n) = \frac{n \cos\left(\frac{\pi}{n}\right)}{1 + \cos\left(\frac{\pi}{n}\right)},$$

but the value of $\Theta(C_n)$ is not known.

The theta number versus Delsarte's bound. Let G be a graph whose adjacency matrix can be written as $\sum_{i \in M} A_i$, where $M \subseteq \{1, \ldots, N\}$ and $A_0, A_1, \ldots, A_N$ are 0/1 symmetric matrices forming an *association scheme*; that is, $A_0 = I$, $\sum_{i=0}^N A_i = J$, there exist scalars $p_{ij}^k \, (i, j, k = 1, \ldots, N)$ such that $A_i A_j = A_j A_i = \sum_{k=0}^N p_{ij}^k A_k$. As the matrices $A_0, \ldots, A_N$ commute, they have a common basis of eigenvectors and therefore positive semidefiniteness of a matrix $X := \sum_{i=0}^N x_i A_i$ can be expressed by a linear system of inequalities in $x_1, \ldots, x_N$. Therefore, one finds that the theta numbers $\vartheta(G)$, $\vartheta'(G)$ can be computed by solving a linear programming problem. Based on this, Schrijver (1979) shows that $\vartheta'(G)$ coincides with a linear programming bound introduced earlier by Delsarte (1973).

These ideas have been extended to general semidefinite programs by Goemans and Rendl (1999).

5 Semidefinite relaxation for the max-cut problem

We present here results dealing with the basic semidefinite relaxation of the cut polytope and its application to designing good approximation algorithms for the max-cut problem.

Given a graph $G = (V, E)$, the *cut* $\delta(S)$ induced by a vertex set $S \subseteq V$ is the set of edges with exactly one endpoint in S. Given edge weights $w \in \mathbb{Q}^E$, the *max-cut problem* consists of finding a cut $\delta(S)$ whose weight $w(\delta(S)) := \sum_{ij \in \delta(S)} w_{ij}$ is maximum. Let $\mathrm{mc}(G, w)$ denote the maximum weight of a cut in G. A comprehensive survey about the max-cut problem can be found in Poljak and Tuza (1995). The max-cut problem is one of the basic NP-hard problems studied by Karp (1972). Moreover, it cannot be approximated with an arbitrary precision; namely, Håstad (1997) shows that for $\rho > \frac{16}{17} = 0.94117$ there is no ρ-approximation algorithm for max-cut if $P \neq \mathrm{NP}$. [A ρ-approximation algorithm is an algorithm that returns in polynomial time a cut whose weight is at least ρ times the maximum weight of a cut; ρ being called the *performance ratio* or *guarantee*.] On the other hand,

Goemans and Williamson (1995) prove a 0.878-approximation algorithm for max-cut that will be presented in Section 5.3 below.

5.1 The basic linear relaxation

As before, the *cut polytope* CUT(G) is the polytope in $\mathbb{R}^E$ defined as the convex hull of the vectors $z^S \in \{\pm 1\}^E$ for $S \subseteq V$, where $z^S_{ij} = -1$ if and only if $|S \cap \{i,j\}| = 1$. The weight of the cut $\delta(S)$ can be expressed as $\frac{1}{2}\sum_{ij\in E} w_{ij}(1 - z^S_{ij})$. Hence the max-cut problem is the problem of optimizing the linear objective function

$$\frac{1}{2}\sum_{ij\in E} w_{ij}(1 - z_{ij}) \tag{72}$$

over CUT(G). The *circuit inequalities*:

$$\sum_{ij\in F} x_{ij} - \sum_{ij\in E(C)\setminus F} x_{ij} \geq 2 - |C|, \tag{73}$$

where C is a circuit in G and F is a subset of $E(C)$ with an odd cardinality, are valid for CUT(G) as they express the fact that a cut and a circuit must have an even intersection. Together with the bounds $-1 \leq x_{ij} \leq 1 (ij \in E)$ they define the *metric polytope* MET(G). Thus CUT(G) $\subseteq$ MET(G); moreover, the only ± 1 vectors in MET(G) are the cut vectors $z^S (S \subseteq V)$. An inequality (73) defines a facet of CUT(G) if and only if C is a chordless circuit in G while an inequality $\pm x_{ij} \leq 1$ is facet defining if and only if ij does not belong to a triangle (Barahona and Mahjoub (1986)). Hence the metric polytope MET(K_n) is defined by the $4\binom{n}{3}$ *triangle inequalities*:

$$x_{ij} + x_{ik} + x_{jk} \geq -1, \quad x_{ij} - x_{ik} - x_{jk} \geq -1 \tag{74}$$

for all triples $i, j, k \in \{1, \ldots, n\}$. Therefore, one can optimize any linear objective function over MET(K_n) in polynomial time. The same holds for MET(G), since MET(G) is equal to the projection of MET(K_n) on the subspace $\mathbb{R}^E$ indexed by the edge set of G (Barahona (1993)). The inclusion CUT(G) $\subseteq$ MET(G) holds at equality if and only if G has no K_5-minor (Barahona and Mahjoub (1986)). Therefore, the max-cut problem can be solved in polynomial time for the graphs with no K_5-minor (including the planar graphs).

The polytope

$$Q(G) := \left\{ x \in [-1,1]^E | \sum_{ij \in E(C)} x_{ij} \geq 2 - |C| \quad \text{for all odd circuits } C \text{ in } G \right\}$$

contains the metric polytope MET(G) and its ± 1-vectors correspond to the bipartite subgraphs of G. Therefore, the max-cut problem for nonnegative weights can be reformulated as the problem of maximizing (72) over the ± 1-vectors in $Q(G)$. A graph G is said to be *weakly bipartite* when all the vertices of $Q(G)$ are ± 1-valued. It is shown in Grötschel and Pulleyblank (1981) that one can optimize in polynomial time a linear objective function over $Q(G)$. Therefore, the max-cut problem can be solved in polynomial time for weakly bipartite graphs with nonnegative edge weights. Guenin (2001) characterized the weakly bipartite graphs as those graphs *containing no odd* K_5-minor (they include the graphs with no K_5-minor, the graphs having two nodes covering all odd circuits, etc.), settling a conjecture posed by Seymour (1977). (See Schrijver (2002) for a shorter proof.) Poljak (1991) shows that, for nonnegative edge weights, one obtains in fact the same optimum value when optimizing (72) over MET(G) or over $Q(G)$.

Let met(G, w) denote the optimum value of (72) maximized over $x \in$ MET(G). When all edge weights are equal to 1, we also use the notation met(G) in place of met(G, w) (and analogously mc(G) in place of mc(G, w)). How well does the polyhedral bound met(G, w) approximate the max-cut value mc(G, w)? In order to compare the two bounds, we assume that all edge weights are nonnegative. Then,

$$\text{met}(G, w) \leq w(E) = \sum_{ij \in E} w_{ij} \quad \text{and} \quad \text{mc}(G, w) \geq \frac{1}{2} w(E).$$

(To see the latter inequality, consider an optimum cut $\delta(S)$ and the associated partition $(S, V \backslash S)$. Then, for every node $i \in V$, the sum of the weights of the edges connecting i to the opposite class of the partition is greater than or equal to the sum of the weights of the edges connecting i to nodes in the same class, since otherwise moving i to the other class would produce a heavier cut.) Therefore,

$$\frac{\text{mc}(G, w)}{\text{met}(G, w)} \geq \frac{1}{2}.$$

In fact, the ratio $\frac{mc(G,w)}{met(G,w)}$ tends to $\frac{1}{2}$ for certain classes of graphs (cf. Poljak (1991), Poljak and Tuza (1994)) which shows that in the worst case the metric polytope does not provide a better approximation than the trivial relaxation of CUT(G) by the cube $[-1, 1]^E$.

5.2 The basic semidefinite relaxation

The max-cut problem can be reformulated as the following integer quadratic program:

$$\begin{aligned} \mathrm{mc}(G, w) = \max \quad & \frac{1}{2}\sum_{ij\in E} w_{ij}(1 - x_i x_j) \\ \text{s.t.} \quad & x_1, \ldots, x_n \in \{\pm 1\}. \end{aligned} \tag{75}$$

For $x \in \{\pm 1\}^n$, the matrix $X := xx^T$ is positive semidefinite with all diagonal elements equal to one. Thus relaxing the rank one condition on X, we obtain the following *semidefinite relaxation* for max-cut:

$$\begin{aligned} \mathrm{sdp}(G, w) := \max \quad & \frac{1}{2}\sum_{ij\in E} w_{ij}(1 - x_{ij}) \\ \text{s.t.} \quad & x_{ii} = 1 \ \forall i \in \{1, \ldots, n\} \\ & X = (x_{ij}) \succeq 0. \end{aligned} \tag{76}$$

The set

$$\mathcal{E}_n := \{X = (x_{ij})_{i,j=1}^n \mid X \succeq 0 \quad \text{and} \quad x_{ii} = 1 \ \forall i \in \{1, \ldots, n\}\} \tag{77}$$

is the basic semidefinite relaxation of the cut polytope $\mathrm{CUT}(K_n)$. More precisely,

$$x \in \mathrm{CUT}(K_n) \Rightarrow \mathrm{mat}(x) \in \mathcal{E}_n \tag{78}$$

where $\mathrm{mat}(x)$ is the $n \times n$ symmetric matrix with ones on its main diagonal and x_{ij} as off-diagonal entries.

The quantity $\mathrm{sdp}(G, w)$ can be computed in polynomial time (with an arbitrary precision). The objective function in (76) is equal to $\frac{1}{4}\langle L_w, X\rangle$, where $L_w = (l_{ij})$ is the *Laplacian* matrix defined by $l_{ii} := w(\delta(i))$ and $l_{ij} := -w_{ij}$ for $i \neq j$ (assigning weight 0 to non edges). Hence, the dual of the semidefinite program (76) is

$$\frac{1}{4}\min\left\{\sum_{i=1}^n y_i \mid \mathrm{diag}(y) - L_w \succeq 0\right\} \tag{79}$$

and there is no duality gap (since I is a strictly feasible solution to (76)). Set $s = \frac{1}{n}y^T e$ and $u = se - y$; then $u^T e = 0$ and $\mathrm{diag}(y) - L_w = sI - \mathrm{diag}(u) - L_w \succeq 0$ if and only if $\lambda_{\max}(L_w + \mathrm{diag}(u)) \leq s$. Therefore, (79) can be rewritten as the following eigenvalue optimization problem:

$$\frac{n}{4}\min\left\{\lambda_{\max}(L_w + \mathrm{diag}(u)) \mid \sum_{i=1}^n u_i = 0\right\}; \tag{80}$$

this eigenvalue upper bound for max-cut had been introduced and studied earlier by Delorme and Poljak (1993a,b). One can also verify directly that (80) is an upper bound for max-cut. Indeed, for $x \in \{\pm 1\}^n$ and $u \in \mathbb{R}^n$ with $\sum_i u_i = 0$, one has:

$$w(\delta(S)) = \frac{1}{4} x^T L_w x = \frac{1}{4} x^T (L_w + \mathrm{diag}(u)) x = \frac{n}{4} \frac{x^T (L_w + \mathrm{diag}(u)) x}{x^T x}$$

which is less than or equal to $\frac{n}{4} \lambda_{\max}(L_w + \mathrm{diag}(u))$ by the Rayleigh principle. The program (80) can be shown to have a unique minimizer u (when $w \neq 0$); this minimizer u is equal to the null vector, for instance, when G is vertex transitive, in which case the computation of the semidefinite bound amounts to an eigenvalue computation (Delorme and Poljak (1993a)). Based on this, one can compute the semidefinite bound for unweighted circuits. Namely, $\mathrm{mc}(C_{2k}) = \mathrm{sdp}(C_{2k}) = 2k$ and $\mathrm{mc}(C_{2k+1}) = 2k$ while $\mathrm{sdp}(C_{2k+1}) = \frac{2k+1}{4}(2 + 2\cos(\frac{\pi}{2k+1}))$. Hence,

$$\frac{\mathrm{mc}(C_5)}{\mathrm{sdp}(C_5)} = \frac{32}{25 + 5\sqrt{5}} \sim 0.88445;$$

the same ratio is obtained for some other circulant graphs (Mohar and Poljak (1990)).

Much research has been done for evaluating the *integrality ratio* $\frac{\mathrm{mc}(G,w)}{\mathrm{sdp}(G,w)}$ and for comparing the polyhedral and semidefinite bounds. Poljak (1991) proved the following inequality relating the two bounds:

$$\frac{\mathrm{met}(G,w)}{\mathrm{sdp}(G,w)} \geq \frac{32}{25 + 5\sqrt{5}} \quad \text{for any graph } G \text{ and } w \geq 0. \tag{81}$$

Therefore, the inequality

$$\frac{\mathrm{mc}(G,w)}{\mathrm{sdp}(G,w)} \geq \frac{32}{25 + 5\sqrt{5}} \tag{82}$$

holds for any weakly bipartite graph (G, w) with $w \geq 0$. The bound (82) remains valid for unweighted line graphs and the better bound $\frac{8}{9}$ was proved for the complete graph K_n with edge weights $w_{ij} := b_i b_j$ (given $b_1, \ldots, b_n \in \mathbb{R}_+$) or for Paley graphs (Delorme and Poljak (1993a)). Moreover, the integrality ratio is asymptotically equal to 1 for the random graphs $G_{n,p}$ (p denoting the edge probability) (Delorme and Poljak (1993a)).

Goemans and Williamson (1995) proved the following bound for the integrality ratio:

$$\frac{\mathrm{mc}(G,w)}{\mathrm{sdp}(G,w)} \geq \alpha_0 \quad \text{for any graph } G \text{ and } w \geq 0, \tag{83}$$

where $0.87856 < \alpha_0 < 0.87857$ and α_0 is defined by

$$\alpha_0 := \min_{0<\theta\leq\pi} \frac{2}{\pi}\frac{\theta}{1-\cos\theta}. \tag{84}$$

Moreover, they present a randomized algorithm producing a cut whose expected weight is at least $\alpha_0 \cdot \mathrm{sdp}(G, w)$; their result will be described in the next subsection.

Until recently, no example was known of a graph having a worst integrality ratio than C_5 and it had been conjectured by Delorme and Poljak (1993a) that $\frac{32}{25+5\sqrt{5}}$ is the worst possible value for the integrality ratio. Feige and Schechtman (2001, 2002) disproved this conjecture and proved that the worst case value for the integrality ratio $\frac{mc(G,w)}{sdp(G,w)}$ is *equal* to the Goemans–Williamson quantity α_0; we will come back to this result later in this section.

5.3 The Goemans–Williamson randomized approximation algorithm for max-cut

The randomized approximation algorithm of Goemans and Williamson (1995) for max-cut goes as follows; its analysis will need the assumption that the edge weights are nonnegative.

(1) *The semidefinite optimization phase*: Solve the semidefinite program (76). Let $X = (x_{ij})$ be an optimum solution and let $v_1, \ldots, v_n \in \mathbb{R}^d$ (for some $d \leq n$) such that $x_{ij} = v_i^T v_j$ for all $i, j \in \{1, \ldots, n\}$.

(2) *The random hyperplane rounding phase*: Generate a random unit vector r and set $S := \{i \mid v_i^T r \geq 0\}$. Then, $\delta(S)$ is the randomized cut returned by the algorithm.

The hyperplane H_r with normal r cuts the space into two half-spaces and an edge ij belongs to the cut $\delta(S)$ if and only if the vectors v_i and v_j do not belong to the same half-space. Hence the probability that an edge ij belongs to $\delta(S)$ is equal to $\frac{\arccos(v_i^T v_j)}{\pi}$ and the expected weight $E(w(S))$ of the cut $\delta(S)$ is equal to

$$\begin{aligned} E(w(S)) &= \sum_{ij\in E} w_{ij}\frac{\arccos(v_i^T v_j)}{\pi} \\ &= \sum_{ij\in E} w_{ij}\frac{1-v_i^T v_j}{2}\cdot\frac{2}{\pi}\frac{\arccos(v_i^T v_j)}{1-v_i^T v_j} \geq \alpha_0 \cdot \mathrm{sdp}(G, w). \end{aligned}$$

The last inequality holds if we assume that $w \geq 0$. As $E(w(S)) \leq \mathrm{mc}(G, w)$, we find

$$\frac{\mathrm{mc}(G, w)}{\mathrm{sdp}(G, w)} \geq \frac{E(w(S))}{\mathrm{sdp}(G, w)} \geq \alpha_0 > 0.87856. \tag{85}$$

As a biproduct of the analysis, we obtain the following trigonometric reformulation for max-cut with $w \geq 0$:

$$\begin{aligned} \mathrm{mc}(G, w) = \ & \max \quad \textstyle\sum_{ij \in E} w_{ij} \frac{\arccos(v_i^T v_j)}{\pi} \\ & \text{s.t.} \quad v_1, \ldots, v_n \text{ unit vectors in } \mathbb{R}^n. \end{aligned} \tag{86}$$

Mahajan and Ramesh (1995) have shown that the above randomized algorithm can be derandomized, therefore giving a deterministic α_0-approximation algorithm for max-cut. Let us stress that until then the best known approximation algorithm was the simple random partition algorithm (which assigns a node to either side of the partition independently with probability $\frac{1}{2}$) with a performance ratio of $\frac{1}{2}$.

As mentioned above, the integrality ratio $\frac{mc(G, w)}{sdp(G, w)}$ is equal to α_0 in the worst case. More precisely, Feige and Schechtman (2001, 2002) show that for every $\epsilon > 0$ there exists a graph G (unweighted) for which the ratio is at most $\alpha_0 + \epsilon$. The basic idea of their construction is as follows. Let θ_0 denote the angle where the minimum in the definiton of

$$\alpha_0 = \min_{0 < \theta \leq \pi} \frac{2}{\pi} \frac{\theta}{1 - \cos\theta}$$

is attained; $\theta_0 \sim 2.331122$ is the nonzero root of $\cos\theta + \theta \sin\theta = 1$. Let $[\theta_1, \theta_2]$ be the largest interval containing θ_0 satisfying

$$\theta \in [\theta_1, \theta_2] \Rightarrow \frac{2}{\pi} \frac{\theta}{1 - \cos\theta} \leq \alpha_0 + \epsilon.$$

Distribute n point $v_1, \ldots, v_n$ uniformly on the unit sphere S^{d-1} in $\mathbb{R}^d$ and let G be the graph on n nodes where there is an edge ij if and only if the angle between v_i and v_j belongs to $[\theta_1, \theta_2]$. Applying the random hyperplane rounding phase to the vectors $v_1, \ldots, v_n$, the above analysis shows that the expected weight of the returned cut satisfies

$$\frac{E(w(S))}{\mathrm{sdp}(G)} \leq \alpha_0 + \epsilon.$$

The crucial part of the proof consists then of showing that for some suitable choice of the dimension d and of the distribution of the n points on the sphere S^{d-1} the expected weight $E(w(S))$ is not far from the max-cut value mc(G).

Nesterov (1997) shows the weaker bound:

$$\frac{E(w(S))}{\mathrm{sdp}(G,w)} \geq \frac{2}{\pi} \sim 0.63661 \tag{87}$$

for the larger class of weight functions w satisfying $L_w \succeq 0$. (Note indeed that $L_w \succeq 0$ if $w \geq 0$.) Hence, the GW rounding technique applies to a larger class of instances at the cost of obtaining a weaker performance ratio. Cf. Section 6.1 for more details.

The above analysis of the GW algorithm shows that its performance guarantee is at least α_0. Karloff (1999) shows that it is, in fact, equal to α_0. For this, he constructs a class of graphs G (edge weights are equal to 1) for which the ratio $\frac{E(w(S))}{\mathrm{sdp}(G,w)}$ can be made arbitrarily close to α_0. (The graphs constructed by Feige and Schechtman (2002) display the same behavior; the construction of Karloff has however a simpler proof.) These graphs are the Johnson graphs $J(m, \frac{m}{2}, b)$ for m even, $b \leq \frac{m}{12}$ having the collection of subsets of $\{1, \ldots, m\}$ of cardinality $\frac{m}{2}$ as node set and two nodes being adjacent if their intersection has cardinality b. An additional feature of these graphs is that $\mathrm{mc}(G,w) = \mathrm{sdp}(G,w)$. Hence, one of the problems that the Karloff's example emphasizes is that although the semidefinite program already solves the max-cut problem at optimality, the GW approximation algorithm is not able to recognize this fact and to take advantage of it for producing a better cut. As a matter of fact, recognizing whether $\mathrm{sdp}(G,w) = \mathrm{mc}(G,w)$ for given weights w is an NP-complete problem (Delorme and Poljak (1993b), Laurent and Poljak (1995)).

Goemans and Williamson (1995) show that their algorithm behaves, in fact, better for graphs having $\frac{\mathrm{sdp}(G,w)}{w(E)} \geq \frac{85}{100}$ (and thus for graphs having very large cuts). To express their result, set $h(t) := \frac{1}{\pi}\arccos(1-2t)$, $t_0 := \frac{1-\cos\theta_0}{2} \sim 0.84458$, where $\theta_0 \sim 2.331122$ is the angle at which the minimum in the definition of $\alpha_0 = \min_{0<\theta\leq\pi} \frac{2}{\pi}\frac{\theta}{1-\cos\theta}$ is attained. Then, $\frac{h(t_0)}{t_0} = \alpha_0$ and it follows from the definition of α_0 that $h(t) \geq \alpha_0 t$ for $t \in [0,1]$. Further, set

$$\alpha_{GW}(t) := \frac{h(t)}{t} \quad \text{if } t \in [t_0, 1] \quad \text{and} \quad \alpha_{GW}(t) := \alpha_0 \quad \text{if } t \in [0, t_0].$$

One can verify that the function $\tilde{h}(t) := \alpha_{GW}(t)t$ is convex on $[0,1]$ and $\tilde{h} \leq h$. From this it follows that

$$\frac{E(w(S))}{\mathrm{sdp}(G,w)} \geq \alpha_{\mathrm{GW}}(A), \quad \text{where} \quad A := \frac{\mathrm{sdp}(G,w)}{w(E)}. \tag{88}$$

Indeed, setting $y_{ij} := \frac{1-v_i^T v_j}{2}$, we have:

$$\frac{E(w(S))}{w(E)} = \sum_{ij\in E} \frac{w_{ij}}{w(E)} h(y_{ij}) \geq \sum_{ij\in E} \frac{w_{ij}}{w(E)} \tilde{h}(y_{ij}) \geq \tilde{h}\left(\sum_{ij\in E} \frac{w_{ij}}{w(E)} y_{ij}\right)$$
$$= \tilde{h}(A) = \alpha_{\mathrm{GW}}(A) \cdot A$$

which implies (88). Therefore, the performance guarantee of the GW algorithm is at least $\alpha_{\mathrm{GW}}(A)$ which is greater than α_0 when $A > t_0$ and tends to 1 as A tends to 1. Extending Karloff's result, Alon and Sudakov (2000) construct (unweighted) graphs G for which $\mathrm{mc}(G, w) = \mathrm{sdp}(G, w)$ and $\frac{E(w(S))}{\mathrm{sdp}(G,w)} = \alpha_{\mathrm{GW}}(A)$ for any $A = \frac{\mathrm{sdp}(G,w)}{w(E)} \geq t_0$, which shows that the performance guarantee of the GW algorithm is equal to $\alpha_{\mathrm{GW}}(A)$. For the remaining values of A, $\frac{1}{2} \leq A < t_0$, Alon, Sudakov, and Zwick (2002) construct graphs satisfying $\mathrm{mc}(G, w) = \mathrm{sdp}(G, w)$ and $\frac{E(w(S))}{\mathrm{sdp}(G,w)} = \alpha_0$ which shows that the analysis of Goemans and Williamson is also tight in this case.

5.4 How to improve the Goemans–Williamson algorithm?

There are several ways in which one can try to modify the basic algorithm of Goemans and Williamson in order to obtain an approximation algorithm with a better performance ratio.

Adding valid inequalities. Perhaps the most natural idea is to strengthen the basic semidefinite relaxation by adding inequalities valid for the cut polytope. For instance, one can add all triangle inequalities; denote by $\mathrm{sdp}'(G, w)$ the optimum value of the semidefinite program obtained by adding the triangle inequalities to (76). The new integrality ratio $\frac{\mathrm{mc}(G,w)}{\mathrm{sdp}'(G,w)}$ is equal to 1 for graphs with no K_5-minor (thus for C_5). For K_5 (with edge weights 1) it is equal to $\frac{24}{25} = 0.96$. However this is not the worst case; Feige and Schechtman (2002) construct graphs for which the new integrality ratio is no better than roughly 0.891.

On the other hand, the example of Karloff shows that the GW randomized approximation algorithm applied to the tighter semidefinite relaxation does not have a better performance guarantee. The same remains true if we would add to the semidefinite relaxation *all* inequalities valid for the cut polytope (because the Karloff's graphs satisfy $\frac{E(w(S))}{\mathrm{sdp}(G,w)} \sim \alpha_0$ while $\mathrm{mc}(G, w) = \mathrm{sdp}(G, w)$!). Therefore, in order to improve the performance guarantee, besides adding some valid inequalities, a new rounding technique will be needed. We now present two ideas along these lines: the first from Feige, Karpinski, and Langberg (2000a) uses triangle inequalities and adds a "local search" phase to the GW algorithm, the second from Zwick (1999) can be seen as a mixing of the hyperplane rounding technique and the basic random algorithm.

Adding valid inequalities and a local search phase. Feige, Karpinski and Langberg (2000a) have presented an approximation algorithm for max-cut with a better performance guarantee for graphs with a bounded maximum degree Δ (edge weights are assumed to be equal to one). Their algorithm has two new features: triangle inequalities are added to the basic semidefinite relaxation (also some triangle equalities in the case $\Delta = 3$) and an additional "greedy" phase is added after the GW hyperplane rounding phase.

Given a partition $(S, V\backslash S)$, a vertex v belonging, say, to S, is called *misplaced* if it has more neighbours in S than in $V\backslash S$; then the cut $\delta(S\backslash\{v\})$ has more edges than the cut $\delta(S)$. One of the basic ideas underlying the FKL algorithm is that, if $(S, V\backslash S)$ is the partition produced by the hyperplane rounding phase and if all angles $\arccos(v_i^T v_j)$ are equal to θ_0 (which implies $E(w(S)) = \alpha_0 \cdot \text{sdp}(G, w)$), then there is a positive probability (depending on Δ alone) of finding a misplaced vertex in the partition and, therefore, one can improve the cut.

In the case $\Delta = 3$ the FKL algorithm goes as follows. In the first step one solves the semidefinite program (76) to which have been added all triangle inequalities as well as the triangle equalities $x_j + x_{ik} + x_{jk} = -1$ for all triples (i, j, k) for which ij, $ik \in E$ (such equality is indeed valid for a maximum cut for, if not, the vertex i would be misplaced). Then the hyperplane rounding phase is applied to the optimum matrix X, producing a partition $(S, V\backslash S)$. After that comes an additional *greedy phase*: if the partition $(S, V\backslash S)$ has a misplaced vertex v, move it to the other side of the partition and repeat until no misplaced vertex can be found. If at some step there are several misplaced vertices, we move the misplaced vertex v for which the ratio between the number of edges gained in the cut by moving v and the number of triples (i, j, k) with $ij, ik \in E$ and i misplaced destroyed by this action, is maximal.

It is shown in Feige, Karpinski and Langberg (2000a) that the expected weight of the final partition returned by the FKL algorithm satisfies

$$E(w(S)) \geq 0.919 \cdot \text{sdp}(G, w). \tag{89}$$

For regular graphs of degree 3, one can show an approximation ratio of 0.924 and, for graphs with maximum degree Δ, a ratio of $\alpha_0 + \frac{1}{2^{33}\Delta^4}$. Note that, when $\Delta \geq 4$, one cannot incorporate the triangle equality $x_{ij} + x_{ik} + x_{jk} = -1$ (with ij, $ik \in E$) as it is no longer valid for maximum cuts.

Recently, Halperin, Livnat, and Zwick (2002) gave an improved approximation algorithm for max-cut in graphs of maximum degree 3 with performance guarantee 0.9326. Their algorithm has an additional preprocessing phase (which converts the input graph into a cubic graph satisfying some additional property) and performs the greedy phase in a more global manner; moreover, it applies to a more general problem than max-cut.

Mixing the random hyperplane and the basic random rounding techniques. We saw above that the performance guarantee of the GW algorithm is greater than α_0 for graphs with large cuts (with weight at least 85% of the total weight of edges). Zwick (1999) presents a modification of the GW algorithm which, on the other hand, has a better performance guarantee for graphs having *no* large cuts.

Note that the simple randomized algorithm, which constructs a partition $(S, V\backslash S)$ by assigning a vertex with probability $\frac{1}{2}$ to either side of the partition, produces a cut with expected weight $\frac{w(E)}{2}$ and thus its performance ratio is

$$\alpha_{\mathrm{rand}}(A) := \frac{1}{2A} \quad \text{where} \quad A = \frac{\mathrm{sdp}(G, w)}{w(E)}.$$

Note, moreover, that this algorithm is equivalent to applying the hyperplane rounding technique to the standard unit vectors $e_1, \ldots, e_n$, with the identity matrix as Gram matrix. As $\alpha_{\mathrm{rand}}(A) \geq \alpha_{\mathrm{GW}}(A)$ when $\frac{1}{2} \leq A \leq \frac{1}{2\alpha_0} \sim 0.569113$, Zwick's idea is to make a "mix" of the hyperplane rounding and basic random algorithms. For this, if X is the optimum matrix obtained when solving the basic semidefinite program (76), set

$$X' := (\cos^2 \gamma_A)X + (\sin^2 \gamma_A)I$$

where $\gamma_A \in [0, \pi]$ is suitably chosen. Namely, if $A \geq t_0$ then $\gamma_A := 0$ and if $\frac{1}{2}A \leq t_0$, then solve the following equations for c and t:

$$\frac{\arccos(c(1-2t)) - \arccos c}{t} = \frac{2c}{\sqrt{1 - c^2(1-2t)^2}},$$

$$\frac{1 - \frac{t}{A}}{\sqrt{1 - c^2}} = \frac{1 - 2t}{\sqrt{1 - c^2(1-2t)^2}}$$

(there is a unique solution c_A, t_A such that $0 \leq c_A \leq 1$ and $\frac{3}{4} \leq t_A \leq t_0$) and set $\gamma_A := \arccos(\sqrt{c_A})$. Note that γ_A tends to $\frac{\pi}{2}$ as A tends to $\frac{1}{2}$. Then a randomized cut $\delta(S)$ is produced by applying the hyperplane rounding phase to the modified matrix X'. Zwick shows that

$$\frac{E(w(S))}{\mathrm{sdp}(G, w)} \geq \alpha_{\mathrm{rot}}(A) \quad \text{for any graph } G \text{ and } w \geq 0 \tag{90}$$

where $\alpha_{\mathrm{rot}}(A) := \alpha_{\mathrm{GW}}(A)$ for $A \geq t_0$ and, setting $h_c(t) := \frac{\arccos(c(1-2t))}{\pi}$,

$$\alpha_{\mathrm{rot}}(A) := \left(\frac{1}{A} - \frac{1}{t_A}\right) h_{c_A}(0) + \frac{1}{t_A} h_{c_A}(t_A)$$

for $\frac{1}{2} \leq A \leq t_0$. The new performance guarantee is at least $\alpha_{\text{rot}}(A)$, which is greater than $\alpha_{\text{rand}}(A)$ and $\alpha_{\text{GW}}(A)$ when $A < t_0$. For instance, $\alpha_{\text{rot}}(A) \geq 0.88$ if $A \leq 0.75$, $\alpha_{\text{rot}}(A) \geq 0.91$ if $A \leq 0.6$. Alon, Sudakov and Zwick (2002) show that the analysis is tight; for this they construct graphs having $\text{mc}(G, w) = \text{sdp}(G, w)$ and $\frac{E(w(S))}{\text{sdp}(G,w)} = \alpha_{\text{rot}}(A)$ for any $\frac{1}{2} \leq A \leq t_0$.

Inapproximability results. Summarizing, the best performance guarantee of an approximation algorithm for max-cut (with nonnegative weights) known so far is $\alpha_0 \sim 0.87856$. In fact, $\frac{16}{17} \sim 0.94117$ is the best performance guarantee that one can hope for. Indeed, Håstad (1997) shows that, for any $\epsilon > 0$, there is no $(\frac{16}{17} + \epsilon)$-approximation algorithm for max-cut if $P \neq NP$. Berman and Karpinski (1998) show that it is NP-hard to approximate max-cut in cubic graphs beyond the ratio of 0.997 (while there is an 0.932-approximation algorithm as we saw above).

On the positive side, Arora, Karger, and Karpinski (1995) show that the max-cut problem has a polynomial time approximation scheme (that is, an $(1-\epsilon)$-approximation algorithm for any $\epsilon > 0$) when restricted to dense graphs, that is, graphs with $O(n^2)$ edges. De la Vega (1996) described independently a randomized approximation scheme for max-cut in graphs with minimum degree cn for some constant $c > 0$.

We have seen in Section 3.6 several techniques permitting to construct semidefinite relaxations of the cut polytope refining the basic one. Thus a natural and very interesting question is whether some of them can be used for proving a better integrality ratio (better than the Goemans–Williamson bound α_0) and for designing an approximation algorithm for max-cut with an improved performance ratio. The most natural candidate to consider might be the Lasserre relaxation $Q_1(K_n)$ (defined using (47) and (48)) or its subset, the Anjos–Wolkowicz relaxation F_n (defined using (47)).

6 Applications of semidefinite programming and the rounding hyperplane technique to other combinatorial optimization problems

The method developed by Goemans and Williamson for approximating the max-cut problem has been applied and generalized to a large number of combinatorial optimization problems. Summarizing, their method consists of the following two phases:

(1) *The semidefinite optimization phase*, which finds a set of vectors $v_1, \ldots, v_n$ providing a Cholesky factorization of an optimum solution to the SDP program relaxing the original combinatorial problem.
(2) *The random hyperplane rounding phase*, which constructs a solution to the original combinatorial problem by looking at the positions of the vectors v_i with respect to some random hyperplane.

The basic method of Goemans and Williamson may have to be modified in order to be applied to some other combinatorial problems. In the first phase, one has to choose an appropriate SDP relaxation of the problem at hand and, in the second phase, one may have to adapt the rounding procedure. For instance, if one wants to approximate graph coloring and max k-cut problems, one should consider more general partitions of the space using more than one random hyperplane. One may also have to add an additional phase permitting to modify the returned solution; for instance, to turn the returned cut into a bisection if one wants to approximate the bisection problem. It turns out that the analysis of the extended approximation algorithms is often more complicated than that of the basic GW algorithm; it sometimes needs the evaluation of certain integral formulas that are hard to evaluate numerically.

In this section we present approximation algorithms based on these ideas for the following problems: general quadratic programming problems, maximum bisection and k-cut problems, coloring, stable sets, MAX SAT, and maximum directed cut problems.

Of course, the above is not an exhaustive list of the problems for which semidefinite programming combined with randomized rounding permits to obtain good approximations. There are other interesting problems, that we could not cover here, to which these techniques apply; this is the case, e.g., for scheduling (see Skutella (2001)).

6.1 *Approximating quadratic programming*

We consider here the Boolean quadratic programming problem:

$$\begin{aligned} m^*(A) := \max \quad & x^T A x \\ \text{s.t.} \quad & x \in \{\pm 1\}^n \end{aligned} \tag{91}$$

where A is a symmetric matrix of order n, and its natural SDP relaxation:

$$\begin{aligned} s^*(A) := \max \quad & \langle A, X \rangle \\ \text{s.t.} \quad & X_{ii} = 1\,(i = 1, \ldots, n) \\ & X \succeq 0. \end{aligned} \tag{92}$$

Obviously, $m^*(A) \le s^*(A)$. How well does the semidefinite bound $s^*(A)$ approximate $m^*(A)$? Obviously $m^*(A) = s^*(A)$ when all off-diagonal entries of A are nonnegative. We saw in Section 5.3 that $\frac{m^*(A)}{s^*(A)} \ge \alpha_0$ (the GW ratio from (84)) in the special case when A is the Laplacian matrix of a graph; that is, when $Ae = 0$ and $A_{ij} \le 0$ for all $i \ne j$. (Note that these conditions imply that $A \succeq 0$.) Nesterov (1997) studies the quality of the SDP relaxation for general A. When $A \succeq 0$ he shows the lower bound $\frac{2}{\pi}$ for the ratio $\frac{m*(A)}{s*(A)}$ and, based on this, he gives upper bounds for the relative accuracy $s^*(A) - m^*(A)$ for indefinite A. the basic step consists in giving a trigonometric reformulation of the problem (91), analogous to the trigonometric reformulation (86) for max-cut.

Proposition 15. *Given a symmetric matrix* A,

$$\begin{array}{rl} m^*(A) = \max & \dfrac{2}{\pi}\langle A, \arcsin(X)\rangle \\ \text{s.t.} & X_{ii} = 1 \ (i = 1, \ldots, n) \\ & X \succeq 0 \end{array} \tag{93}$$

setting $\arcsin(X) := (\arcsin(x_{ij}))_{i,j=1}^n$. *Moreover,* $m^*(A) \geq \frac{2}{\pi} s^*(A)$ *if* $A \succeq 0$.

Proof. Denote by μ the maximum of the program (93). Let x be an optimum solution to the program (91) and set $X := xx^T$. Then X is feasible for (93) with objective value $\frac{2}{\pi}\langle A, \arcsin(X)\rangle = \langle A, xx^T\rangle = m^*(A)$, which shows that $m^*(A) \leq \mu$. Conversely, let X be an optimum solution to (93) and let $v_1, \ldots, v_n$ be vectors such that $X_{ij} = v_i^T v_j$ for all i,j. Let r be a random unit vector. Then the expected value of $\text{sign}(r^T v_i)\text{sign}(r^T v_j)$ is equal to

$$1 - 2\,\text{prob}(\text{sign}(r^T v_i) \neq \text{sign}(r^T v_j)) = 1 - 2\frac{\arccos(v_i^T v_j)}{\pi} = \frac{2}{\pi}\arcsin(v_i^T v_j).$$

Therefore, the expected value E_A of $\sum_{i,j} a_{ij}\text{sign}(r^T v_i)\text{sign}(r^T v_j)$ is equal to $\frac{2}{\pi}\sum_{i,j} a_{ij}\arcsin(v_i^T v_j) = \frac{2}{\pi}\langle A, \arcsin(X)\rangle = \mu$. On the other hand, $\sum_{i,j} a_{ij}\text{sign}(r^T v_i)\text{sign}(r^T v_j) \leq m^*(A)$, since the vector $(\text{sign}(r^T v_i))_{i=1}^n$ is feasible for (91) for any unit vector r. This implies that $E_A \leq m^*(A)$ and thus $\mu \leq m^*(A)$. Assume $A \succeq 0$. Then, $\langle A, \arcsin(X)\rangle = \langle A, \arcsin(X) - X\rangle + \langle A, X\rangle \geq \langle A, X\rangle$, using the fact that $\arcsin(X) - X \succeq 0$ if $X \succeq 0$. Hence, $m^*(A) \geq \frac{2}{\pi} s^*(A)$ if $A \succeq 0$. □

Let $m^*(A)$ (resp. $s_*(A)$) denote the optimum value of the program (91) (resp. (92)) where we replace maximization by minimization. Applying the duality theorem for semidefinite programming, we obtain:

$$s^*(A) = \min(e^T y \mid \text{diag}(y) - A \succeq 0), \tag{94}$$

$$s_*(A) = \max(e^T z \mid A - \text{diag}(z) \succeq 0). \tag{95}$$

For $0 \leq \alpha \leq 1$, set $s_\alpha := \alpha s^*(A) + (1 - \alpha)s_*(A)$.

Lemma 16. *For* $\alpha := \frac{2}{\pi}$, $s_*(A) \leq m_*(A) \leq s_{1-\alpha} \leq s_\alpha \leq m^*(A) \leq s^*(A)$.

Proof. We show the inequality $m_*(A) \leq s_{1-\alpha}(A)$, that is, $s^*(A) - m_*(A) \geq \frac{2}{\pi}(s^*(A) - s_*(A))$. Let y (resp. z) be an optimum solution to (94) (resp. (95)).

Then,

$$s^*(A) - m_*(A) = e^T y + m^*(-A) = m^*(\mathrm{diag}(y) - A) \geq \frac{2}{\pi} s^*(\mathrm{diag}(y) - A)$$

by Proposition 15, since $\mathrm{diag}(y) - A \succeq 0$. To conclude, note that $s^*(\mathrm{diag}(y) - A) = e^T y + s^*(-A) = e^T y - s_*(A) = s^*(A) - s_*(A)$. The inequality $s_\alpha(A) \leq m^*(A)$ can be shown similarly. □

The above lemma can be used for proving the following bounds on the relative accuracy $m^*(A) - s_\alpha$.

Theorem 17. *Set* $\alpha := \frac{2}{\pi}$ *and* $\beta := \frac{\alpha^2 + 2\alpha - 1}{3\alpha - 1}$. *Then,*

$$\frac{m^*(A) - s_\alpha}{m^*(A) - m_*(A)} \leq \frac{\pi}{2} - 1 < \frac{4}{7} \quad \textit{and} \quad \frac{|m^*(A) - s_\beta(A)|}{m^*(A) - m_*(A)} \leq \frac{\pi - 2}{6 - \pi} < \frac{2}{5}.$$

The above results can be extended to quadratic problems of the form:

$$\max \quad x^T A x \quad \text{subject to } [x]^2 \in F$$

where F is a closed convex set in $\mathbb{R}^n$ and $[x]^2 := (x_1^2, \ldots, x_n^2)$. See Tseng (2003), Chapter 13 in Wolkowicz, Saigal and Vandenberghe (2000), Ye (1999), Zhang (2000) for further results. Inapproximability results are given in Bellare and Rogaway (1995).

6.2 Approximating the maximum bisection problem

The *maximum weight bisection problem* is a variant of the max-cut problem where one wants to find a cut $\delta(S)$ such that $|S| = \frac{n}{2}$ (a *bisection* or *equicut*) (n being assumed even) having maximum weight. This is an NP-hard problem, for which no approximation algorithm with a performance ratio $> \frac{16}{17}$ exists unless $P = NP$ (Håstad (1997)). Polynomial time approximation schemes are known to exist for this problem over dense graphs (Arora, Karger and Karpinski (1995)) and over planar graphs (Jansen, Karpinski, and Lingas (2000)).

Extending the Goemans–Williamson approach to max-cut, Frieze and Jerrum (1997) gave a randomized 0.651-approximation algorithm for the maximum weight bisection problem. Ye (2001) improved the performance ratio to 0.6993 by combining the Frieze–Jerrum approach with some rotation argument applied to the optimum solution of the semidefinite relaxation. Halperin and Zwick (2001a) further improved the approximation ratio to 0.7016 by strengthening the SDP relaxation with the triangle inequalities. Details are given below.

Given a graph $G=(V,E)$ ($V=\{1,\ldots,n\}$) and edge weights $w \in \mathbb{R}^E_+$, the maximum weight bisection problem reads:

$$\begin{array}{ll} \max & \frac{1}{2}\sum_{ij\in E} w_{ij}(1-x_ix_j) \\ \text{s.t.} & \sum_{i=1}^{n} x_i = 0 \\ & x_1,\ldots,x_n \in \{\pm 1\}. \end{array} \tag{96}$$

A natural semidefinite relaxation is:

$$\begin{array}{ll} W^* := \max & \frac{1}{2}\sum_{ij\in E} w_{ij}(1-X_{ij}) \\ \text{s.t.} & X_{ii} = 1 \, (i \in V) \\ & \langle J, X\rangle = 0 \\ & X \succeq 0 \end{array} \tag{97}$$

The Frieze–Jerrum approximation algorithm.

(1) *The SDP optimization phase*: Solve the SDP (97), let X be an optimum solution and let $v_1,\ldots,v_n$ be vectors such that $X_{ij} = v_i^T v_j$ for all i,j.

(2) *The random hyperplane rounding phase*: Choose a random unit vector r and define the associated cut $\delta(S)$ where $S := \{i \in V \mid r^T v_i \geq 0\}$.

(3) *Constructing a bisection*: Without loss of generality, assume that $|S| \geq \frac{n}{2}$. For $i \in S$, set $W(i) := \sum_{j \notin S} w_{ij}$. Order the elements of S as $i_1,\ldots,i_{|S|}$ in such a way that $W(i_1) \geq \cdots \geq W(i_{|S|})$ and define $\tilde{S} := \{i_1,\ldots,i_{\frac{n}{2}}\}$.

Then $\delta(\tilde{S})$ is a bisection whose weight satisfies

$$w(\delta(\tilde{S})) \geq \frac{n}{2|S|} w(\delta(S)). \tag{98}$$

Consider the random variables $W := w(\delta(S))$ and $C := |S|(n-|S|)$; W is the weight of the cut $\delta(S)$ in G while C is the number of pairs $(i,j) \in V^2$ that are cut by the partition $(S, V\backslash S)$ (that is, the cardinality of the cut $\delta(S)$ viewed as cut in the complete graph K_n). The analysis of the GW algorithm from Section 5.3 shows the following lower bounds for the expected value $E(W)$ and $E(C)$:

$$E(W) \geq \alpha_0 W^*, \tag{99}$$

$$E(C) \geq \alpha_0 C^* \tag{100}$$

where $C^* := \frac{n^2}{4}$. Define the random variable

$$Z := \frac{W}{W^*} + \frac{C}{C^*}. \tag{101}$$

Then, $Z \leq 2$ and $E(Z) \geq 2\alpha_0$.

Lemma 18. *If $Z \geq 2\alpha_0$ then $w(\delta(\tilde{S})) \geq 2(\sqrt{2\alpha_0} - 1)W^*$.*

Proof. Set $w(\delta(S)) = \lambda W^*$ and $|S| = \sigma n$. Then, $Z = \lambda + 4\sigma(1-\sigma) \geq 2\alpha_0$, implying $\lambda \geq 2\alpha_0 - 4\sigma(1-\sigma)$. Using (98), we obtain that

$$w(\delta(\tilde{S})) \geq \frac{n}{2|S|} w(\delta(S)) = \frac{\lambda W^*}{2\sigma} \geq W^* \frac{2\alpha_0 - 4\sigma(1-\sigma)}{2\sigma} \geq 2(\sqrt{2\alpha_0} - 1)W^*.$$

(The last inequality being a simple verification.) □

As $E(Z) \geq 2\alpha_0$, the strategy employed by Frieze and Jerrum in order to find a bisection satisfying the conclusion of Lemma 18 is to repeat the above steps 2 and 3 of the algorithm N times, where N depends on some small $\epsilon > 0$ ($N = \lceil \frac{1}{\epsilon} \ln \frac{1}{\epsilon} \rceil$) and to choose as output bisection the heaviest among the N bisections produced throughout the N runs. Then, with high probability, the largest among the variables Z produced throughout the N runs will be greater than or equal to $2\alpha_0$. Therefore, it follows from Lemma 18 that the weight of the output bisection is at least $(2(\sqrt{2\alpha_0} - 1) - \epsilon)W^*$. For ϵ small enough, this shows a performance ratio of 0.651.

Ye (2001) shows an improved approximation ratio of 0.6993. For this, he modifies the Jerrum–Frieze algorithm in the following way. Instead of applying the random hyperplane rounding phase to the optimum solution X of (97), he applies it to the modified matrix $\rho X + (1-\rho)I$, where ρ is a parameter to be determined. This operation is analogous to the "outward rotation" used by Zwick (1999) for the max-cut problem and mentioned in Section 5.4.

The starting point is to replace relations (99) and (100) by

$$E(W) \geq \alpha W^* \quad \text{and} \quad E(C) \geq \gamma C^* \tag{102}$$

where $\alpha = \alpha(\rho)$ and $\gamma = \gamma(\rho)$ are lower bounds to be determined on the ratios $\frac{E(W)}{W^*}$ and $\frac{E(C)}{C^*}$, respectively. In fact, the following choices can be made for α, γ:

$$\alpha(\rho) := \min_{-1 \leq x < 1} \frac{2}{\pi} \frac{\arccos(\rho x)}{1 - x}, \tag{103}$$

$$\gamma(\rho) := \min_{-1 \leq x < 1} \frac{2}{\pi} \frac{\arccos(\rho x) - x \arccos \rho}{1 - x}. \tag{104}$$

Indeed,

$$E(W) = \frac{1}{2} \sum_{ij \in E} w_{ij} \frac{2}{\pi} \arccos(\rho X_{ij}) \geq \alpha(\rho) W^*.$$

By the definition of $\gamma(\rho)$, $\frac{2}{\pi}\arccos(\rho x) \geq (1 - x)\gamma(\rho) + \frac{2}{\pi} x \arccos \rho$ for $x \in [-1, 1]$. Therefore,

$$\begin{aligned} E(C) &= \frac{1}{4} \sum_{i \neq j \in \{1,\ldots,n\}} \frac{2}{\pi} \arccos(\rho X_{ij}) \\ &\geq \frac{1}{4} \gamma(\rho) \sum_{i \neq j} (1 - X_{ij}) + \frac{1}{2\pi} \arccos \rho \sum_{i \neq j} X_{ij} \\ &= \frac{n^2}{4} \gamma(\rho) - \frac{\arccos \rho}{2\pi} n. \end{aligned}$$

For n large enough, the linear term can be ignored and the result follows.

Modify the definition of Z from (101) as

$$Z := \frac{W}{W^*} + \tau \frac{C}{C^*} \quad \text{where} \quad \tau := \frac{\alpha}{2\beta} \left(\frac{1}{\sqrt{1 - \beta}} - 1 \right).$$

The proof of Lemma 18 can be adapted to show that, if $Z \geq \alpha + \tau\gamma$, then

$$E(w(\tilde{S})) \geq \frac{\alpha}{1 + \sqrt{1 - \beta}} W^*.$$

For $\rho = 0.89$, one can compute that $\alpha(\rho) \geq 0.8355$, $\gamma(\rho) \geq 0.9621$, and $\frac{\alpha}{1+\sqrt{1-\beta}} > 0.6993$. Therefore, this shows that Ye's algorithm is a 0.6993-approximation algorithm.

Halperin and Zwick (2001a) can improve the performance ratio to 0.7016. They achieve this by adding one more ingredient to Ye's algorithm;

namely, they strengthen the SDP relaxation (97) by adding the triangle inequalities:

$$X_{ij} + X_{ik} + X_{jk} \geq -1, \quad X_{ij} - X_{ik} - X_{jk} \geq -1$$

for distinct $i, j, k \in \{1, \ldots, n\}$. Although triangle inequalities had already been used earlier by some authors to obtain better approximations (e.g., in Feige, Karpinski nd Langberg (2000a) for the max-cut problem in bounded degree graphs as mentioned in Section 5.4), they were always analyzed from a local point of view (e.g., in the above mentioned example, in a local search phase, searching for misplaced vertices). In contrast, Halperin and Zwick are able to make a global analysis of the contribution of triangle inequalities. Namely, they show that the function $\gamma(\rho)$ from (104) can be replaced by

$$\gamma'(\rho) := \min_{-1 \leq x \leq -\frac{1}{3}} \frac{1}{\pi}\left(\arccos(\rho x) + \frac{3(x+1)}{4}\arccos\left(-\frac{\rho}{3}\right) + \frac{1-3x}{4}\arccos\rho\right),$$

which enables them to demonstrate a better performance ratio (using appropriate values for the parameters ρ and τ). (Note the $\gamma'(\rho) > \gamma(\rho)$ for $0 < \rho < 1$.)

Let us give a flavor of how the function $\gamma'(\rho)$ comes up. The goal is to find a lower bound for the ratio $\frac{E(C)}{C^*} = \frac{4}{\pi n^2}\sum_{1 \leq i < j \leq n} \arccos(\rho X_{ij})$. Let A (resp. B, C) denote the set of pairs ij for which $X_{ij} < -\frac{1}{3}$ (resp. $-\frac{1}{3} \leq X_{ij} \leq 0$, $0 \leq X_{ij} \leq 1$). By the triangle inequalities, the graph on $\{1, \ldots, n\}$ with edge set A is triangle free, which implies that $|A| \leq \frac{n^2}{4}$. Thus the optimum value of the following nonlinear program is a lower bound for $\frac{E(C)}{C^*}$:

$$\begin{aligned}
\min \quad & \frac{4}{\pi n^2}\sum_{i<j} \arccos(\rho z_{ij}) \\
\text{s.t.} \quad & \sum_{i<j} z_{ij} = -\frac{n}{2} \\
& -1 \leq z_{ij} \leq 1 \, (i < j) \\
& \left|\{ij \mid z_{ij} < -\tfrac{1}{3}\}\right| \leq \frac{n^2}{4}.
\end{aligned}$$

Halperin and Zwick show then that the above minimum can be expressed in closed form as $\gamma'(\rho)$.

Feige, Karpinski, and Langberg (2000b) design a 0.795-approximation algorithm for the maximum bisection problem restricted to regular graphs. One of their key results is the following: given a cut $\delta(S)$ in a regular graph G, one can efficiently construct a bisection $\delta(S')$ whose weight is at least 0.9027 $w(\delta(S))$. Hence, if we start with the cut $\delta(S)$ given as output of the

Goemans–Williamson algorithm, then this gives an approximation algorithm with performance ratio $0.9027 \cdot 0.878 \sim 0.793$; a further improvement is demonstrated in Feige, Karpinski and Langberg (2000b).

Extensions to variations of the bisection problem. The following variations of the bisection problem have been studied in the literature: (i) the *maximum* $\frac{n}{2}$-*vertex cover problem*, (ii) the *maximum* $\frac{n}{2}$-*dense subgraph problem*, (*iii*) the *maximum* $\frac{n}{2}$-*uncut problem*, which ask for a subset $S \subseteq V$ of size $\frac{n}{2}$ maximizing the total weight of the edges incident to S, contained in S, contained in S or its complement, respectively. Halperin and Zwick (2001a) treat these three problems (together with the maximum bisection problem as well as some directed analogues) in a unified framework and they can show the best approximation ratios known up to today, namely, 0.8452 for problem (i), 0.6221 for problem (ii), and 0.6436 for problem (iii).

6.3 *Approximating the max k-cut problem*

Given a graph $G=(V,E)$, edge weights $w \in \mathbb{R}_+^E$ and an integer $k \geq 2$, the *max k-cut problem* asks for a partition $\mathcal{P}=(S_1,\ldots,S_k)$ of V whose weight $w(\mathcal{P}) := \sum_{1 \leq h < h' \leq k} \sum_{ij \in E | i \in S_h, j \in S_{h'}} w_{ij}$ is maximum. The set of edges whose end nodes belong to distinct classes of the partition is a *k-cut*, denoted as $\delta(S_1,\ldots,S_k)$. For $k=2$, we find the max-cut problem. For any $k \geq 2$, the max k-cut problem is NP-hard; moreover, there can be no polynomial time approximation algorithm for it with performance ratio $1-\frac{1}{239k}$, unless P = NP (Kann, Khanna, Lagergren, and Panconesi (1997)).

A simple heuristic for max k-cut is to partition V randomly into k sets. As the probability that two nodes fall in the same class is $\frac{1}{k}$, the expected weight of the k-cut produced in this way is $\sum_{ij \in E} w_{ij}(1-\frac{1}{k}) \geq w(E)\,(1-\frac{1}{k})$ and, therefore, the simple random partition heuristic has a performance guarantee of $1-\frac{1}{k}$.

Frieze and Jerrum (1997) present an approximation algorithm for max k-cut with performance guarantee α_k satisfying

(i) $\alpha_k > 1-\frac{1}{k}$ and $\lim_{k\to\infty} \frac{\alpha_k-(1-\frac{1}{k})}{(2k^{-2}\ln k)} = 1$,

(ii) $\alpha_2 = \alpha_0 \geq 0.878567$ (recall (84)), $\alpha_3 \geq 0.832718$, $\alpha_4 \geq 0.850304$, $\alpha_5 \geq 0.874243$, $\alpha_{10} \geq 0.926642$, $\alpha_{100} \geq 0.990625$.

In particular, the Frieze–Jerrum algorithm has a better performance guarantee than the simple random heuristic.

One can model the max k-cut problem on a graph $G=(V,E)\,(V=\{1,\ldots,n\})$ by having n variables $x_1,\ldots,x_n$ taking one of k possible values. For $k=2$ the 2 possible values are ± 1 and for $k \geq 2$ one can choose as possible values a set of k unit vectors $a_1,\ldots,a_k \in \mathbb{R}^{k-1}$ satisfying

$$a_i^T a_j = -\frac{1}{k-1} \quad \text{for} \quad 1 \leq i \neq j \leq k.$$

(Such vectors exist since the matrix $\frac{k}{k-1}I_k - \frac{1}{k-1}J_k$ is positive semidefinite.) Hence the max k-cut problem can be formulated as

$$\begin{aligned} \mathrm{mc}_k(G,w) := \max \quad & \frac{k-1}{k}\sum_{ij\in E} w_{ij}(1 - x_i^T x_j) \\ \text{s.t.} \quad & x_1,\ldots,x_n \in \{a_1,\ldots,a_k\} \end{aligned} \tag{105}$$

and the following is a semidefinite relaxation of (105):

$$\begin{aligned} \mathrm{sdp}_k(G,w) := \max \quad & \frac{k-1}{k}\sum_{ij\in E} w_{ij}(1 - X_{ij}) \\ \text{s.t.} \quad & X_{ii} = 1 \qquad (i \in V) \\ & X_{ij} \geq -\frac{1}{k-1} \qquad (i \neq j \in V) \\ & X \succeq 0. \end{aligned} \tag{106}$$

The Frieze–Jerrum approximation algorithm for max k-cut.

(1) Solve (106) to obtain unit vectors $v_1,\ldots,v_n$ satisfying $v_i^T v_j \geq -\frac{1}{k-1} (i,j \in V)$ and $\mathrm{sdp}_k(G,w) = \frac{k-1}{k}\sum_{ij\in E} w_{ij}(1 - v_i^T v_j)$.
(2) Choose k independent random vectors $r_1,\ldots,r_k \in \mathbb{R}^n$. (This can be done by chosing their kn components as independent random variables from the standard normal distribution with mean 0 and variance 1.)
(3) Partition V into $S_1,\ldots,S_k$ where S_h consists of the nodes $i \in V$ for which $v_i^T r_h = \max_{h'=1,\ldots,k} v_i^T r_{h'}$. (Break ties arbitrarily as they occur with probability 0.)

When $k=2$ the algorithm reduces to the Goemans–Williamson algorithm for max-cut. Given two unit vectors $u, v \in \mathbb{R}^n$, the probability that $\max_{1\leq h\leq k} u^T r_h$ and $\max_{1\leq h\leq k} v^T r_h$ are both attained by the same vector within $r_1,\ldots,r_k$ depends only on the angle between u and v, i.e., on $\rho := u^T v$, and it is equal to $k \cdot \mathrm{prob}\,(u^T r_1 = \max_{1\leq h\leq k} u^T r_h$ and $v^T r_1 = \max_{1\leq h\leq k} v^T r_h)$; denote this probability as $kI(\rho)$. Then the expected weight of the k-cut $\delta(S_1,\ldots,S_k)$ produced by the Frieze–Jerrum algorithm is equal to

$$\begin{aligned} \sum_{ij\in E} w_{ij}\,\mathrm{prob}(ij \in \delta(S_1,\ldots,S_k)) &= \sum_{ij\in E} w_{ij}(1 - kI(v_i^T v_j)) \\ &= \textstyle\sum_{ij\in E} w_{ij}\left(\frac{k}{k-1}\,\frac{1-kI(v_i^T v_j)}{1 - v_i^T v_j}\right)\left(\frac{k-1}{k}(1 - v_i^T v_j)\right) \geq \alpha_k\, \mathrm{sdp}_k(G,w), \end{aligned}$$

setting

$$\alpha_k := \min_{-\frac{1}{k-1}\leq\rho<1} \frac{k}{k-1}\,\frac{1-kI(\rho)}{1-\rho}. \tag{107}$$

For $k=2$, $\alpha_2=\alpha_0$ can be computed exactly. For $k\geq 3$, the evaluation of α_k is more complicated and relies on the computation of the function $I(\rho)$ which can be expressed as multiple integral. Using a Taylor series expansion for $I(\rho)$, Frieze and Jerrum could show the lower bonds for α_k mentioned at the beginning of this subsection.

For $k=3$, de Klerk, Pasechnik, and Warners (2004) give a closed form expression for $I(\rho)$ which enables them to show that

$$\alpha_3 = \frac{7}{12} + \frac{3}{4\pi^2}\arccos^2(-1/4).$$

Thus $\alpha_3 > 0.836008$ (instead of the lower bound 0.832718 of Frieze and Jerrum). Goemans and Williamson (2001) find the same expression for α_3 using another formulation for max 3-cut based on complex semidefinite programming.

De Klerk, Pasechnik and Warners (2004) prove a better lower bound for α_k for small $k\geq 3$. For instance, they show that $\alpha_4\geq 0.857487$ (instead of 0.850304). For this they present another approximation algorithm for max k-cut (equivalent to the Frieze–Jerrum algorithm for the graphs G with $\vartheta(\bar{G})\leq k$) which enables them to reformulate the function $I(\rho)$ in terms of the volume of a spherical simplex and do more precise computations.

The minimum k-cut problem is also studied in the literature, in particular, because of its applications to frequency assignment (see Eisenblätter (2001, 2002)). Whereas good approximation algorithms exist for the maximum k-cut problem, the minimum k-cut problem cannot be approximated within a ratio of $O(|E|)$ unless P=NP. Semidefinite relaxations are nevertheless used in practice for deriving good lower bounds for the problem (see Eisenblätter (2001, 2002)).

6.4 *Approximating graph coloring*

Determining the chromatic number of a graph is a hard problem. Lund and Yannakakis (1993) show that there is a constant $\epsilon>0$ for which there exists no polynomial algorithm which can color any graph G using at most $n^{\epsilon}\chi(G)$ colors unless P=NP. Khanna, Linial, and Safra (2000) show that it is not possible to color a 3-colorable graph with 4 colors in polynomial time unless P=NP.

On the positive side, Wigderson (1983) shows that it is possible to color in polynomial time a 3-colorable graph with $3\lceil\sqrt{n}\rceil$ colors and, more generally, a k-colorable graph with $2kn^{1-\frac{1}{k-1}}$ colors; we will come back to this result later in this section. Later Blum (1994) gives a polynomial time algorithm coloring a 3-colorable graph with $O(n^{\frac{3}{8}}\log^{\frac{8}{5}} n)$. Using semidefinite programming and randomized rounding, Karger, Motwani, and Sudan (1998) present a randomized polynomial time algorithm which colors a 3-colorable graph with maximum degree Δ with $O(\Delta^{\frac{1}{3}}\sqrt{\log\Delta}\log n)$ or $O(n^{\frac{1}{4}}\sqrt{\log n})$ colors and, more generally, a k-colorable graph with $O(\Delta^{1-\frac{2}{k}}\sqrt{\log\Delta\log n})$ or $O(n^{1-\frac{3}{k+1}}\sqrt{\log n})$ colors. This result was later refined by Halperin, Nathaniel, and Zwick (2001), who proved that a k-colorable graph with maximum degree Δ can be colored in randomized polynomial time with $O(\Delta^{1-\frac{2}{k}}(\log\Delta)^{\frac{1}{k}}\log n)$. Further coloring results can be found in Blum and Karger (1997), Halldórsson (1993), Halperin, Nathaniel and Zwick (2001).

In what follows we present some of these results. We first prove a weaker version of the Karger–Motwani–Sudan result, namely, how to find a $O(n^{0.387})$ coloring for a 3-colorable graph. This enables us to introduce the basic tools used in Karger, Motwani and Sudan (1998): vectors k-coloring, k-semicoloring, hyperplane rounding, and a result of Wigderson (1983). Then we describe the Halperin–Nathaniel–Zwick algorithm for finding a $O(\Delta^{\frac{1}{3}}(\log\Delta)^{\frac{1}{3}}\log n)$-coloring of a 3-colorable graph with maximum degree Δ. (For simplicity in the exposition we only treat the case $k=3$.) This result is based on a new randomized rounding technique introduced in Karger, Motwani and Sudan (1998), using the standard n-dimensional normal distribution (instead of the distribution on the unit sphere) and vector projections. We finally describe the $O(n^{1-\frac{3}{k+1}}\sqrt{\log n})$-coloring algorithm for k-colorable graphs of Karger, Motwani, and Sudan.

Vector coloring. The first step in the Karger–Motwani–Sudan algorithm consists in solving a semidefinite relaxation for the coloring problem. We saw in Sections 4.2 and 4.4 that the theta number $\vartheta(\bar{G})$ and its variations $\vartheta'(\bar{G})$ and $\vartheta^+(\bar{G})$ constitute lower bounds for the chromatic number of G. Karger, Motwani, and Sudan consider the SDP program (67) defining $\vartheta'(\bar{G})$ as a SDP relaxation for the coloring problem and they introduce the notion of vector coloring. A *vector k-coloring* of G is an assignment of vectors $v_1, \ldots, v_n$ to the nodes of G such that $v_i^T v_j \leq -\frac{1}{k-1}$ for every edge $ij \in E$. Then the *vector chromatic number* $\chi_v(G)$ is defined as the smallest $k \geq 2$ for which there exists a vector k-coloring. By the discussion above, $\chi_v(G) = \vartheta'(\bar{G})$. If in the definition of vector coloring one requires that the inequalities $v_i^T v_j \leq -\frac{1}{k-1}$ hold at equality for all edges, then we obtain the *strict vector chromatic number* which coincides with $\vartheta(\bar{G})$. More strongly, one can consider the *strong vector chromatic number* $\vartheta^+(\bar{G})$ which is defined by requiring $v_i^T v_j = -\frac{1}{k-1}$ for all edges and $v_i^T v_j \geq -\frac{1}{k-1}$ for all nonedges. Therefore, the vector chromatic number is less than or equal to the strict vector chromatic number, which in turn is less than or equal to the strong vector chromatic number, which is a lower bound for the chromatic number (recall (69)).

Let us point out that the gap between the chromatic number and all these vector chromatic numbers can be arbitrarily large. Karger, Motwani and Sudan (1998) construct a class of graphs having $\chi_v(G)=3$ while $\chi(G) \geq n^{0.0113}$. Feige (1997) shows that for all $\epsilon > 0$ there exist families of graphs with $\chi(G) \geq \vartheta(\bar{G})n^{1-\varepsilon}$ and Charikar (2002) proves an analogous result for the strong vector chromatic number.

Semicoloring. The hard part in the Karger–Motwani–Sudan algorithm consists of constructing a good proper coloring from a vector k-coloring. There are two steps: first construct a semicoloring and then from it a proper coloring. A *k-semicoloring* of a graph on n nodes is an assignment of k colors to at least half of the nodes in such a way that no two adjacent nodes receive the same color. This is a useful notion, as an algorithm for semicoloring yields an algorithm for proper coloring.

Lemma 19. *Let $f: \mathbb{Z}_+ \to \mathbb{Z}_+$ be a monotone increasing function. If there is a randomized polynomial time algorithm which $f(i)$-semicolors every i-vertex subgraph of graph G, then this algorithm can color G with $O(f(n)\log n)$ colors. Moreover, if there exists some $\epsilon > 0$ such that $f(i) = O(i^\epsilon)$ for all i, then the algorithm can color G with $f(n)$ colors.*

Proof. We show how to color any p-vertex subgraph H of G. By assumption one can semicolor H with $f(p)$ colors. Let S denote the set of nodes of H that have not been colored; then $|S| \le \frac{p}{2}$. One can recursively color the subgraph of H induced by S using a new set of colors.

Let $c(p)$ denote the maximum number of colors that the above algorithm needs for coloring an arbitrary p-vertex subgraph of G. Then,

$$c(p) \le c\left(\frac{p}{2}\right) + f(p).$$

This recurrence relation implies that $c(p) = O(f(p)\log p)$. Moreover, if $f(p) = p^\epsilon$, one can easily verify that $c(p) = O(f(p))$. □

In view of Lemma 19, we are now left with the task of transforming a vector k-coloring into a good semicoloring.

Coloring a 3-colorable graph with $O(n^{0.387})$-colors.

Theorem 20. *Every vector 3-colorable graph G with maximum degree Δ has a $O(\Delta^{\log_3 2})$-semicoloring which can be constructed in polynomial time with high probability.*

Proof. Let $v_1, \ldots, v_n \in \mathbb{R}^n$ be unit vectors forming a vector 3-coloring of G, i.e., $v_i^T v_j \le -\frac{1}{2}$ for all edges $ij \in E$; this means that the angle between v_i and v_j is at least $\frac{2\pi}{3}$ for all edges $ij \in E$. Choose independently N random hyperplanes. This induces a partition of the space $\mathbb{R}^n$ into 2^N regions and one colors the nodes of G with 2^N colors depending in which region their associated vectors v_i are located. Then the probability that an edge is monochromatic is at most 3^{-N} and thus the expected number of monochromatic edges is at most $|E|3^{-N} \le \frac{1}{2}n\Delta 3^{-N}$. By Markov's inequality, the probability that the number of monochromatic edges is more than twice the expected number is at most $\frac{1}{2}$. After repeating the process t times, we find with probability $\ge 1 - \frac{1}{2^t}$

a coloring of G for which the number of monochromatic edges is at most $n\Delta 3^{-N}$. Setting $N := 2 + \lceil \log_3 \Delta \rceil$, we have $n\Delta 3^{-N} \leq \frac{n}{4}$. As the number of nodes that are incident to a monochromatic edge is $\leq \frac{n}{2}$, we have found a semicoloring using $2^N \leq 8\Delta^{\log_3 2}$ colors. □

As $\log_3 2 < 0.631$, Theorem 20 and Lemma 19 imply a coloring with $O(n^{0.631})$ colors. This is yet weaker than Wigderson's $O(\sqrt{n})$-coloring algorithm. In fact, the result can be improved using the following idea of Wigderson.

Theorem 21. *There is a polynomial time algorithm which, given a 3-colorable graph G and a constant $\delta \leq n$, finds an induced subgraph H of G with maximum degree $\Delta_H < \delta$ and a $\frac{2n}{\delta}$-coloring of $G \backslash H$.*

Proof. If G has a node v of degree $\geq \delta$, color the subgraph induced by $N(v)$ with two colors and delete $\{v\} \cup N(v)$ from G. We repeat this process using two new colors at each deleted neighborhood and stop when we arrive at a graph H whose maximum degree is less than δ. □

Applying Theorem 21 with $\delta = \sqrt{n}$ and the fact that a graph with maximum degree Δ has a $(\Delta + 1)$-coloring, one finds Wigderson's polynomial algorithm for coloring a 3-colorable graph with $3\lceil\sqrt{n}\rceil$ colors. More strongly, one can prove:

Theorem 22. *A 3-colorable graph can be colored with $O(n^{0.387})$ colors by a polynomial time randomized algorithm.*

Proof. Let G be a 3-colorable graph. Applying Theorem 21 with $\delta := n^{0.613}$, we find an induced subgraph H of maximum degree $\Delta_H < \delta$ and a coloring of $G \backslash H$ using $\frac{2n}{\delta} = O(n^{0.387})$ colors. By Theorem 20 and Lemma 19, H can be colored with $O(\delta^{\log_3 2}) = O(n^{0.387})$ colors. This shows the result. □

Improved coloring algorithm using "rounding via vector projections". In order to achieve the better $O(\Delta^{\frac{1}{3}}(\log \Delta)^{\frac{1}{3}} \log n)$-coloring algorithm for a 3-colorable graph, one has to improve Theorem 20 and to show how to construct in randomized polynomial time a $O(\Delta^{\frac{1}{3}}(\log \Delta)^{\frac{1}{3}})$-semicoloring. (Indeed, the desired coloring follows then as a direct application of Lemma 19.) For this, Karger, Motwani, and Sudan introduced another randomized technique for constructing a semicoloring from a vector coloring whose analysis has been refined by Halperin, Nathaniel and Zwick (2001) and is presented below. The main step consists of proving the following result.

Theorem 23. *Let G be a vector 3-colorable graph on n nodes with maximum degree Δ. Then an independent set of size $\Omega\left(\frac{n}{\Delta^{\frac{1}{3}}(\log \Delta)^{\frac{1}{3}}}\right)$ can be found in randomized polynomial time.*

Indeed if Theorem 23 holds, then one can easily construct a $O(\Delta^{\frac{1}{3}}(\log \Delta)^{\frac{1}{3}})$-semicoloring. For this, assign one color to the nodes of the independent set found in Theorem 23 and recurse on the remaining nodes. One can verify that after $O(\Delta^{\frac{1}{3}}(\log \Delta)^{\frac{1}{3}})$ recursive steps, one has properly colored at least half of the nodes; that is, one has constructed a $O(\Delta^{\frac{1}{3}}(\log \Delta)^{\frac{1}{3}})$-semicoloring.

We now turn to the proof of Theorem 23. Let $v_1, \ldots, v_n$ be unit vectors forming a vector 3-coloring of G (i.e., $v_i^T v_j \leq -\frac{1}{2}$ for all edges ij) and set $c := \sqrt{\frac{2}{3}\ln\Delta - \frac{1}{3}\ln\ln\Delta}$. Choose a random vector r according to the standard n-dimensional normal distribution; this means that the components $r_1, \ldots, r_n$ of r are independent random variables, each being distributed according to the standard normal distribution.

Set $I := \{i \in \{1, \ldots, n\} | r^T v_i \geq c\}$, $n' := |I|$, and let m (resp., m') denote the number of edges of G (resp. the number of edges of G contained in I). Then an independent set $J \subseteq I$ can be obtained by removing one vertex from each edge contained in I; thus $|J| \geq n' - m'$. Intuitively there cannot be too many edges within I. Indeed the vectors assigned to the endpoints of an edge are rather far apart since their angle is at least $\frac{2\pi}{3}$, while the vectors assigned to the vertices in I should all be close to r since they have a large inner product with r. The proof consists of showing that the expected value of $n' - m'$ is equal to

$$\Omega\left(\frac{n}{\Delta^{1/3}(\log\Delta)^{1/3}}\right).$$

The expected size of I is

$$E(n') = \sum_{i=1}^{n} \text{prob}(v_i^T r \geq c) = n \cdot \text{prob}(v_1^T r \geq c)$$

and the expected number of edges contained in I is

$$E(m') = \sum_{ij \in E} \text{prob}(v_i^T r \geq c \text{ and } v_j^T r \geq c) = m \cdot \text{prob}(v_1^T r \geq c \text{ and } v_2^T r \geq c)$$

where v_1 and v_2 denote two unit vectors satisfying $v_1^T v_2 \leq -\frac{1}{2}$. The following properties of the standard n-dimensional normal distribution will be used (see Karger, Motwani and Sudan (1998)).

Lemma 24. *Let u_1 and u_2 be unit vectors and let r be a random vector chosen according to the standard n-dimensional normal distribution. Let $N(x) = \int_x^\infty \phi(y)dy$ denote the tail of the standard normal distribution, where $\phi(x) = \frac{1}{\sqrt{2\pi}}\exp(-\frac{x^2}{2})$ is its density function.*

(i) *The inner product $r^T u_1$ is distributed according to the standard normal distribution. Therefore,* $\text{prob}(u_1^T r \geq c) = N(c)$.

(ii) *If u_1 and u_2 are orthogonal, then $u_1^T r$ and $u_2^T r$ are independent random variables.*

(iii) $(\frac{1}{x} - \frac{1}{x^3})\phi(x) \leq N(x) \leq \frac{1}{x}\phi(x)$ *for $x > 0$.*

It follows from Lemma 24 (i) that $E(n') = n \cdot N(c)$. We now evaluate $E(m')$. As before, v_1 and v_2 are two unit vectors such that $v_1^T v_2 \leq -\frac{1}{2}$. Since the probability $P_{12} := \text{prob}(v_1^T r \geq c \text{ and } v_2^T r \geq c)$ is a monotone increasing function of $v_1^T v_2$, it attains its maximum value when $v_1^T v_2 = -\frac{1}{2}$. We can therefore assume that $v_1^T v_2 = -\frac{1}{2}$. Karger, Motwani and Sudan (1998) show the upper bound $N(2c)$ for the probability P_{12} and, using a refinement of their method, Halperin, Nathaniel and Zwick (2001) prove the sharper bound $N(\sqrt{2}c)^2$.

Lemma 26. *If v_1 and v_2 are unit vectors such that $v_1^T v_2 = -\frac{1}{2}$, then* $\text{prob}(v_1^T r \geq c \text{ and } v_2^T r \geq c) \leq N(\sqrt{2}c)^2$.

Proof. Let r' denote the orthogonal projection of r on the plane spanned by v_1 and v_2. Then r' follows the standard 2-dimensional normal distribution and $v_i^T r' = v_i^T r$ for $i = 1, 2$. Hence we can work in the plane; Fig. 2 will help visualize the argument. Write r' as $r' = \alpha \cdot c v_1 + \beta \cdot c(v_1 + 2v_2)$ for some scalars α, β. As v_1 is orthogonal to $v_1 + 2v_2$, we find that $v_1^T r' \geq c$ if and only if $\alpha \geq 1$; that is, if r' belongs to the half-plane lying above the line $(D_1 A B_1)$ (see Fig. 2). Hence the probability P_{12} is equal to the probability that r' falls within the wedge defined by the angle $\angle B_1 A B_2$ (this is the shaded area in Fig. 2). Karger, Motwani and Sudan (1998) bound this probability by the probability that r' lies on the right side of the vertical line through A, which is equal to $\text{prob}((v_1 + v_2)^T r' \geq 2c)$ and thus to $N(2c)$ (since $v_1 + v_2$ is a unit vector). The better bound $N(\sqrt{2}c)^2$ can be shown as follows. Let u_1, u_2 be orthogonal unit vectors in the plane forming each the angle $\frac{\pi}{4}$ with $v_1 + v_2$. Denote by E_i the intersection point of the line through the origin parallel to u_i with the line through A perpendicular to u_i. One can easily verify that E_i is at distance $\sqrt{2}c$ from the origin. Now one can bound the probability P_{12} by the probability that r' falls within the wedge defined by the angle $\angle C_1 A C_2$. The latter probability is just $\text{prob}(u_1^T r' \geq \sqrt{2}c \text{ and } u_2^T r' \geq \sqrt{2}c)$ which (by Lemma 24 (i) (ii)) is equal to $N(\sqrt{2}c)^2$. □

We can now conclude the proof of Theorem 23. Lemma 26 implies that $E(m') \leq m \cdot N(\sqrt{2}c)^2$. As $m \leq \frac{n\Delta}{2}$, we obtain that

$$E(n' - m') \geq n \cdot N(c) - \frac{n\Delta}{2} N(\sqrt{2}c)^2 = n\left(N(c) - \frac{\Delta}{2} N(\sqrt{2}c)^2\right).$$

Using Lemma 24 (iii) we find that

$$\frac{N(c)}{N(\sqrt{2}c)^2} \geq \frac{\left(\frac{1}{c} - \frac{1}{c^3}\right)\frac{1}{\sqrt{2\pi}} e^{-\frac{c^2}{2}}}{\frac{1}{4c^2\pi} e^{-2c^2}} = 2\left(1 - \frac{1}{c^2}\right)\sqrt{2\pi} c e^{\frac{3}{2}c^2}.$$

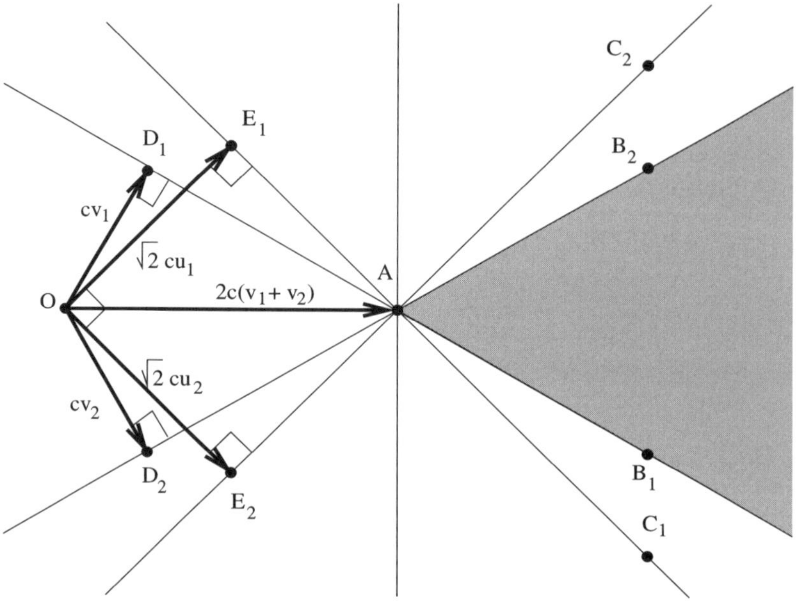

Fig. 2.

As $c = \sqrt{\frac{2}{3}\ln\Delta - \frac{1}{3}\ln\ln\Delta}$, we have $e^{\frac{3}{2}c^2} = \frac{\Delta}{\sqrt{\ln\Delta}}$. One can verify that

$$2\left(1-\frac{1}{c^2}\right)\sqrt{2\pi}ce^{\frac{3}{2}c^2} > \sqrt{2\pi}ce^{\frac{3}{2}c^2} > \Delta.$$

(This holds for Δ large enough. However, one can color G with $\Delta + 1$ colors in polynomial time (using a greedy algorithm) and thus find a stable set of size at least $\frac{n}{\Delta+1}$ which is $\Omega\left(\frac{n}{\Delta^{\frac{1}{3}}(\log\Delta)^{\frac{1}{3}}}\right)$for bounded Δ.) This shows that $N(c) > \Delta \cdot N(\sqrt{2}c)^2$. Therefore, $E(n' - m') \geq \frac{n}{2}N(c)$, and, using again Lemma 24 (iii),

$$E(n' - m') \geq \frac{n}{2}\left(\frac{1}{c} - \frac{1}{c^3}\right)\frac{1}{\sqrt{2\pi}}e^{\frac{-c^2}{2}} = \Omega\left(\frac{n}{\Delta^{\frac{1}{3}}(\log\Delta)^{\frac{1}{3}}}\right).$$

This concludes the proof of Theorem 23.

We mention below the k-analogue of Theorem 23, whose proof is similar. The analogue of Lemma 26 is that the probability P_{12} is bounded by

$$N\left(\sqrt{\frac{k-1}{k-2}}c\right)^2, \text{ where } c = \sqrt{\left(1-\frac{2}{k}\right)(2\ln\Delta - \ln\ln\Delta)}.$$

Theorem 27. *Let G be a vector k-colorable graph* ($k \geq 2$) *on n nodes with maximum degree* Δ. *Then an independent set of size* $\Omega\left(\frac{n}{\Delta^{1-\frac{2}{k}}(\log \Delta)^{\frac{1}{k}}}\right)$ *can be found in randomized polynomial time.*

Feige, Langberg, and Schechtman (2002) show that this result is in some sense best possible. They show that, for all $\epsilon > 0$ and $k > 2$, there are infinitely many graphs G that are vector k-colorable and satisfy $\alpha(G) \leq \frac{n}{\Delta^{1-\frac{2}{k}-\epsilon}}$, where n is the number of nodes and Δ is the maximum degree satisfying $\Delta > n^{\delta}$ for some constant $\delta > 0$.

The $O(n^{1-\frac{3}{k+1}}\sqrt{n})$-coloring algorithm of Karger–Motwani–Sudan for vector k-colorable graphs. As before, it suffices to show that one can find in randomized polynomial time an independent set of size

$$\Omega\left(\frac{n^{\frac{3}{k+1}}}{\sqrt{\log n}}\right) = \Omega\left(\frac{n}{n^{1-\frac{3}{k+1}}\sqrt{\log n}}\right)$$

in a vector k-colorable graph. (Indeed, using recursion, one can then find in randomized polynomial time a semicoloring using $O(n^{1-\frac{3}{k+1}}\sqrt{\log n})$ colors and thus, using Lemma 19, a coloring using the same number of colors.) The result is shown by induction on k. Suppose the result holds for any vector $(k-1)$-colorable graph. Set $\Delta_k(n) := n^{\frac{k}{k+1}}$ and let G be a vector k-colorable graph on n nodes. We distinguish two cases.

Suppose first that G has a node u of degree greater than $\Delta_k(n)$ and consider a subgraph H of G induced by a subset of $\Delta_k(n)$ nodes contained in the neighbourhood of u. Then H is vector $(k-1)$-colorable (easy to verify; see Karger, Motwani and Sudan (1998)). By the induction assumption, we can find an independent set in H (and thus in G) of size

$$\Omega\left(\frac{\Delta_k(n)^{\frac{3}{k}}}{\sqrt{\log \Delta_k(n)}}\right) = \Omega\left(\frac{n^{\frac{3}{k+1}}}{\sqrt{\log n}}\right).$$

Suppose now that the maximum degree Δ of G is less than or equal to $\Delta_k(n)$. It follows from Theorem 27 that we can find an independent set in G of size

$$\Omega\left(\frac{n}{\Delta_k(n)^{1-\frac{2}{k}}\sqrt{\log \Delta_k(n)}}\right) = \Omega\left(\frac{n^{\frac{3}{k+1}}}{\sqrt{\log n}}\right).$$

This concludes the proof.

6.5 Approximating the maximum stable set and vertex cover problems

The stable set problem. Determining the stability number of a graph is a hard problem. Arora, Lund, Motwani, Sudan, and Szegedy (1992) show the existence of a constant $\epsilon > 0$ for which there is no polynomial time algorithm permitting to find a stable set in a graph G of size at least $n^{-\epsilon}\alpha(G)$ unless P = NP. We saw in Section 4.2 that the theta number $\vartheta(G)$ is a polynomially computable upper bound for $\alpha(G)$ which is tight for perfect graphs, in which case a maximum cardinality stable set can be found in polynomial time. For general graphs, the gap between $\alpha(G)$ and $\vartheta(G)$ can be arbitrarily large. Indeed, Feige (1997) shows that, for all $\epsilon > 0$, there is a family of graphs for which $\vartheta(G) > n^{1-\epsilon}\alpha(G)$. The proof of Feige is nonconstructive; Alon and Kahale (1998) gave the following constructive proof for this result.

Theorem 28. *For every $\epsilon > 0$ one can construct a family of graphs on n nodes for which $\vartheta(G) \geq (\frac{1}{2} - \epsilon)n$ and $\alpha(G) = O(n^{\delta})$ where $0 < \delta < 1$ is a constant depending on ϵ.*

Proof. Given integers $0 < s < q$, let G_{qs} denote the graph on $n = \binom{2q}{q}$ nodes corresponding to all subsets A of $Q := \{1, \ldots, 2q\}$ with cardinality $|A| = q$, where A, B are adjacent if $|A \cap B| = s$. We begin with evaluating the theta number of G_{qs}. For every vertex A of G_{qs}, set $d_A := (x+1)\chi^A - \chi^Q$, where x is the largest root of the quadratic polynomial $sx^2 - 2(q-s)x + s = 0$. Then, $d_A^T d_B = 0$ for all adjacent A, B. Therefore, the vectors $v_A := \frac{d_a}{\|d_A\|}$ form an orthonormal representation of $\bar{G}_{qs}$. Setting $d := \frac{1}{\sqrt{2q}}(1, \ldots, 1)^T$ and using the definition from Theorem 12, we obtain:

$$\vartheta(G_{qs}) \geq \sum_A (d^T v_A)^2 = n\frac{(x-1)^2}{2(x^2+1)} = \frac{n}{2}\frac{q-2s}{q-s}.$$

In order to evaluate the stability number of G_{qs}, one can use the following result of Frankl and Rödl (1987): For every $\gamma > 0$, there exists $0 < \delta < 1$ for which $\alpha(G_{qs}) \leq n^{\delta}$ if $\gamma q < s < (1-\gamma)q$.

We now indicate how to choose the parameters q, s in order to achieve the conclusion of the theorem. Let $\epsilon > 0$ be given. Define s as the largest integer for which $s < \frac{q}{2}$ and $\frac{q-2s}{2(q-s)} > \frac{1}{2} - \epsilon$(i.e., $s < \frac{2q\epsilon}{1+2\epsilon}$). Choose γ such that $0 < \gamma < \frac{s}{q}$. Then $\gamma q < s < (1-\gamma)q$ and thus $\alpha(G_{qs}) \leq n^{\delta}$ for some $0 < \delta < 1$ by the Frankl–Rödl result. □

On the positive side, Alon and Kahale (1998) show the following two results; we present the second one without proof.

Theorem 29. *Let G be a graph on n nodes, $k \geq 3$, $m \geq 1$ be integers. If $\vartheta(G) \geq \frac{n}{k} + m$, then an independent set of cardinality $\Omega(m^{\frac{3}{k+1}}\log^{-\frac{1}{2}} m)$ can be found in randomized polynomial time.*

Proof. Using the definition of $\vartheta(G)$ from Theorem 12, there exist unit vectors $d, v_1, \ldots, v_n$ where $v_1, \ldots, v_n$ form an orthonormal representation of $\bar{G}$. These vectors can be found in polynomial time since, as the proof of Theorem 12 shows, they can be computed from an optimum solution to the SDP program (58). Order the nodes in such a way that $(d^T v_1)^2 \geq \cdots \geq (d^T v_n)^2$. As $\vartheta(G) \geq \frac{n}{k} + m$ and $(d^T v_i)^2 \leq 1$ for all i, we have $(d^T v_m)^2 \geq \frac{1}{k}$. Let H denote the subgraph of G induced by the nodes $1, \ldots, m$. Then, $v_1, \ldots, v_m$ is an orthonormal representation of $\bar{H}$, the complementary graph of H. Using the definition of the theta number from Theorem 14, we deduce that

$$\vartheta(\bar{H}) \leq \max_{i=1,\ldots,m} \frac{1}{(d^T v_i)^2} \leq k.$$

Therefore, H has a vector k-coloring. Applying the Karger–Motwani–Sudan results from the preceding subsection, one can find in randomized polynomial time a $O(m^{1-\frac{3}{k+1}}\sqrt{\log m})$ coloring of H. Then the largest color class in this coloring has cardinality $\Omega(m^{\frac{3}{k+1}}\log^{-\frac{1}{2}} m)$. □

Theorem 30. *If G is a graph on n nodes such that $\vartheta(G) > Mn^{1-\frac{2}{k}}$ for an appropriate absolute constant M, one can find in polynomial time a stable set in G of cardinality k.* □

Halperin, Nathaniel and Zwick (2001) show the following extension of Theorem 29.

Theorem 31. *Let G be a graph on n nodes that contains an independent set of size at least $\frac{n}{\alpha}$, where $\alpha \geq 1$, and set $k := \lfloor \alpha \rfloor$. Then an independent set of G of size $\tilde{\Omega}(n^{f(\alpha)})$ can be found in randomized polynomial time, where*

$$f(\alpha) = \frac{1}{k} \cdot \frac{\alpha(\alpha-1)}{\alpha(\alpha-k) + \frac{k^2-1}{3}}$$

(the notation $\tilde{\Omega}$ meaning that logarithmic factors are hidden). In particular, $f(\alpha) = 1$ for $1 \leq \alpha \leq 2$, $f(\alpha) = \frac{\alpha}{2(\alpha-1)}$ for $2 \leq \alpha \leq 3$, and $f(k) = \frac{3}{k+1}$ for every integer $k \geq 1$.

See, e.g., Halldórsson (1998, 1999), Halperin (2002) for further results.

The vertex cover problem. We now turn to the vertex cover problem. A subset $X \subseteq V$ is a *vertex cover* if every edge is adjacent to a node in X; that is, if $V \backslash X$ is a stable set. Denote by $vc(G)$ the minimum cardinality of a vertex cover in G. Thus $vc(G) = n - \alpha(G)$ and determining $vc(G)$ is therefore an NP-hard problem.

It is well known that $vc(G)$ can be approximated within a factor of 2 in polynomial time. An easy way to see it is to take a maximal matching M; then the set C of vertices covered by M forms a vertex cover such that

$vc(G) \le |C| = 2|M| \le 2 \cdot vc(G)$. Alternatively, this can be seen using an LP relaxation of the problem. Indeed, consider the LP problem:

$$\begin{aligned} lp(G) := \min \quad & \sum_{i \in V} x_i \\ \text{s.t.} \quad & x_i + x_j \ge 1 \quad (ij \in E) \\ & 0 \le x_i \le 1 \quad (i \in V) \end{aligned} \tag{108}$$

which is a linear relaxation of the vertex cover problem:

$$\begin{aligned} vc(G) := \min \quad & \sum_{i \in V} x_i \\ \text{s.t.} \quad & x_i + x_j \ge 1 \quad (ij \in E) \\ & x_i \in \{0, 1\} \quad (i \in V). \end{aligned} \tag{109}$$

Obviously, $lp(G) \le vc(G)$. Moreover, $vc(G) \le 2 \cdot lp(G)$; indeed, given an optimum solution x to (108), the set $X := \{i \in V | x_i \ge \frac{1}{2}\}$ is a vertex cover whose cardinality satisfies $|I| \le 2 \cdot lp(G)$.

On the negative side, it is known that the minimum vertex cover problem cannot be approximated in polynomial time within any factor smaller than $10\sqrt{5} - 21 \sim 1.36067$ if P $\neq$ NP (Dinur and Safra (2002)). The existence of a polynomial time approximation algorithm for the vertex cover problem with performance ratio $2 - \varepsilon$ remains, however, open for any $\varepsilon > 0$. Kleinberg and Goemans (1998) propose to use the following semidefinite relaxation of the problem (109):

$$\begin{aligned} sd(G) := \min \quad & \sum_{i=1}^{n} \frac{1 + v_0^T v_i}{2} \\ \text{s.t.} \quad & (v_0 - v_i)^T (v_0 - v_j) = 0 \quad (ij \in E) \\ & v_0, v_1, \ldots, v_n \quad \text{unit vectors.} \end{aligned} \tag{110}$$

They show that this semidefinite bound $sd(G)$ is equal to the obvious lower bound $n - \vartheta(G)$ for $vc(G)$, where $\vartheta(G)$ is the theta number bounding $\alpha(G)$. To see it, consider the matrix $X = (x_{ij})_{i,j=0}^n$ where $x_{ij} = v_i^T v_j$ and $v_0, \ldots, v_n$ satisfy (110); then X is constrained to be positive semidefinite with an all ones diagonal and to satisfy $1 + x_{ij} - x_{0i} - x_{0j} = 0$ for all edges ij of G. If we define the matrix $Y = (y_{ij})_{i,j=1}^n$ by

$$y_{ij} = \frac{1}{4}(1 + x_{ij} - x_{0i} - x_{0j}) \quad \text{for} \quad i, j = 1, \ldots, n,$$

then the objective function in (110) reads $n - \sum_{i=1}^n y_{ii}$ and X is feasible for (110) if and only if Y satisfies $Y - \text{diag}(Y)\text{diag}(Y)^T \succeq 0$ and $y_{ij} = 0 (ij \in E)$; that is, if the vector $(y_{ii})_{i=1}^n$ belongs to the theta body TH(G). (We use the definition of $\vartheta(G)$ from Theorem 11. See Laurent, Poljak and Rendl (1997) for details on the above $X \to Y$ mapping.)

A first observation is that this SDP bound is at least as good as the LP bound; namely,

$$sd(G) = n - \vartheta(G) \geq lp(G).$$

To see it, use the definition from Theorem 12. Let d be a unit vector and $v_1, \ldots, v_n$ an orthonormal representation of $\bar{G}$ such that $\vartheta(G) = \sum_{i \in V}(d^T v_i)^2$. Set $x_i := 1 - (d^T v_i)^2$ $(i \in V)$. Then x is a feasible solution to the program (108) which shows that $lp(G) \leq \sum_i x_i = n - \vartheta(G)$.

Kleinberg and Goemans (1998) construct a class of graphs G for which the ratio $\frac{vc(G)}{n-\vartheta(G)}$ converges to 2 as n goes to infinity, which shows that no improvement is made by using SDP instead of LP. (In fact, the class of graphs constructed in Theorem 28 displays the same behavior.) They also propose to strengthen the semidefinite program (110) by adding to it the constraints

$$(v_0 - v_i)^T (v_0 - v_j) \geq 0 \quad (ij \in \bar{E});$$

the new semidefinite bound can be verified to be equal to $n - \vartheta'(G)$, where $\vartheta'(G)$ is the sharpening of $\vartheta(G)$ introduced in Section 4.4. Charikar (2002) shows that the new integrality gap $\frac{vc(G)}{n-\vartheta'(G)}$ can again be made arbitrarily close to 2.

Improved approximation algorithms exist for graphs with bounded maximum degree Δ. Improving on earlier results, Halperin (2002) shows that, for graphs with maximum degree Δ, the semidefinite relaxation (110) together with suitable randomized rounding permits to derive an approximation algorithm for the minimum vertex cover problem with performance ratio $2 - (1 - o(1))\frac{2 \ln \ln \Delta}{\ln \Delta}$ for large Δ. We sketch this result below.

Halperin's algorithm is based on the following observation. Given a scalar $x \geq 0$, the set $C := (i \in \{1, \ldots, n\}; \mid v_0^T v_i \geq -x\}$ is a vertex cover. Note that for $x = 0$, we have $|C| \leq 2 \cdot sd(\mathrm{G})$ and thus this gives again a 2-approximation algorithm. Moreover, if J is an independent set contained in the set $S_2 := \{i \in \{1, \ldots, n\} \mid -x \leq v_0^T v_i < x\}$, then the set $C \backslash J$ is still a vertex cover. When x is small, nodes in S_2 correspond to vectors v_i that are approximately orthogonal to v_0 and thus the endpoints of an edge contained in S_2 correspond to approximately opposite vectors. Hence the set S_2 is likely to contain few edges and thus a large independent set J; therefore, the set $C \backslash J$ is likely to be a small vertex cover.

More precisely, Halperin defines $x = \Theta(\frac{\ln \ln \Delta}{\ln \Delta})$ and the sets $S_1 := \{i \in \{1, \ldots, n\} \mid v_0^T v_i \geq x\}$ and $S_2 = \{i \in \{1, \ldots, n\} \mid -x \leq v_0^T v_i < x\}$ as above (thus $C = S_1 \cup S_2$). Then, $|S_1| \leq \frac{2}{x+1} sd(G)$ and $|S_2| \leq \frac{2}{1-x} sd(G)$. A large independent set J can be found in S_2 using the "rounding via vector projections" technique from Karger, Motwani and Sudan (1998), exposed earlier in Section 6.4. Indeed, if ij is an edge contained in S_2, then $v_i^T v_j = v_0^T v_i + v_0^T v_j - 1 < 2x - 1$. Hence, the subgraph of G induced by S_2 has a vector k-coloring for $k = \frac{2(1-x)}{1-2x}$. Therefore, Theorem 27 can be used for finding a large independent set in S_2. These facts yield the desired performance ratio; see Halperin (2002) for details.

As mentioned above, no polynomial time approximation algorithm is known for the vertex cover problem having a performance ratio $2 - \varepsilon$ with $\varepsilon > 0$. In fact, no tractable linear relaxation is known for (109), having an integrality gap lower than 2. Arora, Bollabás, and Lovász (2002) initiate a more systematic approach for proving nonexistence of tighter relaxations. They show an integrality gap of $2 - o(1)$ for some fairly general families of LP relaxations of (109). A first family consists of the LP relaxations in which each constraint has at most ϵn variables. A second family involves LP relaxations in which each constraint has defect at most ϵn; the defect of an inequality $a^T x \geq b$ being $2b - \sum_i a_i$. A third family consists of the LP relaxations obtained after O(1) iterations of the Lovász–Schrijver N operator applied to the LP in (108). It is an open question whether an analog result holds for the N_+ operator.

6.6 *Approximating MAX SAT*

An instance of the MAX SAT problem in the Boolean variables $x_1, \ldots, x_n$ is composed of a collection $\mathcal{C}$ of clauses C with nonnegative weights w_C associated to them. Each *clause* C is of the form $z_1 \vee \cdots \vee z_k$ where each z_j is either a variable x_i or its negation $\bar{x}_i$ (called a *literal*); k is its *length* and C is satisfied if at least one of the literals $z_1, \ldots, z_k$ is assigned value 1 (if a variable x_i is assigned value 1 then its negation $\bar{x}_i$ is assigned value 0 and vice versa). The MAX SAT problem consists of finding an assignment of 0/1 value to the variables $x_1, \ldots, x_n$ so that the total weight of the satisfied clauses is maximized. Given an integer $k \geq 1$, the MAX kSAT problem is the special instance of MAX SAT where each clause has length at most k and MAX EkSAT is the instance where all clauses have length exactly k; an instance of MAX SAT is said to be *satisfiable* if there is an assignment of the x_i's satisfying all its clauses.

The MAX SAT and MAX kSAT problems are NP-hard. Moreover, Håstad (1997) proved that, for any $\epsilon > 0$, there is no $(\frac{7}{8} + \epsilon)$-approximation algorithm for MAX SAT, unless P = NP; his result also holds when restricted to satisfiable instances of MAX E3SAT. Håstad (1997) also proved that, for any $\epsilon > 0$, there is no $(\frac{21}{22} + \epsilon)$-approximation algorithm for MAX 2SAT unless P = NP.

A $\frac{3}{4}$-approximation algorithm for MAX SAT. The first approximation algorithm for MAX SAT is the following $\frac{1}{2}$-approximation algorithm due to

Johnson (1974). Given $p_i \in [0, 1]$ ($i = 1, \ldots, n$), set independently and randomly each variable x_i to 1 with probability p_i. Then the probability that a clause $C := \vee_{i \in I_C^+} x_i \vee \vee_{i \in I_C^-} \bar{x}_i$ is satisfied is equal to $1 - \prod_{i \in I_C^+}(1 - p_i) \prod_{i \in I_C^-} p_i$. If we set all p_i's to $\frac{1}{2}$, then the total expected weight $\hat{W}_1$ of satisfied clauses satisfies:

$$\hat{W}_1 = \sum_{C \in \mathcal{C}} w_C \left(1 - \frac{1}{2^{k_C}}\right) \geq \frac{1}{2} \sum_{C \in \mathcal{C}} w_C$$

where k_C is the length of clause C. Therefore, this gives a randomized $\frac{1}{2}$-approximation algorithm for MAX SAT or a $(1 - 2^{-k})$-approximation algorithm for instances MAX SAT where all clauses have length $\geq k$ (thus with performance ratio $\frac{3}{4}$ for MAX E2SAT and $\frac{7}{8}$ for MAX E3SAT); it can be derandomized using the method of conditional probabilities.

Goemans and Wiliamson (1994) give an improved $\frac{3}{4}$-approximation algorithm using linear programming. Consider the integer programming problem:

$$\begin{array}{lll} \max & \sum_{C \in \mathcal{C}} w_C z_C & \\ \text{s.t.} & z_C \leq \sum_{i \in I_C^+} y_i + \sum_{i \in I_C^-} (1 - y_i) \ (C \in \mathcal{C}) & \\ & 0 \leq z_C \leq 1 & (C \in \mathcal{C}) \\ & y_i \in \{0, 1\} & (i = 1, \ldots, n) \end{array} \tag{111}$$

and let Z_{LP}^* denote the optimum value of its linear programming relaxation obtained by relaxing the condition $y_i \in \{0, 1\}$ by $0 \leq y_i \leq 1$. If (y, z) is an optimum solution to (111), letting $x_i = 1$ if and only if $y_i = 1$, then clause C is satisfied precisely when $z_C = 1$; hence (111) solves the MAX SAT problem. The GW approximation algorithm goes as follows. First, solve the LP relaxation of (111) and let (y, z) be an optimum solution to it. Then, apply the Johnson's algorithm using the probabilities $p_i := y_i$; that is, set x_i to 1 with probability y_i. Setting $\beta_k := 1 - (1 - \frac{1}{k})^k$ and using the fact[4] that $1 - \prod_{i \in I_C^+}(1 - y_i) \prod_{i \in I_C^-} y_i \geq \beta_{k_C} z_C$, we find that the expected weight $\hat{W}_2$ of satisfied clauses satisfies:

$$\hat{W}_2 = \sum_{C \in \mathcal{C}} w_C \left(1 - \prod_{i \in I_C^+}(1 - y_i) \prod_{i \in I_C^-} y_i\right) \geq \sum_{C \in \mathcal{C}} w_C z_C \beta_{k_C}.$$

As β_k is a monotone decreasing function of k, this gives a randomized β_k-approximation algorithm for instances of MAX SAT where all clauses have at most k literals; thus a $(1 - \frac{1}{e})$ approximation algorithm for MAX SAT, since $\lim_{k \to \infty}(1 - \frac{1}{k})^k = \frac{1}{e}$.

[4] The proof uses the arithmetic/geometric mean inequality: $\frac{a_1 + \cdots + a_n}{n} \geq (a_1 \ldots a_n)^{\frac{1}{n}}$ for any nonnegative numbers $a_1, \ldots, a_n$.

In order to obtain the promised $\frac{3}{4}$ performance ratio, it suffices to combine the above two algorithms. For this, note that $\frac{1}{2}(1 - \frac{1}{2^k} + \beta_k) \geq \frac{3}{4}$ for all $k \geq 1$. Therefore, $\frac{1}{2}(\hat{W}_1 + \hat{W}_2) \geq \frac{3}{4} Z^*_{LP}$. Hence the following is a $\frac{3}{4}$-approximation algorithm for MAX SAT: with probability $\frac{1}{2}$, use the probabilities $p_i := \frac{1}{2}$ for determining the variables x_i and, with probability $\frac{1}{2}$, use instead the probabilities $p_i := y_i$.

Other $\frac{3}{4}$-approximation algorithms for MAX SAT are given by Goemans and Williamson (1994). Instead of setting $x_i = 1$ with probability y_i, they set $x_i = 1$ with probability $f(y_i)$ for some suitably chosen function $f(\cdot)$.

Better approximation algorithms can be obtained using semidefinite relaxations instead of linear ones combined with adequate rounding techniques, as we now see.

The Goemans–Williamson α_0-approximation algorithm for MAX 2SAT and their 0.7554-approximation algorithm for MAX SAT. Using a semidefinite relaxation for MAX SAT instead of a linear one and the hyperplane rounding technique, one can show a better approximation algorithm. It is convenient to introduce the new Boolean variables $x_{n+i} = \bar{x}_i$ for $i = 1, \ldots, n$. Then a clause C can be expressed as a disjunction $C = \vee_{i \in I_C} x_i$, of the variables $x_1, \ldots, x_{2n}$, with $I_C \subseteq \{1, \ldots, 2n\}$. It is also convenient to work with ± 1 variables v_i (instead of $y_i \in \{0,1\}$) and to introduce an additional ± 1 variable v_0, the convention being to set x_i to 1 if $v_i = -v_0$ and to 0 if $v_i = v_0$. Hence the formulation (111) of MAX SAT can be rewritten as

$$\begin{array}{lll} \max & \sum_{C \in \mathcal{C}} w_C z_C & \\ \text{s.t.} & z_C \leq \sum_{i \in I_C} \frac{1 - v_0 \cdot v_i}{2} & (C \in \mathcal{C}) \\ & 0 \leq z_C \leq 1 & (C \in \mathcal{C}) \\ & v_i \cdot v_{n+i} = -1 & (i = 1, \ldots, n) \\ & v_0, v_1, \ldots, v_{2n} \in \{\pm 1\}. & \end{array} \tag{112}$$

For each clause $C = x_i \vee x_j$ of length 2, one can add the constraint:

$$z_C \leq 1 - \left(\frac{1 + v_0 \cdot v_i}{2}\right)\left(\frac{1 + v_0 \cdot v_j}{2}\right) = \frac{3 - v_0 \cdot v_i - v_0 \cdot v_j - v_i \cdot v_j}{4} \tag{113}$$

which, in fact, implies the constraint $z_C \leq \frac{1 - v_0 \cdot v_i}{2} + \frac{1 - v_0 \cdot v_j}{2}$.

Let (SDP) denote the semidefinite relaxation of the program (112) augmented with the constraints (113) for all clauses of length 2, which is obtained by introducing a matrix variable $X = (X_{ij})_{i,j=0}^{2n} \succeq 0$ and replacing each product $v_i \cdot v_j$ by X_{ij}. In other words, this amounts to replacing the constraint $v_0, \ldots, v_{2n} \in \{\pm 1\}$ by the constraint $v_0, \ldots, v_{2n} \in S^n$, S^n being the unit sphere in $\mathbb{R}^{n+1}$ (the product $v_i \cdot v_j$ meaning then the inner product $v_i^T v_j$).

Goemans and Williamson (1995) show that their basic α_0-approximation algorithm for max-cut extends to MAX 2SAT. Namely, solve the relaxation (SDP) and let $v_0, \ldots, v_n$ be the optimum unit vectors solving it; select a random unit vector r and let H_r be the hyperplane with normal vector r; set x_i to 1 if the hyperplane H_r separates v_0 and v_i and to 0 otherwise. Let θ_{ij} denote the angle (v_i, v_j). Then the probability prob(v_0, v_i) that the clause x_i is satisfied is equal to the probability that H_r separates v_0 and v_i and thus

$$\text{prob}\,(v_0, v_i) = \frac{\theta_{0i}}{\pi};$$

the probability prob(v_0, v_i, v_j) that the clause $x_i \vee x_j$ is satisfied is equal to the probability that a random hyperplane separates v_0 from at least one of v_i and v_j which can be verified to be equal to

$$\text{prob}\,(v_0, v_1, v_j) = \frac{1}{2\pi}(\theta_{0i} + \theta_{0j} + \theta_{ij})$$

using the inclusion/exclusion principle. Therefore, for a clause $C = x_i \vee x_j$, we have

$$\frac{\text{prob}(v_0, v_i, v_j)}{z_C} \geq \frac{2}{\pi}\,\frac{\theta_{0i} + \theta_{0j} + \theta_{ij}}{3 - \cos\theta_{0i} - \cos\theta_{0j} - \cos\theta_{ij}} \geq \alpha_0,$$

where $\alpha_0 \simeq 0.87856$ is the Goemans–Williamson ratio from (84). The above relation also holds when $i = j$, i.e., when C is a clause of length 1, in which case one lets prob(v_0, v_i, v_j) = prob(v_0, v_i). Hence the expected total weight of satisfied clauses is greater than or equal to α_0 times the optimum value of the relaxation (SDP); this gives therefore an α_0-approximation algorithm for MAX 2SAT.

This improved MAX 2SAT algorithm leads to a slightly improved 0.7554-approximation algorithm for general MAX SAT. For this, one considers the following three algorithms: (1) set x_i to 1 independently with probability $p_i := \frac{1}{2}$; (2) set x_i to 1 independently with probability $p_i := \frac{1 - v_0^T v_i}{2}$; (3) select a random hyperplane H_r and set x_i to 1 if H_r separates v_0 and v_i (the v_i's being the optimum vectors to the relaxation (SDP)). One chooses algorithm (i) with probability q_i where $q_1 = q_2 = 0.4785$ and $q_3 = 1 - q_1 - q_2 = 0.0430$. Then the expected weight of the satisfied clauses is at least

$$\sum_{C|k_C \leq 2} w_C z_C \left(\frac{3}{2} q_1 + q_3 \cdot \alpha_0\right) + \sum_{C|k_C \geq 3} w_C z_C \cdot q_1 \left(1 - \frac{1}{2^k} + 1 - \left(1 - \frac{1}{k}\right)^k\right)$$

which can be verified to be at least $0.7554 \cdot \sum_C w_C z_C$. A refinement of this algorithm is given by Goemans and Williamson (1994) with an improved performance ratio 0.7584.

The improved Feige–Goemans 0.931-approximation algorithm for MAX 2SAT. Feige and Goemans (1995) show an improved performance ratio of about 0.931 for MAX 2SAT. For this, they strengthen the semidefinite relaxation (SDP) by adding to it the triangle inequalities:

$$X_{0i} + X_{0j} + X_{ij} \geq -1, \quad X_{0i} - X_{0j} - X_{ij} \geq -1, \quad -X_{0i} - X_{0j} + X_{ij} \geq -1 \tag{114}$$

for all $i, j \in \{1, \ldots, 2n\}$. Moreover, they replace the vectors $v_0, v_1, \ldots, v_n$ (obtained from the optimum solution to the strengthened semidefinite program) by a new set of vectors $v'_0, \ldots, v'_n$ obtained by applying some *rotation* to the v_i's. Then the assignment for the Boolean variables x_i are generated from the v'_i using as before the hyperplane rounding technique.

Let us explain how the vectors v'_i are generated from the v_i's. Let f: $[0, \pi] \to [0, \pi]$ be a continuous function such that $f(0) = 0$ and $f(\pi - \theta) = \pi - f(\theta)$. As before, θ_{ij} denotes the angle (v_i, v_j). The vector v_i is rotated in the plane spanned by v_0 and v_i until it forms an angle of $f(\theta_{0i})$ with v_0; the resulting vector is v'_i. If $v_i = v_0$ then $v'_i = v_i$. Moreover, let $v'_{n+i} = -v'_i$ for $i = 1, \ldots, n$. Let θ'_{ij} be the angle (v'_i, v'_j). Then $\theta'_{0i} = f(\theta_{0i})$ and Feige and Goemans (1995) show the following equation permitting to express θ'_{ij} in terms of θ_{ij}:

$$\cos\theta'_{ij} = \cos\theta'_{0i}\cos\theta'_{0j} + \frac{\cos\theta_{ij} - \cos\theta_{0i}\cos\theta_{0j}}{\sin\theta_{0i}\sin\theta_{0j}}\sin\theta'_{0i}\sin\theta'_{0j}. \tag{115}$$

The probability that the clause $x_i \vee x_j$ is satisfied is now equal to

$$\text{prob}\,(v_0, v'_i, v'_j) = \frac{\theta'_{0i} + \theta'_{0j} + \theta'_{ij}}{2\pi}$$

while the contribution of this clause to the objective function of the semidefinite relaxation is

$$z_C \leq \frac{3 - \cos\theta_{0i} - \cos\theta_{0j} - \cos\theta_{ij}}{4}.$$

The performance ratio of the approximation algorithm using a rotation function f is, therefore, at least

$$\beta(f) := \min \frac{2}{\pi} \cdot \frac{\theta'_{01} + \theta'_{02} + \theta'_{12}}{3 - \cos\theta_{01} - \cos\theta_{02} - \cos\theta_{12}}$$

where the minimum is taken over all θ_{01}, θ_{02}, $\theta_{12} \in [0, \pi]$ for which $\cos\theta_{01}$, $\cos\theta_{02}, \cos\theta_{12}$ satisfy the triangle inequalities (114). Recall that $\theta'_{0i} = f(\theta_{0i})$ and relation (115) permits to express θ'_{12} in terms of θ_{01}, θ_{02}, and θ_{12}.

Feige and Goemans (1995) used a rotation function of the form

$$f_\lambda(\theta) = (1-\lambda)\theta + \lambda\frac{\pi}{2}(1-\cos\theta) \tag{116}$$

and, for the choice $\lambda = 0.806765$, they claim the lower bound 0.93109 for $\beta(f)$. Proving a correct evaluation of $\beta(f)$ is a nontrivial task, since the minimization program defining $\beta(f)$ is too complicated to be handled analytically. Zwick (2000) makes a detailed and rigorous analysis enabling him to prove a performance ratio of 0.931091 for MAX 2SAT.

The Matuura–Matsui 0.935-approximation algorithm for MAX 2SAT. Matuura and Matsui (2001b) designed an approximation algorithm for MAX 2SAT with performance ratio 0.935. As in the Feige–Goemans algorithm, their starting point is to use the semidefinite relaxation (SDP') of MAX 2SAT obtained from (112) by adding the constraints (113) for the clauses of length 2 and the triangle inequalities (114); they fix v_0 to be equal to $(1, 0, \ldots, 0)^T$. Let $v_1, \ldots, v_n$ be the unit vectors obtained from an optimum solution to the program (SDP'). No rotation is applied to the vectors v_i as in the Feige–Goemans algorithm. The new ingredient in the algorithm of Matuura–Matsui consists of selecting the random hyperplane using a distribution function f on the sphere which is skewed towards v_0 and uniform in any direction orthogonal to v_0, instead of a uniform distribution.

Let $\mathcal{F}_n$ denote the set of functions $f: S^n \to \mathbb{R}_+$ satisfying $\int_{S^n} f(v)dv = 1$, $f(v) = f(-v)$ for all $v \in S^n$, and $f(u) = f(v)$ for all u, $v \in S^n$ such that $u^T v_0 = v^T v_0$. Let $f \in \mathcal{F}_n$ and let the random unit vector r be now chosen according to the distribution function f. Then, $\text{prob}(v_i, v_j | f)$ denotes the probability that the clause $x_i \vee x_j$ is satisfied, i.e., as before, the probability that $\text{sign}(r^T v_0) \neq \text{sign}(r^T v_i)$ or $\text{sign}(r^T v_0) \neq \text{sign}(r^T v_j)$. Let P denote the linear subspace spanned by v_0, v_i, v_j and let $\hat{f}$ denote the distribution on S^2 obtained by projecting onto P; that is, $\hat{f}(v') := \int_{T(v')} f(v)dv$, where $T(v')$ is the set of all $v \in S^n$ whose projection on P is parallel to v'. Then the new approximation ratio of the algorithm is equal to

$$\alpha_{\hat{f}} := \min \frac{\text{prob}(v_i, v_j | \hat{f})}{\frac{1}{4}(3 - v_0^T v_i - v_0^T v_j - v_i^T v_j)}$$

where the minimum is taken over all $v_i, v_j \in S^2$ which together with $v_0 = (1, 0, 0)^T$ have their pairwise inner products satisfying the triangle inequalities (114).

The difficulty consists of constructing a distribution function $f \in \mathcal{F}_n$ for which $\alpha_{\hat{f}}$ is large. Matuura and Matsui (2001) show the following. The function

$$g(v) := \cos^{1/1.3}(\theta) \quad \text{for all } v \in S^2 \quad \text{with } |v_0^T v| = \cos\theta, \tag{117}$$

is a distribution function on S^2 belonging to $\mathcal{F}_2$; it satisfies $\alpha_g \geq 0.935$ (this is proved numerically); and there exists $f \in \mathcal{F}_n$ for which $\hat{f} = g$.

The Lewin–Livnat–Zwick 0.940-approximation algorithm for MAX 2SAT. Lewin, Livnat, and Zwick (2002) achieve this improved performance ratio by combining the skewed hyperplane rounding technique exploited by Matuura and Matsui (2001b) with the pre-rounding rotation phase used by Feige and Goemans (1995).

The Karloff–Zwick $\frac{7}{8}$-approximation algorithm for MAX 3SAT. Karloff and Zwick (1997) present an approximation algorithm for MAX 3SAT whose performance ratio they conjecture to be equal to $\frac{7}{8} = 0.875$, thus the best possible since Håstad (1997) proved the nonexistence of an approximation algorithm with performance ratio $> \frac{7}{8}$ unless P = NP. Previous algorithms were using a reduction to the case of MAX 2SAT; for instance, Trevisan, Sorkin, Sudan, and Williamson (1996) give a 0.801-approximation algorithm for MAX 3SAT using the Feige-Goemans 0.931 result for MAX 2SAT. Karloff and Zwick do not make such a reduction but consider instead the following direct semidefinite relaxation for MAX 3SAT:

$$\begin{array}{ll}
\max & \displaystyle\sum_{i,j,k \in \{1,\ldots,2n\}} w_{ijk} z_{ijk} \\
\text{s.t.} & z_{ijk} \leq \text{relax}\,(v_0, v_i, v_j, v_k) \\
& v_i \cdot v_{n+i} = -1 \quad (i = 1, \ldots, n) \\
& v_0, \ldots, v_{2n} \in S^n,\, z_{ijk} \in \mathbb{R},
\end{array}$$

where z_{ijk} is a scalar attached to the clause $x_i \vee x_j \vee x_k$ and

$$\begin{aligned}
\text{relax}(v_0, v_i, v_j, v_k) := \min\Bigg(& 1 - \frac{(v_0 + v_i)^T (v_j + v_k)}{4}, \\
& 1 - \frac{(v_0 + v_j)^T (v_i + v_k)}{4}, 1 - \frac{(v_0 + v_k)^T (v_i + v_j)}{4}, 1 \Bigg).
\end{aligned}$$

Note indeed that when the v_i's are ± 1 scalars, then relax (v_0, v_i, v_j, v_k) is equal to 0 precisely when $v_0 = v_i = v_j = v_k$ which corresponds to setting all variables x_i, x_j, x_k to 0 and thus to the clause $x_i \vee x_j \vee x_k$ not being satisfied.

Denote again by $\text{prob}(v_0, v_i, v_j, v_k)$ the probability that $x_i \vee x_j \vee x_k$ is satisfied and set

$$\text{ratio}(v_0, v_i, v_j, v_k) := \frac{\text{prob}(v_0, v_i, v_j, v_k)}{\text{relax}(v_0, v_i, v_j, v_k)}.$$

For a clause of length 1 or 2 (obtained by letting $j=k=0$ or $k=0$), it follows from the analysis of the GW algorithm that $\text{ratio}(v_0, v_i, v_j, v_k) \geq \alpha_0 > \frac{7}{8}$. For clauses of length 3, the analysis is technically much more involved and requires the computation of the volume of spherical tetrahedra as we now see.

Clearly, $\text{prob}(v_0, v_i, v_j, v_k)$ is equal to the probability that the random hyperplane H_r separates v_0 from at least one of v_i, v_j, v_k and thus to

$$1 - 2 \cdot \text{prob}(r^T v_h \geq 0 \; \forall h = 0, i, j, k).$$

We may assume without loss of generality that v_0, v_i, v_j, v_k lie in $\mathbb{R}^4$ and, since we are only interested in the inner products $r^T v_h$, we can replace r by its normalized projection on $\mathbb{R}^4$ which is then uniformly distributed on the sphere S^3. Define

$$T(v_0, v_i, v_j, v_k) := \{r \in S^3 \mid r^T v_h \geq 0 \; \forall h = 0, i, j, k\}.$$

Then,

$$\text{prob}(v_0, v_i, v_j, v_k) = 1 - 2 \cdot \frac{\text{vol}(T(v_0, v_i, v_j, v_k))}{\text{vol}(S^3)}$$

where $\text{vol}(\cdot)$ denotes the 3-dimensional spherical volume. As $\text{vol}\,(S^3) = 2\pi^2$, we find that

$$\text{prob}(v_0, v_i, v_j, v_k) = 1 - 2 \cdot \frac{\text{vol}(T(v_0, v_i, v_j, v_k))}{\pi^2}.$$

When the vectors v_0, v_i, v_j, v_k are linearly independent, $T(v_0, v_i, v_j, v_k)$ is a spherical tetrahedron, whose vertices are the vectors $v'_0, v'_i, v'_j, v'_k \in S^3$ satisfying $v_h^T v'_h > 0$ for all h and $v_{h_1}^T v'_{h_2} = 0$ for all distinct h_1, h_2. That is,

$$T(v_0, v_i, v_j, v_k) = \left\{ \sum_{h=0,i,j,k} \alpha_h v'_h \middle| \alpha_h \geq 0, \sum_h \alpha_h = 1 \right\}.$$

Therefore, evaluating the quantity ratio (v_0, v_i, v_j, v_k) and thus the performance ratio of the algorithm relies on proving certain inequalities about volumes of spherical tetrahedra.

Karloff and Zwick (1997) show that $\text{prob}(v_0, v_i, v_j, v_k) \geq \frac{7}{8}$ whenever $\text{relax}(v_0, v_i, v_j, v_k) = 1$, which shows a performance ratio $\frac{7}{8}$ for satisfiable instances of MAX 3SAT. Their proof is computer assisted as it involves one computation carried out with Mathematica. Zwick (2002) can prove the performance ratio $\frac{7}{8}$ for general MAX 3SAT. Although his proof is again computer assisted, it can however be considered as a rigorous proof since it is carried out using a new system called RealSearch, written by Zwick, which involves only interval arithmetic (instead of floating point arithmetic). We refer to Zwick's paper for an interesting presentation and discussion.

Further extensions. Karloff and Zwick (1997) describe a procedure for constructing strong semidefinite relaxations for general constraint satisfaction problems and thus for MAX kSAT. Halperin and Zwick (2001b) study approximation algorithms for MAX 4SAT using the semidefinite relaxation provided by the Karloff–Zwick recipe. The analysis of the classic hyperplane rounding technique necessitates now the evaluation of the probability prob$(v_0, \ldots, v_4)$ that a random hyperplane separates v_0 from at least one of $v_1, \ldots, v_4$. Luckily, using the inclusion/exclusion formula, this probability can be expressed in terms of the probabilities prob(v_i, v_j) and prob(v_i, v_j, v_k, v_l) that were considered above. In this way, Halperin and Zwick can show a performance ratio of 0.845173 for MAX 4SAT, thus below the target ratio of $\frac{7}{8}$. They study in detail a variety of other possible rounding strategies which enable them to obtain some improved performance ratios, like 0.8721.

Asano and Williamson (2000) present an improved approximation algorithm for MAX SAT with performance ratio 0.7846. For this, they use a new family of approximation algorithms extending the $\frac{3}{4}$-approximation algorithm of Goemans and Williamson (1994) (presented earlier in this section) combined with the semidefinite approach for MAX 2SAT and MAX 3SAT of Karloff and Zwick (1997) and Feige and Goemans (1995).

Further work related to defining stronger semidefinite relaxations for the satisfiability problem can be found, e.g., in Anjos (2004), de Klerk, Warners, and van Maaren (2000), Warners (1999).

6.7 *Approximating the maximum directed cut problem*

Given a directed graph $G=(V, A)$ and weights $w \in \mathbb{Q}_+^A$ associated to its arcs, the *maximum directed cut problem* asks for a directed cut $\delta^+(S)$ of maximum weight where, for $S \subseteq V$, the *directed cut* (or *dicut*) $\delta^+(S)$ is the set of arcs ij with $i \in S$ and $j \notin S$. This problem is NP-hard, since the max-cut problem in a undirected graph H reduces to the maximum dicut problem in the directed graph obtained by replacing each edge of H by two opposite arcs. Moreover, no approximation algorithm for the maximum dicut problem exists having a performance ratio $> \frac{12}{13}$ unless P = NP (Håstad (1997)).

The simple random partition algorithm (which assigns each node to S independently with probability $\frac{1}{2}$) has a performance ratio $\frac{1}{4}$. Goemans and Williamson (1995) show that their basic approximation algorithm for max-cut can be extended to the maximum dicut problem with performance ratio 0.79607. Feige and Goemans (1995) prove an improved performance ratio of 0.859. These algorithms use the same ideas as the algorithms for MAX 2SAT presented in the same papers. Before presenting them, we mention a simple

$\frac{1}{2}$-approximation algorithm of Halperin and Zwick (2001c) using a linear relaxation of the problem; this algorithm can in fact be turned into a purely combinatorial algorithm.

A $\frac{1}{2}$-approximation algorithm by Halperin and Zwick. Consider the following linear program:

$$\begin{array}{lll} \max & \sum_{ij\in A} w_{ij}z_{ij} & \\ \text{s.t.} & z_{ij} \leq x_i & (ij \in A) \\ & z_{ij} \leq 1 - x_j & (ij \in A) \\ & 0 \leq x_i \leq 1 & (i \in V). \end{array} \tag{118}$$

If we replace the linear constraint $0 \leq x \leq 1$ by the integer constraint $x \in \{0,1\}^V$ then we obtain a formulation for the maximum dicut problem; the dicut $\delta^+(S)$ with $S = \{i \mid x_i = 1\}$ being an optimum dicut. Halperin and Zwick (2001c) show that the program (118) has a half-integer optimum solution. To see it, note first that (118) is equivalent to the program:

$$\begin{array}{lll} \max & \sum_{ij\in A} w_{ij}z_{ij} & \\ \text{s.t.} & z_{ij} + z_{jk} \leq 1 & (ij \in A, jk \in A) \\ & 0 \leq z_{ij} \leq 1 & (ij \in A). \end{array} \tag{119}$$

Indeed, if (z, x) is feasible for (118), then z is feasible for (119); conversely, if z is feasible for (119) then (z, x) is feasible for (118), where $x_i := \max_{ij\in A} z_{ij}$ if $\delta^+(i) \neq \emptyset$ and $x_i := 0$ otherwise. Now, the constraints in (119) define in fact the fractional stable set polytope of the line graph of G (whose nodes are the arcs, with two arcs being adjacent if they form a path in G). Since the vertices of the fractional stable set polytope are half-integral, it follows that (119) and thus (118) has a half-integral optimum solution (x, z). Then one constructs a directed cut $\delta^+(S)$ by putting node $i \in V$ in S with probability x_i. The expected weight of $\delta^+(S)$ is at least $\frac{1}{2}w^Tz$. Therefore, this gives a $\frac{1}{2}$-approximation algorithm. Moreover, this algorithm can be made purely combinatorial since a half-integral solution can be found using a bipartite matching algorithm (see Halperin and Zwick (2001c)).

The Goemans–Williamson 0.796-approximation algorithm. One can alternatively model the maximum dicut problem in the following way. Given $v_0, v_1, \ldots, v_n \in \{\pm 1\}$ and $S := \{i \in \{1, \ldots, n\} \mid v_i = v_0\}$, the quantity

$$\frac{1}{4}(1 + v_0 \cdot v_i)(1 - v_0 \cdot v_j) = \frac{1}{4}(1 + v_0 \cdot v_i - v_0 \cdot v_j - v_i \cdot v_j)$$

is equal to 1 if $ij \in \delta^+(S)$ and to 0 otherwise. Therefore, the following program solves the maximum dicut problem:

$$\begin{array}{ll} \max & \sum_{ij \in A} w_{ij} \frac{1}{4}(1 + v_0 \cdot v_i - v_0 \cdot v_j - v_i \cdot v_j) \\ \text{s.t.} & v_0, v_1, \ldots, v_n \in \{\pm 1\} \end{array} \qquad (120)$$

Let (SDP) denote the relaxation of (120) obtained by replacing the condition $v_0, v_1, \ldots, v_n \in \{\pm 1\}$ by the condition $v_0, v_1, \ldots, v_n \in S^n$ and let z_{sdp} denote its optimum value. Goemans and Williamson propose the following analog of their max-cut algorithm for solving the maximum dicut problem: Solve (SDP) and let $v_0, \ldots, v_n$ be an optimum solution to it; select a random unit vector r and let $S := \{i \in \{1, \ldots, n\} \mid \text{sign}(v_0 \cdot r) = \text{sign}(v_i \cdot r)\}$. Let θ_{ij} denote the angle (v_i, v_j). Then the expected weight $E(S)$ of the dicut $\delta^+(S)$ is equal to

$$E(S) = \sum_{ij \in A} w_{ij} \frac{1}{2\pi}(-\theta_{0i} + \theta_{0j} + \theta_{ij}).$$

In order to bound $\frac{E(S)}{z_{sdp}}$, one has to find lower bounds for the quantity

$$\frac{2}{\pi} \frac{-\theta_{0i} + \theta_{0j} + \theta_{ij}}{1 + \cos\theta_{0i} - \cos\theta_{0j} - \cos\theta_{ij}}.$$

Goemans and Williamson show the lower bound

$$\beta := \min_{0 \le \theta < \arccos(-1/3)} \frac{2}{\pi} \frac{2\pi - 3\theta}{1 + 3\cos\theta} > 0.79607.$$

for it. Therefore, the above algorithm has performance ratio $\beta > 0.79607$.

The Feige–Goemans approximation algorithm. Feige and Goemans (1995) propose an improved approximation algorithm for the maximum dicut problem analog to their improved approximation algorithm for MAX 2SAT. Namely, strengthen the semidefinite program (SDP) by adding to it the triangle inequalities (114); replace the vectors $v_0, \ldots, v_n$ obtained as optimum solution of the strengthened SDP program by a new set of vectors $v'_0, \ldots, v'_n$ obtained by applying some rotation function to the v_i's; generate from the v'_i's the directed cut $\delta^+(S)$ where $S := \{i \in \{1, \ldots, n\} \mid \text{sign}(v'_0 \cdot r) = \text{sign}(v'_i \cdot r)\}$. Thus one should now find lower bounds for the quantity

$$\frac{2}{\pi} \frac{-\theta'_{0i} + \theta'_{0j} + \theta'_{ij}}{1 + \cos\theta_{0i} - \cos\theta_{0j} - \cos\theta_{ij}}.$$

Using the rotation function f_λ from (16) with $\lambda = \frac{1}{2}$, Feige and Goemans claim a performance ratio of 0.857. Zwick (2000) makes a detailed analysis of their algorithm enabling him to show a performance ratio of 0.859643 (using an adequate rotation function).

The Matuura–Matsui 0.863-approximation algorithm. Matuura and Matsui (2001a) propose an approximation algorithm for the maximum directed cut problem with performance ratio 0.863. Analogously to their algorithm for MAX 2SAT presented in the previous subsection, it relies on solving the semidefinite relaxation strengthened by the triangle inequalities (114) and applying the random hyperplane rounding phase using a distribution on the sphere which is skewed towards v_0 and uniform in any direction orthogonal to v_0. As a concrete choice, they propose to use the distribution function on S^2:

$$g(v) = \cos^{1/1.8}(\theta) \quad \text{for all } v \in S^2 \text{ with } |v_0^T v| = \cos\theta \tag{121}$$

which can be realized as projection of a distribution on S^n and permits to show an approximation ratio of 0.863. (Compare (121) with the function g from (117) used for MAX 2SAT.)

The Lewin–Livnat–Zwick 0.874-approximation algorithm. Analogously to their improved algorithm for MAX 2SAT, Lewin, Livnat, and Zwick (2002) achieve this improved performance guarantee by combining the ideas of first suitably rotating the vectors obtained as solutions of the semidefinite program and of then using a skewed distribution function for choosing the random hyperplane.

7 Further Topics

7.1 Approximating polynomial programming using semidefinite programming

We come back in this section to the problem of approximating polynomial programs using semidefinite programming, which was already considered in Section 3.8. We present here the main ideas underlying this approach. They use results about representations of positive polynomials as sums of squares and moment sequences. Sums of squares will again be used in the next subsection for approximating the copositive cone. We then mention briefly some extensions to the general problem of testing whether a semialgebraic set is empty.

Polynomial programs, sums of squares of polynomials, and moment sequences. Consider the following *polynomial programming problem*:

$$\min g(x) \quad \text{subject to} \quad g_\ell(x) \geq 0 \quad (\ell = 1, \ldots, m) \tag{122}$$

where g, g_ℓ are polynomials in $x=(x_1,\ldots,x_n)$. This is a very general problem which contains linear programming (when all polynomials have degree one) and 0/1 linear programming (since the integrality condition $x_i\in\{0,1\}$ can be expressed as the polynomial equation: $x_i^2-x_i=0$). We mentioned in Section 3.8 that, under some technical assumption, the problem (122) can be approximated (getting arbitrarily close to its optimum) by the sequence of semidefinite programs (56). This result, due to Lasserre (2001a), relies on the fact that certain positive polynomials can be represented as sums of squares of polynomials. This idea of using sums of squares of polynomials for approximating polynomial programs has been introduced by Shor (1987a,b, 1998) and used by several other authors including Nesterov (2000) and Parrilo (2000, 2003); it seems to yield a more powerful method than other existing algebraic methods, see Parrilo and Sturmfels (2003) for a comparison.

We would like to explain briefly here the main ideas underlying this approach. For simplicity, consider first the unconstrained problem:

$$p^* := \min g(x) \quad \text{subject to } x\in\mathbb{R}^n \tag{123}$$

where $g(x)=\sum_{\alpha\in S_{2d}} g_\alpha x^\alpha$ is a polynomial of even degree $2d$; here S_k denotes the set of sequences $\alpha\in\mathbb{Z}_+^n$ with $|\alpha|:=\sum_{i=1}^n\alpha_i\le k$ for any integer k. One can assume w.l.o.g. that $g(0)=g_0=0$. In what follows the polynomial $g(x)$ is identified with its sequence of coefficients $g=(g_\alpha)_{\alpha\in S_{2d}}$. Obviously, (123) can be rewritten as

$$p^* = \max\lambda \quad \text{subject to } g(x)-\lambda\ge 0\ \forall x\in\mathbb{R}^n. \tag{124}$$

Testing whether a polynomial is nonnegative is a hard problem, since it contains the problem of testing whether a matrix is copositive (see the next subsection). Lower bounds for p^* can be obtained by considering sufficient conditions for the polynomial $g(x)-\lambda$ to be nonnegative $\mathbb{R}^n$. An obvious such sufficient condition being that $g(x)-\lambda$ be a sum of squares of polynomials. Therefore,

$$p^*\ge\max\lambda \quad \text{subject to } g(x)-\lambda \text{ is a sum of squares.} \tag{125}$$

Testing whether a polynomial $p(x)$ is a sum of squares of polynomials amounts to testing feasibility of a semidefinite program (cf. e.g., Powers and Wörmann (1998)). Indeed, say $p(x)$ has degree $2d$, and let $z:=(x^\alpha)_{\alpha\in S_d}$ be the vector consisting of all monomials of degree $\le d$. Then one can easily verify that $p(x)$ is a sum of squares if and only if $p(x)=z^TXz$ (identical polynomials) for some positive semidefinite matrix X. For $\gamma\in S_{2d}$, set

$$B_\gamma := \sum_{\alpha,\beta\in S_d|\alpha+\beta=\gamma} E_{\alpha,\beta},$$

where $E_{\alpha,\beta}$ is the elementary matrix with all zero entries except 1 at positions (α, β) and (β, α).

Proposition 32. *A polynomial $p(x)$ of degree $2d$ is a sum of squares of polynomials if and only if the following semidefinite program:*

$$X \succeq 0, \langle B_\gamma, X\rangle = p_\gamma \quad (\gamma \in S_{2d}) \tag{126}$$

is feasible, where X is of order $\binom{n+d}{d}$ and with $\binom{n+2d}{2d}$ equations.

Proof. As

$$z^T X z = \sum_{\alpha,\beta \in S_d} X_{\alpha,\beta} x^{\alpha+\beta} = \sum_{\gamma \in S_{2d}} x^\gamma \left(\sum_{\substack{\alpha,\beta \in S_d \\ \alpha+\beta=\gamma}} X_{\alpha,\beta} \right) = \sum_{\gamma \in S_{2d}} x^\gamma \langle B_\gamma, X\rangle,$$

$p(x) = z^T X z$ for some $X \succeq 0$ (which is equivalent to $p(x)$ being a sum of squares) if and only if the system (126) is feasible. □

Note that the program (126) has a polynomial size for fixed n or d. Based on the result from Proposition 32, one can reformulate the lower bound for p^* from (125) as

$$\begin{aligned} p^* \geq \max\ & \lambda & = \max\ & -\langle B_0, X\rangle \\ \text{s.t.}\ & g(x) - \lambda \text{ is a sum of squares} & \text{s.t.}\ & \langle B_\gamma, X\rangle = g_\gamma\ (\gamma \in S_{2d} \setminus \{0\}). \end{aligned} \tag{127}$$

One can alternatively proceed in the following way for finding lower bounds for p^*. Obviously,

$$p^* = \min_\mu \int g(x) d\mu(x) \tag{128}$$

where the minimum is taken over all probability measures μ on $\mathbb{R}^n$. Define a sequence $y = (y_\alpha)_{\alpha \in S_{2d}}$ to be a *moment sequence* if $y_\alpha = \int x^\alpha d\mu(x)$ $(\alpha \in S_{2d})$ for some nonnegative measure μ on $\mathbb{R}^n$. Hence, (128) can be rewritten as

$$p^* = \min \sum_\alpha g_\alpha y_\alpha \quad \text{s.t. } y \text{ is a moment sequence and } y_0 = 1. \tag{129}$$

Lower bounds for p^* can be obtained by replacing the condition that y be a moment sequence by a necessary condition for it. An obvious such necessary condition is that the moment matrix $M_d^{\mathbb{Z}}(y) = (y_{\alpha+\beta})_{\alpha,\beta\in S_d}$ (recall (54)) be positive semidefinite. Thus we find the following lower bound for p^*:

$$p^* \geq \min g^T y \quad \text{subject to } M_d^{\mathbb{Z}}(y) \succeq 0 \quad \text{and} \quad y_0 = 1. \tag{130}$$

Note that the constraint in (130) is precisely condition (56) (when there are no constraints $g_\ell(x) \geq 0$). Since $M_d^{\mathbb{Z}}(y) = B_0 y_0 + \sum_{\gamma\in S_{2d}\setminus\{0\}} B_\gamma y_\gamma$, the semidefinite programs in (130) and in (127) are in fact dual of each other, which reflects the duality existing between the theories of nonnegative polynomials and of moment sequences.

The lower bound from (127) is equal to p^* if $g(x) - p^*$ is a sum of squares; this holds for $n = 1$ but not in general if $n \geq 2$. In general one can estimate p^* asymptotically by a sequence of SDP's analogous to (127) if one assumes that an upper bound R is known a priori on the norm of a global minimizer x of $g(x)$, in which case

$$p^* = \min g(x) \quad \text{subject to } g_1(x) := R - \sum_{i=1}^n x_i^2 \geq 0.$$

Indeed, one can then use a result of Putinar (1993) (quoted in Theorem 33 below) and conclude that, for any $\epsilon > 0$, the polynomial $g(x) - p^* + \epsilon$ is *positive* on $F := \{x \mid g_1(x) \geq 0\}$ and thus can be decomposed as $p(x) + p_1(x)g_1(x)$ for some polynomials $p(x)$ and $p_1(x)$ that are sums of squares. Testing for the existence of such decomposition where $2t \geq \max$ (deg p, deg$(p_1 g_1)$) can be expressed as a SDP program analog to (127). Its dual (analog to (130)) reads:

$$p_t^* := \min g^T y \quad \text{subject to } M_t(y) \succeq 0, \quad M_{t-1}(g_1 * y) \succeq 0, \quad y_0 = 1.$$

Putinar's result permits to show the asymptotic convergence of p_t^* to p^* when t goes to infinity.

Theorem 33. (Putinar (1993)) *Let $g_1, \ldots, g_m$ be polynomials and set $F := \{x \in \mathbb{R}^n | g_1(x) \geq 0, \ldots, g_m(x) \geq 0\}$. Assume that F is compact and that there exists a polynomial u satisfying* (i) *the set $\{x \in \mathbb{R}^n \mid u(x) \geq 0\}$ is compact and* (ii) *u can be decomposed as $u_0 + \sum_{\ell=1}^m u_\ell g_\ell$ for some polynomials $u_0, \ldots, u_m$ that are sums of squares. Then every polynomial $p(x)$ which is positive on F can be decomposed as $p = p_0 + \sum_{\ell=1}^m p_\ell g_\ell$ for some polynomials $p_0, \ldots, p_m$ that are sums of squares.*

The above reasoning extends to the general program (122) if the assumption of Theorem 33 holds. This is the case, e.g., if the set $\{x \mid g_\ell(x) \geq 0\}$ is compact for one of the polynomials defining F. Then, Putinar's result permits to claim that, for any $\epsilon > 0$, the polynomial

$g(x) - p^* + \epsilon$ can be decomposed as $p(x) + \sum_{\ell=1}^{m} p_\ell(x) g_\ell(x)$ for some polynomials $p(x)$, $p_\ell(x)$ that are sums of squares. Based on this, one can derive the asymptotic convergence to p^* of the minimum of $g^T y$ taken over all y satisfying (56) when t goes to ∞. In the 0/1 case, when the constraints $x_i^2 - x_i = 0$ ($i = 1, \ldots, n$) are part of the system defining F, there is in fact finite convergence in n steps (Lasserre (2001b)) (see Section 3).

Semidefinite programming and the Positivstellensatz. Consider the following system:

$$\begin{aligned} f_j(x) &\geq 0 \quad (j = 1, \ldots, s) \\ g_k(x) &\neq 0 \quad (k = 1, \ldots, t) \\ h_\ell(x) &= 0 \quad (\ell = 1, \ldots, u) \end{aligned} \tag{131}$$

where all f_j, g_k, h_ℓ are polynomials in the real variable $x = (x_1, \ldots, x_n)$. The complexity of the problem of testing feasibility of this system has been the object of intensive research. Tarski (1951) showed that this problem is decidable and since then a number of other algorithms have been proposed, in particular, by Renegar (1992) and Basu, Pollack, and Roy (1996).

We saw in Proposition 32 that testing whether a polynomial is a sum of squares can be formulated as a semidefinite program. Parrilo (2000) showed that the general problem of testing infeasibility of the system (131) can also be formulated as a semidefinite programming problem (of very large size). This is based on the following result of real algebraic geometry, known as the "Positivstellensatz". The Positivstellensatz asserts that for a system of polynomial (in)equalities, either there is a solution in $\mathbb{R}^n$, or there is a polynomial identity giving a certificate that no real solution exists. This gives therefore a common generalization of Hilbert's "Nullstellensatz" (in the complex case) and Farkas' lemma (for linear systems).

Theorem 34. (Stengle (1974), Bochnak, Coste and Roy (1987)) *The system* (131) *is infeasible if and only if there exists polynomials f, g, h of the form*

$$\begin{aligned} f(x) &= \sum_{S \subseteq \{1,\ldots,s\}} p_S \left(\prod_{j \in S} f_j \right) \quad \text{where all } p_S \text{ are sums of squares} \\ g(x) &= \prod_{k \in K} g_k \quad \text{where } K \subseteq \{1, \ldots, t\} \\ h(x) &= \sum_{\ell=1}^{u} q_\ell h_\ell \quad \text{where all } q_\ell \text{ are polynomials} \end{aligned}$$

satisfying the equality $f + g^2 + h = 0$.

Bounds are known a priori for the degrees of the polynomials in the Positivstellensatz which make it possible to test infeasibility of the system (131) via semidefinite programming. However, these bounds are very large (triply exponential in n). Practically, one can use semidefinite programming for searching for infeasibility certificates of bounded degree.

7.2 Approximating combinatorial problems using copositive programming

We have seen throughout this chapter how semidefinite programming can be used for approximating combinatorial optimization problems. The idea of using the copositive cone and its dual, the cone of completely positive matrices, instead of the positive semidefinite cone has also been considered; cf., e.g., Bomze, Dür, de Kleck, Roos, Quist and Terlaky (2000), Quist, de Klerk, Roos, and Terlaky (1998). We present below some results of de Klerk and Pasechnik (2002) showing how the stability number of a graph can be computed using copositive relaxations.

Let us first recall some definitions. A symmetric matrix M of order n is *copositive* if $x^TMx \geq 0$ for all $x \in \mathbb{R}^n_+$ and M is *completely positive* if $M = \sum_{i=1}^k u_i u_i^T$ for some nonnegative vectors $u_1, \ldots, u_k$. Let $\mathcal{C}_n$ denote the set of symmetric copositive matrices of order n; its dual cone $\mathcal{C}_n^*$ is the set of completely positive matrices. Hence,

$$\mathcal{C}_n^* \subseteq \mathrm{PSD}_n = \mathrm{PSD}_n^* \subseteq \mathcal{C}_n.$$

Testing whether a matrix M is copositive is a co-NP-complete problem (Murty and Kabadi (1987)).

Let $G = (V, E)$ $(V = \{1, \ldots, n\})$ be a graph and consider its theta number $\vartheta(G)$, defined by

$$\vartheta(G) = \max \langle J, X \rangle \quad \text{s.t.} \quad X_{ij} = 0 \ (ij \in E), \mathrm{Tr}(X) = 1, X \succeq 0 \tag{132}$$

(same as definition (58)). Then, $\vartheta(G)$ is an upper bound for the stability number of G, since for any stable set S in G, the matrix $X_S := \frac{1}{|S|}\chi^S(\chi^S)^T$ is feasible for the semidefinite program (132). Note that X_S is in fact completely positive. Therefore, one can define a tighter upper bound for $\alpha(G)$ by replacing in (132) the condition $X \succeq 0$ by the condition $X \in \mathcal{C}_n^*$. Letting A denote the adjacency matrix of G, we obtain:

$$\begin{array}{llll} \alpha(G) \leq \max & \langle J, X \rangle & \leq \min & \lambda \\ \text{s.t.} & \mathrm{Tr}X = 1 & \text{s.t.} & \lambda I + yA - J \in \mathcal{C}_n \\ & X_{ij} = 0 \ (ij \in E) & & \lambda, y \in \mathbb{R} \\ & X \in \mathcal{C}_n^* & & \end{array} \tag{133}$$

where the right most program is obtained from the left most one using cone-LP duality. Using the following formulation for $\alpha(G)$ due to Motzkin and Straus (1965):

$$\frac{1}{\alpha(G)} = \min x^T(A+I)x \quad \text{subject to } x \geq 0 \text{ and } \sum_{i=1}^{n} x_i = 1,$$

one finds that the matrix $\alpha(G)(I+A)-J$ is copositive. This implies that the optimum value of the right most program in (133) is at most $\alpha(G)$. Therefore, equality holds throughout in (133). This shows again that copositive programming in not tractable.

Parrilo (2000) proposes to approximate the copositive cone using sums of squares of polynomials. For this, note that a matrix M is copositive if and only if the polynomial

$$g_M(x) := \sum_{i,j=1}^{n} M_{ij} x_i^2 x_j^2$$

is nonnegative on $\mathbb{R}^n$. Therefore, an obvious sufficient condition for M to be copositive is that $g_M(x)$ be a sum of squares or, more generally, that the polynomial $g_M(x)(\sum_{i=1}^n x_i^2)^r$ be a sum of squares for some integer $r \geq 0$. A theorem of Pólya asserts that, conversely, if M is strictly copositive (i.e., $x^T M x > 0$ for all $x \in \mathbb{R}^n_+ \setminus \{0\}$), then $g_M(x)(\sum_{i=1}^n x_i^2)^r$ has nonnegative coefficients and thus is a sum of squares for some r. Powers and Reznick (2001) give some upper bound for this integer r (depending only on M).

Let $\mathcal{K}_n^r$ denote the set of symmetric matrices M of order n for which $g_M(x)(\sum_{i=1}^n x_i^2)^r$ is a sum of squares. Thus

$$\text{PSD}_n \subseteq \mathcal{K}_n^0 \subseteq \cdots \subseteq \mathcal{K}_n^r \subseteq \mathcal{C}_n.$$

We saw in the preceding subsection that testing whether a polynomial is a sum of squares can be solved via the semidefinite program (126). Therefore one can test membership in $\mathcal{K}_n^r$ via semidefinite programming. For instance, Parrilo (2000) shows that

$$M \in \mathcal{K}_n^0 \Leftrightarrow M = P + N \quad \text{for some } P \succeq 0, N \geq 0.$$

Moreover, $M \in \mathcal{K}_n^1$ if and only if the following system:

$$\begin{aligned} M - X^{(i)} &\succeq 0 \quad (i=1,\ldots,n) \\ X_{ii}^{(i)} &= 0 \quad (i=1,\ldots,n) \\ X_{ii}^{(j)} + 2X_{ij}^{(i)} &= 0 \quad (i \neq j = 1,\ldots,n) \\ X_{jk}^{(i)} + X_{ik}^{(j)} + X_{ij}^{(k)} &\geq 0 \quad (1 \leq i < j < k \leq n) \end{aligned}$$

has a solution, where $X^{(1)}, \ldots, X^{(n)}$ are symmetric $n \times n$ matrices (Parrilo (2000) and Bomze and de Klerk (2002)).

Replacing in (133) the condition $\lambda I + yA - J \in \mathcal{C}_n$ by the condition $\lambda I + yA - J \in \mathcal{K}_n^r$, one can define the parameter

$$\vartheta^r(G) := \min \lambda \quad \text{subject to } \lambda I + yA - J \in \mathcal{K}_n^r.$$

Using the bound of Powers and Reznick (2001), de Klerk and Pasechnik (2002) show that

$$\alpha(G) = \lfloor \vartheta^r(G) \rfloor \quad \text{if } r \geq \alpha^2(G).$$

The same conclusion holds if we replace $\mathcal{K}_n^r$ by the cone $\mathcal{C}_n^r$ consisting of the matrices M for which $g_M(x)\left(\sum_{i=1}^n x_i^2\right)^r$ has only nonnegative coefficients. Bomze and de Klerk (2002) give the following characterization for the cone $\mathcal{C}_n^r$:

$$\begin{aligned} \mathcal{C}_n^r = \{ & M \text{ symmetric } n \times n \mid x^T M x - x^T \text{diag}(M) \geq 0 \\ & \text{for all } x \in \mathbb{Z}_+^n \text{ with } \sum_{i=1}^n x_i = r + 2 \}. \end{aligned} \tag{134}$$

It is also shown in de Klerk and Pasechnik (2002) that $\vartheta^0(G) = \vartheta'(G)$, the Schrijver parameter from (65); $\vartheta^1(G) = \alpha(G)$ if G is an odd circuit, an odd wheel or their complement, or if $\alpha(G) = 2$. It is conjectured in de Klerk and Pasechnik (2002) that $\vartheta^{\alpha(G)-1}(G) = \alpha(G)$.

Bomze and de Klerk (2002) extend these ideas to standard quadratic optimization problems, of the form:

$$p^* := \min x^T Q x \quad \text{s.t.} \quad x \in \Delta := \{x \in \mathbb{R}_+^n \mid e^T x = 1\} \tag{135}$$

where Q is a symmetric matrix. Problem (135) is equivalent to any of the following dual problems:

$$\begin{aligned} p^* &= \min \langle Q, X \rangle \quad \text{s.t.} \quad \langle J, X \rangle = 1, X \in \mathcal{C}_n^* \\ &= \max \lambda \quad \text{s.t.} \quad Q - \lambda J \in \mathcal{C}_n, \lambda \in \mathbb{R}. \end{aligned} \tag{136}$$

If we replace in (136), the cone $\mathcal{C}_n$ by its subcone $\mathcal{C}_n^r$ (defined above), we obtain a lower bound p^r for p^*. Setting $\bar{p} := \max_{x \in \Delta} x^T Q x$, we have that $p^r \leq p^* \leq \bar{p}$. Bomze and de Klerk (2002) show the following inequality about the quality of the approximation p^r:

$$p^* - p^r \leq \frac{1}{r+1}(\bar{p} - p^*).$$

Using the characterization of C_n^r from (134), the bound p^r can be expressed as

$$p^r = \frac{r+2}{r+1}\left(\min_{x\in\Delta(r)} x^T Q x - \frac{1}{r+2} x^T \operatorname{diag} Q\right),$$

where $\Delta(r)$ is the grid approximation of Δ consisting of the points $x \in \Delta$ with $(r+2)x \in \mathbb{Z}_+^n$. Thus, the minimum value $p_{\Delta(r)}$ of $x^T Q x$ over $\Delta(r)$ satisfies:

$$p^r \le p^* \le p_{\Delta(r)} \le \bar{p}.$$

Bomze and de Klerk (2002) prove that

$$p_{\Delta(r)} - p^* \le \frac{1}{r+2}(\bar{p} - p^*).$$

Therefore, the grid approximation of Δ by $\Delta(r)$ provides a polynomial time approximation scheme for the standard quadratic optimization problem (135).

An extension leading to a PTAS for the optimization of polynomials of fixed degree d over the simplex Δ can be found in de Klerk, Laurent, and Parrilo (2004).

8 Semidefinite programming and the quadratic assignment problem

Quadratic problems in binary variables are the prime source for semidefinite models in combinatorial optimization. The simplest form, unconstrained quadratic programming in binary variables, corresponds to Max-Cut, and was described in detail in Section 5.

Assuming that the binary variables are the elements of a permutation matrix leads to the *Quadratic Assignment Problem* (*QAP*). Formally, QAP consists in minimizing

$$\operatorname{Tr}(AXB + C)X^T \tag{137}$$

over all permutation matrices X. One usually assumes that A and B are symmetric matrices of order n, while the linear term C is an arbitrary matrix of order n. There are many applications of this model problem, for instance in location theory. We refer to the recent monograph (Cela (1998)) for a description of published applications of QAP in Operations Research and combinatorial optimization.

The cost function (137) is quadratic in the matrix variable X. To rewrite this we use the vec-operator and (9). This leads to

$$\operatorname{Tr} AXBX^T = \langle \operatorname{vec}(X), \operatorname{vec}(AXB)\rangle = x^T(B \otimes A)x, \tag{138}$$

because B is assumed to be symmetric. We can therefore express QAP equivalently as

$$\min\{x^T(B \otimes A)x + c^T x \colon x = \mathrm{vec}(X),\ X \text{ permutation matrix}\}.$$

Here, $c = \mathrm{vec}(C)$. To derive semidefinite relaxations of QAP we follow the generic pattern and linearize by introducing a new matrix for xx^T, leading to the study of

$$P = \mathrm{conv}(xx^T : x = \mathrm{vec}(X),\ X \text{ permutation matrix}\}.$$

In section 3, we observed that any $Y \in P$ must satisfy the semidefiniteness condition (20), which in our present notation amounts to

$$Z = \begin{pmatrix} 1 & z^T \\ z & Y \end{pmatrix} \succeq 0,\ \mathrm{diag}(Y) = z.$$

The first question is to identify the smallest subcone of semidefinite matrices that contains P.

We use the following parametrization of matrices having row and column sums equal to e, the vector of all ones, see Hadley, Rendl, and Wolkowicz (1992).

Lemma 35. (Hadley, Rendl and Wolkowicz (1992)) *Let V be an $n \times (n-1)$ matrix with $V^T e = 0$ and rank$(V) = n-1$. Then*

$$\begin{aligned} \mathcal{E} &:= \{X \in \mathbb{R}^{n \times n} \colon Xe = X^T e = e\} \\ &= \left\{\frac{1}{n} ee^T + VMV^T \colon M \in \mathbb{R}^{(n-1)\times(n-1)}\right\} =: \mathcal{E}'. \end{aligned}$$

Proof. Let $Z = \frac{1}{n} ee^T + VMV^T \in \mathcal{E}'$. Then $Ze = Z^T e = e$, because $V^T e = 0$, hence $Z \in \mathcal{E}$. To see the other inclusion, let $V = QR$ be the QR-decomposition of V, i.e., $Q^T Q = I$, $QQ^T = I - \frac{1}{n} ee^T$ and rank$(R) = n-1$. Let $X \in \mathcal{E}$ and set $M := R^{-1} Q^T X Q (R^{-1})^T$. Then $\frac{1}{n} ee^T + VMV^T = X \in \mathcal{E}'$. □

We use this parametrization and define

$$W := \left(\frac{1}{n} e \otimes e,\ V \otimes V\right).$$

V can be any basis of $e^{\perp}$, as in the previous lemma. We can now describe the smallest subcone containing P.

Lemma 36. *Let* $Y \in P$. *Then there exists a symmetric matrix* R *of order* $(n-1)^2+1$, *indexed from* 0 *to* $(n-1)^2$, *such that*

$$R \succeq 0,\ r_{00}=1,\ Y = WRW^T.$$

Proof. (see also Zhao, Karisch, Rendl, and Wolkowicz (1998)) We first look at the extreme points of P, so let X be a permutation matrix. Thus we can write X as $X = \frac{1}{n}ee^T + VMV^T$, for some matrix M. Let $m = \text{vec}(M)$. Then, using (9),

$$x = \text{vec}(X) = \frac{1}{n} e \otimes e + (V \otimes V)m = Wz,$$

with $z = \binom{1}{m}$. Now $xx^T = Wzz^TW^T = WRW^T$, with $r_{00}=1$, $R \succeq 0$. The same holds for convex combinations formed from several permutation matrices.□

To see that the set

$$\hat{P} := \left\{ Y\colon \exists\, R \quad \text{such that } Y = WRW^T, z = \text{diag}(Y), \quad \begin{pmatrix} 1 & z^T \\ z & Y \end{pmatrix} \succeq 0 \right\} \tag{139}$$

is indeed the smallest subcone of positive semidefinite matrices containing P, it is sufficient to provide a positive definite matrix $\hat{R}$, such that $W\hat{R}W^T \in P$. In Zhao, Karisch, Rendl and Wolkowicz (1998) it is shown that

$$\hat{R} = \begin{pmatrix} 1 & 0 \\ 0 & \frac{1}{n^2(n-1)}(nI_{n-1} - E_{n-1}) \otimes (nI_{n-1} - E_{n-1}) \end{pmatrix} \succ 0$$

gives

$$W\hat{R}W^T = \frac{1}{n!} \sum_{X \in \Pi} (xx^T),$$

the barycenter of P. Here

$$V = \begin{pmatrix} I_{n-1} \\ -e_{n-1}^T \end{pmatrix}$$

has to be used in the definition of W.

Eliminating Y leaves the matrix variable R and n^2+1 equality constraints, fixing the first row equal to the main diagonal, and setting the first element equal to 1.

Thus we arrive at the following basic SDP relaxation of QAP:

$$(QAP_{R1}) \quad \begin{array}{ll} \min & \operatorname{Tr}(B \otimes A + \operatorname{Diag}(c))Y \\ \text{such that } & Y = WRW^T \in \hat{P},\ r_{00} = 1. \end{array} \tag{140}$$

It is instructive to look at $W\hat{R}W^T$ for small values of n. For $n=3$ we get

$$W\hat{R}W^T = \frac{1}{6}\left(\begin{array}{ccc|ccc|ccc} 2 & 0 & 0 & 0 & 1 & 1 & 0 & 1 & 1 \\ 0 & 2 & 0 & 1 & 0 & 1 & 1 & 0 & 1 \\ 0 & 0 & 2 & 1 & 1 & 0 & 1 & 1 & 0 \\ \hline 0 & 1 & 1 & 2 & 0 & 0 & 0 & 1 & 1 \\ 1 & 0 & 1 & 0 & 2 & 0 & 1 & 0 & 1 \\ 1 & 1 & 0 & 0 & 0 & 2 & 1 & 1 & 0 \\ \hline 0 & 1 & 1 & 0 & 1 & 1 & 2 & 0 & 0 \\ 1 & 0 & 1 & 1 & 0 & 1 & 0 & 2 & 0 \\ 1 & 1 & 0 & 1 & 1 & 0 & 0 & 0 & 2 \end{array}\right).$$

The zero pattern in this matrix is not incidental. In fact, any $X \in P$ will have entries equal 0 at positions corresponding to $x_{ij}x_{ik}$ and $x_{ji}x_{ki}$ for $j \neq k$. This corresponds to the off-diagonal elements of the main diagonal blocks, and the main-diagonal elements of the off diagonal blocks. To express these constraints, we introduce some more notation, and index the elements of matrices in P alternatively by $P = (p_{(i,j),(k,l)})$ for i, j, k, l between 1 and n.

Hence we can strengthen the above relaxation by asking that

$$y_{rs} = 0 \quad \text{for } r = (i,j),\ s = (i,k), \quad \text{or } r = (j,i),\ s = (k,j),\ j \neq k.$$

We collect all these equations in the constraint $G(Y) = 0$. Adding it to (140) results in a stronger relaxation. In Zhao, Karisch, Rendl and Wolkowicz (1998) this model is called the "Gangster model." Aside from n^2+1 equality constraints from the basic model, we have $O(n^3)$ equations in this extended model. This amounts to serious computational work, but results in a very strong lower bound for QAP.

$$(QAP_{R2}) \quad \begin{array}{ll} \min & \operatorname{Tr}(B \otimes A + \operatorname{Diag}(c))Y \\ \text{such that } & Y = WRW^T \in \hat{P},\ r_{00} = 1,\ G(Y) = 0. \end{array} \tag{141}$$

Finally, one can include the constraints $y_{rs} \geq 0$ for all r, s, leading to

$$(QAP_{R3}) \quad \begin{array}{ll} \min & \operatorname{Tr}(B \otimes A + \operatorname{Diag}(c))Y \\ \text{such that } & Y = WRW^T \in \hat{P},\ r_{00} = 1,\ G(Y) = 0,\ Y \geq 0. \end{array} \tag{142}$$

The resulting SDP has $O(n^4)$ constraints and cannot be solved in a straightforward way by interior point methods for problems of interesting size ($n \geq 15$).

The Anstreicher–Brixius bound. Anstreicher and Brixius (2001) and Anstreicher, Brixius, Goux, and Linderoth (2002) have recently achieved a breakthrough in solving several instances of QAP which could not be solved by previous methods. The size of these instances ranges from $n=20$ to $n=36$. The key to this breakthrough lies in the use of a bound for QAP that is both "fast" to compute, and gives "good" approximations to the exact value of QAP. This bounding procedure combines orthogonal, semidefinite, and convex quadratic relaxations in a nontrivial way, starting from the Hoffman–Wielandt inequality, Theorem 5.

A simple way to derive this bound goes as follows. We use the parametrization

$$X = \frac{1}{n} ee^T + VYV^T \tag{143}$$

from Lemma 35, and assume in addition that $V^T V = I_{n-1}$. Substituting this into the cost of function of QAP results in

$$\begin{aligned} \operatorname{Tr}(AXB + C)X^T = \operatorname{Tr} \hat{A} Y \hat{B} Y^T + \operatorname{Tr}\left(\hat{C} + \frac{2}{n} V^T A e e^T B V \right) Y^T \\ + \frac{1}{n^2} s(A)s(B) + \frac{1}{n} s(C), \end{aligned} \tag{144}$$

where $\hat{A} = V^T AV$, $\hat{B} = V^T BV$, $\hat{C} = V^T CV$, and $s(M) := e^T Me = \sum_{ij} m_{ij}$. The condition $V^T V = I$ implies that X in (143) is orthogonal if and only if Y is. Hadley, Rendl and Wolkowicz (1992) use this to bound the quadratic term in Y by the minimal scalar product of the eigenvalues of $\hat{A}$ and $\hat{B}$, see Theorem 5. Anstreicher and Brixius (2001) use this observation as a starting point and observe that for any symmetric matrix $\hat{S}$, and any orthogonal Y, one has

$$0 = \operatorname{Tr} \hat{S}(I - YY^T) = \operatorname{Tr} \hat{S} - \operatorname{Tr} \hat{S} Y I Y^T = \operatorname{Tr} \hat{S} - \operatorname{Tr}(I \otimes \hat{S})(yy^T).$$

This results in the following identity, true for any orthogonal Y and any symmetric $\hat{S}$, $\hat{T}$:

$$\operatorname{Tr} \hat{A} Y \hat{B} Y^T = \operatorname{Tr}(\hat{S} + \hat{T}) + \operatorname{Tr}(\hat{B} \otimes \hat{A} - I \otimes \hat{S} - \hat{T} \otimes I)(yy^T). \tag{145}$$

We use $\hat{Q} = \hat{B} \otimes \hat{A} - I \otimes \hat{S} - \hat{T} \otimes I$, $\hat{D} = \hat{C} + \frac{2}{n} V^T Aee^T BV$ and substitute this into (144) to get

$$\mathrm{Tr}(AXB + C)X^T = \mathrm{Tr}(\hat{S} + \hat{T}) + y^T \hat{Q} y + \hat{d}^T y + \frac{1}{n^2} s(A)s(B) + \frac{1}{n} s(C), \tag{146}$$

This relation is true for any orthogonal X and Y related by (143) and symmetric $\hat{S}$, $\hat{T}$. It is useful to express the parts in (146) containing Y by the orthogonal matrix X. To do this we use the following identity:

$$\begin{aligned} 0 &= \mathrm{Tr}\, \hat{S}(I - V^T V) = \mathrm{Tr}\, \hat{S}(I - V^T XX^T V) = \mathrm{Tr}\, \hat{S} - \mathrm{Tr}(V\hat{S}V^T)XIX^T \\ &= \mathrm{Tr}\, \hat{S} - \mathrm{Tr}(I \otimes V\hat{S}V^T)(xx^T). \end{aligned}$$

Hence, for any orthogonal X, and any symmetric $\hat{S}$, $\hat{T}$ we also have

$$\mathrm{Tr}(AXB + C)X^T = \mathrm{Tr}(\hat{S} + \hat{T}) + x^T Q x + c^T x. \tag{147}$$

Here $Q = B \otimes A - I \otimes (V\hat{S}V^T) - (V\hat{T}V^T) \otimes I$. Comparing (146) and (147) we note that

$$y^T \hat{Q} y + \hat{d}^T y + \frac{1}{n^2} s(A)s(B) + \frac{1}{n} s(C) = x^T Q x + c^T x.$$

It should be observed that Q and $\hat{Q}$ above depend on the specific choice of $\hat{S}$, $\hat{T}$. Anstreicher and Brixius use the optimal solution $\hat{S}$, $\hat{T}$ from Theorem 6 and observe that dual feasibility yields $\hat{Q} \succeq 0$. Therefore the above problem is a convex quadratic programming problem. We denote its optimal solution as the Anstreicher–Brixius bound ABB(A, B, C).

$$\begin{aligned} ABB(A, B, C) := \mathrm{Tr}(\hat{S} + \hat{T}) + \min\{x^T Q x + c^T x \colon x \\ = \mathrm{vec}(X),\ X \text{ doubly stochastic}\}. \end{aligned}$$

The interesting observation here is that $\hat{S}$, $\hat{T}$ are obtained as a by-product of the Hoffman–Wielandt inequality, and that the resulting matrix Q is positive semidefinite over the set of doubly stochastic matrices (as a consequence of Theorem 6). These facts imply that the Anstreicher–Brixius bound is tractable.

To give a flavor of the quality of these bounds, we provide the following computational results on the standard test sets from Nugent, Vollman, and Ruml (1968). These data sets have the following characteristics. The linear term C is equal to 0. The matrix B represents the rectilinear cell distance of a rectangular array of cells, hence there is some symmetry in these data. In case of $n = 12$, the resulting rectangular cell array has the following form:

1	2	3	4
5	6	7	8
9	10	11	12

We observe that the distance matrix B would not change, if the following cell array would have been used:

4	3	2	1
8	7	6	5
12	11	10	9

Mathematically speaking, there exist several permutation matrices X, such that $B = XBX^T$. Exploiting all these symmetries, it is sufficient to consider only the subproblems where the cells 1, 2, 5, 6 are assigned to some fixed location, say 1. All other permutations can be obtained by exploiting the automorphisms inherent in B.

We denote these subproblems by nug12.1, nug12.2, nug12.5, nug12.6 in Table 1. The instance $n = 15$ has a distance matrix B corresponding to a 5×3 rectangular grid, leading to subproblems nug 15.1, nug 15.2, nug 15.3, nug 15.6, nug 15.7, nug 15.8. The optimal values for these instances are contained in the column labeled "exact." These values can be computed routinely for $n \approx 15$. The biggest instance $n = 30$ was only recently solved to

Table 1.
Semidefinite relaxations and optimal value for some instances from the Nugent collection of test data. The column labeled QAP_{R3} gives lower estimates of the bound computed by the bundle method

Problem	Exact	QAP_{R2}	QAP_{R3}	ABB
nug12	578	529.3	552.1	482
nug12.1	586	550.7	573.6	–
nug12.2	586	550.6	571.3	–
nug12.5	578	551.8	572.2	–
nug12.6	600	555.8	578.8	–
nug15	1150	1070.5	1106.1	996
nug15.1	1150	1103.4	1131.6	–
nug15.2	1168	1116.3	1147.8	–
nug15.3	1164	1120.9	1148.4	–
nug15.6	1166	1113.6	1144.9	–
nug15.7	1182	1130.3	1161.9	–
nug15.8	1184	1134.1	1162.2	–
nug20	2570	2385.6	2441.9	2254
nug30	6124	5695.4	5803.2	5365

optimality, see Anstreicher, Brixius, Goux and Linderoth (2002). The computational results for QAP_{R3} are from the dissertation of Sotirov (2003). It is computationally infeasible to solve this relaxation by interior points. Sotirov uses the bundle method to get approximate solutions of QAP_{R3}. Hence the values are only lower estimates of the true bound. The values of QAP_{R2} were obtained by Sotirov and Wolkowicz[5] by making use of the NEOS distributed computing system. The bounds are obtained using interior point methods. The computational effort to get these values is prohibitively big. A more practical approach consists in using bundle methods to bargain computational efficiency against a slight decrease in the quality of the bound. Finally, the values of the Anstreicher–Brixius bound *ABB* are from Anstreicher and Brixius (2001).

These results indicate that the SDP models in combination with bundle methods may open the way to improved Branch and Bound approaches to solve larger QAP instances.

9 Epilogue: semidefinite programming and algebraic connectivity

An implicit message of all the preceeding sections is that semidefinite programming relaxations have a high potential to significantly improve on purely polyhedral relaxations. This may give the wrong impression that semidefinite programming is a universal remedy to improve upon linear relaxations. This is in principle true, if we assume that some sort of semidefiniteness constraint is *added* to the polyhedral model.

If a model based on semidefinite programming is used *instead* of a linear model, it need not be true that the semidefinite model dominates the linear one. We conclude with an illustration of this perhaps not quite intuitive statement.

We consider the Traveling Salesman Problem (TSP), i.e., the problem of finding a shortest Hamiltonian cycle in an edge weighted graph. This problem is well known to be NP-hard, and has stimulated research since the late 1950's.

We need to recall some notation from graph theory. For an edge weighted graph, given by its weighted adjacency matrix X, with $X \geq 0$, $\mathrm{diag}(X) = 0$ (setting to 0 the entries corresponding to nonedges), we consider vertex partitions $(S, V \setminus S)$ of its vertex set V and define

$$X(S, V \setminus S) := \sum_{i \in S, j \notin S} x_{ij}$$

to be the weight of the cut, given by S. The *edge connectivity* $\mu(X)$ of X is defined as

$$\mu(X) := \min\{X(S, V \setminus S) : S \subseteq V, 1 \leq |S| \leq |V| - 1\}.$$

[5] Personal communication, 2001.

The polyhedral approach to TSP is based on approximating the convex hull of all Hamiltonian cycles by considering all two-edge connected graphs. Formally, this amounts to optimizing over the following set:

$$\{X: 0 \leq x_{ij} \leq 1, \operatorname{diag}(X) = 0, Xe = 2e, \mu(X) = 2\}. \tag{148}$$

Even though there are $O(2^n)$ linear constraints defining this (polyhedral) set, it is possible to optimize over it in polynomial time, by using the ellipsoid method (because the separation problem amounts to a minimum capacity cut problem, which can thus be solved in polynomial time). It is also interesting to note that no combinatorial algorithm of provably polynomial running time exists for optimizing a linear function over this set.

Recently, Cvetcović, Canglavić, and Kovačevič-Vujčić (1999) have proposed a model where 2-edge connectivity is replaced by the algebraic connectivity, leading to an SDP relaxation.

Fiedler (1973) introduces the algebraic connectivity of a graph, given by its weighted adjacency matrix $X \geq 0$, $\operatorname{diag}(X) = 0$, as follows. Let $L(X) := D - X$ be the Laplacian matrix corresponding to X, where $D := Diag(Xe)$, the diagonal matrix having the row sums of X on its main diagonal. Since $De = Xe$, it is clear that 0 is an eigenvalue of $L(X)$ corresponding to the eigenvector e. Moreover $X \geq 0$ implies by the Gersgorin disk theorem, that all eigenvalues of $L(X)$ are nonnegative, i.e., $L(X)$ is positive semidefinite in this case. Fiedler observed that the second smallest eigenvalue $\lambda_2(L(X)) = \min_{\|u\|=1, u^T e=0} u^T L(X) u$ is equal to 0 if and only if X is the adjacency matrix of a disconnected graph, otherwise $\lambda_2(L(X)) > 0$. Note also that $\lambda_2(L(X))$ is concave in X. Fiedler therefore denotes $\alpha(X) := \lambda_2(L(X))$ as the algebraic connectivity of the graph, given by the adjacency matrix X. It is not difficult to calculate $\alpha(C_n)$, the algebraic connectivity of a cycle on n nodes,

$$\alpha(C_n) = 2\left(1 - \cos\left(\frac{2\pi}{n}\right)\right) =: h_n$$

The concavity of $\alpha(X)$ therefore implies that

$$\alpha(X) \geq h_n$$

for any convex combination X of Hamiltonian cycles. We also note that the Taylor expansion of $\cos(x)$ gives $h_n \leq \frac{4\pi^2}{n^2}$. Cvetcović, Cangalvić and Kovačevič-Vujčić (1999) propose to replace the polyhedral constraints $\mu(X) \geq 2$ by the nonlinear condition $\alpha(X) \geq h_n$, which can easily be shown to be equivalent to the semidefiniteness constraint

$$L(X) + ee^T - h_n I \succeq 0$$

on X. Replacing edge connectivity by algebraic connectivity in (148) leads to optimizing over

$$\{X: 0 \leq x_{ij} \leq 1, \operatorname{diag}(X) = 0, Xe = 2e, L(X) + ee^T - h_n I \succeq 0\}. \tag{149}$$

This looks like a reasonable bargain, as we replace $O(2^n)$ linear constraints by a single semidefiniteness constraint. The crucial question of course is whether we can say anything about the relative strength of the two relaxations. Since $L(X) + ee^T \succeq 0$ it is clear that

$$\lambda_{\min}(L(X) + ee^T - h_n I) \geq -h_n \geq -\frac{4\pi^2}{n^2}.$$

Therefore the semidefiniteness constraint in (149) is nearly satisfied for any $X \geq 0$ as the dimension increases. We can say even more. Any matrix X feasible for (148) satisfies $\alpha(X) \geq h_n$, see Fiedler (1972) and the handbook Wolkowicz et al. (2000), Chapter 12 for further details. In other words, the simple semidefinite relaxation given by (149) is dominated by the polyhedral edge connectivity model (148).

10 Appendix: surveys, books and software

Semidefinite Programming has undergone a rapid development in the last decade. We close with some practical information on semidefinite programming in connection with recent books, surveys, software, and web-sites. The references given here are by no means complete and reflect our personal taste. We apologize for any possible omissions.

Books and Survey papers. The proceedings volume (Pardalos and Wolkowicz (1998)) presents one of the first collection of papers devoted to semidefinite programming in connection with combinatorial optimization. The handbook by Wolkowicz, Saigal and Vandenberghe (2000) is currently a prime source for nearly all aspects of semidefinite optimization. It contains contributions from leading experts in the field, covering in 20 chapters algorithms, theory and applications. With nearly 900 references, it also reflects the state of the art up to about the year 1999. We also refer to de Klerk (2002) for a recent monograph on semidefinite programming, featuring also the development up to 2002.

The survey paper by Vandenberghe and Boyd (1996) has set the stage for many algorithmic and theoretical developments, that were to follow in the last few years. The surveys given by Lovász (2003) and Goemans (1997) focus on the interplay between semidefinite programming and NP-hard combinatorial optimization problems. We also refer to Rendl (1999) and Todd (2001) for surveys focusing on algorithmic aspects and also the position of semidefinite programming in the context of general convex programming.

Software. The algorithmic machinery to solve semidefinite programs is rather sophisticated. It is therefore highly appreciated that many researchers offer their software to the scientific community for free use. The following two packages are currently considered state-of-the-art to deal with general semidefinite problems.

SEDUMI: http://fewcal.kub.nl/software/sedumi.html
SDPT3: http://www.math.nus.edu.sg/mathtohkc/sdpt3.html

Both packages use Matlab as the working horse and implement interior-point methods. The following package is written in C, and contains also specially tailored subroutines to compute the ϑ function.

CSDP: http://www.nmt.edu/~borchers/csdp.html

For large-scale problems, where interior-point methods are out of reach, the spectral bundle approach may be a possible alternative:

SBMethod: http://www-user.tu-chemnitz.de/~helmberg/SBMethod.html

Finally, we mention the NEOS Server, where SDP problem instances can be solved through the internet. NEOS offers several solvers and allows the user to submit the data in several formats. It can be found at

http://www-neos.mcs.anl.gov/neos/

Web-sites. Finally, we refer to the following two web-sites, which have been maintained over a long period of time, so we expect them to survive also in the future.

The optimization-online web-site maintains an electronic library of technical reports in the field of optimization. A prominent part covers semidefinite programming and combinatorial optimization.

http://www.optimization-online.org

The semidefinite programming web-site maintained by C. Helmberg contains up-to-date information on various activities related to semidefinite programming (conferences, workshops, publications, software, people working in the field, etc.)

http://www-user.tu-chemnitz.de/~helmberg/semidef.html

The web-site

`http://plato.asu.edu/topics/problems/nlores.html#semidef`

maintained by H. Mittelmann summarizes further packages for semidefinite programming, and also provides benchmarks, comparing many of the publically available packages on a substantial list of problem instances.

Acknowledgments

We thank a referee for his careful reading and his suggestions that helped improve the presentation of this chapter. Supported by ADONET, Marie Curie Research Training Network MRTN-CT-2003-504438.

Note added in Proof

This chapter was completed at the end of 2002. It reflects the state of the art up to 2002. The most recent developments are not covered.

References

Aguilera, N. E., S. M. Bianchi, G. L. Nasini (2004). Lift and project relaxations for the matching polytope and related polytopes. *Discrete Applied Mathematics* 134, 193–212.

Aguilera, N. E., M. S. Escalante, G. L. Nasini (2002a). The disjunctive procedure and blocker duality. *Discrete Applied Mathematics*, 121, 1–13.

Aguilera, N. E., M. S. Escalante, G. L. Nasini (2002b). A generalization of the perfect graph theorem under the disjunctive index. *Mathematics of Operations Research* 27, 460–469.

Alfakih, A. (2000). Graph rigidity via Euclidean distance matrices. *Linear Algebra and its Applications* 310, 149–165.

Alfakih, A. (2001). On rigidity and realizability of weighted graphs. *Linear Algebra and its Applications* 325, 57–70.

Alfakih, A., A. Khandani, H. Wolkowicz (1999). Solving Euclidean distance matrix completion problems via semidefinite programming. *Computational Optimization and Applications* 12, 13–30.

Alfakih, A., H. Wolkowicz (1998). On the embeddability of weighted graphs in Euclidean spaces. Technical Report, CORR 98-12, Department of Combinatorics and Optimization, University of Waterloo. Available at http://orion.math.uwaterloo.ca/˜hwolkowi/.

Alizadeh, F. (1995). Interior point methods in semidefinite programming with applications in combinatorial optimization. *SIAM Journal on Optimization* 5, 13–51.

Alon, N., N. Kahale (1998). Approximating the independence number via the ϑ-function. *Mathematical Programming* 80, 253–264.

Alon, N., B. Sudakov (2000). Bipartite subgraphs and the smallest eigenvalue. *Combinatorics, Probability and Compuiting* 9, 1–12.

Alon, N., B. Sudakov, U. Zwick (2002). Constructing worst case instances for semidefinite programming based approximation algorithms. *SIAM Journal on Discrete Mathematics* 15, 58–72. [Preliminary version in *Proceedings of 12th ACM-SIAM Symposium on Discrete Algorithms*, pages 92–100, 2001.]

Anjos, M. F. (2001). *New Convex Relaxations for the Maximum Cut and VLSI Layout Problems*. PhD thesis, University of Waterloo.

Anjos, M. (2004). An improved semidefinite programming relaxation for the satisfiability problem. *Mathematical Programming*.

Anjos, M. F., H. Wolkowicz (2002a). Strengthened semidefinite relaxations via a second lifting for the max-cut problem. *Discrete Applied Mathematics* 119, 79–106.

Anjos, M. F., H. Wolkowicz (2002b). Geometry of semidefinite Max-Cut relaxations via ranks. *Journal of Combinatorial Optimization* 6, 237–270.

Anstreicher, K., N. Brixius (2001). A lower bound for the Quadratic Assignment Problem based on Convex Quadratic Programming. *Mathematical Programming* 89, 341–357.

Anstreicher, K., N. Brixius, J.-P. Goux, J. Linderoth (2002). Solving large quadratic assignment problems on computational grids. *Mathematical Programming B* 91, 563–588.

Anstreicher, K., H. Wolkowicz (2000). On Lagrangian relaxation of quadratic matrix constraints. *SIAM Journal on Matrix Analysis and its Applications* 22, 41–55.

Arora, S., B. Bollobás, L. Lovász (2002). Proving integrality gaps without knowing the linear program. In *Proceedings of the 43rd IEEE Symposium on Foundations of Computer Science*, IEEE Computer Science Press, Los Alamitos, CA.

Arora, S., D. Karger, M. Karpinski (1995). Polynomial time approximation schemes for dense instances of NP-hard problems. In *Proceedings of the 27th Annual ACM Symposium on Theory of Computing*, ACM, New York, pp. 284–293.

Arora, S., C. Lund, R. Motwani, M. Sudan, M. Szegedy (1992). Proof verification and intractability of approximation problems. In *Proceedings of the 33rd IEEE Symposium on Foundations of Computer Science*, IEEE Computer Science Press, Los Alamitos, CA, pp. 14–23.

Asano, T., D. P. Williamson. Improved approximation algorithms for MAX SAT. In *Proceedings of 11th ACM-SIAM Symposium on Discrete Algorithms*, pp. 96–115.

Balas, E. (1979). Disjunctive programming. *Annals of Discrete Mathematics* 5, 3–51.

Balas, E., S. Ceria, G. Cornuéjols (1993). A lift-and-project cutting plane algorithm for mixed 0–1 programs. *Mathematical Programming* 58, 295–324.

Ball, M. O., W. Liu, W. R. Pulleyblank (1989). Two terminal Steiner tree polyhedra, in: B. Tulkens, H. Tulkens (eds.), *Contributions to Operations Research and Economics*, MIT Press, Cambridge, MA, pp. 251–284.

Barahona, F. (1993). On cuts and matchings in planar graphs. *Mathematical Programming* 60, 53–68.

Barahona, F. (1982). On the computational complexity of Ising spin glass models. *Journal of Physics A, Mathematical and General* 15, 3241–3253.

Barahona, F. (1983). The max-cut problem on graphs not contractible to K_5. *Operations Research Letters* 2, 107–111.

Barahona, F., A. R. Mahjoub (1986). On the cut polytope. *Mathematical Programming* 36, 157–173.

Barahona, F., A. R. Mahjoub (1994). Compositions of graphs and polyhedra. II: stable sets. *SIAM Journal on Discrete Mathematics* 7, 359–371.

Barvinok, A. I. (1993). Feasibility testing for systems of real quadratic equations. *Discrete and Computational Geometry* 10, 1–13.

Barvinok, A. I. (1995). Problems of distance geometry and convex properties of quadratic maps. *Discrete and Computational Geometry* 13, 189–202.

Barvinok, A. I. (2001). A remark on the rank of positive semidefinite matrices subject to affine constraints. *Discrete and Computational Geometry* 25, 23–31.

Basu, S., R. Pollack, M.-F. Roy (1996). On the combinatorial and algebraic complexity of quantifier elimination. *Journal of the Association for Computing Machinery* 43, 1002–1045.

Bellare, M., P. Rogaway (1995). The complexity of approximating a nonlinear program. *Mathematical programming* 69, 429–441.

Berge, C. (1962). Sur une conjecture relative au problème des codes optimaux. Communication, 13ème assemblée générale de l'URSI, Tokyo.

Berman, P., M. Karpinski (1998). On some tighter inapproximability results, further improvements. *Electronic Colloquium on Computational Complexity*, Report TR98-065.

Bienstock, D., M. Zuckerberg (2004). Subset algebra lift operators for 0 – 1 integer programming. *SIAM Journal on Optimization* 15, 63–95.

Blum, A. (1994). New approximation algorithms for graph coloring. *Journal of the Association for Computing Machinery* 41, 470–516. [Preliminary version in *Proceedings of the 21st Annual ACM Symposium on Theory of Computing*, ACM, New York, pages 535–542, 1989 and in *Proceedings of the 31st IEEE Symposium on Foundations of Computer Science*, IEEE Computer Science Press, Los Alamitos, CA, pages 554–562, 1990.]

Blum, A., D. Karger (1997). An $\tilde{O}(n^{3/14})$-coloring algorithm for 3-colorable graphs. *Information Processing Letters* 61, 49–53.

Bochnak, J., M. Coste, M.-F. Roy (1987). *Géométrie Algébrique Réelle*, Springer-Verlag.

Bockmayr, A., F. Eisenbrand, M. Hartmann, A. S. Schulz (1999). On the Chvátal rank of polytopes in the 0/1 cube. *Discrete Applied Mathematics* 98, 21–27.

Bomze, I. M., M. Dür, E. de Klerk, C. Roos, A. J. Quist, T. Terlaky (2000). On copositive programming and standard quadratic optimization problems. *Journal of Global Optimization* 18, 301–320.

Bomze, I. M., E. de Klerk (2002). Solving standard quadratic optimization problems via linear, semidefinite and copositive programming. *Journal of Global Optimization* 24, 163–185.

Borwein, J. M., H. Wolkowicz (1981). Regularizing the abstract convex program. *Journal of Mathematical Analysis and Applications* 83, 495–530.

Bourgain, J. (1985). On Lipschitz embedding of finite metric spaces in Hilbert space. *Israel Journal of Mathematics* 52, 46–52.

Caprara, A., A. N. Letchford (2003). On the separation of split cuts and related inequalities. *Mathematical Programming Series B* 94, 279–294.

Cela, E. (1998). *The Quadratic Assignment Problem: Theroy and Algorithms*, Kluwer Academic Publishers, USA.

Ceria, S. (1993). *Lift-and-Project Methods for Mixed 0-1 Programs*. PhD dissertation, Graduate School of Industrial Administration, Carnegie Mellon University, US.

Ceria, S., G. Pataki (1998). Solving integer and disjunctive programs by lift-and-project, in: R. E. Bixby, E. A. Boyd, R. Z. Rios-Mercato (eds.), IPCO VI, *Lecture Notes in Computer Science* 1412, 271–283.

Charikar, M. (2002). On semidefinite programming relaxations for graph colouring and vertex cover. In *Proceedings of 13th ACM-SIAM Symposium on Discrete Algorithms*, pp. 616–620.

Chudnovsky, M., N. Robertson, P. Seymour, R. Thomas (2002). The strong perfect graph theorem. To appear in *Annals of Mathematics*.

Chvátal, V. (1973). Edmonds polytopes and a hierarchy of combinatorial problems. *Discrete Mathematics* 4, 305–337.

Chvátal, V. (1975). On certain polytopes associated with graphs. *Journal of Combinatorial Theroy B* 18, 138–154.

Chvátal, V., W. Cook, M. Hartman (1989). On cutting-plane proofs in combinatorial optimization. *Linear Algebra and its Applications* 114/115, 455–499.

Cook, W., S. Dash (2001). On the matrix-cut rank of polyhedra. *Mathematics of Operations Research* 26, 19–30.

Cook, W., R. Kannan, A. Schrijver (1990). Chvátal closures for mixed integer programming problems. *Mathematical Programming* 47, 155–174.

Cornuéjols, G., Y. Li (2001a). Elementary closures for integer programs. *Operations Research Letters* 28, 1–8.

Cornuéjols, G., Y. Li (2001b). On the rank of mixed 0-1 polyhedra, in: K. Aardal, A. M. H. Gerards (eds.), IPCO 2001, *Lecture Notes in Computer Science* 2081, 71–77.

Cornuéjols, G., Y. Li (2002). A connection between cutting plane theory and the geometry of numbers. *Mathematical Programming A* 93, 123–127.

Crippen, G. M., T. F. Havel (1988). *Distance Geometry and Molecular Conformation*, Research Studies Press, Taunton, Somerset, England.

Cvetković, D., M. Cangalvic', V. Kovačevič-Vujčić (1999). Semidefinite programming methods for the symmetric traveling salesman problem, In *Proceedings of the 7th International IPCO Conference*, Graz, Austria, pp. 126–136.

Dash, S. (2001). *On the Matrix Cuts of Lovász and Schrijver and their Use in Integer Programming*. PhD thesis, Rice University.

Dash, S. (2002). An exponential lower bound on the length of some classes of branch-and-cut proofs, in: W. J. Cook, A. S. Schulz (eds.), IPCO 2002, *Lecture Notes in Computer Science* 2337, 145–160.

Delorme, C., S. Poljak (1993a). Laplacian eigenvalues and the maximum cut problem. *Mathematical Programming* 62, 557–574.

Delorme, C., S. Poljak (1993b). Combinatorial properties and the complexity of a max-cut approximation. *European Journal of Combinatorics* 14, 313–333.

Delorme, C., S. Poljak (1993c). The performance of an eigenvalue bound on the max-cut problem in some classes of graphs. *Discrete Mathematics* 111, 145–156.

Delsarte, P. (1973). An algebraic approach to the association schemes of coding theory. *Philips Research Reports Supplements* , No. 10.

Deza, M., M. Laurent (1997). *Geometry of Cuts and Metrics*, Springer-Verlag.

Dinur, I., S. Safra (2002). The importance of being biased, In *Proceedings of the 34th Annual ACM Symposium on Theory of Computing*, ACM, New York, pp. 33–42.

Duffin, R. J. (1956). Infinite Programmes, in: H. W. Kuhn, A. W. Tucker (eds.), *Linear Inequalities and Related Systems*, Annals of Mathematicals, Studies Vol. 38, Princeton University Press, pp. 157–170.

Eisenblätter, A. (2001). *Frequency Assignment in GSM Networks: Models, Heuristics, and Lower Bounds*. PhD Thesis, TU Berlin, Germany, Available at ftp://ftp.zib.de/pub/zib-publications/books/PhD_eisenblaetter.ps.Z.

Eisenblätter, A. (2002). The semidefinite relaxation of the k-partition polytope is strong, in: W. J. Cook, A. S. Schulz (eds.), IPCO 2002, *Lecture Notes in Computer Science* 2337, pp. 273–290.

Eisenbrand, F. (1999). On the membership problem for the elementary closure of a polyhedron. *Combinatorica* 19, 299–300.

Eisenbrand, F., A. S. Schulz (1999). Bounds on the Chvátal rank of polytopes in the 0/1 cube, in: G. Cornuéjols et al. (eds.), IPCO 1999, *Lecture Notes in Computer Science* 1610, 137–150.

Feige, U. (1997). Randomized graph products, chromatic numbers, and the Lovász ϑ-function. *Combinatorica* 17, 79–90. [Preliminary version in *Proceedings of the 27th Annual ACM Symposium on Theory of Computing*, ACM, New York, pp. 635–640, 1995.]

Feige, U. (1999). Randomized rounding of semidefinite programs – variations on the MAX CUT example. Randomization, Approximation, and Combinatorial Optimization, Proceedings of Random-Approx'99. *Lecture Notes in Computer Science* 1671, 189–196, Springer-Verlag.

Feige, U., M. Goemans (1995). Approximating the value of two prover proof systems, with applications to MAX 2SAT and MAX DICUT. In *Proceedings of the 3rd Israel Symposium on the Theory of Computing and Systems*, ACM, New York, pp. 182–189.

Feige, U., M. Karpinski, M. Langberg (2000a). Improved approximation of max-cut on graphs of bounded degree. *Electronic Colloquium on Computational Complexity*, Report TR00-021.

Feige, U., M. Karpinski, M. Langberg (2000b). A note on approximating max-bisection on regular graphs. *Electronic Colloquium on Computational Complexity*, Report TR00-043.

Feige, U., R. Krauthgamer (2003). The probable value of the Lovász–Schrijver relaxations for maximum independent set. *SIAM Journal on Computing* 32, 345–370.

Feige, U., M. Langberg, G. Schechtman (2002). Graphs with tiny vector chromatic numbers and huge chromatic numbers. In *Proceedings of the 43rd Annual IEEE Symposium on Foundations of Computer Science*, IEEE Computer Science Press, Los Alamitos, CA.

Feige, U., G. Schechtman (2001). On the integrality ratio of semidefinite relaxations of MAX CUT. In *Proceedings of the 33rd Annual ACM Symposium on Theory of Computing*, ACM, New York, 433–442.

Feige, U., G. Schechtman (2002). On the optimality of the random hyperplane rounding technique for MAX CUT. *Random Structures and Algorithms* 20, 403–440.

Fiedler, M. (1972). Bounds for eigenvalues of doubly stochastic matrices. *Linear Algebra and its Applications* 5, 299–310.

Fiedler, M. (1973). Algebraic connectivity of graphs. *Czechoslovak Mathematical Journal* 23, 298–305.

Frankl, P., V. Rödl (1987). Forbidden intersections. *Transactions of the American Mathematical Society* 300, 259–286.

Frieze, A., M. Jerrum (1997). Improved approximation algorithms for MAX k-CUT and MAX BISECTION. *Algorithmica* 18, 67–81. [Preliminary version in *Proceedings of the 4th International IPCO Conference,* Copenhagen, *Lecture Notes in Computer Science*, 920, 1–13, 1995.]

Fujie, T., M. Kojima (1997). Semidefinite programming relaxation for nonconvex quadratic programs. *Journal of Global Optimization* 10, 367–380.

Fulkerson, D. R. (1972). Anti-blocking polyhedra. *Journal of Combinatorial Theory B* 12, 50–71.

Garey, M. R., D. S. Johnson, L. Stockmeyer (1976). Some simplified NP-complete graph problems. *Theoretical Computer Science* 1, 237–267.

Goemans, M. X. (1997). Semidefinite programming in combinatorial optimization. *Mathematical Programming* 143–161.

Goemans, M., F. Rendl (1999). Semidefinite programs and association schemes. *Computing* 63, 331–340.

Goemans, M. X., L. Tunçel (2001). When does the positive semidefiniteness constraint help in lifting procedures? *Mathematics of Operations Research* 26, 796–815.

Goemans, M. X., D. P. Williamson (1994). New 3/4-approximation algorithms for the maximum satisfiability problem. *SIAM Journal on Discrete Mathematics* 7, 656–666.

Goemans, M. X., D. P. Williamson (1995). Improved approximation algorithms for maximum cuts and satisfiability problems using semidefinite programming. *Journal of the Association for Computing Machinery* 42, 1115–1145. [Preliminary version in *Proceedings of the 26th Annual ACM Symposium on Theory of Computing*, ACM, New York, pp. 422–431, 1994.]

Goemans, M. X., D. P. Williamson (2001). Approximation algorithms for MAX-3-CUT and other problems via complex semidefinite programming. In *Proceedings of the 33rd Annual ACM Symposium on Theory of Computing*, ACM, New York, pp. 443–452.

Grigoriev, D., E. A. Hirsch, D. V. Pasechnik (2002). Complexity of semi-algebraic proofs. *Lecture Notes in Computer Science* 2285, 419–430.

Grigoriev, D., E. de Klerk, D. V. Pasechnik (2003). Finding optimum subject to few quadratic constraints in polynomial time. Preprint, Extended abstract available at http://www.thi.informatik.uni-frankfurt.de/~dima/misc/qp-ea.ps

Grone, R., C. R. Johnson, E. M. Sá, H. Wolkowicz (1984). Positive definite completions of partial Hermitian matrices. *Linear Algebra and its Applications* 58, 109–124.

Grötschel, M., L. Lovász, A. Schrijver (1988). *Geometric Algorithms and Combinatorial Optimization*, Springer-Verlag, Berlin, New York,

Grötschel, M., W. R. Pulleyblank (1981). Weakly bipartite graphs and the max-cut problem. *Operations Research Letters* 1, 23–27.

Gruber, G., F. Rendl. (2003). Computational experience with stable set relaxations. *SIAM Journal on Optimization,* 13, 1014–1028.

Guenin, B. (2001). A characterization of weakly bipartite graphs. *Journal of Combinatorial Theory B* 81, 112–168.

Hadley, S. W., F. Rendl, H. Wolkowicz (1992). A new lower bound via projection for the quadratic assignment problem. *Mathematics of Operations Research* 17, 727–739.

Halldórsson, M. M. (1993). A still better performance guarantee for approximate graph coloring. *Information Processing Letters* 45, 19–23.

Halldórsson, M. M. (1998). Approximations of independent sets in graphs, in: K. Jansen, J. Rolim (eds.), APPROX '98, *Lecture Notes in Computer Science* 1444, 1–14.

Halldórsson, M. M. (1999). Approximations of weighted independent sets and hereditary subset problems, in: T. Asano et al. (eds.), COCOON '99, *Lecture Notes in Computer Science* 1627, 261–270.

Halperin, E. (2002). Improved approximation algorithms for the vertex cover problem in graphs and hypergraphs. *SIAM Journal on Computing* 31, 1608–1623. [Preliminary version in *Proceedings of 11th ACM-SIAM Symposium on Discrete Algorithms*, pp. 329–337, 2000.]

Halperin, E., D. Livnat, U. Zwick (2002). MAX-CUT in cubic graphs. In *Proceedings of 13th ACM-SIAM Symposium on Discrete Algorithms* pp. 506–513.

Halperin, E., R. Nathaniel, U. Zwick (2001). Coloring k-colorable graphs using relatively small palettes. In *Proceedings of 12th ACM-SIAM Symposium on Discrete Algorithms* pp. 319–326.

Halperin, E., U. Zwick (2001a). A unified framework for obtaining improved approximations algorithms for maximum graph bisection problems, in: K. Aardal, A. M. H. Gerards (eds.), IPCO 2001, *Lecture Notes in Computer Science* 2081, 210–225.

Halperin, E., U. Zwick (2001b). Approximation algorithms for MAX 4-SAT and rounding procedures for semidefinite programs. *Journal of Algorithms* 40, 184–211. [Preliminary version in *Proceedings of the 7th conference on Integer Programming and Combinatorial Optimization,* Graz, Austria, pp. 202–217, 1999.]

Halperin, E., U. Zwick (2001c). Combinatorial approximation algorithms for the maximum directed cut problem, In: *Proceedings of 12th ACM-SIAM Symposium on Discrete Algorithms* pp. 1–7.

Håstad, J. (1997). Some optimal inapproximability results. In *Proceedings of the 29th Annual ACM Symposium on the Theory of Computing*, ACM, New York, pp. 1–10. [Full version in *Electronic Colloquium on Computational Complexity*, Report TR97-037.]

Helmberg, C., F. Rendl, R. J. Vanderbei, H. Wolkowicz (1996). An interior-point method for semidefinite programming. *SIAM Journal on Optimization* 6, 342–361.

Helmberg, C., F. Rendl, R. Weismantel (2000). A semidefinite programming approach to the quadratic knapsack problem. *Journal of Combinatorial Optimization* 4, 197–215.

Hill, R. D., S. R. Waters (1987). On the cone of positive semidefinite matrices. *Linear Algebra and its Applications* 90, 81–88.

Hoffman, A. J., H. W. Wielandt (1953). The variation of the spectrum of a normal matrix. *Duke Mathematical Journal* 20, 37–39.

Horn, R. A., C. R. Johnson (1985). *Matrix Analysis*, Cambridge University Press.

Jansen, K., M. Karpinski, A. Lingas (2000). A polynomial time approximation scheme for MAX-BISECTION on planar graphs. *Electronic Colloquium on Computational Complexity*, Report TR00-064.

Johnson, C.R. (1990). Matrix completion problems: a survey, in: C. R. Johnson (ed.), *Matrix Theory and Applications*, Volume 40 of *Proceedings of Symposia in Applied Mathematics*, American Mathematical Society, Providence, Rhode Island, pp. 171–198.

Johnson, D. (1974). Approximation algorithms for combinatorial problems. *Journal of Computer and System Sciences* 9, 256–278.

Johnson, C. R., B. Kroschel, H. Wolkowicz (1998). An interior-point method for approximate positive semidefinite completions. *Computational Optimization and Applications* 9, 175–190.

Kann, V., S. Khanna, J. Lagergren, A. Panconesi (1997). On the hardness of approximating MAX k-CUT and its dual. *Chicago Journal of Theoretical Computer Science* 2.

Karger, D., R. Motwani, M. Sudan (1998). Approximate graph colouring by semidefinite programming. *Journal of the Association for Computing Machinery* 45, 246–265. [Preliminary version in *Proceedings of 35th IEEE Symposium on Foundations of Computer Science*, IEEE Computer Science Press, Los Alamitos, CA, pages 2–13, 1994.]

Karloff, H. (1999). How good is the Goemans–Williamson max-cut algorithm? *SIAM Journal on Computing* 29, 336–350.

Karloff, H., U. Zwick (1997). A 7/8-approximation algorithm for MAX 3SAT? In *Proceedings of the 38th Annual IEEE Symposium on Foundations of Computer Science*, IEEE Computer Science Press, Los Alamitos, CA, pp. 406–415.

Karp, R. M. (1972). Reducibility among combinatorial problems. In *Complexity of Computer Computations*, Plenum Press, New York, pp. 85–103.

Khachiyan, L., L. Porkolab (1997). Computing integral points in convex semi-algebraic sets. In *38th Annual Symposium on Foundations of Computer Science*, IEEE Computer Science Press, Los Alamitos, CA, pp. 162–171.

Khachiyan, L., L. Porkolab (2000). Integer optimization on convex semialgebraic sets. *Discrete and Computational Geometry* 23, 207–224.

Khanna, S., N. Linial, S. Safra (2000). On the hardness of approximating the chromatic number. *Combinatorica* 20, 393–415. [Preliminary version in *Proceedings of the 2nd Israel Symposium on Theory and Computing Systems*, IEEE Computer Society Press, Los Alamos, CA, pp. 250–260, 1993.]

Kleinberg, J., M. X. Goemans (1998). The Lovász theta function and a semidefinite programming relaxation of vertex cover. *SIAM Journal on Discrete Mathematics* 11, 196–204.

de Klerk, E. (2002). *Aspects of Semidefinite Programming: Interior Point Algorithms and Selected Applications*, Kluwer.

de Klerk, E., M. Laurent, P. Parrilo (2004). A PTAS for the minimization of polynomials of fixed degree over the simplex. Preprint.

de Klerk, E., D. V. Pasechnik (2002). Approximation of the stability number of a graph via copositive programming. *SIAM Journal on Optimization* 12, 875–892.

de Klerk, E., D. V. Pasechnik, J. P. Warners (2004). Approximate graph colouring and MAX-*k*-CUT algorithms based on the theta-function. *Journal of Combinatorial Optimization* 8, 267–294.

de Klerk, E., J. P. Warners, H. van Maaren (2000). Relaxations of the satisfiability problem using semidefinite programming. *Journal of Automated Reasoning* 24, 37–65.

Knuth, D. E. (1994). The sandwich theorem. *Electronic Journal of Combinatorics* 1, 1–48.

Kojima, M., S. Shindoh, S. Hara (1997). Interior-point methods for the monotone semidefinite linear complementarity problem in symmetric matrices. *SIAM Journal on Optimization* 7, 86–125.

Kojima, M., L. Tunçel (2000). Cones of matrices and successive convex relaxations of nonconvex sets. *SIAM Journal on Optimization* 10, 750–778.

Lasserre, J. B. (2000). Optimality conditions and LMI relaxations for $0-1$ programs. Technical Report N. 00099, LAAS, Toulouse.

Lasserre, J. B. (2001a). Global optimization with polynomials and the problem of moments. *SIAM Journal on Optimization* 11, 796–817.

Lasserre, J. B. (2001b). An explicit exact SDP relaxation for nonlinear $0-1$ programs, in: K. Aardal, A. M. H. Gerards (eds.), IPCO 2001, *Lecture Notes in Computer Science* 2081, 293–303. [See also: An explicit equivalent positive semidefinite program for nonlinear 0-1 programs. *SIAM Journal on Optimization* 12, 756–769, 2002.]

Lasserre, J. B. (2002). Semidefinite programming vs. LP relaxations for polynomial programming. *Mathematics of Operations Research* 27, 347–360.

Laurent, M. (1997). The real positive semidefinite completion problem for series-parallel graphs. *Linear Algebra and its Applications* 252, 347–366.

Laurent, M. (1998a). A connection between positive semidefinite and Euclidean distance matrix completion problems. *Linear Algebra and its Applications* 273, 9–22.

Laurent, M. (1998b). A tour d'horizon on positive semidefinite and Euclidean distance matrix completion problems, in: P. Pardalos, H. Wolkowicz (eds.), *Topics in Semidefinite and Interior-Point Methods*, Vol. 18 of the Fields Institute for Research in Mathematical Science, Communication Series, Providence, Rhode Island, pp. 51–76.

Laurent, M. (2000). Polynomial instances of the positive semidefinite and Euclidean distance matrix completion problems. *SIAM Journal on Matrix Analysis and its Applications* 22, 874–894.

Laurent, M. (2001a). On the sparsity order of a graph and its deficiency in chordality. *Combinatorica* 21, 543–570.

Laurent, M. (2001b). Tighter linear and semidefinite relaxations for max-cut based on the Lovász-Schrijver lift-and-project procedure. *SIAM Journal on Optimization* 12, 345–375.

Laurent, M. (2003a). A comparison of the Sherali–Adams, Lovász–Schrijver and Lasserre relaxations for 0, 1-programming. *Mathematics of Operations Research* 28(3), 470–496.

Laurent, M. (2003b). Lower bound for the number of iterations in semidefinite hierarchies for the cut polytope. *Mathematical of Operations Reaserch* 28(4), 871–883.

Laurent, M. (2004). Semidefinite relaxations for Max-Cut, in: M. Grötschel (ed.), *The Sharpest Cut: The Impact of Manfred Padberg and his Work, MPS-SIAM Series in Optimization 4*, pp. 291–327.

Laurent, M., S. Poljak (1995). On a positive semidefinite relaxation of the cut polytope. *Linear Algebra and its Applications* 223/224, 439–461.

Laurent, M., S. Poljak (1996). On the facial structure of the set of correlation matrices. *SIAM Journal on Matrix Analysis and its Applications* 17, 530–547.

Laurent, M., S. Poljak, F. Rendl (1997). Connections between semidefinite relaxations of the max-cut and stable set problems. *Mathematical Programming* 77, 225–246.

Lenstra, H. W. Jr. (1983). Integer programming with a fixed number of variables. *Mathematics of Operations Research* 8, 538–548.

Lewin, M., D. Livnat, U. Zwick (2002). Improved rounding techniques for the MAX 2-SAT and MAX DI-CUT problems, in: W. J. Cook, A. S. Schulz (eds.), IPCO 2002, *Lecture Notes in Computer Science* 2337, 67–82.

Linial, N., E. London, Yu. Rabinovich (1995). The geometry of graphs and some of its algorithmic consequences. *Combinatorica* 15, 215–245.

Linial, N., A. Magen, A. Naor (2002). Girth and Euclidean distortion. *Geometric and Functional Analysis* 12, 380–394.

Linial, N., M. E. Sachs (2003). On the Euclidean distortion of complete binary trees. *Discrete and Computational Geometry* 29, 19–21.

Lipták, L., L. Tunçel (2003). Stable set problem and the lift-and-project ranks of graphs. *Mathematical Programming Ser. B* 98, 319–353.

Liu, W. (1988). *Extended Formulations and Polyhedral Projection.* PhD thesis, Department of Combinatorics and Optimization, University of Waterloo, Canada.

Lovász, L. (1972). Normal hypergraphs and the perfect graph conjecture. *Discrete Mathematics* 2, 253–267.

Lovász, L. (1979). On the Shannon capacity of a graph. *IEEE Transactions on Information Theory* IT-25, 1–7.

Lovász, L. (1994). Stable sets and polynomials. *Discrete Mathematics* 124, 137–153.

Lovász, L. (2003). Semidefinite programs and combinatorial optimization, in: B. A. Reed, C. L. Sales (eds.), *Recent Advances in Algorithms and Combinatorics*, CMS Books in Mathematics, Springer, pp. 137–194.

Lovász, L., A. Schrijver (1991). Cones of matrices and set-functions and 0-1 optimization. *SIAM Journal on Optimization* 1, 166–190.

Lund, C., M. Yannakakis (1993). On the hardness of approximating minimization problems. In *Proceedings of the 25th Annual ACM Symposium on Theory of Computing*, ACM, New York, pp. 286–293.

Maculan, N. (1987). The Steiner problem in graphs. *Annals of Discrete Mathematics* 31, 185–222.

Mahajan, S., H. Ramesh (1995). Derandomizing semidefinite programming based approximation algorithms. In *Proceedings of the 36th Symposium on Foundations of Computer Science*, IEEE Computer Science Press, Los Alamitos, CA, pp. 162–169.

Matuura, S., T. Matsui (2001a). 0.863-approximation algorithm for MAX DICUT, in: M. Goemans et al. (eds.), APPROX 2001 and RANDOM 2001, *Lecture Notes in Computer Science* 2129, 138–146.

Matuura, S., T. Matsui (2001b). 0.935-approximation randomized algorithm for MAX 2SAT and its derandomization. Technical Report METR 2001–03, University of Tokyo, Available at http://www.keisu.t.u-tokyo.ac.jp/METR.html.

McEliece, R. J., E. R. Rodemich, H. C. Rumsey, Jr. (1978). The Lovász bound and some generalizations. *Journal of Combinatorics and System Sciences* 3, 134–152.

Meurdesoif, P. (2000). Strenghtening the Lovász $\vartheta(\bar{G})$ bound for graph colouring. Preprint, [*Mathematical Programming,* to appear].

Mohar, B., S. Poljak (1990). Eigenvalues and the max-cut problem. *Czechoslovak Mathematical Journal* 40, 343–352.

Monteiro, R. D. C. (1997). Primal-dual path-following algorithms for semidefinite programming. *SIAM Journal on Optimization* 7, 663–678.

Motzkin, T. S., E. G. Straus (1965). Maxima for graphs and a new proof of a theorem of Túran. *Canadian Journal of Mathematics* 17, 533–540.

Murty, K. G., S. N. Kabadi (1987). Some NP-complete problems in quadratic and linear programming. *Mathematical Programming* 39, 117–129.

Nemhauser, G., L. Wolsey (1988). *Integer and Combinatorial Optimization*, John Wiley and Sons, New York.

Nesterov, Y. (1997). Quality of semidefinite relaxation for nonconvex quadratic optimization. CORE Discussion Paper # 9719, Belgium.

Nesterov, Y. (1998). Semidefinite relaxation and nonconvex quadratic optimization. *Optimization Methods and Software* 9, 141–160.

Nesterov, Y. (2000). Squared functional systems and optimization problems, in: J. B. G. Frenk, C. Roos, T. Terlaky, S. Zhang (eds.), *High Performance Optimization*, Kluwer Academic Publishers, pp. 405–440.

von Neumann, J. (1937). Some matrix inequalities and metrization of matrix space. *Tomsk Univ. Rev.* 1, 286–300, (reprinted in: John von Neumann: Collected works, Vol. 4, A. H. Taub ed., MacMillan, 205–219, 1962.).

Nugent, C. E., T. E. Vollman, J. Ruml (1968). An experimental comparison of techniques for the assignment of facilities to locations. *Operations Research* 16, 150–173.

Overton, M. L., R. S. Womersley (1992). On the sum of the largest eigenvalues of a symmetric matrix. *SIAM Journal on Matrix Analysis and its Applications* 13, 41–45.

Pardalos, P. M., H. Wolkowicz (eds.) (1998). Topics in semidefinite programming and interior point methods. *Fields Institute Communications 18*, American Mathematical Society.

Parrilo, P. A. (2000). *Structured Semidefinite Programs and Semialgebraic Geometry Methods in Robustness and Oprimization.* PhD thesis, California Institute of Technology.

Parrilo, P. A. (2003). Semidefinite programming relaxations for semialgebraic problems. *Mathematical Programming Ser. B* 96, 293–320.

Parrilo, P. A., B. Sturmfels (2003). Minimizing polynomial functions, in: S. Basu, L. Gonzalez-Vega (eds.), *Algorithmic and Quantitative Real Algebraic Geometry, DIMACS Series in Discrete Mathematics and Theoretical Computer Science*, Vol. 60.

Pataki, G. (1996). Cone-LP's and semidefinite programs: geometry and a simplex-type methods. in: W. H. Cunningham, S. T. MacCormick, M. Queyranne (eds.), IPCO 1996, *Lecture Notes in Computer Science* 1084, 162–174.

Pataki, G. (1998). On the rank of extreme matrices in semidefinite programs and the multiplicity of optimal eigenvalues. *Mathematics of Operations Research* 23, 339–358.

Poljak, S. (1991). Polyhedral and eigenvalue approximations of the max-cut problem, in: *Sets, Graphs, and Numbers*, Vol. 60 of *Colloquia Mathematica Societatis János Bolyai*, Budapest, Hungary, pp. 569–581.

Poljak, S., F. Rendl (1995). Nonpolyhedral relaxations of graph-bisection problems. *SIAM Journal on Optimization* 5, 467–487.

Poljak, S., Z. Tuza (1994). The expected relative error of the polyhedral approximation of the max-cut problem. *Operations Research Letters* 16, 191–198.

Poljak, S., Z. Tuza (1995). Maximum cuts and largest bipartite subgraphs, in: W. Cook, L. Lovász, P. Seymour (eds.), *Combinatorial Optimization*, Vol. 20 of *DIMACS Series in Discrete Mathematics and Theoretical Computer Science*, American Mathematical Society, Providence, RI, pp. 181–244.

Porkolab, L., L. Khachiyan (1997). On the complexity of semidefinite programs. *Journal of Global Optimization* 10, 351–365.

Powers, V., B. Reznick (2001). A new bound for Pólya's Theorem with applications to polynomials positive on polyhedra. *Journal of Pure and Applied Algebra* 164, 221–229.

Powers, V., T. Wörmann (1998). An algorithm for sums of squares of real polynomials. *Journal of Pure and Applied Algebra* 127, 99–104.

Putinar, M. (1993). Positive polynomials on compact semi-algebraic sets. *Indiana University Mathematics Journal* 42, 969–984.
Quist, A. J., E. de Klerk, T. Roos, C. Terlaky (1998). Copositive relaxation for general quadratic programming. *Optimization Methods and Software* 9, 185–209.
Ramana, M. V. (1997). An exact duality theory for semidefinite programming and its complexity implications. *Mathematical Programming* 77, 129–162.
Ramana, M. V., A. Goldman (1995). Some geometric results in semidefinite programming. *Journal of Global Optimization* 7, 33–50.
Ramana, M. V., L. Tunçel, H. Wolkowicz (1997). Strong duality for semidefinite programming. *SIAM Journal on Optimization* 7, 641–662.
Reed, B. A., A. J. L. Ramirez (2001). *Perfect Graphs*, Wiley.
Rendl, F. (1999). Semidefinite programming and combinatorial optimization. *Applied Numerical Mathematics* 29, 255–281.
Renegar, J. (1992). On the computational complexity and geometry of the first order theory of the reals. *Journal of Symbolic Computation* 13(3), 255–352.
Schrijver, A. (1979). A comparison of the Delsarte and Lovász bounds. *IEEE Transactions on Information Theory* IT-25, 425–429.
Schrijver, A. (1986). *Theory of Linear and Integer Programming*, John Wiley and Sons, New York.
Schrijver, A. (2002). A short proof of Guenin's characterization of weakly bipartite graphs. *Journal of Combinatorial Theory B* 85, 255–260.
Schrijver, A. (2003). *Combinatorial Optimization – Polyhedra and Efficiency*, Springer-Verlag, Berlin.
Seymour, P. D. (1977). The matroids with the max-flow min-cut property. *Journal of Combinatorial Theory B* 23, 189–222.
Sherali, H., W. Adams (1990). A hierarchy of relaxations between the continuous and convex hull representations for zero-one programming problems. *SIAM Journal on Discrete Mathematics* 3, 411–430.
Sherali, H., W. Adams (1994). A hierarchy of relaxations and convex hull representations for mixed-integer zero-one programming problems. *Discrete Applied Mathematics* 52, 83–106.
Sherali, H., W. Adams (1997). *A Reformulation-Linearization Technique (RLT) for Solving Discrete and Continuous Nonconvex Problems*, Kluwer.
Sherali, H., C. H. Tuncbilek (1992). A global optimization algorithm for polynomial programming problems using a reformulation-linearization technique. *Journal of Global Optimization* 2, 101–112.
Sherali, H. D., C. H. Tuncbilek (1997). Reformulation-linearization/convexification relaxations for univariate and multivariate polynomial programming problems. *Operations Research Letters* 21, 1–10.
Shor, N. Z. (1987a). An approach to obtaining global extremums in polynomial mathematical programming problems. *Kibernetika* 5, 102–106.
Shor, N. Z. (1987b). Class of global minimum bounds of polynomial functions. *Cybernetics* 6, 731–734. [Translated from *Kibernetika*, 6, 9–11, 1987.]
Shor, N. Z. (1998). *Nondifferentiable Optimization and Polynomial Problems*, Kluwer Academic Publishers.
Skutella, M. (2001). Convex quadratic and semidefinite programming relaxations in scheduling. *Journal of the Association for Computing Machinery* 48, 206–242.
Sotirov, R. (2003). *Bundle methods in combinatorial optimization.* PhD thesis, University of Klagenfurt.
Stengle, G. (1974). A Nullstellensatz and a Positivstellensatz in semialgebraic geometry. *Mathematische Annalen* 207, 87–97.
Stephen, T., L. Tunçel (1999). On a representation of the matching polytope via semidefinite liftings. *Mathematics of Operations Research* 24, 1–7.
Szegedy, M. (1994). A note on the ϑ number of Lovász and the generalized Delsarte bound. In *Proceedings of the 35th Annual IEEE Symposium on Foundations of Computer Science*, IEEE Computer Science Press, Los Alamitos, CA, pp. 36–39.

Todd, M. J. (1999). A study of search directions directions in primal-dual interior-point methods for semidefinite programming. *Optimization Methods and Software* 11, 1–46.

Todd, M. J. (2001). Semidefinite programming. *Acta Numerica* 10, 515–560.

Trevisan, L., G. B. Sorkin, M. Sudan, D. P. Williamson (1996). Gadgets, approximation, and linear programming. In *Proceedings of the 37th Annual IEEE Symposium on Foundations of Computer Science*, IEEE Computer Science Press, Los Alamitos, CA, pp. 617–626.

Tseng, P. (2003). Further results on approximating nonconvex quadratic optimization by semidefinite programming relaxation. *SIAM Journal on Optimization* 14, 268–283.

Vandenberghe, L., S. Boyd (1996). Semidefinite programming. *SIAM Review* 38, 49–95.

de la Vega, W. F. (1996). MAX-CUT has a randomized approximation scheme in dense graphs. *Random Structures and Algorithms* 8, 187–198.

Warners, J. P. (1999). *Nonlinear Approaches to Satisfiability Problems.* PhD thesis, Technical University Eindhoven.

Wigderson, A. (1983). Improving the performance guarantee for approximate graph colouring. *Journal of the Association for Computing Machinery* 30, 729–735.

Wolkowicz, H., R. Saigal, L. Vandenberghe (eds.) (2000). *Handbook of Semidefinite Programming*, Kluwer.

Yannakakis, M. (1988). Expressing combinatorial optimization problems by linear programs. In *Proceedings of the 29th International IEEE Symposium on Foundations of Computer Science*, IEEE Computer Science Press, Los Alamitos, CA, pp. 223–228.

Yannakakis, M. (1994). On the approximation of maximum satisfiability. *Journal of Algorithms* 17, 475–502.

Ye, Y. (1999). Approximating quadratic programming with bound and quadratic constraints. *Mathematical Programming* 84, 219–226.

Ye, Y. (2001). A 0.699-approximation algorithm for Max-Bisection. *Mathematical Programming* 90, 101–111.

Zhang, Y. (1998). On extending some primal-dual interior-point algorithms from linear programming to semidefinite programming. *SIAM Journal on Optimization* 8, 365–386.

Zhang, S. (2000). Quadratic minimization and semidefinite relaxation. *Mathematical Progamming* 87, 453–465.

Zhao, Q., S. E. Karisch, F. Rendl, H. Wolkowicz (1998). Semidefinite programming relaxations for the Quadratic Assignment Problem. *Journal of Combinatorial Optimization* 2, 71–109.

Zwick, U. (1999). Outward rotations: a tool for rounding solutions of semidefinite programming relaxations, with applications to MAX CUT and other problems. In *Proceedings of the 31st Annual ACM Symposium on Theory of Computing*, ACM, New York, pp. 679–687.

Zwick, U. (2000). Analyzing the MAX 2-SAT and MAX DI-CUT approximation algorithms of Feige and Goemans. Preprint. Available at http://www.math.tau.ac.il/~zwick/.

Zwick, U. (2002). Computer assisted proof of optimal approximability results. In *Proceedings of 13th ACM-SIAM Symposium on Discrete Algorithms* pp. 496–505.

K. Aardal et al., Eds., *Handbooks in OR & MS, Vol. 12*

Chapter 9

Algorithms for Stochastic Mixed-Integer Programming Models

Suvrajeet Sen

MORE Institute, SIE Department, University of Arizona, Tucson, AZ 85721, USA

Abstract

In this chapter, we will study algorithms for both two-stage as well as multi-stage stochastic mixed-integer programs. We present stagewise (resource-directive) decomposition methods for two-stage models, and scenario (price-directive) decomposition methods for multi-stage models. The manner in which these models are decomposed relies not only on the specific data elements that are random, but also on the manner in which the integer (decision) variables interact with these data elements. Accordingly, we study a variety of structures ranging from models that allow randomness in all data elements, to those that allow only specific elements (e.g. right-hand-side) to be influenced by randomness. Since the decomposition algorithms presented here are based on certain results from integer programming, the relevant background is also provided in this chapter.

1 Introduction

Integer Programming (IP), and Stochastic Programming (SP) constitute two of the more vibrant areas of research in optimization. Both areas have blossomed into fields that have solid mathematical foundations, reliable algorithms and software, and a plethora of applications that continue to challenge the current state-of-the-art computing resources. For a variety of reasons, these areas have matured independently. A study of SMIP requires that we integrate the methods of continuous optimization (SP) and those of discrete optimization (IP). With the exception of a joint appreciation for Benders' decomposition (Benders [1962] and Van Slyke and Wets [1969]), the IP and SP communities have, for many years, kept their distance from a large class of stochastic mixed-integer programming (SMIP) models. Indeed, the only class of SMIP models that has attracted its fair share of attention is the one for which Benders' decomposition is applicable without further mathematical developments. Such models are typically two-stage stochastic programs in which the first-stage decisions are mixed-integer, and the second-stage (recourse) decisions are obtained from linear programming (LP) models. Research on other classes of SMIP models is recent; some of the first structural results for integer recourse problems are only about a decade old (e.g. Schultz [1993]). The first algorithms also began to appear around the same time (e.g. Laporte and Louveaux [1993]). As for dissertations, the first in the area appears to be Stougie [1985], and a few of the early notable ones may be Takriti [1994], Van der Vlerk [1995], and Caroe [1998], to name a few.

In the last few years there has been a flurry of activity resulting in rapid growth of the area. This chapter is devoted to algorithmic issues that have a bearing on two focal points. First, we focus on decomposition algorithms because they have the potential to provide scalable approaches for large-scale models. For realistic SP models, the ability to handle a large number of potential scenarios is critical. The second focal point deals with integer recourse models (i.e. the integer variables are associated with recourse decisions in stages two and beyond). These issues are intimately related to IP decomposition which is likely to be of interest to researches in both SP as well as IP. We hope that this chapter will motivate readers to investigate novel algorithms that will be scalable enough to solve practical stochastic mixed-integer programming models.

Problem Setting

A two-stage SMIP model is one in which a subset of both first and second-stage variables are required to satisfy integer restrictions. To state the problem, let $\tilde{\omega}$ denote a random variable used to model data uncertainty in a two-stage model. (We postpone the statement of a multi-stage problem to section 4.) Since SP models are intended for decision-making, a decision vector x must be chosen in such a manner that the consequences of the decisions (evaluated under several alternative outcomes of $\tilde{\omega}$) are accommodated within an optimal choice model. The consequences of the first-stage decisions are measured through an optimization problem (called the recourse problem) which allows the decision-maker to adapt to an observation of the data (random variable). Suppose that an observation of $\tilde{\omega}$ is denoted ω. Then the consequences of choosing x in the face of an outcome ω may be modeled as

$$h(x, \omega) = \operatorname{Min} g(\omega)^\top y \tag{1.1a}$$

$$W(\omega)y \geq r(\omega) - T(\omega)x \tag{1.1b}$$

$$y \geq 0, y_j \text{ integer}, \ j \in J_2, \tag{1.1c}$$

where J_2 is an index set that may include some or all the variables listed in $y \in \Re^{n_2}$. Throughout this chapter, we will assume that all realizations $W(\omega)$ are rational matrices of size $m_2 \times n_2$. Whenever J_2 is non-empty, and $|J_2| \neq n_2$,

(1.1) is said to provide a model with mixed-integer recourse. Although (1.1) is stated as though the random variable influences all data, most applications lead to models that lead to only some data uncertainty, which in turn lead to certain specialized models.

A typical decision-maker used his/her attitude towards risk to order alternative choices of x. In the decision analysis literature, the collection of possible choices are usually a few in number, and for such cases, it is possible to enumerate all the choices. For more complicated decision models, where the choices may be too many to enumerate, one resorts to optimization techniques, and more specifically to stochastic programming.

While several alternative "risk preferences" have been incorporated within SP models recently (see Ogryczak and Ruszczynski [2002], Riis and Schultz [2003], Takriti and Ahmed [2004]), the predominant approach in the SP literature is the "expected value" model. In order to focus our attention on complications arising from integer restrictions on decision variables, we will restrict our study to the "expected value" model. For this setting, the two-stage SMIP model may be stated as follows.

$$\operatorname*{Min}_{x \in X \cap \mathbf{X}} c^\top x + E[h(x, \tilde{\omega})], \tag{1.2}$$

where $\tilde{\omega}$ denotes a random variable defined on a probability space $(\Omega, \mathcal{A}, \mathcal{P})$, X a convex polyhedron, and $\mathbf{X}$ denotes either the set of binary vectors $\mathcal{B}$, or integer vectors $\mathcal{Z}$ or even mixed-integer vectors $\mathcal{M} = \{x \mid x \geq 0, x_j \text{ integer}, j \in J_1\}$, where J_1 is a given index set consisting of some or all of first-stage variables $x \in \Re^{n_1}$. Whenever we refer to the two-stage SMIP problems, we will be referring to (1.1,1.2). Throughout this chapter, we will assume that the random variables have finite support, so that the expectation in (1.2) reduces to a summation.

Within the stochastic programming literature, a realization of $\tilde{\omega}$ is known as a "scenario". As such, the second-stage problem (1.1) is often referred to as a "scenario subproblem." Because of its dependence on the first-stage decision x, the value function $h(\,)$ is referred to as the recourse function. Accordingly, $E[h(\,)]$ is called the expected recourse function of the two-stage model. These two-stage models are said to have a fixed recourse matrix (or simply fixed recourse) when the matrix $W(\omega)$ is deterministic; that is, $W(\omega) = W$. If the matrix $T(\omega)$ is deterministic, (i.e., $T(\omega) = T$), the stochastic program is said to have fixed tenders. When the second-stage problem is feasible for all choices of $x \in \Re^{n_1}$, the model is said to possess the complete recourse property; moreover, if the second-stage problem is feasible for all $x \in X \cap \mathbf{X}$, then it is said to possess the relatively complete recourse property. When the matrix W has the special structure that $W = (I, -I)$, the second-stage decision variables are continuous, and the constraints (1.1b) are equations, then the resulting problem is called a stochastic program with "simple recourse." In this special case, the second-stage variables simply measure the deviation from an uncertain target. The standard news-vendor problem of perishable inventory management is a stochastic program with simple recourse. It turns out that the continuous simple recourse problem is one class of models that is very amenable to accurate solutions (Kall and Mayer [1996]). Moreover as discussed subsequently, these models may be used in connection with methods for the solution of simple integer recourse models.

Algorithmic research in stochastic programming has focused on methods that are intended to accommodate a large number of scenarios so that realistic applications can be addressed. This has led to novel decomposition algorithms, some deterministic (e.g. Rockafellar and Wets [1991], Mulvey and Ruszczynski [1995]), and some stochastic (Higle and Sen [1991], Infanger [1992]). In this chapter we will adopt a deterministic decomposition paradigm. Such approaches are particularly relevant for SMIP because the idea of solving a series of small MIP problems to ultimately solve a large SMIP is computationally appealing. Moreover, due to the proliferation of networks of computers, such decomposition methods are likely to be more scalable than methods that treat the entire SMIP as one large deterministic MIP. Accordingly, this chapter is dedicated to decomposition-based algorithms for SMIP.

In this chapter, we will examine algorithms for both two-stage and multi-stage stochastic mixed-integer programs. In section 2, we will summarize some preliminary results that will have a bearing on the development of decomposition algorithms for SMIP. Section 3 is devoted to two-stage models under alternative assumptions that specify the structure of the model. For each class of models, we will discuss the decomposition method that best suits the structure. Section 4 deals with multi-stage models. We remind the reader that the state-of-the-art in this area is still in a state of flux, and encourage him/her to participate in our exploration to find ways to solve these very challenging problems.

2 Preliminaries for decomposition algorithms

The presence of integer decisions in (1.1) adds significant complications to designing decomposition algorithms for SMIP. In devising decomposition methods for these problems, it becomes necessary to draw upon results from the theory of IP. Most relevant to this study are results from IP duality, value functions, and disjunctive programming. The material in this section relies mainly on the work of Wolsey [1981] for IP duality, Blair and Jeroslow [1982], Blair [1995] for IP/MIP value functions, and Balas [1979] for disjunctive programming. Of course, some of this material is available in Nemhauser and Wolsey [1988]. We will also provide bridges from the world of MIP into that of SMIP. The first bridge deals with the properties of the SMIP recourse function which derive from properties of the MIP value function. These results were obtained by Schultz [1993]. The next bridge is that provided in the framework of Caroe and Tind [1998].

Structural Properties

Definition 2.1. *$f: \Re^n \rightarrow \Re$ is said to be sub-additive if $f(u+v) \leq f(u) + f(v)$. When this inequality is reversed, f is said to be super-additive.*

In order to state some results about the value function of an IP/MIP, we restate (1.1) in a familiar form, without the dependence on the data random variable, or the first-stage decision.

$$h(r) = \text{Min } g^\top y \tag{2.1a}$$

$$Wy \geq r \tag{2.1b}$$

$$y \geq 0, y_j \text{ integer}, \ j \in J_2. \tag{2.1c}$$

Proposition 2.2.

a) *The value function ($h(r)$) associated with (2.1) is non-decreasing, lower semi-continuous, and sub-additive over its effective domain (i.e. over the set of right hand sides for which the value function is finite).*

b) *Consider an SMIP as stated in (1.1,1.2) and suppose that the random variables have finite support. If the effective domain of the expected recourse function $E[h(\ )]$ is non-empty, then it is lower semi-continuous, and sub-additive on its effective domain.*

c) *Assume that the matrix W and the right-hand side vector r are integral, and (2.1) is a pure IP. Let v denote any vector of m_2 integers. Then the value function h is constant over sets of the form*

$$\left\{z \mid v - (1, \ldots, 1)^\top < z \leq v\right\}, \quad \forall v \in \mathcal{Z}^{m_2}.$$

For a proof of part a), please consult chapter II.3 of Nemhauser and Wolsey [1988]. Of course part b) follows from the fact that the expected recourse function is a finite sum of lower semi-continuous and sub-additive functions. And part c) is obvious since W and y have entries that are integers. This theorem is used in Schultz, Stougie, and Van der Vlerk [1998], as well as Ahmed, Tawarmalani and Sahinidis [2004] (see section 3).

For the case in which the random variables in SMIP are continuous, one may obtain continuity of the recourse function, but at a price. The following result requires that the random variables be absolutely continuous, which as we discuss below, is a significant restriction for constrained optimization problems.

Proposition 2.3. *Assume that (1.1) has randomness only in $r(\tilde{\omega})$, and let the probability space of this random variable, denoted $(\Omega, \mathcal{A}, \mathcal{P})$, be such that $\mathcal{P}$*

is absolutely continuous with respect to the Lebesgue measure in $\Re^{m_2}$. Moreover, suppose that the following hold.

a) (Dual feasibility). There exists $\theta \geq 0$ such that $W^\top \theta \leq g$.
b) (Complete recourse). For any choice of r in (2.1), the MIP feasible set is non-empty
c) (Finite expectation). $E[||r(\tilde{\omega})||] < \infty$.

Then, the expected recourse function is continuous.

This result was proven by Schultz [1993]. We should draw some parallels between the above result for SMIP and requirements for differentiability of the expected recourse function in SLP problems. While the latter possess expected recourse functions that are continuous, differentiability of the expected recourse function in SLP problems requires a similar absolute continuity condition (with respect to the Lebesgue measures in $\Re^{m_2}$). We remind the reader that even when a SLP has continuous random variables, the expected recourse function may fail to satisfy differentiability due to the lack of absolute continuity (Sen [1993]). By the same token, the SMIP expected recourse function may fail to be continuous without the assumption of absolute continuity as required above. It so happens that the requirement of absolute continuity (with respect to the Lebesgue measure in $\Re^{m_2}$) is rather restrictive from the point of view of practical optimization models. In order to appreciate this, observe that many practical LP/IP models have constraints that are entirely deterministic; for example, flow conservation/balance constraints often have no randomness in them. Formulations of this type (where some constraints are completely deterministic) fail to satisfy the requirement that the measure $\mathcal{P}$ is absolutely continuous with respect to the Lebesgue measure in $\Re^{m_2}$. Thus, just as differentiability is a luxury for SLP problems, continuity is a luxury for SMIP problems.

IP Duality

We now turn to an application of sub-additivity, especially its role in the theory of valid inequalities and IP duality.

Definition 2.4.

a) Let $\mathcal{S}$ denote the set of feasible points of an MIP such as (2.1). If $y \in \mathcal{S}$ implies $\pi^\top y \geq \pi_0$, then the latter is called a valid inequality for the set $\mathcal{S}$.
b) A monoid is a set M such that $0 \in M$, and if $W_1, W_2 \in M$, then $W_1 + W_2 \in M$.

Theorem 2.5. *Let $Y = \{y \in \Re^n_+ \mid Wy \geq r\}$, and assume that the entries of W are rational. Consider a pure integer program whose feasible set $\mathcal{S} = Y \cap \mathcal{Z}^{n_2}$ is non-empty.*

a) *If F is a sub-additive function defined on the monoid generated by the columns $\{W_j\}_{j=1}^{n_2}$ of W, then*

$$\sum_j F(W_j) y_j \geq F(r)$$

is valid inequality.

b) *Let $\pi^\top y \geq \pi_0$ denote a valid inequality for $\mathcal{S}$. Then, there is a sub-additive non-decreasing function F defined on the monoid generated by the columns W_j of W such that $F(0) = 0$, $\pi_j \geq F(W_j)$ and $\pi_0 \leq F(r)$.*

The reader may consult the book by Nemhauser and Wolsey [1988] for more on sub-additive duality. Given the above theorem, the sub-additive dual of (2.1) is as follows.

$$\max_{F \text{ sub-additive}} \quad F(r) \tag{2.2a}$$

$$\text{s.t } F(W_j) \leq g_j, \quad \forall j \tag{2.2b}$$

$$F(0) = 0. \tag{2.2c}$$

Several standard notions such as strong-duality and complementary slackness hold for this primal-dual pair. Moreover, Gomory's fractional cuts lead to a class of sub-additive functions constructed from using the ceiling operation on coefficients of linear valid inequalities; that is, functions of the form

$$\sum_j \lceil \pi_j \rceil y_j \geq \lceil \pi_0 \rceil,$$

where $\pi^\top y \geq \pi_0$ is a valid inequality for $\mathcal{S}$ (defined in Theorem 2.5). Such functions, which are referred to as Chvatal functions, are sub-additive and provide the appropriate class of dual price functions for the analysis of Gomory's fractional cuts. However, it is important to note that other algorithmic procedures for IP develop other dual price functions. For instance, branch-and-bound (B&B) methods generate non-decreasing, piecewise linear concave functions that provide solutions to a slightly different dual problem. In this sense, IP algorithms differ from algorithms for convex programming for which linear price functions are sufficient. For a more in-depth review of non-convex price functions (sub-additive or others), the reader should refer to Tind and Wolsey [1981]. Because certain algorithms do not necessarily generate sub-additive price functions, Caroe and Tind [1998] state an IP dual problem over a class of non-decreasing functions, which of course, includes the value function of (2.1). Therefore, the dual problem used in Caroe and Tind [1998] is as follows.

$$\max_{F \text{ non-decreasing}} \quad F(r) \tag{2.3a}$$

$$\text{s.t } F(Wy) \leq g^\top y \tag{2.3b}$$

$$F(0) = 0. \tag{2.3c}$$

We are now in a position to discuss the conceptual framework provided in Caroe and Tind [1998]. Their investigation demonstrates that on a conceptual level, it is possible to generalize the structure of Benders' decomposition (or L-shaped method) to decompose SMIP problems. However as noted in Caroe and Tind [1998], this conceptual scheme does not address practical computational difficulties associated with solving first-stage approximations which contain non-convex functions such as Chvatal functions. Nevertheless, the approach provides a conceptual bridge between MIP and SMIP problems.

In order to maintain simplicity in this presentation, we assume that the second-stage problem satisfies the complete recourse property. Assuming that the random variable modeling uncertainty is discrete, with finite support $(\Omega = \{\omega^1, \ldots, \omega^N\})$, a two-stage SMIP may be stated as

$$\text{Min } c^\top x + \sum_{\omega \in \Omega} p(\omega) g(\omega)^\top y(\omega) \tag{2.4a}$$

$$\text{s.t } Ax \geq b \tag{2.4b}$$

$$T(\omega)x + Wy(\omega) \geq r(\omega), \quad \forall \omega \in \Omega \tag{2.4c}$$

$$\left(x, \{y(\omega)\}_{\omega \in \Omega}\right) \geq 0, \, x_j \text{ integer}, \; j \in J_1, \quad \text{and } y_j(\omega) \text{ integer}, \; \forall_j \in J_2. \tag{2.4d}$$

Despite the fact that there are several assumptions underlying (2.4), it is somewhat general from the IP point of view since both the first and second stages allow general integer variables.

Following Caroe and Tind [1998], suppose we wish to apply a resource directive decomposition method, similar to Benders' decomposition. At iteration k of such a method, we solve one second-stage subproblem for each outcome ω, and assuming that we have chosen an appropriate solution method for the second-stage, then we obtain a non-decreasing price function $F^{\omega k}(r(\omega) - T(\omega)x)$ for each outcome $\omega \in \Omega$. Consequently, we obtain a "cut" of the form

$$\eta \geq \sum_{\omega \in \Omega} p(\omega) F^{\omega k}(r(\omega) - T(\omega)x).$$

Hence, as the iterations proceed, one obtains a sequence of relaxed master programs of the following form.

$$\text{Min } c^\top x + \eta \tag{2.5a}$$

$$\text{s.t } Ax \geq b \tag{2.5b}$$

$$\eta \geq \sum_{\omega \in \Omega} p(\omega) F^{\omega t}(r(\omega) - T(\omega)x), \quad t = 1, \ldots, k \tag{2.5c}$$

$$x \geq 0, x_j \text{ integer}, \ j \in J_1. \tag{2.5d}$$

As with Benders' (or L-shaped) decomposition, each iteration augments the first-stage approximation with one additional collection of price functions as shown in (2.5c). The rest of the procedure also mimics Benders' decomposition in that the sequence of objective values of (2.5) generates an increasing sequence of lower bounds, whereas, the subproblems at each iteration provide values used to compute an upper bound. The method stops when the upper and lower bounds are sufficiently close. Provided that the second-stage problems are solved using Gomory's cuts, or B&B, it is not difficult to show that the method must terminate in finitely many steps. Of course, finiteness also presumes that (2.5) can be solved in finite time.

We now visit the question of computational practicality of the procedure outlined above. The main observation is that the first-stage (master program) can be computationally unwieldy because the Chvatal functions arising from Gomory's method and piecewise linear concave functions resulting from B&B are nonconvex and are directly imported into the first-stage minimization [see (2.5c)]. These functions render the first-stage problem somewhat intractable. In section 3, we will discuss methods that will convexify such functions, thus leading to a more manageable first-stage problem.

Disjunctive Programming

Disjunctive programming focuses on characterizing the convex hull of disjunctive sets of the form

$$\mathcal{S} = \cup_{h \in H} S_h, \tag{2.6}$$

where H is a finite index set, and the sets S_h are polyhedral sets represented as

$$S_h = \{y \mid G_h y \geq r_h, y \geq 0\}. \tag{2.7}$$

This line of work originated with Balas [1975], and further developed in Blair and Jeroslow [1978]. Balas [1979] and Sherali and Shetty [1980] provide a comprehensive treatment of the approach, as well as its connections with other approaches for IP. Balas, Ceria and Cornuéjols [1993] provide computational results for such methods under a particular reincarnation called "lift-and-project" cuts.

The disjunction stated in (2.6, 2.7) is said to be in *disjunctive normal form* (i.e., none of the terms S_h contain any disjunction). It is important to recognize that the set of feasible solutions of any mixed-integer (0-1) program can be written as the union of polyhedra as in (2.6, 2.7) above. However, the number of elements in H can be exponentially large, thus making an explicit representation computationally impractical. If one is satisfied with weaker relaxations, then more manageable disjunctions can be stated. For example, the lift-and-project inequalities of Balas, Ceria and Cornuéjols [1993] use conjunctions associated with a linear relaxation together with one disjunction of the form: $y_j \leq 0$ or $y_j \geq 1$, for some $j \in J_2$. (Of course, y_j is assumed to be a binary variable.) For such a disjunctive set, the cardinality of H is two, with one polyhedron containing the inequalities $Wy \geq r, y \geq 0, -y_j \geq 0$ and the other polyhedron defined by $Wy \geq r, y \geq 0, y_j \geq 1$. For binary problems it is customary to include the bound constraint $-y \geq -1$ in $Wy \geq r$. Observe that in the notation of (2.6, 2.7), the matrices G_h differ only by one row, since W is common to both. Since there are only two atoms in the disjunction, it is computationally manageable. Indeed, it is not difficult to see that there is a hierarchy of disjunctions that one may use in developing relaxations of the integer program. Assuming that we have chosen some convenient level within the hierarchy, the index set H is specified, and we may proceed to obtain convex relaxations of the non-convex set. The idea of using alternative relaxations is also at the heart of the reformulation-linearization technique (RLT) of Sherali and Adams [1990].

The following result is known as the disjunctive cut principle. The forward part of this theorem is due to Balas [1975], and the converse is due to Blair and Jeroslow [1978]. In the following, the column vector G_{hj} denotes the jth column of the matrix G_h.

Theorem 2.6. *Let $\mathcal{S}$ and S_h be defined as in (2.6, 2.7) respectively. If $\lambda_h \geq 0$ for all $h \in H$, then*

$$\sum_j \left\{ \operatorname*{Max}_{h \in H} \lambda_h^\top G_{hj} \right\} y_j \geq \operatorname*{Min}_{h \in H} \lambda_h^\top r_h \tag{2.8}$$

is a valid inequality for $\mathcal{S}$. Conversely, suppose that $\pi^\top y \geq \pi_0$ is a valid inequality, and $H^* = \{h \in H \mid S_h \neq \emptyset\}$. Then there exist nonnegative vectors $\{\lambda_h\}_{h \in H^*}$ such that

$$\pi_j \geq \operatorname*{Max}_{h \in H^*} \lambda_h^\top G_{hj}, \quad \text{and} \quad \pi_0 \leq \operatorname*{Min}_{h \in H^*} \lambda_h^\top r_h. \tag{2.9}$$

Armed with this characterization of valid inequalities for the disjunctive set $\mathcal{S}$, we can develop a variety of relaxations of a mixed-integer linear program. The quality of the relaxations will, of course, depend on the choice of disjunction used, and the subset of valid inequalities used in the approximation. In the process of solving a MIP, suppose that we have obtained a solution to some linear relaxation, and assuming that the solution is fractional, we wish to separate it from the set of IP solutions using a valid inequality. Using one or more of the fractional variables to define H, we can state a disjunction such that the IP solutions are a subset of $\mathcal{S} = \cup_{h \in H} S_h$. Theorem 2.6 is useful for developing convexifications of the feasible mixed-integer solutions of the second-stage MIP.

The strongest (deepest) inequalities that one can derive are those that yield the closure of the convex hull of $\mathcal{S}$, denoted clconv($\mathcal{S}$). The following result of Balas [1979] provides an important characterization of the facets of clconv($\mathcal{S}$).

Theorem 2.7. *Let the reverse polar of $\mathcal{S}$, denoted $\mathcal{S}^\#$, be defined as*

$$\mathcal{S}^\# = \left\{ (\pi, \pi_0) | \textit{there are nonnegative vectors } \{\lambda_h\}_{h \in H} \textit{ such that } (2.9) \textit{ is satisfied.} \right\}$$

When π_0 is fixed, we denote the reverse polar by $\mathcal{S}^\#(\pi_0)$. Assume that $\mathcal{S}$ is full dimensional and $S_h \neq \emptyset$ for all $h \in H$. An inequality $\pi^\top y \geq \pi_0$ with $\pi_0 \neq 0$ is a facet of clconv($\mathcal{S}$) if and only if (π, π_0) is an extreme point of $\mathcal{S}^\#(\pi_0)$. Furthermore, if $\pi^\top y \geq 0$ is a facet of cclonv($\mathcal{S}$) then (π, π_0) is an extreme direction of $\mathcal{S}^\#(\pi_0)$ for all π_0.

Balas [1979] observes that for $\pi \neq 0$, if $(\pi, 0)$ is an extreme direction of $\mathcal{S}^\#$, then $\pi^\top y \geq 0$ is either a facet of clconv($\mathcal{S}$) or there exist two facets $(\pi^1)^\top y \geq \pi_0^1$ and $(\pi^2)^\top y \geq \pi_0^2$ such that $\pi = \pi^1 + \pi^2$ and $\pi_0^1 + \pi_0^2 = 0$. In any event, Theorem 2.7 provides access to a sufficiently rich collection of valid inequalities to the permit clconv($\mathcal{S}$) to be obtained algorithmically. The notion of reverse polars will be extensively used in section 3 to develop convexifications of certain non-convex functions, including price functions resulting from B&B methods for the second-stage.

In studying the behavior of sequential cutting plane methods, it is important to recognize that without appropriate safeguards, one may not, in fact, recover the convex hull of the set of feasible integer points (see Jeroslow [1980], Sen and Sherali [1985]). In such cases, the cutting plane method may not converge. We maintain however, that this is essentially a theoretical concern since practical schemes use cutting planes in conjunction with a B&B method, which are of course finitely convergent.

Before closing this section, we discuss a certain special class of disjunctions for which sequential convexification (one variable at a time) does yield the requisite closure of the convex hull of integer feasible points. This class of disjunctions gives rise to facial disjunction sets, which are described next.

A disjunctive set in *conjuctive normal form* may be stated in the form

$$\mathcal{S} = Y \cap_{j \in J} D_j,$$

where Y is a polyhedron, J is a finite index set, and each set D_j is defined by the union of finitely many halfspaces. The set $\mathcal{S}$ is said to possess the facial property for each j, every hyperplane used in the definition of D_j contains some face of Y. It is not difficult to see that a 0-1 MIP is a facial disjunctive program. For these problems Y is a polyhedral set that includes the "box" constraints $0 \leq y_j \leq 1, j \in J_2$, and the disjunctive sets D_j are defined as follows.

$$D_j = \{y \mid y_j \leq 0\} \cup \{y \mid y_j \geq 1\}.$$

Balas [1979] has shown that for sets with the facial property, one can recover the set clconv($\mathcal{S}$) by generating a sequence of convex hulls recursively. Let $j_1, j_2, \ldots$, etc. denote the indices of J_2, and initialize $j_0 = 0$, $Q_0 = Y$. Then

$$Q_{jk} = \text{clconv}(Q_{jk-1} \cap D_{jk}), \tag{2.10}$$

and the final convex hull operation yields clconv($\mathcal{S}$). Thus for a facial disjunctive program, the complete convexification can be obtained by convexifying the set by using disjunctions one variable at a time. As shown in Sen and Higle [2000], this result provides the basis for the convergence of the convex hull of second-stage feasible (mixed-binary) solutions using sequential convexification.

3 Decomposition algorithms for two-stage SMIP: stagewise decomposition

In this section, we study various classes of two-stage SMIP problems for which stagewise (resource-directive) decomposition algorithms appear to be quite appropriate. Recall that we have chosen to focus on the case of two-stage problems with integer recourse (in the second-stage). Our presentation excludes SMIP models in which the recourse function is defined using the LP value function. This is not to suggest that these problems (with integer first-stage, and continuous second-stage) are well solved. Significant challenges do remain, although they are mainly computational. For instance, the stochastic B&B method of Norkin, Ermoliev and Ruszczynski [1998] raises several interesting questions, especially those regarding its relationship with machine learning. By the same token, computational studies (e.g. Verweij et al [2003]) for this class of problems are of great importance. However, such an excursion would detract from our mission to foster a deeper understanding of the challenges associated with integer recourse models.

Much of this presentation revolves around convexification of the value functions of the second-stage IP. This section is divided into the following subsections.

- Simple Integer Recourse Models with Random RHS
- Binary First-stage, Arbitrary Second-stage
- Binary First-stage, 0-1 MIP Second-stage with Fixed Recourse
- Binary First-stage, MIP Second-stage
- Continuous First-stage, Integer Second-stage and Fixed Tenders
- 0-1 MIP in Both Stages with General Random Data

The heading for the subsections below indicate the above classification, and the subheadings identify the solution approach discussed in that subsection.

Simple Integer Recourse Models with Random RHS: Connections with the Continuum

The Simple Integer Recourse (SIR) model is the pure integer analog of the continuous simple recourse model. Unlike the continuous version of the simple recourse model, this version is intended for "news-vendor"-type models of "large-ticket" items. This class of models introduced by Louveaux and Van der Vlerk [1993], has been studied extensively in a series of papers by Klein Haneveld, Stougie and Van der Vlerk [1995, 1996]. We assume that all data elements except the right-hand side are fixed, and that the matrix T has full row rank. Moreover, assume that $g_i^+, g_i^- > 0, i = 1, \ldots, m_2$. Let $r_i(\omega)$ and t_i denote the ith row of $r(\omega)$ and T respectively, and let $\chi_i = t_i x$. Moreover, define a scalar function

$\lceil v \rceil^+ = \max\{0, \lceil v \rceil\}$ and $\lfloor v \rfloor^- = \max\{0, -\lfloor v \rfloor\}$. Then the statement of the SIR model is as follows.

$$\operatorname*{Min}_{x \in X \cap \mathbf{X}} \left\{ c^\top x + \sum_i E\left[g_i^+ \lceil r_i(\tilde{\omega}) - \chi_i \rceil^+ + g_i^- \lfloor r_i(\tilde{\omega}) - \chi_i \rfloor^- \right] \mid \chi = Tx \right\}. \quad (3.1)$$

This relatively simple problem provides a glimpse at some of the difficulties associated with SMIP problems in general. Under the assumptions specified earlier, Klein Haneveld, Stougie and Van der Vlerk [1995, 1996] have shown that whenever $r_i(\tilde{\omega})$ has finite support, and T has full row-rank, it is possible to compute the convex hull of the expected recourse function by using enumeration over each dimension i. We describe this procedure below. However, it is important to note that since the set $X \cap \mathbf{X}$ will not be used in the convexification process, the resulting optimization problem will only provide a lower bound. Further B&B search may be necessary to close the gap.

The expected recourse function in (3.1) has an extremely important property which relates it to its continuous counterpart. Let the ith component of the expected recourse function of the continuous counterpart be denoted $R_i(\chi_i)$, and the ith component of the expected recourse function in (3.1) be denoted $\hat{R}_i(\chi_i)$. That is,

$$\hat{R}_i(\chi_i) = E\left[g_i^+ \lceil r_i(\tilde{\omega}) - \chi_i \rceil^+ + g_i^- \lfloor r_i(\tilde{\omega}) - \chi_i \rfloor^- \right].$$

Then,

$$R_i(\chi_i) \le \hat{R}_i(\chi_i) \le R_i(\chi_i) + \max\{g_i^+, g_i^-\}. \quad (3.2)$$

The next result (also proved by Klein Haneveld, Stougie and Van der Vlerk [1995, 1996]) is very interesting.

Theorem 3.1. *Let $\hat{R}_i^c$ denote any convex function that satisfies (3.2), and let $(\hat{R}_i^c)'_+$ denote its right directional derivative. Then, for $a \in \Re$*

$$P_i(a) = \frac{(\hat{R}_i^c)'_+(a) + g_i^+}{g_i^+ + g_i^-}$$

is a cumulative distribution function (cdf). Moreover, if ϑ_i is a random variable with cdf P_i, then for all $\chi_i \in \Re$,

$$\hat{R}_i^c(\chi_i) = g_i^+ E\left[(\vartheta_i - \chi_i)^+\right] + g_i^- E\left[(\chi_i - \vartheta_i)^+\right] + \frac{g_i^+ c_i^+ + g_i^- c_i^-}{g_i^+ + g_i^-}, \quad (3.3)$$

where $(v)^+ = \max\{0, v\}$, *and* c_i^+, c_i^- *are asymptotic discrepancies between* $\hat{R}_i$ *and* R_i *defined as follows*

$$c_i^+ = \lim_{\chi_i \to \infty} \hat{R}_i(\chi_i) - R(\chi_i), \quad \text{and} \quad c_i^- = \lim_{\chi_i \to -\infty} \hat{R}_i(\chi_i) - R(\chi_i).$$

Note that unlike (3.1), the expectations in (3.3) do not include any ceiling/floor functions. Hence it is clear that if we are able to identify random variables ϑ_i with cdf P_i, then, we may use the continuous counterpart to obtain a tight approximation of the SIR model.

In order to develop the requisite cdf, the authors construct a convex function by creating the convex hull of $\hat{R}_i$. In order to do so, assume that $r_i(\tilde{\omega})$ has finite support $\Omega = \{\omega^1, \ldots, \omega^N\}$. Then, the points of discontinuity of $\hat{R}_i$ can be characterized as $\cup_{\omega \in \Omega}\{r_i(\omega) + \mathcal{Z}\}$, where $\mathcal{Z}$ denotes the set of integers. Moreover, $\hat{R}_i$ is constant in between the points of discontinuity. Consequently, the convex hull of $\hat{R}_i$ can be obtained by using the convex hull of $(\chi_i, \hat{R}_i(\chi_i))$ at finitely many points of discontinuity. This convex hull (in two-space) can be constructed by adopting a method called Graham scan. This method works by first considering a piecewise linear function that joins the points of dicontinuity $(\chi_i, \hat{R}_i(\chi_i))$, and then verifying whether the right directional derivative at a point is greater than the left directional derivative at that point, for only such points can belong to the boundary of the convex hull. Proceeding in this manner, the method constructs the convex hull, and hence the function $\hat{R}_i^c$. Thereafter, the optimization of a continuous simple recourse problem may be undertaken. This procedure then provides a good lower bound to the optimal value of the SIR model. It is important to bear in mind that there is one additional assumption necessary; the matrix T must have full rank so that the convex hull of the (m_2-dimensional) expected recourse function may be obtained by adding all of the elements $\hat{R}_i^c, i = 1, \ldots, m_2$. This lower bounding scheme may also be incorporated within a B&B procedure to find an optimal solution to the problem.

Binary First-stage, Arbitrary Second-stage: First-stage cuts

For SMIP problems studied in this subsection, we use $\mathbf{X} = \mathcal{B}$ (binary vectors) in (1.1,1.2). Laporte and Louveaux [1993] provide valid inequalities that can be applied to a wide class of expected recourse functions, so long as the first-stage decisions are binary. In particular, the second-stage problems admissible under this scheme include all optimization problems that have a known lower bound on expected recourse function. As one might expect, such widely applicable cuts rely mainly on the fact that the first-stage decisions are binary. The algorithmic setting within which the inequalities of Laporte and Louveaux [1993] are used follows the basic outline of Benders' decomposition (or L-shaped method). That is, at each iteration k, we solve one master program, and as many subproblems as there are outcomes of the random variable. Interestingly, despite the non-convexity of value functions of general optimization problems (including MIPs), the valid inequality provided by Laporte and Louveaux [1993] is linear. As shown in the development below, the linearity derives from a property of the binary first-stage variables.

At iteration k, let the first-stage decision x^k be given, and let

$$I_k = \{i \mid x_i^k = 1\}, \quad Z_k = \{1, \dots, n_1\} - I_k.$$

Next define the linear function

$$\delta_k(x) = |I_k| - \left[\sum_{i \in I_k} x_i - \sum_{i \in Z_k} x_i\right].$$

It can be easily seen that when $x = x^k$ (assumed binary), $\delta_k(x) = 0$; whereas, for all other binary vectors $x \neq x^k$, at least one of the components must switch "states." Hence for $x \neq x^k$, we have

$$\left[\sum_{i \in I_k} x_i - \sum_{i \in Z_k} x_i\right] \leq |I_k| - 1, \quad \text{i.e. } \delta_k(x) \geq 1. \tag{3.4a}$$

Next suppose that a lower bound on the expected recourse function, denoted $\bar{h}_\ell$, is available. Let $\bar{h}(x^k)$ denote the value of the expected recourse function for a given x^k. If $\bar{h}(x^k) = \infty$ (i.e. the second-stage is infeasible), then (3.4a) can be used to delete x^k. On the other hand, if $\bar{h}(x^k)$ is finite, then the following inequality is valid.

$$\eta \geq \bar{h}(x^k) - \delta_k(x)\left[\bar{h}(x^k) - \bar{h}_\ell\right]. \tag{3.4b}$$

This is the "optimality" cut of Laporte and Louveaux [1993]. To verify its validity, observe that when $x = x^k$, the second term in (3.4b) vanishes, and hence the master program recovers the value of the expected recourse function. On the other hand, if $x \neq x^k$, then,

$$\delta_k(x)\left[\bar{h}(x^k) - \bar{h}_\ell\right] \geq \left[\bar{h}(x^k) - \bar{h}_\ell\right].$$

Hence, for all $x \neq x^k$, the right-hand side of (3.4b) obeys

$$\bar{h}(x^k) - \delta_k(x)\left[\bar{h}(x^k) - \bar{h}_\ell\right] \leq \bar{h}(x^k) - \bar{h}(x^k) + \bar{h}_\ell = \bar{h}_\ell.$$

It is interesting to observe that the structure of the second-stage is not critical to the validity of the cut. For the sake of expositional simplicity, we state the algorithm of Laporte and Louveaux [1993] under the complete recourse assumption, thus requiring only (3.4b). If this assumption is not satisfied, then one would also include (3.4a) in the algorithmic process. In the following, $\bar{x}$ denotes an incumbent, $\bar{f}$ its objective value, and f_ℓ, f_u are lower and upper bounds, respectively, on the entire objective function. We use the notation $\alpha + \beta x$ to denote the right-hand side of (3.4b).

First-Stage Cuts for SP with Binary First Stage

0. *Initialize.* $k \leftarrow 0$ Let $\epsilon \geq 0$, $x^1 \in X \cap \mathcal{B}$ and $\bar{h}_\ell$ (a lower bound on the expected recourse function) be given. Define $\eta_0(x) = \bar{h}_\ell$, $f_u = \infty$.
1. *Obtain a Cut*
   $k \leftarrow k+1$. Evaluate the second-stage objective value $\bar{h}(x^k)$. Use (3.4b) to define the cut $\alpha + \beta x$.
2. *Update the Piecewise Linear Approx.*
   (a) Define $\eta_k(x) = \text{Max}\{\eta_{k-1}(x), \alpha + \beta x\}$, and $f_k(x) = c^\top x + \eta_k(x)$.
   (b) Update the upper bound (if possible): $f_u \leftarrow \text{Min}\{f_u, f_k(x^k)\}$. If a new upper bound is obtained, $\bar{x} \leftarrow x^k, \bar{f} \leftarrow f_u$.
3. *Solve the Master Problem.* Let $x^{k+1} \in \text{argmin } \{f_k(x) \,|\, x \in X \cap \mathcal{B}\}$.
4. *Stopping Rule.* $f_\ell = f_k(x^{k+1})$. If $f_u - f_\ell \leq \epsilon$, declare $\bar{x}$ as an ϵ–optimum and stop. Otherwise, repeat from 1.

The above algorithm has been stated in a manner that mimics the Kelley-type methods of convex programming (Kelley [1960]) since the L-shaped method of Van Slyke and Wets [1969] is a method of this type. The main distinctions are in step 1 (cut formation), and step 3 (the solution of the master problem) which requires the solution of a binary IP. We note however that there are various other ways to implement these cuts. For instance, if the solution method adopted for the master program is a B&B method, then one can generate a cut at any node (of the B&B tree) at which a binary solution is encountered. Such an implementation would have the benefit of generating cuts during the B&B process at the cost of carrying out multiple evaluations of the second-stage objective during the B&B process. We close this subsection with an illustration of this scheme.

Example 3.2. Consider the following two-stage problem

$$
\begin{aligned}
&\mathrm{Min} - x_1 + 0.25(-2y_1(1) + 4y_2(1)) + 0.75(-2y_1(2) + 4y_2(2))\\
&\quad -3x_1 - 3y_1(1) + 2y_2(1) \geq -4\\
&\quad -5x_1 - 3y_1(2) + 2y_2(2) \geq -8\\
&\quad\ x_1, y_1(1), y_1(2) \in \{0, 1\}, y_2(1), y_2(2) \geq 0.
\end{aligned}
$$

To maintain notational simplicity in this example, we simply use $\omega = \{1, 2\}$, instead of our regular notation of $\{\omega^1, \omega^2\}$. From the above data, it is easily seen that $-2y_1 + 4y_2 \geq -2$ for $y_1 \in \{0, 1\}$ and $y_2 \geq 0$. Hence $\bar{h}_\ell = -2$ is a valid lower bound for the second-stage problems.

0. *Initialization.* $k = 0$, and let $\epsilon = 0$, $x_1^1 = 0$, $\bar{h}_\ell = -2$, $f_u = \infty$, $\eta_0(x) = -2$.

Iteration 1

1. *Obtain a cut.* For the given x_1^1, we solve each second-stage MIP subproblem. We get $y_1(1) = 1$, $y_2(1) = 0$, $y_1(2) = 1$, $y_2(2) = 0$, and $\bar{h}(x_1^1) = -2$. Moreover, $\delta(x_1) = x_1$, so that the cut is $\eta \geq -2 -(x_1)(-2 + 2) = -2$.
2. *Update the Piecewise Linear Approximation.* The upper bound is $f_u = \mathrm{Min}\{\infty, 0 + f_1(0)\} = -2$. The incumbent is $\bar{x}_1 = 0, \bar{f} = -2$.
3. *Solve the Master Program.*

   $$\mathrm{Min}\{-x_1 + \eta \mid \eta \geq -2, x_1 \in \{0, 1\}\}.$$

   $x_1^2 = 1$ solves this problem, and the lower bound $f_\ell = -3$.
4. *Stopping Rule.* Since $f_u - f_\ell > 0$, repeat from step 1.

Iteration 2

1. *Obtain a cut.* For $x_1^2 = 1$ solve each second-stage MIP subproblem. We get $y_1(1) = 0$, $y_1(2) = 1$, $y_2(2) = 0$, yielding $\bar{h}(x_1^2) = -1.5$. Now, $\delta(x_1) = 1 - x_1$, and the cut is $\eta \geq -1.5 - (1 - x_1)(-1.5 + 2) = -2 + 0.5x_1$.
2. *Update the Piecewise Linear Approximation.* The upper bound is $f_u = \mathrm{Min}\{-2, -1 - 1.5\} = -2.5$, hence, $\bar{x}_1 = 1, \bar{f} = -2.5$.
3. *Solve the Master Program.*

   $$\mathrm{Min}\{-x_1 + \eta \mid \eta \geq -2, \eta \geq -2 + 0.5x_1, x_1 \in \{0, 1\}\}.$$

   $x_1^3 = 1$ solves this problem, and the lower bound $f_\ell = -2.5$.
4. *Stopping Rule.* Since $f_u - f_\ell = 0$, the method stops with $\bar{x}_1 = 1$ as the optimal solution.

As in this example, all 2^{n_1} valid inequalities may be generated in the worst case (where n_1 is the number of first-stage binary variables). However, the finiteness of the method is obvious.

Binary First-stage, 0-1 MIP Second-stage with Fixed Recourse: Cuts in both stages

In this subsection we impose the following structure on (1.1,1.2): a fixed recourse matrix, binary first-stage variables, and mixed-integer (binary) recourse decisions. The methodology here is one of sequential convexification of the integer recourse problem. The main motivation for sequential convexification is to avoid the need to solve every subproblem from scratch in each iteration. These procedures will be presented in the context of algorithms that operate within the framework of Benders' decomposition, as in the previous subsection; that is, in iteration k, a first-stage decision, denoted x^k, is provided to the subproblems, which in turn returns an inequality that provides a linear approximation of the expected recourse function. The cuts derived here use disjunctive programming. This approach has been used to solve some rather large server location problem, and the computational results reported in Ntaimo and Sen [2004] are encouraging. Cuts for this class of models can also be derived using the RLT framework, and has appeared in the work of Sherali and Fraticelli [2002].

We start this development with the assumption that by using appropriately penalized continuous variables, the subproblem remains feasible for any restriction of the integer variables y_j, $j \in J_2$. Let x^k be given, and suppose that matrices W_k, $T_k(\omega)$ and $r_k(\omega)$ are given. Initially (i.e. $k=1$) these matrices are simply $W, T(\omega)$ and $r(\omega)$, and recall that in our notation, we include the constraints $-y_j \geq -1, j \in J_2$ explicitly in $Wy \geq r(\omega) - T(\omega)x$. (Similarly, the constraint $-x \geq 1$ is also included in the constraints $x \in X$.) During the course of solving the 0-1 MIP subproblem for outcome ω, suppose that we happen to solve the following LP relaxation.

$$\text{Min} \quad g^\top y \tag{3.5a}$$

$$\text{s.t.} \quad W_k y \geq r_k(\omega) - T_k(\omega)x \tag{3.5b}$$

$$y \in \Re_+^{n_2}. \tag{3.5c}$$

Whenever the solution to this problem is fractional, we will be able to derive a valid inequality that can be used in all subsequent iterations. Let $y^k(\omega)$ denote a solution to (3.5), and let $j(k)$ denote an index $j \in J_2$ for which $y_j^k(\omega)$ is non-integer for one or more $\omega \in \Omega$. To eliminate this non-integer solution, a disjunction of the following form may be used:

$$\mathcal{S}_k(x^k, \omega) = \mathcal{S}_{0,j(k)}(x^k, \omega) \cup \mathcal{S}_{1,j(k)}(x^k, \omega),$$

where

$$\mathcal{S}_{0,j(k)}\left(x^k, \omega\right) = \left\{y \in \Re_+^{n_2} \mid W_k y \geq r_k(\omega) - T_k(\omega)x^k, -y_{j(k)} \geq 0\right\} \qquad (3.6a)$$

$$\mathcal{S}_{1,j(k)}\left(x^k, \omega\right) = \left\{y \in \Re_+^{n_2} \mid W_k y \geq r_k(\omega) - T_k(\omega)x^k, y_{j(k)} \geq 1\right\}. \qquad (3.6b)$$

The index $j(k)$ is referred to as the "disjunction variable" for iteration k. This is precisely the disjunction used in the lift-and-project cuts of Balas, Ceria and Cornuéjols [1993]. To connect this development with the subsection on disjunctive cuts, we observe that $H=\{0, 1\}$. We assume that the subproblems remain feasible for any restriction of the integer variables, and thus both (3.6a) and (3.6b) are non-empty.

Let $\lambda_{0,1}$ denote the vector of multipliers associated with the rows of W_k in (3.6a), and $\lambda_{0,2}$ denote the scalar multiplier associated with the fixed variable $y_{j(k)}$ in (3.6a). Let $\lambda_{1,1}$ and $\lambda_{1,2}$ be similarly defined for (3.6b). Then Theorem (2.6) implies that if $(\pi, \pi_0(\omega), \omega \in \Omega)$ satisfy (3.7), then $\pi^\top y \geq \pi_0(\omega)$ is a valid inequality for $\mathcal{S}_k(x^k, \omega)$.

$$\pi_j \geq \lambda_{0,1}^T W_{jk} - I_j^k \lambda_{0,2} \; \forall j \qquad (3.7a)$$

$$\pi_j \geq \lambda_{1,1}^\top W_{jk} + I_j^k \lambda_{1,2} \; \forall j \qquad (3.7b)$$

$$\pi_0(\omega) \leq \lambda_{0,1}^\top \left(r_k(\omega) - T_k(\omega)x^k\right) \; \forall \omega \in \Omega \qquad (3.7c)$$

$$\pi_0(\omega) \leq \lambda_{1,1}^\top \left(r_k(\omega) - T_k(\omega)x^k\right) + \lambda_{1,2} \; \forall \omega \in \Omega \qquad (3.7d)$$

$$-1 \leq \pi_j \leq 1, \forall_j, -1 \leq \pi_0(\omega) \leq 1, \forall \omega \in \Omega \qquad (3.7e)$$

$$\lambda_{0,1}, \lambda_{0,2}, \lambda_{1,1}, \lambda_{1,2} \geq 0 \qquad (3.7f)$$

where

$$I_j^k = \begin{cases} 0, & \text{if } j \neq j(k) \\ 1, & \text{otherwise.} \end{cases}$$

Remark 3.3. Several objectives have been proposed in the disjunctive programming literature for choosing cut coefficients (Sherali and Shetty [1980]). One possibility for SMIP problems is to maximize the expected value of the depth of cut: $E[\pi_0(\omega)] - E[y^k(\omega)]\pi$. We should note that the optimal objective value of the resulting LP can be zero, which implies that the inequality generated by the LP does not delete some of the fractional points $y^k(\omega)$, $\omega \in \Omega_k$. Here Ω_k denotes those $\omega \in \Omega$ for which $y^k(\omega)$ does not satisfy mixed-integer feasibility. So long as the cut deletes a fractional $y^k(\omega)$ for some ω, we may proceed with the algorithm. However, if we obtain an inequality such that $(\pi^k)^\top y^k(\omega) \geq \pi_0^k(\omega)$, for all $\omega \in \Omega_k$, then one such outcome should be removed from the expectation operation $E[y^k(\tilde{\omega})]$, and this vector should be replaced by a conditional expectation over the remaining vectors $y^k(\omega)$. Since the rest of the LP remains unaltered, the re-optimization should be carried out using a "warm start." Other objective functions can also be used for the cut generation process. For instance, we could maximize the function $\text{Min}_{\omega\in\Omega}\pi_0(\omega) - y^k(\omega)^\top \pi$.

For vectors $x \neq x^k$, the cut may need to be modified in order to maintain its validity. Sen and Higle [2000] show that for any other x, one only needs to modify the right-hand side scalar π_0; in other words, the vector π^k provides valid cut coefficients as long as the recourse matrix is fixed. This result, known as the Common Cut Coefficients (C^3) Theorem, was proven in Sen and Higle [2000], and a general version may be stated as follows.

Theorem 3.4. (The C^3 Theorem). *Consider a 0-1 SMIP with a fixed recourse matrix. For $(x, \omega) \in X \times \Omega$, let $Y(x, \omega) = \{y \in \Re_+^{n_2} \mid Wy \geq r(\omega) - T(\omega)x, y_j \in \{0, 1\}, j \in J_2\}$, the set of mixed-integer feasible solutions for the second-stage mixed-integer linear program. Suppose that $\{C_h, d_h\}_{h\in H}$, is a finite collection of appropriately dimensioned matrices and vectors such that for all $(x, \omega) \in X \times \Omega$*

$$Y(x, \omega) \subseteq \cup_{h\in H}\left\{y \in \Re_+^{n_2} \mid C_h y \geq d_h\right\}.$$

Let

$$\mathcal{S}_h(x, \omega) = \left\{y \in \Re_+^{n_2} \mid W_y \geq r(\omega) - T(\omega)x, C_h y \geq d_h\right\},$$

and let

$$\mathcal{S} = \cup_{h\in H}\mathcal{S}_h(x, \omega).$$

Let $(\bar{x}, \bar{\omega})$ be given, and suppose that $\mathcal{S}_h(\bar{x}, \bar{\omega})$ is nonempty for all $h \in H$ and $\pi^\top y \geq \pi_0(\bar{x}, \bar{\omega})$ is a valid inequality for $\mathcal{S}(\bar{x}, \bar{\omega})$. There exists a function, $\pi_0 : X \times \Omega \to \Re$ such that for all $(x, \omega) \in X \times \Omega, \pi^\top y \geq \pi_0(x, \omega)$ is a valid inequality for $\mathcal{S}(x, \omega)$.

Although the above theorem is stated for general disjunctions indexed by H, we only use $H=\{0,1\}$ in this development. The LP used to obtain the common cut coefficients is known as the C^3LP, and its solution $(\pi^k)^\top$ is appended to W_k in order to obtain W_{k+1}. In order to be able to use these coefficients in subsequent iterations, we will also calculate a new row to append to $T_k(\omega)$, and $r_k(\omega)$ respectively. These new rows will be obtained by solving some other LPs, which we will refer to as RHS-LPs. These calculations are summarized next.

Let $\lambda^k_{0,1}, \lambda^k_{0,2}, \lambda^k_{1,1}, \lambda^k_{1,2} \geq 0$ denote the values obtained from C^3LP in iteration k. Since these multipliers are non-negative, Theorem 2.6 allows us to use these multipliers for any choice of (x,ω). Hence by using these multipliers, the right-hand side function $\pi_0(x,\omega)$ can be written as

$$\pi_0(x,\omega) = \mathrm{Min}\Big\{\big(\lambda^k_{0,1}\big)^\top r_k(\omega) - \big(\lambda^k_{0,1}\big)^\top T_k(\omega)x,\ \big(\lambda^k_{1,1}\big)^\top r_k(\omega) + \lambda^k_{1,2} - \big(\lambda^k_{1,1}\big)^\top T_k(\omega)x\Big\}.$$

For notational convenience, we put

$$\bar{\nu}_0(\omega) = \big(\lambda^k_{0,1}\big)^\top r_k(\omega),\ \bar{\nu}_1(\omega) = \big(\lambda^k_{1,1}\big)^\top r_k(\omega) + \lambda^k_{1,2}$$

and

$$\big[\bar{\gamma}_h(\omega)\big]^\top = \big(\lambda^k_{h,1}\big)^\top T_k(\omega), \qquad h \in \{0,1\},$$

so that

$$\pi_0(x,\omega) = \mathrm{Min}\Big\{\bar{\nu}_0(\omega) - \big[\bar{\gamma}_0(\omega)\big]^\top x,\ \bar{\nu}_1(\omega) - \big[\bar{\gamma}_1(\omega)\big]^\top x\Big\}.$$

Being the minimum of two affine functions, the epigraph of $\pi_0(x,\omega)$ can be represented as the union of the two half-spaces. Hence the epigraph of $\pi_0(x,\omega)$, restricted to the set X will be denoted as $\Pi_X(\omega)$, and represented as

$$\Pi_X(\omega) = \cup_{h\in H} E_h(\omega),$$

where $H=\{0,1\}$ and

$$E_h(\omega) = \big\{(\eta, x) \mid \eta \geq \bar{\nu}_h(\omega) - \bar{\gamma}_h(\omega)^\top x,\ x \in X\big\}. \tag{3.8}$$

Here $X = \{x \in \Re^{n_1} \mid Ax \geq b, x \geq 0\}$, and we assume that the inequality $-x \geq -1$ is included in the constraints $Ax \geq b$. It follows that the closure of the convex hull of $\Pi_X(\omega)$ provides the appropriate convexification of $\pi_0(x, \omega)$. This computational procedure is discussed next.

In the following, we assume that for all $x \in X$, $\eta \geq 0$ in (3.8). As long as X is bounded, there is no loss of generality with this assumption, because the epigraph can be translated to ensure that $\eta \geq 0$. Analogous to the concept of reverse polars (see Theorem 2.7), Sen and Higle [2000] define the epi-reverse polar, denoted $\Pi_X^{\dagger}(\omega)$, as

$$
\begin{aligned}
\Pi_X^{\dagger}(\omega) = \{\sigma_0(\omega) \in \Re, \sigma(\omega) \in \Re^{n_1}, \zeta(\omega) \in \Re \text{ such that}& \\
\text{for } h = 0, 1,\ \exists\, \tau_h \in \Re^{m_1}, \tau_{0h} \in \Re& \\
\sigma_0(\omega) \geq \tau_{0h} \quad \forall h \in \{0, 1\}& \\
\sum_h \tau_{0h} = 1& \\
\sigma_j(\omega) \geq \tau_h^{\top} A_j + \tau_{0h}\bar{\gamma}_{hj}(\omega) \quad \forall h \in \{0, 1\}; j = 1, \ldots, n_1& \\
\zeta(\omega) \leq \tau_h^{\top} b + \tau_{0h}\bar{\nu}_h(\omega) \quad \forall h \in \{0, 1\}& \\
\tau_h \geq 0, \tau_{0h} \geq 0, \quad h \in \{0, 1\}\}.&
\end{aligned}
$$

The term "epi-reverse polar" is intended to indicate that we are using the reverse polar of an epigraph to characterize its convex hull (see Theorem 2.7). Note that the epi-reverse polar allows only those facets of the closure of the convex hull of $\Pi_X(\omega)$ that have a positive coefficient for the variable η. From Theorem 2.7, we can obtain all necessary facets of the closure of the convex hull of $\pi_0(x, \omega)$. We can derive one such facet by solving the following problem, which we refer to as the RHS-LP(ω).

$$
\begin{aligned}
\text{Max} \quad & \zeta(\omega) - \sigma_0(\omega) - \left(x^k\right)^{\top}\sigma(\omega) \\
\text{s.t.} \quad & (\sigma_0(\omega), \sigma(\omega), \zeta(\omega)) \in \Pi_X^{\dagger}(\omega).
\end{aligned} \tag{3.9}
$$

With an optimal solution to (3.9), $(\sigma_0^k(\omega), \sigma^k(\omega), \zeta^k(\omega))$, we obtain $\nu^k(\omega) = \frac{\zeta^k(\omega)}{\sigma_0^k(\omega)}$ and $\gamma^k(\omega) = \frac{\sigma^k(\omega)}{\sigma_0^k(\omega)}$. For each $\omega \in \Omega$, these coefficients are used to update the right-hand-side functions $r_{k+1}(\omega) = [r_k(\omega)^{\top}, \nu^k(\omega)]^{\top}$, and $T_{k+1}(\omega) = [T_k(\omega)^{\top}; \gamma^k(\omega)]^{\top}$.

One can summarize a cutting plane method of the form presented in the previous subsection by replacing step 1 of that method by a new version of step 1 as summarized below. Sen and Higle [2000] provide a proof of convergence of convex hull approximations based on an extension of (2.10). We caution however that as with any cutting plane method, its full benefits can only be realized when it is incorporated within a B&B method. Such a branch-and-cut approach is discussed in the following subsection.

Deriving Cuts for Both Stages

1. *Obtain a Cut*
   $k \leftarrow k+1$.
   (a) (Solve the LP relaxation for all ω). Given x^k, solve the LP relaxation of each subproblem, $\omega \in \Omega$.
   (b) (Solve C^3-LP). Optimize some objective from Remark 3.3, over the set in (3.7). Append the solution $(\pi^k)^\top$ to the matrix W_k to obtain W_{k+1}.
   (c) (Solve RHS-LP(ω) for all ω). Solve (3.9) for all $\omega \in \Omega$, and derive $r_{k+1}(\omega)$, $T_{k+1}(\omega)$.
   (d) (Solve an enhanced LP relaxation for all ω). Using the updated matrices $W_{k+1}, r_{k+1}(\omega), T_{k+1}(\omega)$, solve an LP relaxation for each $\omega \in \Omega$.
   (e) (Benders' Cut). Using the dual multipliers from step (d), derive a Benders' cut denoted $\alpha + \beta x$.

Example 3.5. The instance considered here is the same as that in Example 3.2. While this example illustrates the process of cut formation, it is too small to really demonstrate the benefits that might accrue from adding cuts into the subproblem. A slightly larger instance (motivated by the example in Schultz, Stougie and Van der Vlerk [1998]) which requires a few more iterations, and one that demonstrates the advantages of stronger LP relaxations appears in Sen, Higle and Ntaimo [2002], and Ntaimo and Sen [2004]. As in Example 3.2, we use $\omega = \{1, 2\}$.

Iteration 1

The LP relaxation of the subproblem in iteration 1 (see Example 3.2) provides integer optimal solutions. Hence, for its iteration, we use the cut obtained in Example 3.2 (without using the Benders' cut). In this case, the calculations of this iteration mimic those for iteration 1 in Example 3.2. The resulting value of x_1 is $x_1^2 = 1$.

Iteration 2

In the following, elements of the vector λ_{01} will be denoted λ_{011} and λ_{012}. Similarly, elements of λ_{11} will be denoted λ_{111} and λ_{112}.

1. *Derive cuts for both stages.*

1a) Putting $x_1^2 = 1$, solve the LP relaxation of the subproblems for $\omega = 1, 2$. For $\omega = 1$, we get $y_1(1) = 1/3$ and $y_2(1) = 0$; similarly for $\omega = 2$, we get $y_1(2) = 1$ and $y_2(2) = 0$.

1b) Solve the C^3LP using $E(y_1, y_2) = (0.833, 0)$.

$$
\begin{aligned}
\text{Max} \quad & 0.25\pi_0(1) + 0.75\pi_0(2) - 0.833\pi_1 \\
\text{s.t.} \quad & \pi_1 + 3\lambda_{011} + \lambda_{012} + \lambda_{02} \geq 0 \\
& \pi_1 + 3\lambda_{111} + \lambda_{112} - \lambda_{12} \geq 0 \\
& \pi_2 - 2\lambda_{011} \geq 0 \\
& \pi_2 - 2\lambda_{111} \geq 0 \\
& \pi_0(1) + \lambda_{011} + \lambda_{012} \leq 0 \\
& \pi_0(1) + \lambda_{111} + \lambda_{112} - \lambda_{12} \leq 0 \\
& \pi_0(2) + 3\lambda_{011} + \lambda_{012} \leq 0 \\
& \pi_0(2) + 3\lambda_{111} + \lambda_{112} - \lambda_{12} \leq 0 \\
& -1 \leq \pi_j \leq 1, \forall j, -1 \leq \pi_0(\omega) \leq 1, \forall \omega, \lambda \geq 0.
\end{aligned}
$$

The optimal objective value of this LP is 0.083, and the cut coefficients are $(\pi_1^1, \pi_2^1)^\top = (-1, 1)$, and the multipliers $\lambda_{01}^\top = (0, 0)$, $\lambda_{02} = 1$, whereas, $\lambda_{11}^\top = (0.5, 0)$, $\lambda_{12} = 0.5$.

1c) For $H = \{0,1\}$ we will now compute $\bar{\nu}_h(\omega)$ and $\bar{\gamma}_h(\omega)$ so that the sets $E_h(\omega)$, $h \in H$ can be determined for all ω. Thereafter the union of these sets can be convexified using the RHS-LP (3.9). Using the multipliers $\lambda_{01} = (0, 0)$, $\lambda_{02} = 1$, we obtain $\bar{\nu}_0(1) = 0$, and $\bar{\gamma}_0(1) = 0$. Hence

$$E_0(1) = \{0 \leq x_1 \leq 1 \mid \eta \geq 0\},$$

and similarly by using $\lambda_{11} = (0.5, 0)$, $\lambda_{12} = 0.5$ we have

$$E_1(1) = \{0 \leq x_1 \leq 1 \mid \eta \geq -1.5 + 1.5x_1\}.$$

Clearly, the convex hull of these two sets is $E_1(1)$, and the facet can be obtained using linear programming. In the same manner, we obtain

$$
\begin{aligned}
& E_0(2) = \{0 \leq x_1 \leq 1, \eta \geq 0\}, \quad \text{and} \\
& E_1(2) = \{0 \leq x_1 \leq 1, \eta \geq -3.5 + 2.5x_1\}.
\end{aligned}
$$

Once again the convex hull of these two sets is $E_1(2)$, and the facet can be derived using linear programming. In any event, the matrices are updated as follows: we obtain W_2 by appending the row $(-1,1)$ to W; $r_2(1)$ is obtained by appending the scalar -1.5 to $(r_1(1))^\top = (-4, -1)$, $r_2(2)$ is obtained by appending the scalar -3.5 to $(r_1(2))^\top = (-8, -1)$. Finally we append the "row" 1.5 to $T_1(1)$ to obtain $T_2(1)$, and the "row" 2.5 is appended to $T_1(2)$, and the resultant is $T_2(2)$.

1d) Solve the LP relaxation associated with each of the updated subproblems using $x_1^1 = 1$. Then we obtain the MIP feasible solutions for each subproblem: $y_1(1) = 0$, $y_2(1) = 0$, $y_1(2) = 1$, $y_2(2) = 0$.

1e) The Benders' cut in this instance is $\eta \geq -4.75 + 3.25x_1$.

(Steps 2,3,4). As in Example 3.2, the optimal solution to the first-stage master problem is $x_1^3 = 1$, with a lower bound $f_\ell = -2.5$, and the algorithm stops.

Remark 3.6. In this instance, the Benders' cut for the first-stage is weaker than that obtained in Example 3.2. The benefit however comes from the fact that the Benders' cut requires only LP solves in the second-stage, and that the second-stage LPs are strengthened sequentially. Hence if there was a need to iterate further, the cut-enhanced relaxations could be used. In contrast, the cuts of the previous subsection requires the solution of as many 0-1 MIP instances as there are scenarios.

Binary First-stage, MIP Second-stage: Branch-and-Cut

We continue with the two-stage SMIP models (1.1,1.2), and the methods of this subsection will accommodate general integers in the second-stage. The methods studied thus far have not used the properties of B&B algorithms in any significant way. Our goal for this subsection is to develop a cut that will convey information uncovered during the stage-two B&B process to the first-stage model. This development appears in Sen and Sherali [2002] who refer to this as the D^2-BAC method. While our development proceeds with the fixed recourse assumption, the validity of the cuts are independent of this assumption.

Consider a partial B&B tree generated during a "partial solve" of the second-stage problem. Let $Q(\omega)$ denote the set of nodes of the tree that have been explored for the subproblem associated with scenario ω. We will assume that all nodes of the B&B tree are associated with a feasible LP relaxation, and that nodes are fathomed when the LP lower bound exceeds the best available upper bound. This may be accomplished by introducing artificial variables, if necessary. The D^2-BAC strategy revolves around using the dual problem associated with the LP relaxation (one for each node), and then stating a disjunction that will provide a valid inequality for the first-stage problem.

For any node $q \in Q(\omega)$, let $z_{q\ell}(\omega)$ and $z_{qu}(\omega)$ denote vectors whose elements are used to define lower and upper bounds, respectively, on the second-stage (integer) variables. In some cases, an element of z_{qu} may be $+\infty$, and in this case, the associated constraint may be ignored, implying that the associated dual multiplier is fixed at 0. In any event, the LP relaxation for node q may be written as

$$\begin{aligned} \text{Min} \quad & g^\top y \\ & W_k y \geq r_k(\omega) - T_k(\omega)x \\ & y \geq 0 \\ & y \geq z_{q\ell}(\omega) - y \geq -z_{qu}(\omega), \end{aligned}$$

and, the corresponding dual LP is

$$\begin{aligned} \text{Max} \quad & \theta_q(\omega)^\top [r_k(\omega) - T_k(\omega)x] + \psi_{q\ell}(\omega)^\top z_{q\ell}(\omega) - \psi_{qu}(\omega)^\top z_{qu}(\omega) \\ & \theta_q(\omega)^\top W_k + \psi_{q\ell}(\omega)^\top - \psi_{qu}(\omega)^\top \leq g^\top \\ & \theta_q(\omega) \geq 0, \ \psi_{q\ell}(\omega) \geq 0, \psi_{qu}(\omega) \geq 0, \end{aligned}$$

where the vectors $\psi_{q\ell}(\omega)$, and $\psi_{qu}(\omega)$ are appropriately dimensioned. Note also that we assume that the second-stage constraints include cuts that are similar to those developed in the previous subsection, so that W_k, $r_k(\omega)$, and $T_k(\omega)$ are updated from one iteration to the next.

We now turn our attention to approximating the value function of the second-stage MIP. As noted in section 2, the IP and MIP value functions are complicated objects. Certain convex approximations have been proposed by perturbing the distribution of the random right-hand-side vector (Van der Vlerk [2004]). For problems with a totally unimodular (TU) recourse matrix, this approach provides an optimal solution. For more general recourse matrices, these approximations only provide a lower bound. Consequently, we resort to a different approach for SMIP problems that do not satisfy the TU requirement.

The B&B tree, together with the LP relaxations at these nodes, provide important information that can be used to approximate MIP value functions. The main observation is that the B&B tree embodies a disjunction, and when coupled with the value functions of LP relaxations of each node, we obtain a disjunctive description of an approximation to the MIP value function. By using the disjunctive cut principle, we will then obtain linear inequalities (cuts) that can be used to build value function approximations. In order to do so, we assume that we have a lower bound h_ℓ such that $h(x, \tilde{\omega}) \geq h_\ell$ (almost surely) for all $x \in X$. Without loss of generality, this bound may be assumed to be 0.

Consider a node $q \in Q(\omega)$ and let $(\theta_q^k(\omega), \psi_{q\ell}^k(\omega), \psi_{qu}^k(\omega))$ denote optimal dual multipliers for node q. Then a lower bounding function may be obtained by requiring that $x \in X$ and that the following disjunction holds.

$$\eta \geq \theta_q^k(\omega)^\top [r_k(\omega) - T_k(\omega)x] + \psi_{q\ell}^k(\omega)^\top z_{q\ell}(\omega) - \psi_{qu}^h(\omega)^\top z_{qu}(\omega) \quad \text{for at least one } q \in Q(\omega). \tag{3.10}$$

Note that each inequality in (3.10) corresponds to a second-stage value function approximation that is valid only when the restrictions (on the y-variables) associated with node $q \in Q(\omega)$ hold true. Since any optimal solution of the second-stage must be associated with at least one of the nodes $q \in Q(\omega)$, the disjunction (3.10) is valid. By assumption, we have $\eta \geq 0$. Hence, $x \in X$ and (3.10) leads to the following disjunction:

$$\Pi_X(\omega) = \left\{(\eta, x) \in \cup_{q \in Q(\omega)} E_q^k(\omega)\right\},$$

where

$$E_q^k(\omega) = \left\{(\eta, x) \mid \eta \geq \bar{\nu}_q^k(\omega) - \bar{\gamma}_q^k(\omega)^\top x,\ Ax \geq b,\ x \geq 0,\ \eta \geq 0\right\},$$

with,

$$\bar{\nu}_q^k(\omega) = \theta_q^k(\omega)^\top r_k(\omega) + \psi_{q\ell}^k(\omega)^\top z_{q\ell}(\omega) - \psi_{qu}^k(\omega)^\top z_{qu}(\omega),$$

and

$$\bar{\gamma}_q^k(\omega)^\top = \theta_q^k(\omega)^\top T_k(\omega).$$

The arguments provided above are essentially the same as that used in the previous subsection, although the precise setting is different. In the previous subsection, we convexified the right-hand side function of a valid inequality derived from the disjunctive cut principle. In this subsection, we convexify an approximation of the second-stage value function. Yet, the tools we use are the same. As before, we derive the epi-reverse polar which we denote by $\Pi_X^\dagger(\omega)$.

$$\begin{aligned}
\Pi_X^\dagger(\omega) = \{&\sigma_0(\omega) \in \Re,\ \sigma(\omega), \in \Re^{n_1},\ \zeta(\omega) \in \Re \mid \forall q \in Q(\omega), \\
&\exists\ \tau_q(\omega) \geq 0,\ \tau_{0q}(\omega) \in \Re_+ \\
\text{s.t } &\sigma_0(\omega) \geq \tau_{0q}(\omega)\ \forall q \in Q(\omega) \\
&\sum_{q \in Q(\omega)} \tau_{0q}(\omega) = 1 \\
&\sigma_j(\omega) \geq \tau_q(\omega)^\top A_j + \tau_{0q}(\omega)\bar{\gamma}_{qj}^k(\omega) \quad \forall q \in Q(\omega), j = 1, \ldots, n_1 \\
&\zeta(\omega) \leq \tau_q(\omega)^\top b + \tau_{0q}(\omega)\bar{\nu}_q^k(\omega) \quad \forall q \in Q(\omega) \\
&\tau_q(\omega) \geq 0, \quad \tau_{0q}(\omega) \geq 0 \quad \forall q \in Q(\omega)\}.
\end{aligned} \tag{3.11}$$

As the reader will undoubtedly notice, the number of atoms in the disjunction here depend on the number nodes available from the B&B tree, whereas, the disjunctions of the previous subsection contained exactly two atoms. In any event, the cut is obtained by choosing non-negative multipliers $\tau_{0q}^k(\omega)$, $\tau_q^k(\omega)$ for all q, and then using the "Min" and "Max" operations as follows:

$$\sigma_0^k(\omega) = \operatorname*{Max}_q \tau_{0q}^k(\omega)$$
$$\sigma_j^k(\omega) = \operatorname*{Max}_q \left\{ \tau_q^k(\omega)^\top A_j + \tau_{0q}^k(\omega)\bar{\gamma}_{qj}^k(\omega) \right\} \quad \forall j$$
$$\zeta^k(\omega) = \operatorname*{Min}_q \left\{ \left[\tau_q^k(\omega)\right]^\top b + \tau_{0q}^k(\omega)\bar{\nu}_q^k(\omega) \right\}.$$

These parameters can also be obtained by using an LP of the form (3.9), and the disjunctive cut for any outcome ω is then given by

$$\sigma_0^k(\omega)\eta + \sum_j \sigma_j^k(\omega)x_j \geq \zeta^k(\omega),$$

where the conditions in (3.11) imply that $\sigma_0^k(\omega) \geq \operatorname{Max}_q \tau_{0q}^k(\omega) > 0$. Hence, the epi-reverse polar only allows those facets (of the convex hull of $\Pi_X(\omega)$) that have a positive coefficient for the variable η.

The "optimality cut" to be included in the first-stage master in iteration k is given by

$$\eta \geq E\left[\frac{\zeta^k(\tilde{\omega})}{\sigma_0^k(\tilde{\omega})}\right] - E\left[\frac{\sigma^k(\tilde{\omega})}{\sigma_0^k(\tilde{\omega})}\right]^\top x. \tag{3.12.k}$$

It is obvious that one can also devise a multi-cut method in which the above optimality cut is disaggregated into several inequalities (e.g. Birge and Louveaux [1997]). The following asymptotic result is proved in Sen and Sherali [2002].

Proposition 3.7. *Assume that $h(x, \tilde{\omega}) \geq 0$ wp1 for all $x \in X$. Let the first-stage approximation solved in iteration k be*

$$\operatorname{Min}\left\{c^\top x + \eta \mid \eta \geq 0, x \in X \cap \mathcal{B}, (\eta, x) \text{ satisfies } (3.12.1), \ldots, (3.12.k)\right\}.$$

Moreover, assume that the second-stage subproblem is a mixed-integer linear program whose partial solutions are obtained using a branch-and-bound method in which all LP relaxations are feasible, and nodes are fathomed only when the lower bound (on the second-stage) exceeds the best available upper bound (for the second-stage). Suppose that there exists an iteration K such that for

$k \geq K$, the branch-and-bound method (for each second-stage subproblem) provides an optimal second-stage solution for all $\omega \in \Omega$, thus yielding an upper bound on the two-stage problem. Then the resulting D^2-BAC algorithm provides an optimal first-stage solution.

Continuous First-stage, Integer Second-stage and Fixed Tenders: Branch-and-Bound

With the exception of the SIR models, all others studied thus far were restricted to models in which the first-stage decisions are restricted to be binary. For problems in which the first-stage includes continuous decision variables, but the second-stage has mixed-integer variables, the situation is more complex. For certain special cases however, there are some practical B&B methods. We summarize one such algorithm which is applicable to problems with purely integer recourse, and fixed tenders T (see (1.1, 1.2)). This method is due to Ahmed, Tawarmalani and Sahinidis [2004].

The essential observation in this method is part c) of Proposition 2.2; namely, the value function of a pure IP (with integer W) is constant over hyper-rectangles ("boxes"). Moreover, if the set $X = \{x \mid Ax \geq b, x \geq 0\}$ is bounded, then there are only finitely many such boxes. This observation was first used in Schultz, Stougie and Van der Vlerk [1998] to design an enumerative scheme for first-stage decisions, while the second-stage decisions were obtained using polynomial ideal theory. However, enumeration in multi-dimensional problems needs far greater care, and this is where the work of Ahmed, Tawarmalani and Sahinidis [2004] makes its contribution. The idea is to transform the original two-stage stochastic integer program into a global optimization problem in the space of "tender variables" $\chi = Tx$. The transformed problem is as follows.

$$\operatorname*{Min}_{\chi \in \mathcal{X}} \varphi(\chi),$$

where $\mathcal{X} = \{\chi \mid Tx = \chi, x \in X\}$ and φ is defined as the sum of

$$\phi(\chi) = \operatorname{Min}\{c^\top x \mid Tx = \chi, x \in X\} \quad \text{and} \quad \Phi(\chi) = \sum_{\omega \in \Omega} p(\omega) h(r(\omega) - \chi),$$

where $h(r(\omega) - \chi)$ denotes the value function resulting from the value of a pure IP with right-hand side is $r(\omega) - \chi$ (see (2.1)). Moreover, the recourse matrix W is allowed to depend upon ω. This is one more distinction between the methods of the previous subsections and the one presented here.

Using part c) of Theorem 2.2, the search space of relevance is a collection of boxes of the form $\Pi_{i=1}^{m_2} [\ell_i, u_i)$ that may be used to partition the space of tenders. Not having both ends of each interval in the box requires that lower bounds be computed with some care. Ahmed, Tawarmalani and Sahinidis [2004] provide guidelines so that closed intervals can be used within the optimization calculations. Their method is summarized as follows.

Branch and Bound for Continuous First Stage with Pure Integers and Fixed Tenders in the Second

0. *Initialize.*
   $k \leftarrow 0$.
   a) Rescale the recourse matrices to be integer. Preprocess to find $\epsilon > 0$, so that boxes have the form $\Pi_{i=1}^{m_2}[\ell_i, u_i - \epsilon]$. Since this step (choosing ϵ) is fairly detailed, we refer the reader to Ahmed, Tawarmalani and Sahinidis [2004].
   b) Identify an initial box B^0 such that $\mathcal{X} \subseteq B^0$. Calculate a lower bound φ_ℓ^0, and $y^0(\omega)$ as second-stage solutions during the lower bounding process. If we find $\chi^0 \in \mathcal{X}$ such that $\varphi(\chi^0) = \varphi_\ell^0$, then declare χ^0 as optimal and stop.
   c) Initialize $\mathcal{L}$, the list of boxes, with its sole element B^0, and record φ_ℓ^0, and $y^0(\omega)$. Specify an incumbent solution, which may be NULL, and its value (possibly $+\infty$). The incumbent solution and its value are denoted $\bar{\chi}$ and $\bar{\varphi}$, respectively.
1. *Node Selection and Branching*
   a) If the list $\mathcal{L}$ is empty, then declare the incumbent solution as optimal, unless the latter is NULL, in which case the problem is infeasible.
   b) $k \leftarrow k+1$. Select a box B^k with the smallest lower bound (i.e. $\varphi_\ell^k \leq \varphi_\ell^t, \forall t \in \mathcal{L}$). Remove B^k from the list $\mathcal{L}$. Partition B^k into two boxes by subdividing one edge of the box. Several choices are possible (see below). Denote these boxes as B^+ and B^-.
2. *Bounding*
   a) (Lower Bounding). For each newly created box, B^+, B^-, calculate a lower bound $\varphi_\ell^+, \varphi_\ell^-$ (resp.). Include those boxes in $\mathcal{L}$ for which the lower bounds are less than $\bar{\varphi}$. For each box included in $\mathcal{L}$, record the lower bounds $(\varphi_\ell^+, \varphi_\ell^-)$ as well as associated (non-integer) solutions $y^+(\omega)$ and $y^-(\omega)$. (These second-stage solutions are used for selecting the edge of the box which will be subdivided for partitioning.) Moreover, record χ^+, χ^-, the tenders obtained while solving the lower bounding problems for B^+ and B^- resp.

b) (Upper Bounding). If $\chi^+ \in \mathcal{X}$ and $\varphi(\chi^+)=\varphi_\ell^+$, then update the incumbent solution and value ($\bar{\chi} \leftarrow \chi^+$, $\bar{\varphi} \leftarrow \varphi(\chi^+)$) provided $\varphi(\chi^+) < \bar{\varphi}$. Similarly, if $\chi^- \in \mathcal{X}$ and $\varphi(\chi^-)=\varphi_\ell^-$, then update the incumbent solution and value ($\bar{\chi} \leftarrow \chi^-, \bar{\varphi} \leftarrow \varphi(\chi^-)$) provided $\varphi(\chi^-) < \bar{\varphi}$.

3. *Fathoming*
Remove all those boxes from $\mathcal{L}$ whose recorded lower bounds exceed $\bar{\varphi}$. Repeat from step 1.

There are two important details to be discussed: a) the lower bounding problem, and b) the choice of the edge for subdivision. Given any box B, let ℓ, u denote the vector of the upper and lower bounds for χ admissible to that box. Then, a lower bound on $\varphi(\cdot)$ for $\chi \in B$ can be calculated by evaluating $\Phi(u-\epsilon)$, and minimizing $\phi(\chi)$ over the set $\chi \in B$. The non-decreasing nature of IP value functions (see Proposition 2.2) imply that $\Phi(u-\varepsilon) \leq \Phi(\chi)$, $\forall\, \chi \in B$. Hence the lower bounding scheme is easily justified. It is also worth mentioning that this evaluation can be performed without having any interactions between the stages or the scenarios, and hence is very well suited for parallel and/or distributed computing. Finally, there are several possible choices for subdividing an edge; the one suggested by the authors is analogous to a "most fractional" rule (see Remark 4.2).

0-1 MIP in Both Stages with General Random Data: Branch and Cut

Of all the methods discussed in this section, the one summarized here has the most in common with standard deterministic integer programming. One may attribute this to the fact that in the absence of any special structure associated with the random elements, it is easiest to view the entire SMIP as a very large deterministic MIP. This method was studied by Caroe [1998]. In order to keep the discussion simple, we only present the cutting plane version of the method, which essentially mimics any cutting plane method for MIP. The extension to a branch-and-cut method will be obvious.

Consider the deterministic equivalent problem stated in (2.4) under the assumption that the integer variables are restricted to be binary. Suppose that we solve the LP relaxation of this problem, and we obtain an LP optimum point $(\bar{x}, \bar{y}(\omega), \omega \in \Omega)$. If these vectors satisfy the mixed-integer feasibility requirement, then the method stops. Otherwise, one derives cuts for those $\omega \in \Omega$ for which the pair $\bar{x}, \bar{y}(\omega)$ does not satisfy the mixed-integer feasibility requirement. The new cuts are added to the deterministic equivalent, and the process resumes (by solving the LP relaxation). One could use any cutting plane method to derive the cuts, but Caroe [1998] suggests using the lift-and-project cuts popularized by Balas, Ceria and Cornuéjols [1993].

Given our emphasis on decomposition, the reader has probably guessed that there is some decomposition lurking in the background here. Of course, the reader is right; note that since each cut is in the space of variables $(x, y(\omega))$, the cut coefficients maintain the dual-block angular structure of (2.4). Because the cuts maintain this structure, the solution of the LP relaxation within this method relies on two-stage SLP methods (e.g. L-shaped decomposition). We should observe that unlike the IP decomposition methodology of all the previous subsections, this method relies on SLP decomposition, and as a result, convexification (cutting plane) steps are undertaken only at those iterations at which an SLP optimum is found, and when such an optimum is non-integer. Of course, the method is easily generalized to the branch-and-cut setting.

4 Decomposition algorithms for multi-stage SMIP: scenario decomposition

As with stochastic linear programs (SLP), the Stagewise Decomposition algorithms discussed in the previous section scale well with respect to the number of scenarios in the two-stage case. Indeed for SLP, these algorithms have been extended to the case of arbitrarily many scenarios (e.g. continuous random variables) using sampling in the two-stage case. However, the scalability of stagewise decompositon methods with respect to multiple decision stages may be suspect. In this section we present two scenario decomposition methods for multi-stage SMIP. These methods, based on branch-and-price (B&P) (Lulli and Sen [2002]), and Lagrangian relaxation (Caroe and Schultz [1999]), share a lot in common. Accordingly, we will present one of the methods (B&P) in detail, and then show how B&P can be easily adapted for Langrangian relaxation. We mention another method, a heuristic by Lokketangen and Woodruff [1996] which combines a Tabu search heuristic with progressive hedging. As with the Lagrangian relaxation in IP, scenario decomposition methods allow us to exploit special structure while remaining applicable to a wide class of problems.

A Scenario Formulation and a Branch-and-Price Algorithm

There are several alternative ways in which a multi-stage stochastic programming model can be formulated. We restrict ourselves to modeling discrete random variables which evolve over discrete points in time which we refer to as stages. More general SP models have been treated as far back as Olsen [1976], and more recently by Wright [1994], and Dentcheva and Roemisch [2002]. The latter paper is particularly relevant for those interested in multi-stage SMIP, and there, the reader will also find a more succinct measure theoretic (as well as convex analytic) treatment of the problem. Because we restrict ourselves to discrete random variables, the data evolution process can be described in graph theoretic terms. For this class of models, any possible trajectory of data may be represented as a path that traverses a series of nodes on a graph. Each node is associated with a stage index t, and represents not only the piece of data revealed at stage t, but also the history of data revealed prior to stage t. Thus multi-stage SP models work with "path-dependent" data, as opposed to "state-dependent" data of Markov decision processes. Arcs on this graph represent the process of data (knowledge) discovery with the passage of time (stages). Since a node in stage t represents the entire history until stage t, it (the node) can only have a unique predecessor. Consequently, the resulting graph is a tree referred to as a scenario tree. A complete path from the root of the tree to a leaf node represents a scenario.

Dynamic deterministic models consider only one scenario and note that for such problems one can associate decisions with each node of the scenario. For SP models, this idea is generalized so that decisions can be associated with every node on the scenario tree, and an SP model is one that chooses decisions for each node in such a manner as to optimize some performance measure. While several papers address other measures of performance (e.g. Ogryczak and Ruszcynski [2002], and Rockafellar and Uryasev [2002]), the most commonly studied measure remains the expected value model. In this case, decisions associated with nodes of the tree must be made in such a way that the expected value of decisions on the entire tree is optimized. (Here the expectation is calculated by weighting the cost of decisions at each node by the probability of visiting that node.) There are several equivalent mathematical representations of this problem, one of which is called the scenario formulation. This is the one we pursue here, although other formulations (e.g. the nodal formulation) may be of interest for the other algorithms.

Let the stages in the model be indexed by $t \in \mathcal{T} = \{1, \ldots, T\}$, the collection of nodes of the scenario tree be denoted $\mathfrak{J}$, and let Ω denote the set of all scenarios. By assumption there are finitely many scenarios indexed by ω, and each has a probability $p(\omega)$. Let us associate decisions $x(\omega) = (x_1(\omega), \ldots, x_T(\omega))$ with each scenario $\omega \in \Omega$. The decisions $x_t(\omega)$ are mixed-integer vectors with $j \in J_t$ denoting the index (set) of integer components in stage t. It is important to note that since ω denotes a complete trajectory (for stages in $\mathcal{T} = \{1, \ldots, T\}$), these decision vectors are allowed to be clairvoyant. In other words, $x_t(\omega)$ may use information from the periods $j > t$ because the argument ω is a complete trajectory! Such *clairvoyant decisions are unacceptable* since they violate the requirement that decisions in stage t cannot use data revealed in future stages ($j > t$). One way to impose this non-clairvoyance requirement is to impose the condition that scenarios which share the *same history of data* until node n, must also share the *same history of decisions* until that node. In order to model this requirement, we introduce some additional mixed-integer vectors $z_n, n \in \mathfrak{J}$. Let Ω_n denote a collection of scenarios (paths) that pass through node n. Moreover, define a mapping $\mathcal{H} : \mathcal{T} \times \Omega \to \mathfrak{J}$ such that for any 2-tuple (t, ω), $\mathcal{H}(t, \omega)$ provides that node n in stage t for which $\omega \in \Omega_n$. Then, the non-clairvoyance condition (commonly referred to as non-anticipativity) requires that

$$x_t(\omega) - z_{\mathcal{H}(t,\omega)} = 0 \quad \forall (t,\omega). \tag{4.1}$$

Higle and Sen [2002] refer to this as the "state variable formulation;" there are several equivalent ways to state non-anticipativity requirement (e.g. Rockafellar and Wets [1991], Mulvey and Ruszcznski [1995]). We will also use J_t to index all integer elements of $z_{\mathcal{H}(t,\omega)}$. The ability to directly address the "state variable" (z) eases the exposition (and even computer programming) considerably, and hence we choose this formulation here. Finally, for a given $\omega \in \Omega$, we will use $z(\omega)$ to designate the trajectory of decision states associated with ω.

(4.1) not only ensures the logical dependence of decisions on data, but also frees us up to use data associated with an entire scenario without having to trace it in a stage-by-stage manner. Thus, we will concatenate all stagewise data into vectors and matrices that can be indexed by ω. Thus, the trajectory of cost coefficients associated with scenario ω will be denoted $c(\omega)$, the collection of technology matrices by $A(\omega)$ and the right-hand side by $b(\omega)$. In the following we use $x_{jt}(\omega)$ to denote the jth element of the vector $x_t(\omega)$, a sub-vector of $x(\omega)$. Next define the set

$$X(\omega) = \left\{x(\omega) \mid A(\omega)x(\omega) \geq b(\omega),\, x(\omega) \geq 0,\, x_{jt}(\omega) \text{ integer},\ j \in J_t, \forall t\right\}.$$

Given the above setup, a multi-stage SMIP problem can now be stated as a large-scale MIP of the following form:

$$\text{Min}\left\{ \sum_{\omega\in\Omega} p(\omega)c(\omega)^{\top}x(\omega) \mid x(\omega) \in X(\omega), \text{ and } \left\{x(\omega)\right\}_{\omega\in\Omega} \text{ satisfies (4.1) } \forall \omega \in \Omega \right\}. \tag{4.2}$$

It should be clear that the above formulation is amenable to solution using decomposition because the only constraints that couple the scenarios together are (4.1). For many practical problems, this collection of constraints may be so large that aggregation schemes may be necessary to solve the large practical problems (see Higle, Rayco and Sen [2002]). However, for moderately sized problems, B&P and similar deterministic decomposition schemes are reasonably effective, and perform better than solving the entire deterministic equivalent using state-of-the-art software like CPLEX (Lulli and Sen [2002]). The following exposition assumes familiarity with standard column generation methods (see e.g. Martin [1999]).

The B&P algorithm may be described as one that combines column generation with branch-and-bound (B&B) or branch-and-cut (B&C). For the sake of simplicity, we avoid the inclusion of cuts, although this is clearly do-able. The lower bounding scheme within a B&P algorithm requires the solution of an LP master problem whose columns are supplied by a mixed-integer subproblem. Let e denote an event (during the B&B process) at which the algorithm requires the solution of an LP (master). This procedure will begin with those columns that are available at the time of event e, and then generate further columns as necessary to solve the LP. We will denote the collection of columns available at the start of event e by the set $I^e(\omega)$, and those at the end of the event by $I^e_+(\omega)$. For column generation iterations in the interim (between the start and end of the column generation process) we will simply denote the set of columns by $I^e(\omega)$, and the columns themselves by $\{\{x_i(\omega)\}, i \in I^e(\omega)\}_{\omega \in \Omega}$.

Since the branching phase will impose integrality restrictions on the "state variables" z we use the notation z_ℓ and z_u to denote lower and upper bounds on z for any nodal problem associated with B&P iteration. (As usual, some of the upper bounds in the vector z_u could be $+\infty$.)

Given a collection of columns $\{x_i(\omega), i \in I^e(\omega), \omega \in \Omega\}$, the non-anticipativity constraints (4.1) can be expressed as

$$\sum_{i \in I^e(\omega)} \rho_i x_i(\omega) - z(\omega) = 0, \quad \forall \omega \tag{4.3a}$$

$$z_\ell \leq z(\omega) \leq z_u \quad \forall \omega \tag{4.3b}$$

$$\sum_{i \in I^e(\omega)} \rho_i(\omega) = 1 \quad \forall \omega \tag{4.3c}$$

$$\rho_i(\omega) \geq 0 \quad \forall i, \omega \tag{4.3d}$$

Whenever the above set is empty, we assume that a series of "Phase I" iterations (of the column generation scheme) can be performed for those scenarios for which the columns make it infeasible to satisfy the range restrictions on some element of $z(\omega)$. In this case, a "Phase I" problem is solved for each offending scenario and columns are generated to minimize deviations from the box (4.3b). We assume that whenever (4.3) is infeasible, such a procedure is adopted to render a feasible collection of columns in the master program which is stated as follows.

$$\text{Min}\left\{\sum_{\omega \in \Omega} p(\omega) \sum_{i \in I^e(\omega)} \left[c(\omega)^\top x_i(\omega)\right] \rho_i(\omega) \text{ where } \left\{\rho_i(\omega), i \in I^e(\omega)\right\}_{\omega \in \Omega} \text{ satisfies (4.3)}\right\}. \tag{4.4}$$

Given a dual multiplier estimate $\mu(\omega)$ for the non-anticipativity constraints (4.3a) in the master problem, the subproblem for generating columns for scenario $\omega \in \Omega$ is as follows.

$$D(\mu(\omega), \omega) = \mathrm{Min}\{[p(\omega)c(\omega) - \mu(\omega)]^{\top} x(\omega) \mid x(\omega) \in X(\omega)\}. \tag{4.5}$$

While each iteration of column generation (LP solve) uses a different vector $\mu(\omega)$, we have suppressed this dependence for notational simplicity. In any case, the column generation procedure continues until $D(\mu(\omega), \omega) - \theta(\omega) \geq 0$ $\forall \omega \in \Omega$, where $\theta(\omega)$ is a dual multiplier associated with the convexity constraint (4.3c). Because of the way in which $X(\omega)$ is defined, (4.5) is a deterministic MIP, and one solves as many of these as there are columns generated during the algorithm. As a result, it is best to use the B&P method in situations where (4.5) has some special structure, so that the MIP in (4.5) is solved efficiently. This is the same requirement as in deterministic applications of B&P (e.g. Barnhart et al [1998]). In Lulli and Sen [2002], the structure they utilized for their computational results was the stochastic batch sizing problem. Nevertheless, the B&P method is applicable to the more general problem. The algorithm may be summarized as follows.

Branch and Price for Multi-Stage SMIP

0. *Initialize.*
   a) $k \leftarrow 0$, $e \leftarrow 0$, $I^e = \emptyset$. B^0 denotes a box for which $0 \leq z \leq +\infty$. (The notation I^e includes columns for all $\omega \in \Omega$; the same holds for I^e_+.)
   b) Solve (4.4) and its optimal value is f^0_ℓ, and a solution z^0. If the elements of z^0 satisfy the mixed-integer variable requirements, then we declare z^0 as optimal, and stop.
   c) $I^{e+1} \leftarrow I^e_+, e \leftarrow e+1$. Initialize $\mathcal{L}$, the list of boxes, with its sole element B^0, and record its lower bound f^0_ℓ, and a solution z^0. Specify an incumbent solution, which may be NULL, and its value (possibly $+\infty$). The incumbent solution and its value are denoted $\bar{z}$ and $\bar{f}$ respectively.
1. *Node Selection and Branching*
   a) If the list $\mathcal{L}$ is empty, then declare the incumbent solution as optimal, unless the latter is NULL, in which case the problem is infeasible.

b) $k \leftarrow k+1$. Select a box B^k with the smallest lower bound (i.e. $f_\ell^k \leq f_\ell^v, \forall v \in \mathcal{L}$). Remove B^k from the list $\mathcal{L}$ and partition B^k into two boxes so that z^k does not belong to either box, (e.g. choose the "most fractional" variable in z^k, and create two subproblems by partitioning). Denote these boxes as B^+ and B^-.

2. *Bounding*

a) (Lower Bounding). Let $I^{e+1} \leftarrow I_+^e, e \leftarrow e+1$. For the newly created box B^+, solve the associated LP relaxation (4.4) using column generation. This procedure provides the lower bound f_ℓ^+ and a solution z^+. Let $I^{e+1} \leftarrow I_+^e$, $e \leftarrow e+1$. Now solve the LP relaxation (4.4) associated with B^-, and obtain a lower bound f_ℓ^-, and a solution z^-. Include those boxes in $\mathcal{L}$ for which the lower bounds are less than $\bar{f}$. For each box included in $\mathcal{L}$, associate the lower bounds (f_ℓ^+, f_ℓ^-) as well as associated (non-mixed-integer) solutions z^+ and z^-.

b) (Upper Bounding). If z^+ satisfies mixed-integer requirements and $f_\ell^+ < \bar{f}$, then update the incumbent solution and value $(\bar{z} \leftarrow z^+, \bar{f} \leftarrow f^+)$. Similarly, if z^- satisfies the mixed-integer requirement, then update the incumbent solution and value $(\bar{z} \leftarrow z^-, \bar{f} \leftarrow f^-)$.

3. *Fathoming*
Remove all those boxes from $\mathcal{L}$ whose recorded lower bounds exceed $\bar{f}$. Repeat from step 1.

Remark 4.1. While we have stated the B&P method by using z as the branching variables, it is clearly possible to use branching on the original x variables. This is the approach implemented in Lulli and Sen [2002].

Remark 4.2. The term "most fractional" may be interpreted in the following sense: if a variable z_j has a value $\bar{z}_j$, and which is in the interval $z_{\ell,j} \leq \bar{z}_j \leq z_{u,j}$, then assuming $z_{\ell,j}, z_{u,j}$ are both integers, the measure of integrality that one may use is $\min\{\bar{z}_j - z_{\ell,j}, z_{u,j} - \bar{z}_j\}$. The "most fractional" variable then is the one for which this measure is the largest. Another measure could be based on the "relatively most fractional" index:

$$\min\left\{\frac{\bar{z}_j - z_{\ell,j}}{z_{u,j} - z_{\ell,j}}, \frac{z_{u,j} - \bar{z}_j}{z_{u,j} - z_{\ell,j}}\right\}.$$

Lagrangian Relaxation and Duality

The algorithmic outline of the previous subsection can be easily adapted to use Lagrangian relaxation as suggested in Caroe and Schultz [1999]. The only modification necessary is in step 2a, where the primal LP (4.4) is replaced by a dual. The exact formulation of the dual problem used in Caroe and Schultz [1999] is slightly different from the one we will use because our branching variables are z, whereas, they branch on the $x(\omega)$ variables directly. However, the procedures are essentially the same. We now proceed to the equivalent dual problem that may be used for an algorithm based on the Lagrangian relaxation.

When there are no bounds placed on the "state variables" z (i.e. the root node of the B&B tree), the following dual is equivalent to the Langrangian dual

$$\operatorname*{Max}_{\mu}\left\{\sum_{\omega\in\Omega} D(\mu(\omega),\omega)\mid \sum_{\omega\in\Omega_n}\mu(\omega)=0, \quad \forall n\in\Im\right\} \tag{4.6}$$

where $\mu=\{\mu(\omega)\}_{\omega\in\Omega}$, and $D(\mu(\omega),\omega)$ is the dual function defined in (4.5). It is not customary to include equality constraints for a Lagrangian dual, but for this particular formulation of non-anticipativity, imposing the dual constraints accommodates the coupling variables z implicitly. There are also some interesting probabilistic and economic features that result from re-scaling dual variables in (4.6) (see Higle and Sen [2002]). Nevertheless, (4.6) will suffice for our algorithmic purposes.

Note that as one proceeds with the branch-and-bound iterations, partitioning the space of "state variables" induces different bound on them. In turn, these bound should be imposed on the primal variables in (4.5). Thus, the dual lower bounds are selectively improved to close the duality gap via the B&B process.

We should note that the dual problem associated with any node results in a nondifferentiable optimization problem, and consequently, Caroe and Schultz [1999] suggest that it be solved using subgradient or bundle based methods (e.g. Kiwiel [1990]). While (4.6) is not the unconstrained problem of Caroe and Schultz [1999], the dual constraints in (4.6) have such a special structure that they do not impede any projection based subgradient algorithm.

In addition to their similarities in structure, B&P and Lagrangian relaxation also lead to equivalent convexifications, as long as the same non-anticipativity constraints are relaxed (see Shapiro [1979], Dentcheva and Roemisch [2002]). Nevertheless, these methods have their computational differences. The master problems in B&P are usually solved using LP software which has become extremely reliable and scalable. It is also interesting to note that B&P algorithms also have a natural criterion for curtailing the size of the master program. In particular, note that we can set aside those columns (in the master) that do not satisfy the bound restrictions imposed at any given node. While this is not necessary, it certainly reduces the size of the master problem. Moreover, the primal approach leads to primal solutions from which branching is quite easy. For dual-based methods, primal solution recovery is necessary before good branching schemes (e.g. strong branching) can be devised. However, further computational research is necessary for a comparison of these algorithms.

We close this section with a comment of duality gaps for multi-stage SMIP. Alternative formulations of the dual problem may result in different duality gaps for multi-stage SMIP. For example, Dentcheva and Roemisch [2002] compare duality gaps arising from relaxing nodal constraints (in a nodal SP formulation) with gaps obtained from relaxing non-anticipativity constraints of the scenario formulation. They show that scenario decomposition methods, such as the ones presented in this section, provide smaller duality gaps than nodal decomposition. Results of this nature are extremely important in the design of algorithms for SMIP. And a final word of caution regarding duality gaps is that without using algorithms that ensure the search for a global optimum (e.g. branch-and-bound), it is difficult to guarantee that the duality gap for SMIP vanishes, even if the number of scenarios is infinitely large, as in problems with continuous random variables (see Sen, Higle and Birge [2000]).

5 Conclusions

In this chapter, we have studied several classes of SMIP models. However, there are many more models and applications that call for further research. We provide a brief synopsis of some of these areas.

We begin by noting that the probabilistically constrained problem with discrete random variables has been recognized by several authors as a disjunctive program (e.g. Prekopa [1990], Sen [1992]). These authors treat the problem from alternative view points, one of which may be considered a dual of the other. More recently, Dentcheva, Prekopa and Ruszczynski [2000] have proposed extensions that allow more realistic algorithms than previously studied. Nevertheless, there are several open issues, including models with random technology matrices, multi-stage models with stage-dependent probabilistic constraints, and more. Another area of investigation deals with the application of test sets to the solution of SMIP problems (Schultz, Stougie and Van der Vlerk [1998], Hemmecke and Schultz [2003]). The reader will find more on this topic in the recent survey by Louveaux and Schultz [2003]. Another survey of interest is the one by Klein Haneveld and Van der Vlerk [1999].

In addition to the above methods, SMIP models are also giving rise to new applications and heuristics. Network routing and vehicle routing problems have been studied by Verweij et al [2003], and Laporte, Van Hamme and Louveaux [2002]. Another classic problem that has attracted a fair amount of attention is the stochastic unit-commitment problem (Takriti, Birge and Long [1996], Nowak and Römisch [2000]). Recent applications in supply chain planning have given rise to new algorithms by Alonso-Ayuso et al [2002]. Other related applications include the work on stochastic lot sizing problems (Lokketangen and Woodruff [1996], Lulli and Sen [2002]). It so happens, that all of these applications lead to multi-stage models, which are among the most challenging SMIP problems. Given such complexity, we expect that the study of good heuristics will be of immense value. Papers on multi-stage capacity expansion planning (Ahmed and Sahinidis [2003], MirHassani et al [2000] and others) constitute a step in this direction.

As shown in this chapter, the IP literature has much to contribute to the solution of SMIP problems. Conversely, decomposition approaches studied within the context of SP have the potential to contribute to the decomposition of IP models in general, and of course, SMIP models in particular. As one can surmise, research on SMIP models has picked up considerable steam over the past few years, and we expect this trend to continue. These problems may be characterized as "guard challenge" problems, and we expect modern computer technology to play a major role in the solution of these models. We believe that distributed computing provides the ideal platform for the implementation of decomposition algorithms for SMIP, and expect that vigorous research will overcome this "grand challenge." The reader may stay updated on this progress through the SIP web site http://mally.eco.rug.nl/spbib.html.

Acknowledgments

I am grateful to the National Science Foundation (DMI-9978780, and CISE-9975050) for its support in this line of enquiry. I wish to thank Guglielmo Lulli, George Nemhauser, and an anonymous referee for their thoughtful comments on an earlier version of this chapter. The finishing touches on this work were completed during my stay as an EPSRC Fellow at the CARISMA Center of the Mathematics Department at the University of Brunel, U.K. My host, Gautam Mitra, was instrumental in arranging this visit, and I thank him for an invigorating stay.

References

Ahmed, S., M. Tawarmalani, and N.V. Sahinidis [2004], "A finite branch and bound algorithm for two-stage stochastic integer programs," *Mathematical Programming*, **100**, pp. 355–377.

Ahmed, S., and N.V. Sahinidis [2003], "An approximation scheme for stochastic integer programs arising in capacity expansion," *Operations Research*, **51**, pp. 461–471.

Alonso-Ayuso, A., L.F. Escudero, A. Garin, M.T. Orteno and G. Peres [2003], "An approach for strategic supply chain planning under uncertainty based on stochastic 0-1 programming," *Journal of Global Optimization*, **26**, pp. 97–124.

Balas, E. [1975], "Disjunctive programming: cutting planes from logical conditions," in *Non-linear Programming 2*, (O.L. Mangasarian, R.R. Meyer and S.M. Robinson, eds.), Academic Press, N.Y.

Balas, E. [1979], "Disjunctive programming," *Annals of Discrete Mathematics*, **5**, pp. 3–51.

Balas, E., S. Ceria, and G. Cornuéjols [1993], "A lift-and-project cutting plane algorithm for mixed 0-1 programs," *Mathematical Programming*, **58**, pp. 295–324.

Barnhart, C., E.L. Johnson, G.L. Nemhauser, M.W.P. Savelsberg and P.H. Vance [1998], "Branch-and-Price: Column generation for solving huge integer programs," *Operations Research*, **46**, 316–329.

Benders, J.F. [1962], "Partitioning procedures for solving mixed-variable programming problems," *Numerische Mathematic*, **4**, pp. 238–252.

Birge, J.R. and F. Louveaux [1997], *Introduction to Stochastic Programming*, Springer.

Blair [1980], "Facial disjunctive programs and sequence of cutting planes," *Discrete Applied Mathematics*, **2**, pp. 173–179.

Blair, C. [1995], "A closed-form representation of mixed-integer program value functions," *Mathematical Programming*, **71**, pp. 127–136.

Blair, C. and R. Jeroslow [1978], "A converse for disjunctive constraints," *Journal of Optimization Theory and Applications*, **25**, pp. 195–206.

Blair, C. and R. Jeroslow [1982], "The value function of an integer program," *Mathematical Programming*, **23**, pp. 237–273.

Caroe, C.C. [1998], *Decomposition in Stochastic Integer Programming*. PhD thesis, Institute of Mathematical Sciences, Dept. of Operations Research, University of Copenhagen, Denmark.

Caroe, C.C. and R. Schultz [1999], "Dual decomposition in stochastic integer programming," *Operations Research Letters*, **24**, pp. 37–45.

Caroe, C.C. and J. Tind [1998], "L-shaped decomposition of two-stage stochastic programs with integer recourse," *Mathematical Programming*, **83**, no. 3, pp. 139–152.

Dentcheva, D., A. Prekopa, and A. Ruszczynski [2000], "Concavity and efficient points for discrete distributions in stochastic programming," *Mathematical Programming*, **89**, pp. 55–79.

Dentcheva, D. and W. Roemisch [2002], "Duality gaps in nonconvex stochastic optimization," Institute of Mathematics, Humboldt University, Berlin, Germany (also Stochastic Programming E-Print Series, 2002–13).

Hemmecke, R. and R. Schultz [2003], "Decomposition of test sets in stochastic integer programming," *Mathematical Programming*, **94**, pp. 323–341.

Higle, J.L., B. Rayco, and S. Sen [2002], "Stochastic Scenario Decomposition for Multi-stage Stochastic Programs," Working paper, SIE Department, University of Arizona, Tucson, AZ 85721.

Higle, J.L. and S. Sen [1991], "Stochastic Decomposition: An algorithm for two-stage linear programs with recourse," *Math. of Operations Research*, **16**, pp. 650–669.

Higle, J.L. and S. Sen [2002], "Duality of Multistage Convex Stochastic Programs," to appear in *Annals of Operations Research*.

Infanger, G. [1992], "Monte Carlo (importance) sampling within a Benders' decomposition algorithm for stochastic linear programs," *Annals of Operations Research*, **39**, pp. 69–95.

Jeroslow, R. [1980], "A cutting plane game for facial disjunctive programs," *SIAM Journal on Control and Optimization*, **18**, pp. 264–281.

Kall, P. and J. Mayer [1996], "An interactive model management system for stochastic linear programs," *Mathematical Programming*, **75**, pp. 221–240.

Kelley, J.E. [1960], "The cutting plane method for convex programs," *Journal of SIAM*, **8**, pp. 703–712.

Kiwiel, K.C. [1990], "Proximity control in bundle methods for convex non-differentiable optimization," *Mathematical Programming*, **46**, pp. 105–122.

Klein Haneveld, W.K., L. Stougie, and M.H. van der Vlerk [1995], "On the convex hull of the simple integer recourse objective function," *Annals of Operations Research*, **56**, pp. 209–224.

Klein Haneveld, W.K., L. Stougie, and M.H. van der Vlerk [1996], "An algorithm for the construction of convex hulls in simple integer recourse programming," *Annals of Operations Research*, **64**, pp. 67–81.

Klein Haneveld, W.K. and M.H. van der Vlerk [1999], "Stochastic integer programming: general models and algorithms," *Annals of Operations Research*, **85**, pp. 39–57.

Laporte, G. and F.V. Louveaux [1993], "The integer L-shaped methods for stochastic integer programs with complete recouse," *Operations Research Letters*, **13**, pp. 133–142.

Laporte, G., L. Van Hamme, and F.V. Louveaux [2002], "An integer L-shaped algorithm for the capacitated vehicle routing problem with stochastic demands," *Operations Research*, **50**, pp. 415–423.

Lokketangen, A. and D.L. Woodruff [1996], "Progressive hedging and tabu search applied to mixed integer (0,1) multi-stage stochastic programming," *Journal of Heuristics*, **2**, pp. 111–128.

Louveaux, F.V. and R. Schultz [2003], "Stochastic Integer Programming," *Handbook on Stochastic Programming*, (A. Ruszczynski and A. Shapiro, eds.), pp. 213–264.

Louveaux, F.V. and M.H. van der Vlerk [1993], "Stochastic Programming with Simple Integer Recourse," *Mathematical Programming*, **61**, pp. 301–325.

Lulli, G. and S. Sen [2002], "A Branch and Price Algorithm for Multi-stage Stochastic Integer Programs with Applications to Stochastic Lot Sizing Problems," to appear in *Management Science*.

Martin, R.K. [1999], *Large Scale Linear and Integer Optimization*, Kluwer Academic Publishers.

MirHassani, S.A., C. Lucas, G. Mitra, E. Messina, and C.A. Poojari [2000], "Computational solution of capacity planning models under uncertainty," *Parallel Computing*, **26**, pp. 511–538.

Mulvey, J.M. and A. Ruszczynski [1995], "A new scenario decomposition method for large scale stochastic optimization," *Operations Research*, **43**, pp. 477–490.

Nemhauser, G. and L.A. Wolsey [1988], *Integer and Combinatorial Optimization*, John Wiley and Sons.

Norkin, V.I., Y.M. Ermoaliev, and A. Ruszczynski [1998], "On optimal allocation of indivisibles under uncertainty," *Operations Research*, **46**, no. 3, pp. 381–395.

Nowak, M. and W. Römisch [2000], "Stochastic Lagrangian relaxation applied to power scheduling in a hydro-thermal system under uncertainty," *Annals of Operations Research*, **100**, pp. 251–272.

Ntaimo, L. and S. Sen [2004], "The million variable "march" for stochastic combinatorial optimization, with applications to stochastic server location problems," to appear in *Journal of Global Optimization*.

Ogryczak, W. and A. Ruszczynski [2002], "Dual stochastic dominance and related mean-risk models," *SIAM J. on Optimization*, **13**, pp. 60–78.

Olsen, P. [1976], "Discretization of multistage stochastic programming," *Mathematical Programming*, **6**, pp. 111–124.

Prekopa [1990], "Dual method for a one-stage stochastic programming problem with random RHS obeying a discrete probability distribution," *Zeitschrift fur Operations Research*, **38**, pp. 441–461.

Riis, M. and R. Schultz [2003], "Applying the minimum risk criterion in stochastic recourse programs," *Computational Optimization and Applications*, **24**, pp. 267–288.

Rockafellar, R.T. and R.J.-B. Wets [1991], "Scenario and policy aggregation in optimization under uncertainty," *Mathematics of Operations Research*, **16**, pp. 119–147.

Rockafeller, R.T. and S. Uryasev [2002], "Conditional value-at-risk for general loss distributions," *Journal of Banking and Finance*, **26**, pp. 1443–1471.

Schultz, R. [1993], "Continuity properties of expectation functions in stochastic integer programming," *Mathematics of Operations Research*, **18**, pp. 578–589.

Schultz, R., L. Stougie, and M.H. van der Vlerk [1998], "Solving stochastic programs with integer recourse by enumeration: a framework using Grobner basis reduction," *Mathematical Programming*, **83**, no. 2, pp. 71–94.

Sen, S. [1992], "Relaxations for probabilistically constrained programs with discrete random variables," *Operations Research Letters*, **11**, pp. 81–86.

Sen, S. [1993], "Subgradient decomposition and the differentiability of the recourse function of a two-stage stochastic LP with recourse," *Operations Research Letters*, **13**, pp. 143–148.

Sen, S. and J.L. Higle [2000], "The C^3 theorem and D^2 algorithm for large scale stochastic optimization: set convexification," working paper SIE Department, University of Arizona, Tucson, AZ 85721 (also Stochastic Programming E-print Series 2000-26) to appear in Mathematical Programming (2005).

Sen, S., J.L. Higle, and J.R. Birge [2000], "Duality Gaps in Stochastic Integer Programming," *Journal of Global Optimization*, **18**, pp. 189–194.

Sen S., J.L. Higle and L.A. Ntaimo [2002], "A Summary and Illustration of Disjunctive Decomposition with Set Convexification," *Stochastic Integer Programming and Network Interdiction Models* (D.L. Woodruff ed.), pp. 105, 125, Kluwer Academic Press, Dordrecht, The Netherlands.

Sen, S. and H.D. Sherali [1985], "On the convergence of cutting plane algorithms for a class of nonconvex mathematical programs," *Mathematical Programming*, **31**, pp. 42–56.

Sen, S. and H.D. Sherali [2002], "Decomposition with Branch-and-Cut Approaches for Two-Stage Stochastic Integer Programming" working paper, MORE Institute, SIE Department, University of Arizona, Tucson, AZ (http://www.sie.arizona.edu/SPEED-CS/raptormore/more/papers/dbacs.pdf) to appear in Mathematical Programming (2005).

Shapiro, J. [1979], *Mathematical Programming: Structures and Algorithms*, John Wiley and Sons.

Sherali, H.D. and W.P. Adams [1990], "A hierarchy of relaxations between the continuous and convex hull representations for zero-one programming problems," *SIAM Journal on Discrete Mathematics*, **3**, pp. 411–430.

Sherali, H.D. and B.M.P. Fraticelli [2002], "A modification of Benders' decomposition algorithm for discrete subproblems: an approach for stochastic programs with integer recourse," *Journal of Global Optimization*, **22**, pp. 319–342.

Sherali, H.D. and C.M. Shetty [1980], "*Optimization with Disjunctive Constraints*," Lecture Notes in Economics and Math. Systems, Vol. 181, Springer-Verlag, Berlin.

Stougie, L. [1985], "Design and analysis of algorithms for stochastic integer programming," Ph.D. thesis, Center for Mathematics and Computer Science, Amsterdam, The Netherlands.

Takriti, S. [1994], "On-line solution of linear programs with varying RHS," Ph.D. dissertation, IOE Department, University of Michigan, Ann Arbor, MI.

Takriti, S. and S. Ahmed [2004], "On robust optimization of two-stage systems," *Mathematical Programming*, **99**, pp. 109–126.

Takriti, S., J.R. Birge, and E. Long [1996], "A stochastic model for the unit commitment problem," *IEEE Trans. of Power Systems*, **11**, pp. 1497–1508.

Tind, J. and L.A. Wolsey [1981], "An elementary survey of general duality theory in mathematical programming," *Mathematical Programming*, **21**, pp. 241–261.

van der Vlerk, M.H. [1995], *Stochastic Programming with Integer Recourse*, Thesis Rijksuniversiteit Groningen, Labyrinth Publication, The Netherlands.

van der Vlerk, M.H. [2004], "Convex approximations for complete integer recourse models," *Mathematical Programming*, **99**, pp. 287–310.

Van Slyke, R. and R.J.-B. Wets [1969], "L-Shaped linear programs with applications to optimal control and stochastic programming," *SIAM J. on Appl. Math.*, **17**, pp. 638–663.

Verweij, B., S. Ahmed, A.J. Kleywegt, G. Nemhauser, and A. Shapiro [2003], "The sample average approximation method applied to stochastic routing problems: a computational study," *Computational Optimization and Algorithms*, **24**, pp. 289–334.

Wolsey, L.A. [1981], "Integer programming duality: price functions and sensitivity analysis," *Mathematical Programming*, **20**, pp. 173–195.

Wright, S.E. [1994], "Primal-dual aggregation and disaggregation for stochastic linear programs," *Mathematics of Operations Research*, **19**, pp. 893–908.

K. Aardal et al., Eds., *Handbooks in OR & MS, Vol. 12*

Chapter 10

Constraint Programming

Alexander Bockmayr

Université Henri Poincaré, LORIA, B.P. 239, F-54506 Vandœuvre-lès-Nancy, France
E-mail: Alexander.Bockmayr@loria.fr

John N. Hooker

Tepper School of Business, Carnegie Mellon University, Pittsburgh, PA 15213, USA
E-mail: john@hooker.tepper.cmu.edu

Abstract

Constraint programming (CP) methods exhibit several parallels with branch-and-cut methods for mixed integer programming (MIP). Both generate a branching tree. Both use inference methods that take advantage of problem structure: cutting planes in the case of MIP, and filtering algorithms in the case of CP. A major difference, however, is that CP associates each constraint with an algorithm that operates on the solution space so as to remove infeasible solutions. This allows CP to exploit substructure in the problem in a way that MIP cannot, while MIP benefits from strong continuous relaxations that are unavailable in CP. This chapter outlines the basic concepts of CP, including consistency, global constraints, constraint propagation, filtering, finite domain modeling, and search techniques. It concludes by indicating how CP may be integrated with MIP to combine their complementary strengths.

1 Introduction

A discrete optimization problem can be given a declarative or procedural formulation, and both have their advantages. A declarative formulation simply states the constraints and objective function. It allows one to describe what sort of solution one seeks without the distraction of algorithmic details. A procedural formulation specifies how to search for a solution, and it therefore allows one to take advantage of insight into the problem in order to direct the search. The ideal, of course, would be to have the best of both worlds, and this is the goal of constraint programming.

The task seems impossible at first. A declarative formulation is static, and a procedural formulation dynamic, in ways that appear fundamentally at odds. For example, setting $x=0$ at one point in a procedure and $x=1$ at another point is natural and routine, but doing the same in a declarative model would simply result in an infeasible constraint set.

Despite the obstacles, the constraint programming community has developed ways to weave procedural and declarative elements together. The evolution of ideas passed through logic programming, constraint satisfaction, constraint logic programming, concurrent constraint programming, constraint handling rules, and constraint programming (not necessarily in that order). One idea that has been distilled from this research program is to *view a constraint as invoking a procedure*. This is the basic idea of constraint programming.

1.1 Constraints as procedures

A constraint programmer writes a constraint declaratively but views it as a procedure that operates on the solution space. Each constraint contributes a relaxation of itself to the *constraint store*, which limits the portion of the space that must be searched. The constraints in the constraint store should be easy in the sense that it is easy to generate feasible solutions for them. The overall solution strategy is to find a feasible solution of the original problem by enumerating solutions of the constraint store in a way to be described shortly.

In current practice the constraint store primarily contains very simple *in-domain* constraints, which restrict a variable to a domain of possible values. The domain of a variable is typically an interval of real numbers or a finite set. The latter can be a set of any sort of objects, not necessarily numbers, a fact which lends considerable modeling power to constraint programming.

The idea of treating a constraint as a procedure is a very natural one for a community trained in computer science, because statements in a computer program typically invoke procedures. This simple device yields a powerful tool for exploiting problem structure. In most practical applications, there are some subsets of constraints that have special structure, but the problem as a whole does not. Existing optimization methods can deal with this situation to some extent, for instance by using Benders decomposition to isolate a linear part, by presolving a network flow subproblem, and so forth. However, most methods that exploit special structure require that the entire problem exhibit the structure. Constraint programming avoids this difficulty by associating procedures with highly structured subsets of constraints. This allows procedures to be designed to exploit the properties of the constraints.

Strictly speaking, constraint programming associates procedures with individual constraints rather than subsets of constraints, but this is overcome with the concept of *global constraints*. A global constraint is a single constraint that represents a highly structured set of constraints. An example would be an `all different` constraint that requires that a set of variables take distinct values. It represents a large set of pairwise disequations. A global constraint can be designed to invoke the best known technology for dealing with its particular structure. This contrasts with the traditional approach used in optimization, in which the solver receives the problem as a set of undifferentiated constraints. If the solver is to exploit any substructure in the problem, it must find it, as some commercial solvers find network substructure. Global constraints, by contrast, allow the user to alert the solver to the portions of the problem that have special structure.

How one can solve a problem by applying special-purpose procedures to individual constraints? What links these procedures together? This is where the constraint store comes into play. Each procedure applies a *filtering* algorithm that eliminate some values from the variable domains. In particular, it eliminates values that cannot be part of any feasible solution for that constraint. The restricted domains are in effect in-domain constraints that are implied by the constraint. They become part of the constraint store, which is passed on to the next constraint to be processed. In this way the constraint store "propagates" the results of one filtering procedure to the others.

Naturally the constraints must be processed in some order, and different systems do this in different ways. In constraint logic programming systems like CHIP, constraints are embedded into a logic programming language (Prolog). In programs written for the ILOG Solver, constraints are objects in a C++ program that determines how the constraints are processed. Programs written in OPL Studio have a more declarative look, and the system exerts more control over the processing.

A constraint program can therefore be viewed as a "program" in the sense of a computer program: the statements invoke procedures, and control is passed from one statement to another, although the user may not specify the details of how this is done. This contrasts with mathematical programs, which are not computer programs at all but are fully declarative statements of the problem. They are called programs because of George Dantzig's early application of linear programming to logistics "programming" (planning) in the military. Notwithstanding this difference, a constraint programming formulation tends to look more like a mathematical programming model than a computer program, since the user writes constraints declaratively rather than writing a code to enforce the constraints.

1.2 Parallels with branch and cut

The issue remains as to how to enumerate solutions of the constraint store in order to find one that is feasible in the original problem. The process is analogous to branch-and-cut algorithms for integer programming, as Table 1 illustrates. Suppose that the problem contains variables $x = [x_1, \ldots, x_n]$ with domains $D_1, \ldots, D_n$. If the domains D_j can all be reduced to singletons $\{v_j\}$, and if $v = [v_1, \ldots, v_n]$ is feasible, then $x = v$ solves the problem. Setting $x = v$ in effect solves the constraint store, and the solution of the constraint store happens to be feasible in the original problem. This is analogous to solving the continuous relaxation of an integer programming problem (which is the "constraint store" for such a problem) and obtaining an integer solution.

Table 1.
Comparison of constraint programming search with branch-and-cut

	Constraint Programming	*Branch-and-Cut*
Constraint store (relaxation)	Set of in-domain constraints	Continuous relaxation (linear inequalities)
Branching	Branch by splitting a nonsingleton domain, or by branching on a constraint	Branch on a variable with a noninteger value in the solution of the relaxation
Inference	Reduce variable domains (i.e., add in-domain constraints to constraint store); add nogoods	Add cutting planes to relaxation (which also contains inequalities from the original IP); add Benders or separating cuts*
Bounding	None	Solve continuous relaxation to get bound
Feasible solution is obtained at a node...	When domains are singletons and constraints are satisfied	When solution of relaxation is integral
Node is infeasible...	When at least one domain is empty	When continuous relaxation is infeasible
Search backtracks...	When node is infeasible	When node is infeasible, relaxation has integral solution, or tree can be pruned due to bounding

*Commercial solvers also typically apply preprocessing at the root note, which can be viewed as a rudimentary form of inference or constraint propagation.

If the domains are not all singletons, then there are two possibilities. One is that there is an empty domain, in which case the problem is infeasible. This is analogous to an infeasible continuous relaxation in branch-and-cut. A second possibility is that some domain D_j contains more than a single value, whereupon it is necessary to enumerate solutions of the constraint store by branching. One can branch on x_j by partitioning D_j into smaller domains, each corresponding to a branch. One could in theory continue to branch until all solutions are enumerated, but as in the branch-and-cut, a new relaxation (in this case, a new set of domains) is generated at each node of the branching tree. Relaxations become tighter as one descends into the tree, since the domains start out smaller and are further reduced through constraint propagation. The search continues until the domains are singletons, or at least one is empty, at every leaf node of the search tree.

The main parallel between this process and the branch-and-cut methods is that both involve *branch and infer*, to use the term of Bockmayr and Kasper (1998). Constraint programming infers in-domain constraints at each node of the branching tree in order to create a constraint store (relaxation). Branch and cut infers linear inequalities at each node in order to generate a continuous relaxation. In the latter case, some of the inequalities in the relaxation appear as inequality constraints of the original integer programming problem and so are trivial to infer, and others are cutting planes that strengthen the relaxation.

Another form of inference that occurs in both constraint programming and integer programming is *constraint learning*, also known as the *nogood generation*. Nogoods are typically formulated when a trial solution (or partial solution) is found to be infeasible or suboptimal. They are constraints designed to exclude the trial solution as the search continues, and perhaps other solutions that are unsatisfactory for similar reasons. Nogoods are closely parallel to the integer programming concept of Benders cuts, which are likewise generated when the solution of the master program yields a suboptimal or infeasible solution. They are less clearly analogous to cutting planes, except perhaps separating cuts, which are generated to "cut off" a nonintegral solution.

Constraint programming and integer programming exploit problem structure primarily in the inference stage. Constraint programmers, for example, invest considerable efforts into the design of filters that exploit the structure of global constraints, just as integer programmers study the polyhedral structure of certain problem classes to generate strong cutting planes.

There are three main differences between the two approaches.

- Branch and cut generally seeks an optimal rather than a feasible solution. This is a minor difference, because it is easy to incorporate optimization into a constraint programming solver. Simply impose a bound on the value of the objective function and tighten the bound whenever a feasible solution is found.
- Branch and cut solves a relaxation at every node with little or no constraint propagation, whereas constraint programming relies more on propagation but does not solve a relaxation. (One might say that it "solves" the constraint store in the special case in which the domains are singletons.) In the branch and cut, solution of the relaxation provides a bound on the optimal value that often allows pruning of the search tree. It can also guide branching, as for instance when one branches on a variable with nonintegral value.
- The constraint store is much richer in the case of the branch-and-cut methods, because it contains linear inequalities rather than simply in-domain constraints. Fortunately, the two types of constraint store can be used simultaneously in the hybrid methods discussed below.

1.3 Constraint satisfaction

Issues that arise in domain reduction and branching search are addressed in the *constraint satisfaction* literature, which is complementary to the optimization literature in interesting ways.

Perhaps the fundamental idea of constraint satisfaction is that of a *consistent* constraint set, which is roughly parallel to that of a convex hull description in integer programming. In this context, "consistent" does not mean feasible or satisfiable. It means that the constraints provide a description of the feasible set that is explicit enough to reduce backtracking, where the amount of reduction depends on the type of consistency maintained. In particular, a *strong n-consistency* (where n is the number of variables) eliminates backtracking altogether, and weaker forms of consistency can do the same under certain conditions.

If an integer/linear programming constraint set is a convex hull description, it in some sense provides an explicit description of the feasible set. Every facet of the convex hull of the feasible set is explicitly indicated. One can solve the problem easily by solving its continuous relaxation. There is no need to use a backtracking search such as branch and bound or branch and cut.

In a similar fashion, a strongly n-consistent constraint set allows one to solve the problem easily with a simple greedy algorithm. For each variable, assign to it the first value in its domain that, in conjunction with the assignments already made, violates no constraint. (A constraint cannot be violated until all of its variables have been assigned.) In general, one will reach a point where no value in the domain will work, and it is necessary to backtrack and try other values for previous assignments. However, if the constraint set is strongly n-consistent, the greedy algorithm always works. The constraint set contains explicit constraints that rule out any partial assignment that cannot be completed to obtain a feasible solution.

Weaker forms of consistency that have proved useful include k-consistency ($k < n$), arc consistency, generalized arc consistency, and bound consistency. These are discussed in Section 2 below. The idea of consistency does not seem to have developed in the optimization literature, although cutting planes and preprocessing techniques serve in part to make a constraint set more nearly consistent.

The constraint satisfaction literature also deals with search strategies, variable and value selection in a branching search, and efficient methods for constraint propagation. These are discussed in Section 3.

1.3.1 Hybrid methods

Constraint programming and optimization have complementary strengths that can be profitably combined.

- Problems often have some constraints that propagate well, and others that relax well. A hybrid method can deal with both kinds of constraints.
- Constraint programming's idea of global constraints can exploit substructure in the problem, while optimization methods for highly structured problem classes can be useful for solving relaxations.

- Constraint satisfaction can contribute filtering algorithms for global constraints, while optimization can contribute relaxations for them.

Due to the advantages of hybridization, constraint programming is likely to become established in the operations research community as part of a hybrid method, rather than as a technique to be used in isolation.

The most obvious sort of hybrid method takes advantages of the parallel between constraint solvers and branch-and-cut methods. At each node of the search tree, constraint propagation creates a constraint store of in-domain constraints, and polyhedral relaxation creates a constraint store of inequalities. The two constraint stores can enrich each other, since reduced domains impose bounds on variables, and bounds on variables can reduce domains. The inequality relaxation is solved to obtain a bound on the optimal value, which prunes the search tree as in the branch-and-cut methods. This method might be called a *branch, infer and relax* (BIR) method.

One major advantage of a BIR method is that one gets the benefits of polyhedral relaxation without having to express the problem in an inequality form. The inequality relaxations are generated within the solver by relaxation procedures that are associated with global constraints, a process that is invisible to the user. A second advantage is that solvers can easily exploit the best known relaxation technology. If a global constraint represents a set of traveling salesman constraints, for example, it can generate a linear relaxation containing the best known cutting planes for the problem. Today, much cutting plane technology goes unused because there is no systematic way to apply it in general-purpose solvers. To overcome this problem, the SCIL system (Althaus, Bockmayr, Elf, Kasper, Jünger and Mehlhorn, 2002) transfers the concept of global constraints from constraint programming to integer programming.

Another promising approach to hybrid methods uses a generalized Benders decomposition. One partitions the variables $[x, y]$ and searches over values of x. The problem of finding an optimal value for x is the *master problem.* For each value v enumerated, an optimal value of y is computed on the assumption that $x = v$; this is the subproblem. In classical Benders decomposition, the subproblem is a linear or nonlinear programming problem, and its dual solution yields a Benders cut that is added to the master problem. The Benders cut requires all future values of x enumerated to be better than v. One keeps adding Benders cuts and re-solving until no more Benders cuts can be generated.

This process can be generalized in a way that unites optimization and constraint programming. The subproblem is set up as a constraint programming problem. Its "dual" can be defined as an *inference dual,* which generalizes the classical dual and can be solved in the course of solving the primal with constraint programming methods. The dual solution yields a generalized Benders cut that is added to the master problem. The master problem is formulated and solved as a traditional optimization problem, such as a mixed integer programming problem. In this way the decomposition scheme combines optimization and constraint programming methods.

BIR and generalized Benders decomposition can be viewed as special cases of a general algorithm that enumerates a series of problem restrictions and solves a relaxation for each. In BIR, the leaf nodes of the search tree correspond to restrictions, and their continuous relaxations are solved. In Benders, the subproblems are problem restrictions, and the master problems are relaxations. This provides a basis for a general scheme for integrating optimization, constraint programming and local search methods Hooker (2003).

1.4 Performance issues

A problem-solving technology should be evaluated with respect to modeling power and development time as well as solution speed.

Constraint programming provides a flexible modeling framework that tends to result in succinct models that are easier to debug than mathematical programming models. In addition, its quasi-procedural approach allows the user to provide the solver information on how best to attack the problem. For example, users can choose global constraints that indicate substructure in the model, and they can define the search conveniently within the model specification.

Constraint programming has other advantages as well. Rather than choosing between two alternative formulations, the modeler can simply use both and significantly speed the solution by doing so. The modeler can add side constraints to a structured model without slowing the solution, as often happens in mathematical programming. Side constraints actually tend to accelerate the solution by improving propagation.

On the other hand, the modeler must be familiar with a sizeable lexicon of global constraints in order to write a succinct model, while integer programming models use only a few primitive terms. A good deal of experimentation may be necessary to find the right model and search strategy for an efficient solution, and the process is more an art than a science.

The computational performance of constraint programming relative to integer programming is difficult to summarize. Constraint programming may be faster when the constraints contain only two or three variables, since such constraints propagate more effectively. When constraints contain many variables, the continuous relaxations of integer programming may become indispensible.

Broadly speaking, constraint programming may be more effective for scheduling problems, particularly resource-constrained scheduling problems, or other combinatorial problems (e.g., problems involving disjunctions) for which the integer programming model tends to be large or have a weak continuous relaxation. This is particularly true if the goal is to find a feasible solution or to optimize a min/max objective, such as makespan.

Integer programming may excel on structured problems that define a well-studied polyhedron, such as the traveling salesman problem. Constraint programming may become competitive when such problems are complicated with side constraints, such as time windows in the case of the traveling salesman problem, or when they are part of a larger model.

It is often said that constraint programming is more effective for "highly-constrained" problems, presumably because constraint propagation is better. Yet this can be misleading, since one can make a problem highly constrained by placing a tight bound on a cost function with many variables. Such a maneuver is likely to make the problem intractable for constraint programming.

The recent trend of combining constraint programming and integer programming makes such comparisons less relevant, since the emphasis shifts to how the strengths of the two methods can complement each other. The computational advantage of integration can be substantial. For example, a hybrid method recently solved product configuration problems 300–600 times faster than either mixed integer programming (CPLEX) or constraint programming (ILOG Solver) (Ottosson and Thorsteinsson, 2000). The problems required selecting each component in some product, such as a computer, from a set of component types; thus one might select a power supply to be any of several wattages. The number of components ranged from 16 to 20 and the number of component types from 20 to 30.

In another study, a hybrid method based on Benders decomposition resulted in even greater speedups for machine scheduling (Jain and Grossmann, 2001; Hooker, 2000; Thorsteinsson, 2001; Bockmayr and Pisaruk, 2003). Each job was scheduled on one of several machines, subject to time windows, where the machines run at different speeds and process each job at a different cost. The speedups increase with problem size and reach five to six orders of magnitude, relative to CPLEX and the ILOG Scheduler, for 20 jobs and 5 machines. Section 4.3 discusses this problem in detail, and Section 4.5 surveys other applications of hybrid methods.

2 Constraints

In this section, we give a more detailed treatment of the declarative and procedural aspects of constraint reasoning.

2.1 *What is a constraint?*

A constraint $c(x_1, \ldots, x_n)$ typically involves a finite number of decision variables $x_1, \ldots, x_n$. Each variable x_j can take a value v_j from a finite set D_j, which is called the *domain* of x_j. The constraint c defines a relation

$R_c \subset D_1 \times \cdots \times D_n$. It is satisfied if $(v_1, \ldots, v_n) \in R_c$. A *constraint satisfaction problem* is a finite set $C = \{c_1, \ldots, c_m\}$ of constraints on a common set of variables $\{x_1, \ldots, x_n\}$. It is *satisfiable* or *feasible* if there exists a tuple $(v_1, \ldots, v_n)$ that satisfies simultaneously all the constraints in C. A *constraint optimization problem* involves in addition, an objective function $f(x_1, \ldots, x_n)$ that has to be maximized or minimized over the set of all feasible solutions. Many constraint satisfaction problems are NP-complete.

2.2 Arithmetic versus symbolic constraints

The concept of "constraint" in constraint programming is very general. It includes classical mathematical constraints like linear or nonlinear equations and inequalities, which are often called *arithmetic constraints*. A crucial feature of constraint programming, however, is that it offers in addition a large variety of other constraints, which we call *symbolic constraints*. In principle, a symbolic constraint could be defined by any relation $R \subset D_1 \times \cdots \times D_n$. However, in order to be useful for constraint programming, it should have a natural declarative reading, and efficient filtering algorithms (see Section 2.6). Symbolic constraints that arise by grouping together a number of simple constraints, each on a small number of variables, into a new constraint involving all these variables together, are called *global* constraints. Global constraints are a key concept of constraint programming. On the declarative level, they increase the expressive power. On the operational side, they improve efficiency.

2.3 Global constraints

We next give an overview of some popular global constraints.

Alldifferent. The constraint `alldifferent`$([x_1, \ldots, x_n])$ states that the variables $x_1, \ldots, x_n$ should take pairwise different values (Régin, 1994; Puget, 1998; Mehlhorn and Thiel, 2000). From a declarative point of view, this is equivalent to a system of disequations $x_i \neq x_j$, for all $1 \leq i < j \leq n$. Grouping together these constraints into one global constraint allows one to make more powerful inferences. For example, consider the system $x_1 \neq x_2$, $x_2 \neq x_3$, $x_1 \neq x_3$, with 0–1 variables x_1, x_2, x_3. Each of these constraints can be satisfied individually, they are *locally consistent* in the terminology of Section 2.4. However, given a global view of all constraints together, one may deduce that the problem is infeasible. A variant of this constraint is the symmetric `alldifferent` constraint (Régin, 1999b).

Element. The `element` constraint `element` (i, l, v) expresses that the i-th variable in a list of variables $l = [x_1, \ldots, x_n]$ takes the value v, i.e., $x_i = v$. Consider an assignment problem where m tasks have to be assigned to n machines. In integer programming, we would use mn binary variables x_{ij}

indicating whether or not task i is assigned to machine j. If c_{ij} is the corresponding cost, the objective function is $\sum_{i=1}^{m}\sum_{j=1}^{n} c_{ij}x_{ij}$. In constraint programming, one typically uses m domain variables x_i with domain $D_i = \{1, \ldots, n\}$. Note that $x_i = j$ if and only if $x_{ij} = 1$. Using constraints `element` $(x_i, [c_{i1}, \ldots, c_{in}], c_i)$, with domain variables c_i, the objective function can be stated as $\sum_{i=1}^{m} c_i$.

Cumulative. The `cumulative` constraint has been introduced to model scheduling problems (Aggoun and Beldiceanu, 1993; Caseau and Laburthe, 1997; Baptiste, Pape and Nuijten, 2001; Beldiceanu and Carlsson, 2002). Suppose there are n tasks. Task j has starting time s_j, duration d_j and needs r_j units of a given resource. The constraint

$$\texttt{cumulative}\,([\mathtt{s}_1, \ldots, \mathtt{s}_n], [\mathtt{d}_1, \ldots, \mathtt{d}_n], [\mathtt{r}_1, \ldots, \mathtt{r}_n], \mathtt{l}, \mathtt{e})$$

states that the tasks have to be executed in such a way that the global resource limit l is never exceeded and e is the end of the schedule [see Fig. 1(a)].

Diffn. The constraint

$$\texttt{diffn}([[o_{11}, \ldots, o_{1n}, l_{11}, \ldots, l_{1n}], \ldots, [o_{m1}, \ldots o_{mn}, l_{m1}, \ldots l_{mn}]])$$

states that m rectangles in n-dimensional space should not overlap (Beldiceanu and Contejean, 1994; Beldiceanu and Carlsson, 2001). Here, o_{ij} gives the origin and l_{ij} the length of the rectangle i in dimension j, [see Fig. 1(b)]. Applications of this constraint include resource allocation and packing problems. Beldiceanu, Qi and Thiel (2001) consider nonoverlapping constraints between convex polytopes.

Cycle. The `cycle` constraint allows one to define cycles in a directed graph (Beldiceanu and Contejean, 1994; Caseau and Laburthe, 1997; Bourreau,

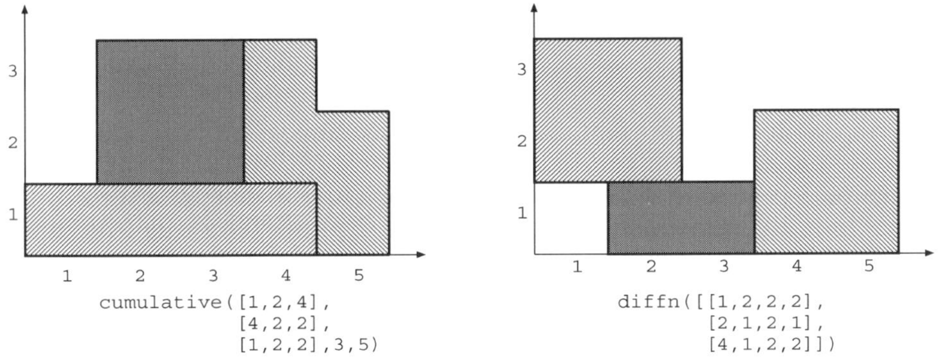

(a) (b)

Fig. 1. (a) Cumulative constraint (b) Diffn constraint.

1999). For each node in the graph, one introduces a variable s_i whose domain contains the nodes that can be reached from node i. The constraint `cycle (k, [s1, ..., sn])` holds if the variables s_i are instantiated in such a way that precisely k cycles are obtained. A typical application of this constraint are vehicle routing problems.

Cardinality. The `cardinality` constraint restricts the number of times a value is taken by a number of variables (Beldiceanu and Contejean, 1994; Régin 1996; Régin and Puget, 1997; Régin, 1999a). Application areas include personnel planning and sequencing problems. An extension of `cardinality` is the sequence constraint that allows one to define complex patterns on the values taken by a sequence of variables (Beldiceanu, Aggoun and Contejean, 1996).

Sortedness. The `sort` constraint `sort` $(x_1, \ldots, x_{n;} y_1, \ldots, y_n)$ expresses that the n-tuple $(y_1, \ldots, y_n)$ is obtained from the n-tuple $(x_1, \ldots, x_n)$ by sorting the elements in nondecreasing order (Bleuzen-Guernalec and Colmerauer, 2000; Mehlhorn and Thiel, 2000). It was introduced in (Older, Swinkels and van Emden, 1995) to model and solve job-shop scheduling problems. Zhou (1997) considered a variant with $3n$ variables that makes explicit the permutation linking the x's and y's.

Flow. The `flow` constraint can be used to model flows in generalized networks (Bockmayr, Pisaruk and Aggoun, 2001). In particular, it can handle conversion nodes that arise when modeling production processes. A typical application area is the supply chain optimization.

This list of global constraints is not exhaustive. Various other constraints have been proposed in the literature, e.g., (Régin and Rueher, 2000; Beldiceanu, 2001). A classification scheme for global constraints that subsumes a variety of the existing constraints (but not all of them) is introduced in Beldiceanu (2000).

2.4 *Local consistency*

From a declarative point of view, a constraint $c(x_1, \ldots, x_n)$ defines a relation on the Cartesian product $D_1 \times \cdots \times D_n$ of the corresponding domains. In general, it is computationally prohibitive to determine directly the tuples $(v_1, \ldots, v_n)$ that satisfy the constraint. Typically, constraint programming systems try to *filter* the domains D_j, i.e., to remove values v_j that cannot occur in a solution.

A constraint $c(x_1, \ldots, x_n)$ is *generalized arc consistent* (Mohr and Masini, 1988) if for any variable x_i and any value $v_i \in D_i$, there exist values $v_j \in D_j$, for all $j \neq i$, such that $c(v_1, \ldots, v_n)$ holds. Generalized arc consistency is a basic concept in constraint reasoning. Stronger notions of consistency have been introduced in the literature, like path consistency, k-consistency, or (i,j)-consistency. Freuder (1985) introduced *(i,j)-consistency* for binary constraints. Given values for i variables, satisfying the constraints on those variables, and given any other j (or fewer) variables, there exist values for those j variables such that the $i+j$ values taken together satisfy all constraints on the $i+j$ variables. With this definition, *k-consistency* is the same as $(k-1,1)$-consistency. *Path consistency* corresponds to 3- resp. (2, 1)-consistency, and *arc consistency* to 2- resp. (1,1)-consistency. *Strong k-consistency* is defined as j-consistency, for all $j \leq k$.

A problem can be made arc consistent by removing inconsistent values from the variable domains, i.e., values that cannot appear in any solution. Achieving k-consistency for $k \geq 3$ requires to remove tuples of values (instead of values) from $D_1 \times \cdots \times D_n$. The corresponding algorithms become rather expensive. Therefore, their use in constraint programming is limited. Recently, consistency notions have been introduced that are stronger than arc consistency, but still use only domain filtering (as opposed to filtering the Cartesian product), see Debruyne and Bessière (2001), Prosser, Stergiou and Walsh (2000).

Bound consistency is a restricted form of generalized arc consistency, where we reason only on the bounds of the variables. Assume that D_j is totally ordered, typically $D_j \subset \mathbb{Z}$. A constraint $c(x_1, \ldots, x_n)$ is *bound consistent* (Puget, 1998) if for any variable x_i and each bound value $v_i \in \{\min(D_i), \max(D_i)\}$, there exist values $v_j \in [\min(D_i), \max(D_i)]$, for all $j \neq i$, such that $c(v_1, \ldots, v_n)$ holds.

Most work on constraint satisfaction problems in the artificial intelligence community has been done on binary constraints. However, recently, the nonbinary case has been receiving more and more attention (Bessière, 1999; Stergiou and Walsh, 1999b; Zhang and Yap, 2000). Bacchus, Chen, van Beek and Walsh (2002) study two transformations from nonbinary to binary constraints, the dual transformation and the hidden (variable) transformation, and formally compare local consistency techniques applied to the original and the transformed problem.

2.5 *Constraint propagation*

In general, a constraint problem contains many constraints. When achieving an arc consistency for one constraint through filtering, other constraints, which were consistent before, may become inconsistent. Therefore, filtering has to be applied repeatedly to constraints that share common variables, until no further domain reduction is possible. This process is called *constraint propagation*.

The classical method for achieving arc consistency is the algorithm AC 3 (Mackworth, 1977b). Consider a constraint satisfaction problem C with unary constraints $c_i(x_i)$ and binary constraints $c_{ij}(x_i, x_j)$, where $i<j$. Let arc(C)

denote the set of all ordered pairs (i, j) and (j, i) such that there is a constraint $c_{ij}(x_i, x_j)$ in C.

```
Algorithm AC-3 (Mackworth77)
  for i ← 1 to n do D_i ← {v ∈ D_i | c_i(v)};
  Q ← {(i,j) | (i,j) ∈ arc(C)};
  while Q not empty do
    select and delete any arc (i,j) from Q;
    if revise(i,j) then Q ← Q ∪ {(k,i) | (k,i) ∈ arc(C), k ≠ i, k ≠ j};
  end while
end
```

The procedure `revise` (i, j) removes all values $v \in D_i$ for which there is no corresponding value $w \in D_j$ such that $c_{ij}(v, w)$ holds. It returns `true` if at least one value can be removed from D_i, and `false` otherwise. If e is the number of binary constraints and d a bound on the domain size, the complexity of AC 3 is $O(ed^3)$.

Various extensions and refinements of the original algorithm AC3 have been proposed. Some of these algorithms achieve the optimal worst case complexity $O(ed^2)$, others have an improved average case complexity;

- AC 4 (Mohr and Henderson, 1986),
- AC 5 (van Hentenryck and Graf, 1992),
- AC 6 (Bessière, 1994),
- AC 7 (Bessière, Freuder and Régin, 1999),
- AC 2000 and AC 2001 (Bessière and Régin, 2001, see also Zhang and Yap, 2001).

Again these papers focus on binary constraints. Extensions to the non-binary case, i.e., generalized arc consistency, are discussed in Mackworth (1977a), Mohr and Masini (1988), Bessière and Régin (1997, 2001).

2.6 Filtering algorithms for global constraints

Local consistency techniques for linear arithmetic constraints may look similar to preprocessing in integer programming. Symbolic constraints in constraint programming, however, come with their own filtering algorithms. These are specific to the constraint and therefore can be much more efficient than the general techniques presented in the previous section. Efficient filtering algorithms are a key reason for the success of constraint programming. They make it possible to embed problem-specific algorithms, e.g., from graph theory or scheduling, into a general purpose solver. We illustrate this on two examples.

2.6.1 *Alldifferent*

First we discuss a filtering algorithm for the `alldifferent` constraint (Régin, 1994). Let $x_1, \ldots, x_n$ be the variables and $D_1, \ldots, D_n$ be the corresponding domains. We construct a bipartite graph G to represent the problem in graph-theoretic terms. For each variable x_j we introduce a node on the left, and for each value $v_j \in D_1 \cup \cdots \cup D_n$ a node on the right. There is an edge between x_i and v_j iff $v_j \in D_i$. Then the constraint `alldifferent` $([x_1, \ldots, x_n])$ is satisfiable iff the graph G has a matching covering all the variables.

Our goal is to remove redundant edges from G. Suppose we are given a matching M in G covering all the variables. Matching theory tells us that an edge $(x, v) \notin M$ belongs to some maximum matching iff it belongs either to an even alternating cycle or an even alternating path starting in a free node. A node is free if it is not covered by M. An alternating path or cycle is a simple path or cycle whose edges alternately belong to M and its complement. We orient the graph by directing all edges in M from values to variables, and all edges not in M from variables to values. In the directed version of G, the first kind of edge is an edge in some strongly connected component, and the second kind of edge is an edge that is reachable from a free node. This yields a linear-time algorithm for removing redundant edges. If no matching M is known, the complexity becomes $O(\sqrt{n} \cdot m)$, where m is the number of edges in G.

Puget (1998) devised an $O(n \cdot \log n)$ algorithm for *bound* consistency of `all different`, a simplified and faster version was obtained in Mehlhorn and Thiel (2000). Stergiou and Walsh (1999a) compare different notions of consistency for `alldifferent`, see also van Hoeve (2001).

2.6.2 *Cumulative*

Next we give a short introduction to constraint propagation techniques for resource constraints in scheduling. There is an extensive literature on this subject. We consider here only the simplest example of a one-machine resource constraint in the non-preemptive case. For a more detailed treatment and a guide to the literature, we refer to Baptiste et al. (2001).

We are given a set of *activities* $\{A_1, \ldots, A_n\}$ that have to be executed on a single *resource* R. For each activity, we introduce three domain variables, $start(A_i)$, $end(A_i)$, $proc(A_i)$, that represent the start time, the end time, and the processing time, respectively. The processing time is the difference between the end and the start time, $proc(A_i) = end(A_i) - start(A_i)$. Given an initial release date r_i and a deadline d_i, activity A_i has to be performed in the time interval $[r_i, d_i - 1]$. During propagation, these bounds will be updated so that they always denote the current *earliest starting time* and *latest end time* of activity A_i.

Different techniques can be applied to filter the domains of the variables $start(A_i)$ and $end(A_i)$ (Baptiste et al., 2001):

Time tables. Maintain bound consistency on the formula $\sum_{i=1}^{n} x(A_i, t) \leq 1$, for any time t. Here $x(A_i, t)$ is a 0–1 variable indicating whether or not activity A_i executes at time t.

Disjunctive constraint propagation. Maintain bound consistency on the formula

$$\left[end(A_i) \leq start(A_j)\right] \vee \left[end(A_j) \leq start(A_i)\right]$$

Edge finding. This is one of the key techniques for resource constraints. Given a set of activities Ω, let r_Ω, d_Ω and p_Ω, respectively, denote the smallest earliest starting time, the largest latest end time, and the sum of the minimal processing times of the activities in Ω. Let $A_i \ll A_j$ mean that A_i executes before A_j, and $A_i \ll \Omega$ (resp. $A_i \gg \Omega$) that A_i executes before (resp. after) all activities in Ω. Then the following inferences can be performed:

$$\begin{aligned}
&\forall \Omega, \forall A_i \notin \Omega\left[d_{\Omega \cup \{A_i\}} - r_\Omega < p_\Omega + p_i\right] \Rightarrow [A_i \ll \Omega] \\
&\forall \Omega, \forall A_i \notin \Omega\left[d_\Omega - r_{\Omega \cup \{A_i\}} < p_\Omega + p_i\right] \Rightarrow [A_i \gg \Omega] \\
&\forall \Omega, \forall A_i \notin \Omega[A_i \ll \Omega] \Rightarrow \left[end(A_i) \leq \min_{\theta \neq \Omega' \subseteq \Omega}(d_{\Omega'} - p_{\Omega'})\right] \\
&\forall \Omega, \forall A_i \notin \Omega[A_i \ll \Omega] \Rightarrow \left[start(A_i) \geq \max_{\theta \neq \Omega' \subseteq \Omega}(r_{\Omega'} + p_{\Omega'})\right]
\end{aligned}$$

Edge-finding reasons on sets of activities. Given n activities, a priori $O(n2^n)$ pairs (A_i, Ω) have to be considered. Carlier and Pinson (1990) present an algorithm that improves the time bounds in $O(n^2)$.

Not-first, not-last. The previous techniques try to determine whether an activity A_i *must* be the first (or the last) within a set of activities $\Omega \cup \{A_i\}$. Alternatively, one may try to find out whether A_i *can* be the first (or last) activity in $\Omega \cup \{A_i\}$. If this is not the case, one may deduce that A_i cannot start before the end of at least one activity in Ω (or that A_i cannot end after the start of at least one activity in Ω) which leads to another set of inference rules.

2.7 Modeling in constraint programming: an illustrative example

To illustrate the variety of models that may exist in constraint programming, we consider the reconstruction of pictures in discrete tomography (Bockmayr, Kasper and Zajac, 1998). While in integer programming we are looking for models that provide tight linear relaxations, constraint programming aims at models that allow for strong filtering and propagation between different constraints.

A *two-dimensional binary picture* is given by a binary matrix $X \in \{0, 1\}^{m \times n}$. Intuitively, a pixel is black iff the corresponding matrix element is 1. A binary picture X is:

- *Horizontally convex*, if the set of 1's in each row is convex, i.e., $x_{ij_1} = x_{ij_2} = 1$ implies $x_{ij} = 1$, for all $1 \leq i \leq m, 1 \leq j_1 < j < j_2 \leq n$.
- *Vertically convex*, if the set of 1's in each column is convex, i.e. $x_{i_1j} = x_{i_2j} = 1$ implies $x_{ij} = 1$, for all $1 \leq i_1 < i < i_2 \leq m, 1 \leq j \leq n$.
- *Connected* or a *polyomino*, if the set of 1's in the matrix is connected with respect to the adjacency relation where each matrix element is adjacent to its two vertical and horizontal neighbors.

Given two vectors $h = (h_1 \ldots, h_m) \in \mathbb{N}^m$, $v = (v_1, \ldots, v_n) \in \mathbb{N}^n$, the *reconstruction problem* of a binary picture from *orthogonal projections* consists in finding $X \in \{0, 1\}^{m \times n}$ such that

- $\sum_{j=1}^{n} x_{ij} = h_i$, for $i = 1, \ldots, m$ (horizontal projections)
- $\sum_{i=1}^{m} x_{ij} = v_j$, for $j = 1, \ldots, n$ (vertical projections)

The complexity of the reconstruction problem depends on the additional properties that are required for the picture (Woeginger, 2001).

	$v+h$ convex	v convex	h convex	No restriction
Connected	P	NP-complete	NP-complete	NP-complete
No restriction	NP-complete	NP-complete	NP-complete	P

2.7.1 0–1 Models

The above properties may be modeled in many different ways. In integer linear programming, one typically uses 0–1 variables x_{ij}. The binary picture $X \subseteq \{0, 1\}^{m \times n}$ with horizontal and vertical projections $h \in \mathbb{N}^m, v \in N^n$ is horizontally convex iff the following set of linear inequalities is satisfied:

$$h_i \cdot x_{ik} + \sum_{j=k+h_i}^{n} x_{ij} \leq h_i, \quad \text{for all } 1 \leq i \leq m, 1 \leq k \leq n.$$

X is vertically convex iff

$$v_j \cdot x_{kj} + \sum_{i=k+v_j}^{m} x_{ij} \leq v_i, \quad \text{for all } 1 \leq k \leq m, 1 \leq j \leq n.$$

The connectivity of a horizontally convex picture can be expressed as follows:

$$\sum_{j=k}^{k+h_i-1} x_{ij} - \sum_{j=k}^{k+h_i-1} x_{i+1j} \leq h_i - 1, \quad \text{for all } 1 \leq i \leq m-1,$$
$$1 \leq k \leq n - h_i + 1.$$

This leads to $O(mn)$ variables and constraints.

2.7.2 Finite domain models

In finite domain constraint programming, 0–1 variables are usually avoided. For each row resp. column in the given $m \times n$-matrix, we introduce a finite domain variable

- $x_i \in \{1, \ldots, n\}$, for all $i = 1, \ldots, m$, resp.
- $y_j \in \{1, \ldots, m\}$, for all $j = 1, \ldots, n$.

If $h = (h_1, \ldots, h_n)$ and $v = (v_1, \ldots, v_m)$ are the horizontal and vertical projections, then $x_i = j$ says that the block of h_i 1's for row i starts at column j. Analogously, $y_j = i$ expresses that the block of v_j 1's for column j starts in row i.

Conditional propagation. To ensure that the values of the variables x_i and y_j are compatible with each other, we impose the constraints

$$x_i \leq j < x_i + h_i \Leftrightarrow y_j \leq i < y_j + v_j, \quad \text{for all } i = 1, \ldots, m, j = 1, \ldots, n.$$

Such constraints may be realized by *conditional propagation* rules of the form `if` C then P, saying that, as soon as the remaining values for the variables satisfy the condition C, the constraints P become active. This models horizontal/vertical projections and convexity. To ensure connectivity, we have to forbid that the block in row $i+1$ ends left of the block in row i or that the block in row $i+1$ starts right of the block in row i. Negating this disjunction yields the linear inequalities

$$x_i \leq x_{i+1} + h_{i+1} - 1 \text{ and } x_{i+1} \leq x_i + h_i, \quad \text{for all } i = 1, \ldots, m-1.$$

The above constraints are sufficient to model the reconstruction problem. However, we may try to improve propagation by adding further constraints, which are redundant from the declarative point of view, but provide additional filtering techniques on the procedural side. *Adding redundant constraints* is a standard technique in constraint programming. Again, there is a problem-dependent tradeoff between the cost of the filtering algorithm and the domain reductions that are obtained.

Cumulative. For example, we may use the `cumulative` constraint. We identify each horizontal block in the image with a task $(x_i, h_i, 1)$ which starts at time x_i, has duration h_i, and requires 1 resource unit. For each column j, we introduce an additional task $(j, 1, m - v_j + 1)$, which starts at time j, has duration 1, and uses $m - v_j + 1$ resource units. These complementary tasks model vertical projections numbers. The capacity of the resource is $m+1$ and all the tasks end before time $n+1$. Thus, the constraint

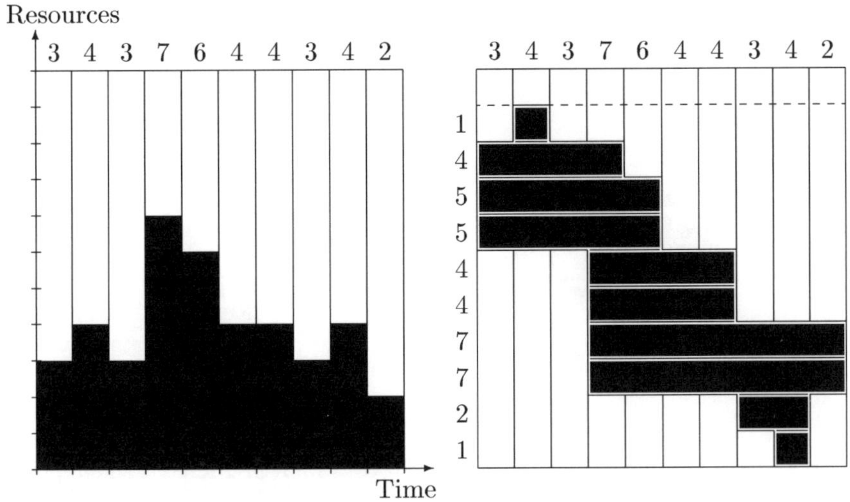

Fig. 2. Cumulative constraint in discrete tomography.

$$\texttt{cumulative}\begin{pmatrix} [x_1, \ldots, x_m, & 1, \ldots, n], \\ [h_1, \ldots, h_m, & 1, \ldots, 1], \\ [1, \ldots, 1, & m - v_1 + 1, \ldots, m - v_n + 1], \\ m + 1, & n + 1 \end{pmatrix}$$

models horizontal/vertical projection numbers, and horizontal convexity, see Fig. 2.

Diffn. Another possibility is to use the `diffn` constraint. Here, we look at polyomino reconstruction as packing of two-dimensional rectangles. We model the problem by an extended version of the `diffn` constraint (Beldiceanu and Contejean, 1994), involving four arguments. In the first argument, we define the rectangles. For each black horizontal block in the picture, we introduce a rectangle

$$R_i = [x_i, i, h_i, 1].$$

with origins (x_i, i) and lengths $(h_i, 1)$, $i = 1, \ldots, m$. To model vertical convexity, we introduce $2n$ additional rectangles

$$S_{1,j} = \left[j, 0, 1, l_{j,1}\right], \quad S_{2,j} = \left[j, m + 1 - l_{j,2}, 1, l_{j,2}\right].$$

which correspond to two white blocks in each column. The variables l_{jk} define the height of these rectangles. To ensure that each white block has a nonzero

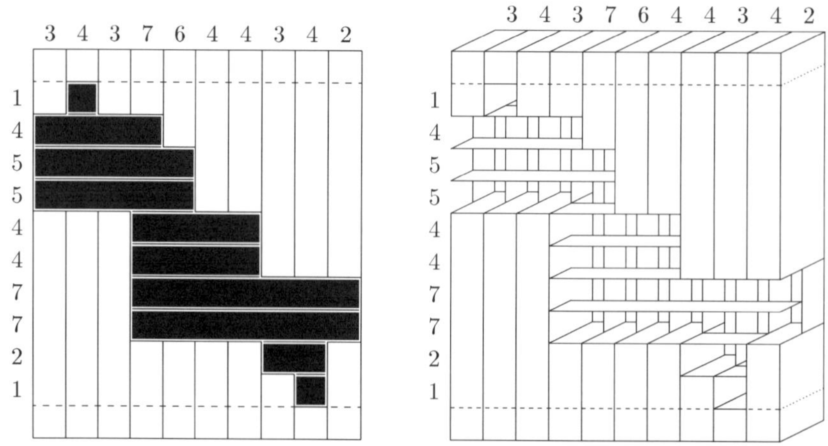

Fig. 3. Two- and three-dimensional `diffn` constraint in discrete tomography.

surface, we introduce two additional rows 0 and $m+1$, see Fig. 3 for an illustration.

The second argument of the `diffn` constraint says that the total number of rows and columns is $m+2$ resp. n. In the third argument, we express that the distance between the two white rectangles in column j has to be equal to v_j. To model connectivity, we state in the fourth argument that each pair of successive rectangles has a contact in at least one position. This is represented by the list $[[1, 2, c_1], \ldots, [m-1, m, c_{m-1}]]$, with domain variables $c_i \geq 1$. Thus, the whole reconstruction problem can be modeled by a *single* `diffn` constraint:

$$\mathtt{diffn}\begin{pmatrix} [R_1, \ldots, R_m, S_{1,1}, \ldots, S_{1,n}, S_{2,1}, \ldots, S_{2,n}], \\ [n, m+2], \\ [[m+1, m+n+1, v_1], \ldots, [m+n, m+2*n, v_n]], \\ [[1, 2, c_1], \ldots, [m-1, m, c_{m-1}]] \end{pmatrix}$$

Note that this model involves only the row variables x_i, not the column variables y_j. It is also possible to use row and column variables simultaneously. This leads to another model based on a single `diffn` constraint in three dimensions, see Fig. 3. Here, the third dimension is used to ensure that row and column variables define the same picture.

3 Search

Filtering algorithms reduce the domains of the variables. In general, this is not enough to determine a solution. Therefore, filtering is typically embedded into a search algorithm. Whenever, after filtering, the domain D of a variable x contains more than one value, we may split D into nonempty subdomains $D = D_1 \cup \cdots \cup D_k, k \geq 2$, and consider k new problems $C \cup \{x \in D_1\}, \ldots, C \cup \{x \in D_k\}$. Assuming $D_i \neq D$, we may apply filtering again in order to get further domain reductions. Alternatively, we may branch on a constraint like $x + y \leq c$ or $x + y \geq c + 1$. By repeating this process, we obtain a *search tree*. There are many different ways to construct and to traverse this tree.

The basic search algorithm in constraint programming is *backtracking*. Variables are instantiated one after the other. As soon as all variables of some constraint have been instantiated, this constraint is evaluated. If it is satisfied, instantiation goes on. Otherwise, at least one variable becomes uninstantiated and a new value is tried.

There are many ways to improve standard backtracking. Following (Dechter, 1992), we may distinguish between look-ahead and look-back schemes. *Look-ahead* schemes are invoked before extending the current partial solution. The most important techniques are strategies for selecting the next variable or value and maintaining local consistency in order to reduce the search space. *Look-back* schemes are invoked when one has encountered a dead-end and backtracking becomes necessary. This includes heuristics how far to backtrack (back-jumping) or what constraints to record in order to avoid that the same conflict rises again later in the search (nogoods) (Dechter, 1990; Prosser, 1993). We focus here on the look-ahead techniques that are widely used in constraint programming. A comprehensive survey on look-back methods can be found in Dechter and Frost (2002). For possible combinations of look-forward and look-back schemes, we also refer to Jussien, Debruyne and Boizumault (2000), Chen and van Beek (2001).

3.1 Variable and value ordering

In many cases, the domains $D_1, \ldots, D_k$ used in splitting are singleton sets that correspond to the different values in the domain D. The process of assigning to the variables their possible values and constructing the corresponding search tree is often called *labeling*. During labeling, two important decisions have to be made:

- In which order should the variables be instantiated (variable selection)?
- In which order should the values be assigned to a selected variable (value selection)?

These orderings may be defined *statically,* i.e. before starting the search, or *dynamically* by taking into account the current state of the search tree. Dynamic variable selection strategies may be the following:

- Choose the variable with the smallest domain ("first fail").

- Choose the variable with the smallest domain that occurs in most of the constraints ("most constrained").
- Choose the variable which has the smallest/largest lower/upper bound on its domain.

Value orderings include:

- Try first the minimal value in the current domain.
- Try first the maximal value in the current domain.
- Try first some value in the middle of the current domain.

Variable and value selection strategies have a great impact on the efficiency of the search, see e.g., Gent, MacIntyre, Prosser, Smith and Walsh (1996), Prosser (1998). Finding good variable or value ordering heuristics is often crucial when solving hard problems.

3.2 Complete search

Whenever we reach a new node of the search tree, typically by assigning a value to a variable, filtering and constraint propagation may be applied again. Depending on the effort we want to spend at the node, we may enforce different levels of consistency.

Forward checking (FC) performs arc consistency between the variable x that has just been instantiated and the uninstantiated variables. Only those values in the domain of an uninstantiated variable are maintained that are compatible with the current choice for x. If the domain of a variable becomes empty, backtracking becomes necessary. Forward-checking for nonbinary constraints is described in Bessière, Meseguer, Freuder and Larrosa (1999), while a general framework for extending forward checking is developed in Bacchus (2000).

Full look-ahead or *Maintaining Arc Consistency* (MAC) performs arc consistency for all pairs of uninstantiated variables (in addition to forward checking), see Sabin and Freuder (1997) for an improved version. *Partial look-ahead* is an intermediate form, where only one direction of each edge in the constraint graph is considered.

Again there is a tradeoff between the effort needed to enforce local consistency and the corresponding pruning of the search tree. For a long time, it was believed that FC or FC with Conflict-Directed Backjumping (CBJ) (Prosser, 1993), together with the first-fail heuristics, is the most efficient strategy for solving constraint satisfaction problems. (Sabin and Freuder, 1994, Bessière and Régin, 1996) argued that MAC is more efficient than FC (or FC-CBJ) on hard problems and justified this by a number of empirical results.

An important issue is *symmetry breaking* during search. Various techniques have been proposed in the literature, we refer to McDonald and Smith (2002), Puget (2002) for some recent work.

3.3 Heuristic search

For many practical problems, complete search methods may be unable to find a solution. In such cases, one may use heuristics in order to guide the search towards regions of the search space that are likely to contain solutions.

Limited discrepancy search (LDS) (Harvey and Ginsberg, 1995) is based on the idea that a heuristic that normally leads to a solution may fail only because a small number of wrong choices are made. To correct these mistakes, LDS searches paths in the tree that follow the heuristic almost everywhere, except in a limited number of cases where a different choice is made. These are called *discrepancies*. *Depth-bounded discrepancy search* (DDS) is a refinement of LDS that biases search to discrepancies high in the tree (Walsh, 1997). It uses an iteratively increasing depth-bound. Discrepancies below this bound are forbidden.

Interleaved depth-first search (IDFS) (Meseguer, 1997) is another strategy to prevent standard depth-first search to fall into mistakes. IDFS searches in parallel several subtrees, called *active*, at certain levels of the trees, called *parallel*. The current active tree is searched depth-first until a leaf is found. If this is a solution, search terminates. Otherwise, the state of the current tree is recorded so that it can be resumed later, and another active subtree is considered. There are two variants of this method. In *Pure* IDFS, all levels are parallel and all subtrees are active. *Limited* IDFS considers a limited number of active subtrees and a limited number of parallel levels, typically at the top of the tree. An experimental comparison of DDS and IDFS can be found in Meseguer and Walsh (1998).

Another possibility to overcome the problem of making wrong choices early in the backtrack search is *randomization* together with *restart* techniques (Gomes, Selman, and Kautz, 1998, Ruan, Horvitz, and Kautz, 2002). Here, a certain amount of randomness is introduced into the search strategy. If the algorithm does not terminate within a given number of backtracks, which is called the *cutoff* value, the run is halted and restarted with a new random seed. Randomized search algorithms can be made complete by using learning techniques that record and consult all nogoods discovered during the search. Combining randomization with learning has been particularly successful in propositional satisfiability solvers (Zhang and Malik, 2002).

3.4 Constraint programming languages and systems

As has been pointed out already in Section 1.1, the term "programming" may have two different meanings, see also Lustig and Puget (2001):

- Mathematical programming, i.e., solving mathematical optimization problems.
- Computer programming, i.e., writing computer programs in a programming language.

Constraint programming makes contributions on both the sides. On the one hand, it provides a new approach to solving discrete optimization problems. On the other hand, the constraint solving techniques are embedded into a high-level programming language so that they become easily accessible even to a nonexpert user.

There are different ways of integrating constraints into a programming language. Early work in this direction was done by Laurière (1978) in the language ALICE. Constraint programming as it is known today first appeared in the form of *constraint logic programming*, with logic programming as the underlying programming language paradigm (Colmerauer, 1987; Jaffar and Lassez, 1987). In logic programming (Prolog), search and backtracking are built into the language. This greatly facilitates the development of search algorithms. Constraint satisfaction techniques have been studied in artificial intelligence since the early 70s. They were first introduced into logic programming in the CHIP system (Dincbas, van Hentenryck, Simonis, Aggoun and Graf, 1988; van Hentenryck, 1989). Puget (1994) showed that the basic concepts of constraint logic programming can also be realized in a C++ environment, which lead to the development of ILOG Solver. Another possible approach is the *concurrent constraint programming* paradigm (cc) (Saraswat, 1993), with systems such as cc(FD) (van Hentenryck, Saraswat and Deville, 1998) or Oz (Smolka, 1995).

The standard way to develop a constraint program is to use the host programming language in order to build the constraint model and to specify the search strategy. In recent years, new declarative languages have been proposed on top of existing constraint programming systems, which allow one to define both the constraints and the search strategy in a very high-level way. Examples include OPL (van Hentenryck, 1999), PLAM (Barth and Bockmayr, 1998), MOSEL (Colombani and Heipcke, 2002), or more specifically for search SALSA (Laburthe and Caseau, 2002). These languages provide high-level algebraic and set notation, similarly to algebraic modeling languages in mathematical programming. In addition to arithmetic constraints, they also support the different symbolic constraints that are typical for constraint programming. Furthermore, they allow the user to specify search procedures in a high-level way.

As an example, we present an OPL model for solving a job-shop scheduling problem (van Hentenryck, Michel, Perron and Régin, 1999), see Fig. 4. Part 1 of the model contains various declarations concerning machines, jobs, tasks, the duration of the tasks, and the resources they require. Part 2 declares the activities and resources of the problem, which are predefined concepts in OPL. In Part 3, symbolic precedence and resource constraints are stated. Finally, the search strategy is specified in Part 4. It uses limited discrepancy search and a ranking of the resources.

While languages such as OPL provide a very elegant modeling and solution environment, particular problems that require specific solution strategies and heuristics may not be expressible in such a high-level framework. In that case,

```
int nbMachines = ...;
range Machines 1..nbMachines;
int nbJobs = ...;
range Jobs 1..nbJobs;
int nbTasks = ...;
range Tasks 1..nbTasks;
Machines resource[Jobs,Tasks] = ...;
int+ duration[Jobs,Tasks] = ...;
int totalDuration = sum(j in Jobs, t in Tasks) duration[j,t];

scheduleHorizon = totalDuration;
Activity task[j in Jobs, t in Tasks](duration[j,t]);
Activity makespan(0);
UnaryResource tool[Machines];

minimize
   makespan.end
subject to {
   forall(j in Jobs)
      task[j,nbTasks] precedes makespan;
   forall(j in Jobs)
      forall(t in 1..nbTasks-1)
         task[j,t] precedes task[j,t+1];
   forall(j in Jobs)
       forall(t in Tasks)
          task[j,t] requires tool[resource[j,t]];
};

search {
   LDSearch() {
      forall(r in Machines ordered by increasing localSlack(tool[r]))
         rank(tool[r]);
   }
}
```

Fig. 4. A job-shop model in OPL (van Hentenryck et al., 1999).

the user has to work directly with the underlying constraint programming system.

We finish this section with a short overview of some current constraint programming systems, see Table 2. While this information has been compiled to the best of our knowledge, we cannot guarantee its correctness and completeness. For a more detailed description, we refer to the corresponding web sites.

4 Hybrid methods

Hybrid methods have developed over the last decade in both the constraint programming and the optimization communities.

Constraint programmers initially conceived hybrid methods as double modeling approaches, in which some constraints are given both a constraint programming and a mixed integer programming formulation. The two formulations are linked and pass domain reductions and/or infeasibility information to each other. Little and Darby-Dowman (1995) were early

Table 2.
Constraint programming systems

System	Availability	Constraints	Language	Web site
B-prolog	Commercial	Finite domain	Prolog	www.probp.com
CHIP	Commercial	Finite domain, Boolean, Linear rational, Hybrid	Prolog, C, C++	www.cosytec.com
Choco	Free	Finite domain	Claire	www.choco-constraints.net
Eclipse	Free for nonprofit	Finite domain, Hybrid	Prolog	www.icparc.ic.ac.uk/eclipse/
GNU Prolog	Free	Finite domain	Prolog	gnu-prolog.inria.fr
IF/Prolog	Commercial	Finite domain Boolean, Linear arithmetic	Prolog	www.ifcomputer.co.jp
ILOG	Commercial	Finite domain, Hybrid	C++, Java	www.ilog.com
NCL	Commercial	Finite domain		www.enginest.com
Mozart	Free	Finite domain	Oz	www.mozart-oz.org
Prolog IV	Commercial	Finite domain, Linear/nonlinear interval arithmetic	Prolog	prologianet.univ-mrs.fr
Sicstus	Commercial	Finite domain, Boolean, linear real/rational	Prolog	www.sics.se/sicstus/

proponents of double modeling, along with Rodošek, Wallace and Hajian (1997) and Wallace, Novello and Schimpf (1997), who adapted the constraint logic programming system ECLiPSe so that linear constraints could be dispatched to commercial linear programming solvers (CPLEX and XPRESS-MP). Double modeling requires some knowledge of which formulation is better for a given constraint, an issue studied by Darby-Dowman and Little (1998) and others. The constraints community also began to recognize the parallel between constraint solvers and mixed integer solvers, as evidenced by Bockmayr and Kasper (1998).

In more recent work, Heipcke (1998, 1999) proposed several variations of double modeling. Focacci, Lodi, and Milano (1999a,b, 2000) and Sellmann (2002) adapted several optimization ideas to a constraint programming context, such as reduced cost variable fixing and Refalo (1999) integrated piecewise linear modeling through "tight cooperation" between constraint propagation and a linear relaxation. ILOG's OPL Studio (van Hentenryck, 1999) and Dash's Mosel system (Colombani and Heipcke, 2002) are commercial modeling systems that can invoke both constraint programming and integer programming solvers and pass a certain amount of information from one to the other.

The mathematical programming community initially conceived hybrid methods as generalizations of branch and cut or a logic-based form of Benders decomposition. Drawing on the work of Beaumont (1990), Hooker (1994) and Hooker and Osorio (1999) proposed mixed logical/linear programming (MLLP) as an extension of mixed integer/linear programming (MILP). Several investigators applied similar hybrid methods to process design and scheduling problems (Cagan, Grossmann and Hooker 1997; Grossmann, Hooker, Raman and Yan, 1994; Pinto and Grossmann, 1997; Raman and Grossmann, 1991, 1993, 1994; Türkay and Grossmann, 1996) and a nonlinear version of the method to truss structure design (Bollapragada, Ghattas and Hooker, 2001). Bockmayr and Pisaruk (2003) develop a branch-and-cut algorithm for mixed integer programs augmented by monotone Boolean constraints that are handled by constraint programming. The key to this approach are separation heuristics that allow one to use constraint programming to detect infeasibility and to generate cutting planes for possibly fractional solutions of the mixed integer program.

The logic-based Benders approach was initially developed for circuit verification by Hooker and Yan (1995) and in general by Hooker (1995, 2000) and Hooker and Ottosson (2003). As noted earlier, Jain and Grossmann (2001) and Hooker (2004) found that the Benders approach can dramatically accelerate the solution of planning and scheduling problems. Hooker (2000) observed that the master problem need only be solved once if a Benders cut is generated for each feasible solution found during its solution. Thorsteinsson (2001) obtained an additional order of magnitude speedup for the Jain and Grossmann problem by implementing this idea, which he called *branch and check*. Benders decomposition has recently generated interest on the constraint programming side, as in the work of Eremin and Wallace (2001).

More recently, Aron, Hooker and Yunes (2004) integrated MIP and CP in a general high-level modeler and solver (SIMPL) that is based on an infer-relax-and-restrict algorithmic framework of which BIR and logic-based Benders are special cases. It searches over problems restrictions that become search tree nodes in BIR and subproblems in Benders, while the relaxations are continuous relaxations in BIR and master problems in Benders.

The double modeling and MLLP methods can, by and large, be viewed as special cases of branch-infer-and-relax, which we examine first. We then take up the Benders approach and present Jain and Grossmann's machine scheduling example. Finally, we briefly discuss continuous relaxations of the common global constraints and survey some further applications.

4.1 Branch, infer and relax

Table 3 summarizes the elements of a branch-infer-and-relax (BIR) method. The basic idea is to combine, at each node of the search tree, the filtering and propagation of constraint programming with the relaxation and cutting plane generation of mixed integer programming.

In its simplest form, a BIR method maintains three main data structures: the original set $\mathcal{C}$ of constraints, a constraint store $\mathcal{S}$ that normally contains

Table 3.
Basic elements of branch-infer-and-relax methods

Constraint store (relaxation)	Maintain a constraint store (primarily in-domain constraints) and create a relaxation at each node of the search tree.
Branching	Branch by splitting a nonsingleton domain, perhaps using the solution of the relaxation as a guide.
Inference	Reduce variable domains. Generate cutting planes for the relaxation as well as for constraint propagation.
Bounding	Solve the relaxation to get a bound.
Feasible solution is obtained at a node...	When search variables can be assigned values that are consistent with the solution of the relaxation, and all constraints are satisfied.
Node is infeasible...	When at least one domain is empty or the relaxation is infeasible.
Search backtracks...	When a node is infeasible, a feasible solution is found at a node, or the tree can be pruned due to bounding.

in-domain constraints, and a relaxation $\mathcal{R}$ that may, for example, contain a linear programming relaxation. The constraint store is itself a relaxation, but for convenience, we refer only to $\mathcal{R}$ as the relaxation.

The problem to be solved is to minimize $f(x,y)$ subject to $\mathcal{C}$ and $\mathcal{S}$. The search proceeds by branching on the *search variables* x, and the *solution variables* y receive values from the solution of $\mathcal{R}$. The search variables are often discrete, but in a continuous nonlinear problem they may be continuous variables with interval domains, and branching may consist of splitting an interval (van Hentenryck, Michel and Benhamou, 1998).

The hybrid algorithm consists of a recursive procedure Search $(\mathcal{C}, \mathcal{S})$ and proceeds as follows. Initially the user calls Search $(\mathcal{C}, \mathcal{S})$ with $\mathcal{C}$ the original set of constraints, and $\mathcal{S}$ containing the initial variable domains. $UB = \infty$ is the initial upper bound on the optimal value. Each call to Search $(\mathcal{C}, \mathcal{S})$ executes the following steps.

(1) *Infer constraints for the constraint store.* Process each constraint in $\mathcal{C}$ so as to reduce domains in $\mathcal{S}$. Cycle through the constraints of $\mathcal{C}$ using the desired method of constraint propagation (Section 3). If no domains are empty, continue to Step 2.

(2) *Infer constraints for the relaxation.* Process each constraint in $\mathcal{C}$ so as to generate a set of constraints to be added to the relaxation $\mathcal{R}$, where $\mathcal{R}$ is initially empty. The constraints in $\mathcal{R}$ contain a subset x' of the variables x and all solution variables y, and they may contain new solution variables u that do not appear in $\mathcal{C}$. Constraints in $\mathcal{R}$ that contain no new variables may be added to $\mathcal{C}$ in order to enhance constraint propagation. Cutting planes, for instance, might be added to both $\mathcal{R}$ and $\mathcal{C}$. Continue to Step 3.

(3) *Solve the relaxation.* Minimize the relaxation's objective function $f(x', y, u)$ subject to $\mathcal{R}$. Let LB be the optimal value that results, with $LB = \infty$ if there is no solution. If $LB < UB$ continue to Step 4.

(4) *Infer post-relaxation constraints.* If desired, use the solution of the relaxation to generate further constraints for $\mathcal{C}$, such as separating cuts, fixed variables based on reduced costs, and other types of nogoods. Continue to Step 5.
(5) *Identify a solution.* If possible, assign some value $\bar{x}$ to x that is consistent with the current domains and the optimal solution $(\bar{x}', \bar{y})$ of the relaxation. If $(x, y) = (\bar{x}, \bar{y})$ is feasible for $\mathcal{C}$, let $UB = LB$, and add the constraint $f(x) < UB$ to $\mathcal{C}$ at all subsequent nodes (to search for a better solution). Otherwise go to Step 6.
(6) *Branch.* Branch on some search variable x_j by splitting its domain D_j into smaller domains $D_{j1}, \ldots, D_{jp}$ and calling Search $(\mathcal{C}, \mathcal{S}_k)$ for $k = 1, \ldots, p$, where $\mathcal{S}_k$ is $\mathcal{S}$ augmented with the in-domain constraint $x_j \in D_{jk}$. One can also branch on a violated constraint.

In Step 3, the relaxation $\mathcal{R}$ can depend on the current variable domains. This allows for a more flexible modeling. For example, it is often convenient to use conditional constraints of the form $g(x) \rightarrow h(y)$ where $\rightarrow$ means "implies." Such a constraint generates the constraint $h(y)$, for $\mathcal{R}$ when and if the search variable domains become small enough to determine that $g(x)$ is satisfied. If $g(x)$ is not determined to be satisfied, no action is taken.

One common occurrence of the conditional constraints is in fixed charge problems, where cy_1 is the variable cost of an activity running at level y_1, and an additional fixed charge d is incurred when $y_1 > 0$. If x_1 is a Boolean variable that is true when the fixed charge is incurred, a skeletal fixed charge problem can be written as:

$$\begin{array}{ll} \text{minimize} & cy_1 + y_2 \\ \text{subject to} & x_1 \rightarrow (y_2 \geq d) \\ & \text{not-}x_1 \rightarrow (y_1 \leq 0) \\ & x_1 \in \{T, F\}, y_1 \in [0, M], y_2 \in [0, \infty) \end{array} \tag{1}$$

where x_1 is the only search variable and y_2 represents the fixed cost incurred. The constraint $y_2 \geq d$ is added to $\mathcal{R}$ when and if x_1 becomes true in the course of the BIR algorithm, and $y_1 \leq 0$ is added when x_1 becomes false.

In practice, the two conditional constraints of (1) should be written as a single global constraint that will be discussed below in Section 4.4:

$$\text{inequality-or}\left(\begin{bmatrix} x_1 \\ \text{not-}x_1 \end{bmatrix}, \begin{bmatrix} y_2 \geq d \\ y_1 \leq 0 \end{bmatrix} \right)$$

The constraint signals that the two conditional constraints enforce a disjunction $(y_2 \geq d) \vee (y_1 \leq 0)$, which can be given a simple and useful continuous relaxation introduced by Beaumont (1990). (The $\vee$ is an inclusive "or.") In this case the relaxation is $dy_1 \leq My_2$, which the inequality-or constraint generates for $\mathcal{R}$ even before the value of x_1 is determined.

4.2 Benders decomposition

Another promising framework for the hybrid methods is a logic-based form of Benders decomposition, a well-known optimization technique (Benders 1962; Geoffrion 1972). The problem is written using a partition $[x, y]$ of the variables:

$$\begin{array}{lll} \text{minimize} & f(x, y) & \\ \text{subject to} & g_i(x, y), & \text{all } i \end{array} \tag{2}$$

The basic idea is to search values of x in a *master problem*, and for each value enumerated solve the *subproblem* of finding an optimal y. Solution of a subproblem generates a *Benders cut* that is added to the master problem. The cut excludes some values of x that can be no better than the value just tried.

The variable x is initially assigned an arbitrary value $\bar{x}$. This gives rise to a *subproblem* in the y variables:

$$\begin{array}{lll} \text{minimize} & f(\bar{x}, y) & \\ \text{subject to} & g_i(\bar{x}, y), & \text{all } i \end{array} \tag{3}$$

Solution of the subproblem yields a *Benders cut* $z \geq B_{\bar{x}}(x)$ that has two properties:

(a) When x is fixed to any given value $\hat{x}$ the optimal value of (2) is at least $B_{\bar{x}}(\hat{x})$.
(b) When x is fixed to $\bar{x}$ the optimal value of (2) is exactly $B_{\bar{x}}(\bar{x})$.

If the subproblem (3) is infeasible, its optimal value is infinite, and $B_{\bar{x}}(\bar{x}) = \infty$. If the subproblem is unbounded, then (2) is unbounded, and the algorithm terminates. How Benders cuts are generated will be discussed shortly.

In the Kth iteration, the *master problem* minimizes z subject to all Benders cuts that have been generated so far.

$$\begin{array}{lll} \text{minimize} & z & \\ \text{subject to} & z \geq B_{x^k}(x), & k = 1, \ldots, K-1 \end{array} \tag{4}$$

A solution $\bar{x}$ of the master problem is labeled x^K, and it gives rise to the next subproblem. The procedure terminates when the master problem has the same optimal value as the previous subproblem (infinite if the original problem is infeasible), or when the subproblem is unbounded. The computation can sometimes be accelerated by observing that (b) need not hold until the last iteration.

To obtain a Benders cut from the subproblem (3), one solves the *inference dual* of (3):

$$\begin{array}{ll} \text{maximize} & v \\ \text{subject to} & (g_i(\bar{x}, y), \text{ all } i) \rightarrow (f(\bar{x}, y) \geq v) \end{array} \tag{5}$$

The inference dual seeks the largest lower bound on the subproblem's objective function that can be inferred from its constraints. If the subproblem has a finite optimal value, clearly its dual has the same optimal value. If the subproblem is unbounded (infeasible), then the dual is infeasible (unbounded).

Suppose that $\bar{v}$ is the optimal value of the subproblem dual ($\bar{v} = -\infty$ if the dual is infeasible). A solution of the dual takes the form of a proof that deduces $f(\bar{x}, y) \geq \bar{v}$ from the constraints $g_i(\bar{x}, y)$. The dual solution proves that $\bar{v}$ is a lower bound on the value of the subproblem (3), and therefore a lower bound on the value z of the original problem (2) when $x = \bar{x}$. The key to obtain a Benders cut is to structure the proof so that it is parameterized by x. Thus if $x = \bar{x}$ the proof establishes the lower bound $\bar{v} = B_{\bar{x}}(\bar{x})$ on z. If x has some other value $\hat{x}$, the proof establishes a valid lower bound $B_{\bar{x}}(\hat{x})$ on z. This yields the Benders cut $z \geq B_{\bar{x}}(x)$.

In a classical Benders decomposition, the subproblem is a linear programming problem, and its inference dual is the standard linear programming dual. The Benders cuts take the form of linear inequalities. Benders cuts can also be obtained when the subproblem is a 0-1 programming problem (Hooker 2000, Hooker and Ottosson 2003).

Logic-based Benders can integrate MIP and CP if one formulates the master problem as an MIP problem, and the subproblem as a CP problem. Constraint programming provides a natural context for generating Benders cuts because it shows that $f(\bar{x}, \bar{y}) = v$ is an optimal value of (3) by providing an infeasibility proof of (3) when $f(\bar{x}, \bar{y}) < v$ is added to the constraint set. This proof can be regarded as a solution of the inference dual.

4.3 Machine scheduling example

A machine assignment and scheduling problem of Jain and Grossmann (2001) illustrates a Benders approach in which the subproblem is solved by constraint programming.

Each job j is assigned to one of several machines i that operate at different speeds. Each assignment results in a processing time d_{ij} and incurs a processing cost c_{ij}. There is a release date r_j and a due date s_j for each job j. The objective is to minimize processing cost while observing release and due dates.

To formulate the problem, let x_j be the machine to which job j is assigned and t_j the start time for job j. It is also convenient to let $[t_j | x_j = i]$ denote the tuple of start times of jobs assigned to machine i, arranged in increasing order of the job number. The problem can be written as:

$$\begin{array}{lll} \text{minimize} & \sum_j c_{x_j j} & (a) \\ \text{subject to} & t_j \geq r_j, \quad \text{all } j & (b) \\ & t_j + d_{x_j j} \leq S_j, \quad \text{all } j & (c) \\ & \texttt{cumulative}\left(\left[t_j | x_j = i\right], \left[d_{ij} | x_j = i\right], e, 1\right), \quad \text{all } i & (d) \end{array} \tag{6}$$

The objective function (a) measures the total processing cost. Constraints (b) and (c) observe release times and deadlines. The `cumulative` constraint (d) ensures that jobs assigned to each machine are scheduled so that they do not overlap. (Recall that e is a vector of ones.)

The problem has two parts: the assignment of jobs to machines, and the scheduling of jobs on each machine. The assignment problem is treated as the master problem and solved with mixed integer programming methods. Once the assignments are made, the subproblems are dispatched to a constraint programming solver to find a feasible schedule. If there is no feasible schedule, a Benders cut is generated.

Variables x go into the master problem and t into the subproblem. If x has been fixed to $\bar{x}$, the subproblem is

$$\begin{array}{l} t_j \geq r_j, \quad \text{all } j \\ t_j + d_{\bar{x}_j j} \leq S_j, \quad \text{all } j \\ \texttt{cumulative}\left(\left[t_j | \bar{x}_j = i\right], \left[d_{ij} | \bar{x}_j = i\right], e, 1\right), \quad \text{all } i \end{array} \tag{7}$$

The subproblem can be decomposed into smaller problems, one for each machine. If a smaller problem is infeasible for some i, then the jobs assigned to machine i cannot all be scheduled on that machine. In fact, going beyond Jain and Grossmann (2001), there may be a subset J of these jobs that cannot be scheduled on machine i. This gives rise to a Benders cut stating that at least one of the jobs in J must be assigned to another machine.

$$\bigvee_{j \in J} (x_j \neq i) \tag{8}$$

Let x^k be the solution of the kth master problem, I^k the set of machines i in the resulting subproblem for which the schedule is infeasible, and J_{ki} the infeasible subset for machine i. The master problem can now be written as:

$$\begin{array}{ll} \text{minimize} & \sum_j c_{x_j j} \\ \text{subject to} & \bigvee_{j \in J_{ki}} (x_j \neq i), \; i \in I^k, \; k = 1, \ldots, K \end{array} \tag{9}$$

The master problem can be reformulated for solution with conventional integer programming technology. Let x_{ij} be a 0-1 variable that is 1 when job j is assigned to machine i. The master problem (9) can be written as:

$$\begin{array}{lll} \text{minimize} & \sum_{i,j} c_{ij}x_{ij} & (a) \\ \text{subject to} & \sum_{j\in J_{ki}} (1-x_{ij}) \geq 1, \quad i \in I^k,\ k=1,\ldots,K & (b) \\ & \sum_j d_{ij}x_{ij} \leq \max_j\{s_j\} - \min_j\{r_j\}, \quad \text{all } i & (c) \\ & x_{ij} \in \{0,1\}, \quad \text{all } i,j & (d) \end{array}$$

Constraints (c) are valid cuts added to strengthen the continuous relaxation. They simply say that the total processing time on each machine must fit between the earliest release time and the latest deadline. Stronger relaxations are available as well. Appropriate Benders cuts are much less obvious when the subproblem is an optimization rather than a feasibility problem, as in minimum makespan and minimum tardiness problems. Hooker (2004) develops effective Benders cuts for these problems and generalizes the subproblem to accommodate cumulative scheduling.

4.4 Continuous relaxations for global constraints

Continuous relaxations for global constraints can accelerate solution by exploiting substructure in a model. Relaxations have been developed for several constraints, although other constraints have yet to be addressed. Relaxations for many of the constraints discussed below are summarized by Hooker (2000, 2002); see also Refalo (2000).

The *inequality-or* constraint, discussed above in the context of fixed charge problems, may be written,

$$\texttt{inequality-or}\left(\begin{bmatrix} x_1 \\ \vdots \\ x_k \end{bmatrix}, \begin{bmatrix} A^1y \geq a^1 \\ \vdots \\ A^ky \geq a^k \end{bmatrix}\right)$$

It requires that x_i be true and $A^iy \geq a^i$ be satisfied for at least one $i \in \{1,\ldots,k\}$. A convex hull relaxation can be obtained by introducing new variables, as shown by Balas (1975, 1979). The well-known "big-M" lifted relaxation is weaker than the convex hull relaxation but requires fewer variables. Hooker and Osorio (1999) discuss how to tighten the big-M relaxation.

A *disjunction of single inequalities*

$$(a^1y \geq \alpha_1) \vee \cdots \vee (a^k \geq \alpha_k)$$

relaxes to a single inequality, as shown by Beaumont (1990). Hooker and Osorio (1999) provide a closed-form expression for a tighter right-hand side.

Cardinality rules provide for more complex logical conditions:

If at least k of $x_1, \ldots, x_m$ are true, then at least ℓ of $y_1, \ldots y_n$ are true.

Yan and Hooker (1999) describe a convex hull relaxation for such rules. Convex hull characterizations and separation routines for disjunctions of monotone polyhedra are given in (Balas, Bockmayr, Pisaruk and Wolsey, 2004).

Piecewise linear functions can easily be given a convex hull relaxation that, when properly used, can result in faster solution than mixed integer programming with specially ordered sets of type 2 (Ottosson, Thorsteinsson and Hooker, 1999). Refalo (1999) shows how to use the relaxation in "tight cooperation" with domain reduction to obtain maximum benefit.

The *alldifferent* constraint can be given a convex hull relaxation as described by Hooker (2000) and Williams and Yan (2001).

The *element* constraint is particularly useful for implementing variable indices. An expression of the form u_y can be encoded by replacing it with the variable z and adding the constraint `element` $(y, (u_1, \ldots, u_n), z)$. Here $u_1, \ldots, u_n$ may be constants or variables. Hooker, Ottosson, Thorsteinsson and Kim (1999) present various relaxations of the element constraint, including a convex hull relaxation when the variables $u_1, \ldots, u_n$ have the same upper bound (Hooker, 2000).

The important *cumulative* constraint has been given three relaxations by Hooker and Yan (2002). One relaxation consists of facet defining inequalities in the special case in which some jobs have identical characteristics.

Lagrangean relaxation can be employed in a hybrid setting. Sellmann and Fahle (2001) use it to strengthen propagation of knapsack constraints in an automatic recording problem. Benoist, Laburthe and Rottembourg (2001) apply it to a traveling tournament problem. It is unclear whether this work suggests a general method for integrating Lagrangean relaxation with constraint propagation.

4.5 Other applications

Hybrid methods have been applied to a number of problems other than those already mentioned. Transportation applications include vehicle routing with time windows (Caseau, Silverstein and Laburthe, 2001; Focacci, Lodi and Milano, 1999b), vehicle routing combined with inventory management (Lau and Liu, 1999), crew rostering (Caprara et al., 1998; Junker, Karisch, Kohl, Vaaben, Fahle and Sellmann, 1999), the traveling tournament problem (Benoist et al., 2001), and the classical transportation problem with piecewise linear costs (Refalo, 1999).

Scheduling applications include machine scheduling (Heipcke, 1998; Raman and Grossmann, 1993), sequencing with setups (Focacci, Lodi and Milano, 1999a), hoist scheduling (Rodošek and Wallace, 1998), employee scheduling (Partouche, 1998), dynamic scheduling (Sakkout, Richards and Wallace, 1998), and lesson timetables (Focacci et al., 1999a). Production scheduling applications include scheduling with resource constraints (Pinto and Grossmann, 1997) and with labor resource constraints in particular (Heipcke, 1999), two-stage process scheduling (Jain and Grossmann, 2001), machine allocation and scheduling (Lustig and Puget, 2001), production flow planning with machine assignment (Heipcke, 1999), scheduling with piecewise linear costs (Ottosson et al., 1999), scheduling with earliness and tardiness costs (Beck, 2001), and organization of a boat party (Hooker and Osorio 1999; Smith, Brailsford, Hubbard and Williams, 1996).

Other areas of application include inventory management (Rodošek et al., 1997), office cleaning (Heipcke, 1999), product configuration (Ottosson and Thorsteinsson, 2000), generalized assignment problems (Darby-Dowman, Little, Mitra and Zaffalon, 1997), multidimensional knapsack problems (Osorio and Glover, 2001), automatic recording of television shows (Sellmann and Fahle, 2001), resource allocation in ATM networks (Lauvergne, David and Boizumault, 2001), and assembly line balancing (Bockmayr and Pisaruk, 2001).

Benders-based hybrid methods provide a natural decomposition for manufacturing and supply chain problems in which resource assignment issues combine with scheduling issues. Recent industrial applications along this line include automobile assembly (Beauseigneur and Noiré, 2003), polypropylene manufacture (Timpe, 2003), and paint production (Constantino, 2003).

References

Aggoun, A., N. Beldiceanu (1993). Extending CHIP in order to solve complex scheduling and placement problems. *Mathl. Comput. Modelling* 17(7), 57–73.

Althaus, E., A. Bockmayr, M. Elf, T. Kasper, M. Jünger, K. Mehlhorn (2002). SCIL-Symbolic constraints in integer linear programming. *10th European Symposium on Algorithms, ESA' 02*, Springer, Rome, LNCS 2461, pp. 75–87.

Aron, I., J. N. Hooker, T. M. Yunes (2004). SIMPL, a system for integrating optimization techniques. CPAIOR 2004, Springer, Cambridge, MA, LNCS 3011.

Bacchus, F. (2000). Extending forward checking. *Principles and Practice of Constraint Programming. CP'2000*, Springer, Singapore, LNCS 1894, pp. 35–51.

Bacchus, F., X. Chen, P. van Beek, T. Walsh (2002). Binary vs. non-binary constraints. *Artificial Intelligence*, 140, 1–37.

Balas, E. (1975). Disjunctive programming: cutting planes from logical conditions. in: O. L. Mangasarian, R. R. Meyer, S. M. Robinson (eds.), *Nonlinear Programming 2*, Academic Press, New York, pp. 279–312.

Balas, E. (1979). Disjunctive programming. *Annals of Discrete Mathematics* 5, 3–51.

Balas, E., A. Bockmayr, N. Pisaruk, L. Wolsey (2004). On unions and dominants of polytopes. *Mathematical Programming*, Ser. A 99, 223–239.

Baptiste, P., C. L. Pape (2000). Constraint propagation and decomposition techniques for highly disjunctive and highly cumulative project scheduling problems. *Constraints* 5(1/2), 119–139.

Baptiste, P., C. L. Pape, W. Nuijten (2001). Constraint-based scheduling. *International Series in Operations Research and Management Science*, Vol. 39, Kluwer.

Barth, P., A. Bockmayr (1998). Modelling discrete optimisation problems in constraint logic programming. *Annals of Operations Research* 81, 467–496.

Beaumont, N. (1990). An algorithm for disjunctive programs. *Europ. J. Oper. Res.* 48, 362–371.

Beauseigneur, M., S. Noiré (2003). Solving the car sequencing problem using combined CP/MIP for PSA Peugeot Citroën, LISCOS Project Summary Meeting, Brussels (28 March 2003).

Beck, C. (2001). A hybrid approach to scheduling with earliness and tardiness costs. *Third International Workshop on Integration of AI and OR Techniques (CPAIOR01)*.

Beldiceanu, N. (2000). Global constraints as graph properties on a structured network of elementary constraints of the same type. *Principles and Practice of Constraint Programming, CP'2000*, Springer, Singapore, LNCS 1894, pp. 52–66.

Beldiceanu, N. (2001). Pruning for the *minimum* constraint family and for the *number of distinct values* constraint family. *Principles and Practice of Constraint Programming, CP'2001*, Springer, Paphos, Cyprus, LNCS 2239, pp. 211–224.

Beldiceanu, N., A. Aggoun, E. Contejean (1996). Introducing constrained sequences in CHIP. *Technical Report*, COSYTEC S.A., Orsay, France.

Beldiceanu, N., M. Carlsson (2001). Sweep as a generic pruning technique applied to the non-overlapping rectangles constraint. *Principles and Practice of Constraint Programming, CP'2001*, Springer, Paphos, Cyprus, LNCS 2239, pp. 377–391.

Beldiceanu, N., M. Carlsson (2002). A new multi-resource cummulatives constraint with negative heights. *Principles and Practice of Constraint Programming, CP'2002*, Springer, Ithaca, NY, LNCS 2470, pp. 63–79.

Beldiceanu, N., E. Contejean (1994). Introducing global constraints in CHIP. *Mathl. Comput. Modelling* 20(12), 97–123.

Beldiceanu, N., G. Qi, S. Thiel (2001). Non-overlapping constraints between convex polytopes. *Principles and Practice of Constraint Programming, CP'2001*, Springer, Paphos, Cyprus, LNCS 2239, pp. 392–407.

Benders, J. F. (1962). Partitioning procedures for solving mixed-variables programming problems. *Numerische Mathematik* 4, 238–252.

Benoist, T., F. Laburthe, B. Rottembourg (2001). Lagrange relaxation and constraint programming collaborative schemes for traveling tournament problems. *Third International Workshop on Integration of AI and OR Techniques (CPAIOR01)*.

Bessière, C. (1994). Arc-consistency and arc-consistency again. *Artificial Intelligence* 65, 179–190.

Bessière, C. (1999). Non-binary constraints. *Principles and Practice of Constraint Programming, CP'99*, Springer, Alexandria, VA, LNCS 1713, pp. 24–27.

Bessière, C., E. Freuder, J.-C. Régin (1999). Using constraint meta-knowledge to reduce arc consistency computation. *Artificial Intelligence* 107, 125–148.

Bessière, C., P. Meseguer, E. C. Freuder, J. Larrosa (1999). On forward checking for non-binary constraint satisfaction. *Principles and Practice of Constraint Programming, CP'99*, Springer, Alexandria, VA, LNCS 1713, pp. 88–102.

Bessière, C., J.-C. Régin (1996). MAC and combined heuristics: two reasons to forsake FC (and CBJ?) on hard problems. *Principles and Practice of Constraint Programming, CP'96*, Springer, Cambridge, MA, LNCS 1118, pp. 61–75.

Bessière, C., J.-C. Régin (1997). Arc consistency for general constraint networks: preliminary results. *15th Intern. Joint Conf. Artificial Intelligence, IJCAI'97, Nagoya, Japan*, Vol. 1, pp. 398–404.

Bessière, C., J.-C. Régin (2001). Refining the basic constraint propagation algorithm. *17th Intern. Joint Conf. Artificial Intelligence, IJCAI'01, Seattle*, Vol. 1, pp. 309–315.

Bleuzen-Guernalec, N., A. Colmerauer (2000). Optimal narrowing of a block of sortings in optimal time. *Constraints* 5(1/2), 85–118.

Bockmayr, A., T. Kasper (1998). Branch and infer: a unifying framework for integer and finite domain constraint programming. *INFORMS Journal on Computing* 10, 287–300.

Bockmayr, A., T. Kasper, T. Zajac (1998). Reconstructing binary pictures in discrete tomography. *16th European Conference on Operational Research, EURO XVI, Bruxelles.*

Bockmayr, A., N. Pisaruk (2001). Solving assembly line balancing problems by combining IP and CP. *Sixth Annual Workshop of the ERCIM Working Group on Constraints, Prague*, http://arXiv.org/abs/cs.DM/0106002.

Bockmayr, A., N. Pisaruk (2003). Detecting infeasibility and generating cuts for MIP using CP. *5th International Workshop on Integration of AI and OR Techniques in Constraint Programming for Combinatorial Optimization Problems, CPAIOR'03*, Montreal, pp. 24–34.

Bockmayr, A., N. Pisaruk, A. Aggoun (2001). Network flow problems in constraint programming. *Principles and Practice of Constraint Programming, CP'2001*, Springer, Paphos, Cyprus, LNCS 2239, pp. 196–210.

Bollapragada, S., O. Ghattas, J. N. Hooker (2001). Optimal design of truss structures by mixed logical and linear programming. *Operations Research* 49, 42–51.

Bourreau, E. (1999). *Traitement de Contraintes sur les Graphes en Programmation par Contraintes*, PhD thesis, L.I.P.N., Univ. Paris 13.

Cagan, J., I. E. Grossmann, J. N. Hooker (1997). A conceptual framework for combining artificial intelligence and optimization in engineering design. *Research in Engineering Design* 49, 20–34.

Caprara, A., F. Focacci, E. Lamma, P. Mello, M. Milano, P. Toth, D. Vigo (1998). Integrating constraint logic programming and operations research techniques for the crew rostering problem. *Software-Practice and Experience* 28, 49–76.

Carlier, J., E. Pinson (1990). A practical use of Jackson's preemptive schedule of solving the job-shop problem. *Annals of Operations Research* 26, 269–287.

Caseau, Y., F. Laburthe (1997). Solving small TSP's with constraints. *14th International Conference on Logic Programming, ICLP'97*, MIT Press, Leuven, pp. 316–330.

Caseau, Y., G. Silverstein, F. Laburthe (2001). Learning hybrid algorithms for vehicle routing problems. *Third International Workshop on Integration of AI and OR Techniques* (*CPAIOR01*).

Chen, X., P. van Beek (2001). Conflict-directed backjumping revisited. *Journal of Artificial Intelligence Research* 14, 53–81.

Colmerauer, A. (1987). Introduction to PROLOG III, *4th Annual ESPRIT Conference, North Holland, Bruxelles*, See also: Comm. ACM 33(1990), 69–90.

Colombani, Y., S. Heipcke (2002). Mosel: en extensible environment for modeling and programming solutions. *4th International Workshop on Integration of AI and OR techniques in Constraint Programming for Combinatorial Optimization Problems, CP-AI-OR'02, Le Croisic, France*, pp. 277–290.

Constantino, M. (2003). Integrated lot-sizing and scheduling of Barbot's paint production using combined MIP/CP, LISCOS Project Summary Meeting, Brussels (28 March 2003).

Darby-Dowman, K., J. Little (1998). Properties of some combinatorial optimization problems and their effect on the performance of integer programming and constraint logic programming. *INFORMS Journal on Computing* 10, 276–286.

Darby-Dowman, K., J. Little, G. Mitra, M. Zaffalon (1997). Constraint logic programming and integer programming approaches and their collaboration in solving an assignment scheduling problem. *Constraints* 1, 245–264.

Debruyne, R., C. Bessière (2001). Domain filtering consistencies. *Journal of Artificial Intelligence Research* 14, 205–230.

Dechter, R. (1990). Enhancement schemes for constraint processing: back jumping, learning, and cutset decomposition. *Artificial Intelligence* 41, 273–312.

Dechter, R. (1992). Constraint networks, in: S. Shapiro (ed.), *Encyclopedia of artificial intelligence*, Vol. 1. Wiley, 276–285.

Dechter, R. (2003). *Constraint Processing*, Morgan Kaufmann.

Dechter, R., D. Frost (2002). Backjump-based backtracking for constraint satisfaction problems. *Artificial Intelligence* 136, 147–188.

Dincbas, M., P. van Hentenryck, H. Simonis, A. Aggoun, T. Graf (1988). The constraint logic programming language CHIP. *Fifth Generation Computer Systems, Tokyo, 1988*, Springer.

Eremin, A., M. Wallace (2001). Hybrid Benders decomposition algorithms in constraint logic programming. *Seventh International Conference on Principles and Practice of Constraint Programming (CP2001)*.

Focacci, F., A. Lodi, M. Milano (1999a). Cost-based domain filtering. *Principles and Practice of Constraint Programming, Lecture Notes in Computer Science*, Vol. 1713, pp. 189–203.

Focacci, F., A. Lodi, M. Milano (1999b). Solving TSP with time windows with constraints. *16th International Conference on Logic Programming*, Las Cruces, NM.

Focacci, F., A. Lodi, M. Milano (2000). Cutting planes in constraint programming: an hybrid approach. *Principles and Practice of Constraint Programming, CP'2000*, Springer, Singapore, LNCS 1894, pp. 187–201.

Freuder, E. C. (1985). A sufficient condition for backtrack-bounded search. *Journal of the Association for Computing Machinery* 32(4), 755–761.

Frühwirth, T., S. Abdennadher (2003). *Essentials of Constraint Programming*, Springer.

Gent, I. P., E. MacIntyre, P. Prosser, B. M. Smith, T. Walsh (1996). An empirical study of dynamic variable ordering heuristics for the constraint satisfaction problem. *Principles and Practice of Constraint Programming, CP'96*, Springer, Cambridge, MA, LNCS 1118, pp. 179–193.

Geoffrion, A. M. (1972). Generalized Benders decomposition. *Journal of Optimization Theory and Applications* 10, 237–260.

Gomes, C. P., B. Selman, H. A. Kautz (1998). Boosting combinatorial search through randomization. *Proc. 15th National Conference of Artificial Intelligence (AAAI'98) and 10th Innovative Applications of Artificial Intelligence Conference (IAAI'98)*, pp. 431–437.

Grossmann, I. E., J. N. Hooker, R. Raman, H. Yan (1994). Logic cuts for processing networks with fixed charges. *Computers and Operations Research* 421, 265–279.

Harvey, W. D., M. L. Ginsberg (1995). Limited discrepancy search. *14th Intern. Joint Conf. Artificial Intelligence, IJCAI'95, Montreal*, Vol. 1, pp. 607–615.

Heipcke, S. (1998). Integrating constraint programming techniques into mathematical programming. *Proceedings, 13th European Conference on Artificial Intelligence*, Wiley, New York, pp. 259–260.

Heipcke, S. (1999). *Combined Modeling and Problem Solving in Mathematical Programming and Constraint Programming*, PhD thesis, Univ. Buckingham.

Hooker, J. N. (1994). Logic-based methods for optimization, in: A. Borning (ed.), *Principles and Practice of Constraint Programming, Lecture Notes in Computer Science*, Vol. 874, Springer, pp. 336–349.

Hooker, J. N. (1995). Logic-based Benders decomposition, *INFORMS National Meeting.*

Hooker, J. N. (2000). *Logic-based Methods for Optimization: Combining Optimization and Constraint Satisfaction*, John Wiley and Sons.

Hooker, J. N. (2002). Logic, optimization and constraint programming. *INFORMS Journal on Computing* 14, 295–321.

Hooker, J. N. (2003). A framework for integrating solution methods, in: H. K. Bhargava, Mong Ye (eds.), *Computational Modeling and Problem Solving in the Networked World* (*Proceedings of ICS 2003*), Kluwer, pp. 3–30.

Hooker, J. N. (2004). A hybrid method for planning and scheduling. *Principles and Practices of Constraint Programming (CP2004)*, Springer, Cambridge, MA, LNCS 3258.

Hooker, J., M. A. Osorio (1999). Mixed logical/linear programming. *Discrete Applied Mathematics* 96–97, 395–442.

Hooker, J. N., G. Ottosson (2003). Logic-based Benders decomposition. *Mathematical Programming* 96, 33–60.

Hooker, J. N., G. Ottosson, E. Thorsteinsson, H.-J. Kim. (1999). On integrating constraint propagation and linear programming for combinatorial optimization. *Proceedings, 16th National Conference on Artificial Intelligence*, MIT Press, Cambridge, MA, pp. 136–141.

Hooker, J. N., H. Yan (1995). Logic circuit verification by Benders decomposition, in: V. Saraswat P. V. Hentenryck (eds.), *Principles and Practice of Constraint Programming: the Newport Papers*, MIT Press, Cambridge, MA, pp. 267–288.

Hooker, J. N., H. Yan (2002). A relaxation for the cumulative constraint, in: P. Van Hentenryck (ed.), *Principles and Practice of Constraint Programming (CP2002), Lecture Notes in Computer Science*, 2470(2002), 686–690.

Jaffar, J., J.-L. Lassez (1987). Constraint logic programming. *Proc. 14th ACM Symp. Principles of Programming Languages*, Munich.

Jain, V., I. E. Grossmann (2001). Algorithms for hybrid MILP/CP models for a class of optimization problems, *INFORMS J. Computing* 13(4), 258–276.

Junker, U., S. E. Karisch, N. Kohl, B. Vaaben, T. Fahle, M. Sellmann (1999). A framework for constraint programming based column generation, in: J. Jaffar (eds.), *Principles and Practice of Constraint Programming, Lecture Notes in Computer Science*, Vol. 1713, Springer, Berlin, 261–274.

Jussien, N., R. Debruyne, P. Boizumault (2000). Maintaining arc consistency within dynamic backtracking. *Principles and Practice of Constraint Programming, CP'2000*, Springer, Singapore, LNCS 1894, pp. 249–261.

Laburthe, F., Y. Caseau (2002). SALSA: A language for search algorithms. *Constraints* 7(3), 255–288.

Lau, H. C., Q. Z. Liu (1999). Collaborative model and algorithms for supporting real-time distribution logistics systems. *CP99 Post-conference Workshop on Large Scale Combinatorial Optimization and Constraints*, 30–44.

Laurière, J.-L. (1978). A language and a program for stating and for solving combinatorial problems. *Artificial Intelligence* 10, 29–127.

Lauvergne, M., P. David, P. Boizumault (2001). Resource allocation in ATM networks: a hybrid approach. *Third International Workshop on the Integration of AI and OR Techniques* (*CPAIOR 2001*).

Little, J., K. Darby-Dowman (1995). The significance of constraint logic programming to operational research, in: M. Lawrence, C. Wilson (eds.), *Operational Research*, pp. 20–45.

Lustig, I. J., J.-F. Puget (2001). Program does not equal program. Constraint programming and its relationship to mathematical programming. *Interfaces*, 31, 29–53.

Mackworth, A. (1977a). On reading sketch maps. *5th Intern. Joint Conf. Artificial Intelligence, IJCAI'77*, Cambridge MA, pp. 598–606.

Mackworth, A. (1977b). Consistency in networks of relations. *Artificial Intelligence* 8, 99–118.

Marriott, K., P. J. Stuckey (1998). *Programming with Constraints*, MIT Press.

McDonald, I., B. Smith (2002). Partial symmetry breaking, *Principles and Practice of Constraint Programming, CP'2002*, Springer, Ithaca, NY, LNCS 2470, pp. 431–445.

Mehlhorn, K., S. Thiel (2000). Faster algorithms for bound-consistency of the sortedness and the alldifferent constraint. *Principles and Practice of Constraint Programming, CP'2000*, Springer, Singapore, LNCS 1894, pp. 306–319.

Meseguer, P. (1997). Interleaved depth-first search. *15th Intern. Joint Conf. Artificial Intelligence, IJCAI'97, Nagoya, Japan*, Vol. 2, pp. 1382–1387.

Meseguer, P., T. Walsh (1998). Interleaved and discrepancy based search. *13th Europ. Conf. Artificial Intelligence, Brighton, UK*, John Wiley and Sons, pp. 229–233.

Mohr, R., T. C. Henderson (1986). Arc and path consistency revisited. *Artificial Intelligence* 28, 225–233.

Mohr, R., G. Masini (1988). Good old discrete relaxation. *Proc. 8th European Conference on Artificial Intelligence*, Pitman Publishers, Munich, FRG, pp. 651–656.

Older, W. J., G. M. Swinkels, M. H. van Emden (1995). Getting to the real problem: experience with BNR prolog in OR. *Practical Application of Prolog, PAP'95*, Paris.

Osorio, M. A., F. Glover (2001). Logic cuts using surrogate constraint analysis in the multidimensional knapsack problem. *Third International Workshop on Integration of AI and OR Techniques* (*CPAIOR01*).

Ottosson, G., E. Thorsteinsson (2000). Linear relaxations and reduced-cost based propagation of continuous variable subscripts. *Second International Workshop on Integration of AI and OR Techniques in Constraint Programming for Combinatorial Optimization Problems, CPAIOR2000*, University of Paderborn.

Ottosson, G., E. Thorsteinsson, J. N. Hooker (1999). Mixed global constraints and inference in hybrid CLP-IP solvers, *CP99 Post-Conference Workshop on Large Scale Combinatorial Optimization and Constraints*, pp. 57–78.

Partouche, A. (1998). *Planification d'horaires de travail*, PhD thesis, Université Paris-Daphiné, U. F. R. Sciences des Organisations.

Pinto, J. M., I. E. Grossmann (1997). A logic-based approach to scheduling problems with resource constraints. *Computers and Chemical Engineering* 21, 801–818.

Prosser, P. (1993). Hybrid algorithms for the constraint satisfaction problem. *Computational Intelligence* 9, 268–299.

Prosser, P. (1998). The dynamics of dynamic variable ordering heuristics. *Principles and Practice of Constraint Programming, CP'98*, Springer, Pisa, LNCS 1520, pp. 17–23.

Prosser, P., K. Stergiou, T. Walsh (2000). Singleton consistencies. *Principles and Practice of Constraint Programming, CP'2000*, Springer, Singapore, LNCS 1894, pp. 353–368.

Puget, J. F. (1994). AC++ implementation of CLP. *Technical report*, ILOG S. A. http://www.ilog.com.

Puget, J.-F. (1998). A fast algorithm for the bound consistency of alldiff constraints. *Proc. 15th National Conference on Artificial Intelligence (AAAI'98) and 10th Conference on Innovative Applications of Aritificial Intelligence (IAAI'98)*, AAAI Press, pp. 359–366.

Puget, J.-F. (2002). Symmetry breaking revisited. *Principles and Practice of Constraint Programming, CP'2002*, Springer, Ithaca, NY, LNCS 2470, pp. 446–461.

Raman, R., I. Grossmann (1991). Symbolic integration of logic in mixed-integer linear programming techniques for process synthesis. *Computers and Chemical Engineering* 17, 909–927.

Raman, R., I. Grossman (1993). Relation between MILP modeling and logical inference for chemical process synthesis. *Computers and Chemical Engineering* 15, 73–84.

Raman, R., I. Grossman (1994). Modeling and computational techniques for logic based integer programming. *Computers and Chemical Engineering* 18, 563–578.

Refalo, P. (1999). Tight cooperation and its application in piecewise linear optimization. *Principles and Practice of Constraint Programming, CP'99*, Springer, Alexandria, VA, LNCS 1713, pp. 375–389.

Refalo, P. (2000). Linear formulation of constraint programming models and hybrid solvers. *Principles and Practice of Constraint Programming, CP'2000*, Springer, Singapore, LNCS 1894, pp. 369–383.

Régin, J.-C. (1994). A filtering algorithm for constraints of difference in CSPs. *Proc. 12th National Conference on Artificial Intelligence, AAAI'94, Seattle*, Vol. 1, pp. 362–367.

Régin, J.-C. (1996). Generalized arc consistency for global cardinality constraint. *Proc. 13th National Conference on Artificial Intelligence, AAAI'96, Protland*, Vol. 1, pp. 209–215.

Régin, J.-C. (1999a). Arc consistency for global cardinality constraints with costs. *Principles and Practice of Constraint Programming, CP'99*, Springer, Alexandria, VA, LNCS, 1713, pp. 390–404.

Régin, J.-C. (1999b). The symmetric alldiff constraint. *Proc. 16th International Joint Conference on Artificial Intelligence, IJCAI'99, San Francisco*, Vol. 1, pp. 420–425.

Régin, J.-C., J.-F. Puget (1997). A filtering algorithm for global cardinality constraints with costs. *Principles and Practice of Constraint Programming, CP'97*, Springer, Linz, Austria, LNCS 1330, pp. 32–46.

Régin, J.-C., M. Rueher (2000). A global constraint combining a sum constraint and difference constraint. *Principles and Practice of Constraint Programming, CP'2000*, Springer, Singapore, LNCS 1894, pp. 384–395.

Rodošek, R., M. Wallace (1998). A generic model and hybrid algorithm for hoist scheduling problems. *Principles and Practice of Constraint Programming (CP98). Lecture Notes in Computer Science*, Vol. 1520, Springer, 385–399.

Rodošek, R., M. Wallace, M. Hajian (1997). A new approach to integrating mixed integer programming and constraint logic programming. *Annals of Operations Research* 86, 63–87.

Ruan, Y., E. Horvitz, H. A. Hautz (2002). Restart policies with dependence among runs: a dynamic programming approach. *Principles and Practice of Constraint Programming, CP'2002*, Springer, Ithaca, NY, LNCS 2470, 573–586.

Sabin, D., E. C. Freuder (1994). Contradicting conventional wisdom in constraint satisfaction. *Principles and Practice of Constraint Programming, PPCP'94*, Springer, Rosario, LNCS 874, pp. 10–20.

Sabin, D., E. C. Freuder (1997). Understanding and improving the MAC algorithm. *Principles and Practice of Constraint Programming, CP'97*, Springer, Linz, Austria, LNCS 1330, pp. 167–181.

Sakkout, L. E., T. Richards, M. Wallace (1998). Minimal perturbance in dynamic scheduling, in: H. Prade (ed.), *Proceedings, 13th European Conference on Artificial Intelligence*, Vol. 48. Wiley, New York, pp. 504–508.

Saraswat, V. A. (1993). Concurrent constraint programming. *ACM Doctoral Dissertation Awards*, MIT Press.

Sellmann, M. (2002). *Reduction Techniques in Constraint Programming and Combinatorial Optimization*, PhD thesis, Univ. Paderborn.

Sellmann, M., T. Fahle (2001). CP-based lagrangian relaxation for a multimedia application. *Third International Workshop on the Integration of AI and OR Techniques* (*CPAIOR 2001*).

Smith, B. M., S. C. Brailsford, P. M. Hubbard, H. P. Williams (1996). The progressive party problem: integer linear programming and constraint programming compared. *Constraints* 1, 119–138.

Smolka, G. (1995). The Oz programming model, in: J. van Leeuwen (ed.), *Computer Science Today: Recent Trends and Developments*, Springer, LNCS 1000.

Stergiou, K., T. Walsh (1999a). The difference all-difference makes. *16th Intern. Joint Conf. Artificial Intelligence, IJCAI'99*, Stockholm, pp. 414–419.

Stergiou, K., T. Walsh (1999b). Encodings of non-binary constraint satisfaction problems. *Proc. 16th National Conference on Artificial Intelligence* (*AAAI'99*) *and 11th Conference on Innovative Applications of Artificial Intelligence (IAAI'99)*, pp. 163–168.

Thorsteinsson, E. S. (2001). Branch-and-check: a hybrid framework integrating mixed integer programming and constraint logic programming. *Seventh International Conference on Principles and Practice of Constraint Programming* (*CP2001*).

Timpe, C. (2003). Solving BASF's plastics production planning and lot-sizing problem using combined CP/MIP, *LISCOS Project Summary Meeting*, Brussels (28 March 2003).

Türkay, M., I. E. Grossmann (1996). Logic-based MINLP algorithms for the optimal synthesis of process networks. *Computers and Chemical Engineering* 20, 959–978.

van Hentenryck, P. (1989). *Constraint Satisfaction in Logic Programming*, MIT Press.

van Hentenryck, P. (1999). *The OPL Optimization Programming Language*, MIT Press. (with contributions by I. Lustig, L. Michel, J.-F. Puget).

van Hentenryck, P., T. Graf (1992). A generic arc consistency algorithm and its specializations. *Artificial Intelligence* 57, 291–321.

van Hentenryck, P., L. Michel, F. Benhaumou (1998). Newton: constraint programming over non-linear constraints. *Science of Programming* 30, 83–118.

van Hentenryck, P., L. Michel, L. Perron, J.-C. Régin (1999). Constraint programming in OPL. *Principles and Practice of Declarative Programming, International Conference PPDP'99*, Springer, Paris, LNCS 1702, pp. 98–116.

van Hentenryck, P., V. Saraswat, Y. Deville (1998). Design, implementation, and evaluation of the constraint language cc(FD), *Journal of Logic Programming* 37(1–3), 139–164.

van Hoeve, W. J. (2001). The alldifferent constraint: a survey. *Sixth Annual Workshop of the ERCIM Working Group on Constraints, Prague*. http://arXiv.org/abs/cs.PL/0105015.

Wallace, M., S. Novello, J. Schimpf (1997). ECLiPSe: a platform for constraint logic programming. *ICL Systems Journal* 12, 159–200.

Walsh, T. (1997). Depth-bounded discrepancy search. *15th Intern. Joint Conf. Artificial Intelligence, IJCAI'97, Nagoya, Japan*, Vol. 2, pp. 1388–1395.

Williams, H. P., H. Yan (2001). Representations of the all-different predicate of constraint satisfaction in integer programming. *INFORMS Journal on Computing* 13, 96–103.

Woeginger, G. J. (2001). The reconstruction of polyominoes from their orthogonal projections. *Information Processing Letters* 77(5–6), 225–229.

Yan, H., J. N. Hooker (1999). Tight representation of logical constraints as cardinality rules. *Mathematical Programming* 85, 363–377.

Zhang, L., S. Malik (2002). The quest for efficient Boolean satisfiability solvers. *18th International Conference on Automated Deduction, CADE-18*, Springer, Copenhagen, LNCS 2392, pp. 295–313.

Zhang, Y., R. H. C. Yap (2000). Arc consistency on *n*-ary monotonic and linear constraints. *Principles and Practice of Constraints Programming, CP'2000*, Springer, Singapore, LNCS 1894, pp. 470–483.

Zhang, Y., Yap, R. H. C. (2001). Making AC-3 an optimal algorithm. *17th Intern. Joint Conf. Artificial Intelligence, IJCAI'01, Seattle*, Vol. 1, pp. 316–321.

Zhou, J. (1997). *Computing Smallest Cartesian Products of Intervals: Application to the Job-Shop Scheduling Problem*, PhD thesis, Univ. de la Méditerranée Aix-Marseille II.

Zhou, J. (2000). Introduction to the constraint language NCL. *Journal of Logic Programming* 45(1–3), 71–103.

Index